Lecture Notes in Computer Science

Edited by G. Goos and J. Hartmanis
Series: I.F.I.P. TC7 Optimization Conferences

41

Optimization Techniques

Modeling and Optimization in the Service of Man

Part 2

Proceedings, 7th IFIP Conference
Nice, September 1975

Edited by Jean Cea

Springer-Verlag
Berlin · Heidelberg · New York

Lecture Notes in Computer Science

Vol. 1: GI-Gesellschaft für Informatik e.V. 3. Jahrestagung, Hamburg, 8.–10. Oktober 1973. Herausgegeben im Auftrag der Gesellschaft für Informatik von W. Brauer. XI, 508 Seiten. 1973.

Vol. 2: GI-Gesellschaft für Informatik e.V. 1. Fachtagung über Automatentheorie und Formale Sprachen, Bonn, 9.–12. Juli 1973. Herausgegeben im Auftrag der Gesellschaft für Informatik von K.-H. Böhling und K. Indermark. VII, 322 Seiten. 1973.

Vol. 3: 5th Conference on Optimization Techniques, Part I. (Series: I.F.I.P. TC7 Optimization Conferences.) Edited by R. Conti and A. Ruberti. XIII, 565 pages. 1973.

Vol. 4: 5th Conference on Optimization Techniques, Part II. (Series: I.F.I.P. TC7 Optimization Conferences.) Edited by R. Conti and A. Ruberti. XIII, 389 pages. 1973.

Vol. 5: International Symposium on Theoretical Programming. Edited by A. Ershov and V. A. Nepomniaschy. VI, 407 pages. 1974.

Vol. 6: B. T. Smith, J. M. Boyle, J. J. Dongarra, B. S. Garbow, Y. Ikebe, V. C. Klema, and C. B. Moler, Matrix Eigensystem Routines – EISPACK Guide. XI, 551 pages. 2nd Edition 1974. 1976.

Vol. 7: 3. Fachtagung über Programmiersprachen, Kiel, 5.–7. März 1974. Herausgegeben von B. Schlender und W. Frielinghaus. VI, 225 Seiten. 1974.

Vol. 8: GI-NTG Fachtagung über Struktur und Betrieb von Rechensystemen, Braunschweig, 20.–22. März 1974. Herausgegeben im Auftrag der GI und der NTG von H.-O. Leilich. VI, 340 Seiten. 1974.

Vol. 9: GI-BIFOA Internationale Fachtagung: Informationszentren in Wirtschaft und Verwaltung. Köln, 17./18. Sept. 1973. Herausgegeben im Auftrag der GI und dem BIFOA von P. Schmitz. VI, 259 Seiten. 1974.

Vol. 10: Computing Methods in Applied Sciences and Engineering, Part 1. International Symposium, Versailles, December 17–21, 1973. Edited by R. Glowinski and J. L. Lions. X, 497 pages. 1974.

Vol. 11: Computing Methods in Applied Sciences and Engineering, Part 2. International Symposium, Versailles, December 17–21, 1973. Edited by R. Glowinski and J. L. Lions. X, 434 pages. 1974.

Vol. 12: GFK-GI-GMR Fachtagung Prozessrechner 1974. Karlsruhe, 10.–11. Juni 1974. Herausgegeben von G. Krüger und R. Friehmelt. XI, 620 Seiten. 1974.

Vol. 13: Rechnerstrukturen und Betriebsprogrammierung, Erlangen, 1970. (GI-Gesellschaft für Informatik e.V.) Herausgegeben von W. Händler und P. P. Spies. VII, 333 Seiten. 1974.

Vol. 14: Automata, Languages and Programming – 2nd Colloquium, University of Saarbrücken, July 29–August 2, 1974. Edited by J. Loeckx. VIII, 611 pages. 1974.

Vol. 15: L Systems. Edited by A. Salomaa and G. Rozenberg. VI, 338 pages. 1974.

Vol. 16: Operating Systems, International Symposium, Rocquencourt 1974. Edited by E. Gelenbe and C. Kaiser. VIII, 310 pages. 1974.

Vol. 17: Rechner-Gestützter Unterricht RGU '74, Fachtagung, Hamburg, 12.–14. August 1974, ACU-Arbeitskreis Computer-Unterstützter Unterricht. Herausgegeben im Auftrag der GI von K. Brunnstein, K. Haefner und W. Händler. X, 417 Seiten. 1974.

Vol. 18: K. Jensen and N. E. Wirth, PASCAL – User Manual and Report. VII, 170 pages. Corrected Reprint of the 2nd Edition 1976.

Vol. 19: Programming Symposium. Proceedings 1974. V, 425 pages. 1974.

Vol. 20: J. Engelfriet, Simple Program Schemes and Formal Languages. VII, 254 pages. 1974.

Vol. 21: Compiler Construction, An Advanced Course. Edited by F. L. Bauer and J. Eickel. XIV. 621 pages. 1974.

Vol. 22: Formal Aspects of Cognitive Processes. Proceedings 1972. Edited by T. Storer and D. Winter. V, 214 pages. 1975.

Vol. 23: Programming Methodology. 4th Informatik Symposium, IBM Germany Wildbad, September 25–27, 1974. Edited by C. E. Hackl. VI, 501 pages. 1975.

Vol. 24: Parallel Processing. Proceedings 1974. Edited by T. Feng. VI, 433 pages. 1975.

Vol. 25: Category Theory Applied to Computation and Control. Proceedings 1974. Edited by E. G. Manes. X, 245 pages. 1975.

Vol. 26: GI-4. Jahrestagung, Berlin, 9.–12. Oktober 1974. Herausgegeben im Auftrag der GI von D. Siefkes. IX, 748 Seiten. 1975.

Vol. 27: Optimization Techniques. IFIP Technical Conference. Novosibirsk, July 1–7, 1974. (Series: I.F.I.P. TC7 Optimization Conferences.) Edited by G. I. Marchuk. VIII, 507 pages. 1975.

Vol. 28: Mathematical Foundations of Computer Science. 3rd Symposium at Jadwisin near Warsaw, June 17–22, 1974. Edited by A. Blikle. VII, 484 pages. 1975.

Vol. 29: Interval Mathematics. Procedings 1975. Edited by K. Nickel. VI, 331 pages. 1975.

Vol. 30: Software Engineering. An Advanced Course. Edited by F. L. Bauer. (Formerly published 1973 as Lecture Notes in Economics and Mathematical Systems, Vol. 81) XII, 545 pages. 1975.

Vol. 31: S. H. Fuller, Analysis of Drum and Disk Storage Units. IX, 283 pages. 1975.

Vol. 32: Mathematical Foundations of Computer Science 1975. Proceedings 1975. Edited by J. Bečvář. X, 476 pages. 1975.

Vol. 33: Automata Theory and Formal Languages, Kaiserslautern, May 20–23, 1975. Edited by H. Brakhage on behalf of GI. VIII, 292 Seiten. 1975.

Vol. 34: GI – 5. Jahrestagung, Dortmund 8.–10. Oktober 1975. Herausgegeben im Auftrag der GI von J. Mühlbacher. X, 755 Seiten. 1975.

Vol. 35: W. Everling, Exercises in Computer Systems Analysis. (Formerly published 1972 as Lecture Notes in Economics and Mathematical Systems, Vol. 65) VIII, 184 pages. 1975.

Vol. 36: S. A. Greibach, Theory of Program Structures: Schemes, Semantics, Verification. XV, 364 pages. 1975.

Vol. 37: C. Böhm, λ-Calculus and Computer Science Theory. Proceedings 1975. XII, 370 pages. 1975.

Vol. 38: P. Branquart, J.-P. Cardinael, J. Lewi, J.-P. Delescaille, M. Vanbegin. An Optimized Translation Process and Its Application to ALGOL 68. IX, 334 pages. 1976.

Vol. 39: Data Base Systems. Proceedings 1975. Edited by H. Hasselmeier and W. Spruth. VI, 386 pages. 1976.

Vol. 40: Optimization Techniques. Modeling and Optimization in the Service of Man. Part 1. Proceedings 1975. Edited by J. Cea. XIV, 854 pages. 1976.

Vol. 41: Optimization Techniques. Modeling and Optimization in the Service of Man. Part 2. Proceedings 1975. Edited by J. Cea. XIII, 852 pages. 1976.

Lecture Notes in
Computer Science

Edited by G. Goos and J. Hartmanis
Series: I.F.I.P. TC7 Optimization Conferences

41

Optimization Techniques

Modeling and Optimization in the Service of Man Part 2

Proceedings, 7th IFIP Conference
Nice, September 8–12, 1975

Edited by Jean Cea

Springer-Verlag
Berlin · Heidelberg · New York 1976

Editor
Jean Cea
Département de Mathématiques
Faculté des Sciences
Parc Valrose
06034 Nice Cedex/France

Library of Congress Cataloging in Publication Data
IFIP Conference on Optimization Techniques, 7th, Nice, 1975.
Optimization techniques.

(Series, I. F. I. P. TC 7 optimization conferences)
(Lecture notes in computer science ; 40-41)
Sponsored by the IFIP Technical Committee on Optimization (TC 7).
1. Mathematical optimization--Congresses.
2. Mathematical models--Congresses. I. Céa, Jean, 1932- II. International Federation for Information Processing. Technical Committee on Optimization (TC 7). III. Title. IV. Series: International Federation for Information Processing. Technical Committee on Optimization (TC 7). Series, I. F. I. P. TC 7 optimization conferences. V. Series: Lecture notes in computer science ; 40-41.
QA402.5.I173 1975 001.4'24 76-9857

AMS Subject Classifications (1970): 49.02, 49A35, 49A40, 49B35, 49B40, 49D10, 49D45, 49G99, 65K05, 90C10, 90C20, 90C30, 90C50, 90C99, 90 D05, 92A15, 93.02, 93B05, 93B10, 93B20, 93B30, 93B35, 93B99, 93C20, 93E05, 93E20, 94.00
CR Subject Classifications (1974): 3.1, 3.2, 3.3, 4.9, 5.1, 5.4

ISBN 3-540-07623-9 Springer-Verlag Berlin · Heidelberg · New York
ISBN 0-387-07623-9 Springer-Verlag New York · Heidelberg · Berlin

Printed in Germany
Printing and binding: Beltz Offsetdruck, Hemsbach/Bergstr.

PREFACE

These Proceedings are based on the papers presented at the 7th IFIP Conference on Optimization Techniques held in Nice, September 8-12, 1975. The Conference was sponsored by the IFIP Technical Committee on Optimization (TC 7) with the cooperation of:

AFCET (Association Française pour la Cybernétique Economique et Technique)

IRIA (Institut de Recherche en Informatique et en Automatique)

SMF (Société Mathématique de France)

Université de Nice, Ville de Nice and Conseil Général des Alpes-Maritimes.

The Conference was devoted to recent advances in optimization techniques and their application to modeling, identification and control of large systems. Major emphasis of the Conference was on the most recent application areas including: environmental systems, socio-economic systems, biological systems.

The Proceedings are divided into two volumes: In the first are collected the papers in which the methodological aspects are emphasized; in the second those dealing with various application areas.

The international Program Committee of the Conference consisted of:

A.V. Balakrishnan (U.S.A.), B. Fraeijs de Veubeke (Belgium), G. Nyiry (Hungary), A. Kalliauer (Austria), L.L. Lions (France), G. Marchuk (USSR), C. Olech (Poland), L.S. Pontryagin (USSR), A. Ruberti (Italy), F. Stoer (RFG), J.H. Westcott (United Kingdom), K. Yajima (Japan).

PART 2

TABLE OF CONTENTS

---

*paper not received

MATHEMATICAL PROGRAMMING

OPTIMAL CONTROL STOCHASTIC

PART 1

TABLE OF CONTENTS

INVITED SPEAKERS

MEDICINE AND BIOLOGY

---

*paper not received

HUMAN ENVIRONMENT (WATER POLLUTION)

HUMAN ENVIRONMENTS (SOCIOLOGY, URBAN SYSTEMS, PHYSICS, CHEMISTRY)

HUMAN ENVIRONMENTS (ENERGY, WORLD MODELS)

OPERATIONAL RESEARCH

** This paper was received during production and has been included at the end of the volume (page 834).

ASSOCIATED SOFTWARE PROBLEMS

ON THE MARGINAL VALUE OF AN ANTAGONISTIC GAME

Joachim Hartung
Institut für Angewandte Mathematik
D-53 Bonn, Wegelerstraße 6

Let X and Y be the sets of strategies of two antagonistic players, and for $t \in [0,T] \subset \mathbb{R}$, $T > 0$,

$$f(x,y,t) : X \times Y \to \mathbb{R}$$

may be a payoff function in the parametric two-person zero-sum game

$$G_t := (X,\ Y,\ f(x,y,t))\ ,\quad t \in [0,T],$$

where it is to maximize over X and to minimize over Y.

Definition 1: If $v(t) = \mathrm{val}(G_t)$ is the value of G_t, $t \in [0,T]$, then the marginal value of the family of games $\{G_t\}_{t \in [0,T]}$ in the point $t = 0$ is defined as

$$v'_+(0) := \lim_{t \to +0} \frac{v(t) - v(0)}{t} .$$

We consider conditions - different from that given for instance in [3], [5] - for the existence of the marginal value, which enable us to derive from the marginal value theorem a method for solving a wide class of constrained games.

Let X, Y be not empty closed convex subsets of real topological linear spaces.

Definition 2: A function $g: X \times Y \to \mathbb{R}$ is 'sup-inf compact', if the level sets $\{x \in X \mid g(x,y) \geq a\}$, $y \in Y$, $a \in \mathbb{R}$, and $\{y \in Y \mid g(x,y) \leq b\}$, $x \in X$, $b \in \mathbb{R}$, are compact.

Let X(t), Y(t) denote the sets of optimal strategies in G_t, $t \in [0,T]$.

Theorem 1: (marginal value theorem)

If (i) $f(x,y,t)$ is concave (convex) in x (in y) for $t \in [0,T]$,

(ii) $f(x,y,0)$ is upper (lower) semicontinuous in x (in y),

(iii) $f(x,y,0)$ has a saddle point on $X \times Y$,

(iv) $f'(x,y,0) := \frac{\partial f(x,y,t)}{\partial t}\Big|_{t=+0}$ exists, and

$$O(1,x) \leq \frac{f(x,y,t)-f(x,y,0)}{t} - f'(x,y,0) \leq O(1,y)\ ,\ \text{for } t \in (0,T],$$

where $O(1,x)$ is independent of y and $O(1,x) \to 0$ for $t \to +0$ and fixed $x \in X$, $O(1,y)$ is independent of x and $O(1,y) \to 0$ for $t \to +0$ and fixed $y \in Y$,

(v) $f'(x,y,0)$ and $f(x,y,t)$ for $t \in (0,T]$ are sup-inf compact on $X \times Y$,

then $v'_+(0)$ exists and

$$v'_+(0) = \max_{x \in X(0)} \min_{y \in Y(0)} f'(x,y,0) = \min_{y \in Y(0)} \max_{x \in X(0)} f'(x,y,0) \quad . \tag{1}$$

Proof:

Because of (i), (iii) and (v) the sets $X(t)$, $Y(t)$ are not empty and there exist the values $v(t)$, for $t \in [0,T]$. Let $x_t \in X(t)$, $y_t \in Y(t)$, $t \in [0,T]$, then we have

$$f(x_o,y_t,t) - f(x_o,y_t,0) \leq v(t) - v(0) \leq f(x_t,y_o,t) - f(x_t,y_o,0) \quad . \tag{2}$$

For $t \in (0,T]$ we get from (iv)

$$\bigwedge_{x \in X} \bigwedge_{y \in Y} O(t,x) + tf'(x,y,0) \leq f(x,y,t) - f(x,y,0) \leq O(t,y) + t \cdot f'(x,y,0) \quad , \tag{3}$$

and with (2)

$$t \cdot f'(x_o,y_t,0) + O(t,x_o) \leq f(x_o,y_t,t) - f(x_o,y_t,0) \leq v(t) - v(0) \leq f(x_t,y_o,t) - f(x_t,y_o,0) \leq t \cdot f'(x_t,y_o,0) + O(t,y_o) \quad . \tag{4}$$

From (v) we get

$$f'(x,y,0) \geq \min_{y \in Y} f'(x,y,0) =: f'(x,y(x),0) > -\infty \tag{5}$$

and

$$f'(x,y,0) \leq \max_{x \in X} f'(x,y,0) =: f'(x(y),y,0) < +\infty \quad ,$$

such that with (4):

$$t \cdot f'(x_o,y(x_o),0) + O(t,x_o) \leq v(t) - v(0) \leq t \cdot f'(x(y_o),y_o,0) + O(t,y_o) \quad , \tag{6}$$

which means that

$$\lim_{t \to +0} v(t) = v(0) \quad . \tag{7}$$

Dividing in (4) by t, (5) yields

(8) $$f'(x_o,y_t,0) + o(1,x_o) \leq f'(x(y_o),y_o,0) + o(1,y_o) \quad , \quad \text{and}$$

$$f'(x_t,y_o,0) + o(1,y_o) \geq f'(x_o,y(x_o),0) + o(1,x_o) \quad ,$$

which by (v) means, that the x_t, y_t are elements of compact sets independent of t. Therefore $\{x_t\}_{t\to+0}$, $\{y_t\}_{t\to+0}$ have accumulation points $\hat{x} \in X$, $\hat{y} \in Y$ and convergent subsequences $\{x_{t_n}\}_{n\in\mathbb{N}} \subset \{x_t\}_{t\to+0}$, $\{y_{t_n}\}_{n\in\mathbb{N}} \subset \{y_t\}_{t\to+0}$, such that

$$\lim_{n\to\infty} x_{t_n} = \hat{x} \quad ,$$

$$\lim_{n\to\infty} y_{t_n} = \hat{y} \quad .$$

By (ii) we get for all $x \in X$

(9) $$f(x,\hat{y},0) \leq \underline{\lim}_{t_n\to+0} f(x,y_{t_n},0) \leq \overline{\lim}_{t_n\to+0} f(x,y_{t_n},0) \quad ,$$

and with (iv) and (5)

(10) $$v(t_n) - f(x,y_{t_n},0) \geq f(x,y_{t_n}) - f(x,y_{t_n},0)$$
$$\geq t_n\cdot f'(x,y(x),0) + o(t_n,x) \quad ,$$

(11) $$\underline{\lim}_{t_n\to+0} (v(t_n) - f(x,y_{t_n},0)) \geq 0 \quad ,$$

and by (7)

(12) $$\overline{\lim}_{t_n\to+0} f(x,y_{t_n},0) \leq v(0) \quad .$$

Because of (9) this gives

(13) $$f(x,\hat{y},0) \leq v(0) \quad \text{for all } x \in X \quad ,$$

i.e. $\hat{y}$ is an optimal strategy in G_o. Analogously you show $\hat{x} \in X(0)$.
From inequality (4) it follows

(14) $$f'(x_o,y_t,0) + o(1,x_o) \leq \frac{v(t) - v(0)}{t}$$
$$\leq f'(x_t,y_o,0) + o(1,y_o) \quad ,$$
$$\text{for all } x_o \in X(0),\ y_o \in Y(0) \quad ,$$

such that

(15) $$\sup_{x\in X(0)} f'(x,y_t,0) + o(1,x) \leq \frac{v(t) - v(0)}{t}$$
$$\leq \inf_{y\in Y(0)} f'(x_t,y,0) + o(1,y) \quad .$$

Since $y \to f'(x,y,0)$ is lower semicontinuous, also $y \to \sup_{x\in X(0)} f'(x,y,0)$ is lower semicontinuous. Similarly, $x \to \inf_{y\in Y(0)} f'(x,y,0)$ is upper semicontinuous.

Let P and Q be the sets of accumulation points of $\{x_t\}_{t\to+0}$ and $\{y_t\}_{t\to+0}$, respectively, and for $p \in P$, $q \in Q$

$$\{x_{t(p)}\}_{t(p)\to+0} \subset \{x_t\}_{t\to+0}\ ,\quad \{y_{t(q)}\}_{t(q)\to+0} \subset \{y_t\}_{t\to+0}$$

may be convergent subsequences such that

$$\lim_{t(p)\to+0} x_{t(p)} = p\ ,\qquad \lim_{t(q)\to+0} x_{t(q)} = q\ .$$

From (15) we get now

$$(16)\qquad \begin{aligned} &\inf_{q\in Q}\ \sup_{x\in X(0)}\ f'(x,q,0)\\ &\leqslant \inf_{q\in Q}\ \underline{\lim}_{t(q)\to+0}\ \sup_{x\in X(0)}\ f'(x,y_{t(q)},0)\\ &\leqslant \underline{\lim}_{t\to+0}\ \sup_{x\in X(0)}\ f'(x,y_t,0)\\ &\leqslant \underline{\lim}_{t\to+0}\ \frac{v(t)-v(0)}{t}\\ &\leqslant \overline{\lim}_{t\to+0}\ \frac{v(t)-v(0)}{t}\\ &\leqslant \overline{\lim}_{t\to+0}\ \inf_{y\in Y(0)}\ f'(x_t,y,0)\\ &\leqslant \sup_{p\in P}\ \overline{\lim}_{t(p)\to+0}\ \inf_{y\in Y(0)}\ f'(x_{t(p)},y,0)\\ &\leqslant \sup_{p\in P}\ \inf_{y\in Y(0)}\ f'(p,y,0)\ . \end{aligned}$$

By (13) we have

$$(17)\qquad P \subset X(0)\ ,\quad Q \subset Y(0)\ ,$$

and (16) yields

$$(18)\qquad \begin{aligned} &\inf_{y\in Y(0)}\ \sup_{x\in X(0)}\ f'(x,y,0)\\ &\leqslant \inf_{y\in Q}\ \sup_{x\in X(0)}\ f'(x,y,0)\\ &\leqslant \underline{\lim}_{t\to+0}\ \frac{v(t)-v(0)}{t}\\ &\leqslant \overline{\lim}_{t\to+0}\ \frac{v(t)-v(0)}{t}\\ &\leqslant \sup_{x\in P}\ \inf_{y\in Y(0)}\ f'(x,y,0)\\ &\leqslant \sup_{x\in X(0)}\ \inf_{y\in Y(0)}\ f'(x,y,0)\ . \end{aligned}$$

On the other hand there holds the sup-inf inequality

(19) $$\sup_{x\varepsilon X(0)} \inf_{y\varepsilon Y(0)} f'(x,y,0) \leqslant \inf_{y\varepsilon Y(0)} \sup_{x\varepsilon X(0)} f'(x,y,0) .$$

Thus $v_+'(0) = \lim_{t\to+0} \frac{v(t) - v(0)}{t}$ exists and by the sup-inf compactness of $f'(x,y,0)$ we get

$$v_+'(0) = \max_{x\varepsilon X(0)} \min_{y\varepsilon Y(0)} f'(x,y,0)$$

$$= \min_{y\varepsilon Y(0)} \max_{x\varepsilon X(0)} f'(x,y,0) .$$

<u>Theorem 2:</u>

Let $x_t \varepsilon X(t)$, $y(t) \varepsilon Y(t)$, $t \varepsilon (0,T]$.

For any accumulation points $\hat{x}$, $\hat{y}$ of $\{x_t\}_{t\to+0}$, $\{y_t\}_{t\to+0}$, respectively, it holds then under the assumptions of Theorem 1:

(20) $$f'(\hat{x},\hat{y},0) = v_+'(0) .$$

<u>Proof:</u>

Let $x_{t_n} \to \hat{x}$, for $t_n \to +0$, then we get from (15) with the results of Theorem 1:

(21) $$\sup_{x\varepsilon X(0)} f'(x,\hat{y},0) \leqslant \underline{\lim}_{t_n\to+0} \sup_{x\to X(0)} f'(x,y_{t_n},0)$$

$$\leqslant v_+'(0) = val(X(0),Y(0),f'(x,y,0)) .$$

Since $\hat{y} \varepsilon Y(0)$ that means, $\hat{y}$ is an optimal strategy in the game $(X(0),Y(0),f'(x,y,0))$. Similarly, $\hat{x}$ is an optimal strategy in this game.

Now we come to an application of the results above to constrained games.

Let U, V be normed real vector spaces, $C \subset U$, $K \subset V$ not empty closed convex cones, $g: X \to U$, $h: Y \to V$ continuous and concave relative to the cones, and $\phi: X \times Y \to \mathbb{R}$ be upper-lower semicontinuous, concave-convex and sup--inf compact.

We consider the constrained game

(22) $$(CG) := (\{x\varepsilon X \mid g(x)\varepsilon C\}, \{y\varepsilon Y \mid h(y)\varepsilon K\}, \phi(x,y)) .$$

With

$$d(g(x),C) := \inf_{c\varepsilon C} \|g(x) - c\| ,$$

$$d(h(y),K) := \inf_{k\varepsilon K} \|h(y) - k\| ,$$

we define for $t \geqslant 0$

(23) $$f(x,y,t) := d(h(y),K) - d(g(x),C) + t\cdot\phi(x,y) .$$

A solution method for (CG) is given by the following

<u>Theorem 3:</u>

Let $\{t_n\}_{n\in N}$ be a positive real nullsequence and (CG) may have admissable strategies.

Then (i) the unconstrained games $(X,Y,f(x,y,t_n))$ have optimal strategies x_{t_n}, y_{t_n} .

(ii) $\{x_{t_n}\}_{n\in N}$, $\{y_{t_n}\}_{n\in N}$ have accumulation points $\hat{x}$, $\hat{y}$ and $\hat{x}$, $\hat{y}$ are optimal strategies for (CG).

(iii) $\lim_{n\to\infty} \{\phi(x_{t_n},y_{t_n}) + \frac{1}{t_n}(d(h(y),K) - d(g(x),C)\} = \phi(\hat{x},\hat{y}) = val(CG)$

<u>Proof:</u>

First we show that $d(g(x),C)$ is continuous and convex in x.
$d(.,C)$ is continuous on U, $g(.)$ is continuous on X, thus $d(g(.),C)$ is continuous on X. g is concave on C, i.e. for $x_1,x_2 \in X$ and $0 \leq \alpha \leq 1$ it holds $g(\alpha x_1 + (1-\alpha)x_2) - \alpha g(x_1) - (1-\alpha)g(x_2) \in C$. Then for all $c \in C$: $g(\alpha x_1 + (1-\alpha)x_2) - \alpha g(x_1) - (1-\alpha)g\ x_2) + c \in C$, which gives

$$\begin{aligned}&\|\alpha g(x_1) + (1-\alpha)g(x_2) - c\| \\ &\quad = \|g(\alpha x_1 + (1-\alpha)x_2) - (g(\alpha x_1 + (1-\alpha)x_2) - \\ &\qquad - \alpha(g(x_1) - (1-\alpha)g(x_2) + c)\| \\ &\quad \geq \inf_{d\in C} \|g(\alpha x_1 + (1-\alpha)x_2) - d\| .\end{aligned}$$

Theorem 3.4 in [6] states, that $d(.,C)$ is convex. Thus we have

$$\begin{aligned}d(g(\alpha x_1 + (1-\alpha)x_2),C) &\leq d(\alpha g(x_1) + (1-\alpha)g(x_2),C) \\ &\leq \alpha d(g(x_1),C) + (1-\alpha)d(g(x_2),C).\end{aligned}$$

It is $f(x,y,0) = d(h(y),K) - d(g(x),C)$. We show $X(0) = \{x \in X \mid g(y) \in C\}$, $Y(0) = \{y \in Y \mid h(y) \in K\}$. Let $x_o \in X(0)$, $y_o \in Y(0)$, then for all $x \in X$, $y \in Y$ it is valid

$$\begin{aligned}d(h(y_o),K) - d(g(x),C) &\leq d(h(y_o),K) - d(g(x_o),C) \\ &\leq d(h(y),K) - d(g(x_o),C) .\end{aligned}$$

It follows $d(h(y_o),K) \leq d(h(y),K)$ for all $y \in Y$.
Since (CG) should have admissible strategies, there exists a $y \in Y$ with $h(y) \in K$. So $d(h(y_o),K) \leq 0$, i.e. $h(y_o) \in K$. Analogously it can be shown that $g(x_o) \in C$, under the assumption that there exists an $x \in X$ with $g(x) \in C$. Now let $\hat{x} \in X$, $\hat{y} \in Y$ such that $g(\hat{x}) \in C$, $h(\hat{y}) \in K$. Then $d(g(\hat{y}),C) = d(h(\hat{y}),K) = 0$ and for all $x \in X$, $y \in Y$ we have

$$-d(g(x),C) \leq 0 \leq d(g(y),K) \quad ,$$

i.e. $\hat{x} \in X(0)$, $\hat{y} \in Y(0)$. Further we have $v(0) = 0$.
$f(x,y,t)$ is sup-inf compact for $t > 0$, because $f'(x,y,0) = \phi(x,y)$ has this property and $f(x,y,0)$ is bounded from above (below) in x (in y). From this fact part (i) of the theorem follows.

We have shown that the assumptions of theorem 1. are fulfilled, such that (ii) and (iii) follow from the theorems 1. and 2., respectively from the proofs.

To solve the problem of finding optimal strategies of the games $(X,Y,f(x,y,t))$, you often have to take algorithms which need for convergency the function $f(.,.,t)$ to be strictly concave-convex, as for example the successive approximation method given in [1]. If f does not possess this property, we can do the following:

Let $\psi(x,y)$ be a strictly concave-convex upper-lower semicontinuous real valued function on $X \times Y$, which is bounded from above (below) in x (in y) by some $a(y) \in \mathbb{R}$ $(b(x) \in \mathbb{R})$. Then we define

$$F(x,y,t) := f(x,y,t) + t^2\psi(x,y) \ .$$

If the conditions of theorem 1. are fulfilled for f, then also for F. We show this for the condition (iv):

If $\quad O_f(1,x) \leq \dfrac{f(x,y,t) - f(x,y,0)}{t} - f'(x,y,0) \leq O_f(1,y) \ , \ t > 0,$

then we define $\quad O_F(1,x) := O_f(1,x) + t\cdot b(x) \ , \ O_F(1,y) := O_f(1,y) + t\cdot a(y)$ and get

$$O_F(1,x) \leq \frac{f(x,y,t) - f(x,y,0)}{t} + t\cdot\psi(x,y) - f'(x,y,0) \leq O_F(1,y).$$

Since $\quad F(x,y,0) = f(x,y,0)$, $F'(x,y,0) = f'(x,y,0)$ this states 1.(iv) for F. Furthermore, let $V(t) := \mathrm{val}(X,Y,F(x,y,t))$, then $V(0) = v(0)$ and $V_+'(0) = v_+'(0)$. Thus in order to compute $v_+'(0)$, instead of $(X,Y,f(x,y,t))$ we can solve the games $(X,Y,F(x,y,t))$, which have unique solutions (for $t > 0$).

Under certain conditions it can be shown that the accumulation points of the corresponding optimal strategies for $t \to +0$ are uniquely determined, such that the whole sequences are converging to optimal strategies of the game $(X(0), Y(0), f'(x,y,0))$.

This kind of regularization is particularly interesting for the above given method for solving constrained games.

References

[1] DEM'JANOV, V.F., Successive Approximation for Finding Saddle Points, Soviet Math. Dokl. 8, 6, 1967.
[2] DUGOŠIJA, Đ. u. J. HARTUNG, Über Stabilität in Zwei-Personen-Null--Summen-Spielen, Matematica Balkanica (Beograd) 4, 1974.
[3] ГОЛЬШТЕЙН, Е.Г., и С.М. МОВШОВИЧ: Непрерывная зависимость от параметра множества решений минимаксной задачи, Экономика и Матем. методы 4, 6, 1968.
[4] HARTUNG, J., Penalty-Methoden für Kontrollprobleme und Open-Loop--Differential-Spiele, to appear in: R. Bulirsch, W. Oettli, J. Stoer (eds.) "Optimierungstheorie und optimale Steuerungen", Lect. Not. Math., Springer, Berlin, Heidelberg, New York, 1975.
[5] MILLS, H., Marginal Values of Matrix Games and Linear Programs, in: H.W. Kuhn, A.W. Tucker (eds.) "Linear Inequalities and Related Systems", Princeton University Press, Princeton, New Jersey, 1956.
[6] VALENTINE, F.A., Convex Sets, McGraw-Hill, New York, 1964.

OPTIMIZATION OF STRUCTURAL ELEMENTS

Jean-Louis ARMAND
Institut de Recherches de la Construction Navale
47, rue de Monceau - 75008 PARIS, FRANCE
also
Ecole Polytechnique, 5 rue Descartes - 75005 PARIS, FRANCE

Abstract

A numerical method for the solution of structural optimization problems involving ordinary differential equations is presented for a simple situation where the constraint is of an aeroelastic nature. The method is adapted from optimal control theory and has proven successful in a number of structural optimization problems. Its extension to two dimensional structures is outlined ; limitation to situations involving plates, however, is emphasized. It is assumed that the instability exhibited by the optimality condition is related to the fact that plates cannot in general achieve global extrema. Suggestions for further research in this area are presented.

1. INTRODUCTION

The general theory of optimization finds a privileged field of application within the framework of Mechanics. Optimization of structural design, in particular, has raised considerable interest in recent years. The available literature concerned with this subject is vast. A comprehensive and organized review up to 1972 can be found in an article by Niordson and Pedersen (1972). Another excellent survey, concerned with constraints of a dynamic nature only, has been provided by Pierson (1972). Since then the flow of new research publications has continued steadily. A new phase seems now to have been reached with the apparition of monographs by Moe and Gisvold (1971), Hemp (1973), Gallagher and Zienkiewicz (1973), Distefano (1974), Majid (1974), among others. Research in structural optimization mainly follows two separate directions. The first approach, sometimes termed structural synthesis, is concerned with the optimization of discrete or discretized structures ; it is cast as a mathematical programming problem of extremizing an objective function subject to several equality or inequality constraints.

For a more complete description, the reader is referred to the papers by Schmit (1971) or Fox (1973), as well as to the texts by Moe and Gisvold (1971) or Gallagher and Zienkiewicz (1973) already mentioned. Although relatively complex structures can be considered, the number of design variables remains, unfortunately, very limited for practical reasons. The second approach is concerned with the optimal design of continuous structural elements, the properties of which are amenable to description by ordinary (one-dimensional structures) or partial (two- and three - dimensional structures) differential equations. A considerable literature exists in this field, although it is systematically ignored by those advocating the mathematical programming approach. This is rather unfortunate, since the continuous approach has the merit of raising basic questions such as the existence, uniqueness and characterization of the solution, questions which cannot be answered by mathematical programming techniques alone. Moreover, the constitution of a library of optimal structural members under a variety of constraints and boundary conditions would provide the practicing designer with efficient ideas in situations where intuition alone is of little help.

The present paper is concerned with the minimum-mass design of elastic structures whose behaviour is described by ordinary differential equations (no mention will be made of so-called plastic design, which is quite a different subject altogether). Despite a large existing literature, very few numerical solutions for structural elements have been presented, and the majority of those are of academic interest only. We present in what follows a numerical solution procedure borrowed from optimal control theory, which has proven quite useful in situations where the conventional variational approach leads to a set of complex nonlinear equations.

2. OPTIMAL DESIGN OF CONTINUOUS STRUCTURAL ELEMENTS

Application of the classical variational calculus to structural optimization dates as far back as Lagrange (1770-1773) who investigated the ideal shape that a column should possess in order to sustain its own weight. In this approach, the resulting equations are unfortunately quite complex and usually nonlinear. It is extremely rare to find a closed-form solution. A relative exception is the class of so-called linear stiffness structures, for which both stiffness and weight are linear in the design variable (cross-sectional area or thickness) ;

this is the case of sandwich construction in particular. The paper by Prager and Taylor (1968) contains an exhaustive treatment of this subject. A tutorial presentation can be found in Dym (1974). Generally, however, the investigator is left with a complex two-point boundary value problem (TPBVP) to solve in the case of one-dimensional elements, or a system of nonlinear partial differential equations in the case of two- or three-dimensional situations.

Application of optimal control theory to structural optimization was independently suggested by Dixon (1967), Ashley and McIntosh (1968) and Haug (1969). It has up to now hardly been realized that the designer has at his disposal the powerful tools which have been developed for the solution of similar problems arising in control theory, described in the book by Bryson and Ho (1969), among others. The idea is to cast the optimization problem in a form amenable to conventional treatment through the introduction of state and design (control) variables. The optimality conditions are then derived through an extension of classical variational methods (a particular case of the more general Pontryagins's Maximum Principle). The resulting system of equations is then solved by an iterative procedure.

A systematic presentation of the above points as well as an evaluation of the various numerical schemes will be found in a paper by Armand (1973), who also presents a review of the field. McIntosh (1974) also described the foundations of the method in a review paper in which comparisons with discrete solutions can be found. The approach presented here will be best described through a simple example where the constraint is of an aeroelastic nature, first suggested by Ashley and McIntosh (1968) and treated in detail by Armand and Vitte (1970), in which results of computation can be compared with the exact analytical solution.

3. MINIMUM - MASS DESIGN OF A RECTANGULAR WING FOR GIVEN TORSIONAL DIVERGENCE SPEED

3.1 Statement of the problem

Consider a cantilever straight wing with elastic axis perpendicular to the free stream. The wing profile is assumed to be constant along the span and characterized by a lift-coefficient slope a_o. The other parameters and variables are defined in Fig. 1. The thickness T of the skin varies along the span.

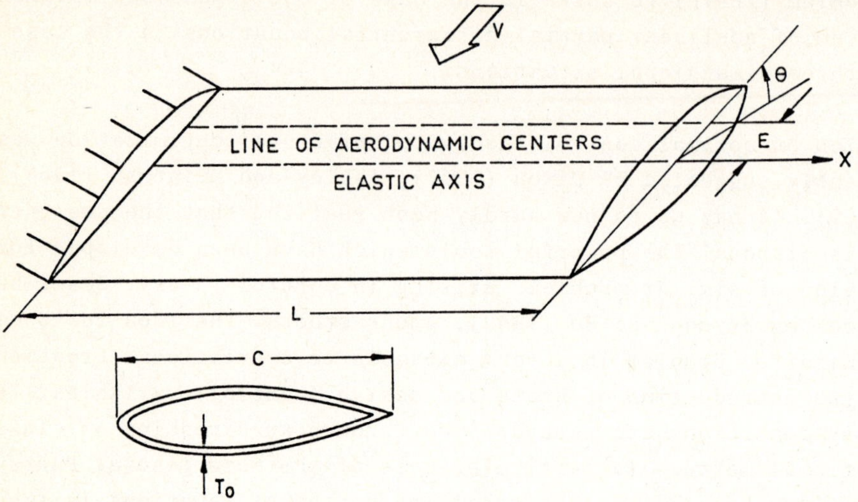

Figure 1 : Unswept cantilever wing with constant chord.

The torsional divergence speed V for this wing is obtained by solving the eigenvalue problem :

$$\frac{d}{dX}\left(GJ\,\frac{d\theta}{dX}\right) + \frac{1}{2}\rho V^2 CEa_o\theta = 0 \qquad (1)$$

$$\theta(0) = 0,$$

$$GJ\frac{d\theta}{dX}\bigg|_{X = L} = 0 \qquad (1')$$

ρ is the free stream density. The reader is referred to Bisplinghoff, Ashley and Halfman (1955) for details.

If we assume that the torsional stiffness of the wing is dominated by the contribution from the skin, then the torsion constant J is directly proportional to the skin thickness (Bredt's formula),

$$J = KT$$

It is possible to rewrite eqs. (1) and (1') in a convenient, non-dimensional form if we introduce that particular wing of constant skin thickness T_o with identical cross-sectional profile and the same torsional divergence speed -the so-called reference wing. We define the dimensionless quantities

$$x = \frac{X}{L}\ , \quad t = \frac{T}{T_o}\ ,$$

and denote by ()' the differentiation with respect to x ; the eigen-value problem (1), (1') is then rewritten as :

$$(t\theta')' + \omega^2\theta = 0\ ,\ \omega = \frac{\pi}{2} \tag{2}$$

$$\theta(0) = 0$$

$$t\theta'\Big|_{x=1} = 0 \tag{2'}$$

The optimization problem is now stated as follows :
Minimize the functional

$$m = \int_0^1 t(x)dx$$

representing the dimensionless skin mass, subject to the constraints (2) and (2').

Eq. (2) is split into a system of two first order differential equations after introducing a new variable s, as follows :

$$\theta' = \frac{s}{t}$$

$$s' = -\omega^2\theta \tag{3}$$

with boundary conditions :

$$\theta(0) = 0 \qquad s(1) = 0 \tag{3'}$$

We introduce unknown Lagrange multipliers λ_θ and λ_s and form the Hamiltonian

$$H = t+\lambda_\theta\frac{s}{t} - \lambda_s\omega^2\theta \tag{4}$$

Necessary conditions for an extremum of m are now derived. The optimality condition is :

$$\frac{\partial H}{\partial t} \equiv 1-\frac{\lambda_\theta s}{t^2} = 0 \tag{5}$$

λ_θ, λ_s are solution of the system of Euler-Lagrange equations :

$$\lambda'_\theta \equiv -\frac{\partial H}{\partial \theta} = \omega^2\lambda_s$$

$$\lambda'_s \equiv -\frac{\partial H}{\partial s} = -\frac{\lambda_\theta}{t} \tag{6}$$

Transversality conditions furnish two additional boundary conditions:

$$\lambda_\theta(1) = 0 \qquad \lambda_s(0) = 0 \tag{6'}$$

Equations (3) and (6) together with eq. (5) and boundary conditions (3') and (6') form a system of 4 first order differential equations in the four unknowns θ, s, λ_θ, λ_s. This system will be solved by resorting to the so-called neighboring extremal methods ; the idea is to construct a nominal solution satisfying the governing differential equations (3), the optimality condition (5) and the Euler-Lagrange equations (6), but not all the boundary conditions (3') and (6').

3.2 The numerical procedure : a transition-matrix algorithm (Newton's method)

The idea is to find the relation between unspecified boundary conditions at one end and specified boundary conditions at the other end. We start by "guessing" the unknown initial conditions $s(0)$ and $\lambda_\theta(0)$, then integrate (3) and (6) forward from 0 to 1, t being determined through (5). We obtain values for $s(1)$ and $\lambda_\theta(1)$, which in general will differ from zero. A "measure" of the variations of the final values of s and λ_θ when the initial boundary conditions are perturbed is provided by the matrix T, the so-called transition matrix

$$T = \begin{bmatrix} \dfrac{\partial s(1)}{\partial s(0)} & \dfrac{\partial s(1)}{\partial \lambda_\theta(0)} \\ \\ \dfrac{\partial \lambda_\theta(1)}{\partial s(0)} & \dfrac{\partial \lambda_\theta(1)}{\partial \lambda_\theta(0)} \end{bmatrix} \tag{7}$$

linking small variations of s and λ_θ at 1 to small variations of the same quantities at 0

$$\begin{bmatrix} \delta s(1) \\ \\ \delta \lambda_\theta(1) \end{bmatrix} = T \begin{bmatrix} \delta s(0) \\ \\ \delta \lambda_\theta(0) \end{bmatrix} \tag{8}$$

To determine T, we begin by computing the first variations $\delta\theta$, δs, $\delta\lambda_\theta$, $\delta\lambda_s$ of the 4 quantities θ, s, λ_θ, λ_s from the system of differential equations defining them ; we obtain :

$$\delta\theta' = \frac{1}{2}\sqrt{\frac{s}{\lambda_\theta}}\left(\frac{\delta s}{s} - \frac{\delta\lambda_\theta}{\lambda_\theta}\right)$$

$$\delta s' = -\omega^2 \delta\theta$$

$$\delta\lambda_\theta' = \omega^2 \delta\lambda_s$$

$$\delta\lambda_s' = \frac{1}{2}\sqrt{\frac{\lambda_\theta}{s}}\left(\frac{\delta s}{s} - \frac{\delta\lambda_\theta}{\lambda_{\dot\theta}}\right) \tag{9}$$

If $\delta s(0)$ is set equal to unity, all other perturbations being initially set to zero, integration of this system (9) together with the initial system (3), (3'), (5), (6), (6') on $[0,1]$ will yield values for $\delta s(1)$ and $\delta\lambda_\theta(1)$ which constitute the first column of T. Similarly, the second column of T is obtained by integrating both systems with, as initial conditions, $\delta\lambda_\theta(0)$ set equal to unity, all other perturbations equal zero, (3') and (5').

If the initial conditions $s(0)$ and $\lambda_\theta(0)$ had been correctly chosen, $s(1)$ and $\lambda_\theta(1)$ would both vanish. We therefore define variations of the final values $\Delta s(1)$ and $\Delta\lambda_\theta(1)$ which will hopefully bring the next solution closer to the desired values of $s(1)$ and $\lambda_\theta(1)$, defined as :

$$\Delta s(1) = -\varepsilon s(1)$$

$$\Delta\lambda_\theta(1) = -\varepsilon\lambda_\theta(1) \qquad 0\leq\varepsilon\leq 1$$

Corresponding variations $\Delta s(0)$ and $\Delta\lambda_\theta(0)$ of initial values are obtained from (8) :

$$\begin{bmatrix}\Delta s(0)\\ \Delta\lambda_\theta(0)\end{bmatrix} = T^{-1}\begin{bmatrix}\Delta s(1)\\ \Delta\lambda_\theta(1)\end{bmatrix}$$

We now take as a new guess :

$$s(0)_{new} = s(0)_{old} + \Delta s(0),$$

$$\lambda_\theta(0)_{new} = \lambda_\theta(0)_{old} + \Delta\lambda_\theta(0)$$

and start the whole process again until $s(1)$ and $\lambda_\theta(1)$ have the specified value zero to the desired accuracy.

With values of $s(0)$ and $\lambda_\theta(0)$ initially equal to 0.3 and 6, respectively, and $\varepsilon = 1$, convergence was very smoothly obtained in 6 iterations with a relative error of 0.04 % on the value of $\dot{t}$ at the root (Fig. 2). The profile obtained is that given by the exact analytical solution (Ashley and McIntosh (1968)).

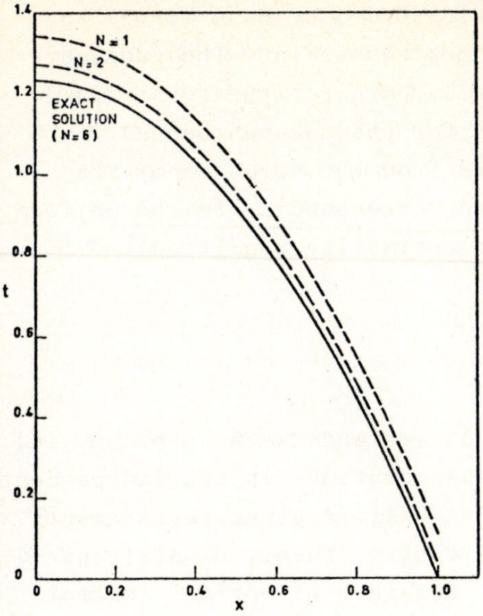

Fig. 2 : A transition-matrix procedure-
no minimal thickness constraint
(N = number of iterations).

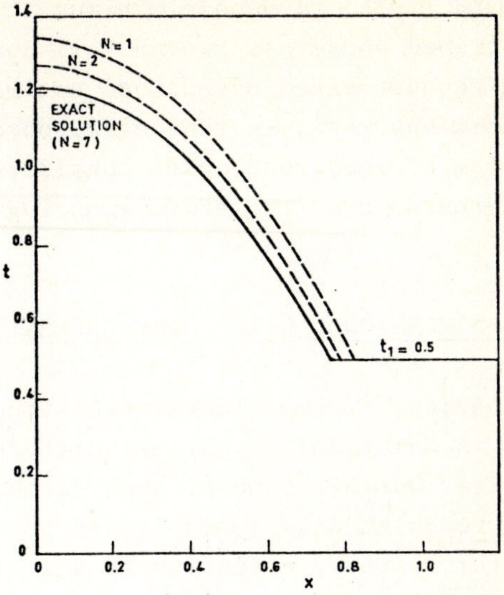

Fig. 3 : A transition-matrix procedure-
minimum thickness constraint

3.3 Optimization with a minimum thickness constraint

The more realistic case of a constraint on the value of the thickness is handled in the easiest fashion. A single statement needs to be added to the computer program, setting t equal to the prescribed value t_{min} everytime its computed value $t = \sqrt{\lambda_\theta s}$ is smaller than t_{min}. Results of the iteration procedure are shown in Fig. 3 for $t_{min} = 0.5$ where they are compared with the exact analytical solution given by Armand and Vitte (1970). The actual distribution was obtained after 7 integrations with a relative error of 0.01 % on t(0).

The problem just described is governed by fairly simple equations, which explains the relative ease with which the numerical solution was iteratively found. Other successful applications of this procedure can be found in the work by Armand and Vitte (1970), or Weisshaar (1970).

A slighty more complex case, for which no closed-form solution exists, is that of a vibrating simply-supported beam, first treated by Niordson (1965), and independently extended by Haug (1969) and Armand (1973) to handle minimum thickness constraint. The procedure described above has proven very powerful, both for the constrained and unconstrained case. Good estimates for the unknown boundary conditions were, in that case, obtained from the solution of the system of equations under consideration corresponding to the uniform reference case (therefore ignoring the optimality condition).

4. TWO-DIMENSIONAL STRUCTURAL ELEMENTS

The behaviour of two-dimensional structural elements such as plates and shells is described by partial differential equations in two independent variables. Optimization of such elements ot satisfy given requirements can be formulated as a problem in optimal control theory of distributed-parameter systems, extension to a spatial domain D of optimal control theory in which only one independent variable, the time t, is considered.

The domain D is the planar image of the spatial structure considered, and is referred to a set of orthonormal axes Ox, Oy.

State and design variables are defined in the same fashion as before ; the deflection (in-plane or out-of-plane) and its derivatives will play the role of state variables, whereas the thickness will in some cases represent a design variable.

For a minimum-mass problem, the functional to be minimized is the surface integral of the thickness over the domain D ; the constraints are in the form of partial differential equations to be satisfied inside the domain, together with equality and/or inequality constraints on functions of the state and/or design variables.

A problem of this type is a special case of a general Mayer-Bolza problem for multiple integrals, which received authoritative treatment from Lurie (1963) in a paper which also hints at possible applications to structural optimization.

A statement of the general optimization problem as well as a derivation of the necessary conditions will be found in Armand (1972).

Applications of the theory to structural optimization problems has been up to now very limited. The plate optimization problem, in particular, has not yet been solved in a satisfactory manner, although many researchers have considered it : reference is made to McIntosh (1974) for a comprehensive review. It shoud be mentioned at this point, however, that numerical problems are not characteristic of two-dimensional plates but also apply to axi-symmetric (one-dimensional) plates as well ; numerical difficulties encountered at the solution stage for situations involving plates, however, undoubtly correspond to problems associated with the non-existence of a global optimum.

The above observations are in accordance with the works of Olhoff (1973) and Mroz (1973), who independently pointed out that the optimization problem in the case of a vibrating plate does not seem to possess a global optimum solution, but instead an infinity of local extrema.

Further research should therefore be pursued in this direction with the tools of control theory, since it is of particular interest to investigate existence and possible uniqueness of solutions to those problems for which an analytical formulation is still possible.

Numerical methods developed in the frame of optimal control of distributed-parameter systems theory and similar to methods used to solve the TPBVP's of classical optimal control have hardly been used yet and will without doubt prove extremely helpful.

A very promising approach is that described by Pierson and Genalo (1974), who applied a gradient projection algorithm to obtain the optimal design of a rectangular panel subject of a flutter speed constraint.

6. ACKNOWLEDGEMENTS

Research described in section 3 was performed while the author was a doctoral student at Stanford University under a National Aeronautics and Space Administration Grant.

The author also wishes to thank Professor Holt Ashley and Dr. S.C. McIntosh, Jr., whose enthusiasm for structural optimization proved contagious.

7. REFERENCES

Armand, J.L. and W.J. Vitte, (1970), "Foundations of Aeroelastic Optimization and Some Applications to Continuous Systems", Report SUDAAR N° 390, Department of Aeronautics and Astronautics, Stanford University, Stanford.

Armand, J.L., (1972),"Applications of the Theory of Optimal Control of Distributed-Parameter Systems to Structural Optimization", NASA CR-2044, June 1972.

Armand J.L., (1973), "Applications of Optimal Control Theory to Structural Optimization : Analytical and Numerical Approach", Proceedings of the IUTAM Symposium on Optimization in Structural Design, Warsaw, Springer-Verlag, Berlin (to be published).

Ashley, Hi. and S.C. McIntosh, Jr., (1968), "Applications of Aeroelastic Constraints in Structural Optimization", Proceedings of the 12 th International Congress of Theoretical and Applied Mechanics, Stanford, Springer-Verlag, Berlin, pp. 100-113.

Bisplinghoff, R.L., Ashley, H. and R.L. Halfman, (1955), "Aeroelasticity", Addison-Wesley Publishing Co., Reading.

Bryson, A.E., Jr. and Y.C. Ho, (1969), "Applied Optimal Control-Optimization, Estimation and Control", Blaisdell Publishing Co., Watham.

Distefano, N., (1974), "Nonlinear Processes in Engineering", Academic Press, New-York.

Dixon, L.C.W., (1967), "Pontryagin's Maximum Principle Applied to the Profile of a Beam", Aeronautical Journal of the Royal Aeronautical Society, 71, pp. 513-515.

Dym, C.L., (1974), "On Some Recent Approaches to Structural Optimization", Journal of Sound and Vibration, 32, pp. 49-70.

Fox, R.L., (1973), "Structural and Mechanical Design Optimization", in Optimization and Design, M. Avriel, M.J. Rijckaert and D.J. Wilde, Eds., Prentice-Hall, Englewood Cliffs, pp. 119-143.

Gallagher, R.H. and O.C. Zienkiewicz, Eds., (1973), Optimum Structural Design : Theory and Applications, John Wiley and Sons, London.

Haug, E.J., Jr., (1969), "Optimal Design of Structural Elements", Lecture Notes, Department of Mechanics and Hydraulics, The University of Iowa, Ames.

Hemp, W.S., (1973), Optimum Structures, Clarendon Press, Oxford.

Lagrange, J.L., (1770-1773), "Sur la Figure des Colonnes", Miscellanea Taurinensia, t.V., in Oeuvres de Lagrange, Tome deuxième, Gauthier-Villars, Paris, 1868, pp. 125-170.

Lurie, K.A., (1963), "The Mayer-Bolza Problem for Multiple Integrals and the Optimization of the Performance of Systems with Distributed Parameters", Applied Mathematics and Mechanics (PMM), Vol. 27, N° 5, March 1963, pp. 1284-1299.

Majid, K.I., (1974), Optimum Design of Structures, Newnes-Butterworths, London.

McIntosh, S.C., Jr., (1974), "Structural Optimization via Optimal Control Techniques : A review", Structural Optimization Symposium AMD-VOL.7, The American Society of Mechanical Engineers, New-York, pp. 49-64

Moe, J. and K.M. Gisvold, (1971), "Optimization and Automated Design of Structures", Report SK/M 21, Division of Ship Structures, Technical University of Trondheim, Trondheim.

Mroz, S., (1973), "Multiparameter Optimal Design of Plates and Shells", Journal of Structural Mechanics, 3, pp. 371-392.

Niordson, F.I., (1965), "On the Optimal Design of a Vibration Beam", Quarterly of Applied Mathematics, 23, pp. 45-53.

Pierson, B.L. and L.J. Genalo, (1974), "Minimum-Weight Design of a Rectangular Flat Panel Subject to a Flutter Speed Constraint", Symposium on Optimization Problems in Engineering and Economics, Naples, Dec. 1974.

Niordson, F.I. and P. Pedersen, (1972), "A Review of Optimal Structural Design", Proceedings of the 13th International Congress of Theoretical and Applied Mechanics, Moscow, Springer-Verlag, Berlin, pp. 264-278.

Olhoff, N., (1973), "On Singularities, Local Optima and Formation of Stiffeners in Optimal Design of Plates", Proceedings of the IUTAM Symposium on Optimization in Structural Design, Warsaw, Springer-Verlag, Berlin (to be published).

Pierson, B.L., (1972), "A survey of Optimal Structural Design Under Dynamic Constraints", International Journal for Numerical Methods in Engineering, 4, pp. 491-499.

Prager, W. and J.E. Taylor, (1968), "Problems of Optimal Structural Design", Journal of Applied Mechanics, 35, pp. 102-106.

Schmit, L.A., Jr., (1971), "Structural Synthesis 1959-1969 : A Decade of Progress", in Recent Advances in Matrix Methods of Structural Analysis and Design, R.H. Gallagher, Y. Yamada and J.T. Oden, Eds., The University of Alabama Press, Huntsville, pp. 565-634.

Weisshaar, T.A., (1970), "An Application of Control Theory Methods to the Optimization of Structures Having Dynamic or Aeroelastic Constraints", Report SUDAAR N° 412, Department of Aeronautics and Astronautics, Stanford University, Stanford.

The policy iteration method for the optimal stopping of a Markov chain with an application

by

K.M. van Hee

0. Summary

In this paper we study the problem of the optimal stopping of a Markov chain with a countable state space. In each state i the controller receives a reward r(i) if he stops the process or he must pay the cost c(i) otherwise. We show that, under the condition that there exists an optimal stopping rule, the policy iteration method, introduced by Howard, produces a sequence of stopping rules for which the expected return converges to the value function. For random walks on the integers with a special reward and cost structure, we show that the policy iteration method gives the solution of a discrete two point boundary value problem with a free boundary. We give a simple algorithm for the computation of the optimal stopping rule.

1. Introduction

Consider a Markov chain $\{X_n \mid n = 0,1,2,\ldots\}$ defined on the probability space $(\Omega,F,\mathbb{P})$. The state space S is countable. We suppose that $\mathbb{P}[X_0 = i] > 0$ for all $i \in S$. Hence $\mathbb{P}_i[A]$, the conditional probability of $A \in F$ given $X_0 = i$, is defined for all $i \in S$.

On S real functions r and c are defined, where r(i) is the reward if the process is stopped in state i and c(i) is the cost if the process goes on. We consider stopping times T (for a definition see [7]). For a nonnegative function g on S we define

$$\mathbb{E}_i[g(X_T)] := \int_{\{T<\infty\}} g(X_T)d\mathbb{P}_i \ .$$

Condition A. Suppose that the reward function r satisfies

$$\mathbb{E}_i[r^+(X_T)] + \mathbb{E}_i[r^-(X_T)] < \infty$$

for all $i \in S$ and all stopping times T.

(Note that: $r^+(i) := \max\{0,r(i)\}$, $r^-(i) := -\min\{0,r(i)\}$).

Let P be the transition matrix of the Markov chain, with components $P(i,j)$ for $i,j \in S$. If the function c on S is integrable for all $\mathbb{P}_i[.]$, we define the function Pc by

$$Pc(i) := \sum_{j\in S} P(i,j)c(j) ,$$

and with induction, if $P^{n-1}c$ is integrable for all $\mathbb{P}_i[.]$

$$P^n c := P(P^{n-1}c) .$$

We call a function c on S a *charge* (see [3]) if

$$\sum_{n=0}^{\infty} P^n|c| < \infty .$$

(Note that for function v and w on S: $v \le w$ if $v(i) \le w(i)$ for all $i \in S$ and $v < w$ if $v(i) < w(i)$ for all $i \in S$. Further $|v|$ is defined by $|v|(i) := |v(i)|$.

Condition B. Either the cost function c is a charge or r and c are nonnegative, both.

Throughout this paper we shall suppose that conditions A and B hold.

We call a function w on S c-*excessive* with respect to the cost function c if

1) $$w \ge -c + Pw$$

2) $$w \ge -\sum_{n=0}^{\infty} P^n c .$$

For a stopping time T the expected return $v_T(i)$, given the starting state i, is defined by

$$v_T(i) := \mathbb{E}_i[r(X_T) - \sum_{n=0}^{T-1} c(X_n)] .$$

The existence of the expected return $v_T(i)$ is guaranteed for all T since $|\mathbb{E}_i[r(X_T)]| < \infty$ for all i and c is either a charge or a nonnegative function. Note that $v_T(i) = -\infty$ is permitted.
The value function v(i) is the supremum over all the stopping times T

$$v(i) := \sup_T v_T(i) .$$

Sometimes we need the following assumption.

Assumption C. There exists an optimal stopping time T^*, i.e. $v_{T^*}(i) = v(i)$ for all $i \in S$.

In the rest of this section we summarize some properties of stopping problems.

1.1. The value function v satisfies the functional equation

$$v(i) = \max\{r(i), -c(i) + \sum_{j \in S} P(i,j).v(j)\}$$

(see [2], [3] or [7]).

1.2. The value function v is the smallest c-excessive function dominating the reward function r (see [2] and [3]).

1.3. If an optimal stopping time exists the entrance time T_Γ in the set $\Gamma := \{i \mid r(i) = v(i)\}$ is optimal (see [2] and [6]).

1.4. If $\sup_{i \in S} |r(i)| < \infty$ and $\inf_{i \in S} c(i) > 0$ then there exists an optimal stopping time (see [2] and [7]).

2. Some preparations

A stopping rule f is a mapping from S to $\{0,1\}$ where $f(i) = 0$ means that the process is stopped in i and $f(i) = 1$ means that the process goes on in state i. The stopping rule f is equivalent with the entrance time T_f in the set $\Gamma_f := \{i \mid f(i) = 0\}$. The expected return under a stopping rule f is indicated by $v_f(i)$.
For a stopping rule f we define

2.1. $D_f := \{i \in S \mid f(i) = 1\}$, the go-ahead set.
$\Gamma_f := S \backslash D_f$, the stopping set.

2.2. P_f is the matrix with components

$$P_f(i,j) := \begin{cases} P(i,j) & \text{if } i \in D_f \\ 0 & \text{otherwise} . \end{cases}$$

2.3. d_f is a function on S with

$$d_f(i) := \begin{cases} r(i) & \text{if } i \in \Gamma_f \\ -c(i) & \text{otherwise} . \end{cases}$$

If assumption C holds, property 1.3 guarantees that the entrance time T_Γ in the set Γ is also optimal. In that case

2.4. $$v(i) = \mathbb{E}_i[r(X_{T_\Gamma}) - \sum_{n=0}^{T_\Gamma - 1} c(X_n)] .$$

According to the stopping time T_Γ we define the stopping rule f_* by

2.5. $$f_*(i) = 0 \quad \text{if and only if } i \in \Gamma .$$

Further let

$$D := S \backslash \Gamma, \ d := d_{f*} \text{ and } \widetilde{P} := P_{f*} .$$

Lemma 1. For each stopping rule f with $v_f \geq r$ we have

1) $|v_f(i)| < \infty$

2) $v_f = \sum_{n=0}^{\infty} P_f^n d_f$

3) $\lim_{n\to\infty} P_f^n |d_f| = 0$ (pointwise convergence)

4) $v_f = d_f + P_f v_f$

5) $\lim_{n\to\infty} P_f^n |v_f| = 0$ (pointwise convergence) .

Proof. If r and c are nonnegative we have

$$0 \leq r(i) \leq v_f(i) \leq \mathbb{E}_i[r(X_{T_f})] < \infty \quad \text{for all } i \in S .$$

Since

$$v_f(i) = \mathbb{E}_i[r(X_{T_f})] - \mathbb{E}_i[\sum_{n=0}^{T_f-1} c(X_n)]$$

we may conclude

$$\mathbb{E}_i[\sum_{n=0}^{T_f-1} |c(X_n)|] < \infty \quad \text{for all } i \in S .$$

Note that if c is a charge this also true. Define:

2.6. $$w_f(i) := \mathbb{E}_i[|r(X_{T_f})|] + \mathbb{E}_i[\sum_{n=0}^{T_f-1} |c(X_n)|] .$$

So we have for both cases of B

$$|v_f(i)| \leq w_f(i) < \infty . \quad \text{(statement 1)}$$

We have the following representation

$$w_f = \sum_{n=0}^{\infty} P_f^n |d_f|$$

(note that $P_f^0(i,j) = 1$ if and only if $i = j$) and in the same way, by absolute convergence,

$$v_f = \sum_{n=0}^{\infty} P_f^n d_f . \quad \text{(statement 2)}$$

Because $w_f < \infty$ we may conclude $P_f^n|d_f| \to 0$ for $n \to \infty$ (statement 3)

$$v_f = \sum_{n=0}^{\infty} P_f^n d_f = d_f + \sum_{n=1}^{\infty} P_f^n d_f .$$

Since

$$\sum_{n=1}^{\infty} P_f^n |d_f|$$

is finite we may change the summation order, hence

$$v_f = d_f + P_f \sum_{n=0}^{\infty} P_f^n d_f = d_f + P_f v_f . \quad \text{(statement 4)}$$

In the same way

$$w_f = d_f + P_f w_f .$$

By iterating this equation we get

$$w_f = \sum_{n=0}^{N} P_f^n d_f + P_f^{N+1} w_f$$

from which it follows that $P_f^n w_f$ tends to 0 if n tends to ∞. Because $|v_f| \le w_f$ we have also

$$\lim_{n\to\infty} P_f^n |v_f| = 0 . \quad \text{(statement 5)} \qquad \square$$

Corollary 1. If C hold we have from 2.4 and lemma 1 that

$$|v(i)| < \infty \text{ for all } i \in S \text{ and } \lim_{n\to\infty} \tilde{P}^n |d| = 0 .$$

Define:

$$w := \sum_{n=0}^{\infty} \tilde{P}^n |d| .$$

By lemma 1 we have

2.7. $$\lim_{n\to\infty} \tilde{P}^n w = 0 .$$

In the next section we study expressions like $P_g^k v_f$, where f and g are stopping rules. We shall give sufficient conditions in lemma 2 for the finiteness of these expressions.

Lemma 2. Let f and g are stopping rules. Suppose $v_f \ge r$. Then $P_g^k |v_f|$ is finite for $k = 1,2,3,\ldots$.

Proof. Let $T := T_f + k$. Using the same arguments as in lemma 1, we derive for c a charge:

$$\mathbb{E}_i[|r(X_T)| + \sum_{n=0}^{T-1} |c(X_n)|] < \infty .$$

Note that

$$\mathbb{E}_i[|r(X_T)| + \sum_{n=0}^{T-1} |c(X_n)|] = \sum_{n=0}^{k-1} P^n|c|(i) + P^k w_f(i)$$

(w_f is defined in 2.6).
Hence

$$|P_g^k v_f| \le P_g^k|v_f| \le P^k|v_f| \le P^k w_f < \infty .$$

Now let r and c be nonnegative.
$P^k v_f$ is defined because $v_f \ge r \ge 0$. Hence $P^k v_f \ge P^k r \ge 0$

$$0 \le P^k v_f(i) = \sum_{j\in S} P^k(i,j)\mathbb{E}_j[r(X_{T_f}) - \sum_{n=0}^{T_f-1} c(X_n)] \le$$

$$\le \sum_{j\in S} P^k(i,j)\mathbb{E}_j[r(X_{T_f})] = \mathbb{E}_i[r(X_T)] < \infty .$$

Define vectors c_f and r_f by

$$c_f(i) := c(i) \text{ if } i \in D_f, \quad r_f(i) := r(i) \text{ if } i \in \Gamma_f$$
$$:= 0 \quad \text{otherwise} \qquad := 0 \quad \text{otherwise} .$$

Note that $|d_f| = r_f + c_f$. It is easy to verify that

$$\sum_{j\in S} P^k(i,j)\mathbb{E}_j[r(X_{T_f})] = P^k \sum_{n=0}^{\infty} P_f^n r_f(i)$$

and

$$\sum_{j\in S} P^k(i,j)\mathbb{E}_j[\sum_{n=0}^{T_f-1} c(X_n)] = P^k \sum_{n=0}^{\infty} P_f^n c_f(i) .$$

Hence $P^k w_f = P^k \sum_{n=0}^{\infty} P_f^n\{r_f + c_f\} < \infty$. Reasoning like before, we see that $P_g^k|v_f| < \infty$.□

3. Policy iteration method

Let f be a stopping rule, such that $\sum_{j\in S} P(i,j)v_f(j)$ is defined. For f we define the *improved stopping rule* g by

3.1. $$g(i) := 0 \quad \text{if } r(i) \ge -c(i) + \sum_{j\in S} P(i,j)v_f(j)$$
$$:= 1 \quad \text{otherwise.}$$

Lemma 3. Let g be the improved stopping rule of f and let $v_f \geq r$. Then

1) $D_g \subset D$

2) $v_f \leq d_g + P_g v_f$.

Proof. We first prove 1).
If $g(i) = 1$ then

$$r(i) < -c(i) + \sum_{j \in S} P(i,j)v_f(j) \leq -c(i) + \sum_{j \in S} P(i,j)v(j) \leq v(i)$$

hence

$$D_g = \{i \mid g(i) = 1\} \subset \{i \mid v(i) > r(i)\} = D .$$

We proceed with 2).
Note that $P_g v_f$ is finite (by lemma 2). Let $i \in D_g$ then $g(i) = 1$, $d_g(i) = -c(i)$, $P_g(i,.) = P(i,.)$ and so

$$r(i) < -c(i) + \sum_{j \in S} P(i,j)v_f(j) = d_g(i) + \sum_{j \in S} P_g(i,j)v_f(j) .$$

Since either

$$v_f(i) = -c(i) + \sum_{j \in S} P(i,j)v_f(j)$$

or $v_f(i) = r(i)$ the statement is true for $i \in D_g$.
If $i \in \Gamma_g$ then $g(i) = 0$, $d_g(i) = r(i)$ and $P_g(i,.) = 0$ and since

$$r(i) \geq -c(i) + \sum_{j \in S} P(i,j)v_f(j)$$

it is true for $i \in \Gamma_g$. □

Lemma 4. Assume C. If g is the improved stopping rule of f and if $v_f \geq r$ then $v_g \geq v_f$.

Proof. From lemma 2 it follows that $P_g^k|v_f|$ exists and is finite for all k. By lemma 3 is $v_f \leq d_g + P_g v_f$. Hence

$$\sum_{k=0}^{N} P_g^k v_f \leq \sum_{k=0}^{N} P_g^k d_g + \sum_{k=1}^{N+1} P_g^k v_f$$

and therefore

$$v_f - P_g^{N+1} v_f \leq \sum_{k=0}^{N} P_g^k d_g .$$

We shall prove that $P_g^N v_f^+ \to 0$ for $N \to \infty$. Consider first the case that $r \geq 0$ and $c \geq 0$. Since $0 \leq r \leq v_f \leq v$ and $D_g \subset D$

$$0 \le P_g^N v_f \le P_g^N v \le \tilde{P}^N v$$

by corollary 1 $\tilde{P}^N v \to$ for $N \to \infty$.

Suppose now that c is a charge:

$$v_f^+ \le v^+ \le w$$

(w is defined in corollary 1) hence

$$P_g^N v_f^+ \le P_g^N w \le \tilde{P}^N w \ .$$

By 2.7

$$\tilde{P}^N w \to 0 \quad \text{for } N \to \infty \ .$$

Therefore

$$v_f \le \sum_{k=0}^{\infty} P_g^k d_g = v_g \ . \qquad \square$$

We define a sequence of stopping rules $\{f_0, f_1, f_2, \ldots\}$ by

3.2. $f_0(i)$ is a stopping rule with $v_{f_0} \ge r$ (for example $f_0(i) = 0$ for all $i \in S$)

f_n is the improved stopping rule of f_{n-1}, $n \ge 1$ (see 3.1).

The method of approximating the optimal stopping rule and its expected return by the sequence 3.2 is called the *policy iteration method*. This method was introduced by Howard [4] for decision processes with a finite state space and discounted rewards. In theorem 1 some properties of the sequence $\{f_0, f_1, f_2, \ldots\}$ are derived. In theorem 2 we study the convergence of v_{f_n} to v.

Most of Howards results carry over to our situation. Call

1) 1) $v_n := v_{f_n}$, 2) $d_n := d_{f_n}$,

3) $D_n := D_{f_n}$, 4) $\Gamma_n := \Gamma_{f_n}$.

Theorem 1. Assume C. The following assertions hold

1) $f_n(i)$ and $v_n(i)$ are nondecreasing in n

2) if $f_n(i_0) < f_{n+1}(i_0)$ then $v_n(i_0) < v_{n+1}(i_0)$.

Proof. It follows from lemma 4 that $v_{n+1} \ge v_n$ for $n \ge 0$, since $v_0 \ge r$. If $f_n(i) = 1$ then

$$r(i) < -c(i) + \sum_{j\in S} P(i,j)v_{n-1}(j) \le -c(i) + \sum_{j\in S} P(i,j)v_n(j), \quad \text{for } n \ge 1$$

hence $f_{n+1}(i) = 1$, which proves assertion 1. Suppose $f_n(i_0) = 0$ and $f_{n+1}(i_0) = 1$, then

$$v_n(i_0) = r(i_0) < -c(i_0) + \sum_{j\in S} P(i,j)v_n(j) \le$$

$$\le -c(i_0) + \sum_{j\in S} P(i_0,j)v_{n+1}(j) = v_{n+1}(i_0) . \qquad \square$$

<u>Theorem 2</u>. Assume C.

1) If, either $v_{n_0} \ge -\sum_{k=0}^{\infty} P^k c$ or $v_{n_0} \ge 0$ for some n_0, then $\lim_{n\to\infty} v_n = v$.
2) If, in addition to 1, $f_n = f_{n+1}$, for some $n \ge n_0$ then v_n is optimal.

<u>Proof</u>. Since $D_n \subset D$ for all n (lemma 3) and since $f_n(i)$ is nondecreasing in n (theorem 1) there exists a set $E \subset S$ such that

$$\lim_{n\to\infty} D_n = E \subset D .$$

And, in the same way, since $v_n(i) \le v(i)$ for all n and since $v_n(i)$ is nondecreasing in n, there exists a function z such that

$$z(i) = \lim_{n\to\infty} v_n(i) .$$

Fix some $i \in E$. For all n sufficiently large is $i \in D_n$ and so:

$$r(i) \le v_n(i) = -c(i) + \sum_{j\in S} P(i,j)v_n(j) \le -c(i) + \sum_{j\in S} P(i,j)v(j) = v(i) .$$

Since $v_n(i) \uparrow z(i)$ we have by monotone convergence

$$-c(i) + \sum_{j\in S} P(i,j)\{v_n(j) - r(j)\} \uparrow -c(i) + \sum_{j\in S} P(i,j)\{z(j) - r(j)\} ,$$

hence

$$z(i) = -c(i) + \sum_{j\in S} P(i,j)z(j) \le v(i) .$$

Fix some $i \in S\backslash E$. For all n it holds that $i \in \Gamma_n$ hence

$$v_n(i) = r(i) \ge -c(i) + \sum_{j\in S} P(i,j)v_n(j)$$

and therefore (again by monotone convergence)

$$z(i) = r(i) \ge -c(i) + \sum_{j\in S} P(i,j)z(j) .$$

So z satisfies the functional equation:

$$z(i) = \max\{r(i), -c(i) - \sum_{j\in S} P(i,j)z(j)\} .$$

Now, suppose $v_{n_0} \geq -\sum_{n=0}^{\infty} P^n c$. Then $z \geq -\sum_{n=0}^{\infty} P^n c$ and since z satisfies the functional equation, z is a c-excessive function dominating r. Because v is the smallest function with this property it must hold that $v = z$. If $v_{n_0} \geq 0$ it must hold that $z \geq 0$ and $v \geq 0$. We now prove that $v = z$ on Γ. Let $i \in \Gamma$:

$$0 \leq v(i) - z(i) \leq r(i) - r(i) = 0 .$$

Let, now $i \in D$:

$$0 \leq v(i) - z(i) \leq \sum_{j\in S} P(i,j)\{v(j) - z(j)\} .$$

Hence $0 \leq v - z \leq \tilde{P}(v - z)$.
Iterating this inequality gives

$$0 \leq v - z \leq \tilde{P}^n(v - z) \leq \tilde{P}^n v \to 0 \quad \text{for } n \to \infty$$

which proves $v = z$. The first assertion is proved.
Suppose $f_n = f_{n+1}$ for some $n \geq n_0$. Then $v_n = v_{n+1}$ and therefore $f_{n+2} = f_{n+1}$. By induction it follows that $z = v_n$ which proves the theorem. □

Lemma 5. Let c be a charge. Let f be the stopping rule defined by $f(i) = 1$ for all $i \in S$ and let g be the improved stopping rule, then

$$v_g \geq v_f \quad \text{and} \quad v_g \geq r .$$

If $v_g = v_f$ then f is optimal.

Proof. Since $v_f = -\sum_{n=0}^{\infty} P^n c$ it holds that Pv_f and $P_g^k v_f$ are finite. Following exactly the proof of lemma 3 we have $v_f \leq d_g + P_g v_f$ and from the proof of lemma 4 it follows, since $P_g^k v_f$ is finite, that

$$v_f - P_g^{n+1} v_f \leq \sum_{k=0}^{n} P_g^k d_g .$$

Note that

$$P_g^n |v_f| \leq P^n |v_f| = P_f^n |v_f| .$$

Since c is a charge:

$$w_f := \sum_{n=0}^{\infty} P^n |c| < \infty .$$

Hence $w_f = |c| + Pw_f$ and therefore $P_f^n w_f$ tends to 0 if n tends to ∞. Because $w_f \geq |v_f|$ we may conclude

$$\lim_{n\to\infty} P_g^n |v_f| = 0 \ .$$

Hence

$$v_f \le \sum_{k=0}^{\infty} P_g^k d_g = v_g \ .$$

If $g(i) = 0$ then $v_g(i) = r(i)$ and if $g(i) = 1$ then

$$r(i) < -c(i) + \sum_{j\in S} P(i,j)v_f(j) = v_f(i) \le v_g(i) \ .$$

Hence $v_g \ge r$.
Now, suppose $v_g = v_f$, then

$$r \le v_f = -c + P_f v_f = -c + Pv_f$$

hence v_f is c-excessive and dominates r. Because $v_f \le v$ and the fact that v is the least function with this property, we have $v = v_f$. □

Corollary 2.

1) If r is nonnegative, we have for $f_0 \equiv 0$ $v_{f_0} \ge r \ge 0$, hence the sequence v_n converges to v.
2) If c is a charge we may start with $f_{-1}(i) := 1$ for all $i \in S$ and try to improve this stopping rule by f_0. If no improvement is possible (i.e. $v_{f_0} = v_{f_{-1}}$) we have already the optimal stopping rule. Otherwise f_0 satisfies

   a) $v_0 = v_{f_0} \ge r$

   b) $v_0 \ge -\sum_{n=0}^{\infty} P^n c$

   hence v_n converges to v.

Counterexamples.

1) There exists a stopping problem satisfying assumptions A, B and C where the policy iteration method does not converge to the optimal stopping rule. Let $S = \{1,2\}$; $r(1) = r(2) = -1$, $c(1) = c(2) = 0$ and $P(1,1) = \alpha = 1 - P(1,2)$, $P(2,2) = \beta = 1 - P(2,1)$. The optimal stopping rule is $f(1) = f(2) = 1$ and $v(1) = v(2) = 0$. The cost function is a charge and $\mathbb{E}_i[|r(X_T)|] \le 1$. Note that $r(1) = \alpha r(1) + (1-\alpha)r(2)$ and $r(2) = \beta r(2) + (1-\beta)r(1)$ so that $r \ge c + Pr$ hence $f_n = f_0 \equiv 0$.
2) There exists a stopping problem satisfying assumptions A and B where the improved policy of f_0 is not at least as good as f_1. Let

$$S = \{0,1,2,3,\ldots\} \cup \{x\}, \qquad 1 > \varepsilon > 0 \ .$$

   For $i = 0,1,2,3,\ldots$:

$$P(i,i+1) = 1 - \varepsilon,\ P(i,x) = \varepsilon,\ r(i) = \frac{1}{(1-\varepsilon)^i},\ c(i) = 0\ .$$

Further:

$$P(x,x) = 1,\ r(x) = 1,\ c(x) = 1\ .$$

Note that r and c are nonnegative both (condition A). We shall examine the stopping time $T_n \equiv n$:

$$v_{T_n}(i) = (1-\varepsilon)^n \frac{1}{(1-\varepsilon)^{i+n}} + \{1 - (1-\varepsilon)^n\}\ .$$

Hence

$$w(i) := \sup_n v_{T_n}(i) = \frac{1}{(1-\varepsilon)^i} + 1\ .$$

This function w satisfies the functional equation

$$w(i) = \max\{r(i), -c(i) + \sum_{j\in S} P(i,j)w(j)\}$$

and $w \geq -\sum_{n=0}^{\infty} P^n c$, hence $w = v$ so that $v(i) < \infty$ from which it follows that $\mathbb{E}_i[|r(X_T)|] < \infty$ for all i and all T (condition B).
For $i = 0,1,2,3,\ldots$:

$$r(i) = \frac{1}{(1-\varepsilon)^i} < (1-\varepsilon)\frac{1}{(1-\varepsilon)^{i+1}} + \varepsilon = -c(i) + \sum_{j\in S} P(i,j)r(j)$$

and $r(x) = 1 > -c(x) + r(x)$.
Hence $f_1(i) = 1$ for $i \in \{0,1,2,3,\ldots\}$ and $f_1(x) = 0$ so that $v_1(i) = 1$ for all i, but $v_0(i) = \frac{1}{(1-\varepsilon)^i} > 1$ for $i = 1,2,3,\ldots$.

4. An application

We shall study in this section the optimal stopping of a random walk on the integers with a special cost and reward structure, to illustrate the computational aspects of the policy iteration method. For simplicity we shall not formulate the results as general as possible.

Definition of the decision process

Consider a random walk on the set of integers (Z). Let the transition matrix P be defined by

4.1. $\qquad P(i,i+1) := p_i,\ P(i,i) := s_i,\ P(i,i-1) = q_i$

with $p_i, q_i > 0$, $s_i \geq 0$ and $p_i + q_i + s_i = 1$. The reward function

4.2. $\qquad 0 \leq r(i) \leq M, \quad i \in Z\ .$

The cost function

4.3. $c(i) \geq \delta > 0, \quad i \in Z$.

Further we assume the existence of integers d, e, $d \leq e$, such that:

4.4. $r(i) < -c(i) + p_i r(i+1) + q_i r(i-1) + s_i r(i)$

if and only if $d \leq i \leq e$. Call $H := \{i \in Z \mid d \leq i \leq e\}$.

Assumption 4.4 says that for $i \in Z \backslash H$ immediately stopping is more profitable than making one more transition. In statistical sequential analysis there are examples of random walks where this assumption is fulfilled in a natural way (compare [5]). In lemma 6 we collect some properties of this process.

Lemma 6. For the sequence of stopping rules $f_0, f_1, f_2, \ldots$ defined in 3.2 with $f_0(i) = 0$ for all $i \in Z$ it holds that

1) there exist numbers $k_n, \ell_n \in Z$ such that

$$D_n = \{i \in Z \mid k_n \leq i \leq \ell_n\}, \quad n = 0,1,2,\ldots .$$

2) $k_n \geq k_{n+1} \geq k_n - 1$ and $\ell_n \leq \ell_{n+1} \leq \ell_n + 1$.

3) for some n f_n is optimal.

Proof. Since $0 \leq r(i) \leq M$ and $c \geq 0$ A and B are satisfied. By 1.4 we know that the entrance time in Γ is optimal, hence the assumption C is fulfilled. By theorem 1 we have $D_n \subset D_{n+1}$ for $n = 0,1,2,3,\ldots$ and by theorem 2 we have $\lim_{n\to\infty} v_n(i) = v(i)$. We shall prove 1 and 2 with induction.

D_0 is empty. It is easy to verify that $f_1(i) = 1$ if and only if $i \in H$, hence $k_1 = d$ and $\ell_1 = e$. Suppose 1 hold for $n = m$. For $i < k_m - 1$ and $i > \ell_m + 1$ it holds that $f_{m+1}(i) = 0$ because $v_m(i) = r(i)$ and $i \in Z\backslash H$. Therefore it can happen only in the points $i = k_m - 1$ and $i = \ell_m + 1$ that $f_{m+1}(i) > f_m(i)$. Since $D_m \subset D_{m+1}$ 1 and 2 are proved. Now the last assertion.

Note that $0 \leq r(i) \leq M$ and $c(i) \geq \delta > 0$ for all $i \in Z$. Choose $1 > \varepsilon > 0$ and a natural number k such that $(1-\varepsilon)k > \frac{M}{\delta}$. Let f be the optimal stopping rule. We shall prove $\mathbb{P}_i[T_f \leq k] \geq \varepsilon$. Suppose the contrary, i.e. let $\mathbb{P}_i[T_f \leq k] < \varepsilon$. Then

$$v_f(i) \leq M - \delta\mathbb{E}_i[T_f] \leq M - \delta(1-\varepsilon)k < 0$$

which is a contradiction.

Hence for all $i \in Z$ Γ must be reachable in at most k steps, so that $D \subset \{i \mid d-k \leq i \leq e+k\}$. Since $D_{n-1} \subset D_n \subset D$ and because D_{n-1} is a proper subset of D_n if $f_{n-1}(i) \neq f_n(i)$ for at least one i we may conclude that $f_{n-1} = f_n$ for some n.

□

Computational aspects

In our case v is the smallest solution of

$$v(i) = \max\{r(i), -c(i) + p_i v(i+1) + s_i v(i) + q_i v(i-1)\} .$$

Because we know the structure of D we may say v is the smallest function x which has the following properties.

For some $k \le d$ and some $\ell \ge e$, $i,k,\ell \in Z$:

1) $x(i) = -c(i) + p_i x(i+1) + s_i x(i) + q_i x(i-1), \quad k \le i \le \ell$

2) $x(i) = r(i), \quad i > \ell,\ i < k$

3) $r(k-1) \ge -c(k-1) + p_{k-1}x(k) + s_{k-1}r(k-1) + q_{k-1}r(k-2)$

$$r(\ell+1) \ge -c(\ell+1) + p_{\ell+1}r(\ell+2) + s_{\ell+1}r(\ell+1) + q_{\ell+1}x(\ell) .$$

This is a *two point boundary value problem* with a *free boundary*. We shall show that for fixed k and ℓ the function x is completely determined by 1 and 2.

Define, for function on Z, the difference operator Δ as usual by

4.5. $\Delta x(i) := x(i+1) - x(i)$.

Consider the difference equation, derivated from 1,

4.6. $p_i \Delta x(i) - q_i \Delta x(i-1) = c(i)$.

Call:

$$z_i := \Delta x(i), \quad a_i := \frac{q_i}{p_i} \quad \text{and} \quad b_i := \frac{c(i)}{p_i} .$$

Hence 4.6 becomes

$$z_i - a_i z_{i-1} = b_i .$$

With induction on m it is easy to verify that for $k \le m \le \ell$

4.7. $z_m = z_{k-1} \prod_{i=k}^{m} a_i + \sum_{i=k}^{m} \{b_i \prod_{j=i+1}^{m} a_j\}$

(an empty product has the value 1, an empty sum the value 0).

Because $x(\ell+1) = r(\ell+1)$ and $x(k-1) = r(k-1)$ it holds that

$$r(\ell+1) - r(k-1) = \sum_{m=k-1}^{\ell} z_m$$

hence

$$4.8. \qquad z_{k-1} = \frac{r(\ell+1) - r(\ell-1) - \sum_{m=k-1}^{\ell} \sum_{i=k}^{m} \{b_i \prod_{j=i+1}^{m} a_j\}}{\sum_{m=k-1}^{\ell} \prod_{i=k}^{m} a_i} .$$

From 4.7 and 4.8 one can compute $z_k, z_{k+1}, \ldots, z$ and even so $x(k), x(k+1), \ldots, x(\ell)$, which shows that the function x is completely determined.

The boundary conditions 3 can be formulated as follows

$$4.9. \qquad \begin{aligned} & z_{k-1} - a_{k-1}\Delta r(k-2) \le b_{k-1} \\ & \Delta r(\ell+1) - a_{\ell+1} z_\ell \le b_{\ell+1} , \end{aligned}$$

which shows that we only have to compute the differences z_k to check 3 and not the function x itself.

It is easy to verify that the sums and products in 4.7 and 4.8 can be computed recursively. We shall formulate an algorithm to compute the optimal stopping rule and the value function v.

Algorithm

1. $k := d$, $\ell := e$,
2. compute z_{k-1} (by 4.8) and z_ℓ (by 4.7), set $i := 0$,
3. if $z_{k-1} - a_{k-1} \cdot \Sigma r(k-2) > b_{k-1}$ then $k := k - 1$ and $i := 1$,
4. if $\Delta r(\ell+1) - a_{\ell+1} z_\ell > b_{\ell+1}$ then $\ell := \ell + 1$ and $i := 1$,
5. if $i = 0$ then goto 6, else goto 2,
6. D is the set $\{i \in Z \mid k \le i \le \ell\}$ and v can be compute by 4.7.

Acknowledgement

The author whishes to express his gratitude to Dr. A. Hordijk for valuable remarks.

Literature

[1] DYNKIN, E.B., JUSCHKEWITSCH, A.A.; Sätze und Aufgaben über Markoffsche Prozesse. Springer-Verslag (1969).

[2] HORDIJK, A., POTHARST, R., RUNNENBURG, J. Th.; Optimaal stoppen van Markov ketens. MC-syllabus19 (1973).

[3] HORDIJK, A.; Dynamic programming and Markov potential theory. MC tract (1974).

[4] HOWARD, R.A.; Dynamic programming and Markov processes. Technology Press, Cambridge Massachusetts (1960).

[5] VAN HEE, K.M., HORDIJK, A.; A sequential sampling problem solved by optimal stopping. MC-rapport SW 25/73 (1973).

[6] VAN HEE, K.M.; Note on memoryless stopping rules. COSOR-notitie R-73-12, T.H. Eindhoven (1974).

[7] ROSS, S.; Applied probability models with optimization applications. Holden-Day (1970).

ALGORITHMES POUR UN PROBLEME INVERSE DISCRET DE STURM-LIOUVILLE.

P. Morel
Université de Bordeaux I
33405 Talence / France

I - Introduction.

On considère sur $[o,\pi]$ l'opérateur différentiel de Sturm-Liouville $L(y) = -y'' + q(x)y = \lambda y$ avec les conditions aux limites $\alpha_1 y(o) + \beta_1 y'(o) = 0$, $\alpha_2 y(\pi) + \beta_2 y'(\pi) = 0$. On appelle problème inverse de valeurs propres la recherche de la fonction q connaissant le spectre de l'opérateur.

Nous voulons obtenir numériquement la fonction q ; c'est donc la version discrétisée de ce dernier problème qui nous intéresse. Après avoir introduit un pas de discrétisation, un maillage ,on obtient un problème matriciel qui légèrement généralisé s'énonce de la manière suivante. On appelle problème (P_s) la recherche d'une matrice n×n diagonale réelle $X = (x_i \delta_{ij}) \in \mathcal{M}_{nn}$ (IR) telle que A étant une matrice n×n donnée symétrique de $\mathcal{M}_{nn}$ (IR) le spectre: Sp (A+X) de A+X soit égal au spectre de la matrice fixée $S = (s_i \delta_{ij}) \in \mathcal{M}_{nn}$ (IR).

On peut faire deux hypothèses ne diminuant en rien la généralité du problème traité. On peut supposer que la diagonale de la matrice A est nulle. En effet X est une solution pour A et S fixées ie. Sp(A+X)=SpS si et seulement si X-Diag A est une solution pour A et S données.
D'autre part soit $t \in \mathbb{R}$ tel que pour $i=1, 2, \ldots, n$ $s_i > t > 0$
Si alors X est telle que :

$$Sp(A+X) = Sp((s_i+t)\,\delta_{ij})$$

alors :

$$X-(t\delta_{ij}) \text{ vérifie } Sp(A+X-(t\delta_{ij})) = Sp(S) ;$$

en d'autres termes on peut supposer que le spectre visé est strictement positif. Nous incluons ces deux hypothèses, non restrictives, dans la formulation du problème (P_S).

2 - Des conditions nécessaires et des conditions suffisantes.

L'étude en dimension 2×2 montre immédiatement que le problème (Ps) ne possède pas toujours de solution. K. Hadeler [2], F. Laborde [4], P. Morel [6, 7] ont donné des conditions nécessaires de plus en plus précises pour que le problème

(Ps) possède des solutions. On montre dans Morel [7] que nécessairement $S = (s_i \delta_{ij})$ doit vérifier :

$$\sum_{i=1}^{n} s_i^2 - \frac{1}{n}(\Sigma s_i)^2 \geqslant \sum_{i,j} a_{ij} a_{ji}$$

ce qui d'une manière équivalente mettant en évidence une nécessaire séparation du spectre visé s'écrit :

$$2n \sum_{i,j} a_{ij} a_{ji} \leqslant \sum_{i,j} (s_i - s_j)^2$$

Dans le cas où A est symétrique $\sum_{i,j} a_{ij} a_{ji} = \sum_{i,j} a_{ij}^2 = \mathrm{tr}\, A^2 = \|A\|_S^2$; $\|A\|_s$ désignant la norme de Schur de A . Sous cette hypothèse cela permet d'affirmer que : l'application $x \to \mu(A+(x\delta_{ij}))$, où $\mu(A+x_i\delta_{ij})$ désigne le vecteur dont les composantes sont les valeurs propres de $A+(x_i\delta_{ij})$ numérotées dans l'ordre non croissant n'est surjective que si A est nulle.

Dans [6], on obtient la condition nécessaire suivante, qui est strictement plus précise que les précédentes

$$(\overset{<}{\mu}(S) | \mu(A)) \leqslant \sum_{i,j} a_{ij}^2 \leqslant (\mu(S) | \mu(A))$$

où $(\overset{<}{\mu}(S) | (\mu(A))$ désigne le produit scalaire entre les vecteurs $\overset{<}{\mu}(S)$ et $\mu(A)$ qui désignent respectivement les valeurs propres de S dans l'ordre croissant et les valeurs propres de A dans l'ordre décroissant.

Ces conditions nécessaires, ne sont suffisantes que si la dimension est inférieure ou égale à 2. Dans Morel [7] on recherche systèmatiquement une localisation de la solution ; cela pour choisir le plus correctement possible une approximation initiale lors de la mise en oeuvre d'un algorithme. Dans cet ordre d'idée citons:

$$\sum_{i=1}^{n} x_i^2 = \sum_{i=1}^{n} s_i^2 - \sum_{i\ j=1}^{n} a_{ij}^2$$

Toutes les conditions suffisantes connues expriment que le spectre visé est suffisamment séparé. Plus précisément que $d(s) = \min_{i \neq j} |s_i - s_j| > f(\lambda_1(A), \ldots, \lambda_n(A))$ où f est une fonction des valeurs propres de A.

Donnons celle de Morel [7] :

$$d(s) = \min_{i \neq j} |s_i - s_j| \geqslant 2^{(1-1/p)} \left(\sum_{i=j}^{n} |\lambda_i(A)|^p\right)^{1/p}$$

Elle recouvre celle de Laborde ($p=+\infty$) et rappelle sans être pourtant identique celle de de Oliviera (p=1) et, la première connue, celle de K. Hadeler (p=2).

Notons que l'on connait également des conditions suffisantes pour un problème analogue à (P_S) mais dans lequel on ne suppose pas que A soit symétrique cf [2, 4, 5, 6, 7].

On peut adapter une démonstration de de Oliveira [1] et de Friedland [9] pour obtenir le résultat d'existence, et en quelque sorte d'unicité suivant.

PROPOSITION 1. - Si $d(s) = \min_{i \neq j} |s_i - s_j| > \ell(A) = \max_{i \neq j} |\lambda_i(A) - \lambda_j(A)|$
alors le problème (P_S) possède
i) $n!$ solutions
ii) une et une seule solution $X = (x_i \delta_{ij})$ vérifiant
$x_1 \geq x_2 \geq \ldots\ldots \geq x_n > 0$

La démonstration est basée sur le théorème du point fixe de Brouwer.

3 - Un algorithme du type des approximations successives et un algorithme du type Newton.

Pour $n \geq 2$ toutes les conditions suffisantes assurant l'existence proviennent de l'application du théorème de Brouwer ;il est donc naturel de rechercher sous quelles conditions l'algorithme des approximations successives sera convergent. Pour montrer qu'un opérateur est une contraction il est classique d'étudier sa dérivée, ce qui entraîne à regarder la dérivabilité de $x \to \mu(A+X)$.

Si x est tel que A+X n'a que des valeurs propres simples alors en ce point $\mu(A+X)$ est de classe C^{∞}, d'après Lancaster [13] et Kato [12]. Le fait d'imposer que A+X n'ait que des valeurs propres simples est assez restrictif mais l'on peut donner des exemples où une telle situation a lieu.

Supposons que A soit symétrique, tridiagonale et que $a_{i-1,i} \neq 0$ $i=2,\ldots,n$ alors d'après Wilkinson [11] page 300,on sait que pour tout $x \in \mathbb{R}^n$ $A+(x_i\delta_{ij})$ n'aura que des valeurs propres simples. Dans ce cas $x \to \mu(A+(x_i\delta_{ij}))$ appartient à $C^{\infty}(\mathbb{R}^n)$. Notons que ce cas correspond exactement à la discrétisation de l'opérateur de Sturm-Liouville.

Supposons que $x \to \mu(A+x_i\delta_{ij})$ soit dans $C^1(\Omega)$. Notons $J(x)$ la valeur en x de la matrice jacobienne de $x \to (\mu(A+X))$; d'après Lancaster [13] on obtient :

$$J(x) = \left[\frac{\partial \mu_i(A+X)}{\partial x_j} \right]_{ij} = (u^2_{ji})_{ij}$$

où $U=(u_{ij})_{ij}$ est la matrice orthogonale qui diagonalise la matrice symétrique $A+X$, ie $A+X=U.\ \mathrm{Diag}\,\mu(A+X).\ U^T$. Il est important de remarquer que $J(x)$ est une matrice doublement stochastique.

Pour les propriétés des matrices doublement stochastiques, on pourra consulter Horn [14], Hardy-Littlewood-Polya [15].

ALG 1 : Un algorithme du type approximations successives : Alg 1

C'est Hadeler [2] qui a obtenu les résultats les plus précis sur l'algorithme des approximations successives :

Alg 1 : $x^{n+1} = x^n + \mu(S) - \mu\,(A+(x_i^n\,\delta_{ij})) \qquad n \geqslant 0$

Reformulons son résultat en introduisant un coefficient de relaxation ω qui assure un meilleur comportement numérique.

PROPOSITION 2. - Soit A appartenant à $\mathfrak{M}_{nn}(\mathbb{R})$, symétrique à diagonale nulle.

Si $\min_{i \neq j} |s_i - s_j| \geqslant 4 \max_i \sqrt{\sum_j a_{ij}^2}$ alors quelque soit $\omega \in]0,1]$ l'application $T : x \to x + \omega(\mu(S) - \mu(A+X))$ est k-lipschitzienne de constante $k \leqslant 13/18$ de la boule $B(s\ d(s)/12)$ dans elle même.

Laborde [4] a démontré également que sous l'hypothèse $\min_{i \neq j} |s_i - s_j| > 2\rho(A)$ $\rho(A)$ rayon spectral de A la solution était un point attractif (cf Ortega [16] page 383) pour les approximations successives.

De fait les conditions de Hadeler, aussi bien que celles de Laborde impliquent que sur la solution $\bar{x}$ le jacobien $J(\bar{x})$ est inversible. Cela donne en quelque sorte la limite de leur résultat car il est facile de construire des exemples pour lesquels une solution existe mais dont le jacobien en ce point n'est pas inversible.

Nous n'avons pas réussi à construire d'exemple pour lequel à la fois

$d(s) > \ell(A)$ et $J(\bar{x})$ non inversible ; mais cette conjecture semble plausible.

L'avantage majeur de cet algorithme est le fait qu'il n'utilise pas les vecteurs propres ; l'unique opération couteuse est l'extraction des valeurs propres de $A+X^n$ ce que l'on réalise par une méthode du type Q. R. avec shift.

Un autre avantage est sa tendance à conserver l'invariant important pour le problème, qu'est la trace. Appelons défaut de trace à l'itération k le nombre :

$$e_k = \sum_{i=1}^{n} x_i^k - \sum_{i=1}^{n} s_i$$

PROPOSITION 3. - Pour tout ω de $]0, 1]$ considérons l'algorithme

$$x^{n+1} = x^n + \omega(s - \mu(A + (x_i^n \delta_{ij}))) \qquad n \geq 0$$

alors si 1/ $e_0 = 0 \Rightarrow \forall k \geq 0 \quad e_k = 0$

2/ $e_0 \neq 0 \Rightarrow \lim_{k\to\infty} e_k = 0$

ALG 2 : Un algorithme du type Newton : Alg 2.

Le problème à résoudre étant essentiellement celui de la résolution d'un système non linéaire, il est naturel d'envisager l'algorithme de Newton.

PROPOSITION 4. - Supposons :

1/ qu'il existe une solution $\bar{x}$ ie $\mu(A+(\bar{x}_i \delta_{ij})) = \mu(S)$

2/ $f : x \to \mu(A+X)$ soit de classe C^1 dans un voisinage Ω de $\bar{x}$

3/ $\forall$ i=1, 2, ... n $\mathcal{R}e\,\rho_i > 0$ ou ρ_i valeurs propres de $J(\bar{x})$

alors $\forall \lambda > 0$ et $\forall \omega \in]0\ 1]$ l'algorithme de Newton

$$\text{alg 2 :} \quad x^{n+1} = x^n - \omega(J(x_n)+\lambda I)^{-1}(\mu(A+(x_i^n \delta_{ij})) - \mu(S))$$

possède $\bar{x}$ comme point d'attraction.

Remarquons que s'il existe une **solution** $\bar{x}$ pour un spectre visé $Sp(s_i \delta_{ij})$ qui est bien séparé alors nécessairement $f : x \to \mu(A+X)$ est de classe C^1 dans un voisinage Ω de $\bar{x}$; la seule hypothèse restante est la 3°. $J(\bar{x}) = (u_{ji}^2)_{ij}$ est une matrice doublement stochastique, ce qui implique d'après le théorème de Gerchgorin que 1 est toujours la valeur de plus grand module d'une part, d'autre part que toutes les autres valeurs propres sont contenues dans la réunion pour $i=1,2,\ldots\ldots,n$ des disques centrés en u_{ii}^2 et de rayon $1-u_{ii}^2$. Tous ces disques seront contenus dans le 1/2 plan $\mathcal{R}e\, z > 0$ dès que $\forall\, i=1\ 2,\ldots\ldots,n \quad u_{ii}^2 > \frac{1}{2}$. Or d'après Laborde [4] cela est réalisé si $\min |s_i - s_j| > 2\rho(A)$. D'où le corollaire

COROLLAIRE 1. - Si A est une matrice symétrique à diagonale nulle et si $\min |s_i - s_j| > 2\rho(A)$ alors

1/ il existe $\bar{x}$ solution de (Ps)

2/ $\bar{x}$ est un point attractif pour l'algorithme de Newton :

$$\text{Alg } 2 : \quad x^{n+1} = x^n - \omega(J(x^n)+\lambda I)^{-1} - (\mu(A+(x_i^n\,\delta_{ij}))-\mu(S)).$$

Une autre façon d'obtenir que tout les disques de centre u_{ii}^2 et de rayon $1-u_{ii}^2$ soient dans $\mathcal{R}e\, z > 0$ et d'imposer que $1-u_{ii}^2 = \sum_{\substack{j=1 \\ j\neq 1}}^{n} u_{ij}^2 < \frac{1}{2}$, puisqu'ils passent tous par le point 1. En adaptant une partie de démonstration de Hadeler [2] on obtient

COROLLAIRE 2. - Si A est symétrique à diagonale nulle et si

$$d(s) = \min_{i\neq j} |s_i - s_j| \geqslant 2\sqrt{3}\ \max_i \sqrt{\sum_j a_{ij}^2} \quad \text{alors}$$

1/ il existe $\bar{x}$ solution de (Ps)

2/ $\bar{x}$ est un point attractif pour l'algorithme de Newton Alg 2.

On peut résumer ces deux corollaires en disant que les conditions qui assurent la convergence des approximations successives, suffisent pour entraîner la convergence de la méthode de Newton.

L'algorithme de Newton nécessite à chaque étape la connaissance de $J(x^n)$ c'est à dire de toutes les valeurs propres et de tous les vecteurs propres de $A+(x_i^n\,\delta_{ij})$. C'est un accroissement de la masse des calculs pour chaque itération, de fait lors des essais numériques nous nous sommes bornés à des matrices tridiagonales symétriques et nous avons employé l'algorithme du type Q.R nommé tq 12 dans Wilkinson-Reinsch [17]. Pour contre partie nous obtenons une convergence très rapide, et le fait assez surprenant que pour des approximations initiales qui sont en normes plus éloignées de la solution, que celles nécessaires à la convergence des approximations successives, nous ayons encore convergence. Ce bon comportement numérique est peut être dû au fait que l'algorithme conserve la trace, ou réduit le défaut de trace.

En effet, on a la :

PROPOSITION 5. - Pour $\omega \in]0\ 2[$ et $\lambda > 0$ considérons l'algorithme

$$x^{n+1} = x^n - \omega(J(x^n) + \lambda I)^{-1}(\mu(A+X^n) - \mu(S))$$

alors

$$e_0 = 0 \Rightarrow \forall k\ e_k = 0$$

$$e_0 \neq 0 \Rightarrow \lim_{k \to \infty} e_k = 0$$

Dans la démonstration on utilise le fait que $J(x^n)$ est une matrice doublement stochastique.

4 - Algorithmes de minimisation.

Dès que l'on sait calculer la dérivée de $x \to \mu(A+(x_i \delta_{ij}))$ il est naturel pour approximer la solution de l'équation $\mu(A+x_i\delta_{ij}) = \mu(S)$ de songer à minimiser $f(x) = \frac{1}{2} \| \mu(A+x_i\delta_{ij}) - \mu(S) \|_2^2$.

Nous ferons l'hypothèse que A est une matrice tridiagonale à diagonale nulle telle que de plus $a_{i,i-1} \neq 0$ $i = 2\ 3 \ldots\ldots, n$; cela pour assurer la dérivabilité de $x \to \mu(A+x_i\delta_{ij})$ en tout $x \in \mathbb{R}^n$.

La fonction $f : x \to f(x) = \frac{1}{2} \|\mu(A+x_i\delta_{ij}) - \mu(S)\|_2^2$ n'est pas convexe, mais elle possède de bonnes propriétés vis à vis d'une méthode de gradient. Par construction f est bornée inférieurement par zéro et il résulte d'un calcul facile que son gradient $\nabla f(x)$ en x vaut :

$$\nabla f(x) = J(x)^T \lfloor \mu(A+x_i\delta_{ij}) - \mu(S) \rfloor$$

où :

$$J(x) = (u_{ji}^2)_{ij}$$

$U = (u_{ij})$ étant la matrice orthogonale qui diagonalise $A+(x_i\delta_{ij})$.

Notons également que : $\lim_{\|x\| \to +\infty} f(x) = +\infty$

Appelons Alg 3 l'algorithme de plus grande descente décrit par

$$\text{Alg 3} : x^{n+1} = x^{n} - \rho_n \nabla f(x^n) \qquad n \geqslant 0$$

Pour assurer la convergence de cet algorithme il reste à faire un choix convergent, au sens de Cea [18], du pas ρ_n.

Notons $\mu(x) = (\mu_1(x), \ldots \mu_n(x))$ le vecteur de $\mathbb{R}^n$ obtenu à partir du vecteur x en renumérotant ses composantes dans l'ordre non croissant. Sur $\mathbb{R}^n \times \mathbb{R}^n$ introduisons après Hardy-Littlewood-Polya la relation $x \mathcal{R} y = x \preccurlyeq y$ qui est vraie si et seulement si :

$$\forall k = 1, 2, \ldots, n-1 \qquad \sum_{i=1}^{k} \mu_i(x) \leqslant \sum_{i=1}^{k} \mu_i(y)$$

$$\sum_{1}^{n} \mu_i(x) = \sum_{1}^{n} \mu_i(y)$$

On a alors le résultat suivant dû à Horn [14].

PROPOSITION 6. - Soit $X = (x_i \delta_{ij})$ fixée. Une condition nécessaire et suffisante pour qu'il existe une matrice réelle symétrique A à diagonale nulle telle que $\mathrm{Sp}\,(A+X) = (s_1, s_2, \ldots, s_n)$ est que $x \preccurlyeq s$.

Notons alors W_s l'ensemble des $x \in \mathbb{R}^n$ tels que $x \preccurlyeq s$ Horn [14] reprenant Hardy-Littlewood-Polya montre que W_s peut encore s'écrire $W_s = \{x \in \mathbb{R}^n \mid x = Ms$ M matrice doublement stochastique$\}$ ce qui prouve que d'après un résultat de Birkoff [23] que W_s est un polyedre convexe compact dont les sommets sont les Ps ; P décrivant l'ensemble des matrices de permutation.

Considérons d'autre part l'orbite $\Theta(S)$ de $S = (s_i \delta_{ij})$, c'est à dire l'ensemble des matrices orthogonalement semblables à S.

$$\Theta(S) = \{B \in \mathcal{M}_m(\mathbb{R}) \mid B = USU^T \qquad U \text{ orthogonale}\}$$

est un ensemble compact, mais non convexe. Il est clair que si x est une solution de (Ps) alors $A + x_i \delta_{ij} \in \Theta(S)$.

Sur l'ensemble des matrices symétriques considérons le produit scalaire $(A\ ,B) = \mathrm{tr}\,(AB)$ et la norme induite, dite norme de Schur $\|A\|_S^2 = \mathrm{tr}\, A^2 = \sum_{i,j=1}^{n} a_{ij}^2$.

L'ensemble des matrices symétriques est alors un espace de Hilbert qui peut se décomposer en somme directe orthogonale entre les matrices diagonales et les matrices symétriques à diagonale nulle. D'après le théorème de Wiedlant-Hoffman [21] la distance de $A+x_i\delta_{ij}$ à $\mathcal{O}(S)$ est donnée par $\|\mu(A+x_i\delta_{ij})-\mu(S)\|_2$ c'est à dire que $f(x)$ représente au facteur 1/2 près le carré de la distance de $A+X$ à $\mathcal{O}(S)$.

Notons $A+W_s$ l'ensemble convexe compact des matrices $A+x_i\delta_{ij}$ où $x\in W_s$. $A+W_s$ est contenu dans un hyperplan passant par A parallèle à l'ensemble des matrices diagonales. Il est clair que résoudre (P_s) c'est trouver un point de l'intersection $\mathcal{O}(S)\cap(A+W_s)$, et qu'une méthode constructive sera l'obtention d'une suite minimisant la distance. Gubin-Polyak-Raik [19], Pierra [20] ont développé des algorithmes de projection successives pour trouver un point de l'intersection de plusieurs convexes ; reprenons cette idée en l'adaptant.

Soit $A^k = A+X^k = A+(x_i^k\,\delta_{ij})$ une matrice de $(A+W_s)$. D'après le théorème de Wiedlandt-Hoffman [21] la distance de A^k à $\mathcal{O}(S)$ est donnée par :

$$\text{dist}\,(A^k,\mathcal{O}(S)) = \min_{B\in\mathcal{O}(S)} \|A^k - B\|_s = \|\mu(A+X^k)-\mu(S)\|_2$$

Car si $A^k = A+X^k = M^k.\ \text{Diag}\,\mu(A+X^k).\ M^{k^T}$, la matrice $B^k = M^k\ \text{Diag}\,\mu(S)\ M^{k^T}$ réalise le minimum de la distance. En d'autres termes, on sait projeter sur $\mathcal{O}(S)$. Notons que $\mathcal{O}(S)$ n'étant pas convexe il peut exister plusieures projections, mais notre façon de procéder en détermine une seule.

La détermination de la projection $A^{k+1} = A+X^{k+1}$ de B^k sur le convexe compact $(A+W_s)$ est particulièrement simple, d'après la proposition 6, il vient :

$$\text{Proj}_{(A+W_s)} B^k = A+\text{Diag}\, B^k$$

car diag $B^k\in W_s$.

On appellera algorithme des projections successives ou Alg 4 l'itération des deux étapes suivantes.

a) $B^k = M^k\ \text{Diag}\,\mu(S)\ M^{k^T}$ si $A^k\ A+X^k = M^k.\ \text{Diag}\,(A+X^k).\ M^{k^T}$

b) $A^{k+1} = A+X^{k+1} = A+\text{Diag}\,B^k$.

On a par construction la proposition suivante :

<u>PROPOSITION 7.</u> - Si $f(x) = \frac{1}{2}\ \|\mu(A+x_i\delta_{ij})-\mu(S)\|_2^2$ alors pour la suite

$\{x^k\}_1^\infty$ fournit par Alg 4 on a $f(x^{k+1}) \leq f(x^k)$

Explicitons le passage de X^n à X^{n+1} dans l'algorithme précédent.

On a :

$$X^{n+1} = X^n + \text{Diag}\,\{M^n(\text{Diag}\,\mu(S) - \text{Diag}\,\mu\,(A+X^n))\,M^{n^T}\}.$$

Cette écriture prouve que l'algorithme Alg 4, de projections successives est l'algorithme de O. Hald [9].

Multiplions à droite chaque terme par e où $e^T=(1,1,1,\ldots\ldots,1)$; il vient :

$$X^{n+1}\,e = x^{n+1} + \text{Diag}\,\{M^n(\text{Diag}\,\mu(S) - \text{Diag}\,\mu(A+X^n)\,M^{n^T}\}\,e$$

d'où

$$x^{n+1} = x^n + J(x^n)^T\,\{\mu(S) - \mu(A+x_i^n\,\delta_{ij})\} = x^n - \nabla f(x^n).$$

Il y a donc coïncidence entre l'algorithme de double projection Alg 4, l'algorithme de Hald, la méthode de plus grande descente Alg 3 avec le choix du pas $\rho_n=1$. C'est la concordance de ces trois méthodes qui va permettre de prouver la convergence.

<u>LEMME.</u> - (Hald [9] page 162). Pour la suite de matrices diagonales obtenues par l'algorithme de Hald [resp Alg 3, Alg 4] on a

$$\sum_{n>0} \|X^{n+1} - X^n\|_s^2 < +\infty$$

Il est connu que ce résultat n'implique pas à lui seul la convergence de la suite $\{x^n\}_1^\infty$. C'est notre interprétation comme méthode de gradient qui permet de conclure.

<u>PROPOSITION 8.</u> - Soit $A \in \mathcal{M}_{nn}(\mathbb{R})$ symétrique à diagonale nulle

Soit $S=(s_i\,\delta_{ij})$ fixée

Alors

1/ il y a coïncidence entre les trois algorithmes de Hald, des projections successives : Alg 4, de gradient à pas fixe $\rho_n=1$: Alg 3

ie. $x^{n+1} = x^n - J(x^n)^T\{\mu(A+x_i^n\,\delta_{ij}) - \mu(S)\}$

2/ tout point adhérant $\bar{x}$ à la suite $\{x^n\}_1^\infty$ est un point stationnaire pour $f(x) = \frac{1}{2}\|\mu(A+x_i\,\delta_{ij}) - \mu(S)\|_2^2$.

De même que l'interprétation de l'algorithme de Hald comme algorithme de descente fournit des variantes, l'interprétation géométrique suscite de même des variantes numériquement intéressantes.

De la relation

$$A+X=M\ \mathrm{Diag}\ \mu(S)\ M^T$$

on déduit en prenant le carré de la norme de Schur des deux membres que

$$\sum_{i=1}^{n} x_i^2 = \sum_{i=1}^{n} s_i^2 - \sum_{i,j=1}^{n} a_{ij}^2 = r^2$$

Si une solution x au problème (Ps) existe alors, nécessairement, $x \in S(o, r) \subset B(o, r)$.
posons

$$A+B(o, r)=\{M \in \mathcal{M}_{nn}(\mathbb{R}) \mid M=A+(x_i\delta_{ij}),\ \sum_{i=1}^{n} x_i^2 \leq r^2\}.$$

Cet ensemble est contenu dans hyperplan parallèle à l'ensemble des matrices diagonales et passant par A.

La solution si elle existe appartient à $(A+B(o, r)) \cap \mathcal{O}(S)$; il est facile d'adapter l'algorithme des projections successives.
Soit :

$$A^k = A+X^k = M^k = M^k.\ \mathrm{Diag}\ \mu(A+X^k).\ M^{k^T}$$

la matrice $C^k=M^k\ \mathrm{Diag}\ \mu(S)M^{k^T}$ réalise le minimum de la distance entre A^k et $\mathcal{O}(S)$. La projection de C^k sur le convexe compact $(A+B(o, r))$ est facile à obtenir ; c'est $A^{k+1} = A+X^{k+1} = A+r \dfrac{\mathrm{Diag}\ C^k}{\|\mathrm{Diag}\ C^k\|_2}$. On itère ces deux projections successives.

L'intérêt dans cet algorithme est la conservation à priori de la norme de la solution.

Il est clair que la suite $\{x^n\}_1^\infty$ obtenue définie un algorithme de descente pour $f(x) = \frac{1}{2} \|\mu(A+x_i\delta_{ij})-\mu(S)\|_2^2$, c'est à dire $f(x^n) \geq f(x^{n+1})$.
De fait cet algorithme n'est rien d'autre que la minimisation de f(x) sous la contrainte $x \in B = B(o, r) = \{x \mid \|x\| \leq r\}$ par une méthode de gradient projeté.

On obtient alors le résultat suivant :

<u>PROPOSITION 9.</u> - Soit A tridiagonale symétrique à diagonale nulle telle que

$a_{i-1,i} \neq 0 \qquad i=2, 3, \dots, n$

alors si

$$K = \sup_{\xi \in \mathbb{R}^n} \|f''(\xi)\| < 2$$

où

$$f(x) = \frac{1}{2} \|\mu'A + x_i \delta_{ij}) - \mu(S)\|_2^2$$

l'algorithme $x^{n+1} = \text{Proj}(x^n - \nabla f(x^n))$ produit une suite dont tous les points adhérants sont des points stationnaires de f.

5 - Essais numériques.

Nous avons fait de nombreuses expériences numériques, aussi nous n'en présentons ici qu'une partie. Dans les essais suivants nous aurons $\omega = 1$ Alg 1, $\omega = 1$ et $\lambda = 0.1/(n^\circ$ de l'itération) dans Alg 2.

Nous avons également multiplié les essais pour étudier le domaine d'attraction d'une solution relativement aux divers algorithmes. L'expérience a révélé un fait assez inhabituel : l'ensemble des approximations initiales pour lesquelles l'algorithme de Newton Alg 2 converge semble plus grand que l'ensemble des approximations initiales assurant la convergence de l'algorithme des approximations successives Alg 1. Naturellement lorsque tous les deux convergent Alg 2 est bien plus rapide que Alg 1.

Pour comparer l'efficacité des divers algorithmes étudiés nous avons construit des problèmes tests à partir de la discrétisation avec un pas $h = 1/(n+1)$, de problèmes de Sturm-Liouville.

$$Ly = y'' + q(x)y = \lambda y$$

$$\alpha_1 y(o) - \beta_1 y'(o) = 0$$

$$\alpha_2 y(1) + \beta_2 y'(1) = 0$$

On approxime $y''(x_i) = y''(ih)$ par $(y_{i-1} - 2y_i + y_{i+1})/h^2$, $y'(o)$ par $(y_1 - y_o)/h$ et $y'(1)$ par $(y_{n+1} - y_n)/h$. Il est alors facile de vérifier que l'approximation de l'opérateur L est la matrice A symétrique tridiagonale ayant des -1 sur les deux codiagonales et $\{C_a + (2 + h^2 q_1), 2 + h^2 q_2, \ldots, 2 + h^2 q_{n-1}, (2 + h^2 q_n) + C_b\}$ comme diagonale, avec $C_a = -\beta_1/(\alpha_1 h + \beta_1)$ et $C_b = -\beta_2/(\alpha_2 h + \beta_2)$. La procédure est alors la suivante. On se donne une fonction q et les constantes $\alpha_1, \alpha_2, \beta_1, \beta_2$ pour $n = 10$ on obtient des matrices 10×10. On calcule alors le spectre de cette matrice par l'algorithme du type Q. R que Wilkinson-Reinsh nomme tq 12 cf (17).

Les programmes pour les calculs des jeux d'essais et les algorithmes Alg 1, Alg 2, Alg 3, Alg 4 sont écrits en Fortran et testés sur IRIS 80 de CII. Pour chaque programme la phase essentielle du calcul des valeurs propres et vecteurs propres est effectuée par le sous programme tq 12.

Nous reproduisons dans les tableaux ci-dessous deux séries d'essais ; l'une est construite à partir d'une fonction $q(x)=x(1-x)$, c'est à dire symétrique sur $[0,1]$, l'autre à partir de $q(x)=1-x$. Pour chaque série nous faisons varier les conditions aux limites.

i) $y(o) = y(1) = 0$

ii) $y'(o) = o$ et $y(1) = 0$

iii) $y(o) = y'(o)$ et $y(1) = y'(1)$

Nous donnons pour diverses itérations l'erreur relative :

$$\| \text{Spectre visé - Spectre obtenu} \|_2 \, / \, \| \text{Spectre visé} \|_2 .$$

L'approximation initiale de la diagonale est pour tous les essais le spectre visé.

Tableau 1 -$q(x) = x(1-x)$; $y(o) = y(1) = 0$

N°	Alg 1	Alg 2	Alg 3	Alg 4
0	0.260	0.260	0.260	0.260
10	$0.351 \; 10^{+1}$	$0.414 \; 10^{-4}$	$0.196 \; 10^{-1}$	$0.174 \; 10^{-1}$
20	diverge	$0.373 \; 10^{-5}$	$0.106 \; 10^{-1}$	$0.940 \; 10^{-2}$
50		$0.153 \; 10^{-5}$	$0.494 \; 10^{-2}$	$0.421 \; 10^{-2}$
80		$0.381 \; 10^{-7}$	$0.322 \; 10^{-2}$	$0.272 \; 10^{-2}$

Tableau 2 $q(x) = x(1-x)$; $y'(o)=0$ $y(1)=0$

N°	Alg 1	Alg 2	Alg 3	Alg 4
0	0.268	0.268	0.268	0.268
10	$0.972 \; 10^{+1}$	0.257	$0.234 \; 10^{-1}$	$0.209 \; 10^{-1}$
20	diverge	$0.159 \; 10^{-9}$	$0.160 \; 10^{-1}$	$0.141 \; 10^{-1}$
50		arrêt	$0.577 \; 10^{-2}$	$0.463 \; 10^{-2}$

Tableau 2 (suite)

N°	Alg 1	Alg 2	Alg 3	Alg 4
80			$0.150\ 10^{-2}$	$0.117\ 10^{-2}$

Tableau 3 $q(x) = x(1-x); y(o) = y'(o); y(1) = y'(1)$

N°	Alg 1	Alg 2	Alg 3	Alg 4
0	0.280	0.280	0.280	0.280
10	$0.112\ 10^{+3}$	0.179	$0.427\ 10^{-1}$	$0.408\ 10^{-1}$
20	diverge	0.128	$0.372\ 10^{-1}$	$0.336\ 10^{-1}$
50		$0.253\ 10^{-5}$	$0.115\ 10^{-1}$	$0.864\ 10^{-2}$
80		$0.669\ 10^{-7}$	$0.386\ 10^{-2}$	$0.351\ 10^{-2}$

Tableau 4 $q(x) = 1-x; y(o) = y(1) = 0$

N°	Alg 1	Alg 2	Alg 3	Alg 4
0	0.259	0.259	0.259	0.259
10	$0.710\ 10^{+1}$	$0.282\ 10^{-4}$	$0.195\ 10^{-1}$	$0.174\ 10^{-1}$
20	diverge	$0.250\ 10^{-5}$	$0.106\ 10^{-1}$	$0.941\ 10^{-2}$
50		$0.584\ 10^{-4}$	$0.494\ 10^{-2}$	$0.422\ 10^{-2}$
80		$0.345\ 10^{-6}$	$0.322\ 10^{-2}$	$0.272\ 10^{-2}$

Tableau 5 $q(x) = 1-x; y'(o) = 0; y(1) = 0$

N°	Alg 1	Alg 2	Alg 3	Alg 4
0	0.268	0.268	0.268	0.268
10	0.976	0.577	$0.235\ 10^{-1}$	$0.408\ 10^{-1}$
20	diverge	0.161	$0.161\ 10^{-1}$	$0.136\ 10^{-1}$
50		$0.196\ 10^{-1}$	$0.573\ 10^{-2}$	$0.863\ 10^{-2}$
80		$0.145\ 10^{-9}$ en N° 60	$0.386\ 10^{-2}$	$0.118\ 10^{-2}$

Tableau **6** $q(x) = 1-x$; $y(o) = y'(o)$; $y(1)=y'(1)$

N°	Alg 1	Alg 2	Alg 3	Alg 4
0	0.279	0.279	0.279	0.279
10	$0.140\ 10^{+2}$	0.314	$0.426\ 10^{-1}$	$0.208\ 10^{-1}$
20	diverge	$0.219\ 10^{-1}$	$0.371\ 10^{-1}$	$0.140\ 10^{-1}$
50		$0.134\ 10^{-5}$	$0.114\ 10^{-1}$	$0.464\ 10^{-2}$
80		$0.157\ 10^{-6}$	$0.386\ 10^{-2}$	$0.118\ 10^{-2}$

A la vue de ces résultats deux remarques au moins s'imposent. La plus importante est que pour l'approximation initiale choisie l'algorithme Alg 1 diverge à chaque fois tandis que les autres convergent ; cela corrobore une remarque déjà faite.

Le seconde remarque consiste en l'opposition entre d'une part les deux algorithmes de minimisation Alg 3 et Alg 4 et d'autre part l'algorithme de Newton Alg 2; Alg 3 et Alg 4 donnent des résultats très similaires avec un très léger avantage pour Alg 4. Mais pour ces deux méthodes la convergence bien que très régulière est aussi très lente ; on n'arrive pas en 80 itérations à dépasser le seuil d'une erreur relative en 10^{-3}, ce qui dans notre cas assure 10 pour cent d'erreur sur la plus petite composante et 0,25 pour cent d'erreur sur la plus grande composante du spectre visé. Par contre Alg 2 donne à chaque fois une erreur relative de l'ordre de 10^{-7}, ce qui assure sept chiffres caractéristiques exacts pour toutes les composantes du spectre visé.

La régularité des algorithmes 3 et 4 et la rapidité de l'algorithme de Newton incite à étudier un algorithme pour le problème inverse des valeurs propres qui aurait ces deux excellentes propriétés.

BIBLIOGRAPHIE

[1] De OLIVEIRA G. - Note on inverse characteristic problem. Numer. Math Vol 15, (1970), 339-341.

[2] HADELER K. P. - Ein inverses Eigen wert problem. Linear algebra and its appl. Vol 1, (1968), 83-101.

[3] HADELER K. P. - Newton-Verfahren für inverse Eigenwertaufgen. Num. Math. Vol 12, (1968), 35-39.

[4] LABORDE F. - Sür un problème inverse de valeurs propres. CRAS tome 268, (1969), 153-156.

[5] CHATELIN -LABORDE F. - Thèse Mathodes numériques de calcul de valeurs propres et vecteurs propres d'un opérateur linéaire. Grenoble 1971.

[6] MOREL P. - A propos d'un problème inverse de valeurs propres. GRAS tome 277, (1973), 125-128.

[7] MOREL P. - Sur le problème inverse des valeurs propres. Numer. Math 23, (1974), 83-94.

[8] HALD O. - On discrete and numerical inverse Sturin-Liouville problems, Uppsala University, Dep. of Computer Sciences, Report 42, 1972.

[9] FRIEDLAND S. - Matrices with prescribed off diagonal elements, Israel J. of Math. Vol 11, (1972), 184-189.

[10] FRIEDLAND S. - Inverse eigenvalue problems. A paraître.

[11] WILKINSON - The algebric eigenvalue problem. Oxford University Press (1965).

[12] KATO T. - Perturbation theory of linear operators. Springer Verlag (1966).

[13] LANCASTER P. - On eigenvalues of matrices dependent on a parametor. Numer. Math Vol 6, (1964), 377-387.

[14] HORN A. - Doubly stochastic matrices and the diagonal of a rotation matrice. Amer. J. Math. Vol 76, (1954), 620-630.

[15] HARDY-LITTLEWOOD-POLYA - Inegalities, Cambridge University Press (1948).

[16] ORTEGA-RHEINBOLDT - Iterature solution of non linear equations in several variables, Academic Press (1970).

[17] WILKINSON- REINSCH - Linear algebra. Handbook for compotations. Springer Verlag.

[18] CEA J. - Optimisation, théorie et algorithmes. Dunod 1971.

[19] GUBIN-POLYAK-RAIK - The method of projections for fonding the common point of convex sets. USSR Comp. Math and Math Phys. Vol 6, (1967), 1-24.

[20] PIERRA G. - Sur le croissement de méthodes de descente. CRASS t 277, (1973) 1071-1074.

[21] WIELANDT-HOFFMAN - The variation of the spectrum of a normal matrix, Doke J. of Math Vol 20, (1953) , 37-39.

[22] GOLSTEIN A. - Constructive real analysis. Harper International Edition (1967).

ETUDE DE PROBLEMES D'OPTIMAL DESIGN

François MURAT et Jacques SIMON

Université Paris VI et CNRS
Laboratoire Associé 189
Analyse Numérique

Tour 55.65 - 5ème étage
4 place Jussieu - PARIS (5°)

On expose ici les grandes lignes d'un travail sur l'optimal design. On travaillera sous des hypothèses fortes pour simplifier l'exposition. Un article plus complet, à paraitre, donnera le détail des démonstrations (qui sont seulement esquissées ici) ainsi que d'autres résultats et la bibliographie.

Les données d'un problème d'optimal design sont sensiblement différentes de celles d'un problème de contrôle habituel, et l'objet de notre travail est de définir un cadre dans lequel on puisse étendre des méthodes et obtenir des résultats classiques en théorie du contrôle.

1. POSITION DU PROBLEME

On appelle "optimal design", ou "contrôle par un domaine géometrique", la recherche d'un domaine (ie ouvert connexe de R^N), réalisant le minimum d'une fonctionnelle donnée définie sur un ensemble de domaines.

On exposera ici les méthodes et les résultats obtenus en se référant à l'exemple suivant, relatif au problème de Neumann non homogène. Etant donnés

(1) $f \in H^1(R^N)$, $g \in H^2(R^N)$, $z \in H^2(R^N)$

(2) D un domaine de R^N, variété à bord de classe $W^{2,\infty}$, et de frontière ∂D,

on définit l'état u_D du système de façon unique par :

(3) $$\begin{cases} -\Delta u_D + u_D = f & \text{dans } D \\ \dfrac{\partial u_D}{\partial n} = g & \text{sur } \partial D \\ u_D \in H^2(D) & \end{cases}$$

et la fonction coût J par

(4) $$J(D) = \int_{\partial D} |u_D - z|^2 \, ds$$

Etant donnée une famille $\mathcal{D}_{ad}$ de domaines vérifiant l'hypothèse (2) on cherche un élément D_o minimisant J sur $\mathcal{D}_{ad}$, i.e. tel que

(5) $$D_o \in \mathcal{D}_{ad} \quad \text{et} \quad J(D_o) \leq J(D) \qquad \forall\, D \in \mathcal{D}_{ad}$$

On se pose 3 problèmes classiques en optimisation :

i) Etablir l'existence d'un domaine optimal D_o

ii) Donner des propriétés des éléments optimaux

iii) Trouver une méthode de calcul d'un élément optimal.

Dans les problèmes de contrôle habituels, le contrôle appartient en général à un sous ensemble d'un espace vectoriel normé, et on utilise des résultats de compacité pour résoudre i), et des résultats de dérivabilité de la fonction coût par rapport au contrôle pour résoudre ii) et iii).

Une des difficultés des problèmes d'optimal design est qu'il n'existe pas dans l'arsenal classique de l'analyse d'ensemble de domaines muni d'une structure d'espace vectoriel normé. Nous allons définir des espaces de domaines munis d'une structure métrique complète et d'une "structure différentielle". Dans ces espaces nous obtiendrons l'existence d'un domaine optimal D_o (quand la famille $\mathcal{D}_{ad}$ est compacte), et la dérivabilité de la fonction $D \to J(D)$ (ce qui conduit aux conditions nécessaires d'optimalité du 1er ordre et à des méthodes de descente).

2. DEFINITION ET PROPRIETES METRIQUES DES ESPACES DE DOMAINES

Etant donné un ouvert connexe Q de R^N (on ne suppose pas de régularité sur Q) et k un entier, $k \geq 1$, on définit un espace de domaines homéomorphes à Q par :

$$\mathcal{D}_Q^{k,\infty} = \{ D \mid D = T(Q) \quad , \quad T \in \mathcal{T}^{k,\infty} \}$$

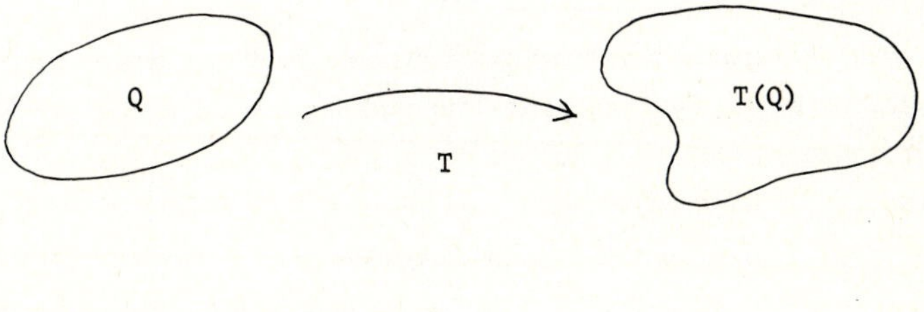

où $\mathcal{T}^{k,\infty}$ est l'espace de bijections régulières de R^N défini ainsi :

$$\mathcal{T}^{k,\infty} = \{ T \mid T \text{ bijection de } R^N \text{ sur lui-même}$$
$$T-I \in W^{k,\infty}(R^N,R^N) \quad \text{et} \quad T^{-1}-I \in W^{k,\infty}(R^N,R^N) \}$$

L'espace $\mathcal{T}^{k,\infty}$ est donc un ensemble de perturbations bornées, à dérivées bornées, de l'identité I de R^N. On munit par exemple $W^{k,\infty}(R^N,R^N)$ de la norme :

$$\|\varphi\|_{W^{k,\infty}(R^N,R^N)} = \underset{x\in R^N}{\text{Sup.ess.}} \{ \sum_{0\leq|\alpha|\leq k} \left| D^\alpha \varphi(x) \right|^2_{R^N} \}^{1/2}$$

et on définit sur $\mathcal{D}_Q^{k,\infty} \times \mathcal{D}_Q^{k,\infty}$ l'application :

$$d_{k,\infty}(D_1,D_2) = \inf_{\substack{T(D_1)=D_2 \\ T \in \mathcal{T}^{k,\infty}}} \{ \|T-I\|_{W^{k,\infty}(R^N,R^N)} + \|T^{-1}-I\|_{W^{k,\infty}(R^N,R^N)} \}$$

L'application $d_{k,\infty}$ est presque une distance, mais elle ne vérifie pas l'inégalité triangulaire. On a cependant les résultats suivants :

<u>THEOREME 1</u>

i) Il existe une constante positive η_k telle que l'application $\hat{d}_{k,\infty} = \inf \{ \sqrt{d_{k,\infty}} , \eta_k \}$ soit une distance sur $\mathcal{D}_Q^{k,\infty}$.

ii) L'espace $\mathcal{D}_Q^{k,\infty}$ muni de cette métrique est complet .

iii) Si $k \geq 2$ et si Q est borné, l'injection de $\mathcal{D}_Q^{k,\infty}$ dans $\mathcal{D}_Q^{k-1,\infty}$ est compacte : Plus précisemment de toute suite $\{D_n\}$ de $\mathcal{D}_Q^{k,\infty}$ telle que $d_{k,\infty}(D_1,D_n) \leq$ cste on peut extraire une sous suite qui converge dans $\mathcal{D}_Q^{k-1,\infty}$ vers un élément $D_o \in \mathcal{D}_Q^{k,\infty}$. ■

3. <u>UN RESULTAT D'EXISTENCE D'UN DOMAINE OPTIMAL</u>

Utilisant les espaces de domaines munis de leur structure métrique introduits au paragraphe 2, on obtient pour l'exemple que nous avons choisi de traiter, le résultat d'existence :

THEOREME 2

Soit Q un ouvert de R^N borné, variété à bord de classe $W^{2,\infty}$. Etant donné $\mathcal{D}_{ad}$ sous ensemble borné de $\mathcal{D}_Q^{2,\infty}$, fermé pour la topologie induite par $d_{1,\infty}$, il existe un domaine D_o réalisant le minimum de J sur $\mathcal{D}_{ad}$, i.e. vérifiant (5). ■

Principe de la démonstration

On utilise le théorème 1 iii) et on se ramène au domaine fixe Q par changement de variable. ■

4. DERIVATION DANS LES ESPACES DE DOMAINES $\mathcal{D}_Q^{k,\infty}$

La notion de dérivée Fréchet s'introduit généralement dans un espace affine normé A d'espace vectoriel sous jacent E, et utilise l'addition, application de A × E dans A. Nous allons étendre la notion de dérivée Fréchet en faisant jouer à $W^{k,\infty}(R^N,R^N)$ le rôle d'espace vectoriel sous jacent à $\mathcal{D}_Q^{k,\infty}$, et à l'application définie par

$$(6) \qquad (D,\Theta) \to (I+\Theta)(D)$$

le rôle de l'addition. Cela est loisible car si Θ est assez petit dans $W^{k,\infty}(R^N,R^N)$, $(I+\Theta)$ appartient à $\mathcal{T}^{k,\infty}$ et $(I+\Theta)(D)$ est donc un élément de $\mathcal{D}_Q^{k,\infty}$ si $D \in \mathcal{D}_Q^{k,\infty}$. De plus l'application $\Theta \to (I+\Theta)(D)$ transforme toute base de voisinages de O dans $W^{k,\infty}(R^N,R^N)$ en une base de voisinages de D dans $\mathcal{D}_Q^{k,\infty}$. Cela nous conduit à poser les définitions :

DEFINITION 1

Une application J de $\mathcal{D}_Q^{k,\infty}$ dans R est dérivable au point $D_o \in \mathcal{D}_Q^{k,\infty}$ si l'application $\Theta \to J^\star(\Theta) = J((I+\Theta)(D_o))$ est Fréchet dérivable (au sens usuel) de $W^{k,\infty}(R^N,R^N)$ dans R au point O. On dira que la dérivée de J au point D_o est

$$\frac{\partial J}{\partial D}(D_o) = \frac{\partial J^\star}{\partial \Theta}(0) \in \mathcal{L}_c(W^{k,\infty}(R^N,R^N);R) \qquad ■$$

DEFINITION 2

Une application G de R dans $\mathcal{D}_Q^{k,\infty}$ est dérivable au point r_o s'il existe une application ${}^\star G$ de R dans $W^{k,\infty}(R^N,R^N)$ telle que

$G(r) = (I+{}^\star G(r))(G(r_o))$ pour tout r dans un voisinage de r_o

${}^\star G(r_o) = 0$

${}^\star G$ est dérivable (au sens usuel) en r_o.

On dira alors que G admet comme dérivée au point r_o

$$\frac{\partial^{\star} G}{\partial r}(r_o) \in \mathcal{L}_c(R;W^{k,\infty}(R^N,R^N)) \equiv W^{k,\infty}(R^N,R^N) \qquad \blacksquare$$

Ces définitions entrainent un certain nombre de propriétés analogues à celles des fonctions dérivables au sens usuel ; Ainsi une application dérivable est localement lipschitzienne, et la composée de 2 applications dérivables est dérivable. Par contre "l'addition" définie par (6) n'est pas injective, et la "soustraction" correspondante est multivoque, ce qui entraine les propriétés spécifiques suivantes :

PROPRIETE 1

Soit J une application de $\mathcal{D}_Q^{k,\infty}$ dans R, dérivable en un point $D_o \in \mathcal{D}_Q^{k,\infty}$ que l'on suppose ouvert borné de R^N, variété à bord de classe C^1 ; Si Θ_1 et Θ_2 sont des éléments de $W^{k,\infty}(R^N,R^N)$ tels que

$$\Theta_1-\Theta_2 \in C^k(R^N,R^N)$$

$$<n,\Theta_1> = <n,\Theta_2> \quad \text{sur } \partial D_o$$

(n désignant la normale extérieure et $<,>$ le produit scalaire dans R^N, de sorte que $<n,\Theta> = \sum_{i=1}^{N} n_i\, \Theta_i$), alors

$$\frac{\partial J}{\partial D}(D_o)\cdot\Theta_1 = \frac{\partial J}{\partial D}(D_o)\cdot\Theta_2$$

Cette première propriété peut se résumer simplement en disant que la dérivée $\frac{\partial J}{\partial D}(D_o).\ \Theta$ ne dépend de Θ presque que par la valeur de sa trace normale $<n,\Theta>$ sur ∂D_o.

De la définition 2, il résulte qu'une application G dérivable en r_o peut admettre plusieurs dérivées en ce point. Si l'on note $\frac{\partial G}{\partial r}(e_o) \subset W^{k,\infty}(R^N,R^N)$ l'ensemble de ces dérivées, on peut presque caractériser cet ensemble. En effet :

PROPRIETE 2

Soit G une application de $\mathcal{D}_Q^{k,\infty}$ dans R, dérivable en un point r_o, et soit $\gamma \in \frac{\partial G}{\partial r}(r_o)$; on suppose $D_o = G(r_o)$ ouvert borné de R^N, variété à bord de classe C^1. Alors

$$\frac{\partial G}{\partial r}(r_o) \subset \{\gamma+\eta \mid \eta \in W^{k,\infty}(R^N,R^N)\ ,\ <n,\eta> = 0 \text{ sur } \partial D_o\}$$

D'autre part si $\gamma_1 \in \frac{\partial G}{\partial r}(r_o)$ appartient à $C^k(R^N,R^N)$, on a

$$\{ \gamma_1+\eta_1 \mid \eta_1 \in W^{k,\infty}(R^N,R^N) \cap C^k(R^N,R^N) \; <n,\eta_1> = 0 \text{ sur } \partial D_o \} \subset \frac{\partial G}{\partial r}(r_o). \quad ■$$

Ces propriétés s'interprètent géométriquement en observant que, quand Θ_1 et Θ_2 sont voisins de 0 dans $W^{k,\infty}(R^N,R^N)$, $(I+\Theta_1)(D_o)$ et $(I+\Theta_2)(D_o)$ définissent le même domaine au second ordre près, "si et seulement si" $<n,\theta_1> = <n,\Theta_2>$ sur ∂D_o.

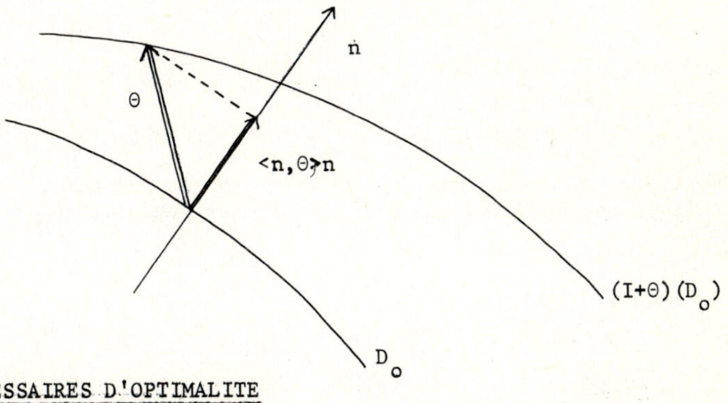

5. CONDITIONS NECESSAIRES D'OPTIMALITE

Les espaces de domaines $\mathcal{D}_Q^{k,\infty}$ étant munis de la "structure différentielle" définie au paragraphe 4, on peut donner des conditions nécessaires d'optimalité générales :

THEOREME 3

Soit $\mathcal{D}_{ad}$ un sous ensemble de $\mathcal{D}_Q^{k,\infty}$ et J une fonction de $\mathcal{D}_{ad}$ dans R. Si D_o réalise le minimum de J sur $\mathcal{D}_{ad}$ et si J est dérivable au point D_o, alors pour tout Θ appartenant à l'espace tangent à $\mathcal{D}_{ad}$ en D_o, on a

$$\frac{\partial J}{\partial D}(D_o) \cdot \theta \geqslant 0 \tag{7}$$

où l'espace tangent à $\mathcal{D}_{ad}$ en D_o est l'ensemble des $\Theta \in W^{k,\infty}(R^N,R^N)$, tels qu'il existe une application G de $[0,1]$ dans $\mathcal{D}_Q^{k,\infty}$, dérivable en zéro, telle que $G(0)=D_o$, $G(h) \in \mathcal{D}_{ad}$ et $\theta \in \frac{\partial G}{\partial h}(0)$. ■

L'espace tangent à $\mathcal{D}_{ad}$ en D_o contient en particulier l'ensemble des $\Theta \in W^{k,\infty}(R^N,R^N)$ tels que $(I+h\Theta)(D_o)$ soit dans $\mathcal{D}_{ad}$ pour h assez petit.

6. DERIVABILITE D'UNE INTEGRALE SUPERFICIELLE PAR RAPPORT A SON DOMAINE D'INTEGRATION

Avant d'énoncer le théorème de dérivabilité de la fonctionnelle associée au problème de Neumann par (4), donnons un théorème de dérivabilité d'une fonctionnelle plus simple définie par une intégrale superficielle, et qui éclaire le résultat et le principe de la démonstration.

On suppose Q ouvert de R^N borné, variété à bord de classe $W^{2,\infty}$. Alors tout $D \in \mathcal{D}_Q^{2,\infty}$ présente les mêmes propriétés. Soit $\tilde{f} \in W^{2,1}(R^N,R)$; on définit la fonctionnelle $\tilde{J}$ par

(8) $$\tilde{J}(D) = \int_{\partial D} \tilde{f}(s)\cdot ds \qquad \forall\, D \in \mathcal{D}_Q^{2,\infty}$$

THEOREME 4

L'application $\tilde{J}$ définie par (8) est dérivable de $\mathcal{D}_Q^{2,\infty}$ dans R en tout point $D_o \in \mathcal{D}_Q^{2,\infty}$ et l'on a

(9) $$\forall\, \tau \in W^{2,\infty}(R^N,R^N) \qquad \frac{\partial \tilde{J}}{\partial D}(D_o)\cdot\tau = \int_{\partial D_o} \langle n,\tau\rangle \left(\frac{\partial \tilde{f}}{\partial n} + Hf\right)$$

où n désigne la normale extérieure, $\frac{\partial \tilde{f}}{\partial n}$ la dérivée normale de $\tilde{f}$ et H la courbure moyenne de la variété ∂D_o. ■

Principe de la démonstration

D'après la définition 1, il faut montrer que l'application $\tilde{J}^\star$ définie par

$$\tilde{J}^\star(\Theta) = \int_{\partial(I+\Theta)(D_o)} \tilde{f}(s)\cdot ds$$

est dérivable en zéro. Pour cela on effectue le changement de variable $(I+\Theta)$.
On montre que :

$$\tilde{J}^\star(\Theta) = \int_{\partial D_o} \tilde{f}\circ(I+\Theta)\cdot |\det(I+\Theta)'|\cdot|{}^t(I+\Theta)'^{-1}n|\, ds$$

où $(I+\Theta)'$ designe la matrice dérivée de $(I+\Theta) \in C^1(R^N,R^N)$.

Dans cette expression, on montre la dérivabilité de chacun des trois termes du produit par rapport à Θ, ce qui démontre la dérivabilité de $\tilde{J}$ et donne la valeur de la dérivée :

$$\begin{cases} \forall\, \tau \in W^{2,\infty}(R^N,R^N) \\ \dfrac{\partial \tilde{J}}{\partial D}(D_o)\cdot\tau = \displaystyle\int_{\partial D_o} \{ <\text{grad}\ \tilde{f},\tau> + \tilde{f}(\text{div}\ \tau - <(\tau)'n,n> \}ds \end{cases}$$

Une intégration par parties sur ∂D_o, analogue à la formule de Stokes sur un ouvert, conduit alors à (9). ∎

7. RESULTAT DE DERIVABILITE POUR LE PROBLEME DE NEUMANN

THEOREME 5.

On suppose Q ouvert borné de R^N, variété à bord de classe $W^{3,\infty}$. Alors la fonction coût définie par (4) est dérivable de $\mathcal{D}_Q^{3,\infty}$ dans R en tout point $D_o \in \mathcal{D}_Q^{3,\infty}$ et sa dérivée est donnée par :

(10) $$\forall\, \Theta \in W^{3,\infty}(R^N,R^N) \qquad \frac{\partial J}{\partial D}(D_o)\cdot\Theta = \int_{\partial D_o} <n,\Theta> \cdot F(u_{D_o},p_{D_o}) \cdot ds$$

où u_{D_o} est l'état du système défini par (3), où p_{D_o} est l'état adjoint défini par :

(11) $$\begin{cases} -\Delta p_{D_o} + p_{D_o} = 0 & \text{dans } D_o \\ \dfrac{\partial p_{D_o}}{\partial n} = 2(u_{D_o}-z) & \text{sur } \partial D_o \\ p_{D_o} \in H^2(D_o) \end{cases}$$

et où F est la fonctionnelle définie par

(12) $$F(u,p) = -<\text{grad}\ u,\text{grad}\ p> + p(H.g + \frac{\partial g}{\partial n} + f-u) + (u-z)(H.(u-z)+4g-2\,\frac{\partial z}{\partial n})$$

(H est la courbure moyenne de la variété ∂D_o). ∎

Remarquons que, comme l'indique la propriété 1, la dérivée ne dépend de Θ que par sa trace normale $<n,\Theta>$.

Ce résultat de dérivabilité permet d'obtenir immédiatement les conditions nécessaires d'optimalité du premier ordre, grâce au théorème 3. Il permet également d'envisager l'emploi de méthodes de descente pour le calcul d'un domaine optimal.

Principe de la démonstration du théorème 5.

Comme dans la démonstration du théorème 4, on effectue le changement de variable $(I+\Theta)$ dans l'intégrale superficielle $J^{\star}(\Theta) = J((I+\Theta)(D_o))$.

On est alors conduit à démontrer la dérivabilité par rapport à Θ d'un certain nombre de termes, et notamment la dérivabilité en zéro de l'application :

$$\Theta \rightarrow (u_{(I+\Theta)(D_o)}) \circ (I+\Theta) \quad \text{de} \quad W^{3,\infty}(R^N,R^N) \text{ dans } H^2(D_o)$$

ce qui constitue un résultat de dérivabilité de l'état du système par rapport au domaine. Pour cela, on effectue le changement de variable $(I+\Theta)$ dans l'équation (3) relative à $(I+\Theta)(D_o)$ et on utilise le théorème de dérivabilité de la fonction implicite.

On termine la démonstration par des intégrations par parties sur D_o et ∂D_o, ce qui conduit à (10). ■

UNE FORMULE DE HADAMARD DANS DES PROBLEMES D'OPTIMAL DESIGN

A. Dervieux
B. Palmerio

Université de Nice
Institut de Mathématiques et Sciences Physiques
Parc Valrose
06034 NICE Cedex - France

Nous nous intéressons à des problèmes d'optimal design qui peuvent être formulés ainsi

Problème : Minimiser la fonctionnelle coût $j(\Omega)$ pour Ω appartenant à $\mathcal{U}_{ad}$; où $\mathcal{U}_{ad}$ est une famille de domaines Ω de $\mathbb{R}^n$.

Pour résoudre numériquement un tel problème une direction d'approche naturelle consiste à chercher la dérivée de la fonctionnelle j par rapport au domaine Ω. Hadamard [9] en 1908 calcula la dérivée de la plus petite valeur propre de certains opérateurs par rapport au domaine géométrique.

Nous étudions ici l'optimal design de systèmes régis par des problèmes aux limites elliptiques bien posés (1). En particulier nous nous referrerons au long de cet article au problème modèle suivant :

Problème modèle : Minimiser dans un ensemble $\mathcal{U}_{ad}$ la fonctionnelle

(0.1) $$j(\Omega) = \tfrac{1}{2}\|z(\Omega) - z_d\|^2_{L^2(\Omega_\circ)}$$

où $z(\Omega)$ est la solution de l'équation d'état dans Ω

(0.2) $$\begin{cases} z(\Omega) \in H^1(\Omega) \quad \text{et} \quad \forall\ \varphi \in H^1(\Omega) \\ \displaystyle\int_\Omega \left\{\nabla z(\Omega)\cdot\nabla\varphi + z(\Omega)\,\varphi - f\varphi\right\} dv = 0 \end{cases}$$

et $\mathcal{U}_{ad}$ est un ensemble d'ouverts vérifiant

$$\forall\, \Omega \in \mathcal{U}_{ad}\ , \quad \Omega_\circ \subset \Omega \subset B$$

(1) Pour des problèmes aux valeurs propres, le lecteur pourra se réferrer à [12] et [7] et à la bibliographie de ces travaux.

$\Omega_\circ$ et B étant deux ouverts fixes de $\mathbb{R}^n$ et

$$z_d \in L^2(\Omega_\circ) \ : \ f \in L^2(B)$$

Le plan que nous suivrons est le suivant

1 Formulation du problème

1.1 Repérage du domaine
1.2 Notations
1.3 Calcul formel du gradient

2 Transport sur un ouvert fixe : dérivabilité

3 Prolongement à un ouvert fixe : formule de Hadamard

4 Exemples

5 Conclusion

1 Formulation du Problème

1.1 Repérage du domaine

Nous utiliserons un repérage classique pour ce genre de problèmes (cf. [3]) ; soit k un entier supérieur ou égal à 2 , B un ouvert borné de $\mathbb{R}^n$ on pose

$$\mathcal{I}_k = \left\{ T \text{ difféomorphisme de B dans B tel que } T \text{ et } T^{-1} \in \left[C^k(\bar{B})\right]^n \right\} \tag{1.1}$$

Soit Ω un ouvert fixe de fermeture incluse dans B, de classe C^k (cf. par exemple [11]) ; on s'intéresse aux domaines images de Ω par certains éléments de $\mathcal{I}_k$: plus précisément l'ensemble $\mathcal{U}_{ad}$ des domaines admissibles est de la forme

$$\mathcal{U}_{ad} = \left\{ \Omega_T = T(\Omega) \ ; \ T \in \mathcal{I}_{ad} \right\} \tag{1.2}$$

où $\mathcal{I}_{ad}$ désigne l'ensemble des difféomorphismes admissibles $\mathcal{I}_{ad} \subset \mathcal{I}_k$.

1.2 Notations

Il sera plus pratique désormais de considérer le difféomorphisme comme nouveau contrôle et de formuler le problème général de la façon suivante :

(P) Minimiser dans $\mathcal{I}_{ad}$ la fonctionnelle

$$j(T) = J(T, z(T)) \tag{1.3}$$

où $z(T)$ est la solution d'un problème aux limites dans Ω_T noté ainsi

$$\begin{cases} z(T) \in Z_T \\ \Phi(T, z(T)) = 0 \text{ dans } Q'_T \end{cases} \tag{1.4}$$

et avec

$$\begin{cases} J : \mathcal{I}_{ad} \times Z_T \longrightarrow \mathbb{R} \\ \Phi : \mathcal{I}_{ad} \times Z_T \longrightarrow Q'_T \end{cases} \tag{1.5}$$;

Z_T et Q'_T sont des espaces de Banach dépendant du contrôle T : J et Φ ne sont pas des applications.

Cas du problème modèle : On pose pour T appartenant à $\mathcal{I}_{ad}$

$$Z_T = H^1(\Omega_T) \quad Q_T = H^1(\Omega_T)$$

$$\forall \varphi \in Z_T \quad , \quad \Psi \in Q_T$$

$$(1.6)\quad \begin{cases} J(T,\varphi) = \frac{1}{2}\|\varphi - z_d\|^2_{L^2(\Omega_o)} \\ \left[\Phi(T,\varphi),\Psi\right]_{Q'_T \times Q_T} = \\ \int_{\Omega_T} \left\{ <\nabla\varphi\ ,\ \nabla\Psi> + \varphi\Psi - f\Psi \right\} dv \end{cases}$$

Le symbole $<\ ,\ >$ désigne, sauf mention contraire, le produit scalaire de $\mathbb{R}^n$.

1.3 Calcul formel du gradient

En dérivant formellement l'équation (1.4) on obtient

$$(1.7)\qquad \frac{\partial \Phi}{\partial T} + \frac{\partial \Phi}{\partial \varphi} \cdot \frac{dz}{dT}(T) = o$$

introduisons l'état adjoint suivant

$$(1.8)\quad \begin{cases} p(T) \in Q_T \quad \text{et} \quad \forall\, \varphi \in Z_T \\ \left[\frac{\partial \Phi}{\partial \varphi}(T, z(T)) \cdot \varphi\ ,\ p(T)\right] = \frac{\partial J}{\partial \varphi}(T, z(T)) \cdot \varphi \end{cases}$$

et utilisons la dérivation composée

$$(1.9)\qquad \frac{dj}{dT}(T) = \frac{\partial J}{\partial T}(T, z(T)) + \frac{\partial J}{\partial \varphi}(T, z(T)) \cdot \frac{dz}{dT}(T)$$

nous obtenons

$$(1.10)\qquad \frac{dj}{dT}(T) = \frac{\partial J}{\partial T}(T, z(T)) - \left[\frac{\partial \Phi}{\partial T}(T, z(T)),\ p(T)\right].$$

Le but de cet article est de présenter un contexte général où ce calcul formel est justifié. Nous mettrons aussi en évidence dans quelques exemples que le deuxième membre de (1.10) est une intégrale de surface du même type que celles introduites par J.Hadamard.

Mais tout d'abord nous nous intéresserons à la dérivabilité de la fonctionnelle j : c'est l'objet du paragraphe **2** .

2 Transport sur un ouvert fixe. Dérivabilité

Soit T un élément de $\mathcal{J}_{ad}$, θ et δ

$$\theta \in [\, C^k(\overline{B})\,]^n$$

$$\delta > o$$

tels que

$$(2.1)\qquad T + t\,\theta \in \mathcal{J}_{ad} \quad \forall\ t \in I = [o, \delta[\quad ;$$

Dans ce qui suit T désignera sans ambiguité possible tantôt un élément fixe vérifiant la condition (2.1) tantôt une variable décrivant l'ensemble $\mathcal{J}_{ad}$.

Nous allons étudier la dérivabilité au sens de Gâteaux de la fonctionnelle j dans la direction θ. La principale difficulté du contexte précédent provient du fait que l'espace des fonctions d'état varie avec le contrôle. Une idée naturelle pour se ramener à des espaces fixes consiste à transporter l'équation d'état sur l'ouvert fixe Ω en utilisant le difféomorphisme T ; d'où l'hypothèse

(H 1) Il existe un espace de Banach Z et un espace de Banach réflexif Q de fonctions définies sur Ω tels que pour tout élément T de $\mathcal{J}_{ad}$ l'application $\varphi \to \varphi \circ T$ est un isomorphisme de Z_T sur Z (resp. Q_T sur Q).

Considérons l'état transporté

$$\tilde{z}(T) = z(T) \circ T \tag{2.2}$$

c'est la solution du système

$$\begin{cases} \tilde{z}(T) \in Z \\ \tilde{\Phi}(T,\tilde{z}(T)) = 0 \text{ dans } Q' \end{cases} \tag{2.3}$$

où l'application $\tilde{\Phi}$ est définie de la manière suivante

$$\begin{cases} \tilde{\Phi} : \mathcal{J}_{ad} \times Z \to Q' \\ \left[\tilde{\Phi}(T,\varphi), \Psi\right] = \left[\Phi(T,\varphi \circ T^{-1}), \Psi \circ T^{-1}\right] \end{cases} \tag{2.4}$$

et le problème (P) (cf. le paragraphe 1.2 équivaut à $(\tilde{P})$

$(\tilde{P})$ Minimiser dans $\mathcal{J}_{ad}$

$$j(T) = \tilde{J}(T,\tilde{z}(T))$$

où $\tilde{z}(T)$ est la solution de (2.3) et avec

$$\begin{cases} \tilde{J} : \mathcal{J}_{ad} \times Z \to \mathbb{R} \\ \tilde{J}(T,\varphi) = J(T,\varphi \circ T^{-1}) \end{cases} \tag{2.5}$$

Dans le <u>cas du problème modèle</u> on aura

$$Z = H^1(\Omega) \qquad Q = H^1(\Omega)$$

$$\tilde{J}(T,\varphi) = \tfrac{1}{2}\|\varphi \circ T^{-1} - z_d\|^2_{L^2(\Omega_o)}$$

$$\left[\tilde{\Phi}(T,\varphi),\Psi\right]_{[H^1(\Omega)]' \times H^1(\Omega)} =$$

$$\int_\Omega \left\{< DT^{-1*}\nabla\varphi, DT^{-1*}\nabla\Psi> + \varphi\Psi - f \circ T\,\Psi\right\} |DT|\, dv$$

On voit dans cet exemple que le contrôle T intervient uniquement dans les coefficients et le deuxième membre de l'équation d'état. Nous montrerons la dérivabilité de j à partir de celle de l'application

$$\begin{cases} \mathcal{J}_{ad} \to Z \\ T \to \tilde{z}(T) \end{cases} \tag{2.6}$$

celle-ci est une conséquence du théorème des fonctions implicites, d'où l'introduction d'hypothèses de régularité sur les applications $\tilde{\Phi}$ et $\tilde{J}$

(H 2) (i) L'application suivante est continue

$$I \times Z \to Q'$$
$$(t, \varphi) \to \tilde{\Phi}(T + t\,\theta\,, \varphi)\,.$$

(ii) Elle est dérivable par rapport à φ et sa dérivée partielle $\partial\tilde{\Phi}/\partial\varphi$ est continue par rapport au couple $(t\,,\varphi)$.

(iii) Elle est dérivable en $(o,\ \tilde{z}(T))$ et l'on a

$$\frac{\partial\tilde{\Phi}}{\partial\varphi}\,(o,\tilde{z}(T)) \in \mathrm{Isom}\,(Z\,, Q')\,.$$

(H 3) L'application

$$I \times Z \to \mathbb{R}$$
$$(t,\varphi) \to \tilde{J}(T + t\,\theta\,, \varphi)$$

est dérivable en $(o,\ \tilde{z}(T))$.

Proposition 2.1 - On se place sous les hypothèses (H 1) à (H 3) ; alors

(i) L'application (2.6) est G-différentiable en T dans la direction θ

$$\tilde{z}\,'(T, \theta) =$$

$$(2.7)\qquad -\left[\frac{\partial\tilde{\Phi}}{\partial\varphi}(T, \tilde{z}(T))\right]^{-1} \circ \left[\frac{d}{dt}(\tilde{\Phi}(T + t\,\theta\,,\ \tilde{z}(T))\right]_{t=o}$$

(ii) La fonctionnelle j est G-différentiable en T dans la direction θ

$$j'\,(T\,, \theta) =$$

$$(2.8)\qquad \frac{d}{dt}\left[\tilde{J}(T+t\,\theta\,,\ \tilde{z}(T))\right]_{t=o} - \frac{d}{dt}\left[\tilde{\Phi}(T+t\,\theta,\ \tilde{z}(T))\,,\ \tilde{p}(T)\right]_{t=o}$$

où $\tilde{p}(T)$ est la solution du système adjoint suivant

$$(2.9)\qquad \begin{cases} \tilde{p}(T) \in Q \\ \left[\dfrac{\partial\tilde{\Phi}}{\partial\varphi}\,(T\,,\tilde{z}(T))\right]^{*} .\ \tilde{p}(T) = \dfrac{\partial\tilde{J}}{\partial\varphi}\,(T\,, \tilde{z}(T))\ . \end{cases}$$

Cette proposition est une adaptation du théorème des fonctions implicites (cf. par exemple [13] p. 283). L'état adjoint est introduit comme dans [2] .

Application au problème modèle.

Les hypothèses de la proposition précédente sont des hypothèses de régularités sur les données et le contrôle. Les conditions suivantes sont suffisantes pour leur vérification

$$(2.10)\qquad \begin{cases} f \in H^1\ (B) \\ k = 2 \quad \text{dans } (2.1) \end{cases}$$

L'état adjoint $\tilde{p}(T)$ est la solution du système

$$(2.11)\quad \begin{cases} \tilde{p}(T) \in H^1(\Omega) \text{ et } \forall \varphi \in H^1(\Omega) \\ \int_\Omega \Big\{ < DT^{-1*} \nabla \tilde{p}(T) \,,\, DT^{-1*} \nabla \varphi > + \tilde{p}(T)\varphi \\ \varphi(\tilde{z}(T) - z_d \circ T) \Big\} |\det T| \, dv = 0 \end{cases}$$

et la formule (2. 8) se met sous la forme suivante

$$(2.12)\qquad j'(T,\theta) = \int_\Omega \Big\{ <G_0, \theta>_{\mathbb{R}^n} + <G_1, D\theta>_{\mathbb{R}^{n^2}} \Big\} \, dv$$

en posant

$$\int_\Omega <G_0, \theta> dv = \int_\Omega <\nabla f \circ T \,,\, \theta> \tilde{p}(T) \; |DT| \, dv$$
$$- \int_{\Omega_0} (\tilde{z}(T) \circ T^{-1} - z_d) <\nabla \tilde{z}(T) \circ T^{-1} \,,\, DT^{-1} \circ T^{-1}.\theta > dv$$

$$\int_\Omega < G_1, D\theta > dv =$$
$$- \int_\Omega \Big\{ < DT^{-1*} \nabla \tilde{z}(T) \,,\, DT^{-1*} \nabla \tilde{p}(T) > + \tilde{z}(T)\, \tilde{p}(T)$$
$$- f \circ T \;\tilde{p}(T) \Big\} \frac{d}{dt} |D(T + t\theta)|_{t=0} \; dv$$
$$+ \int_\Omega \Big\{ < \big[DT^{-1}.\, D\theta .DT^{-1}\big]^* \nabla \tilde{z}(T) \,,\, DT^{-1*} \nabla \tilde{p}(T) >$$
$$+ < DT^{-1*} \nabla \tilde{z}(T) \,,\, \big[DT^{-1}.\, D\theta .\, DT^{-1}\big]^* \nabla \tilde{p}(T) \Big\} |DT| \; dv \,.$$

Du point de vue optimisation il est utile de se placer dans un contexte hilbertien : soit G une solution du problème aux limites

$$\begin{cases} G \in [H^1(B)]^k \\ ((G, \theta))_{[H^1(\Omega)]^n} = \int_\Omega \Big\{ <G_0 \,,\, \theta> + <G_1 \,,\, D\theta> \Big\} dv \;; \end{cases}$$

si G est suffisamment régulière, G est une direction de plus grande descente pour le problème **(P)** .

Cette expression de gradient est peu commode à utiliser et nous allons mettre en évidence une expression plus simple de ce gradient.

3 Prolongement à un ouvert fixe . Formule de Hadamard

Comme annoncé au paragraphe 1.3 nous nous proposons d'exprimer le gradient de la fonctionnelle j non plus à partir des applications $\tilde{\Phi}$ et $\tilde{J}$ mais à partir des correspondances Φ et J ou plus exactement à l'aide d'application $\bar{\Phi}$, $\bar{J}$ prolongeant ces correspondances. Pour cela on introduit deux espaces de Banach Z_B et Q_B de fonctions définies sur B vérifiant l'hypothèse

(H 5) (i) L'application

$$\varphi \longrightarrow \varphi \circ T$$

est un automorphisme de Z_B (resp. Q_B)

ii) Il existe deux opérateurs linéaires continus et à images denses

$$R_1 \; : Z_B \longrightarrow Z$$

$$R_2 \; : Q_B \longrightarrow Q \quad .$$

De ces opérateurs, qui seront dans les exemples des opérateurs de restriction, on déduit par transport

$$R_1^T \; : Z_B \longrightarrow Z_T$$

$$R_2^T \; : Q_B \longrightarrow Q_T$$

définis par

(3.1) $$R_i^T \varphi = \left[R_i(\varphi \circ T\right] \circ T^{-1} \qquad \text{pour } i = 1, 2$$

Soient $\bar{\Phi}$ et $\bar{J}$ les prolongés de Φ et J dans le sens suivant

(3.2) $$\bar{\Phi} \; : \mathcal{I}_{ad} \times Z_B \longrightarrow Q_B'$$
$$\left[\bar{\Phi}(T, \varphi), \Psi\right] = \left[\Phi(T, R_1^T \varphi), R_2^T \Psi\right]$$

(3.3) $$\bar{J} \; : \mathcal{I}_{ad} \times Z_B \longrightarrow \mathbb{R}$$
$$\bar{J}(T, \varphi) = J(T, \mathbb{R}_1^T \varphi)$$

Cas du problème modèle. On pose

$$Z_B = Q_B = H^1(B)$$

et pour $i = 1, 2$

$$R_i \; : H^1(B) \longrightarrow H^1(\Omega)$$
$$\varphi \longrightarrow \varphi|_{\Omega}$$

alors on a

$$\forall \varphi \quad , \Psi \in H^1(B)$$

(3.4) $$[\bar{\Phi}(T, \varphi), \Psi] = \int_{\Omega_T} \left\{< \nabla\varphi, \nabla\Psi > + \varphi\,\Psi - f\,\Psi\right\} dv$$

(3.5) $$\bar{J}(T, \varphi) = \tfrac{1}{2} \|\varphi - z_d\|^2_{L^2(\Omega_\circ)} \qquad \blacksquare$$

Dans cet exemple, des problèmes concernant la régularité des fonctions d'état (direct et adjoint) se posent si l'on veut dériver $\tilde{\Phi}$ par rapport à T ; dans la formulation générale ces problèmes se retrouvent dans la difficulté de déduire de la dérivabilité de $\tilde{\Phi}$ (resp. $\tilde{J}$) celle de $\bar{\Phi}$ (resp. $\bar{J}$), difficulté qui provient du transport par le difféomorphisme (cf. les relations (2.4) et (2.5)). C'est pourquoi nous allons introduire deux espaces de fonctions plus régulières que les fonctions standard des espaces Z_B et Q_B ; les fonctions d'état devront appartenir à ces espaces ; en fait, dans la plupart des exemples on profite d'un supplément de régularité consécutif à la vérification de l'hypothèse (H 2) .

(H 6) Il existe deux sous espaces Z_r et Q_r de Z_B et Q_B (respectivement) tels que l'application

$$(3.6) \qquad t \to \varphi \circ (T + t\,\theta)$$

soit dérivable en o dans Z_B (resp. Q_B) pour tout élément φ de Z_r (resp. Q_r).

<u>Proposition 3.1</u> - On se place sous les hypothèses (H 1) à (H 6) ; on suppose en outre que les systèmes suivants

$$(3.7) \qquad \begin{cases} \bar{z}(T) \in Z_B \\ \bar{\Phi}(T, \bar{z}(T)) = o \quad \text{dans } Q'_B \end{cases}$$

$$(3.8) \qquad \begin{cases} \bar{p}(T) \in Q_B \;, \; \forall \varphi \in Z_B \\ \left[\frac{\partial \bar{\Phi}}{\partial \varphi}(T, \bar{z}(T)) \,.\, \varphi \,,\; \bar{p}(T)\right]_{Q'_B \times Q_B} = \frac{\partial \bar{J}}{\partial \varphi}(T, z(T)) \,.\, \varphi \end{cases}$$

admettent des solutions qui vérifient les conditions de régularité

$$(3.9) \qquad \bar{z}(T) \in Z_r \qquad \bar{p}(T) \in Q_r \,.$$

Alors le gradient (2.8) se met sous la forme

$$(3.10) \qquad j'(T, \theta) = \frac{d}{dt}\left[\bar{J}(T + t\,\theta, \bar{z}(T))\right]_{t=o} - \frac{d}{dt}\left[\bar{\Phi}(T + t\,\theta, \bar{z}(T)), \bar{p}(T)\right]_{t=o} .$$

<u>Proposition 3.2</u> - On se place sous les hypothèses (H 1) à (H 6) et on suppose qu'il existe un opérateur de prolongement $\mathcal{P}$ linéaire continu de Z dans Z_B

$$\mathcal{P} : Z \to Z_B$$

$$R_1 \circ \mathcal{P} = Id_Z \qquad ;$$

soit H un espace de Banach contenant Z_B tel que l'application

$$I \times Z_B \to H$$

$$(t, \varphi) \to \varphi \circ [T + t\,\theta]^{-1}$$

soit dérivable en (o, φ) pour tout élément φ de Z_B .

Alors il existe pour tout élément T de $\mathcal{I}_{ad}$ un prolongement $\bar{z}(T)$ de $z(T)$ dans Z_B dans le sens suivant

$\bar{z}(T)$ est une solution de (3.7), $R_1^T\, \bar{z}(T) = z(T)$ et tel que l'application

$$\mathcal{I}_{ad} \to H$$

$$T \to \bar{z}(T)$$

admet pour Gâteaux-dérivée en T dans la direction θ la solution du système

$$\frac{\partial\bar{\Phi}}{\partial\varphi}(T,\bar{z}(T)) . \bar{z}'(T,\theta) = -\frac{d}{dt}\bar{\Phi}(T+t\theta,\ \bar{z}(T))_{t=o} .$$

Remarque 3.1

Le calcul formel de dérivation composée de paragraphe 1.3 peut être entièrement justifié via la proposition 3.2 mais sous des hypothèses plus fortes que celles de la proposition 3.1 qui n'est donc pas une conséquence de la proposition 3.2 .

4 Exemples

4.1 Quelques lemmes techniques

Nous verrons plus loin que les calculs de gradients relatifs à un certain nombre de problèmes se ramènent grâce à la proposition 3.1 à l'utilisation de quelques formules de dérivation connues en Mécanique des Milieux continus (cf. par exemple [8]).

On conserve les notations des paragraphes précédents, et on pose

$$V = \theta \circ T^{-1}$$

et pour t appartenant à l'intervalle I

$$\Omega_t = (T + t\ \theta)\Omega$$

On peut énoncer les lemmes suivants

Lemme 4.1 Soit u un élément de $W^{1,1}(\mathbb{R}^2)$ (1) alors

$$\frac{d}{dt}\int_{\Omega_t} u\, dv = \int_{\partial\Omega_t} u \ <\vec{n}, V> \ d\sigma$$

où $\vec{n}$ désigne un vecteur unitaire normal à $\partial\Omega_t$ et dirigé vers l'extérieur de Ω_t .

Lemme 4.2 Soit u un élément de $W^{2,1}(\mathbb{R}^2)$ alors

$$\frac{d}{dt}\int_{\partial\Omega_t} u\, d\sigma = \int_{\partial\Omega_t}\Big[<\nabla u, V> + u <DV . \vec{\tau}, \vec{\tau}>\Big] d\sigma$$

où $\vec{\tau}$ est un vecteur tangent à $\partial\Omega_t$.

Remarque 4.1 - La formule précédente peut se mettre sous une forme plus simple en faisant intervenir la courbure de la frontière $\partial\Omega_t$ (cf. [11]).

Lemme 4.3 Soient u un élément de $H^3(\mathbb{R}^n)$ (2) , et w un élément de $H^2(\mathbb{R}^n)$ alors

$$\frac{d}{dt}\int_{\partial\Omega_t} u \ \frac{\partial v}{\partial n}\ d\sigma = \int_{\partial\Omega_t}\{<\nabla u,\ \nabla v> + u\,\Delta\, v\} <\vec{n}, V> \ d\sigma$$

Le lemme 4.3 est une conséquence de la formule de Green et du lemme 4.1 .

---

(1) $W^{m,P}(\Omega) = \{u \in L^P(\Omega)\ ;\forall\alpha\in\mathbb{N}^n,\ |\alpha| \le m \quad D^{\alpha} u \in L^P(\Omega)\} \quad \forall\ \Omega\subset\mathbb{R}^n$.

(2) $H^s(\Omega) = W^{s,2}(\Omega)$

4.2 Résultats

Pour plus de simplicité on se bornera à définir pour chaque exemple une application $\bar{\Phi}$ vérifiant les hypothèses de la proposition 3.1 ainsi que les hypothèses de régularité requises pour les données et pour l'ouvert ; dans la plupart des cas ces hypothèses sont plus que suffisantes pour garantir l'existence des expressions obtenues ; on peut envisager d'améliorer ces hypothèses par un raisonnement par densité (cf. [4]). On montre par ailleurs (cf. [5]) l'existence de formules analogues (intégrales sur le bord variable) sous des hypothèses beaucoup moins restrictives.

Exemple n°1 (Problème modèle) -

On suppose que

$$\mathcal{J}_{ad} \subset \mathcal{J}_2 \quad \text{(cf. (1.1))}$$

et

$$f \in H^1(B) \quad ;$$

on introduit l'état adjoint suivant

$$\left\{ \begin{array}{ll} -\Delta p + p = \chi_{\Omega_0} (z - z_d) & \text{dans} \quad \Omega_T \quad (^1) \\ \dfrac{\partial p}{\partial n} = 0 & \text{sur} \quad \partial\Omega_T \end{array} \right.$$

et le gradient est obtenu sous la forme suivante

$$j'(T, \theta) = -\int_{\partial\Omega_T} \{ <\nabla z, \nabla p> + zp - fp \} <\vec{n}, V> \quad d\sigma$$

Remarque 4.2 - L'application de la proposition 3.2 nous donnerait le résultat suivant :

Proposition - Il existe un prolongement $\bar{z}(T)$ de $z(T)$ dans $H^1(B)$, G-dérivable par rapport à T à valeurs dans $L^2(B)$ (dans $H^k(B)$ en se plaçant dans un contexte suffisamment régulier).

On montre aussi sans supplément de régularité que $z(T)|_{\Omega_0}$ est dérivable à valeur dans $H^1(\Omega_0)$.

Exemple n°2

On suppose

$$\mathcal{J}_{ad} \subset \mathcal{J}_4$$

$$f \in H^1(B) \; ; \; z_d \in H^1(B)$$

On veut calculer le gradient de

$$j(T) = \tfrac{1}{2} \| z - z_d \|^2_{L^2(\Omega_T)}$$

avec

---

(1) χ_{Ω_0} : fonction caractéristique de Ω_0 .

$$\left\{\begin{array}{ll} \Delta^2 z + z = f & \text{dans } \Omega_T \\ \Delta z = o & \\ \frac{\partial}{\partial n} \Delta z = o & \Big\} \text{ sur } \Omega_T \end{array}\right.$$

On applique la proposition 3.1 à l'application

$$\bar{\Phi} : \mathcal{J}_{ad} \times H^2(B) \rightarrow \left[H^2(B)\right]'$$

$$\left[\bar{\Phi}(T, \varphi), \Psi\right] = \int_{\Omega_T} \left\{\Delta\varphi \, \Delta\Psi + \varphi\Psi - f\,\Psi\right\} dv$$

Introduisons l'état adjoint

$$\left\{\begin{array}{ll} \Delta^2 p + p = z - z_d & \text{dans } \Omega_T \\ \Delta p = o & \\ \frac{\partial}{\partial n} \Delta p = o & \Big\} \text{ sur } \partial\Omega_T \end{array}\right.$$

on obtient

$$j'(T, \theta) = \int_{\partial\Omega_T} \left\{\tfrac{1}{2}(z - z_d)^2 - z p + f p\right\} <\vec{n}, V> d\sigma$$

<u>Exemple n°3</u>

On suppose

$$\mathcal{J}_{ad} \subset \mathcal{J}_5$$
$$f \in H^1(B)$$

l'équation d'état est la suivante

$$\left\{\begin{array}{ll} \Delta^2 z = f & \text{dans } \Omega_T \\ z = o & \\ \frac{\partial z}{\partial n} = o & \Big\} \text{ sur } \partial\Omega_T \end{array}\right.$$

la fonctionnelle est identique à celle de l'exemple 2 et l'on prend pour application $\bar{\Phi}$

$$\bar{\Phi} : \mathcal{J}_{ad} \times H^4(B) \rightarrow \left[L^2(B) \times H^1(B) \quad H^1(B)\right]'$$

$$\forall (T, \varphi, \Psi_1, \Psi_2, \Psi_3,) \in \mathcal{J}_{ad} \times H^4(B) \times L^2(B \times \left[H^1(B)\right]^2$$

$$\left[\bar{\Phi}(T, \varphi), (\Psi_1, \Psi_2, \Psi_3)\right] =$$

$$\int_{\Omega_T} (\Delta^2 z - f)\, \Psi_1 \, dv + \int_{\partial\Omega_T} (z\, \Psi_2 + \frac{\partial z}{\partial n} \Psi_3)\, d\sigma .$$

On en déduit l'équation adjointe

$$\left\{\begin{array}{ll} \Delta^2 p_1 = z - z d & \text{dans } \Omega_T \\ p_1 = \frac{\partial p_1}{\partial n} = o & \\ p_2 = \frac{\partial}{\partial n} \Delta p_1 & \text{sur } \partial\Omega_T \\ p_3 = - \Delta p_1 & \end{array}\right.$$

et le gradient

$$j'(T, \theta) = \int_{\partial\Omega_T} \left\{\tfrac{1}{2}(z - z_d)^2 + \Delta z \quad \Delta p\right\} <\vec{n}, V> d\sigma$$

Cette méthode s'applique aussi au contrôle de l'interface d'un système gouverné par un problème de transmission défini dans l'ouvert B tout entier (cf. [6]) .

5 CONCLUSION

La méthode que nous avons présentée s'applique aux problèmes de contrôle par un domaine de systèmes régis par des problèmes bien posés suffisamment réguliers, linéaires ou non linéaires. Elle se généralise au cas de problèmes de valeurs propres ; on retrouve alors les formules de Hadamard.

On peut aussi adapter cette méthode à différents choix de paramétrage, et l'utilisation de ces formules dans des algorithmes du type gradient acceléré donne des résultats encourageants (cf [4] pour des résultats numériques) .

REFERENCES

[1] J.CEA, A. GIOAN, J.MICHEL

[2] G.CHAVENT, Thèse, Paris, 1970

[3] R.COURANT and D.HILBERT, Methods of Mathematical Physics, Vol I, Interscience, New York

[4] A.DERVIEUX, B.PALMERIO, Thèses de 3ème Cycle, Nice, 1974

[5] A.DERVIEUX, B.PALMERIO, Communication aux Journées sur le Contrôle, l'Identification ou la Localisation d'un domaine géométrique, Nice, Octobre 1974.

[6] A.DERVIEUX, B.PALMERIO, C.R.A.S., 1975, t.280, pp.1697-1700 et 1761-1764

[7] A.DERVIEUX, B.PALMERIO, B.ROUSSELET, à paraître

[8] P.GERMAIN, Mécanique des Milieux continus, Masson et Cie, Paris, 1962

[9] J.HADAMARD, Mémoire sur le problème d'analyse relatif à l'équilibre des plaques élastiques encastrées, oeuvre de Jacques Hadamard, C.N.R.S., Paris, 1968

[10] J.L.LIONS, E.MAGENES, Problèmes aux limites non homogènes, Dunod, Paris, 1968

[11] F.MURAT, J.SIMON, Communication à ce congrès

[12] B.ROUSSELET, Communication à ce congrès

[13] L.SCHWARTZ, Cours d'Analyse, Hermann, Paris, 1967.

PROBLEMES INVERSES DE VALEURS PROPRES

B. ROUSSELET

Département de Mathématiques - Université de Nice
Parc Valrose - 06034 NICE

On s'intéresse à des problèmes d'optimisation où le contrôle est un domaine Ω de $\mathbb{R}^n$ (" optimal design " cf DERVIEUX - PALMERIO [5] et MURAT - SIMON [15]) et où la fonctionnelle fait intervenir les valeurs propres d'un problème aux limites au lieu de la solution d'un problème bien posé (par exemple elliptique).

L'origine de ce genre de questions remonte à RAYLEIGH (1877) [16] en passant par HADAMARD [10] , GARABEDIAN [8], KAC [12] .

On s'attachera surtout à dégager la mise en oeuvre d'un algorithme après avoir précisé le problème inverse et la régularité des fonctionnelles.

I . POSITION DU PROBLEME ; REGULARITE DES FONCTIONNELLES

Dans toute la suite, on désignera par L l'ensemble des ouverts bornés, connexes et " lipschitziens " de $\mathbb{R}^n$ (CHENAIS [4])

(1) Le problème direct

Soit $V = H^1(\Omega)$ où $\Omega \in L$ et le problème de valeurs propres variationnel :

$$(1_\Omega) \quad \begin{cases} \forall v \in V \\ a_\Omega(u, v) = \lambda\, b_\Omega(u, v) \qquad (u \neq 0) \end{cases}$$

avec par exemple :

$$\begin{cases} a_\Omega(u, v) = \int_\Omega (A(x)\, \nabla u(x) | \nabla v(x))\, dx + \int_\Omega B(x)\, u(x)\, v(x)\, dx \\ \qquad + \int_{\partial\Omega} \ell(x)\, A(x)\, u(x)\, v(x)\, d\sigma \\ b_\Omega(u, v) = \int_\Omega C(x)\, u(x)\, v(x)\, dx \end{cases}$$

et les <u>hypothèses</u> : $(B, L, C) \in (L^\infty(\mathbb{R}^n))^3$ $C > 0$ presque partout

$A \in W^{1,\infty}(\mathbb{R}^n)$ vérifiant, avec $0 < \alpha \leq M$:

$$\forall x \in \mathbb{R}^n \quad \forall \xi \in \mathbb{R}^n \quad \alpha\|\xi\|^2 \leq (A(x)\xi, \xi) \leq M\|\xi\|^2 .$$

Dans ces conditions, les valeurs propres de (1_Ω) forment un ensemble dénombrable discret de nombres positifs, elles sont de plus toutes de multiplicité finie. KATO [13] , GOULAOUIC [9] ; (on les désigne par $(\lambda_n(\Omega))_{n \in N}$) .

REMARQUE . - pour un type de problème direct de valeurs propres non autoadjoint avec transmissions cf DERVIEUX - PALMERIO - ROUSSELET [6] .

(2) Problèmes **inverses**

Soient $\Lambda_N(\Omega) = (\lambda_k(\Omega))_{k \leq N}$ et $M_N = (\mu_k)_{k \leq N}$ (où pour tout k, $\mu_k \in \mathbb{R}_+$) et :

$$L \xrightarrow{J_N} \mathbb{R}_+$$
$$\Omega \longrightarrow J_N(\Lambda_N(\Omega), M_N)$$

on s'intéresse à la famille de

Problèmes d'optimisation

Trouver $\Omega_o \in L$ tel que pour tout $\Omega \in L$ (resp. tout $\Omega \in L$ et " voisin " de Ω_1 donné)

$$J(\Omega_o) \leq J(\Omega)$$

REMARQUE . - pour l'étude de la différentiabilité, on supposera que $(\Lambda_N, M_N) \longrightarrow J_N(\Lambda_N, M_N)$ est $\mathcal{C}^1$.

EXEMPLES . -

(1) $$J_N(\Omega) = \sum_{i \leq N} |\lambda_i(\Omega) - \mu_i|^2$$

(2) $$J_N(\Omega) = \sum_{i \leq N} \left|\frac{1}{\lambda_i(\Omega)} - \frac{1}{\mu_i}\right|^2$$

(3) $$J_N(\Omega) = \sum_{i \leq N} \left|\frac{\lambda_{i+1}(\Omega)}{\lambda_i(\Omega)} - \frac{\mu_{i+1}}{\mu_i}\right|^2$$

(3) Propriétés des fonctionnelles

La continuité et la différentiabilité résultent de la remarque ci-dessus et des propriétés analogues des valeurs propres :

(a) Continuité

PROPOSITION . - Si l'on munit l'ensemble L de la topologie définie par le système fondamental de voisinages :

$$V(\Omega^*, \eta) = \left\{ \Omega = (I+F)(\Omega^*) \;\middle|\; \begin{array}{l} F \in W^{1,\infty}(\mathbb{R}^n, \mathbb{R}^n) \\ \|F\|_{1,\infty} < \eta \end{array} \right\} \cap L$$

l'application $\Omega \longrightarrow \lambda_n(\Omega)$ de L dans $\mathbb{R}_+$ est continue.

La démonstration repose sur la classique caractérisation variationnelle des valeurs propres de problèmes autoadjoints (" quotient de Rayleigh ") après transport sur un ouvert fixe.

(b) Gateaux - différentiabilité

PROPOSITION . - Soit $F \in W^{1,\infty}(R_n, \mathbb{R}^n)$, $\phi_\varepsilon = I + \varepsilon F$ $(\varepsilon \in \mathbb{R})$, $\Omega_\varepsilon = \phi_\varepsilon(\Omega)$, $\lambda_n(\Omega)$ une valeur propre simple de (1_Ω) , la limite suivante existe :

$$\lim_{\varepsilon \to o} \frac{\lambda_n(\Omega_\varepsilon) - \lambda_n(\Omega)}{\varepsilon}$$ et on la note $\lambda_n'(\Omega, F)$.

La démonstration se fait par changement de variable : ce qui ramène à un domaine fixe avec un opérateur dépendant de ε et on utilise alors KATO [13] Ch VII où l'hypothèse analytique n'est pas essentielle.

REMARQUE . - Calcul explicite donne d'abord une expression assez peu maniable faisant intervenir des intégrales portant sur Ω . Si les fonctions sont assez régulières, on peut exprimer cette dérivée à l'aide d'intégrales de bord (par application de formules de Green) :

$$\lambda_n'(\Omega, F) = A_1 + A_2 \quad \text{où}$$

$$A_1 = \int_{\partial\Omega} (+ A\ |\nabla u|^2 + Bu^2 - \lambda cu^2)(F|\nu)\ d\sigma$$

$$A_2 = \int_{\partial\Omega} \left\lfloor (\nabla(LAu^2)|F) + \ell Au^2(\nabla^* F - (DF\nu|\nu))\right\rfloor d\sigma$$

REMARQUE . - dans le cas de l'équation de Helmholtz (en remarquant que dans ce cas $\nabla u = \frac{\partial u}{\partial \nu}$ sur le bord : ici $V = H_o^1(\Omega)$)

$$\lambda_n'(\Omega, F) = -\int_{\partial\Omega} \left|\frac{\partial u}{\partial \nu}\right|^2 (F|\nu)\ d\sigma \qquad \text{(" formule de Hadamard ")} .$$

Dans le cas d'une valeur propre multiple, l'expression est plus compliquée ; indiquons que pour l'équation de Helmholtz on trouve que si $(\lambda_{n+j}(\Omega))_{j=1,\dots p}$ est une valeur propre d'ordre p , $(\lambda'_{n+J}(\Omega, F))_{j=1,\dots,p}$ sont les valeurs propres de la matrice B de terme général

$$B_{ij} = -\int_{\partial\Omega} \frac{\partial u_{n+i}}{\partial \nu} \frac{\partial u_{n+j}}{\partial \nu} (F.\nu)\ d\sigma$$

où $(u_{n+i})_{i=1,\dots,p}$ sont les fonctions propres associées aux $(\lambda_{n+i})_{i=1,\dots,p}$.

REMARQUE . - Dans la pratique numérique, il est impossible de distinguer une valeur propre multiple de valeurs propres très rapprochées. Cependant (cf MICHELETTI [14]) : pour tout $\epsilon > 0$, il existe $\Omega^* \in V_\epsilon^{(3)}(\Omega)$ pour lesquels toutes les valeurs propres de (1_Ω) sont simples.

Ceci rend légitime l'utilisation de la formule concernant les valeurs propres simples dans les applications numériques.

II . TECHNIQUES DE MISE EN OEUVRE

REMARQUE . - Un algorithme de recherche de domaine optimal devra générer successivement plusieurs domaines approchés dans lesquels on devra résoudre un problème du type (1_Ω) ; les problèmes et les techniques de mise en oeuvre en résultent ; on se limite naturellement à la dimension 2 .

(1) Eléments finis et domaines variables

Problème : trianguler automatiquement et économiquement les domaines générés par l'algorithme sans intervention du programme (travail en batch-processing).

Deux solutions :

. on se restreint à des domaines Ω image d'un domaine de référence $\hat{\Omega}$ que l'on triangule au départ $(\hat{\mathcal{T}}_h)$; la triangulation de $\Omega = \mathcal{T}_h = \phi(\hat{\mathcal{T}}_h)$.

. on triangule d'abord un domaine de référence $\hat{\Omega}$, d'où un domaine de référence approché $\hat{\Omega}_h$; on travaille directement sur des domaines " discrets " Ω_h et à chaque étape on construit un difféomorphisme ϕ_h transportant la triangulation.

Exemple de triangulation de référence :

Si l'on prend pour $\hat{\Omega}$ le disque unité et des éléments finis de type $\mathcal{P}_1$, la triangulation suivante est très satisfaisante : les noeuds sont définis par

$z_{pq} = \rho_p e^{i\theta_q}$ où :

$$\begin{cases} p = 0, \ldots, n \\ q = 1, \ldots, q_m(p) \quad \text{avec} \quad q_m = 6p \\ \rho_p = ph \qquad \theta_q = q \dfrac{2\pi}{q_m} \end{cases}$$

h est le " pas de la triangulation " .

REMARQUE . - Pour certains problèmes où l'on prévoit d'obtenir des ouverts " à coins " il sera commode de prendre pour $\hat{\Omega}$ un triangle, un carré etc ...

. Transport de la triangulation

– Pour éviter d'aplatir les angles dans le transport, on choisit $\emptyset$ conforme dans $\bar{\hat{\Omega}}$.

– si l'on travaille directement sur les domaines " discrets " on pourra prendre $\emptyset_h$ holomorphe dans un voisinage de $\hat{\Omega}$ et telle que : $|\emptyset_h'(z_k)| > \epsilon_c$ pour tout $z_k \in \hat{\mathcal{C}}_h$.

REMARQUE . – Si $\hat{\theta}_m$ est le plus petit angle de $\hat{\mathcal{C}}_h$ et si $\alpha < \hat{\theta}_m$ et $c_m = \mathrm{Inf}\left(\frac{1}{R(\Gamma_{2h},A_{1,h})} + \frac{1}{R(\Gamma_{3,h},A_{1,h})}\right)$ où l'Inf est pris sur tous les noeuds $A_{i,h} \in \mathcal{C}_h$

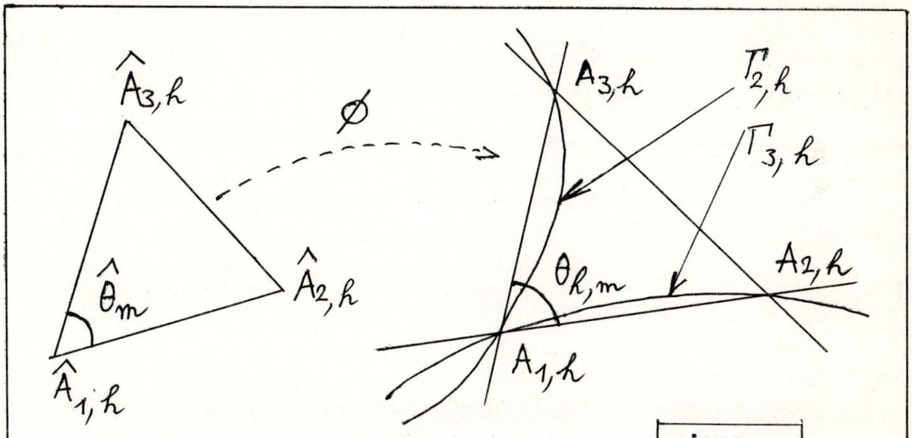

FIGURE 1

$R(\Gamma_{j,h}\ A_{i,h})$ désignant le rayon de courbure (arithmétique) de l'arc Γ_j en $A_{i,h}$ (cf figure [1]) alors si $h \le 2c_m(\hat{\theta}_m - \alpha)$, on a : $\theta_{h,m} \ge \alpha$.

(2) Le Problème discret

. pour la formation des matrices on a utilisé la méthode de formation progressive des coefficients (J. CEA) [3,2]

. le problème de valeurs propres discret : $Ax = \lambda Bx$ A et B sont de g grandes matrices creuses définies positives : il suffit de calculer les plus petites valeurs propres ; les méthodes directes (Jacobi, QR) sont à éviter pour des raisons de mémoire ; on utilise une méthode d'itération simultanée généralisant la méthode de la puissance cf JENNINGS [11] BATHE [1] , ce dernier aimablement indiqué par GLOWINSKI .

(3) Choix de la famille de domaines (dans les applications numériques) .

notations : $\mathcal{H}(\bar{\Omega})$: fonctions holomorphes au voisinage de Ω .

$\mathrm{Conf}\,(\bar{\Omega})$: transformation conformes dans un voisinage de Ω .

$\mathrm{Conf}_{\mathscr{C}_h,\varepsilon}(\bar{\Omega}) = \{ \phi \in \mathcal{H}(\bar{\Omega}) \mid \forall z_k \in \mathscr{C}_h,\ |\phi'(z_k)| > \varepsilon \}$

conformément à l'esprit des éléments finis on considèrera des familles de l'un des types suivants :

(a) $L_N = L_\cap \{ \phi(\hat{\Omega}) \mid \phi \in \mathrm{Conf}(\bar{\hat{\Omega}})_\cap \mathbb{P}_N \}$ avec $\mathbb{C}_{N-1}[z] \subsetneq \mathbb{P}_N \subset \mathbb{C}_N[z]$ désigne les polynômes de degré au plus N ; les " vraies inconnues " sont ici les coefficients (ou bien les racines) des polynômes,

(b) $L_{N,\mathscr{C}_h,\varepsilon} = \{ \phi(\hat{\Omega}_h) \mid \phi \in \mathrm{Conf}_{\mathscr{C}_h,\varepsilon}(\overline{\hat{\Omega}_h})_\cap \mathbb{P}_N \}$; dans ce cas, on choisit $\mathbb{P}_N$ tel que les noeuds de $\partial\hat{\Omega}_h$ forment un ensemble $\mathbb{P}_N$ - unisolvant, les " vraies inconnues " sont ici les coordonnées des noeuds de $\partial\Omega_h$, .

(4) Expressions du gradient

Au niveau de la mise en oeuvre, on considère la fonctionnelle comme ne dépendant que de $2N$ paramètres <u>réels</u> (les parties réelles et imaginaires des " vraies inconnues ") ; on note $J'(a,b) = (G(a)|b)$. Pour des raisons de simplicité, on se limite à l'équation de Helmoltz et à une valeur propre simple : le calcul se fait alors à partir de (2) .

(a) J exprimé à partir des coefficients

$a = ((\alpha_p,\beta_p))_{p \le P}$ où $(\alpha_p + i\beta_p)$ sont les coefficients du polynôme ϕ .

Les composantes du gradient au point a sont :

$G(a) = (\gamma_1, \delta_1, \dots, \gamma_q, \delta_q, \dots, \gamma_P, \delta_P)$ où

$$\left\{\begin{aligned} \gamma_q &= \sum_{p \le Q} \gamma_{qp}^{(1)} \alpha_p + \gamma_{qp}^{(2)} \beta_p \\ \delta_p &= \sum_{p \le P} \delta_{qp}^{(1)} \alpha_p + \alpha_p + \delta_{qp}^{(2)} \beta_p \end{aligned}\right.$$

avec

$$\left\{\begin{aligned} \gamma_{qp}^{(1)} &= \int_0^{2\pi} K_p(\theta) \cos((p-q)\theta)d\theta \ ; \quad \gamma_{qp}^{(2)} = -\int_0^{2\pi} K_p(\theta) \sin((p-q)\theta)d\theta \\ \delta_{qp}^{(1)} &= \int_0^{2\pi} K_p(\theta) \sin((p-q)\theta)d\theta \ ; \quad \delta_{qp}^{(2)} = \int_0^{2\pi} K_p(\theta) \cos((p-q)\theta)d\theta \end{aligned}\right.$$

où

$$K_p(\theta) = \sum_{n \le N} 2p \frac{\partial J}{\partial \lambda_n} \left(\frac{\partial u_n}{\partial \nu} (\phi(e^{i\theta}))\right)^2$$

(b) J exprimé à partir des zéros :

$a = ((\xi_p, \eta_p))_{1 \le p \le P}$ où $\zeta_p = \xi_p + i\eta_p$ sont les zéros du polynôme $\emptyset$.

Les composantes du gradient au point a sont :

$$G(a) = (\gamma_1, \delta_1, \dots, \gamma_q, \delta_q, \dots \gamma_P, \delta_P) \text{ où :}$$

$$\begin{cases} \gamma_q = \int_0^{2\pi} K(\theta) \, Re(L_{aq}(\theta)) d\theta \\ \delta_q = \int_0^{2\pi} K(\theta) \, Im(L_{aq}(\theta)) d\theta \end{cases}$$

avec

$$K(\theta) = \sum_{n \le N} \frac{\partial J}{\partial \lambda_n} \left(\frac{\partial u_n}{\partial \nu} (\emptyset(e^{i\theta})) \right)^2$$

$$L_{aq}(\theta) = \frac{P_a(re^{i\theta})}{re^{-i\theta} - \bar{\zeta}_q} P'_a(re^{i\theta}) e^{i\theta}$$

P_a désignant le polynôme unitaire dont les zéros sont $(\zeta_p)_{1 \le p \le P}$.

(c) J exprimé à partir des coordonnées des noeuds

$a = (z_k)_{1 \le k \le K}$ où les z_k sont les coordonnées (complexes) des noeuds de $\partial\Omega_h$; l'utilisation d'une formule de quadrature :

$$-\int_{\partial\Omega} \left(\frac{\partial u}{\partial \nu}\right)^2 (F|\nu) d\sigma = -\sum \omega_k \left(\frac{\partial u}{\partial \nu}(z_k)\right)^2 (c_k \nu_{kx} + d_k \nu_{ky})$$

avec $F_k = (c_k, d_k)$, donne directement pour la gradient :

$$G(a) = (\gamma_1, \delta_1, \dots, \gamma_q, \delta_q, \dots \gamma_P, \delta_P) \text{ avec :}$$

$$\begin{cases} \gamma_q = -\sum_{n,k} \frac{\partial J}{\partial \lambda_n} \omega_k \left(\frac{\partial u_n}{\partial \nu} (z_k) \right)^2 \nu_{kx} \\ \delta_q = -\sum_{n,k} \frac{\partial J}{\partial \lambda_n} \omega_k \left(\frac{\partial u_n}{\partial \nu} (z_k) \right)^2 \nu_{ky} \end{cases}$$

III . MISE EN OEUVRE

(1) Algorithme

Il s'agit de minimiser sur une partie non convexe de $\mathbb{R}^{2N}$ une fonctionnelle continue et différentiable avec les difficultés suivantes :

. très gros coût pour évaluer la fonctionnelle : résolution d'un problème de valeurs propres discret avec de grandes matrices creuses $(Ax = \lambda Bx)$ via un programme d'éléments finis qui forme ces matrices.

. la fonctionnelle est <u>non convexe</u> : le choix de ρ est délicat (cf BENSASSON [2])

. il est délicat de s'assurer la <u>conformité</u> de la transformation ; on a utilisé la Théorème de Rouché (DIEUDONNE [7])

Ceci nous a amené à choisir une méthode de descente très simple (gradient) (CEA [3,1]) avec un algorithme sophistiqué pour le choix du ρ .

<u>REMARQUE</u> . - On divise le temps d'exécution par 2 à 4 en ajustant les valeurs propres les unes après les autres et ceci améliore aussi un peu la précision.

(2) <u>Les essais numériques</u>

Ils ont été réalisés sur l'IBM 370-168 du C.I.R.C.E. et demandent selon les cas de 1 à 10 mn de temps C.P.U. pour l'exécution en F.T.X.

Des raisons de coût et de temps nous ont ainsi obligé à nous limiter à des éléments finis de type P_1 comprenant seulement une centaine de noeuds et à n'ajuster que cinq valeurs propres. Cela a entraîné une imprécision dans l'évaluation de la fonctionnelle et du gradient qui explique les difficultés d'un certain nombre d'essais.

Les <u>résultats numériques</u> exécutés en prenant comme " vrais paramètres " les coefficients de $\emptyset$ révèlent que l'algorithme se " bloque " parfois pour deux raisons :

- soit parcequ'on n'arrive plus à déformer le domaine en conservant la conformité de $\emptyset$

- soit parceque avec un gradient non négligeable, on n'arrive pas à trouver de ρ qui fasse sensiblement décroître la fonctionnelle

. la première est dûe au choix de la famille de domaines très liés aux difféomorphismes : ce blocage devrait disparaître avec les coordonnées des noeuds frontières comme " vraies inconnues " .

. la première est dûe d'une part à la nature non convexe de la fonctionnelle et par ailleurs au manque de précision des calculs.

<u>Cependant</u>

- dans des situations à valeurs propres simples, on a fréquemment une division de la fonctionnelle par 100 à 200.

- dans le cas de valeurs propres " multiples " on a en général une division par 30 à 40 même en " partant loin du domaine souhaité ", ce qui est d'autant plus intéressant si l'on remarque que les <u>problèmes directs</u> sont alors essentiellement <u>mal conditionnés</u>.

BIBLIOGRAPHIE

[1] BATHE Solution methods for large generalized eigenvalue problems in tructural engineering, University of California (1971)

[2] BENSASSON Optim, une subroutine en Fortran IV à allocation dynamique de mémoire, pour résoudre les problèmes d'optimisation non linéaires (1974) (preprint)

[3] [, 1] CEA Optimisation, théorie et algorithme (Dunod)

[, 2] CEA Approximation variationnelle des problèmes aux limites. Ann. Inst. Fourier, 14, 2 (1964)

[4] CHENAIS On the existence of a solution in a domain identification problem (J. Math. Anal. Appli. (1975))

[5] DERVIEUX - PALMERIO Hadamard's variational formulas in optimum design problems (Communication à ce même Congrès)

[6] DERVIEUX - PALMERIO - ROUSSELET Dessin optimal d'une cellule de réacteur nucléaire (à paraître)

[7] DIEUDONNE Calcul infinitésimal (Hermann)

[8] GARABEDIAN and SCHIFFER Variational problems in the theory of elliptic partial differential equations. J. of Rat. Mechanics and Anal. 2, 137 - 171 (1953)

[9] GOULAOUIC C.I.M.E. (1973, Third Session); Théorie spectrale de problèmes aux limites irréguliers ; applications

[10] HADAMARD Mémoire sur le problème d'analyse relatif à l'équilibre des plaques élastiques encastrées. Oeuvres de Jacques Hadamard, C.N.R.S. Paris (1968)

[11] JENNINGS and ORR Application of the simultaneous iteration method to undamped vibration problems. Internat. J. for Num. Methods in Engineering 3, 13 - 24 (1971)

[12] KAC Can one hear the shape of a drum, A. Math. Monthly (1964)

[13] KATO Perturbation theory for linear operators (Springer : Die Grundlehren der mathematischen Wissenschaften)

[14] MICHELETTI Perturbazione dello spretto di un operatore di tipo variazionale in relazione ad una variazione del campo. Annali di Matematica pura ed applicate. Vol XCVII (1973)

[15] MURAT - SIMON Etude de quelques problèmes d'optimal design (communication à ce même Congrès)

[16] RAYLEIGH The theory of sound (New York Dover Publications)

A DECOMPOSITION TECHNIQUE IN INTEGER LINEAR PROGRAMMING

S. Giulianelli
CSSCCA - CNR

M. Lucertini
Istituto di Automatica
Università di Roma
Via Eudossiana, 18 - 00184 Roma

1. INTRODUCTION

The size of linear integer programming problems that can be successfully solved is generally not very large and only a relatively small number of integer variables can be considered.

In fact the normally used packages are conceived for about 150-300 variables.

The techniques normally used to solve larger integer or mixed integer problems using a branch and bound search method, are based on the "penalty" approach and the choice of suitable lower and upper bounds for the optimal value of objective function. In order to obtain such bounds the Gomory's group theoretic methods together with Lagrange multipliers have been used in many works |1,2,3,4,5,6,7,8,13,15,16|.

The solution procedure proposed in the present work makes use of a decomposition technique that generates a number of subproblems of the original one.

Let the problem be written as:

$$\begin{aligned} &\text{minimize } z, \\ &z = c^T x \\ &\text{s.t.: } Ax = b \\ &x \geq 0, \text{ integer} \end{aligned} \qquad (1)$$

where A is a matrix of rank m of order $m \times n$, $(m<n)$, x and c are n-vectors and b is an m-vector. Further, let A be partitioned as B and N, B being the optimal linear programming basis. Vectors x and c are similarly partitioned into x_B, x_N, c_B and c_N, respectively. Without loss of generality, assume that all the coefficient of A and b are integer. (This is equivalent to assuming that A and b consist of rational numbers).

Expression (1) may be written as follows:

$$\begin{aligned}
&\text{minimize } z,\\
&z = c_B^T x_B + c_N^T x_N\\
&\text{s.t.: } Bx_B + Nx_N = b\\
&x_B, x_N \geq 0, \text{ integers}
\end{aligned} \qquad (2)$$

where B is of order m×m and nonsingular N is of order m×(n-m), c_B and x_B are of order m×1, and c_N and x_N are of order (n-m)×1.

Consider the linear programming problem (2) in the updated form (3):

$$\begin{aligned}
&\text{minimize } z,\\
&z = (c_N^T - c_B^T B^{-1}N)x_N + c_B^T B^{-1}b\\
&\text{s.t.: } x_B + B^{-1}Nx_N = B^{-1}b\\
&x_B, x_N \geq 0
\end{aligned} \qquad (3)$$

The optimal conditions of a linear programming problem, $c_N^T - c_B^T B^{-1}N \geq 0$, must be satisfied, and the non integer optimum is $x_B = B^{-1}b$ and $x_N = 0$. In all but trivial cases, $B^{-1}b$ will not be all-integer.

Therefore the strategy for finding an integer optimum will be to examine certain solutions of the set $x_N \geq 0$ and integer.

2. THE GROUP THEORETIC APPROACH

There are three problems in examining the solutions of the set $\{x_N \geq 0$, integer$\}$ in general:

1) $x_N \geq 0$ and integer are not sufficient to assure that x_B will be integer;
2) $x_N \geq 0$ and integer are not sufficient to assure that the inequalities $x_B \geq 0$ will be satisfied;
3) $x_N \geq 0$ and integer are not sufficient to assure the optimality of an integral solution to (3).

When a solution $x_N \geq 0$ and integer overcomes these three problems simultaneously such x_N determines an optimal integer solution of (3).

The first problem can be resolved by adding the constraints |11, 17|

$$\sum_{j \varepsilon N} (a_{ij} - \lfloor a_{ij} \rfloor) x_j \equiv (b_i - \lfloor b_i \rfloor) \pmod{1} \qquad \forall i=1,\ldots,m \quad (4)$$

where $a \equiv b \pmod{c}$ means that a and b are congruent modulo c, or that a and b differ by an integer multiple of c (i.e., $a-b=r\,c$, r integer). In addition, a_{ij} are the updated matrix coefficients of (3) and $\lfloor a_{ij} \rfloor$ is the largest integer not larger than a_{ij}. (Note that no component of x_B appears in (4)).

In other words, satisfaction of constraints (4) assures us that for x_N, only integer values of x_B will be considered, and the objective function categorizes the optimal solution, therefore solving problem 3. Thus, if we could solve the following problem, we would overcome problems 1 and 3.

$$\begin{aligned} &\text{minimize } z \\ &z = (c_N^T - c_B^T B^{-1} N) x_N + c_B^T B^{-1} b \\ &\text{s.t.: } B^{-1} N x_N \equiv B^{-1} b \pmod{1} \\ &x_N \geq 0, \text{ integer} \end{aligned} \qquad (5)$$

or

$$\begin{aligned} &\text{minimize } z \\ &z = \tilde{c}^T x_N \\ &\text{s.t.: } Dx_N \equiv p \pmod{1} \\ &x_N \geq 0, \text{ integer} \end{aligned} \qquad (6)$$

where:

$$\tilde{c}^T = [c_N^T - c_B^T B^{-1} N]$$

$$D = B^{-1} N$$

$$p = B^{-1} b$$

From (6) follows that:

$$x_B = B^{-1} b - B^{-1} N x_N \qquad (7)$$

It is usually possible to eliminate some of the constraints of (4). Any constraints which can be shown to be congruent modulo one to other equations or congruent modulo one to linear combinations are redundant and may be dropped.

The constraints that cannot be deleted are generating constraints for the group, and are sufficient to admit only valid solutions to the group of constraints. Thus, when the group is cyclic, there is only one constraint necessary to solve the group problem |9,10,11,12,17,18|.

Nevertheless in many real cases the number of the constraints and especially the number of integer variables in (6) is too large for an efficient solution.

In the following a procedure is proposed to formulate two or more I.L.P. problems in a fewer number of variables that can be solved indipendently. The optimal solution obtained is obviously the same of problem (6).

3. DECOMPOSITION TECHNIQUE

Consider the I.L.P. problem written in the last form (6). Let

g.c.d.{·} $\stackrel{\Delta}{=}$ greater common divisor of the set of integer numbers {·}

l.c.m.{·} $\stackrel{\Delta}{=}$ least common multiple of the set of integer numbers {·}

THEOREM - If there exist a column partition of matrix D (by reordering rows and columns of D)

$$D \stackrel{\Delta}{=} |D_1 \vdots D_2|$$

and two positive integers $\{k_1, k_2\}$ such that

$$\text{i)} \quad k_1 D_2 \equiv 0 \pmod{1} \qquad (8)$$
$$k_2 D_1 \equiv 0 \pmod{1}$$

$$\text{ii)} \quad \text{For each } i=1,2 \ \exists\ (l_i, m_i) \text{ with } (d_{l_i m_i} \varepsilon D_i) \qquad (9)$$
$$\text{such that } (k_i d_{l_i m_i} \not\equiv 0 \pmod{1}) \ i=1,2$$

$$\text{iii)} \quad \text{g.c.d.}\{k_1, k_2\} = 1 \qquad (10)$$

Then the optimal solution of (6) is the same of the optimal solution of the following block diagonal form problem (reordering rows of $\tilde{c}$, x_N, P, according to D):

$$\text{minimize } z$$
$$z = \tilde{c}^T x_N$$
$$\text{s.t. } k_i D_i \, x_N^{(i)} \equiv k_i p \pmod{1} \qquad i=1,2 \qquad (11)$$
$$x_N \geq 0 \text{ integer}$$

where $x_N^T \stackrel{\Delta}{=} [x_N^{(1)^T} \vdots x_N^{(2)^T}]$ is a partition of the x_N vector according to the column partition of matrix D.

<u>Proof</u> - Let us consider the sets

$$\Delta \stackrel{\Delta}{=} \{x_N | Dx_N \equiv p(\text{mod.}1),\ x_N \geq 0,\ \text{integer}\} \tag{12}$$

$$\Delta_{k_i} \stackrel{\Delta}{=} \{x_N | k_i Dx_N \equiv k_i\, p\ (\text{mod.}1),\ x_N \geq 0,\ \text{integer}\} \tag{13}$$

$i=1,2$

k_i positive integer

Since

$$\begin{array}{l} (x_N \varepsilon \Delta) \Rightarrow x_B \text{ is an integer vector} \\ (x_N \varepsilon \Delta k_i) \Rightarrow k_i x_B \text{ is an integer vector} \end{array} \tag{14}$$

$i=1,2$

we can write

$$\Delta \subseteq \bigcap_{i=1}^{2} \Delta_{k_i} \tag{15}$$

Further, since (8) and (9), hold, we can write

$$\Delta_{k_i} = \{x_N | k_i D_i x_N^{(i)} \equiv k_i\, p\ (\text{mod.}1),\ x_N \geq 0 \text{ integer}\} \tag{16}$$

$i=1,2$

From (14) and (7) follows:

$$(x_N \ \varepsilon\ \Delta_{k_1} \bigcap \Delta_{k_2}) \Rightarrow \begin{cases} k_1\ x_B = h_1 \\ k_2\ x_B = h_2 \end{cases} \tag{17}$$

where h_1 and h_2 are positive integer m-vectors.

Then

$$x_B = \frac{h_1}{k_1} = \frac{h_2}{k_2} \tag{18}$$

and

$$k_2 h_1 = k_1 h_2 \tag{19}$$

From (19) it derives that k_1 divides each component of $k_2 . h_1$.

Then k_1 and k_2 being relatively prime, for hypothesis (10), it follows

$$h_1 = k_1\, q \quad , \quad h_2 = k_2\, q$$

with q positive integer m-vector.

Therefore

$$(x_N \in \bigcap_{i=1}^{2} \Delta_{k_i}) \quad x_B \text{ is an integer m-vector} \tag{20}$$

or

$$\Delta = \bigcap_{i=1}^{2} \Delta_{k_i} \tag{21}$$

Hence the theorem is proved.

REMARK - The problem (11), with respect to the vector x_N, is in a block diagonal form and then it can be solved with respect to each $x_N^{(i)}$ indipendently.

$$\begin{aligned} &\text{minimize } z_i \\ &z_i = \tilde{c}^{(i)^T} x_N^{(i)} \\ &\text{s.t. } k_i D_i x_N^{(i)} \equiv k_i p \ (\text{mod. } 1) \\ &x_N^{(i)} \geq 0 \text{ integer} \end{aligned} \tag{22}$$

where $\tilde{c}^T = |\tilde{c}^{(1)^T} \vdots \tilde{c}^{(2)^T}|$.

Since the components of D are rational numbers a value of k_i that satisfies (8) can be found as follows. Let:

$$d_j \triangleq j^{th} \text{ column of D} \qquad j=1,2,\ldots,(n-m) \tag{23}$$

$$\begin{aligned} &\phi_j \triangleq \{\text{set of column indices of } D_j\} \qquad j=1,2 \\ &\Gamma \triangleq \{\text{set of column indices of D}\} \end{aligned} \tag{24}$$

we can define the following linear programming problems in a single va riable

$$\begin{aligned} &\text{minimize } y_j \\ &d_j y_j \equiv 0 \ (\text{mod. } 1) \qquad j=1,2,\ldots,(n-m) \\ &y_j \geq 0 \end{aligned} \tag{25}$$

The optimum solution y_j^* of each problem is integer.

If $\{k_1, k_2\}$ are relatively prime, with:

$$k_1 = \text{l.c.m.} \{y_j^* | j \in \phi_2\}$$
$$k_2 = \text{l.c.m.} \{y_j^* | j \in \phi_1\} \quad (26)$$

the conditions (8), (9) and (10) of the previous theorem are satisfacted, and than we can solve the problem (11).

COROLLARY - If in the previous theorem the constraint (10) is dropped

i.e. $\quad \text{g.c.d.} \{k_1, k_2\} = \hat{k} \quad (27)$

with $\hat{k}$ positive integer

the solution of problem (11) is such that

$\hat{k}\, x_B$ is an integer vector

<u>Proof</u> - If:

$$k_2 = k_2' \, \hat{k} \quad (28)$$

from (15) we can write:

$$k_2' \, \hat{k} \, h_1 = k_1 \, h_2 \quad (29)$$

and then:

$$\hat{k} \, h_1 = k_1 \, q \quad (30)$$

therefore the vector $\hat{k}\, x_B$ is an integer vector equal to q.

REMARK - The problem (11) with the hypothesis of the previous corollary is a relaxation of problem (6). In many cases the optimal solution is such that x_B become an integer vector.

4. DECOMPOSITION ALGORITHM

The decomposition procedure consists first in solving the problems (25) for $j=1,2,\ldots,(n-m)$. In order to found a partition of the set Γ, if exists, that satisfies the conditions (8), (9) and (10) with $\{k_1,k_2\}$ given by (26) we can use the following algorithm (Φ is the empty set):

Algorithm:

1. Set: $\phi_1 = \phi_2 = \Omega = \phi$, $\Lambda = \{j \mid j \varepsilon \Gamma\}$

2. Take a $t \varepsilon \Lambda$, remove t from Λ, add t to the set ϕ_1

3. $\forall i \varepsilon \Lambda$ calculate:

$$g_i = \text{g.c.d. } \{y_i^*, y_t^*\}$$

   If $g_i \neq 1$ remove i from Λ and add i to Ω

4. If $\Lambda = \phi$ go to 7 otherwise go to 5

5. If $\Omega = \phi$ go to 7 otherwise go to 6

6. Take a $t \varepsilon \Omega$, remove t from Ω, add t to the set ϕ_1. Go to 3

7. $\phi_2 = \{i \mid i \varepsilon \Gamma,\ i \not\varepsilon \phi_1\}$ Stop.

Since:

$$(\text{g.c.d.}\{y_i^*,y_j^*\}=1;\ \forall i \varepsilon \phi_1,\ \forall j \varepsilon \phi_2) \Rightarrow (\text{g.c.d.}\{k_1,k_2\}=1)$$

with k_1 and k_2 calculated by (26) the partition obtained with this algorithm satisfies the conditions of the previous theorem.

REMARK - In the previous algorithm it is sufficient to consider only the different values of y_j^*.

REMARK - The subproblem defined by the set ϕ_1 cannot be further decomposed. On the other hand the same decomposition procedure can be applied to the subproblem defined by the set ϕ_2.

5. CONCLUSIONS

In this work, using the group theoretical approach we point out some conditions on the $B^{-1}N$ matrix, often verified in practice, that make it possible to transform the system of linear congruences (constraints of problem 2) in a block diagonal form. In some cases, using this procedure, the number of constraints can increase with respect to the number of constraints of problem 2. However, the problem can be solved indipendently for the variables associated with each block.

This procedure leads to the indipendent solution of a number of subproblems in a smaller number of variables.

In the worst case each subproblem requires the same number of constraints as the original problem, but generally this number is smaller.

REFERENCES

|1| J.J.H. FORREST, J.P.H. HIRST, J.A. TOULIN, Practical solution of large mixed integer programming problems with UMPIRE. Management Science, vol. 20, n. 5, January 1974.

|2| G. MITRA, Investigation of some branch and bound algorithms for (0-1) mixed integer linear programming. Mathematical Programming 4, pp. 155-170, 1973.

|3| M. SHAW, Review of computational experience in solving large mixed integer programming problems. pp. 406-412. Applications of Mathematical Programming Techniques, English Universities Press, London 1970.

|4| A.M. GEOFFRION, G.W. GRAVES, Multicommodity distribution system design by Benders decomposition. Management Science, vol.20, n. 5, January 1974.

|5| G. GALLO, E. MARTINO, B. SIMEONE, Group optimization algorithms and some numerical results via a branch and bound approach. Giornate AICA su "Tecniche di Simulazione ed Algoritmi". Milano, Nov. 1972.

|6| J.F. SHAPIRO, Dynamic programming algorithms for the integer programming problem I: the integer programming problem viewed as a knapsack-type problem. Operation Research, 16 January 1968.

|7| J.F. SHAPIRO, Group theoretic algorithms for the integer programming problem II: extension to a general algorithm. Operation Research 16, September 1968.

|8| J.A. TOMLIN, Branch and bound methods for integer and non-convex programming. Integer and non-linear programming, cap. 21, North-Holland, Amsterdam 1970.

|9| L.A. WOLSEY, Extensions of the group theoretic approach in integer programming. Management Science, vol. 18, n. 1, September 1971.

|10| S. ZIONTS, Linear and integer programming, Prentice-Hall, 1974.

|11| T.C. HU, Integer programming and network follows. Addison-Wesley Publishing Company, 1969.

|12| R.E. GOMORY, Some polyhedra related to combinatorial problems. Linear Algebra and Its Applications, n. 2, 1969.

|13| D.E. BELL, Improved bounds for integer programs: a supergroup approach. Research Memorandum of IIASA, November 1973.

|14| H. GREENBERG, Integer programming. Academic Press, 1971.

|15| A.M. GEOFFRION, Lagrangean relaxtion and its uses in integer programming. Western Management Science Institute, Working Paper n. 195.

|16| M.L. FISHER, J.F. SHAPIRO, Constructive duality in integer programming. Massachusetts Institute of Technology. Working Paper OK 008-72, April 1972.

|17| R.S. GARFINKEL, G.L. NEMHAUSER, Integer programming. John Wiley and Sons, 1972.

|18| G.S. MOSTOW J.H SAMSON, I.P. MEYER, Fundamental structure of algebra. Mc Graw-Hill, New York, 1963.

AN INTEGRATED THEORY OF PROBLEMS AS AN ALGEBRAIC BASE FOR COMPLEXITY UNDERSTANDING AND AUTOMATIC PROBLEM SOLVING.

Giovanni Guida - Dino Mandrioli - Amedeo Paci - Marco Somalvico

Milan Polytechnic Artificial Intelligence Project - MP-AI Project
Milan, Italy.

1. INTRODUCTION

The studies about the theory of problem solving, to which belong the results that we shall present in this paper, are intended to achieve the following main goals :

- a rather precise understanding of the human behaviour in problem solving activity;
- a clear definition of what we mean by an Automatic Problem Solver (APS);
- the formulation of an Abstract Theory of Problem-Solving which can clearly point out the theoretical possibilities and the limits of an Automatic Problem Solver;
- a proposal of an efficient structure of an Automatic Problem Solver which can perform the three basic activities of selection, search and learning;
- the formulation of a Theory of Problems which can be helpful as a theoretical base in the design of an Automatic Problem Solver;
- further investigations about Automatic Problem Solvers as non-deterministic interpreters of an high-level representation language and as automatic programmers;
- implications of the developped theories on fields of actual interest; e.g., Industrial Robotics, Computer Aided Medical Diagnosis, Intelligent Data Base Management Systems, Question Answering Systems, etc.

These studies are being developped at the Milan Polytechnic Artificial Intelligence Project since three years.

The purpose of this paper is to present some new results achieved by the authors in the above mentioned fields of research. The paper is divided into two Parts.
In part A, which is constituted by the Sections II and III, an Integrated Theory of Problems in presented in a formal way.
In Part B, which is constituted by the Sections IV and V, some implications of the theory and important directions of future research are described.

More in detail the paper is devoted to present :

- in Section II, the basic definitions and results of the formalization of the State-Space Approach to Problem-Solving;
- in Section III, the basic definitions and results of the formalization of the Problem-Reduction Approach to Problem-Solving.
- in Section IV, a first cut Theory of Complexity, which constitutes an unformal base for the definition of a measure of complexity;
- in Section V, our basic point of view on Problem-Solving and a detailed unformal structure of an Automatic Problem Solver.

Part A - The Formal Theory

In this first part the formalization of the classical State-Space and Problem-Reduction Approaches to Problem-Solving [6], [9] is presented.
This formalization is achieved by means of an algebraic tool strictly connected with the Theory of Graphs [1], [3] and with the Theory of the AND/OR Graphs. Because of the unitary way in which the two different approaches have been considered during the formalization, the outcoming theory can in fact be called an Integrated Theory of Problems.

II. STATE-SPACE : FORMALIZATION

In this Section we present some basic definitions and results of the formalization of the State Space Approach to Problem-Solving. A further investigation of these topics may be found in [4]. The Algebraic Theory of Automata [3] and the Theory of Graphs [1] were both taken into account in setting up the following theory.

The prooves of the theorems of this Section are omitted for the sake of brevity and can be found by the reader in [4].

We first present some basic definitions of the Theory of Graphs which shall be useful for the following investigations.

Definition 2.1

A (directed, labeled) graph is a triple $G = (V,A,R)$ where :

- $V = \{v_o, \ldots, v_{n-1}, \ldots\}$ is a set of elements called the vertices of G;
- $A = \{a_o, \ldots, a_{m-1}, \ldots\}$ is a set of elements called the labels of G;
- $R = \{R_{a_o}, \ldots, R_{a_{m-1}}, \ldots\}$ is a set of functions from V into V.

□

Definition 2.2

An arc of a graph $G = (V,A,R)$ is a couple of vertices $u=(v_o,v_1)$ such that :

$$(\exists a)\ ((a \in A) \wedge (v_o\, R_a = v_1)) \qquad (2.1)$$

the vertex v_o is called the initial vertex and the vertex v_1 is called the final vertex.

□

Definition 2.3

A loop is an arc $u = (v_o, v_1)$ such that $v_o = v_1$.

□

Definition 2.4

A path of a graph $G = (V,A,R)$ from a vertex $v_i \in V$ to a vertex $v_f \in V$ is a finite sequence of vertice $\mu = (v_i \ldots, v_1, \ldots, v_{k-1}, v_f)$

such that :

$$(\exists x)\ ((x \in A^{+}) \wedge (x = a_{i_1} \dots a_{ik}) \wedge$$

$$(v_1 = v_1 R_{a_{i_1}}) \wedge \dots \wedge (v_f = v_{k-1} R_{a_{i_k}}) \qquad (2.2)$$

where:

$$A^{+} = A^{*} \ \{\varepsilon\}$$

The string x is called a generating string of μ. Moreover we say that the path μ has length k.

□

We can now present the basic definitions of the State-Space Approach.

Definition 2.5

A (deterministic) problem schema M is a triple $M = (S, \Sigma, \Gamma)$ where :

- $S = \{s_o, s_1, \dots, s_{n-1}, \dots\}$ is a set of elements called the states of M;
- $\Sigma = \{\sigma_o, \sigma_1, \dots, \sigma_{m-1}, \dots\}$ is a set of elements called the inputs of M;
- $\Gamma = \{\gamma\sigma_o, \gamma\sigma_1, \dots \gamma\sigma_{m-1}, \dots\}$ is a set of functions of S into S called the operators of M.

□

Definition 2.6

A (deterministic) problem $\tilde{P}$ is a quintuple $\tilde{P} = (S, \Sigma, \Gamma, i, f)$: where (S, Σ, Γ) is a (deterministic) problem schema, and :

- $i \in S$ is called the initial state;
- $f \in S$ is called the final state.

□

Definition 2.7

A (deterministic) extended problem P is a quintuple $P=(S, \Sigma, \Gamma, I, F)$ where (S, Σ, Γ) is a (deterministic) problem schema and :

- $I \subseteq S$ is called the set of the initial states;
- $F \subseteq S$ is called the set of the final states.

□

Definition 2.8

A solution of the problem $P = (S, \Sigma, \Gamma, i, f)$ is a string :

$$x = \sigma_{i_1} \sigma_{i_2} \ldots \sigma_{i_k} \in \Sigma^* \qquad (2.3)$$

such that :

$$i \gamma_x = f \qquad (2.4)$$

where :

$$\gamma_x = \gamma_{\sigma_{i_1}} \gamma_{\sigma_{i_2}} \cdots \gamma_{\sigma_{i_k}} \qquad (2.5)$$

(i.e. γ_x is made up by the composition of operators), and γ_ε is the identify function on S, if ε is the null string.

□

Definition 2.9

A solution of the extended problem $\tilde{P} = (S, \Sigma, \Gamma, I, F)$ is a string $x \in \Sigma^*$ such that :

$$(\exists i)\ (\exists f)\ ((i \in I) \wedge (i \gamma_x = f) \wedge (f \in F)) \qquad (2.6)$$

□

Definition 2.10

The solution set of a (extended) problem P ($\tilde{P}$) is the set $X_P \subseteq \Sigma^*$ ($X_{\tilde{P}} \subseteq \Sigma^*$) which contains all the solutions of P ($\tilde{P}$).

□

We outline that the solution set X_P of a problem P is not necessarily finite. We are now able to introduce some initial formal properties of these notions.

Theorem 2.1. Given an extended problem $\tilde{P} = (S, \Sigma, \Gamma, I, F)$ we have :

$$X_{\tilde{P}} = \bigcup_{P_i \in E_{\tilde{P}}} X_{P_i} \qquad (2.7)$$

where :

$$E_{\tilde{P}} = \left\{ P_i \mid P_i = (S, \Sigma, \Gamma, i, f) \wedge (i \in I) \wedge (f \quad F) \right\} \qquad (2.8)$$

□

Although this theorem states a close relation between the solution of an extended problem and the solution of a set of problem, there is no indication about the methods of how to "reduce" in a general case the solution of an extended problem $\tilde{P}$ to the solutions of the set of problems $E_{\tilde{P}}$.

In fact this "reduction" is closely related to the search strategy adopted in the problem solving process.

Conversely, we want to focus our interest in the following pages only on problems and their properties.

We now outline explicitely the close relation existing between the definition of problem and the definition of graph.

Definition 2.11

The underlying graph of the problem $P = (S, \Sigma, \Gamma, i, f)$ is the graph $G = (S, \Sigma, \Gamma)$.

□

We assume as well known the concept of length of a string x that we shall denote by $l(x)$.
We outline that if $x \in \Sigma^*$ is the solution of a problem $P = (S, \Sigma, \Gamma, i, f)$ $l(x)$ is exactly the number of expansions required to obtain the solution x, i.e. the number of times that an operator $\gamma_\sigma \in \Gamma$ has to be applied.

Definition 2.12

A (k-step) solution of a problem $P = (S, \Sigma, \Gamma, i, f)$ is a solution $x \in \Sigma^*$ of P such that $l(x) = K$.

□

Definition 2.13

The (k-step) solution sequence generated by the (k-step) solution $x = \sigma_{i_1} \sigma_{i_2} \dots \sigma_{i_k}$ of the problem P $(S, \Sigma, \Gamma, i, f)$ is the sequence of states:

$$G_x^{(k)} = (i, s_1, s_2, \ldots, s_{k-1}, f) \tag{2.9}$$

such that :

$s_1 = i \gamma \sigma_{i_1}$

$s_2 = s_1 \gamma \sigma_{i_2}$

.
.
.

$$s_{k-1} = s_{k-2} \gamma \sigma_{i_{k-1}} \tag{2.10}$$

$f = s_{k-1} \gamma \sigma_{i_k}$

□

It is evident that a (k-step) solution sequence of a problem is in fact a path of length k from i to f.
Let us now introduce the basic definition of cost.

Definition 2.14

A cost function c on a set Σ is a measure function of Σ^* into R_+ (set of real non negative numbers) such that :

$$c(xy) = c(x) + c(y) \qquad (\forall x)\ (\forall y)\ ((x \in \Sigma^*) \wedge (y \in \Sigma^*)) \tag{2.11}$$

□

Theorem 2.2 . A cost function c on a set Σ is completely determined by its restriction to Σ.

□

Definition 2.15

Given a cost function c on Σ , the simple cost C of a solution $x \in X_p$ of a problem $P = (S, \Sigma, \Gamma, i, f)$ is defined as :

$$C(x) = c(x) \qquad (\forall x)\ (x \in X_p) \tag{2.12}$$

□

Definition 2.16

Given a cost function K on $S \times \Sigma$ the composite cost K of a solution $x \in X_p$ of a problem $P = (S, \Sigma, \Gamma, i, f)$ is defined as :

$$K(x) = K(\tilde{x}) \qquad (2.19)$$

where :

$$\tilde{x} = (i, \sigma_{i_1})(S_1, \sigma_{i_2}) \dots (S_{q-1}, \sigma_{i_q}) \qquad (2.14)$$

iff :

$$x = \sigma_{i_1} \sigma_{i_2} \dots \sigma_{i_q}$$

and

$$G_x^{(k)} = (i, S_1, S_2, S_2, \dots, S_{q-1}, f) \qquad (2.16)$$

is the (q-step) solution sequence generated by the (q-step) solution x of P.

□

Definition 2.17

A solution $\bar{x} \in \Sigma^*$ of a problem $P = (S, \Sigma, \Gamma, i, f)$ is minimal iff :

$$l(\bar{x}) = \min_{x \in X_p} \{ l(x) \} \qquad (2.17)$$

□

Definition 2.18

A solution $\bar{x} \in \Sigma^*$ of a problem $P = (S, \Sigma, \Gamma, i, f)$ is simply (compositely) optimal iff :

$$C(\bar{x}) = \min_{x \in X_p} \{ C(x) \} \; (K(\bar{x})) = \min_{x \in X_p} \{ K(x) \}) \qquad (2.18)$$

□

We outline that in general neither the existence nor the uniqueness of a simply (compositely) optimal o r minimal solution of a problem can be proved.

III. PROBLEM-REDUCTION : FORMALIZATION

In this Section we present some basic definitions and results of the formalization of the Problem- Reduction Approach to Problem-Solving. The adopted formalism is congruent with that one used in the view setting up an unitary, integrated theory.

We first present some basic definitions of the Theory of Graphs which shall be useful for the following investigations.

Definition 3.1

A (directed, labeled) AND/OR graph is a quadruple $\Delta = (V,A,R,W)$ where :

- $V = \{ v_o, \ldots, v_{n-1}, \ldots \}$ is a set of elements called the vertices of ;
- $A = \{ a_o, \ldots, a_{m-1}, \ldots \}$ is a set of elements, called the labels of ;
- $R = \{ R_{a_o}, \ldots, R_{a_{m-1}}, \ldots \}$ is a set of functions from V into V;
- $W = \{ w_o, \ldots, w_{u-1}, \ldots \}$ is a set of elements called the AND/OR constraints of Δ and associated to the vertices.,

where for each $i = 0, \ldots, n-1, \ldots$ associated to the vertex v_i, we have:

$$w_i = \{ d_{i_1}, d_{i_2}, \ldots, d_{i_{k_i}} \} , \qquad K_i \geqslant 0 \qquad (3.1)$$

and for each $j = i_1, i_2, \ldots i_{k_i}$:

$$d_j = a_{J_1} \, a_{J_2} \ldots a_{J_l} \in A^* \qquad (3.2)$$

such that :

$$\bar{d}_j \subseteq \xi_i \qquad (3.3)$$

where :

$$\bar{d}_J = \{ x \mid x \text{ is an element of } d_J \} \subseteq A \qquad (3.4)$$

$$\xi_i = \{a_s \mid (a_s \in A) \wedge (v_i \text{ is an element of the domain of } R_{a_s})\} \subseteq A \qquad (3.5)$$

We note that :

$$d_j = \varepsilon \qquad \text{iff : } \bar{d}_J = \emptyset \qquad (3.6)$$

□

Definition 3.2

A (directed, labeled) canonical AND/OR graph Ω is an AND/OR graph $\Delta = (V,A,R,W)$ such that for each $i = 0,\dots, n-1,\dots$ we have :

$$- |w_i| = 1 , \qquad (3.7)$$

or

$$- \ w_i = \{d_1, d_2,\dots, d_j,\dots, d_{k_i}\}$$

and for each $j = 1,\dots, k_i$:

$$l(d_j) = 1 \qquad (3.8)$$

If condition (3.7) is met we say that the corresponding vertex v_i is an AND-vertex, if condition (3.8) is met we say that the corresponding vertex v_i is an OR-vertex.

□

It is obvious that the set V of the vertices of a canonical AND/OR graph Ω is partitioned into the following two subsets :

- $\bar{V} = \{v \mid (v \in V) \wedge (v \text{ is an AND-vertex of } \Omega)\}$
- $\tilde{V} = \{v \mid (v \in V) \wedge (v \text{ is an OR-vertex of } \Omega)\}$

Moreover an algorithm can be easyly defined which allows to construct for each (directed, labeled) AND/OR graph Δ a (directed, labeled) ca nonical AND/OR graph Ω which is "equivalent" to Δ. (the meaning of the word equivalent is now left to the intuition of the reader).
We shall denote by the symbol $\sqcup$ the concatenation of sequence; i.e., the associative, non commutative, operation which associates to each ordered pair of sequences A_1, A_2 the sequence $A = A_1 \sqcup A_2$ containing exactly the elements of A_1 followed by the elements of A_2.

Definition 3.3

An AND/OR path froma vertex v_i to a set of vertices $V_f = \{ v_{f_1}, v_{f_2}, \ldots, v_{f_h} \}$ of and AND/OR graph $\Delta = (V,A,R,W)$ is a finite sequence $\rho = (b_o, b_1, \ldots, b_{v_1})$ of finite sequences of vertices of Δ such that:

$b_o = (v_i)$

$b_1 = (v_1^1, v_2^1, \ldots, v_{p_1}^1)$

.

.

$$b_q = (v_1^q, v_2^q, \ldots, v_{r_1}^q, v_{r_1+1}^q, \ldots, v_{r_1+r_2}, \ldots, v_{r_1+r_2+\ldots+r_{p_{q-1}}=p_q}^q) \qquad (3.9)$$

.

.

$b_k = (\hat{v}_1^k, \ldots, \hat{v}_{h=p_k}^k)$

where :

$$- b_1 = (v_1^1, v_2^1, \ldots, v_{p_1}^1) \quad \text{iff :} \qquad (3.10)$$

$$(\exists d_j)((d_j \in w_o) \wedge (d_j = a_{j_1} a_{j_2} \ldots a_{j_{p_1}}) \wedge$$
$$(v_o R_{a_{j_1}} = v_1^1) \ (v_o R_{a_{j_2}} = v_2^1) \wedge \ldots \wedge (v_o R_{a_{j_{p_1}}} = v_{p_1}^1)) \qquad (3.11)$$

- for $q = 2,3,\ldots,k$

$$b_q = \bigsqcup_1^{p_{q-1}} \bar{b}_i \qquad (3.12)$$

where :

$$\bar{b}_i = (\hat{v}_i^{q-1}) \quad \text{only if :} \qquad (3.13)$$

$$v_i^{q-1} \in V_f \qquad (3.14)$$

$$\bar{b}_i = (v^q_{r_1+\ldots+r_{i-1}+1}, v^q_{r_1+\ldots+r_{i-1}+2}, \ldots, v^q_{r_1+\ldots+r_{i-1}+r_i}) \qquad (3.15)$$

only if :

$$(\exists d_j)((d_j \in w_i^{q-1}) \wedge (d_j = a_{j_1} a_{j_2} \dots a_{j_{r_i}}) \wedge$$

$$(v_i^{q-1} Ra_{j_1} = v^q_{r_1+\dots+r_{i-1}+1}) \wedge \dots \wedge \qquad (3.16)$$

$$(v_i^{q-1} Ra_{j_{r_i}} = v^q_{r_i+\dots+r_{i-1}+r_i}))$$

(where w_i^{q-1} denotes the $w \in W$ associated to the vertex v_i^{q-1})

- b_{k-1} contains at least one vertex which has not the superscript $\wedge$

- b_k is a permutation of the elements of V_f.

Moreover we say that the path ρ has <u>length</u> K-1.

□

We can now present the basic definitions of the Problem-Reduction Approach. We outline that in our formalization we have defined separately the syntactic aspects and the semantic ones ,in order to obtain a better evidence of the theory. In fact the semantics is presented in Definitions 3.4 and 3.5, and the syntax in Definitions 3.6, 3.7, 3.9 and 3.10.

We first introduce in a formal and general way, the concept of relations between problems, on which our formalization of the Problems-Reduction Approach is based. All the possible relations existing between problems which one of any interest for us can be based on the "comparison" of their solution sets as in the following definitions :

<u>Definition 3.4</u>

An <u>implicant</u> of a problem P is a couple $L = (\Pi, \Psi)$ where :

- $\Pi = \{P_1, P_2, \dots, P_k\}$ is a finite set of problems;
- Ψ is a mapping of $X_{P_1} \times X_{P_2} \times \dots \times X_{P_k}$ into X_P.

We will write :

$$\Pi \xrightarrow{\Psi} P \qquad (3.17)$$

If Ψ is a mapping onto X_p, then L is said a full implicant of P. In this case, we will write :

$$\Pi \xrightarrow{\Psi} P \tag{3.18}$$

□

Some important and usual particular cases of this general definition can be found by the reader in [4].

Definition 3.5

An implicance schema is a couple $\Xi = (\mathfrak{P}, \Theta)$ where :

- $\mathfrak{P} = \{P_o, P_1, \ldots, P_{n-1}, \ldots\}$ is a set of problems;
- $\Theta = \{\Lambda_o, \Lambda_1, \ldots, \Lambda_{m-1}, \ldots\}$ is a set such that $|\Theta| \leq |\mathfrak{P}|$

and for each $j = 0, 1, \ldots, m-1, \ldots$:

$$\Lambda_j = \{L_o, L_1, \ldots, L_{p_j}\} \tag{3.19}$$

is a finite set of implicants of the problem $P_j \in \mathfrak{P}$ such that for each $i = 0, 1, \ldots, p_j$;

$$\Pi_i \subseteq \mathfrak{P} \tag{3.20}$$

□

Definition 3.6

A reduction schema N is a triple $N = (\mathfrak{P}, \Sigma, \Gamma)$ where

- $\mathfrak{P} = \{P_o, P_1, \ldots, P_{n-1}, \ldots\}$ is a set of problems;
- $\Sigma = \{\sigma_o, \sigma_1, \ldots, \sigma_{m-1}, \ldots\}$ is a set of elements, called the inputs of N;
- $\Gamma = \{\gamma_{\sigma_o}, \gamma_{\sigma_1}, \ldots, \gamma_{\sigma_{m-1}}, \ldots\}$ is a set of function of P into P, called the operators of N.

□

Definition 3.7

A constraint on a reduction schema $N = (\mathfrak{P}, \Sigma, \Gamma)$ is a set $Y = \{y_o, y_1, \ldots, y_{n-1}, \ldots\}$

where for each $i = 0, \ldots, n-1, \ldots$, associated to the Problem P_i, we have :

$$y_i = \{ e_1, e_2, \ldots, e_{k_i} \} \qquad k_i \geq 0 \tag{3.21}$$

where for each $J = 1,2,\ldots, k_i$:

$$e_J = \sigma_{j_1} \sigma_{j_2} \ldots \sigma_{j_{l_j}} \in \Sigma^* \tag{3.22}$$

such that :

$$\bar{e}_j \subseteq T_i \tag{3.23}$$

where :

$$\bar{e}_j = \{ x \mid x \text{ is an element of } e_J \} \subseteq \Sigma \tag{3.24}$$

$$T_i = \{ \sigma_s \mid (\sigma_s \in \Sigma) \wedge (P_i \text{ is an element of the domain of } \gamma_{\sigma_s}) \} \tag{3.25}$$

we assume that :

$$e_j = \varepsilon \qquad \text{iff :} \qquad \bar{e}_j = \emptyset \tag{3.26}$$

□

Definition 3.8

A constraint Y on a reduction schema $N = (\mathfrak{P}, \Sigma, \Gamma)$ is natural iff there exists an implicance schema $= (\mathfrak{P}, \Theta)$ such that, if we assume :

$$\mathfrak{P} = \{ P_o, P_1, \ldots, P_{n-1}, \ldots \},$$
$$\Theta = \{ \Lambda_o, \Lambda_1, \ldots, \Lambda_{m-1}, \ldots \}, \tag{3.27}$$
$$Y = \{ y_o, \ldots, y_{n-1}, \ldots \}$$

we have :

- $|\Theta| = |Y|$ (3.28)
- for each $i = 0,1,\ldots, m-1,\ldots, n-1,\ldots$:

$$y_i = \{ e_1, e_2, \ldots, e_{k_i} \} \tag{3.29}$$

and for each $j = 1,2,\ldots, k_i$

$$e_j = \sigma_{j\,1}\ \sigma_{j\,2} \ldots\ \sigma_{j\,l_j} \tag{3.30}$$

iff :

$$\Lambda_i = \{ L_1, L_2, \ldots, L_{k_i} \} \tag{3.31}$$

and

$$\Pi_j = \bigcup_{1}^{l_j}{}_s \{ P_i\ \gamma \sigma_{j_s} \} \tag{3.32}$$

□

Definition 3.9

A (deterministic) reduction problem schema is a quadruple $M = (\mathfrak{P}, \Sigma, \Gamma, Y)$ where $(\mathfrak{P}, \Sigma, \Gamma)$ is a reduction schema and Y is a constraint on $(\mathfrak{P}, \Sigma, \Gamma)$.

□

Definition 3.10

A (deterministic) reduction problem is a sextuple $Z = (\mathfrak{P}, \Sigma, \Gamma, Y, P_i, \mathfrak{P}_f)$ where $(\mathfrak{P}, \Sigma, \Gamma, Y)$ is a reduction problem schema and :

- $P_i \in \mathfrak{P}$ is called the initial problem;
- $\mathfrak{P}_f \subseteq \mathfrak{P}$ is called the set of the final problems.

□

We now outline explicitely the close relation existing between the definition of reduction problem and the definitions of graph and AND/OR graph.

Definition 3.11

The underlying graph of the reduction problem $Z = (\mathfrak{P}, \Sigma, \Gamma, Y, P_i, \mathfrak{P}_f)$ is the graph $G = (\mathfrak{P}, \Sigma, \Gamma)$

□

Definition 3.12

The underlying AND/OR graph of the reduction problem $Z = (\mathfrak{P}, \Sigma,$

Γ, Y, $P_i, \mathfrak{P}_f$) is the AND/OR graph $\Delta = (\mathfrak{P}, \Sigma, \Gamma, Y)$.

□

It is now clear that, as we have already pointed out, the formalism of the reduction problem schemata (or equivalently of the AND/OR graphs) is just the syntax of the implicance schemata, which on the other side are the semantics of the reduction problems schemata. We outline that the same reduction problem schema could have several "interpretations", i.e. associated implicance schemata, and viceversa.

Differently from the method followed in Section II for introducing the concept of solution, we are now going to define first the solution sequence and then the solution of a reduction problem.

Definition 3.13

A (k-step) solution sequence of a reduction problem $Z = (\mathfrak{P}, \Sigma, \Gamma, Y, P_i, \mathfrak{P}_f)$ is an AND/OR path ρ of the underlying AND/OR graph $(\mathfrak{P}, \Sigma, \Gamma, Y)$ from the vertex P_i to the set of vertices $V_f = P_{f_1}, P_{f_2}, \ldots, P_{f_h}$ such that :

$$V_f \subseteq \mathfrak{P}_f \tag{3.33}$$

and

$$l(\rho) = k \tag{3.34}$$

□

Definition 3.14

The (k-step) solution generated by the (k-step) solution sequence $\rho = (b_o, b_1, \ldots, b_{k+1})$ of the problem $Z = (\mathfrak{P}, \Sigma, \Gamma, Y, P_i, \mathfrak{P}_f)$ is the string $x = \mathcal{L}_1 \mathcal{L}_2 \ldots \mathcal{L}_k$ of finite sequences $\mathcal{L}_j$ of strings $\mathfrak{g}_i \in \Sigma^*$ such that if :

$$b_o = (P_i)$$
$$b_1 = (P^1_1, P^1_2, \ldots, P^1_{p_1}) \tag{3.35}$$

.
.
.

$$b_q = (p_1^q,\ p_2^q, \ldots,\ p_{r_1}^q,\ p_{r_1+1}^q, \ldots,\ p_{r_1+r_2}^q, \ldots\ p_{r_1+r_2+\ldots+r_{p_{q-1}} = p_q}^q)$$

.

.

.

$$b_{k+1} = (p_1^{k+1},\ \ldots,\ p_{p_{k+1}}^{k+1})$$

we have :

$$\mathcal{L}_1 = (\mathcal{G}^1{}_1)$$

$$\mathcal{L}_2 = (\mathcal{G}^2{}_1, \mathcal{G}^2{}_2, \ldots, \mathcal{G}^2{}_{p_1}) \qquad (3.36)$$

.

.

.

$$\mathcal{L}_q = (\mathcal{G}^q{}_1, \mathcal{G}^q{}_2,\ \ldots\ , \mathcal{G}^q{}_{p_{q-1}})$$

.

.

.

$$\mathcal{L}_k = (\mathcal{G}^k{}_1,\ \ldots, \mathcal{G}^k{}_{p_k})$$

where :

- $\mathcal{G}^1{}_1$ is exactly the d_j referred to in (3.11)

- for $q = 2,3,\ldots,\ k$:

$$\mathcal{L}_q = \bigsqcup_1^{p_{q-1}}{}_i\ \mathcal{D}_i \qquad (3.37)$$

where :

$$\mathcal{D}_i = (\varepsilon) \qquad (3.38)$$

iff in (3.12) $\bar{b}_i$ is obtained as in (3.13)

$$\mathcal{D}_i = (d_j) \qquad (3.39)$$

where d_j is exactly the string referred to in (3.16) iff in (3.12) b_i is obtained as in (3.15).

Moreover we will call the set $F_x = \{ \hat{p}_1^{k+1}, \ldots, \hat{p}_{p_{k+1}}^{k+1} \} \subseteq \mathcal{P}_f$ the covering set determined by the (k-step) solution x of Z.

□

Definition 3.15

A solution x of a reduction problem 2 = $(\mathcal{P}, \Sigma, \Gamma, Y, P_i, \mathcal{P}_f)$ is a string of finite sequences of strings of Σ^* such that x is a (k-step) solution of Z geenerated by a (k-step) solution sequence ρ of Z for some nonnegative integer k.

□

Definition 3.16

The solution set of a reduction problem Z = $(\mathcal{P}, \Sigma, \Gamma, Y, P_i, \mathcal{P}_f)$ is a set X_Z such that :

$$X_Z = \{ x \mid x \text{ is a solution of } Z \} \tag{3.40}$$

□

Definition 3.17

The total length λ of x solution x = $\mathcal{L}_1 \ldots \mathcal{L}_k$ of a reduction problem Z = $(\mathcal{P}, \Sigma, \Gamma, Y, P_i, \mathcal{P}_f)$ is defined as :

$$\lambda(x) = \sum_{1}^{k}{}_i \sum_{1}^{p_{i-1}}{}_j \, 1 \, (\mathcal{L}_j^i) \tag{3.41}$$

□

We outline that $\lambda(y)$ is exactly the number of expansions required to obtain the solution x, i.e. the number of times that an operator $\gamma_\sigma \in \Gamma$ had to be applied.

We introduce now the important concept of cost of a solution of a reduction problem.

Definition 3.18

Given a cost function c on Σ the simple cost C of a solution x = $\mathcal{L}_1, \ldots, \mathcal{L}_k$ of a reduction problem Z = $(\mathcal{P}, \Sigma, \Gamma, Y, P_1, \mathcal{P}_f)$ is defined as :

$$C(x) = \sum_{1}^{k}{}_j \sum_{o}^{p_{j-1}}{}_i \, c(\mathcal{G}_i^j) \tag{3.42}$$

Definition 3.19

A solution $\bar{x}$ of a reduction problem $Z = (\mathcal{P},\Sigma,\Gamma, Y, P_i, \mathcal{P}_f)$ is minimal iff :

$$\lambda(\bar{x}) = \min_{x \in X_Z} \{ \lambda(x) \} \qquad (3.43)$$

□

Definition 3.20

A solution x of a reduction problem $Z = (\mathcal{P},\Sigma,\Gamma, Y, P_i, \mathcal{P}_f)$ is simply optimal iff :

$$C(x) = \min_{x \in X_Z} \{ C(x) \} \qquad (3.44)$$

□

It is then possibly to define the composite cost of a solution x and a compositely optimal solution x in a fully analogous way as that one followed in Section II. We omit these definitions for the sake of brevity.

Definition 3.21

We say that a problem $P = (S,\Sigma,\Gamma, i, f)$ is solved iff a couple $T = (P, X)$ is given where :

$$X \subseteq X_P \qquad (3.45)$$

and

$$X \neq \emptyset \qquad (3.46)$$

If $X = X_p$ we say that P is fully solved.

□

We conclude this Section by pointing out the close relation existing between the two different approaches presented in Section II and III.

Theorem 3.1. Given a reduction problem $Z = (\mathcal{P},\Sigma,\Gamma, Y, P_i, \mathcal{P}_f)$ and a solution $x \in X_Z$ of Z if all the final problems of the covering set F_s determined by x are solved and if Y is a natural constraint on $(\mathcal{P},\Sigma,\Gamma)$ it is possible to construct a solution $y \in X_{P_i}$ of P_i.

Proof

The proof of this Theorem is directly obtained from Definition 3.8, 3.13, 3.18, and 3.23.
The algorithm for building up y is also easily defined.

□

Part B - Topics for Future Research

In this second part some important directions for further investigation are presented. Informal bases for the definition of a measure of complexity inside the above presented Integrated Theory of Problems and for the design of an Automatic Problem Solver are described in detail.

IV - TOWARD A MEASURE OF COMPLEXITY

In this Section we present some preliminary concepts with the aim of setting up an unformal base for future definition of a measure of complexity inside the above presented Integrated theory of Problems.

The intuitive concept of complexity can be better specified by splitting it into the following three components :

- complexity of a solution;
- complexity of the algorithm used for finding out a solution;
- complexity of a representation of a problem.

Let us now investigate these three different concepts in more detail. The complexity of a solution is determined foundamentally by its simple (composite) cost and by its length. In fact the cost takes into account the complexity of the application of each operator necessary to build up the solution, and the length takes into account the number of an operators which must be applied, i.e. it gives an estimate of the time spent for the construction of the solution. This first type of complexity, which we shall call solution complexity

is of course related to a solution of a problem in a given representation, i.e. it can be useful for compairing different solutions of a some representation of a problem.
The complexity of the algorithm used for finding out the solution must take into account the time spent for the research, the memory occupation, and the number of unsuccessful attempts done in the such activity.
This type of complexity, which we shall call algorithm complexity it therefore related to an algorithm followed in the search of a given solution of a problem in a given representation.
We can then define for each couple (solution, algorithm) of a given representation of a problem a global complexity which takes into account both the solution complexity and the algorithm complexity and gives a quite precise estimate of the complexity of the global solution process of a problem.
Before proceeding further, we note that, as we have already outlined in Section III, the solution of a reduction problem $Z = (\mathfrak{P}, \Sigma, \Gamma, Y, P_i, \mathfrak{P}_f)$ is conceptually different from the solution of P_i, which is in fact the problem that we wish to solve. However the solution of P_i can be constructed from the solution of Z if the conditions of Theorem 3.1 hold. In this case, which we call the Ibrid Approach, it is clear that the global complexity of the solution of P_i must take into account the global complexity of the Problem-Reduction part, the solution complexity of the State-Space part, and the complexity of the construction of the solution of P_i, once the solution of Z is found.

We conclude this Section by briefly investigating the third type of complexity we mentioned above, i.e. the complexity of a representation of a problem, which we shall call the representation complexity. If we refer to a given intuitive problem and to a set of different representations of it, inside the same approach or inside different approaches (i.e., State-Space, Problem-Reduction, or Ibrid), we want to define e measure of complexity which clearly points out which representation is the most suitable (natural) for the given intuitive problem.

The representation complexity must therefore take into account an "average value" of the global complexities of all the possible couples (solution, algorithm) of a given representation of a problem.

V - TOWARD THE DESIGN OF AN AUTOMATIC PROBLEM SOLVER

In this Section we present our basic point of view on Problem - Solving and first-cut, unformal, design of an Automatic Problem Solver. The standpoint of our considerations is the following one: computer science is an experimental discipline which is centered around an unitary and global goal : man-computer interaction in solving problems. This interaction can be syntetically represented as a path connecting the following basic concepts:

- intuitive problem,
- represented problem,
- solved problem.

The path itself is constituted by the three activities of formalization, automatic resolution, and matching.
These concepts are graphically illustrated in Figure 1 and they will be widely investigated in the whole Section.

The intuitive problem is an entity which independently faces the man and can be viewed as an undefined and unlimited source of information.

From it, through the activity of formalization, the man operates an extraction of a finite and precisely described amount of information, namely the represented problem. This information is chosen as valuable and sufficient in order to provide, through mechanical, or interactive, computation, the construction of the solved problem (i.e., the solution of the problem).

It is clear that in principle the solution construction can be performed in an artificial and completely automatic way by means of a general purpose computational tool that we shall call Automatic Problem Solver.

On the other hand, the activity of formalization belongs exclusi-

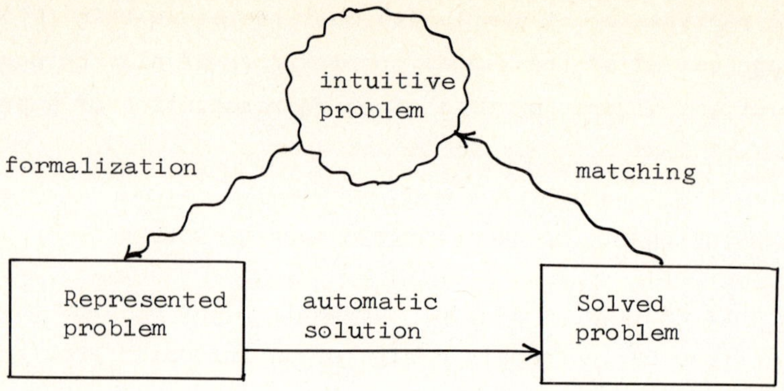

Figure 1 - Problem-Solving activity.

vely to the man.

Let us now investigate in more detail these outlined concepts.

The formalization activity, performed by the man, provides, as it has been previously exposed, the represented problem, as an artificial object which is obtained from the intuitive problem.

The invention of the represented problem consists in the precise description of a finite quantity of information which the man formulates by means of the observation of two distinct entities, namely: the intuitive problem, and the Automatic Problem Solver.
The invention of the represented problem requires that the man performs two basically different activities, in its formalization process.

The first activity is devoted to the specification of the methods and ways which "tune" the Automatic Problem Solver, considered as an originally general purpose tool, into a well precised special purpose tool which is oriented by the semantic domain from which the intuitive problem is originated.

The information described by consequence of this first activity, is called control information and it is the first part of the information contained in the represented problem.

The second activity is dedicated to the selection from the intuitive problem of a finite quantity of information which is considered by the man as useful, and, hopenfully, efficient and sufficient, in order to allow the special Problem Solver to achieve its goal of providing an automatic solution of the problem.

The information described by consequence of this second activity, is called problem information and it is the second part of the information contained in the represented problem.

It is conceptually important to observe that both the two previously described activities are done by the man with the conscience of being faced by ignorances of two different types, namely :

- what part of the problem information is actually relevant to the computer and shall be utilized in order to solve the problem;
- what is the actual way in which this relevant information shall be processed in order to construct the solved problem.

These two types of ignorances are useful to point out two functions performed by the automatic problem solver, which are intended to give artificial answers to these ignorances.

The first function, which is devoted to produce an automatic answer to the first type of ignorance, consists in an appropriate selection of one part of the information contained in the represented problem, and considered, by the Automatic Problem Solver, as useful and relevant for its activity of solving the problem.

This activity is performed by a first part of the Automatic Problem Solver, called Selector, as it is shown in Figure 2, where all the block-structure of an Automatic Problem Solver is illustrated.

Therefore we will call global represented problem the input of the Selector and selected represented problem the output of the Selector.

The second function, which is devoted to produce an automatic answer to the record type of ginorance, consists in a skillful search of the cooperation process, embracing the already selected information, which essentially makes up the solution algorithm and, thus, yields the solution of the problem.

This activity is performed by a second part of the automatic problem solver, called Searcher, as it is shown in Figure 2.

Therefore, while the input of the Searcher will be the selected represented problem, the output of the Searcher will be the solved

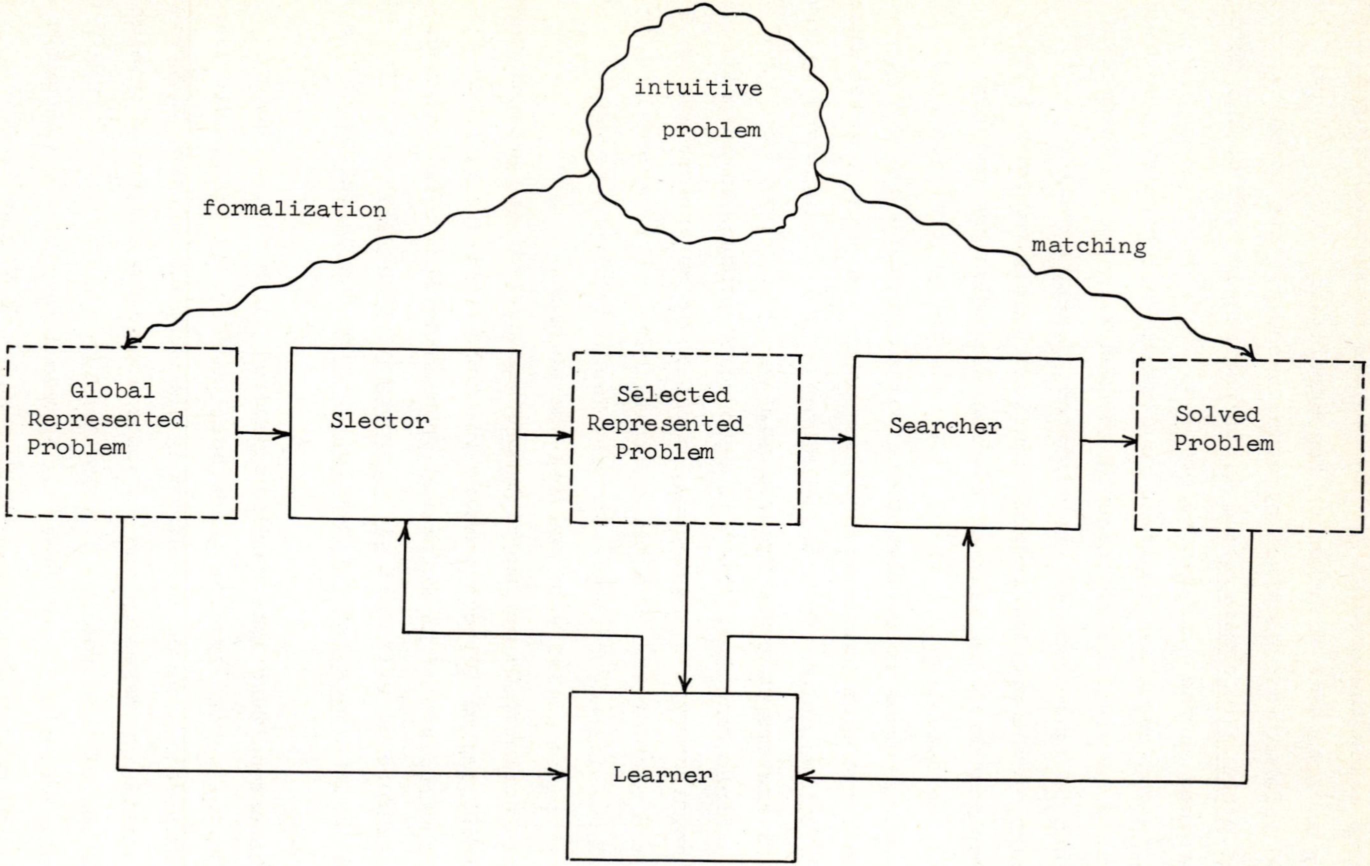

Figure 2 - Structure of an Automatic Problem Solver.

problem.

As it has been previously illustrated, the control information is the information which enables the man to specify the special configuration of the problem solver oriented towards a particular semantic domain.

Therefore, by means of the control information, the structures of the Selector and Searcher are completely specified.

This definition of the structure of the Selector and of the Searcher by the man can be considered just as an initial specification which, during the ongoing solution process, can possibly be changed and improved.

This modification and enhancing activity is the typical activity of learning which is able to provide a dynamic evolution of the structure of the Selector and of the Searcher.

This selfchanging activity is performed by a third part of the automatic problem solver, called Learner, as it is shown in Figure 2.

Therefore, the inputs of the Learner are constituted by the global represented problem, by the selected represented problem, and by the solved problem.

The outputs of the Learner are the automatically constructed and modifiable specifications of the Selector and of the Searcher.

Thus, the kernel of an automatic problem solver appears to be an artificial metasystem which is initialized by the man as an initial system, and, afterwards, can evolve itself, in a way appropriate to enhance its artificial performances in solving problems.

Therefore learning can be viewed as the ability of selfawereness of the whole automatic problem solver.

Whichever has been the method followed by the man in performing its formalization task for the construction of the represented problem it is necessary for him to choose an appropriate formalism apted both to provide a "good" represented problem and to catalyze a "valid" artificial activity for the automatic problem solver.

We can rightfully call such a formalism the Representation Language, which the man needs for cooperating with computer.

While the classic programming languages have been conceived to communicate to the computer algorithms, the representation languages can be conceived to channel to the computer represented problems.

Therefore we can also look at an Automatic Problem Solvers as at the Interpreter of the Representation Language in which the represented problem have been communicated to the computer, and, moreover, as at an Automatic Programmer.

Let us now present a first-cut design of an Automatic Problem Solver and a detailed description ofits mode of operation.

These concepts are graphically illustrated in Figure 3.

The control information and the problem information will constitute, expressed in an appropriate Representation Language, che Control-base and the Problem-base.

The automatic problem solver acts on these two bases of information as an interpreter and can perform the three basic activities of selection, search, and learning.

Its activities are controlled and organized by a monitor systhem. We now examine in detail the above outlined concepts, which are graphically illustrated in Figure 3. Let us begin with the Problem-Base.

The problem-base contains all informations on the problem to be solved (P) and its environment, which the man thinks as sufficient for the solution of P. In fact the problem-base can be built up as a set which contains: P (possibly many different representations), implicants of P, solved problems, other problems and reduction problems (having possibly P as their initial problem), simple and/or composite costs functions for the problems and the reduction problems of the problem-base, etc. The problem-base is first submitted to an ordering process which gives to the information contained in it an hierarchic ordering (e.g., a discrimination net or a tree) with the purpose of allowing an easier and more efficient exploration of it, whenever it is required by the monitor system.

The basic carachteristic of the problem-base is that it constitutes a dynamic set. In fact whenever the automatic problem solver achieves some useful results during its search activity, this is inserted

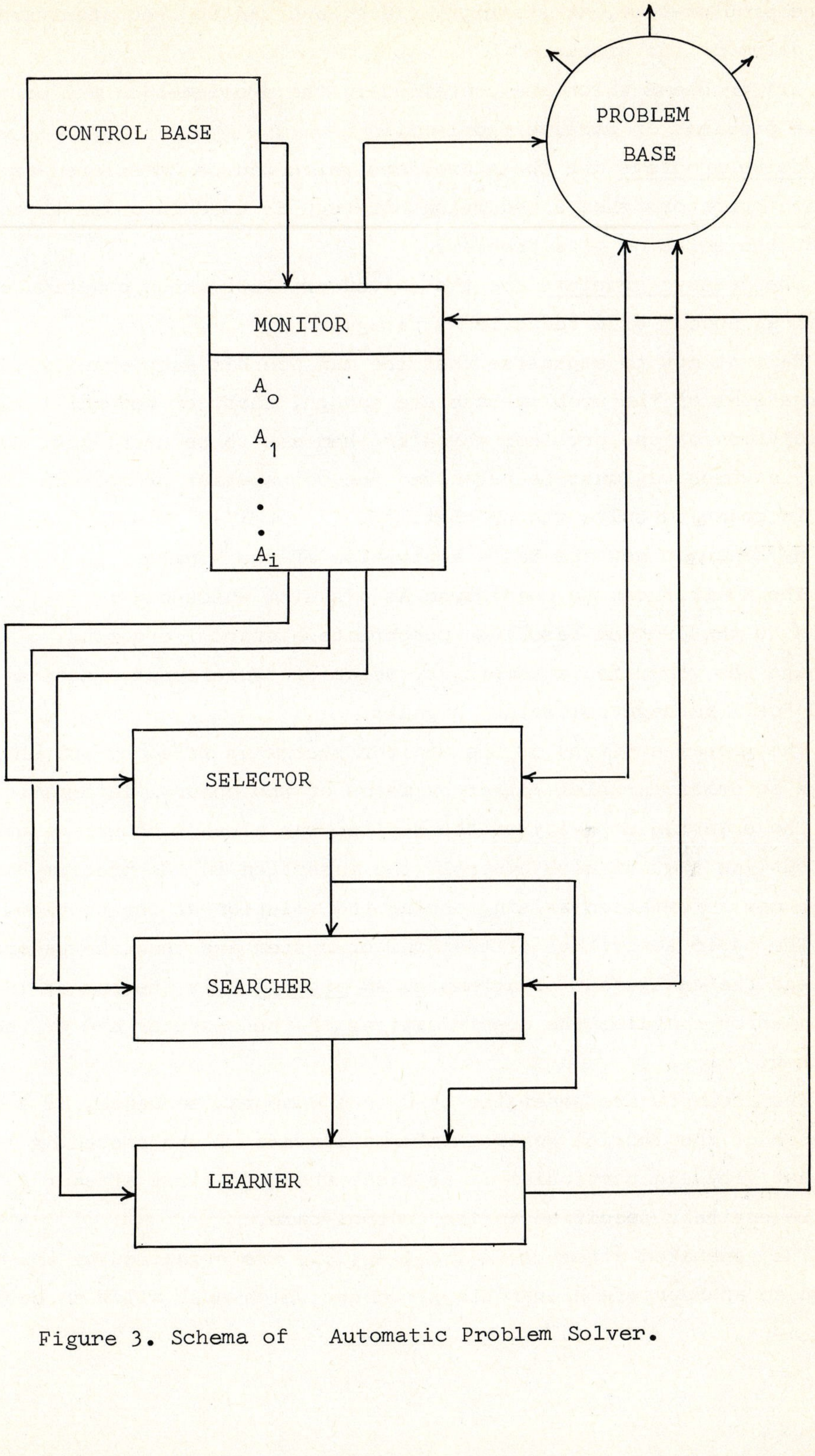

Figure 3. Schema of Automatic Problem Solver.

in the problem-base, at its right place, and can be used afterwards as a datum of the problem.

All problems which are contained in the problem-base may be either active problems or passive problems.
The active problems are those problems which can be considered as reduction operators when a reduction strategy is used in order to construct the solution of a problem.

The passive problems are the solved problems which constitute the "terminal nodes" of a reduction strategy.

We want now to emphasize that the man doesn't know exactly if the informations of the problem-base are enough, short or redundant for the solution of the problem; which of them are to be used; and, moreover, in which way must be organized the cooperation process between them in order to solve the problem.

We describe now the basic activities of the Monitor.

The Monitor can be considered as a system which can :

- give to the problem-base its appropriate hierarchic ordering;
- menage the generated attempts, by means of "interrupt", "activate" and "call garbage collector" signals.

The global activity of the Monitor system is fixed by the user of the automatic problem solver by means of the control information.

The ordering activity of the Monitor was already shown; we will only outline that it also controls the insertion in the problem-base of the new information arising during the solution of the problem.

The basic activities of the Monitor system are then the generation and the management of attempts. An attempt A is the set of information which contains the specifications of the selector and of the Searcher.

The attempts are generated by G in a temporal sequence, as a consequence of the initial control information and of the preceding learning activity. In particular it is clear that the first attempt A_o must be entirely specified in the control-base.

The generated attempts, A_o, A_1, A_2,..., are organized by the monitor in an appropriate hierarchic structure which must allow an easier management.

The attempts must be managed in such a way that they can be considered as coroutines: old attempts can be activated and new attempts cen be interrupted whenever it is useful. The attempt management is done by means of the control information. In Particular the "activate" and "interrupt" signals are generated on the basis of the computational effort done up to a certain point estimated by means of appropriate complexity measures defined in the control-base.
The attempt management must also provide a garbage-collector which distroies the old unuseful attempts, whenever it is necessary.

We can now examine in detail the content of the control-base. The control-base contains all informations that the man can draw from the intuitive problem and from his knowledge of the operating way of the artificial automatic problem solver in order to initialize the metasystem and to control its dynamic development.

In other words, the problem-solver, considered as a metasystem, is an artificial entity, existing outside the user, which has been provided by the artificial intelligence scientist as a general purpose tool.

However, the ingenuity and creativity of the user, can be exploited in order to "tune" such general purpose tool in the direction of the semantic domain from which the problem arises.

Hence, such initializstion (or specification) of the general purpose metasystem constitutes a way of defining a special purpose system. In particular the control-base must contain :

- the complete definition of the Monitor system and of the Learner;
- the first attempt A_o.

Now, we can precisely define the basic activities of the Selector, of the Searcher, and of the Learner.

The Selector can be considered as a system which can act on the problem-base and select: either a set of states to be expanded and a set of inputs for executing the expansions, if an expansion attempt is active at that time; or a set of passive problems which must take part to a reduction operation, and an set of active problems for performing the reduction, if a reduction attempt is active at that time. Of course a bidirectional exchange of informations between the Selec-

tor and the problem-base is provided.
The Searcher can be considered as a system which can act on that part of th. problem base selected by the Selector and perform the expansion or reduction operation.
Of course a bidirectional exchange of information between the Searcher and the problem-base is also provided.
The Learner can be considered as a system which can provide the Monitor with all the requested information for a correct and efficient generation and menagement of attempts taking into account all the "past experience" done by the Selector, the Searcher and the Monitor itself.
The mode of operation of the whole system is then now quite easy to explain. By means of the activity of the Selector and of the Searcher the problem-base is incrementally expanded, i.e. new passive (solved) problems are constructed by means of the already existing ones and of the active problèms. This process continues untill the problem P, that we wish to solve, becomes also a solved one.
The efficiency of the system is provided by means of an adeguate generation and menagement of attempts, i.e. by the activity of the Monitor.

REFERENCES

1. Berge, C. "Théorie des Graphes et Ses Applications. Dunod, Paris, 1958.

2. Coray G. "Additive Features in Positional Games". ORME-IP-IRIA-NATO, Nato Advanced Study Institute on "Computer oriented learning, processes, Procédures informatique d'apprentissage". Bonas (Gers), France, August-September, 1974. Imprimé en France, IRIA, Domaine de Voluceau.

3. Ginzburg, A. "Algebraic Theory of Automata". Academic Press. Inc. New York. 1968.

4. Guida G., Mandrioli D., Paci A., and Somalvico M. "A Formal Framework for an Unitary Approachto the Theory of Problem-Solving". Proceedings IIASA Workshop on Artificial Intelligence - Question Answering Systems. Laxenburg Schloss, Austria. June, 1975.

5. Hartmanis, J. and Stearns, R.E. "Algebraic Structure Theory of Sequential Machines". Prentice-Hall, "Inc. Englewood Cliffs, N.Y. 1966.

6. Nilsson, N.J. "Problem-Solving Methods in Artificial Intelligence" Mc Graw-Hill Book Company, New York. 1971

7. Pohl,I. "Bi-Directional and Heuristic Search in Path Problems". SLAC Report N. 104. Stanford Linear Accelerator Center. Stanford University. Stanford, California, May, 1969.

8. Sangiovanni Vincentelli, A. and Somalvico M. "State-Space Approach in Problem-Solving Optimization". Politecnico di Milano. Istituto di Elettrotecnica ed Elettronica. Laboratorio di Calcolatori. Relazione Interna n. 73-15. MEMO-MP-AIM-12. May, 1973.

9. Minker J., and Vanderburg G.J. "State-Space, Problem-Reduction, and Theorem Proving-Some Relationships", Communication of the ACM, February 1975, Volume 18, Number 2, pp.107-115.

CHOIX D'UNE BASE DANS L'APPROXIMATION D'UNE FONCTION

J. LEMAIRE - M. MORIOU - J. POUGET
Département Informatique
Institut Universitaire de Technologie
95, avenue de Fabron 06041 NICE CEDEX - FRANCE

SUMMARY :

Many models of physical, chemical and biological processes using the description of flow between cells lead to identifying the output of the type $t \longmapsto s(t) = \sum_{i=1}^{n} a_i \, e^{-b_i t}$ $(b_i > 0)$. These particular problems are thoroughly studied in papers and various methods are proposed. After a detailed study of these various methods which consist in determining the parameters n, $\{a_i\}$, $\{b_i\}$, we have reached the following conclusions :

1°) that the adjustment of experimental results by multi exponential function may often be obtained with very good approximation.

2°) that the adjustment of these results is but slightly sensitive to the variation of the parameters n, $\{a_i\}$, $\{b_i\}$.

The interpretation of that last point is quite explicit in the case when the method used leads to solving a linear system. This one is always ill conditionned.

3°) that very few methods allow to approach an over estimation of the error on the various coefficients the approximation.

This last point is particularly dangerous in the case when the parameters a_i and b_i have physical significance particularly in medecine. We have defined for certain methods the function bases for which the conditionning was best, i.e. for which the errors on the coefficients a_i were of the same magnitude order.

In the same way we have defined a spectral method using a moment method and allowing a calculus of the error of approximation.

SOMMAIRE :

De nombreux modèles de processus physiques, chimiques ou biologiques utilisant la notion de flux entre des cellules conduisent à identifier des sorties du type $t \longmapsto s(t) = \sum_{i=1}^{n} a_i \, e^{-b_i t}$ $(b_i > 0)$. Ces problèmes sont parfaitement étudiés dans diverses publications et des méthodes variées sont proposées. Après une étude détaillée de ces diverses méthodes qui consistent à déterminer les paramètres n, $\{a_i\}$, $\{b_i\}$ nous sommes arrivés aux conclusions suivantes ;

1°) que les ajustements de résultats expérimentaux par des fonctions multiexponentielles peuvent souvent être obtenus avec une très bonne approximation.

2°) que l'ajustement de ces résultats est très peu sensible aux variations des paramètres n, $\{a_i\}$, $\{b_i\}$. L'interprétation de ces derniers points est tout à fait claire dans le cas où la méthode utilisée conduit à résoudre un système linéaire. Celui-ci est toujours mal conditionné.

3°) que très peu de méthodes permettent d'obtenir une estimation de l'erreur sur les différents coefficients de l'approximation. Ce dernier point est particulièrement dangereux dans le cas où les paramètres $\{a_i\}$ et $\{b_i\}$ ont une signification physique, particulièrement en médecine. Nous avons défini pour certaines méthodes les fonctions de base pour lesquelles le conditionnement est le meilleur, c'est-à-dire pour lesquelles les erreurs sur les coefficients a_i ont le même ordre de grandeur. Nous avons défini une méthode spectrale utilisant une méthode de moments et permettant un calcul de l'erreur sur l'approximation.

I. - INTRODUCTION

De nombreux modèles de processus physiques, chimiques ou biologiques utilisant la notion de flux entre des cellules ou d'autres modèles conduisent à identifier des systèmes dont les sorties sont du type :

$$t \longmapsto s(t) = \sum_{i=1}^{n} a_i \; e^{-b_i t} \qquad \text{(I-1)}$$

où $a_i \in \mathbb{R}$ et $b_i \in \mathbb{R}^+$ (somme d'exponentielles réelles).

En toute généralité, on a à trouver dans un espace fonctionnel une fonction appartenant à un sous-ensemble qui n'est pas un sous espace vectoriel, approchant au mieux, selon un critère donné, une fonction numérique résultant d'une mesure expérimentale.

Dans le cas particulier, on est conduit à déterminer une base d'un sous espace définie à partir de n fonctions indépendantes du type e^{-bt}($b \in \mathbb{R}^+$), ainsi que les composantes de l'approximation dans cette base.

Ce problème particulier fait partie des problèmes classiques d'identification de systèmes ainsi que la recherche de la dimension de ces systèmes. A ce sujet, un certain nombre de questions sont soulevées.

I-1) LE NOMBRE CONSIDERABLE DE METHODES DECRITES DANS LES PUBLICATIONS POUR AJUSTER DES SORTIES EXPERIMENTALES PAR DES SOMMES D'EXPONENTIELLES REELLES

On peut donner plusieurs raisons à cela :

a) en général, le but final n'est pas d'identifier au mieux la sortie mais le système qui engendre cette sortie. Ce sont en effet les coefficients qui définissent le système qui ont des implications pratiques fondamentales (exemple en médecine dans le cas de l'analyse multicompartimentale). Comme les systèmes numériques à résoudre sont souvent mal conditionnés, on est ainsi poussé à rechercher diverses méthodes.

b) le sentiment, dans certains cas, qu'une méthode plus complexe, ou plus raffinée, devrait conduire à de meilleurs résultats pour l'identification. Cela peut d'ailleurs être justifié dans certains cas, mais ne l'est absolument pas pour d'autres méthodes qui, en plus, ne permettent pas d'aborder un calcul d'erreur.

I-2) LES PROBLEMES D'INDETERMINATION SUR LES COEFFICIENTS

Ces problèmes sont essentiellement dus à deux raisons :

a) les erreurs de mesures expérimentales,

b) surtout les problèmes de mauvais conditionnement des problèmes d'analyse numérique abordés.

La première question peut en partie être résolue par des méthodes statistiques. La seconde question dépend de la nature même des systèmes découlant des modèles choisis. C'est ainsi qu'il serait vain de chercher à identifier deux coefficients b_1 et b_2 correspondant à un modèle du type :

$$t \longmapsto s(t) = a \quad e^{-(b_1+ b_2)t} \qquad \text{(I-2)}$$

ou a_1 et a_2 correspondant à un modèle du type :

$$t \longmapsto s(t) = (a_1+ a_2)\ e^{-bt} \qquad \text{(I-3)}$$

et pourtant b_1, b_2 ou a_1, a_2 peuvent avoir des significations concrètes différentes dans les modèles choisis.

Il est important de signaler que ces deux problèmes ont une implication directe sur la détermination de l'ordre de grandeur de l'erreur commise sur les coefficients qui détermine le système.

A titre d'exemple citons le cas particulier suivant.

Soit le modèle :

$$t \longmapsto s(t) = e^{-t}+ 3.e^{-1.1t} \qquad \text{(I-4)}$$

On se donne les valeurs :

$\sigma(i) = s(i) \qquad i = 1,2..,20$

La recherche d'un modèle utilisant l'écart

$$\delta(\sigma-s) = \sum_{i=1}^{20} |\sigma(i) - s(i)|^2 \qquad (I-5)$$

peut conduire aux deux modèles s_1 et s_2 suivants :

$$s_1(t) = 0.023 + 0.0006\, e^{-6.45t} + 3.997\, e^{-1.075t}$$
$$s_2(t) = 0.001 + 0.017\, e^{-1.747t} + 3.982\, e^{-1.072t} \qquad (I-6)$$

Ceux-ci, bien que très différents, conduisent aux écarts :

$$\sum_{i=1}^{20} |\sigma(i) - s_1(i)|^2 = 0.38 \times 10^{-7}$$
$$\sum_{i=1}^{20} |\sigma(i) - s_2(i)|^2 = 0.12 \times 10^{-8} \qquad (I-7)$$

I-3) QUELS SONT LES PROBLEMES ABORDES DANS NOTRE TRAVAIL ?

Nous nous sommes particulièrement intéressés à deux problèmes :

1° Après une étude des diverses méthodes d'identification , nous décrivons une méthode permettant d'aborder un calcul d'erreur.

2° Dans le cas particulier de méthodes basées sur l'approximation en moyenne quadratique nous avons recherché les bases qui correspondent au cas de meilleurs conditionnement des systèmes numériques à résoudre.

II. - EXEMPLE DE SITUATION PHYSIQUE CONDUISANT A CE MODELE

Un exemple simple de problème conduisant à de tels modèles est la suivante |42| :

Une substance est répartie entre n compartiments de contenance fixe Q_1, Q_2, Q_1, Q_2,...Q_n (au moins pendant un petit intervalle de temps). Ceux-ci échangent leur contenance avec une vitesse k_{ij} du compartiment i vers le compartiment j (cf fig 1).

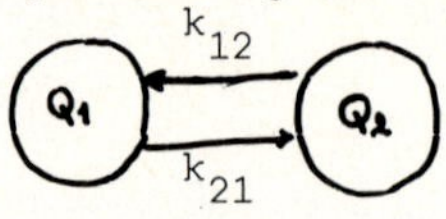

fig.1

Le problème consiste à déterminer les k_{ij}.

Des situations diverses peuvent ainsi être modélisées :

- En Biologie : l'étude des échanges de l'eau entre le corps humain et le milieu ambiant peut être modélisée avec deux compartiments

1° le corps humain,

2° le milieu ambiant.

- En Chimie : les compartiments peuvent avoir une signification chimique telle que "les ions bicarbonate" dans l'étude de la dilution du gaz carbonique ou "un groupe fonctionnel" dans une molécule organique.

Pour déterminer les k_{ij} on utilise un <u>traceur</u> isotope radioactif en général. On obtient ainsi dans le cas de deux compartiments le système suivant :

$$\frac{d_{q1}(t)}{dt} = k_{1\,2}\, q_2 - k_{2\,1}\, q_1$$

$$\frac{d_{q2}(t)}{dt} = k_{2\,1}\, q_1 - k_{1\,2}\, q_2 \qquad \text{(II-1)}$$

où $q_1(t)$ et $q_2(t)$ désignent respectivement les quantités de traceurs dans les compartiments 1 et 2 à l'instant t.

En désignant par :

$$a_i = \frac{q_i}{Q_i} \qquad \text{(II-2)}$$

la concentration ou "activité du traceur" i", on obtient en tenant compte des conditions initiales ($a_2(0) = 0$) et du système (II-1) le système :

$$a_1 = X_1 + X_2\, e^{-(k_{12} + k_{21})t}$$

$$a_2 = X_3 + X_4\, e^{-(k_{12} + k_{21})t}$$

$$a_1(0) = X_1 + X_2 \qquad \text{(II-3)}$$

$$a_2(0) = X_3 + X_4$$

Il est ainsi possible d'obtenir à partir du système (II-3)

$$a_1 = \frac{a_1(0)}{k_{12} + k_{21}} \left(k_{12} - k_{21}\, e^{-(k_{12}+k_{21})t}\right)$$

$$a_2 = \frac{Q_1 k_{21}\, a_1(0)}{Q_2(k_{12}+k_{21})} \left(1 - e^{-(k_{12}+k_{21})t}\right) \qquad \text{(II-4)}$$

On mesure expérimentalement les valeurs de $a_1(t)$ et $a_2(t)$ (cf figure 2)

Ensuite on procède à une estimation de Q_1, Q_2, k_{12}, k_{21} à l'aide de (II-4) et de :

$$Q_1 = \frac{q_1(0)}{a_1(0)}$$

Les diverses méthodes utilisées pour obtenir une estimation des paramètres précités sont décrites dans le paragraphe qui suit.

L'exemple donné montre l'importance qui peut être attachée à la détermination précise des paramètres qui interviennent dans le modèle.

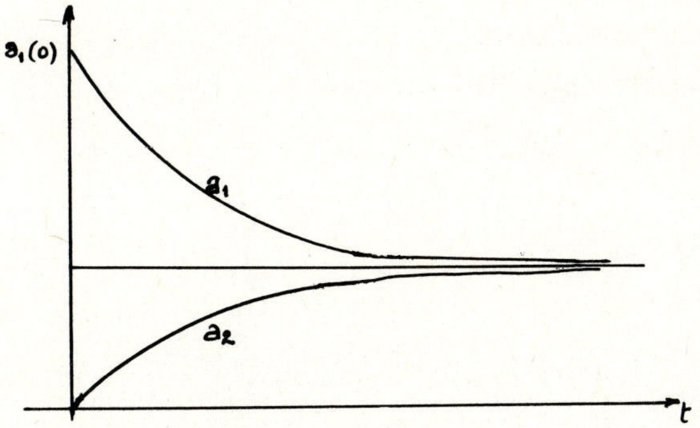

Fig. 2

III. - PRINCIPALES METHODES UTILISEES POUR L'ESTIMATION DES PARAMETRES D'UN MODELE MULTIEXPONENTIEL

III-1 Méthodes asymptotiques (peeling method) |40|, |41|

Ces méthodes utilisent les propriétés asymptotiques des fonctions

$$t \longmapsto s(t) = \sum_{i=1}^{n} a_i\, e^{-b_i t} \qquad \text{(III-1)}$$

$$\text{où } b_1 > b_{n-1} > \dots b_1 > 0$$

lorsque t tend vers + ∞

$$\text{Log } |s(t)| \sim \text{Log } |a_1| - b_1 t \qquad \text{(III-2)}$$

....

$$\text{Log } \left|s(t) - \sum_{i=1}^{p-1} a_i\, e^{-b_i t}\right| \sim \text{Log}|a_p| - b_p t$$

Ayant observé $\sigma(t)$ $(t > 0)$, on réalise une suite d'ajustements linéaires asymptotiques :

$\text{Log } |\sigma(t)| \sim \text{Log } |a_1| - b_1 t$ donne des estimations de a_1 et b_1 (III-3)

....

$\text{Log } \left|\sigma(t) - \sum_{i=1}^{p-1} a_i\, e^{-b_i t}\right| \sim \text{Log}|a_p| - b_p t$ donne des estimations de a_p et b_p

|40| expose un traitement automatique de cette méthode ; |41| présente une technique analogue faisant intervenir $\sigma(t)$ et $\sigma'(t)$ pour éviter les logarithmes.

III-2 Méthode des moindres carrés

Etant donnée une norme sur $l^2|0,1,\dots N|$ ou plus généralement sur l'espace $\mathcal{F}$ des fonctions numériques réelles définies sur $[0, +\infty|$, notée $||.||$ on cherche les solutions du problème :

$$\min_{\substack{n, a_1, a_2 \dots a_n \\ b_1, b_2 \dots b_n}} ||\sigma - s|| \qquad \text{(III-4)}$$

De façon générale, les auteurs utilisent des méthodes de gradients |2| , |9|... ou des méthodes de régression linéaire partielle utilisant la linéarité par rapport aux variables $a_1, a_2, \dots a_n$ |7| , |49|

III-3 Approximation différentielle |25| , |34|...

Ces méthodes utilisent les propriétés différentielles de $s(t) = \sum_{i=1}^{n} a_i\, e^{-b_i t}$ qui est solution d'une équation différentielle :

$$\alpha_o\, s^{(n)}(t) + \alpha_1\, s^{(n-1)}(t) + \dots + \alpha_n\, s(t) = 0 \qquad \text{(III-5)}$$

En utilisant les valeurs expérimentales σ on obtient des estimations des paramètres $\alpha_o, \alpha_1, \dots, \alpha_n$.

Les racines de l'équation caractéristiques de l'équation différentielle fournissent des estimations des paramètres $b_1, b_2 \ldots b_n$.

Une régression linéaire permet ensuite d'obtenir les estimations de $a_1, a_2, \ldots a_n$. D'autres méthodes utilisent des fonctions modulatrices qui permettent de remplacer le calcul des dérivées $\sigma^{(k)}$ par les convolutions $\sigma * \phi^{(k)}$.

III-4 <u>Approximation par des équations récurrentes</u> |36| |37|...

La fonction s est solution de l'équation

$$\alpha_0 D^n s + \ldots \alpha_n s = 0 \qquad \text{(III-6)}$$

où D est l'opérateur différence

$$s(t) \longmapsto Ds(t) = s(t-\tau)$$

Les méthodes utilisées sont alors semblables à celles du paragraphe précédent. La méthode de PRONY |72| est un exemple de telles méthodes.

III-5 <u>Méthodes spectrales</u>

Ces méthodes sont basées sur le fait que s peut s'interpréter comme la transformée de LAPLACE d'une mesure atomique μ. On a en effet

$$s(t) = \sum_{i=1}^{n} a_i e^{-b_i t} = \int_0^\infty e^{-bt} d\mu(b) \qquad \text{(III-7)}$$

$$\text{où } \mu = \sum_{i=1}^{n} a_i \delta_{bi} \qquad \text{(III-8)}$$

Le problème consiste donc à inverser la transformation de LAPLACE après régularisation de la mesure μ.

Diverses méthodes sont proposées :

a) <u>Utilisation de la transformation de FOURIER</u> :

Dans le cas où

$$d\mu(b) = g(b)\, db \qquad \text{(III-9)}$$

on a la formule

$$g(e^{-y}) = \mathcal{F} \frac{\mathcal{F}(e^t s(t))}{\mathcal{F}(e^{t-e^t})} \qquad \text{(III-10)}$$

GARDNER |64| utilise directement cette formule.

SCHLESINGER |65| utilise la transformation de FOURIER rapide.

b) <u>Utilisation des quadratures de GAUSS</u> |66|

Ce sont les méthodes qui utilisent la formulation classique de la transformation de LAPLACE en utilisant pour calculer les quadratures les meilleures quadratures de GAUSS.

c) Approximation de LEGENDRE

Une méthode décrite dans le chapitre IV permet d'obtenir une approximation de la régularité de la mesure μ par des polynômes de LEGENDRE.

IV - APPLICATION D'UNE METHODE DE MOMENTS A L'IDENTIFICATION DES COEFFICIENTS D'UNE SOMME D'EXPONENTIELLES REELLES

IV-1 INTRODUCTION : Dans le paragraphe précédent nous avons mentionné plusieurs auteurs dont les méthodes consistent à remplacer l'estimation des paramètres de s : n, $a_1,\ldots a_n$, $b_1,\ldots b_n$ par ceux de μ , transformée inverse de LAPLACE de s. Leurs méthodes souvent complexes ne sont pas accompagnées de calcul d'erreurs. Nous présentons ici une méthode particulièrement simple du point de vue numérique permettant en outre de tels calculs.

IV-2 DESCRIPTION DE LA METHODE

Notations :

a) Informations sur la fonction à approcher :

$$\sigma_k = \sigma(kh) \; ; \; k = 0,1\ldots N \text{ avec } h \text{ réel} > 0$$

b) Fonction régularisante ϕ de la mesure μ :

$$\phi : \mathbb{R} \to \mathbb{R}^+, \text{ telle que supp } \phi \subset [0,\zeta] \text{ avec } \int_0^\zeta \phi(t)\, dt = 1$$

c) Changement de variable $x = e^{-bh}$:

On pose Φ telle que $|\mu * \phi|\,(b) = h\, \Phi(e^{-bh}) = h\, \Phi(x)$ (IV-1)

Soit $\{b_i\}$ une suite de nombres positifs i = 1,2,..,n, on posera ;

$$x_i = e^{-b_i h} \qquad \text{(IV-2)}$$

d) De plus, on pose : $\Psi(x) = \int_0^x \Phi(t)\, dt$ (IV-3)

$\mathcal{L}$: opérateur "transformée de LAPLACE" et

$$(f,g) = \int_0^1 f(t)\, g(t)\, dt \text{ et } ||f|| = (f,f)^{1/2} \qquad \text{(IV-4)}$$

Principe de la méthode : Comme on suppose que s(t) est de la forme :

$$s(t) = \sum_{i=1}^{n} a_i\, e^{-b_i t}$$

on a :

$$\Phi(x) = \frac{1}{h} \sum_{i=1}^{n} a_i\, \phi\left(\frac{1}{h} \text{Log} \frac{x_i}{x}\right) \qquad \text{(IV-5)}$$

L'étude du nombre "n", de la localisation "$[x_i\, e^{-\zeta h}, x_i]$" et de l'amplitude "$a_i x_i\, \mathcal{L}[\phi](h)$" des sauts de Ψ, permet d'estimer les paramètres n, a_i et b_i de s si ζ est assez petit. Dans cette étude, on remplace Ψ inconnue par sa meilleure approximation polynomiale de degré <N dans l'espace $L^2|0,1|$: $S_N(\psi)$.

Calcul de $S_N(\psi)$: Suivant |68|, il suffit de connaître les N+1 premiers moments de ψ.

Lemme 1 : On démontre facilement le résultat suivant :

$$m_k = \int_0^1 x^k\, \Psi(x)\, dx = \frac{1}{k+1} \left[\sigma_1 \mathcal{L}[\phi](h) - \sigma_{k+2}\, \mathcal{L}[\phi](h(k+2))\right] \qquad \text{(IV-6)}$$

pour k = 0,...n.

Soient alors $L_m(x) = P_m(1-2x) = \sum_{k=o}^{m} \lambda_{m,k}\, x^k$

$m = 0,1\ldots N$ où P_m est le polynôme de LEGENDRE de degré m |69|. La propriété d'orthogonalité

$$(L_N, L_m) = \frac{\delta_{n,m}}{2n+1} \quad \text{permet d'obtenir :}$$

$$S_N(\psi)(x) = \sum_{m=o}^{N} (2m+1)\,(L_m,\psi)\,L_m(x) \qquad \text{(IV-7)}$$

$L_m(x)$ et $\lambda_{m,n}$ sont calculés à l'aide des formules de récurrence |69| :

$$L_o(x) = 1 \qquad L_1(x) = 1 - 2x \qquad \text{(IV-8)}$$

$$mL_m(x) = (2m-1)(1-2x)\,L_{m-1}(x) - (m-1)\,L_{m-2}(x)$$

et

$$(L_m,\psi) = \sum_{k=o}^{m} \lambda_{m,k}\, m_k \qquad \text{(IV-9)}$$

IV-3 CALCUL D'ERREUR

Lemme 2 : si $\phi(b) = \frac{e^{bh}h}{e^{\zeta h}-1}$ dans $[0,\zeta]$

$$||S_N(\psi) - \psi||^2 < \frac{N+1}{(8(N+1)^2-2)(e^{\zeta h}-1)} \left(\sum_{i=1}^{n} |a_i| x_i^{1/2} \right)^2 \qquad \text{(IV-10)}$$

Ce lemme résulte de la formule de PARSEVAL et du fait que :

$$(\psi, L_m)^2 = (\Phi, L^*_m)^2 < ||\Phi||^2\,||L^*_m||^2 \qquad \text{(IV-11)}$$

où |69| : $L^*_m(x) = \int_o^x L_m(t)\,dt = \frac{L_{m-1}(x) - L_{m+1}(x)}{2(2m+1)}$ (IV-12)

Ce choix de ϕ fournit d'ailleurs une majoration minimum pour $||\Phi||$.

IV-4 EXEMPLE

Si on part de l'exemple suivant :

$$\sigma(t) = 10\, e^{-0,9t} + 5\, e^{-0,2t} \qquad \text{(IV-13)}$$

avec les données :

$$h = 1 \qquad \zeta = 0,3 \qquad N = 10$$

on obtient par le calcul $x_1 = 0,406$ $x_2 = 0,818$

$x_1\, e^{-\zeta h} = 0,301$ $x_2\, e^{-\zeta h} = 0,606$

On peut comparer avec les résultats de la figure.

IV-5 CONCLUSION

Présentant les mêmes inconvénients que les méthodes exposées dans |64| |65| |66| : imprécision sur la détermination des sauts de ψ, détection impossible des exponentielles pour lesquelles $|a_i| x_i$ est petit; celle-ci se révèle cependant plus simple du point de vue numérique et permet un calcul d'erreur. Ce dernier permet de mesurer l'effet du produit ζh, l'importance de $a_i x_i$ et, a posteriori, d'apprécier une erreur dans $L^2|0,1|$ si l'on remplace les paramètres de s par leurs estimations.

Un prolongement naturel consiste à appliquer autant de fois qu'il est possible cette méthode sur le même modèle lorsqu'on dispose d'un grand nombre de valeurs $\sigma(t)$, et à interpréter ces résultats de manière statistique. Cette technique paraît d'autant plus appropriée que σ est entachée d'erreur.

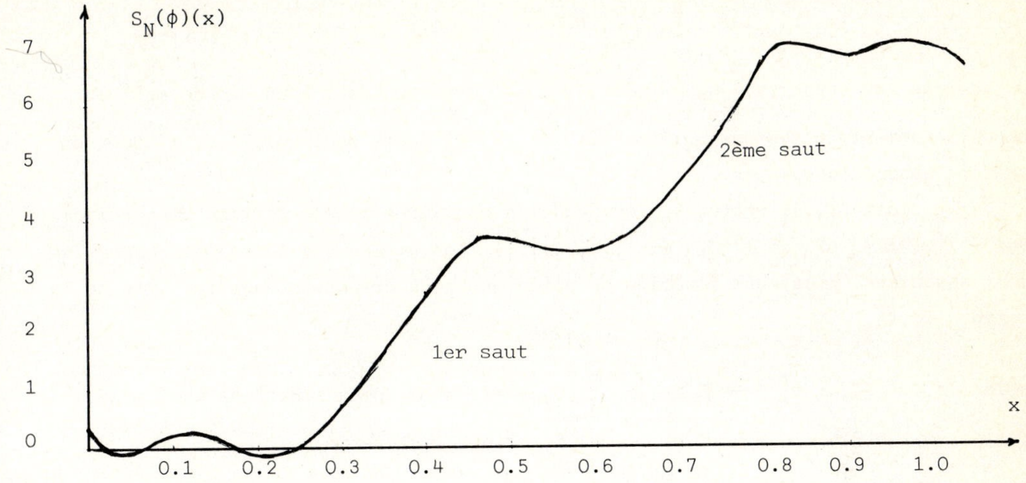

V. - ETUDE DU MEILLEUR CONDITIONNEMENT DANS LE CAS D'UNE APPROXIMATION EN MOYENNE QUADRATIQUE

V-1 En vue d'approcher une fonction numérique définie sur $[0,+\infty[$ ou $\mathbb{N}$ par une somme d'exponentielles réelles

$$s(t) = \sum_{i=1}^{n} a_i \, e^{-b_i t} \qquad \text{(V-1)}$$

On cherche les rapports $\frac{b_1}{b_2}$, $\frac{b_1}{b_3}$, ... $\frac{b_1}{b_n}$ correspondant à un cas de meilleur conditionnement du système qui fournit les coefficients optimum a_1^* ... a_n^* pour un critère quadratique.

Soit $(.,.)_1$ et $(.,.)_2$ les produits scalaires usuels définis respectivement sur $L^2([0, +\infty[\, p)$ et $l^2(\mathbb{N})$ et $||.||_1, ||.||_2$ les normes qui leur sont respectivement associées. p est une fonction de poids que nous prendrons dans la suite de la forme

$$t \longmapsto p(t) = e^{-2\alpha t} \qquad \text{(V-2)}$$

soient $\sigma_1 : [0, +\infty[\longrightarrow \mathbb{R}$ et $\sigma_2 : |0, +\infty| \longrightarrow \mathbb{R}$ deux fonctions numériques.

Enfin, soit $n \in \mathbb{N}^*$ $0 < b_n < b_{n-1} \ldots < b_1$ et e_{b_j} la fonction

$$e_{b_j} : t \in \mathbb{R}^+ \longmapsto e^{-b_j t} \qquad j \in \{1,\ldots n\}$$

Déterminer $a_1^*, a_2^*, \ldots a_n^*$ minimisant :

$(a_1, a_2, \ldots a_n) \in \mathbb{R}^n$ $\quad ||\sigma_i - \sum_{j=1}^{n} a_j e_{b_j}||_i$ $\quad i \in \{1,2\}$ équivaut à chercher les solutions du système

$$M_i (b_1, b_2, \ldots b_n) A = D \qquad \text{(V-3)}$$

où $$^tA = [a_1 \;\; a_2, \ldots a_n] \qquad \text{(V-4)}$$

$$^tD = [(e_{b_1}, \sigma_i), \ldots (e_{b_n}, \sigma_i)] \qquad \text{(V-5)}$$

M_i étant une matrice n,n

$$M_i(b_1, b_2, \ldots b_n) = [(e_{b_p}, e_{b_q})] \quad 1 < p,\, q < n \qquad \text{(V-6)}$$

avec

$$(e_{b_p}, e_{b_q}) = \frac{1}{b_p + b_q - 2\alpha} \quad \text{si } i = 1 \qquad \text{(V-7)}$$

$$(e_{b_p}, e_{b_q}) = \frac{1}{1 - e^{-(b_p + b_q)}} \quad \text{si } i = 2 \qquad \text{(V-8)}$$

V-2 Critère de meilleur conditionnement

M_i est une matrice définie positive et par suite, il existe une matrice orthogonale S telle que

$$^tS \, M_i \, S = \Delta = \text{diag}\,(\lambda_1, \lambda_2, \ldots, \lambda_n) \qquad \text{(V-9)}$$

les λ_i étant positifs et représentant les valeurs propres de M_i.

On a entre l'erreur δA sur A et δD sur D la relation

$$C.\,\delta A = \delta D \qquad \text{(V-10)}$$

Compte-tenu des propriétés de S, on en déduit facilement l'inégalité

$$\frac{||\delta A||}{||A||} < \frac{\max\limits_{j=1,n} \lambda_j}{\min\limits_{j=1,n} \lambda_j} \; \frac{||\delta D||}{||D||} \qquad \text{(V-11)}$$

$||\,.\,||$ représentant une norme classique sur $\mathbb{R}^n$.

Le meilleur conditionnement du système (V-3) correspond donc aux matrices M_i pour lesquelles $\frac{\max \lambda_j}{\min \lambda_j}$ est minimum, i.e. le plus voisin de 1. Ceci peut s'interpréter en faisant intervenir l'ellipsoïde d'erreur sur les a_i qui, dans ce cas, est de la forme

$$^tX\, M_i\, X = \varepsilon \qquad \text{(V-12)}$$

Les valeurs propres $\lambda_1, \lambda_2, \ldots, \lambda_n$ sont liées aux longueurs α_i des demi-axes de l'ellipsoïde précédent par les relations suivantes :

$$\frac{1}{\alpha j^2} = \frac{\lambda_j}{\varepsilon} \qquad \text{(V-13)}$$

Il s'agit donc de déterminer les valeurs des coefficients b_j pour lesquels $\frac{\max \lambda_j}{\min \lambda_j}$ est minimum.

Dans le cas n = 2 MALENGE et POUGET |70| ont montré que la situation optimum est obtenue pour i = 2 lorsque $\frac{b_1+\alpha}{b_2+\alpha} = 3 + 2\sqrt{2}$.

Pour les valeurs de $n > 2$, nous avons recherché un critère qui donne une approche de la condition optimale précédente et pour lequel un calcul théorique est possible. Nous avons ainsi choisi de chercher les solutions du problème

$$\min_{b_1, b_2 \ldots, b_n} \sum_{j=1}^{n} \lambda_j(b_1, b_2 \ldots b_n)$$

sous la contrainte (V-14)

$$\prod_{j=1}^{n} \lambda_j(b_1, b_2, \ldots, b_n) = C^{ste}$$

soit :

$$\min_{b_1, b_2 \ldots b_n} \left[\text{trace } M_i(b_1, b_2 \ldots, b_n)\right] \qquad \text{(V-15)}$$

$$\det M_i(b_1 \ldots b_n) = C^{ste}$$

Ce choix s'inspire de la remarque suivante :

si $K = \{\lambda = (\lambda_1, \lambda_2, \ldots \lambda_n)\ ;\ \lambda_j > 0 \text{ et } \prod_{j=1}^{n} \lambda_j = C \neq 0\}$

et si $\lambda^{(m)} \in K$ vérifie $\lim\limits_{m \to \infty} \sum_{j=1}^{n} \lambda_j^{(m)} = \min\limits_{\lambda \in K} \sum_{j=1}^{n} \lambda_j$

alors $\lambda_j^{(m)}$ tend vers $\sqrt[m]{C}$ lorsque $m \to +\infty$.

V-3 Mise en équation du problème (V-15) - Résolution

Les solutions de ce problème figurent parmi celles de

$$\lambda \frac{\partial}{\partial b_j} \text{ trace } M_i = \frac{\partial}{\partial b_j} \text{ Log det } M_i \qquad \text{(V-16)}$$

Cas i = 1 :

$$\text{trace } M_1(b_1,\dots b_n) = \sum_{j=1}^{n} \frac{1}{2(b_j+\alpha)} \qquad \text{(V-17)}$$

$$\det M_1(b_1,\dots b_n) = \frac{1}{2^n \sum_{j=1}^{n} b_j} \prod_{i<j} \left| \frac{b_i-b_j}{b_i+b_j+2\alpha} \right|^2 \qquad \text{(V-18)}$$

si $\rho_j = \dfrac{b_1+\alpha}{b_j+\alpha}$ $j \in \{2,3\dots n\}$ le système (V-16) implique :

$$1 - \frac{1}{\rho_j} + \frac{4}{1-\rho_j} + \sum_{\substack{k=2\\k\neq j}}^{n} \frac{4\rho_k(\rho_j^2-1)}{(1-\rho_k^2)(\rho_j^2-\rho_k^2)} = 0 \qquad \text{(V-19)}$$

$j = 2,3,\dots n$

système dont les solutions $\rho_2^*,\dots\rho_n^*$ peuvent être estimées par la méthode de NEWTON-RAPHSON.

	ρ_2^*	ρ_3^*	ρ_4^*	ρ_5^*
n=2	$3+2\sqrt{2}$			
n=3	5.77	22.15		
n=4	5.76	21.77	68.23	
n=5	5.76	21.74	66.76	184.18

(V-20)

Cas i = 2 :

Si $u_j = e_{b_j}$ $j \in \{1,2,\dots,n\}$

$$\text{trace } M_2(b_1,b_2,\dots,b_n) = \sum_{j=1}^{n} \frac{1}{1-u_j^2} \qquad \text{(V-21)}$$

$$\det M_2(b_1,b_2,\dots b_n) = \frac{1}{\prod_{j=1}^{n} (1-u_j^2)} \prod_{k<j} \left(\frac{u_k-u_j}{u_k+u_j}\right)^2 \qquad \text{(V-22)}$$

Le système (V-16) implique alors

$T_1(u_1,\dots,u_n) = T_j(u_1,\dots u_n)$ $j = 2,\dots,n$

ou $$T_j(u_1,u_2,\dots u_n) = \frac{1-u_j^2}{u_j} \sum_{k\neq j} \frac{1}{u_j-u_k} + \frac{u_k}{1-u_k u_j} + 1 - u_j^2 \qquad \text{(V-23)}$$

Là encore, la méthode de NEWTON-RAPHSON permet d'estimer les solutions $b_j^* = -\text{Log } u_j^*$
$j=1,\ldots n$

	b_5^*	b_4^*	b_3^*	b_2^*	b_1^*
n=2				0.016	0.1
n=3			0.0040	0.016	0.1
n=4		0.0013	0.0041	0.016	0.1
n=5	0.00047	0.0013	0.0041	0.016	0.1

V-4 Conclusion

Les résultats numériques mettent en évidence le mauvais conditionnement des systèmes rencontrés. Le caractère est évident lorsque les coefficients $b_1, b_2 \ldots b_n$ sont voisins ou d'ordres de grandeurs très différents. Cependant, même dans le cas de conditionnement optimum considéré ici, ces ordres de grandeurs deviennent très vite différents dès que n>3. Ceci a pour conséquence de rendre certaines exponentielles très vite négligeables devant d'autres ; leur identification devient pratiquement impossible. On remarque aussi la convergence des solutions des systèmes (V-19) et (V-23) lorsque n augmente.

L'utilisation de la norme $||.||$ permet de choisir la valeur de α la meilleure lorsque l'on a déjà des estimations des valeurs des coefficients b_j $j \in \{1,\ldots,n\}$.

Pour toutes les méthodes décrites dans III ces résultats permettent de donner une information sur la qualité de l'identification.

B I B L I O G R A P H I E

METHODES DIRECTES - Norme $||.||_2$

|1| SPEARMAN SIMULTANEOUS ESTIMATION FOR A COMPARTMENTAL MODEL
J.J.BEAUCHAMP and R.G. CORNELL - Technometrics Vol. 11, n° 3 August 1969

|2| QUELQUES ASPECTS DES PROBLEMES D'AJUSTEMENT - LA RECRESSION EXPONENTIELLE
Université Claude-Bernard de Lyon - J. ESTEVE - juillet 1971

|3| SHORT COMMUNICATIONS - Investigation of the Effect of Data Error on the Determination of Physiological Parameters by Means of Compartmental Analysis
G.L. ATKINS - Biometric J. 127, 437-438 - 1972

|4| NON-LINEAR LEAST SQUARES ESTIMATION
H.O. HARTLEY and A. BOOKER - Ann. Math. St. p.638-650 - 1965

|5| ASYMPTOTIC PROPERTIES OF NON-LINEAR LEAST SQUARES ESTIMATORS
R.I. JENNRICH - Ann. Math. St. p. 633-643 - 1969

|6| A SIGNIFICANCE TEST FOR EXPONENTIAL REGRESSION
E.S. KEEPING - Ann. Math. St. p. 180-198 - 1951

|7| AN EFFICIENT METHOD FOR MULTIEXPONENTIAL FITTING WITH A COMPUTER
A. LEMAITRE - J.P. MALENGE - Computers and Biomedical Research MS 297 - 1971

|8| MAXIMUM LIKELIHOOD ESTIMATORS OF PARAMETERS IN DOUBLE EXPONENTIAL REGRESSION
S. LIPTON - C. McGILCHRIST - Biometrics p.144-151 - 1963

|9| UNI-KLINIK - STEEPEST DESCENT OPTIMISATION by B. MARREN
ECC Report n° 140 - Electronic Associates Inc. European Computation Center-1972

|10| INVESTIGATION OF THE EFFECT OF DATA ERROR IN THE ANALYSIS OF BIOLOGICAL TRACER DATA FROM THREE COMPARTMENT SYSTEMS
J. MYHILL - J. Theoret. Biol. 23, 218-231 - 1968

|11| A SIMPLE METHOD FOR FITTING AN ASYMPTOTIC REGRESSION CURVE
H.D. PATTERSON - Biometrics p. 323-329 - Sept. 1956

|12| A METHOD OF MAXIMUM-LIKELIHOOD ESTIMATION
F.S.G. RICHARDS - J.R. Stat. Soc. B23, p 469-476 - 196

|13| A COMPARISON OF CONTINUOUS DISTRIBUTIONS OF PARAMETERS OF EXPONENTIAL DECAY CURVES
R.G. ROSSING and M.B. DANFORD - Biometrics, 24 p.117-133 1968

|14| THE APPLICATION OF THE METHOD OF MAXIMUM LIKELIHOOD TO THE ANALYSIS OF TRACER KINETIC DATA - T. SANDOR, M.F. CONROY and N.K. HOLLENBERG - Math. Biosc.9 p 149-159 - 1970

|15| A NOTE ON FITTING MULTIEXPONENTIAL FUNCTIONS OF TIME TO EXPERIMENTAL DATA
M.W. SIMPSON-MORGAN - Mathematical Biosciences 5 (1969), 195-199

|16| ASYMPTOTIC REGRESSION - W.L. STEVENS - Biometrics 1951, p. 247-267

|17| THE SPECTRUM OF A LINEAR NETWORK : A SIMPLE APPROACH TO THE n COMPARTMENT PROBLEM
M.E. TURNER - Biometrics 1964, p. 827-831

|18| A SIGNIFICANCE TEST AND ESTIMATION IN THE CASE OF EXPONENTIAL REGRESSION
D.S. VILLARS - Ann. of Math. Stat. 1947 p.596-600

METHODES DIRECTES Norme $||.||_\infty$

|19| VECTOR-VALUED APPROXIMATION AND ITS APPLICATION TO FITTING EXPONENTIAL DECAY CURVES - G. G. BELFORD - Mathematics of Computation, vol. 28, n° 125 1974

|20| UBER DIE VORZEICHENSTRUKTUR DER EXPONENTIALSUMMEN D. BRAESS - Journal of Approximation Theory 3, 101-113 - 1970

|21| EXPONENTIAL CHEBYSHEV APPROXIMATION ON FINITE SUBSETS OF |0, 1| B.H. ROSMAN - Math. of Computation, vol. 25, n° 115 p. 575-577 - 1971

|22| ON THE CONTINUITY OF THE SET-VALUED EXPONENTIAL METRIC PROJECTION E. SCHMIDT - Journal of Approximation Theory 7, 36-40 - 1973

METHODES DES EQUATIONS DIFFERENTIELLES

|23| STEADY STATE MULTIPLICITY AND STABILITY IN AN ADIABATIC CONTROLLED CYCLED STIRRED TANK REACTOR - J. AUSIKAITIS - A.J. ENGEL - AIChE J.(vol 20, n° 2) 1974

|24| L'ESTIMATION DES PARAMETRES DES MODELES DE SIMULATION C.BEAUGAS - G. COMYN - Rev. Informatique Méd. n° 4 Décembre 1970

|25| A USEFUL APPROXIMATION TO e^{-t^2} by R. BELLMAN, B.G. KASHEF and R. VASUDEVAN Mathematics of Computation, vol. 26 n° 117, p. 233-235 - 1972

|26| ON THE RELIABILITY OF PARAMETER ESTIMATES IN A SET OF SIMULTANEOUS NONLINEAR DIFFERENTIAL EQUATIONS - G. EMIG and L.H. HOSTEN - Chemical Engineering Science Vol. 29 pp 475-483 - Pergamon Press - 1974

|27| A NEW METHOD OF PARAMETER ESTIMATION IN LINEAR SYSTEMS G.R. GAVALAS - AIChE Journal Vol. 19, n° 2 pp 214-222 - 1973

|28| A METHOD OF EXPONENTIAL CURVE FITTING BY NUMERICAL INTEGRATION S.D. FOSS - Biometrics pp 815-821 - Décembre 1970

|29| A COMPUTATIONAL ALGORITHM FOR SEQUENTIAL ESTIMATION R.J. HANSON and P. DYER - The Computer Journal pp 285-290 vol. 14 n° 3 - 1970

|30| NON-LINEAR ESTIMATION THEORY - J.H. SEINFELD - Industrial and Engineering Chem. Vol. 62, n° 1 - pp 32-42 - 1970

|31| A PRACTICAL FILTER FOR SYSTEMS WITH UNKNOWN PARAMETERS T. SOEDA, T. YOSHIMURA - Transactions of the ASME pp 396-401 - Dec. 1973

|32| A HYBRID METHOD FOR THE SOLVING OF OPTIMAL CONTROL PROBLEMS IN A REAL TIME ENVIRONMFNT - A. STEVEN - Electronic Associates Inc. European Comp.Center 1971

|33| FINITE-DIMENSIONAL SENSOR ORBITS AND OPTIMAL NONLINEAR FILTERING J. TING-HO LO - I.E.E.E. Trans. on Inf. Theory Vol. IT-18 n° 5 Sept. 1972

|34| ANALYSE DES COURBES DE DECROISSANCE MULTIEXPONENTIELLES PAR LA METHODE DES FONCTIONS MODULATRICES-APPLICATION A LA FLUORESCENCE B. VALEUR et J. MOIREZ - Journal de Chimie-Physique 70, n° 3 - 1973

|35| A NEW METHOD FOR THE ESTIMATION OF PARAMETERS IN DIFFERENTIAL EQUATIONS B. VAN DEN BOSCH and L. HELLINCKX - AIChE Journal Vol.20, n° 2 pp 250 - 1974

METHODE DES EQUATIONS RECURRENTES

|36| SIMULTANEOUS ESTIMATION BY PARTIAL TOTALS FOR COMPARTIMENTAL MODELS
J. J. BEAUCHAMP - R.G. CORNELL - American Stat. Assoc. J. pp 573-583 - 1968

|37| APPROXIMATION OF SIGNALS BY REAL EXPONENTIALS
P.Y. KESKAR - G.V. LAGO - Dept. of Electrical Eng. University of Missouri Columbia, Missouri 65201

|38| BIOLOGICAL PROBLEMS INVOLVING SUMS OF EXPONENTIAL FUNCTIONS OF TIME : A MATHEMATICAL ANALYSIS THAT REDUCES EXPERIMENTAL TIME
D.H. PARSONS - Mathematical Biosciences 2, 123-128 - 1968

|39| RANDOM DIFFERENCE EQUATIONS OCCURRING IN ONE-COMPARTMENT MODELS
V.R. RAO UPPULURI, P.I. FEDER, L.R. SHENTON
Mathematical Biosciences - pp 143-171 - 1967

PEELING METHOD

|40| MANCINI-PILO - A COMPUTER PROGRAM FOR MULTIEXPONENTIAL FITTING BY THE PEELING METHOD - Comp. Biom. Res (1970) 3 p. 1-14

|41| PERL - A METHOD FOR CURVE-FITTING BY EXPONENTIAL FUNCTIONS
Intern. Journal of Applied Radiation and Isotopes (1960) 8 p. 212-222

ANALYSE MULTICOMPARTIMENTALE

|42| G.L. ATKINS - MODELES A COMPARTIMENTS MULTIPLES POUR LES SYSTEMES BIOLOGIQUES
Gauthier-Villars - 1973

|43| ESTIMATION PROCEDURES FOR CONSECUTIVE FIRST ORDER IRREVERSIBLE REACTIONS
R.C. BAILEY - G.S. EADIE - F. H. SCHMIDT - Biometrics 30, 67-75 - 1974

|44| INVARIANTS IN EXPERIMENTAL DATA ON LINEAR KINETICS AND THE FORMULATION OF MODELS - M. BERMAN - R. SCHOENFELD - Journal of Applied Physics Vol. 27, n° 11 november 1956 - pp 1361-1370

|45| STOCHASTIC MODELS FOR THE DISTRIBUTION OF RADIOACTIVE MATERIAL IN A CONNECTED SYSTEM OF COMPARTMENTS - S.R. BERNARD, L.R. SHENTON, V.R. RAO UPPULURI
Proc. 5th BERKELEY Symp. - Math. Stat. Prob. IV - 481-510

|46| MODELS FOR THE INTERPRETATION OF EXPERIMENTS USING TRACER COMPOUNDS
J. CORNFIELD, J. STEINFELD, S.W. GREENHOUSE - Biometrics 16, pp 212-234 1960

|47| THE DEAD SPACE IN A COMPARTMENTAL LUNG MODEL
J.W. EVANS - D.G. CANTOR - J.R. NORMAN - Bulletin of Math. Biophysics Vol.29 pp 711-718 - 1967

|48| THE LOGARITHMIC CONVEXITY OF THE WASHOUT FUNCTION IN TRACER KINETICS
R.B. GUNN, C.S. PATLAK, J.Z. HEARON - Mathematical Biosciences 4 (1969) 1-6

|49| BALANCE STUDIES ON COMPARTMENTAL SYSTEMS WITH STOCHASTIC INPUTS
J.A. JACQUEZ - F.J. MATHER - J. Theoret. Biol. (1966) 11, 446-458

|50| STOCHASTIC COMPARTMENTAL ANALYSIS : MODEL AND LEAST SQUARES ESTIMATION FROM TIME SERIES DATA - J.H. MATIS - H.O. HARTLEY Biometrics 27, 77-102 - 1971

|51| MULTI-COMPARTMENTAL ANALYSIS IN STEADY STATE AS A STOCHASTIC PROCESS
J. H MATIS - M.W. CARTER - Actabiometrica - 1970

|52| ESTIMATION PROCEDURES FOR CONSECUTIVE FIRST ORDER REVERSIBLE REACTIONS
F.H. SCHMIDT - R.C. BAILEY - Biometry Dept. Prepint n° 28 - October 1973
Emory University - ATLANTA, Georgia 30322

|53| GENERALIZED KINETIC REGRESSION ANALYSIS : HYPERGEOMETRIC KINETICS
M.E. TURNER, R.J. MONROE, L.D. HOMER - Biometrics p 406-428 - 1963

EXEMPLES D'APPLICATION

|54| THE EXPERIMENTAL MEASUREMENT AND THEORETICAL ANALYSIS OF FLOW-SYSTEM TRACER CURVES BY A LINE-SAMPLING METHOD
BANG MO KIM - T.R. HARRIS - Chemical Engineering Science 1973, Vol.28 p1653-1659

|55| NEW METHOD FOR RAPID AND ACCURATE MEASUREMENT OF DECAY TIME CONSTANT
A.BERNALTE - J. LE PAGE - The Review of Scientific Instruments Vol 40 n° 1 - p 71-74 - 1969

|56| EVALUATION OF PHOTOLUMINESCENCE LIFETIMES
J.N. DEMAS - A.W. ADAMSON - The Journal of Phys.Chem. Vol.75, n° 16 - 1971

|57| EXPERIMENTAL VERIFICATION OF A 2-COMPARTMENT MASSTRANSFER MODEL FOR WHOLE BLOOD IN A HAEMODIALYSER - J.D.S.GAYLOR - Med and Biol.Engi. p 90-99 - 1974

|58| A STOCHASTIC MODEL FOR THE REABSORPTION OF ISOTONIC SALINE BY THE TUBULES IN THE KIDNEY - R.J. HARDY, Ph.D. and E.J. WEINMAN - Biometrics 30, 33-40 - 1974

|59| THE DECONVOLUTION OF FLUORESCENCE DECAY CURVES - A NON-METHOD FOR REAL DATA
A.E.W. KNIGHT - B.K. SELINGER - Spectrochimica Acta Vol.27-A p 1223-1234-1971

|60| NEWMAN'S CHAMBER MODELS IN THE STUDY OF BLOOD CIRCULATION THROUGH THE LIVER AND HEART - M. VALENTINUZZI - M.E. VALENTINUZZI - Bulletin of Mathematical Biology - Vol. 35 - 1973

|61| PROGRAMME APPROCHANT PAR UNE COURBE PLURIEXPONENTIELLE DES DONNEES EXPERIMENTALES DISCRETES OBTENUES CHEZ L'HOMME PAR DETECTION RADIOACTIVE
G. VALLEE, J.P. BELIN, C. BEAUGAS - Symposium d'Informatique Médicale de TOULOUSE - mars 1968 - Rapports I.R.I.A.-ROCQUENCOURT

|62| EXPLOITATION D'UN PROGRAMME D'ANALYSE DE LA CINETIQUE DES TRACEURS
Problèmes posés. Exemple du radio-calcium chez l'homme - G. VALLEE, C.BEAUGAS, M. IZEMBART, J.P. MASSIN - Revue Informatique Médicale Vol.2 n° 3 - 1971

METHODES SPECTRALES

|63| METHOD FOR THE ANALYSIS OF MULTICOMPONENT EXPONENTIAL DECAY CURVES
D.G.GARDNER - J.C. GARDNER - G. LAUSH - M.W. MEINKE
Journal of Chemical Physics Vol. 31 n° 4 978-986 - 1959

|64| RESOLUTION OF MULTICOMPONENT EXPONENTIAL DECAY CURVES USING FOURIER TRANSFORMS
D.G. GARDNER - Ann. NEW-YORK Acad. of Sc. Vol.108 Art.1 p.195-203 - 1963

|65| FIT TO EXPERIMENTAL DATA WITH EXPONENTIAL FUNCTIONS USING THE FAST FOURIER TRANSFORM - J.SCHLESINGER - Nuclear Instrument and methods n°106 pp 503-508-1973

|66| NUMERICAL INVERSION OF THE LAPLACE TRANSFORM
J.R.E. BELLMAN - Elsevier 1966

DIVERS

|67| APPLIED ANALYSIS
LANCZOS - Prentice Hall - 1956

|68| MOMENT PROBLEME FÜR EIN ENDLICHES INTERVALL
HANSDORFF - Mathematishe Zeitschrift 16 - pp 220-248 - 1923

|69| CONSTRUCTIVE FUNCTION THEORY
I.P. NATANSON - Vol II - Ungar 1965

|70| APPROXIMATION EN MOYENNE QUADRATIQUE D'UNE FONCTION PAR UNE SOMME D'EXPONENTIELLES. CAS DU MEILLEUR CONDITIONNEMENT
J.P. MALENGE - J. POUGET - CR Acad. Sci. Paris T.275 - n° 25 2° semestre 1972

|71| NUMERICAL INVERTING OF MATRICE OF HIGH ORDER
J. VON NEUMANN and H.H. GOLDSTINE - Proc. Am. Math. Soc. 2 - p.188-202 - 1951

|72| INTRODUCTION TO NUMERICAL ANALYSIS
HILDEBRAND - Mac Graw-Hill (1956)

IMPLEMENTATION NUMERIQUE EN FILTRAGE OPTIMAL NON-LINEAIRE : ALGORITHMES PARALLELES ET COMPARAISON AVEC D'AUTRES SOLUTIONS

F. LEVIEUX
I.R.I.A.
Bt. 13
Domaine de Voluceau
B.P. n° 5
78150 Le Chesnay
France

I. INTRODUCTION

Dans une précédente étude (LEVIEUX [5]) on a montré, que l'on pouvait construire des filtres non-linéaires stables par approximation de l'équation d'évolution de la densité conditionnelle due à KUSHNER [2]. En utilisant des travaux plus récents (ZAKAI [12], ROZOVSKII [10]), on prouve que les schémas d'approximation précédents se prètent aisément à une implémentation sur calculateur parallèle, et conservent sous cette forme leurs propriétés de stabilité et de convergence.

On montre l'efficacité d'une telle solution sur une application à un problème d'interférométrie pour l'observation de rayonnements astronomiques en fréquences millimétriques (LAMBLA [3]). Cette solution est comparée, avec des algorithmes déjà connus, sur le plan des performances et de la fiabilité d'utilisation. On précise ensuite les avantages associés à l'implémentation de cet algorithme, ainsi que le type de problèmes pour lesquels ces avantages sont déterminants.

II. QUELQUES RESULTATS MATHEMATIQUES

II. 1. Présentation du problème et notations.

Soit $X(t,\omega)$ un processus de diffusion, vectoriel de dimension n, vérifiant l'équation :

$$(2.1)\quad X(t,\omega) = X_0(\omega) + \int_0^t f(X(s,\omega),s)ds + \int_0^t g(X(s,\omega),s)db_1(s,\omega).$$

Les fonctions f et g sont respectivement vectorielles et matricielles de dimension n ; b_1 est un mouvement brownien unitaire. Soit f_i les composantes de f et g_{ij} les composantes de la matrice gg'. Enfin désignons par $A(t)$ l'opérateur de Fokker Planck associé à l'équation (2.1) et défini par :

$$(2.2)\quad A(t). = \sum_{i=1}^{n} \frac{\partial[f_i(x,t).]}{\partial x_i} - \frac{1}{2}\sum_{ij=1}^{n} \frac{\partial^2[\sigma_{ij}(x,t).]}{\partial x_i\, \partial x_j}\ .$$

Introduisons les hypothèses suivantes (cf.LEVIEUX [4]) :

(H1) les coefficients $f_i(x,t)$, $\sigma_{ij}(x,t)$ pour $i,j = 1 \dots n$, sont éléments de $C^{n_1}\{(0,T)\ C^{n_0}(R^n)\}$ avec n_0 = Entier $(\frac{n}{2}+1)$ et n_1 = Entier $(\frac{n_0}{2}+1)$.

(H2) la fonction f_i et σ_{ij} sont bornées et Hölderiennes sur $R^n \times (0,T)$ ∎

(H3) il existe $\lambda,\mu > 0$ tel que, quelque soit φ élément de $L^2\{(0,T)\ ;\ V\}$:

$$(2.3)\quad < A(t)\varphi,\varphi >_{VV'} + \lambda\, \|\varphi\|_H^2 \geq \mu\, \|\varphi\|_V^2\ ,$$

où $V \equiv H^1(R^n)$ $H \equiv H' \equiv L^2(R^n)$ et $< , >_{VV'}$ désigne le produit de dualité entre V et son dual V' ∎

(H4) $p_0(x)$ densité de la loi $X_0(\omega)$ est élément de $H^{n_0}(R^n)$ ∎

(H5) il existe un prolongement $\bar{A}(t)$ de l'opérateur $A(t)$ pour $t < 0$ tel que $p_0(x)$ vérifie :

$$(2.4)\quad \int_{R^n} p_0(x)dx = 1$$

$$(2.5)\quad \exists\delta > 0,\ 0 \leq p_0(x) \leq \delta\rho(x,0)$$

(2.6) $\frac{d\rho}{dt}(x,t) + \bar{A}(t)\rho(x,t) = 0$

(2.7) $\exists\tau_0 > 0,\ \rho(x, -\tau_0) = \delta(x)$ ∎

Le processus $x(t,\omega)$ défini par (2.1) et les hypothèses (H1) à (H5) est observé selon une loi physique décrite par la relation :

(2.8) $z(t,\omega) = \int_0^t H(X(s,\omega),s)ds + db_2(s,\omega),$

où b_2 est un mouvement brownien scalaire unitaire et H une fonction instantannée sur $R^n \times (0,T)$, vérifiant l'hypothèse suivante :

(H6) $H(x,t)$ est élément de $L^\infty[R^n \times (0,T)] \cap C^0[R^n \times (0,T)]$ et $\frac{\partial H}{\partial x}$, $\frac{\partial H}{\partial t}$ sont éléments de $L^2[R^n \times (0,T)] \cap L^\infty[R^n \times (0,T)]$.

Suivant ici l'exposé de FUJISAKI, KALLIANPUR, KUNITA [1], on introduit le processus innovation $b(t,\omega)$ associé à $z(t,\omega)$, qui sous les hypothèses présentes, est un mouvement brownien unitaire. Ce processus est défini par la relation :

(2.9) $b(t,\omega) = z(t,\omega) - \int_0^t E\{H[x(s,\omega),s]/B(z_0^s)\}ds,$

où $B(z_0^t)$ désigne la plus petite σ-algèbre contenant les événements de la forme $\alpha \leq z(s,\omega) \leq \beta$ quelque soit $\alpha,\beta \in R$ et $s \in [0,T]$.

Soit $A^*(t)$ l'opérateur adjoint de l'opérateur A. Une fonction scalaire $\varphi(x)$ est dite élément de l'espace $\mathcal{D}(A^*)$ si et seulement si les conditions suivantes sont satisfaites :

(2.10) $E\{\int_0^t [A^*(s)\varphi[x(s,\omega)]]^2 ds\} < +\infty.$

Le processus $M_t(\varphi)$ défini par :

(2.11) $M_t(\varphi) = \varphi[x(t,\omega)] - E\{\varphi[x_0(\omega)]\} + \int_0^t A^*s\varphi[x(s,\omega)]ds$

est une martingale mesurable sur $B(b_1{}_0^t)$ quelque soit φ dans $\mathcal{D}(A^*)$.

Introduisons B_t l'opérateur de $\mathcal{D}(R^n)$ (+) dans $\mathcal{D}'(R^n)$ défini comme suit : B^*_t

(+) $\mathcal{D}(R^n)$ est l'ensemble des fonctions indéfiniment dérivable à support compact.

désignant l'adjoint de B_t, l'expression :

$$(2.12) \quad C_t = M_t(\varphi) \times b_2(t,\omega) - \int_0^t B^*_s \, \varphi(x(s,\omega))ds$$

est une martingale adaptée aux σ-algèbres $B([b_1]_0^t, [b_2]_0^t)$. On montre (voir réf.[11]) que toute fonction φ appartenant à $\mathcal{D}(A^*)$, l'opérateur B* est défini de façon unique sur la relation (2.12) et, par dualité, l'opérateur B.

Le résultat suivant (théorème 4.1 de la réf.[5]) permet de trouver une relation différentielle liant l'espérance mathématique de la fonction $\varphi[x(t,\omega)]$ conditionnée par les observations passées $z(s,\omega)$ avec $s < t$, ce qui est l'objet même du filtrage récursif.

Théorème II.1.

Sous les hypothèses (H1) à (H6), toute fonction φ élément de $\mathcal{D}(A^*)$ vérifie la relation :

$$(2.13) \quad E\{\varphi[x(t,\omega)]/B(z_0^+)\} + \int_0^t E\{A^*_s \, \varphi[x(s,\omega)]/B(z_0^s)\}ds = E\{\varphi[x_0(\omega)]\} +$$

$$\int_0^t [E\{\varphi[x(s,\omega)]H[x(s,\omega),s]/B(z_0^s)\} - E\{\varphi[x(s,\omega)]/B(z_0^s)\} \times E\{H[x(s,\omega),s]/B(z_0^s)\} +$$

$$E\{B^*_s \, \varphi[x(s,\omega)]/B(z_0^s)]db(s,\omega) \quad \blacksquare$$

Pour tenir compte de corrélations éventuelles entre b_1 et b_2, introduisons la fonction vectorielle $S(x(t,\omega),t)$, vérifiant (H6), par la relation formelle :

$$(2.14) \quad E\{db_2(t,\omega)db_1(t,\omega)\} = S(x(t,\omega),t)dt.$$

Notons $p(x,t;\omega)$ la densité de probabilité de la loi $x(t,\omega)$ conditionnée par les réalisations passées $B(z_0^t)$. On peut montrer que, au sens des distributions dans $R^n x(0,T)$, cette densité vérifie la relation :

$$(2.14) \quad p(x,t;\omega) + \int_0^t A_s p(x,s;\omega)ds = p_0(x) + \int_0^t [p(x,s;\omega)[H(x,s)$$

$$- \int_{R^n} H(\alpha,s)p(\alpha,s;\omega)d\alpha] - B_s p(x,s;\omega)]db(s,\omega)$$

A_t et B_t se déduisant de A^*_t et B^*_t par dualité. En particulier B_t prend la forme :

$$(2.15)\quad B_t p(x,t;\omega) = \sum_{i=1}^{n} \frac{\partial}{\partial x_i} [(g(x,t)S(x;t))_i p(x,t;\omega)]$$

où $(.\ )_i$ désigne la $i^{\text{ème}}$ composante du produit matriciel gS (FUJISAKI, KALLIANPUR, KUNITA [1]).

Il faut remarquer, que la relation (2.14) permet effectivement de caractériser les variations de la fonction p au cours du temps, ce qui n'était pas le cas de la relation (2.13) pour la variable $E\{\varphi/B(z_0^t)\}$.

On peut simplifier formellement la relation (2.14) en effectuant le changement de variable suivant (ZAKAI [12], ROZOVSKII [10]). Soit $\tilde{p}$ une fonction solution des relations :

$$(2.16)\quad \tilde{p}(x,t;\omega) = p(x,t;\omega) \int_{R^n} \tilde{p}(x,t;\omega)dx,$$

$$(2.17)\quad \tilde{p}(x,0;\omega) = p_0(x).$$

On peut montrer que, parmi les différentes solutions des relations (2.16), (2.17), il en existe une qui vérifie la relation différentielle :

$$(2.18)\quad \tilde{p}(x,t;\omega) + \int_0^t A_s\tilde{p}(x,s;\omega)ds = p_0(x) + \int_0^t [\tilde{p}(x,s;\omega)H(x,s) - B_s\tilde{p}(x,s;\omega)]dz(s,\omega).$$

On remarquera que l'équation (2.18) est linéaire par rapport à la seule variable $\tilde{p}$ et aux conditions initiales p_0, ce qui sera de grande importance pour la conception d'algorithmes d'approximation de $\tilde{p}$ (et donc de p).

<u>II. 2. Deux résultats d'existence et d'unicité</u>.

Introduisons l'hypothèse suivante :

(H7) Quelque soit $x \in R^n$ et $t \in [0,T]$ il existe un réel positif α_1, tel que, au sens des matrices définies positives :

$$(2.19)\quad I_n - S(x,t)\, S'(x,t) \geq \alpha_1\, I_n.$$

Utilisant un résultat abstrait du à PARDOUX [9], on démontre le théorème suivant :

Théorème II.2.

Sous les hypothèses (H1) à (H7), il existe une solution unique à l'équation (2.18) dans l'espace $L^2\{\Omega x(0,T);V\} \cap L^2\{\Omega,\mu;C^0[(0,T);H]\}$. De plus cette solution est presque partout positive sur $R^n x(0,T)x\Omega$ ∎

On doit montrer ensuite, quelles sont les liaisons existantes entre le problème du filtrage non-linéaire d'une part, la solution de l'équation (2.18) d'autre part. Ceci ne peut se faire, qu'en introduisant la fonction $\bar{p}(x,t;\omega)$ par la relation :

$$(2.20)\quad \bar{p}(x,t;\omega) = \frac{\tilde{p}(x,t;\omega)}{\int_{R^n} \tilde{p}(x,t;\omega)dx}$$

et en montrant que $\bar{p}(x,t;\omega)$ est solution de l' équation donnant la densité de probabilité conditionnelle :

$$(2.21)\quad p(x,t;\omega) + \int_0^t A_s p(x,s;\omega)ds = p_0(x) + \int_0^t [p(x,s;\omega)[H(x,s) - \int_{R^n} H(\alpha,s)p(\alpha,s;\omega)d\alpha] - B_s p(x,s;\omega)]db(s,\omega).$$

Pour cela, on procède, en trois étapes (LEVIEUX [5]) : montrer que la relation (2.21) admet une solution unique de carré sommable ; montrer que l'expression (2.20) est presque toujours définie et correspond à un opérateur indéfiniment différentiable opérant sur la fonction $\tilde{p}$; enfin, appliquer le lemme de ITO en dimension infinie pour identifier les deux membres de la relation (2.20).

Théorème II.3.

Sous les hypothèses (H1) à (H7), il existe une solution unique dans $L^2\{\Omega x(0,T);V\} \cap L^2\{\Omega,\mu;C^0[(0,T);H]\}$ à l'équation (2.21). Cette solution est liée à celle de l'équation (2.18) par la relation (2.20) pour presque tout ω ∎

III. APPROXIMATION DE LA SOLUTION DE L'EQUATION (2.18)

III. 1. Rappel et extension des résultats connus.

Soit h un réel positif destiné à tendre vers zéro. Soit N la partie entière de T/h et h,k,x(t) la fonction caractéristique de l'intervalle $[hk,h(k+1)]$. Introduisons la fonction suivante $p_h(x,t)$ dite semi-discrétisée en temps de la fonction $p(x,t)$ par la relation :

$$(3.1) \quad p_h(x,t) = \sum_{h=0}^{n} X_{h,k}(t)p_h(x,k).$$

La fonction $p_h(x,k)$ est donnée par la récurrence :

$$(3.2) \quad p_h(x,0) = p_0(x)$$

$$(3.3) \quad p_h(x,k+1) - p_h(x,k) + h\,A_h(k)p_h(x,k+1) = \varphi[p_h(x,k)],$$

où φ est un second membre dépendant de l'équation dont on cherche à approximer la solution. Si il s'agit de l'équation (2.18), alors φ est donné par :

$$(3.4) \quad \varphi[p_h(x,k)] = B_h[k,p(x,k)][z(k+1,\omega) - z(k,\omega)].$$

S'il s'agit de l'équation (2.21), alors φ est donné par la relation :

$$(3.5) \quad \varphi[p_h(x,k)] = [p_h(x,k)[H(x,k) - \int_{R^n} H(x,k)p_h(x,k)dx] - B_h(k,p_h(x,k))]$$
$$[z[(k+1)h,\omega] - h\int_{R^n} H(x,k)p_h(x,k)dx].$$

Nous allons montrer le résultat suivant, qui est une extension directe des théorèmes 1.3 et 2.1 chapitre 3 et 2.2 chapitre 5 de la référence [5].

Théorème III.1.

Sous les hypothèses (H1) à (H7), la fonction $p_h(x,t)$ définie par les relations (3.1) (3.2) et (3.5) converge fortement dans $L^2\{\Omega x(0,T)xR^n\}$ vers la solution de l'équation (2.21). ∎

III. 2. Convergence d'un schéma d'approximation implicite et décomposable.

Soit I un ensemble d'indices de cardinal fini et **i** l'indice courant. Introduisons une famille de fonctions $\xi_i(x)$, $i \in I$, chacune vérifiant les hypothèses (H4) et (H5). Enfin, notons λ_i un ensemble de réels positifs ou nuls vérifiant la relation :

$$(3.6) \quad \sum_{i \in I} \lambda_i = 1.$$

Soit $\tilde{p}(x,k)$ la suite de fonction définie de la façon suivante :

$$(3.7) \quad \xi_i(x,0) = \xi_i(x)$$

$$(3.8) \quad \xi_i(x,k+1) - \xi_i(x,k) + hA_h\xi_i(x,k+1) = H(x,k)\xi_i(x,k)[z[(k+1)h,\omega] - z(kh,\omega)].$$

$$(3.9) \quad \tilde{p}(x,k) = \sum_{i \in I} \lambda_i \xi_i(x,k).$$

Enfin, notons $p(x,k)$ la suite de fonction définie par la relation :

$$(3.10) \quad p(x,k) = \frac{\tilde{p}(x,k)}{\int_{R^n} \tilde{p}(x,k)dx}$$

Théorème III.2.

Sous les hypothèses (H1) à (H7), la fonction $p_h(x,t)$ définie par les relations (3.1) et (3.6) à (3.10) converge fortement dans $L^2\{\Omega x(0,T)xR^n\}$ vers la solution de l'équation (2.21). ∎

La démonstration de ce théorème est détaillée dans LEVIEUX [7].

III. 3. Implémentation parallèle du schéma d'approximation précédent.

Les conditions imposées aux fonctions $\xi_i(x)$ sont vérifiées par n'importe quelle base d'éléments finis usuelle. Le théorème III.2. implique la convergence d'un algorithme calculant la solution de l'équation (2.21) par normalisation dans $L^1(R^n)$. Le temps affecté à cette normalisation est largement compensé par la simplicité de l'équation (2.18). Une implémentation parallèle de cet algorithme est possible en utilisant la linéarité de l'opérateur qui fait passer d'une condition initiale à la solution correspondante de (2.18). Supposons que l'on dispose de processeurs S_i capables de calculer la fonction $\xi_i(x,k+1)$ connaissant les fonctions $\xi_i(x,k)$ et $z((k+1)h,\omega)$. Il est possible de calculer la valeur de $\tilde{p}(x,k+1)$ à partir de $\tilde{p}(x,k)$ par la procédure suivante :

- effectuer la décomposition de $\tilde{p}(x,k)$ sur une base de fonctions $\xi_i(x,k)$;

- calculer en parallèle les fonctions $\xi_i(x,k+1)$ à l'aide des processeurs S_i ;
- reconstituer $\tilde{p}(x,k+1)$ à partir des $\xi_i(x,k+1)$ et du résultat de la première étape ;
- incrémenter le temps de h et recommencer.

La difficulté d'une telle implémentation réside dans le choix convenable des fonctions de base ξ_i. Elles peuvent varier depuis une simple fonction "chapeau" jusqu'à une combinaison de gaussiennes (voir réf.[11]). Leur choix est un problème spécifique de l'application traitée et ne peut suivre de règles générales.

Cette méthode a été tentée en simulation sur plusieurs exemples, dont celui que nous présentons dans la section IV. La stabilité du schéma est la même que dans une solution non-parallèle. La seconde étape du calcul, qui est la plus longue dans une implémentation classique, devient la plus courte. Il n'est cependant pas possible de chiffrer exactement le gain de temps, qui est toujours important, mais dépend du nombre et de la complexité des fonctions $\xi_i(x)$.

IV. APPLICATION NUMERIQUE A UN PROBLEME D'INTERFEROMETRE A DEUX ANTENNES

IV. 1. Description du dispositif.

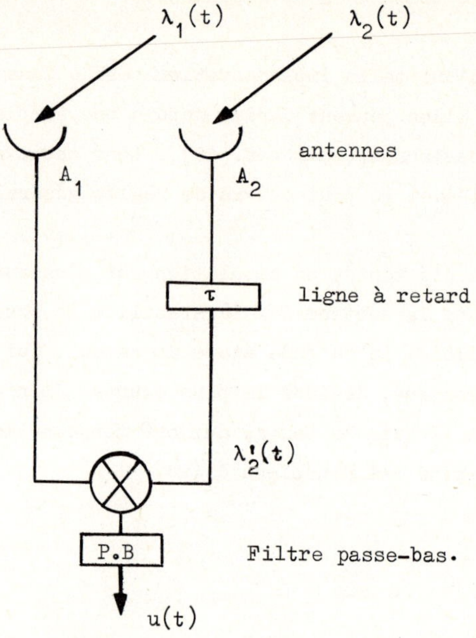

Figure 1.

On dispose de deux antennes A_1 et A_2 avec lesquelles on capte des signaux radio-astronomiques $\lambda_1(t)$, $\lambda_2(t)$. Ces antennes sont réglées sur une fréquence Ω et un retard τ est placé à la sortie de l'antenne A_2. Les signaux observés dans l'intervalle de temps $[0,T]$ sont :

$$\lambda_1(t) = A \cos[\Omega t + \varphi_1(t)],$$

$$\lambda_2'(t) = A \cos[\Omega(t-\tau) + \varphi_2(t)].$$

On fait passer $\lambda_1(t)$ et $\lambda_2'(t)$ dans un multiplieur, puis dans un filtre passe-bas, et on obtient le signal :

$$u(t) = \frac{A^2}{2} \cos[-\Omega\tau + \varphi_2(t) - \varphi_1(t)].$$

Les signaux $\lambda_1(t)$ et $\lambda_2(t)$ étant supposés provenir de la même source, la différence de phase $\varphi_2(t) - \varphi_1(t)$ correspond à la différence de marche de l'onde observée entre les deux antennes, qui ne dépend que de la géométrie du système et de la direction de la source émissive. Compte tenu de la rotation de la terre, cette différence de marche est supposée proportionnelle au temps, soit :

$$\varphi_2(t) - \varphi_1(t) = \Psi_0 + \omega_0 t.$$

En pratique, la fréquence $\omega_0/2\pi$ varie de 0 à 15 Hz. Le signal u(t) a donc la forme suivante :

$$u(t) = \frac{A^2}{2} \cos[\omega_0 t - \Omega\tau + \Psi_0].$$

En prenant pour τ une fonction linéaire du temps, $\tau = \alpha t$, le signal u(t) devient :

$$(4.1) \quad u(t) = a \cos(\omega t + \Psi_0), \text{ avec } \omega = \omega_0 - \Omega\alpha \text{ et } a = \frac{A^2}{2}.$$

On dispose donc à présent d'un signal sinusoïdal u(t) dont on peut faire varier la fréquence en jouant sur le paramètre α. On remarquera que u(t) devient une fonction constante du temps lorsque : $\alpha = \omega_0/\Omega$.

En pratique, les phases $\varphi_1(t)$ et $\varphi_2(t)$ sont soumises à de faibles fluctuations que nous modéliserons sous la forme :

$$\varphi_2(t) - \varphi_1(t) = \Psi_0 + \omega_0 t + \varphi(t),$$

où $\varphi(t)$ est une fonction aléatoire de moyenne nulle sur l'intervalle d'observation.

D'autre part, au signal observé à la sortie du filtre passe-bas s'ajoutent des perturbations dues au trajet de l'onde entre la source et les antennes. On observera donc à la sortie du filtre le signal :

$$(4.2) \quad y(t) = a \cos[\omega t + \Psi_0 + \varphi(t)] + b(t),$$

où b(t) peut être considéré comme un bruit blanc gaussien stationnaire sur $[0,T]$. Le niveau de ce bruit est très élevé : le rapport de la densité spectrale de puissance du bruit à la puissance moyenne du signal peut atteindre 15 db.

IV. 2. Résultats et comparaison avec le filtre Kalman-Bucy étendu.

Des essais comparatifs des filtres non-linéaires (NL) et Kalman-Bucy étendus

(KBE) ont été effectués sur ordinateur IRIS-80. Le bruit dynamique a été pris très faible ($g = 10^{-4}$) en raison du problème d'interférométrie traité. Des tests statistiques ont été effectués en fonction du niveau de bruit d'observation, du pas de discrétisation spatiale, et du pas de discrétisation temporelle. Les résultats présentés ont été obtenus par statistique sur 500 échantillons par niveau de bruit.

Nous prendrons dans toute la suite comme définition du rapport signal sur bruit la quantité :

$$R = \frac{\text{puissance moyenne du signal}}{\text{densité spectrale du bruit}} = \frac{a^2}{2\,r^2}$$

IV. 3. Résultats concernant la mise en œuvre du filtre ; choix des pas de discrétisations spatiale et temporelle.

Régime transitoire et régime permanent (cf.Fig.6).

x_0 étant la valeur autour de laquelle la phase x fluctue faiblement, la réponse du filtre NL, c'est-à-dire la courbe $\hat{x}(t)$ a toujours grossièrement la structure présentée sur la figure 6.

On y distingue un régime transitoire de durée TT, et un régime permanent dans l'intervalle (TT, TM), TM étant la durée totale du filtrage. On appelle erreur permanente e_p, l'erreur quadratique moyenne en régime permanent.

Influence du pas de discrétisation spatiale.

Pour un niveau de bruit donné, l'influence du pas de discrétisation spatiale est représentable par la courbe de l'erreur permanente e_p en fonction du nombre de points en espace NPT. Quel que soit le niveau du bruit cette courbe a l'allure de la figure 2.

Par "points" on peut entendre, soit l'une des fonctions de base $\xi_i(x,k)$, le résultat du calcul étant la fonction $\xi_i(x,k+1)$, soit les nœuds d'un réseau de discrétisation classique, les fonctions ξ_i ayant pour support un de ces nœuds ou un ensemble connexe pris parmi eux.

Le temps de calcul étant proportionnel à NPT, on a intérêt à prendre la valeur NPT_{opt} au début du palier de la courbe $e_p(NPT)$. Les tests montrent que NPT_{opt} ne dépend que du rapport signal à bruit r. On choisira donc un pas de discrétisation spatiale à l'aide de la courbe NPT_{opt} (R) représentée par la figure 3.

On remarque que le nombre de points d'espace requis diminue lorsque le bruit d'observation augmente.

Influence du pas de discrétisation temporelle.

Les essais montrent que l'influence du pas de discrétisation temporelle est analogue à celle du pas de discrétisation spatiale. Quel que soit le niveau de bruit, 20 points par période semble être une bonne discrétisation. En dessous de 20 points, le filtre est trop imprécis, et au-dessus de 20 points, on ne gagne que très peu en précision.

IV. 4. Résultats globaux en fonction du niveau de bruit d'observation.

Le principal résultat est la stabilité du filtre NL en bruit fort, par opposition au filtre KBE qui diverge fréquemment dès que le bruit devient important.

Bruit faible, $R > 3$ dB.

Lorsque la discrétisation spatiale est trop grossière on observe pour le filtre NL de fortes oscillations haute fréquence autour de la bonne valeur x_0 de la phase (cf.Fig.7). Pour remédier à cela, on est amené à prendre une discrétisation spatiale très fine (100 points par période) ; mais le temps de calcul machine est alors très élevé : 50 fois celui du filtre KBE.

On pourrait également remédier à cet inconvénient en plaçant à la sortie du filtre NL un filtre passe-bas. Le filtre KBE est lui très stable en bruit faible, et précis. Il est donc préférable, dans ce cas, de choisir le filtre KBE qui donne une bonne précision, une bonne stabilité et un temps de calcul machine faible par rapport à celui du filtre NL.

Bruit moyen, $0 < R < 3$ dB (cf.Fig.8).

Le filtre KBE présente souvent des divergences et ne sort pas de son régime transitoire. Le pourcentage de divergence est de 10 à 15 % des essais effectués. Lorsque le filtre KBE converge, les deux filtres ont une précision équivalente (Fig.8). Notons d'autre part, que le temps de calcul du filtre NL est de l'ordre de 7 à 10 fois celui du filtre KBE. Le choix de l'un des deux filtres est donc ici moins évident en raison de l'opposition temps de calcul - risque de divergence.

Bruit fort, $R < 0$ dB (cf.Fig.9).

Le filtre KBE devient ici inutilisable en raison du risque très important de divergence (60 % pour R = 15 dB). Le filtre NL est stable et son temps de calcul est toujours 7 à 10 fois celui du filtre KBE. Le filtre NL sera donc choisi incontestablement dans le cas de bruit fort.

Précision du filtre NL.

La précision obtenue par le filtre NL est mesurable par la variance moyenne de l'erreur permanente pour un niveau de bruit donné, que nous estimons par :

$$V = \text{Moyenne sur l'ensemble des échantillons à bruit donné de } \int_{TT}^{TM} [x(t) - \hat{x}(t)]^2 dt.$$

La courbe V(R) est représentée par la figure 4. Remarquons qu'il est difficile de bien définir le régime transitoire du filtre. Ici nous avons pris une définition sévère de TT ; il reste donc encore une partie de transitoire dans (TT, TM) d'où une mesure pessimiste de V. Notons également qu'en bruit fort, les divergences détériorent la précision moyenne du filtre KBE.

Durée des transitoires.

La durée moyenne TT du régime transitoire donne l'ordre de grandeur du temps que met le filtre à accrocher la bonne valeur. Ce temps est mesuré en périodes temporelles $2\,\pi/\omega$. Il varie avec le niveau de bruit suivant la courbe de la figure 5.

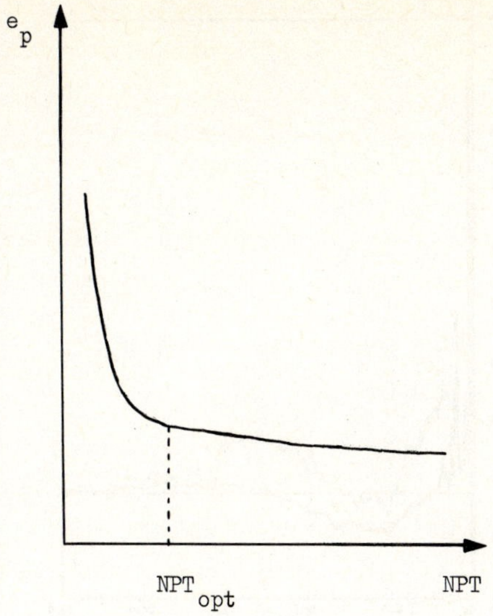

<u>Fig.2</u> : Allure de la courbe $e_p(NPT)$ à niveau de bruit donné.

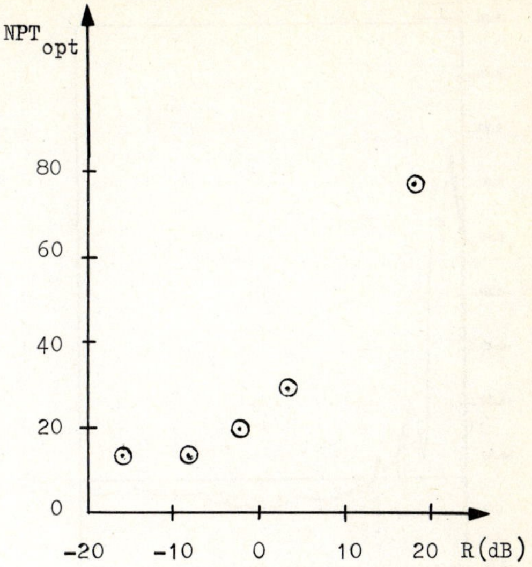

<u>Fig.3</u> : NPT_{opt} en fonction du rapport signal à bruit R.

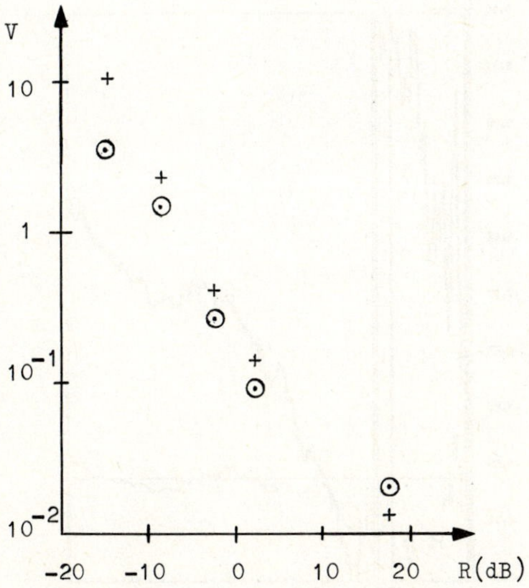

O Filtre NL + Filtre KBE

<u>Fig.4</u> : Précision des filtres.

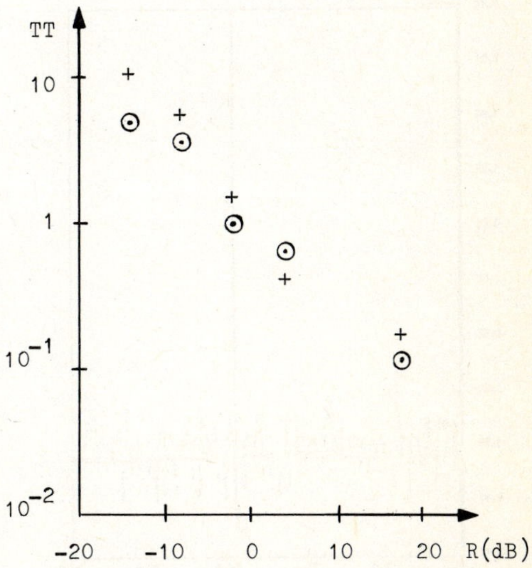

O Filtre NL + Filtre KBE

<u>Fig.5</u> : Durée des transitoires.

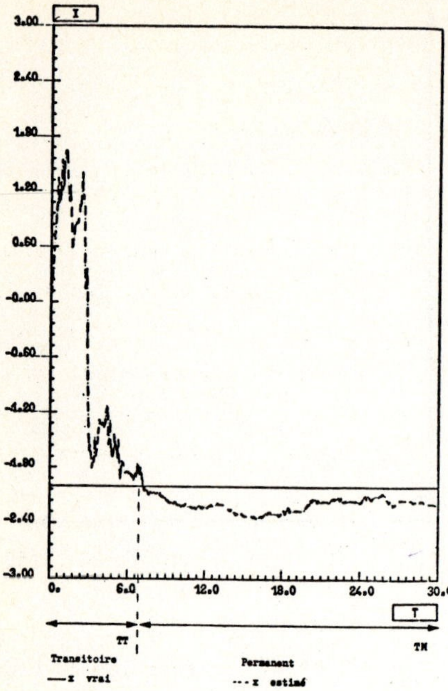

Fig.6 : Régime transitoire et permanent.

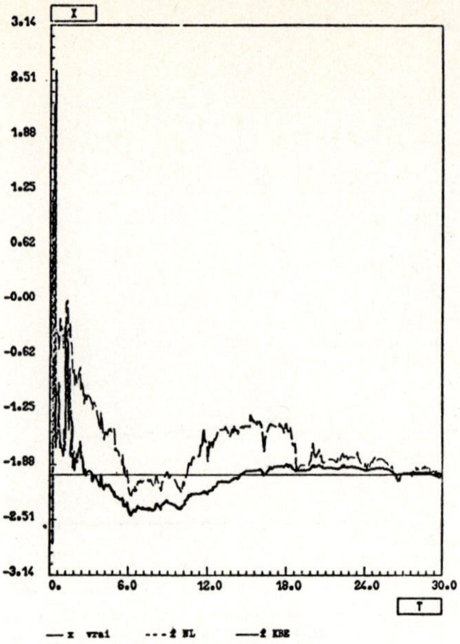

Fig.8 : Exemple de filtrage en bruit moyen.

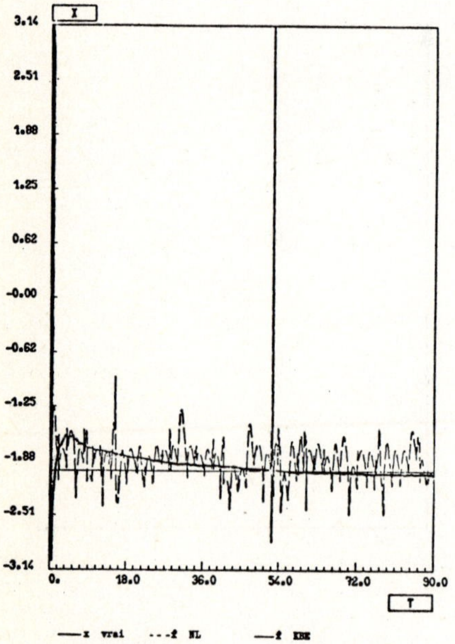

Fig.7 : Oscillations du filtre NL en bruit faible.

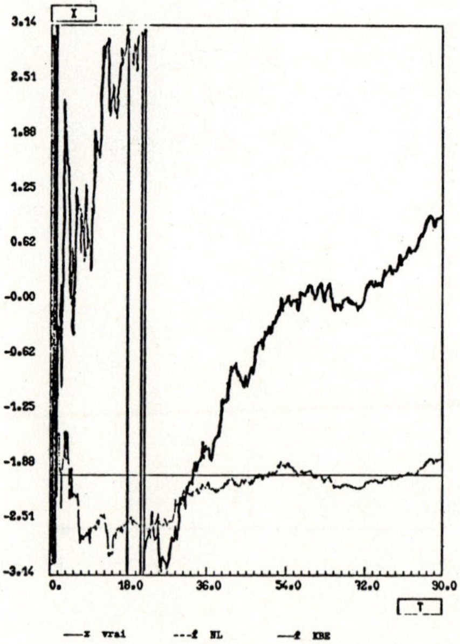

Fig.9 : Divergence du filtre KBE en bruit fort.

V. CONCLUSION

Dans cette étude, on a montré qu'on pouvait simplifier les algorithmes d'approximation de l'équation d'évolution de KUSHNER, sans perdre les propriétés de stabilités et de précision, qui en font l'intérêt. On a d'autre part confirmé, à la suite des travaux de ZAKAI [12] et ROZOWSKII [10], que l'évolution de la densité de probabilité conditionnelle du filtrage non-linéaire récursif est caractérisée par une équation bilinéaire aux dérivés partielles. Pour n suffisamment réduit (inférieur à 4), on a pu montrer que les méthodes par éléments finis, conduisaient à des algorithmes numériques stables et très performants par rapport aux solutions connues. Plusieurs applications en cours confirment, sur le plan pratique, les espoirs que ces résultats fondamentaux avaient fait naître (LAMBLA [3]). En particulier, cette méthode est la seule solution envisageable, lorsque le niveau des bruits perturbateurs est important par rapport aux non-linéarités du dispositif physique de mesures.

VI. REFERENCES

[1] FUJISAKI, KALLIANPUR, KUNITA, Stochastic differential equations for the non-linear filtering problem. Osaka, Journal of Mathematic n° 9 (1972), pp.19-40.

[2] H.J. KUSHNER, Dynamical Equations for Optimal non-linear filtering, Journal of Differential Equations, Vol. 3 (1967), pp.179-190.

[3] J.L. LAMBLA, Application de méthodes récursives de filtrage non-linéaire à l'estimation de la phase d'un signal sinusoïdal fortement bruité, Colloque National sur le Traitement du Signal et ses Applications (GRETSI) Nice 16-21 Juin 1975.

[4] F. LEVIEUX, Un théorème d'existence et d'unicité de la solution d'une équation intégro-différentielle stochastique. C.R. Acad. Sciences. Paris. Ser. A. t. 277 (1973), pp.281-284.

[5] F. LEVIEUX, Filtrage non-linéaire et analyse fonctionnelle. Rapport LABORIA n° 57 (1974).

[6] F. LEVIEUX, Recursive non-linear filtering : Theorical approach, numerical analysis and applications. A paraître Applied Mathematics and Optimization (1975).

[7] F. LEVIEUX, Rapport LABORIA à paraître (1975).

[8] J.L. LIONS, Contrôle optimal de systèmes gouvernés par des équations aux dérivées partielles, Dunod-Gauthier Villars (1968).

[9] E. PARDOUX, Thèse à la faculté des Sciences de Paris (1975).

[10] РОЗОВСКИЙ Б.Л. СТОХАСТИЧЕСКИЕ ДИФФЕРЕНЦИАЛЬНЫЕ УРАВНЕНИЯ ЧАСТНЫХ ПРОИЗВОДНЫХ, ВОЗНИКАЮЩИЕ В ЗАДАЧАХ НЕЛИНЕЙНОЙ ФИЛЬТРАЦИИ - В МОСКОВСКОМ МАТЕМАТИЧЕСКОМ ОБЩЕСТВЕ - ТОМ XXVII (1972), pp. 213-214.

[11] H.W. SORENSON, D.L. ALSPACH, Approximation of density functions by a sum of gaussians for non-linear bayesian estimation, 1st Symposium on non-linear estimation theory and its application, San Diego, Calif. (1970), Western Periodicals, Co, pp.19-31.

[12] M. ZAKAI, On the optimal filtering of diffusion processes. Z. Wahrscheinlichkeitstheorie verw. Gel. Vol. 11 (1969), pp.230-243.

METHODES DE DECOMPOSITION APPLIQUEES
AUX PROBLEMES DE CONTROLE IMPULSIONNEL

S. Maurin
IRIA-LABORIA
Domaine de Voluceau
78150 Le Chesnay (France)

INTRODUCTION

Les problèmes de contrôle impulsionnel interviennent dans un très grand nombre de situations en gestion scientifique des stocks cf. A.F. Veinott Jr [1]. Utilisant la programmation dynamique, A. Bensoussan et J.L. Lions [1] ont introduit pour l'étude de ces problèmes de nouvelles inégalités aux dérivées partielles, les inéquations quasi-variationnelles (I.Q.V).

S'agissant d'une extension des inéquations variationnelles introduites dans J.L. Lions et G. Stampacchia [1], il semble naturel d'appliquer aux I.Q.V les méthodes de décomposition introduites dans G.I. Marchuck [1] et N.N. Yanenko [1]. puis étendues dans Lions-Temam [1] et Bensoussan-Lions-Temam [1] aux problèmes variationnels.

Le but de ce travail est d'adapter ces méthodes aux inéquations quasi-variationnelles stationnaires.

Après un bref rappel du problème économique de gestion de stock étudié puis de la formulation en I.Q.V (§1), nous indiquons comment adapter les méthodes de décomposition récurrentes et parallèles (§2), nous étudions l'aspect numérique, précision et temps de Calcul, en fonction des paramètres dans deux cas de décomposition du second membre (§3).

Enfin nous simulons au §4 en temps partagé sur IRIS 80 CII la politique optimale de commande pour un exemple tridimensionnel stationnaire et nous visualisons l'évolution du stock sur écran de visualisation Tecktronix.

Ce travail est le prolongement d'un travail présenté en Juin 1974 au Congrès International sur la Théorie du Contrôle, les Méthodes Numériques et la Modélisation des Systèmes Informatiques à l'IRIA dans Goursat-Maurin [1].

§. 1. Rappel du problème théorique.

1.1. Un exemple économique.

Nous nous intéressons à la résolution du problème suivant. Nous renvoyons à Bensoussan-Lions [1] et [2] à Lions [1] et Goursat-Maarek [1] et à la bibliographie de ces ouvrages pour des exemples plus généraux.

Soit à gérer sur un intervalle $[0,T]$ un stock de q produits.

On suppose que l'état du système à l'instant s est un vecteur $y(s)$ d'un ensemble fermé $\bar{\Omega} \subset R^q$ où Ω est un ouvert régulier de R^q.

La demande est donnée par l'équation différentielle stochastique: cf. J.I. Gikhman-A.V. Skorokhod [1].

(1.1) $$D(t,t+dt) = \mu(t)\, dt + \sigma(t)\, [w(t+dt) - w(t)].$$

sur l'intervalle $[t, t+dt]$;

ou $\mu(t) = \mu_1(t), \mu_2(t), \dots, \mu_q(t) \in R^q.$

$\sigma(t) = \sigma_1(t), \sigma_2(t), \dots, \sigma_q(t) \in R^q.$

et $w(s)$ est un processus de Wiener de R^q.

Si à l'instant t $y(t) = x \in R^q$

On définit alors une politique de commande

$$v_{xt} = \{\theta^1_{xt}, \xi^1_{xt}, \dots, \theta^i_{xt} \dots, \theta^{Mxt}_{xt}, \xi^{Mxt}_{xt}\}$$

où θ^i_{xt} sont les Mxt instants de commandes des quantités

$$\xi^i_{xt} = (\xi^{1i}_{xt}, \dots, \xi^{qi}_{xt}) \in R^q$$

$y(s)$ obéit alors à l'équation stochastique.

(1.2) $$y(s) = x - \int_t^s \mu(\tau)d\tau - \int_t^s \sigma(\tau)dw(\tau) + \sum_{j=1}^{M_{xt}} \xi^j_{xt}\, \delta_{(s-\theta^j_{xt})} .$$

où δ représente le symbole de Dirac

La fonction coût de la politique v_{xt} est définie par :

(1.3) $$J(v_{xt}) = E_{xt}\Big[\int_t^T f(y(\tau),\tau)d\tau + k\, N_{xt} + \sum_{i=1}^{q} k_i\, N^i_{xt}\ \Big]$$

$f(x,t)$ représente le coût de gestion ou de rupture de stock

k le coût de commandes groupées

k_i le coût de commande du $i^{\text{ème}}$ produit

N_{xt} le nombre de commandes groupées

N^i_{xt} le nombre de commandes du $i^{\text{ème}}$ produit seulement.

On cherche alors v_{xt} telle que

(1.4) $$J(v_{xt}) \leq J(v^*_{xt}) \quad \forall \ v^*_{xt}$$

On pose

(1.5) $$u(x,t) = J(v_{xt}).$$

1.2. <u>Caractérisation de u(x,t)</u>.

<u>Cas évolutif</u> :

On montre dans Bensoussan-Lions [1] et [2] qu'il existe une politique optimale de commande v_{xt} et que $u(x,t)$ est caractérisée par l'inéquation quasi variationnelle suivante :

(1.6) $$< -\frac{\partial u}{\partial t}, v - u >_{\mathcal{L}^2(\Omega)} + a(u,v-u) \geq < f, v - u >_{\mathcal{L}^2(\Omega)}$$

$$\forall \ v \in H^1(\Omega)^{(1)} \quad v \leq M\,u\,(x,t).$$
$$u \leq M(u)$$
$$u(x,T) = 0 \ .$$

où $a(u,v)$ est définie par

(1.7) $$a(u,v) = \sum_{i=1}^{q} \int_\Omega \left[\frac{\sigma_i^2}{2} \frac{\partial u}{\partial x_i}\frac{\partial v}{\partial x_i} + \mu_i \frac{\partial u}{\partial x_i} v\right].dx$$

pour $u, v \in H^1(\Omega)$

et M par

(1.8) $$M(u) = \inf\,(M_o(u), M_1(u),\dots,M_q(u))$$

avec pour $u \in \mathcal{L}^\infty(\Omega)$

$$M_o(u) = k + \inf_{\substack{x \in \Omega \\ x+\xi \in \Omega \\ \xi \geq 0 \quad \xi \in R^q}} u(x+\xi) \ .$$

$$M_i(u) = k_i + \inf_{\substack{x \in \Omega \\ x+\xi_i\vec{e}_i \in \Omega \\ \xi_i \geq 0}} u(x+\xi_i\vec{e}_i) \qquad i=1 \text{ à } q \ .$$

(1) Pour la définition de $H^1(\Omega)$ on se référera à Lions-Magènes [1].

Cas stationnaire :

en introduisant un coefficient d'actualisation α, pour $T = +\infty$, en supposant que μ , σ, et f sont indépendants de t, alors dans (1.5) en posant $u(x,o) = u(x)$

On montre que u satisfait à

(1.9) $$a(u, v-u) \geq \langle f, v-u \rangle_{\mathcal{L}^2(\Omega)}$$

$$u \leq M(u)$$

$$v \leq M(u) \quad v \in H^1(\Omega)$$

où $a(u,v)$ dans (1.7) est changée en

(1.10) $$a(u,v) = \sum_{i=1}^{q} \int \frac{\sigma_1^2}{2} \frac{\partial u^2}{\partial x_i} \frac{\partial v}{\partial x_i} + \mu_i \frac{\partial u}{\partial x_i} v + \int_\Omega \alpha\, u\, v\,.$$

si $f \geq 0$ et $f \in \mathcal{L}^\infty(\Omega)$ on montre que (1.9) admet une seule solution $u \geq 0$.

Bensoussan-Goursat-Lions [1] L. Tartar [1] Th.Laetsch [1].

Nous cherchons donc des algorithmes de décomposition pour résoudre le problème stationnaire.

§. 2. Algorithmes de décomposition.

Nous en citons 2 ; pour d'autres algorithmes et les démonstrations nous renvoyons à Maurin [1].

Notations : (Temam [1]).

Ω est un ouvert régulier de R^q. (cf. ouvrage cité de Lions-Magènes pour la définition d'un ouvert régulier).

(2.1) $$V_i = \{u \in \mathcal{L}^2(\Omega) \quad \frac{\partial u}{\partial x_i} \in \mathcal{L}^2(\Omega)\} \quad i = 1 \text{ à } q .$$

(2.2) $$a_i(u,v) = \int_\Omega \frac{\sigma_i^2}{2} \frac{\partial u}{\partial x_i} \frac{\partial v}{\partial x_i} dx + \int_\Omega \mu_i \frac{\partial u}{\partial x_i} v\, dx + \int_\Omega \alpha_i\, u\, v\, dx.$$

$$= \langle A_i u,v \rangle_{V_i' \times V_i} \qquad \forall\ u, v \in V_i \times V_i$$

où V_i' est le dual de V_i et $\langle\ \rangle_{V_i' \times V_i}$ le produit scalaire de la dualité.

$$\sum_{i=1}^{q} \alpha_i = \alpha .$$

On suppose que

(2.3) $$a_i(u,u) \geq \beta_i \|u\|^2_{V_i} \qquad \text{avec } \beta_i > 0 .$$

(2.4) $$f = \sum_{i=1}^{q} f_i \quad f_i \geq 0 \quad f_i \in \mathcal{L}^\infty(\Omega) .$$

La décomposition de f est supposée fixe ou variable au sens de St Pierre [1].

(2.5) $$u_o = \sup \frac{f}{\alpha} \ . \ \Delta t \text{ positif petit et M donné par (1.8)} .$$

Algorithme récurrent :

On résoud successivement les q équations suivantes :

Etape 1 : Trouver $u^{1/q+1} \in V_1$ tel que

(2.6) $$\left\langle \frac{u^{1/q+1} - u_o}{\Delta t}, v \right\rangle_{\mathcal{L}^2(\Omega)} + a_1(u^{1/q+1}, v) = \langle f_1, v \rangle_{\mathcal{L}^2(\Omega)} \qquad \forall\ v \in V_1 .$$

Etape 2 : Trouver $u^{2/q+1} \in V_2$ tel que

(2.7) $$\left\langle \frac{u^{2/q+1} - u^{1/q+1}}{\Delta t}, v \right\rangle_{\mathcal{L}^2(\Omega)} + a_2(u^{2/q+1}, v) = \langle f_2, v \rangle_{\mathcal{L}^2(\Omega)} \qquad \forall\ v \in V_2 .$$

<u>Etape i</u> : Trouver $u^{i/q+1} \in V_i$ tel que

(2.8) $$< \frac{u^{i/q+1} - u^{i-1/q+1}}{\Delta t}, v >_{\mathcal{L}^2(\Omega)} + a_i(u^{i/q+1}, v) = < f_i, v >_{\mathcal{L}^2(\Omega)} \quad \forall\ v \in V_i.$$

<u>Etape q</u> : Trouver $u^{q/q+1} \in V_q$ tel que

(2.9) $$< \frac{u^{q/q+1} - u^{q-1/q+1}}{\Delta t}, v >_{\mathcal{L}^2(\Omega)} + a_q(u^{q/q+1}, v) = < f_q, v >_{\mathcal{L}^2(\Omega)}$$

<u>Etape q+1</u> :

(2.10) $$u^1 = \inf\ (u^{q/q+1}, M(u^{q/q+1}))$$

puis par récurrence sur r, supposant connaître u^r on résoud successivement

<u>Etape r(q+1)</u>:

Trouver $u^{r+1/q+1} \in V_1$ tel que

(2.11) $$< \frac{u^{r+1/q+1} - u^r}{\Delta t}, v >_{\mathcal{L}^2(\Omega)} + a_1(u^{r+1/q+1}, v) = < f_1, v >_{\mathcal{L}^2(\Omega)} \quad \forall\ v \in V_1.$$

<u>Etape r(q+1) + 2</u> :

Trouver $u^{r+2/q+1} \in V_2$ tel que

(2.12) $$< \frac{u^{r+2/q+1} - u^{r+1/q+1}}{\Delta t}, v >_{\mathcal{L}^2(\Omega)} + a_2(u^{r+2/q+1}, v) = < f_2, v >_{\mathcal{L}^2(\Omega)}$$

$$\forall\ v \in V_2.$$

<u>Etape r(q+1) + i</u> :

Trouver $u^{r+i/q+1} \in V_i$ tel que

(2.13) $$< \frac{u^{r+i/q+1} - u^{r+i-1/q+1}}{\Delta t}, v >_{\mathcal{L}^2(\Omega)} + a_i(u^{r+i/q+1}, v) = < f_i, v >_{\mathcal{L}^2(\Omega)}$$

$$\forall\ v \in V_i.$$

<u>Etape r(q+1)+q</u> :

Trouver $u^{r+q/q+1} \in V_q$ tel que

(2.14) $$< \frac{u^{r+q/q+1} - u^{r+q-1/q+1}}{\Delta t}, v >_{\mathcal{L}^2(\Omega)} + a_q(u^{r+q/q+1}, v) = < f_q, v >_{\mathcal{L}^2(\Omega)} \quad \forall\, v \in V_q.$$

Etape (r+1)(q+1) :

(2.15) $$u^{r+1} = \inf (u^{r+q/q+1}, M(u^{r+q/q+1}))$$

et ainsi de suite.

On démontre alors le

Théorème I : lorsque $r \to \infty$

$u^{r+i/q+1}$ tend en décroissant vers $u_{\Delta t}^{i/q+1}$ dans V_i fort $i=1$ à q.

u^r tend en décroissant vers $u_{\Delta t}^1$ dans $\mathcal{L}^2(\Omega)$ fort ou $u_{\Delta t}^{i/q+1}$ $i=1,2,\ldots,q,q+1$ sont solutions de

(2.16)
$$\begin{gathered}\sum_{i=1}^{q} a_i(u_{\Delta t}^{i/q+1}, v - u_{\Delta t}^{i/q+1}) \geq \sum_{i=1}^{q} \langle f_i, v - u_{\Delta t}^{i/q+1} \rangle_{\mathcal{L}^2(\Omega)} \\ v \leq M(u_{\Delta t}^{q/q+1}) \quad v \in H^1(\Omega) \\ u_{\Delta t}^1 \leq M(u_{\Delta t}^{q/q+1}) \\ \sum_{i=2}^{q+1} \| u_{\Delta t}^{i/q+1} - u_{\Delta t}^{i-1/q+1} \|_{\mathcal{L}^2}^2 \leq C\,\Delta t .\end{gathered}$$

Algorithme parallèle :

On peut adapter l'algorithme précédent en remplaçant dans (2.13) $u^{r+i-1/q+1}$ par u^r $\forall\, r \in N$ $\forall\, i=1,2,\ldots,q+1$ et dans (2.15) $u^{r+q/q+1}$ par $\frac{1}{q}\sum_{i=1}^{q} u^{r+i/q+1}$.

Nous proposons un autre algorithme : J.L. Liens [1].

On résoud successivement les q I.Q.V

Etape 1 : Trouver $u^{1/q+1} \in V_1$ tel que

(2.17) $$\left\langle \frac{u^{1/q+1} - u_0}{\Delta t}, v - u^{1/q+1} \right\rangle_{\mathcal{L}^2(\Omega)} + a_1(u^{1/q+1}, v - u^{1/q+1}) \geq \langle f_1, v - u^{1/q+1} \rangle_{\mathcal{L}^2(\Omega)}$$

$$u^{1/q+1} \leq M_1(u^{1/q+1})$$

$$v \leq M_1(u^{1/q+1}) \quad v \in V_1 .$$

Etape 2 : Trouver $u^{2/q+1} \in V_2$ tel que

(2.18) $$< \frac{u^{2/q+1} - u_o}{\Delta t}, v - u^{2/q+1} >_{\mathcal{L}^2(\Omega)} + a_2(u^{2/q+1}, v - u^{2/q+1}) \geq < f_2, v - u^{2/q+1} >_{\mathcal{L}^2(\Omega)}$$

$$u^{2/q+1} \leq M_2(u^{2/q+1})$$

$$v \leq M_2(u^{2/q+1}) \qquad v \in V_2 .$$

Etape i : Trouver $u^{i/q+1} \in V_i$ tel que

(2.19) $$< \frac{u^{i/q+1} - u_o}{\Delta t}, v - u^{i/q+1} >_{\mathcal{L}^2(\Omega)} + a_i(u^{i/q+1}, v - u^{i/q+1}) \geq < f_i, v - u^{i/q+1} >_{\mathcal{L}^2(\Omega)}$$

$$u^{i/q+1} \leq M_i(u^{i/q+1}) \qquad v \leq M_i(u^{i/q+1}) \qquad v \in V_i$$

Etape q : Trouver $u^{q/q+1} \in V_q$ tel que

(2.20) $$< \frac{u^{q/q+1} - u_o}{\Delta t}, v - u^{q/q+1} >_{\mathcal{L}^2(\Omega)} + a_q(u^{q/q+1}, v - u^{q/q+1}) \geq < f_q, v - u^{q/q+1} >_{\mathcal{L}^2(\Omega)}$$

$$u^{q/q+1} \leq M_q(u^{q/q+1})$$

$$v \leq M_q(u^{q/q+1}) \qquad v \in V_q .$$

Etape q+1 :

(2.21) $$u^1 = \inf\left(\frac{1}{q} \sum_{i=1}^{q} u^{i/q+1} , M_o\left(\frac{1}{q} \sum_{i=1}^{q} u^{i/q+1}\right)\right)$$

puis par récurrence sur r en supposant connaître u_r, on résoud successivement

Etape r(q+1) + 1 : Trouver $u^{r+1/q+1} \in V_1$ tel que

(2.22) $$< \frac{u^{r+1/q+1} - u_r}{\Delta t}, v - u^{r+1/q+1} >_{\mathcal{L}^2(\Omega)} + a_1(u^{r+1/q+1}, v - u^{r+1/q+1}) \geq < f_1, v - u^{r+1/q+1} >_{\mathcal{L}^2(\Omega)}$$

$$u^{r+1/q+1} \leq M_1(u^{r+1/q+1})$$

$$v \leq M_1(u^{r+1/q+1}) \qquad v \in V_1 .$$

Etape r(q+1) + 2 : Trouver $u^{r+2/q+1} \in V_2$ tel que

$$(2.23) \qquad \left\langle \frac{u^{r+2/q+1} - u^r}{\Delta t}, v-u^{r+2/q+1} \right\rangle_{\mathscr{L}^2(\Omega)} + a_2(u^{r+2/q+1}, v-u^{r+2/q+1}) \geq \langle f_2, v-u^{r+2/q+1} \rangle_{\mathscr{L}^2(\Omega)}$$

$$u^{r+2/q+1} \leq M_2(u^{r+2/q+1})$$

$$v \leq M_2(u^{r+2/q+1}) \qquad v \in V_2 .$$

Etape r(q+1) + i : Trouver $u^{r+i/q+1} \in V_i$ tel que

$$(2.24) \qquad \left\langle \frac{u^{r+i/q+1} - u^r}{\Delta t}, v-u^{r+i/q+1} \right\rangle_{\mathscr{L}^2(\Omega)} + a_i(u^{r+i/q+1}, v-u^{r+i/q+1}) \geq \langle f_i, v-u^{r+i/q+1} \rangle_{\mathscr{L}^2(\Omega)}$$

$$u^{r+i/q+1} \leq M_i(u^{r+i/q+1})$$

$$v \leq M_i(u^{r+i/q+1}) \qquad v \in V_i .$$

Etape r(q+1)+q : Trouver $u^{r+q/q+1} \in V_q$ tel que

$$(2.25) \qquad \left\langle \frac{u^{r+q/q+1} - u^r}{\Delta t}, v-u^{r+q/q+1} \right\rangle_{\mathscr{L}^2(\Omega)} + a_q(u^{r+q/q+1}, v-u^{r+q/q+1}) \geq \langle f_q, v-u^{r+q/q+1} \rangle_{\mathscr{L}^2(\Omega)}$$

$$u^{r+q/q+1} \leq M_q(u^{r+q/q+1})$$

$$v \leq M_q(u^{r+q/q+1}) \qquad v \in V_q .$$

Etape (r+1)(q+1)

$$(2.26) \qquad u^{r+1} = \inf\left(\frac{1}{q} \sum_{i=1}^{q} u^{r+i/q+1} , M_o\left(\frac{1}{q} \sum_{i=1}^{q} u^{r+i/q+1}\right)\right)$$

On démontre alors le

Théorème II : lorsque $r \to \infty$ $u^{r+i/q+1}$ tendent en décroissant vers $u_{\Delta t}^{i/q+1}$ dans V_i fort $i = 1$ à q , u^r tend en décroissant vers $u_{\Delta t}^1$ dans $\mathscr{L}^2(\Omega)$ fort où les $u_{\Delta t}^{i/q+1}$ $i=1$ à $q+1$ sont solutions de

$$(2.27) \qquad \sum_{i=1}^{q} a_i(u_{\Delta t}^{i/q+1}, v - u_{\Delta t}^{i/q+1}) \geq \sum_{i=1}^{q} \langle f_i, v - u_{\Delta t}^{i/q+1} \rangle_{\mathscr{L}^2(\Omega)}$$

$$v \leq M_o(\frac{1}{q}\sum_{i=1}^{q} u_{\Delta t}^{i/q+1})$$

$$v \leq M_i(u_{\Delta t}^{i/q+1}) \qquad i=1 \text{ à } q$$

$$u_{\Delta t}^{1} \leq M_o(\frac{1}{q}\sum_{i=1}^{q} u_{\Delta t}^{i/q+1})$$

$$u_{\Delta t}^{i/q+1} \leq M_i(u_{\Delta t}^{i/q+1})$$

$$\sum_{i=1}^{q} \| u_{\Delta t}^{i/q+1} - u_{\Delta t}^{1} \|^2_{\mathcal{L}^2(\Omega)} \leq C\,\Delta t$$

§. 3. Résolution numérique.

On montre dans Bensoussan-Lions [2] que le problème (1.9) est équivalent au suivant dans Ω.

trouver u telle que

$$(3.1)\qquad \begin{cases} A\,u \leq f \\ u \leq M(u) \\ (Au-f)(u-M(u)) = 0 \\ \left.\dfrac{\partial u}{\partial \vec{v}}\right|_{\Gamma} = 0 . \end{cases}$$

où $\left.\dfrac{\partial u}{\partial \vec{v}}\right|_{\Gamma}$ représente la dérivée normale de u dirigée vers l'intérieur de Ω.

et Au est donné par

$$(3.2)\qquad Au = \sum_{i=1}^{q} A_i\, u .$$

avec

$$(3.3)\qquad A_i u = -\frac{\sigma_i^2}{2}\frac{\partial^2 u}{\partial x_i^2} + \mu_i \frac{\partial u}{\partial x_i} + \alpha_i\, u .$$

On discrétise alors le problème (3.1) par les différences finies. Soit h le pas de discrétion.

On note :

(3.4)

A_{ih} la discrétisée de A_i

A_h la discrétisée de A

f_h la discrétisée de f

f_{ih} la discrétisée de f_i

$u_h^o(i_1,i_2,\ldots,i_q)$ la solution discrétisée de l'IQV obtenue par relaxation avec projection au point $i_1,\ldots,i_q$ du maillage de discrétisation de l'ouvert Ω. Bensoussan-Lions [1] .

$u_h^r(i_1,i_2,\ldots,i_q)$ la solution discrétisée obtenue par l'algorithme de décomposition après r étapes de projection au même point $i_1,i_2,\ldots,i_q$.

On pose

$$\text{EPS} = \frac{1}{N_h} \sum_{i_1} \sum_{i_2} \sum_{\cdots} \sum_{i_q} \left| u_h^o(i_1,\ldots,i_q) - u_h^r(i_1,\ldots,i_q) \right| \tag{3.5}$$

ou N_h est le nombre de points de discrétisation de Ω.

On résoud donc par récurrence le système linéaire suivant :

$$(\Delta t\, A_{ih} + I)\, u_h^{r+i/q+1} = f_{ih}\, \Delta t + u_h^{r+i-1/q+1} \tag{3.6}$$

pour $i = 1$ à q .

On utilise un algorithme décrit dans N.N. Yanenko [1] puis on projette

$$u_h^{r+1} = \inf\, (u_h^{r+q/q+1},\ M(u_h^{r+q/q+1})) \tag{3.7}$$

L'étude numérique a été faite de deux points de vue

1. <u>Résolution de (3.1)</u>.

On a étudié la convergence de l'algorithme récurrent en calculant la solution stationnaire de (3.1) puis la quantité EPS.

On a étudié les variations de EPS en fonction de Δt dans les deux cas de décomposition fixe ou variable du second membre.
La décomposition variable utilisée est la suivante

$$(3.8)\qquad \begin{aligned} f_{1h} &= f_h - A_h u_h^r + A_{1h}\, u_h^r\,. \\ f_{2h} &= A_{2h}\, u_h^r \\ &\ \vdots \\ f_{qh} &= A_{qh}\, u_h^r \end{aligned}$$

2. <u>Recherche des surfaces</u> $\Gamma_1\ \Gamma_{10}\ \Gamma_{11} \ldots \Gamma_{1q}$ et $\Sigma_o\ \Sigma_1 \ldots \Sigma_q$. définies par les relations suivantes

Γ_1 = frontière entre les deux régions Ω_1 et Ω_2

suivantes

$$(3.9)\qquad \begin{aligned} \Omega_1 &= \{\, x \quad u(x) = Mu(x) \,\} \subset \Omega \\ \Omega_2 &= \{\, x \quad u(x) < Mu(x) \,\} \subset \Omega \\ \Gamma_{10} &= \{\, x \in \Gamma_1 \quad u(x) = M_o u(x) \,\}\,. \end{aligned}$$

$$\Gamma_{1i} = \{ x \in \Gamma_1 \quad u(x) = M_i\, u(x) \} \qquad i=1 \text{ à } q .$$

$$\Sigma_o = \{ x \quad u(x) = \inf_{\substack{\xi \in \mathbb{R}^q \\ x \in \Omega \\ x+\xi \in \Omega}} u(x + \xi) \} .$$

$$\Sigma_i = \{ x \quad u(x) = \inf_{\substack{x \in \Omega \\ x+\xi_i \vec{e}_i \in \Omega \\ \xi_i \in R}} u(x + \xi_i\, \vec{e}_i) \} \qquad i=1 \text{ à } q .$$

En effet on montre que la donnée de ses surfaces est suffisante à la connaissance de la politique optimale de commande du problème de gestion défini au §.1.

Nous définissons cette politique optimale en vue de la simulation du processus au §.4.

On a donc recherché ces surfaces par l'algorithme de décomposition, les exemples numériques montrant qu'elles deviennent très vite stationnaires.

De manière analogue à 1 on a étudié les variations des temps de Calcul en fonction de Δt et dans les deux cas de décomposition du second membre f.

L'exemple étudié est l'exemple tridimensionnel suivant :

On prend

$$f(x_1, x_2, x_3) = 4|x_1| + 4|x_2| + 4\,|x_3| .$$

$$Mu(x) = k + \inf_{\substack{x \in \Omega \\ x+\xi \in \Omega \\ \xi \geq 0}} u(x+\xi) \qquad k = 1$$

$$\sigma_1^2 = \sigma_2^2 = \sigma_3^2 = 1.6/\pi^2$$

$$\mu_1 = \mu_2 = \mu_3 = 2/\pi$$

$$\alpha = 0.4$$

$$\Omega = [-0.7 \;,\; +0.7]^3$$

$$u_o = 21$$

$$h = 0.1 .$$

On a donc 3375 points de discrétisation à chaque étape. On résoud donc 225 fois 3 systèmes linéaires 15 × 15 relatifs à chaque direction d'espace.

Convergence du coût Décomposition fixe

Δt	Nb d'itérations	Temps exécution	E P S =	Compilation
$\frac{1}{25}$	650	1 mn 40 sec	0,171	2 sec 60
$\frac{1}{50}$	1200	3 mn	0,083	"
$\frac{1}{100}$	2200	6 mn 10 sec	0,042	"

Convergence du coût Décomposition variable

Δt	Nb d'itérations	Temps exécution	E P S =	Compilation
$\frac{1}{25}$	650	2 mn 30 sec	0,048	7 sec 80
$\frac{1}{50}$	1300	5 mn 17 sec	0,020	"
$\frac{1}{100}$	2400	9 mn 30 sec	0,010	"

Recherche de la frontière libre   Décomposition fixe

Δt	Nb d'itérations	Temps exécution	Compilation
$\frac{1}{25}$	25	2 s 40	4 sec
$\frac{1}{50}$	45	5 s 20	"
$\frac{1}{100}$	85	9 s 50	"

Recherche de la frontière libre   Décomposition variable

Δt	Nb d'itérations	Temps exécution	Compilation
$\frac{1}{25}$	25	4 s 50	11 sec
$\frac{1}{50}$	45	8 s	"
$\frac{1}{100}$	85	15 s 11	"

Figure II

les résultats sont présentés sur les tableaux des figures I et II. Les temps d'exécution et de compilation de la figure I sont ceux d'un IBM 370-168 , ceux de la figure II d'un IBM 360-91.

<u>Conclusion</u> : Sur l'exemple étudié on observe les résultats suivants (Figure III)

en ce qui concerne la première étude, on note une dépendance linéaire de la précision par rapport à Δt.

On obtient une précision quatre fois plus grande en utilisant la décomposition variable du second membre.

en ce qui concerne la seconde étude, la décomposition variable n'apporte pas d'amélioration sensible.

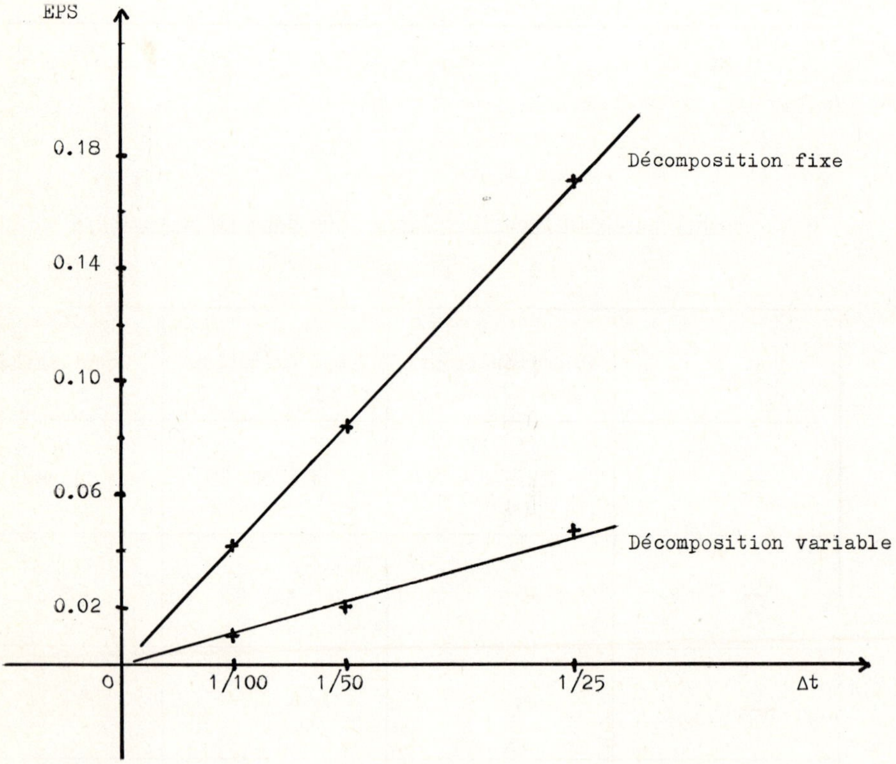

<u>FIGURE III</u>

<u>Variation de EPS en fonction de Δt</u>.

§. 4. Simulation :

Supposons pouvoir simuler la demande définie en (1.1). C'est ce que fait effectivement un programme de bibliothèque ; alors si l'on connaît les surfaces définies en (3.9) on montre que la politique optimale est définie de la manière suivante.

Soit $x = x_1, \ldots, x_q$ l'état initial du stock .

Laissons évoluer le stock suivant la demande, soit $y(s)$ l'état du stock à l'instant s.

Si $y(s) \notin \Gamma_1$ on ne commande pas.

Si $y(s) \in \Gamma_{10}$ on fait une commande groupée $\xi_1, \xi_2, \ldots, \xi_q = \xi \in R^q$ telle que $y(s) + \xi \in \Sigma_0$ puis on laisse évoluer le stock.

Si $y(s) \in \Gamma_{1i}$ $i = 1$ à q pour un i (les exemples numériques montrent que les Γ_{1i} n'intersectent pas) on commende ξ_i du i^e produit telle que $y(s) + \xi_i \vec{e_i} \in \Sigma_i$ et on laisse évoluer le stock.

Nous avons en temps partagé sur IRIS 80 d'une part visualisé les différentes surfaces définies en (3.9) d'autre part simulé l'évolution d'un stock et visualisé la politique optimale relative à chaque produit sur un écran de visualisation Tektronix.

L'organigramme utilisé est le suivant

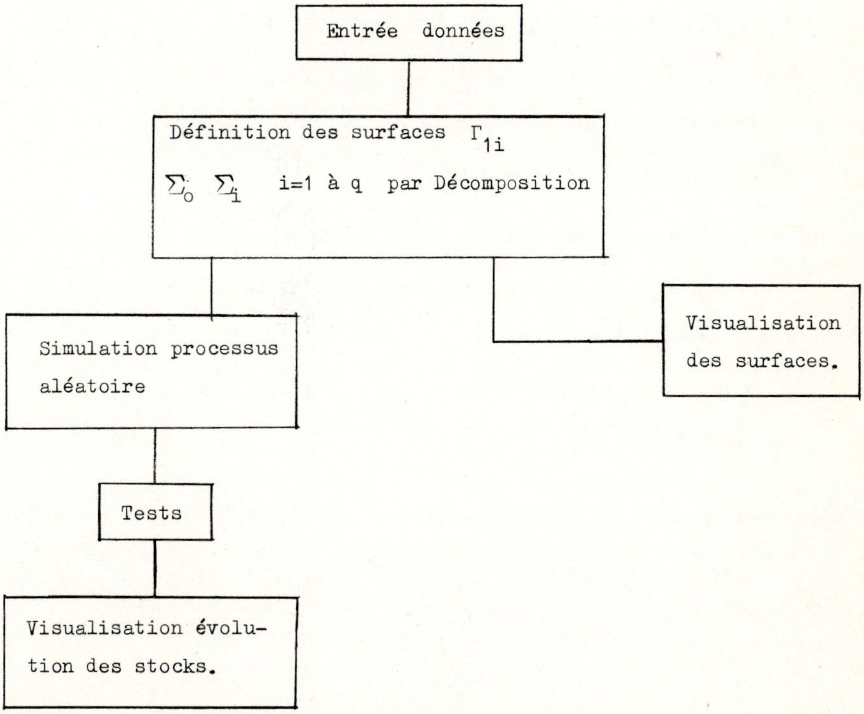

Les données sont entrées sur console en temps partagé, elles sont les suivantes :

- bornes inférieures de l'ouvert Ω pris égal à 1 cube.
- valeurs des coûts $k, k_1, \dots, k_q$.
- nombre d'itérations de l'algorithme.
- valeur du pas de temps Δt.
- coefficients de pénalisation dans le coût de stockage.
- coefficients $\sigma_1^2, \dots, \sigma_q^2$, $\mu_1, \dots, \mu_q$, α de l'IQV.

• le processus est simulé par un programme de bibliothèque.

• la visualisation utilise le software graphique Fortran 2D ou 3D E Saltel [1] projet Lemaire IRIA-LABORIA.

On obtient sur écran de visualisation Tektronix soit, les surfaces Γ_{1i} et Σ_i, soit les diagrammes simultanés de l'évolution dans le temps de chacun des trois produits.

L'exemple étudié est le suivant : $q = 3$

$$\Omega = [\,-0.7 \quad +0.7\,]^3 \qquad \sigma_1^2 = 0.0625 \quad \mu_1 = 2.5$$

$$\Delta t = 1/25 \qquad \sigma_2^2 = 0.25 \quad \mu_2 = 5$$

$$\alpha = 0.4 \qquad \sigma_3^2 = 0.16 \quad \mu_3 = 4 .$$

les coefficients du processus aléatoire sont donc :

$$\sigma_1^2\,\Delta t = 0.0025 \qquad \mu_1\,\Delta t = 0.1$$

$$\sigma_2^2\,\Delta t = 0.01 \qquad \mu_2\,\Delta t = 0.2$$

$$\sigma_3^2\,\Delta t = 0.0064 \qquad \mu_3\,\Delta t = 0.16$$

le coût de gestion est égal à :

$$f(x_1, x_2, x_3) = f_1(x_1) + f_2(x_2) + f_3(x_3) \quad \text{avec}$$

$$f_1(x_1) = 4\,|x_1|$$

$$f_2(x_2) = 10\,|x_2| \quad \text{si } x_2 \leq 0 \qquad f_2(x_2) = 4\,|x_2| \quad x_2 \geq 0$$

$$f_3(x_3) = 4\,|x_3|$$

M définie comme en (1.8) avec

$$k = 1 \quad k_1 = 0.5 \quad k_2 = 0.8 \quad k_3 = 0.2 .$$

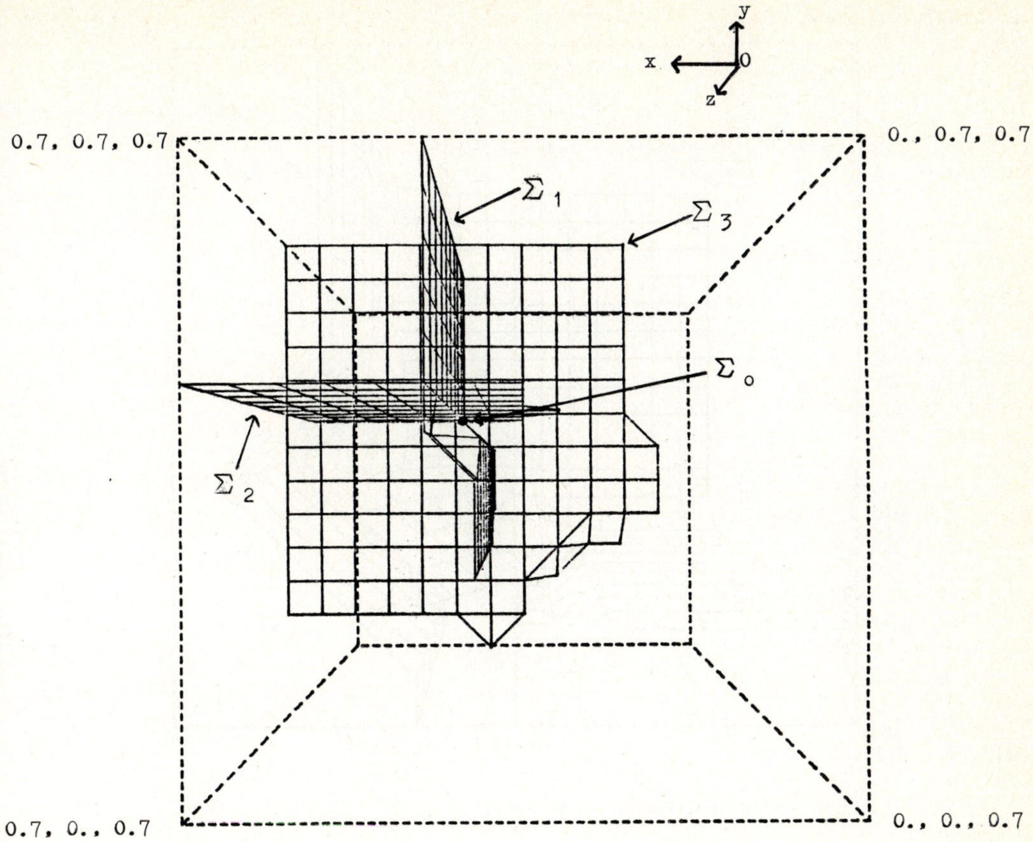

Σ_1 = { minima dans la direction 0_x }

Σ_2 = { minima dans la direction 0_y }

Σ_3 = { minima dans la direction 0_z }

Σ_0 minimum absolu.

Figure IV

Surfaces des minima de u.

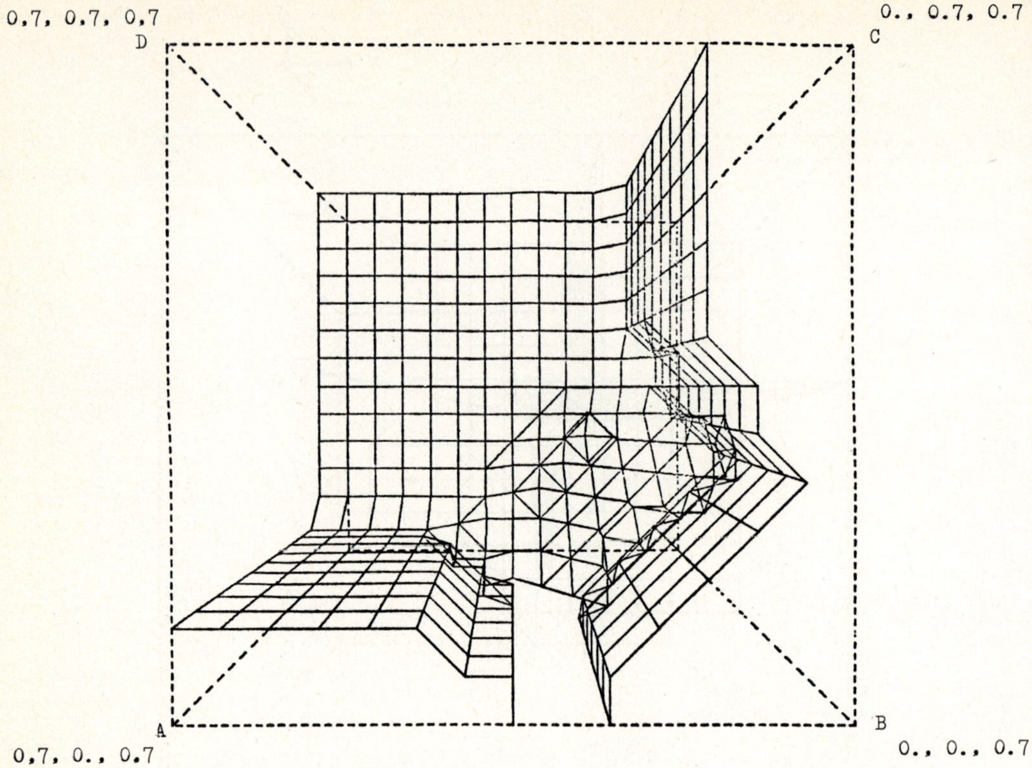

$\sigma_1 = \sigma_2 = \sigma_3 = 0.16$

$\mu_1 = \mu_2 = \mu_3 = 0.7$

$\alpha = 0.4$

M défini comme en 1.8

FIGURE V

Frontière entre $\Omega_1 = \{x\ u(x) = M(u)(x)\}$ et $\Omega_2 = \{x\ u(x) < M(u)(x)\}$.

On s'est arrêté sur l'exemple étudié après dix itérations de l'algorithme de décomposition chaque itération prenant environ 10 secondes en temps réel sur IRIS 80 CII.

On a alors obtenu les figures IV en ce qui concerne les surfaces Σ_i les figures V en ce qui concerne la surface Γ_1 (on peut bien sûr visualiser seulement les Γ_{1i}) et la figure VI en ce qui concerne la simulation.

Precisons que la figure VI est la reproduction d'une photographie de l'écran de visualisation prise à l'IRIA.

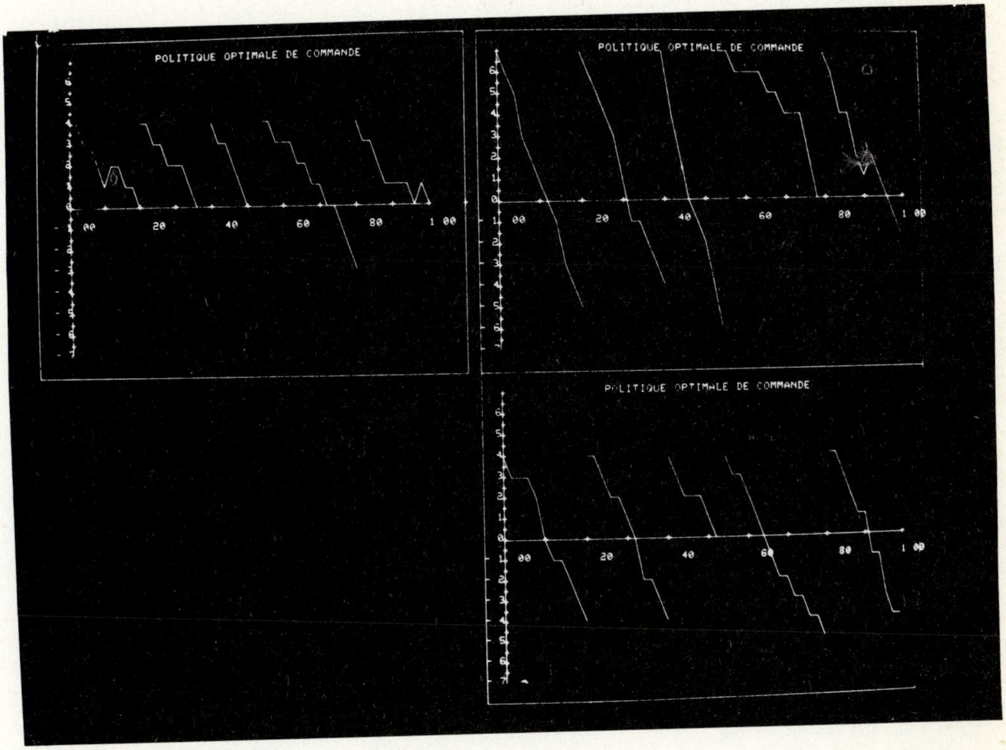

Figure VI

Simulation de gestion de stock + visualisation.

Unité de temps = 1 an

Unité de quantité de produits = 1 tonne

Les discontinuités de l'évolution des stocks définissent les instants de commande de chacun des trois produits.

Bibliographie

A. Bensoussan - J.L. Lions C.R.A.S.

[1] 1) t 276 (1973) pp 1189-1192
2) t 276 (1973) pp 1333-1338
3) t 278 (1974) pp 675-679
4) t 278 (1974) pp 747-751

[2] Temps optimal, Contrôle impulsionnel et application (ouvrage en préparation).

A. Bensoussan - M. Goursat - J.L. Lions C.R.A.S.

[1] t 276 pp 1279-1284

A. Bensoussan - J.L. Lions - R. Temam

[1] Cahier IRIA N° 11 Juin 1972 p 5-189 .

J. Cea et R. Glowinsky

[1] Sur des méthodes d'optimisation par relaxation.

II. Gikman - A.V. Skorohod

[1] Stochastic differential equations.
Springer Verlag Berlin Heidelberg
New York 1972 .

B. Glowinsky - J.L. Lions - R. Tremolières

[1] Livre Dunod Paris 1974 .

M. Goursat - Maarek

[1] Rapport Laboria 1974.

M. Goursat - S. Maurin

[1] Communication au Congrès International sur la théorie du Contrôle, les méthodes Numériques et la Modélisation des sytèmes Informatiques.
IRIA Juin 1974.

C. Leguay

[1] Thèse 3e cycle 1975.

J.L. Lions

[1] On the numerical approximation of problems of impulse control.
IFIP Novosibirsk Juillet 1974.

J.L. Lions - E. Magenes

[1] Problèmes aux limites non homogènes.
Dunod Paris 1968 Tome I.

J.L. Lions - G. Stampacchia

[1] Comm. Pure and Applied Math. XX (1967) pp 493-519 .

J.L. Lions - R. Temam

[1] C.R.A.S. 263 (1966) pp 563-565 .

G.I. Marchuk

[1] **Numerical** methods on Meteorology
A. Colin Paris 1969

S. Maurin

[1] Rapport Laboria 1975 .

E. Saltel

[1] Thèse 3e cycle Utilisation de la generation dynamique d'images et des procédures graphiques dans le Software Metavisu III .

P. St Pierre

[1] Thèse 3e cycle

L. Tartar

[1] C.R.A.S. t 278 pp 1193-1196

R. Temam

[1] Thèse Paris 1967

A.F. Veinott Jr

[1] Management Science 12(11) 1966 pp 745-777

N.N. Yanenko

[1] Méthodes à pas fractionnaires
A. Colin 1968.

A MIXT RELAXATION ALGORITHM APPLIED TO QUASI-VARIATIONNAL INEQUATIONS

J.C. MIELLOU
Faculté des Sciences et des Techniques
La Bouloie - Route de Gray
25030 - BESANCON CEDEX

Introduction -

In the first paragraph we consider a class of finite dimensionnal quasi-variationnal inequalities problems, for which we recall a "Theoretical" algorithm allowing the approximation of a maximal inequalities. Then we recall an heuristic formulation of a method which mixt point relaxation method and the "Theoretical" algorithm mentionned above.

In the second paragraph we introduce in a more general frame (Block-relaxation), a precise formulation and set convergence results for this mixt method. For this purpose we use a notion of "delays" previously introduced by CHAZAN and MIRANKER, in the formulation of multiprocessor relaxation algorithms.

§ I - BACKGROUND : definitions, a finite dimensional I.Q.V. problem ; a "Theoretical" algorithm ; First formulation of a mixt GAUSS-SEIDEL algorithm. Let $\beta \in N$, $E=R^\beta$, $K=R^\beta_+$ -

We note $\leq$ the partial order relation induced on E by the cone K.

We consider the application $\mathcal{A}$:

$$v = \{v_1,\dots,v_k,\dots,v_\beta\} \in R^\beta \xrightarrow{\mathcal{A}} \mathcal{A}(v) = \{\mathcal{A}_1(v),\dots,\mathcal{A}_\ell(v),\dots,\mathcal{A}_\beta(v)\} \in R^\beta.$$

For $k \in \{1,\dots,\beta\}$ let $\ell_k = \{0,\dots,0,1,0,\dots,0\}$ $\forall k,l \in \{1,\dots\beta\}$ we consider the function $\forall t \in R \xrightarrow{\varphi_{k,l}} \varphi_{k,l}(t) = \mathcal{A}_l(v+te_k)$.

Definition 1 -

The application $\mathcal{A}$ *is an M-application (or M-function) if :*
For $k \neq l$ $\quad t \rightarrow \varphi_{k,l}(t)$ *is antitone and moreover :*

$$\mathcal{A}(u) \leq \mathcal{A}(v) \Longrightarrow u \leq v.$$

Definition 2 -

The application $\mathcal{A}$ *is an L-application if :*

$\exists \{\lambda_1,\dots,\lambda_\beta\} \in K$ *such that* $\mathcal{A}+\Lambda$ *be an M-application from* R^β *onto* R^β *where is defined by :*

$$v = \{v_1,\dots,v_l,\dots,v_\beta\} \xrightarrow{\Lambda} \Lambda v = \{\lambda_1 v_1,\dots,\lambda_l v_l,\dots,\lambda_\beta v_\beta\}.$$

Let $\{u^p\}$ be a sequence of vectors ($\forall p \in N$ $u^p \in R^\beta$), we note the fact that $\{u^p\}$ is isotone (resp. antitone) and converges towards u, by $u^p \uparrow u$ (resp. $u^p \downarrow u$).

Définition 3 -

Let a mapping $g : R^\beta \to R^\beta$ be isotone, g is 0-half-continuous on the right (resp. the left) if :

(1.1) $u^p \downarrow u$ (resp $u^p \uparrow u$) $g(u^p) \downarrow g(u)$ (resp $g(u^p) \uparrow g(u)$).

Remark - Definition 1 is due to W.C. RHEINBOLDT, property (1.1) that we call here 0-half-continuity is due to BENSOUSSAN-GOURSAT-LIONS.

we consider now the hypothesis :

(1.2) Let $\mathcal{A} : R^\beta \to R^\beta$ be a continuous mapping which is an L-application ; and let $f \in R^\beta$.

(1.3) Let Φ $K \to K$ be isotone for the order induced on R^β by K, and 0-half-continuous on the right on K.

(1.4) $\forall w' \in K$ Let $\mathcal{U}_{(w')} = \{v \in K | v \leq \Phi(w')\}$.

We formulate the I.Q.V. problem - (I.Q.V. for quasi-variationnal inequalities)

$$(1.5)_a \begin{cases} \text{Find } u \in \mathcal{U}_{(u)} \text{ such that :} \\ (\mathcal{A}(u), v-u) \geq (f, v-u) \quad \forall v \in \mathcal{U}_{(u)} \end{cases}$$

(where (,) notes the usual scalar product on R^β).

Other formulation of problem $(1.5)_a$: $\forall w' \in K$, we mark $\Phi(w') = \{\ldots, \Phi^l_{w'}, \ldots\}$ for $l \in \{1,\ldots,\beta\}$.

Let $\Psi_{[0,\Phi^l_{w'}]}$ is the indicatrix function of the segment $[0, {}^l_{w'}]$, and let $\partial\Psi_{[0,\Phi^l_{w'}]}$ be the sub-differential of $\Psi_{[0,\Phi^l_{w'}]}$.

$$(1.6) \begin{cases} \text{Then } \forall w' \in K, \ \forall l \in \{1,\ldots,\beta\} \text{ we consider the multiapplication } M^{w'}_l \\ \text{which :} \\ \forall u_l \in R \xrightarrow{M^{w'}_l} M^{w'}_l(u_l) = \partial\Psi_{[0,\Phi^l_{w'}]}(u_l) \subset R \\ \text{and we note :} \\ u = \{u_1,\ldots,u_l,\ldots,u_\beta\} \in R^\beta \xrightarrow{M^{w'}} M^{w'}_{(u)} = \{\ldots, M^{w'}_l(u_l), \ldots\},\ l \in \{1,\ldots,\beta\} \end{cases}$$

($M^{w'}$ is an operator of diagonal type).

The problem $(1.5)_a$ can now be formulated in the following manner :

$$(1.5)_b \begin{cases} \text{Find } u \in D(M^u) \text{ such that} \\ 0 \in M^u(u) + \mathcal{A}(u) - f. \end{cases}$$

A "Theoretic" algorithm for the approximation of a solution of problem $(1.5)_{a/\text{or } b/}$

Starting from a convenient initial vector u^o, we consider a sequence of I.V. problems (for Variationnal Inequalities) :

$(1.7)_a$ $(\mathcal{A}(u^{p+1}), v-u^{p+1}) \geq (f, v-u^{p+1})$ $\forall v \in \mathcal{U}(u^p)$, p=0,1,...

These kind of algorithm has been proposed and studied by BENSOUSSAN, GOURSAT, LIONS in the continuous case (A being then an elliptic second order, partial differential operator).

We can also formulate the problems $(1.7)_a$ in the following manner.

$(1.7)_b$ $\quad 0 \in M^{u^p}(u^{p+1}) + \mathcal{A}(u^{p+1}) - f \qquad p = 0,1,\ldots$

<u>First formulation of a mixt GAUSS-SEIDEL algorithm</u> -

BENSOUSSAN - LIONS (see also the work of COMMINCIOLI and alt for analogous free boundary problems) had proposed for the resolution of each problem $(1.5)_{a/\text{ or } b/}$ the point GAUSS-SEIDEL method with projection, which can be formulated by : let $u^{o,p} = u^p$ we approximate u^{p+1} by a sequence $\{u^{q,p}\}$ defined by :

$$(1.8)_a \quad \begin{cases} \forall q \in N, \ \forall l \in \{1,\ldots,\beta\} \quad \mathcal{A}_l(u_1^{q+1,p},\ldots,u_{l-1}^{q+1,p},\tilde{u}_l^{q+1,p}, u_{l+1}^{q,p},\ldots,u_l^{q,p}) = f_l \\ u_l^{q+1,p} = \operatorname{Proj}_{[0,\Phi^l_{u^p}]} \tilde{u}_l^{q+1,p} \end{cases}$$

or equivently :

$$(1.8)_b \quad \forall q \in N, \ \forall l \in \{1,\ldots,\beta\} \ 0 \in M_l^{u^p}(u_l^{q+1,p}) + \mathcal{A}_l(u_1^{q+1,p},\ldots,u_l^{q+1,p},u_{l+1}^{q,p},\ldots) - f_l .$$

For q "great" we replace in $(1.8)_{a/\text{ or } b/}$ u_p by $u^{q,p}$ and start again, in (1.8) a/ or b/, with $u^{o,p+1} = u^{q,p}$.

Here it must be observed that $u^{o,p+1} \neq u^{p+1}$ so we cannot by the use of GAUSS-SEIDEL algorithm for the approximation of I.V. subproblems obtain exactly the "theoretical" algorithm (1.7).

§ II - <u>FORMULATION OF A MIXT RELAXATION ALGORITHM USING A NOTION OF DELAYS</u> -

Our main interest is now : To give a precise formulation of a mixt algorithm allowing the association of the "Theoretical" algorithm (1.7), and relaxation methods (namely under-relaxation and GAUSS-SEIDEL) ; moreover we place ourselves in a little more general frame, than a bove : that is to say, block-relaxation-methods.

<u>Subproblems associated to problem (1.5)</u> -

$$(2.1) \quad \begin{cases} \text{Let } \alpha \in N \text{ such that } \alpha \leq \beta, \text{ and a family of integers } \{\beta_1,\ldots,\beta_i,\ldots,\beta_\alpha\} \\ \text{such that } \sum_{i=1}^{\alpha} \beta_i = \beta . \end{cases}$$

$$(2.2) \quad \begin{cases} \forall i \in \{1,\ldots,\alpha\} \quad E_i = R^{\beta_i} \text{ and } \leq_i \text{ is the relation of partial order indu-} \\ \text{ced by the cone } K_i = R_+^{\beta_i} \text{ then } E = \prod_{i=1}^{\alpha} E_i \ ; \ K = \prod_{i=1}^{\alpha} K_i . \end{cases}$$

$\forall w' \in E, \quad \forall i \in \{1,\ldots,\alpha\} \qquad$ Let :

$$\mathcal{U}^i_{(w')} = \{v_i \in K_i \subset E_i \,|\, v_i \leq_i \Phi^i_{w'} = \{\ldots,\Phi^l_{w'},\ldots\} \text{ for } l \in \{\beta_1+\ldots+\beta_{i-1}-1,\ldots,\beta_1+ +\ldots+\beta_i\}.$$

we note moreover :

$$\forall i \in \{1,\ldots,\alpha\} \quad u_i \in E_i \xrightarrow{M_i^{w'}} M_i^{w'}(u_i) = \{\ldots,M_l^{w'}(u_l),\ldots\} \text{ , for}$$

$$l \in \{\beta_1+\ldots+\beta_{i-1}+1,\ldots,\beta_1+\ldots+\beta_i\}.$$

$\forall u = \{u_1,\ldots,u_i,\ldots,u_\alpha\} \in \prod_{i=1}^{\alpha} E_i$

Let $\mathcal{A}(u) = \{\mathcal{A}_1(u),\ldots,\mathcal{A}_j(u),\ldots,\mathcal{A}_\alpha(u)\} \in \prod_{j=1}^{\alpha} E_j$

$\forall w = \{w_1,\ldots,w_j,\ldots,w_\alpha\} \in \prod_{j=1}^{\alpha} E_j, \quad \forall u_j \in E_i$

Let $a_{i,w}(u_i) = \mathcal{A}_i(w_1,\ldots,w_{i-1},u_i,w_{i+1},\ldots,w_\alpha)$

Let $\textcircled{H} = \{\theta_1,\ldots,\theta_j,\ldots,\theta_\alpha\} \in \prod_{j=1}^{\alpha} [0,+\infty[$ such that :

$$(2.3) \qquad \forall i \in \{1,\ldots,\alpha\} \quad \theta_j \geq \max_{\beta_1+\ldots+\beta_{j-1}<l\leq\beta_1+\ldots+\beta_j} \lambda_l$$

where $\{\lambda_1,\ldots,\lambda_\beta\}$ has been introduced in <u>definition 2</u>.

We consider also $\Omega = \{\omega_1,\ldots,\omega_j,\ldots,\omega_\alpha\} \in \prod_{j=1}^{\alpha} [0,1]$ such that :

$\forall j \in \{1,\ldots,\alpha\}$ either $\omega_j = 1$, or $d_j \in R$, $d_j > 0$ such that

$$(2.4) \qquad \forall w \in R^\beta, \quad \forall v_j^1, v_j^2 \in R^{\beta_j} \quad \text{with} \quad v_j^1 \underset{j}{\leq} v_j^2.$$

$$a_{j,w}(v_j^2) - a_{j,w}(v_j^1) \underset{j}{\leq} d_j (v_j^2 - v_j^1) \text{ and } (1-\omega_j)d_j \leq \theta_j.$$

we consider now the subproblems :

$$(2.5)_a \begin{cases} \forall w, w' \in K, \ \forall j \in \{1,\ldots,\alpha\} \quad \text{find } u_j \in \mathcal{U}^j_{(w')} \text{ such that :} \\ (a_{j,w}(\omega_j u_j+(1-\omega_j)w_j) + \theta_j(u_j-w_j), v_j-u_j)_j \geq (f_j, v_j-u_j) \ \forall v_j \in \mathcal{U}^j_{(\bar{w}')} \\ \text{where } (,)_j \text{ is the usual scalar product on } E_j = R^{\beta_j}). \end{cases}$$

which can also be formulated by :

$\forall w,w' \in K \quad \forall j \in \{1,\ldots,\alpha\}$ find $u_j \in D(M_j^{w'})$ such that :

$$(2.5)_b \qquad f_j \in M_j^{w'}(u_j) + a_{j,w}(\omega_j u_j+(1-\omega_j)w_j) + \theta_j(u_j-w_j)$$

and we wright :

$$(2.5)_c \begin{cases} \forall j \in \{1,\ldots,\alpha\} \quad G_{j,\omega_j,\theta_j}(w,w') = u_j \\ G_{\Omega,\textcircled{H}}(w,w') = \{u_1,\ldots,u_j,\ldots,u_\alpha\}. \end{cases}$$

<u>Proposition 1</u> -

The hypothesis (1.2), (1.3), (1.4), (2.1), (2.2), (2.3) being satisfied, $w, w' \in K \xrightarrow{G_{\Omega,\textcircled{H}}} G_{\Omega,\textcircled{H}}(w,w')$ *is well defined (i.e. ∃ a unique solution of each subproblem* $(2.5)_{a/\text{ or }b/}$*), is an isotone mapping of each of its two arguments w, w', and is continuous relatively to the first argument w, and half-continuous on the right relatively to the second argument w'. Moreover* $u \in K$ *is a solution of the I.Q.V. problem* $(1.5)_{a/\text{ or }b/}$ *iff u is a fixed point of the application :*

$$w \xrightarrow{F} F(w) = G_{\Omega,\textcircled{H}}(w,w)$$

Mixt algorithm associating "theoretical" algorithm (1.7) and relaxation methods -

For the sake of simplicity we suppose that we want to perform r relaxation iterations for the approximation of the solution of each I.V. occuring in the "Theoretical" algorithm $(1.7)_a$. u^o being conveniently choiced (the precise choice shall begiven in proposition 2/) we introduce now three formulation of the same following mixt algorithm :

let $h(p) = p \bmod(\alpha) + 1$; $k(p) = p \bmod(r\alpha)$.

$$(2.6)_a \begin{cases} \text{If} \quad j \neq h(p) \qquad u_j^{p+1} = u_j^p \\ \text{If} \quad j = h(p) \qquad \text{find } u_j^{p+1} \in \mathcal{U}^j(u^{p-k(p)}) \\ (\mathcal{A}_{j,u^p}(\omega_j u_j^{p+1} + (1-\omega_j)u_j^p) + \theta_j(u_j^{p+1} - u_j^p),\ v_j - u_j^{p+1})_j \geq \\ \qquad\qquad (f_j,\ v_j - u_j^{p+1})_j \quad \forall v_j \in \mathcal{U}^j_{(u^{p-k(p)})} \end{cases}$$

$$(2.6)_b \begin{cases} \text{If} \quad j \neq h(p) \qquad u_j^{p+1} = u_j^p \\ \text{If} \quad j = h(p) \qquad \text{find } u_j^{p+1} \in D(M_j^{u^{p-k(p)}}) \text{ such that} \\ f_j \in M_j^{u^{p-k(p)}}(u_j^{p+1}) + \mathcal{A}_{j,u^p}(\omega_j\, u_j^{p+1} + (1-\omega_j)\, u_j^p) + \theta_j(u_j^{p+1} - u_j^p) \end{cases}$$

and :

$$(2.6)_c \begin{cases} \text{if} \quad j \neq h(p) \quad u_j^{p+1} = u_j^p \\ \text{if} \quad j = h(p) \quad u_j^{p+1} = G_{j,\omega_j,\theta_j}(u^p,\ u^{p-k(p)}). \end{cases}$$

Proposition 2 -

The hypothesis are the some that those of proposition 1, and let $u^o \in K$, be such that $f \leq \mathcal{A}(u^o)$, then the sequence $\{u^p\}$ produced by algorithm $(2.6)_{a/\ b/\ or\ c/}$ is such that $u^p \downarrow u$ maximal solution of problem $(1.5)_{a/\ or\ b/}$, and fixed point of $w \xrightarrow{F} F(w) = G_{\Omega,\, \Theta}(w,w)$.

Remarks - Il we cancel the constraints of our problem and :

- take $\alpha = 1$ and so $\beta_1 = \beta$ we can give an interpretation of algorithm (2.6) by the scheme :

$$\frac{u^{p+1} - u^p}{k} + \mathcal{A}(\omega u^{p+1} + (1-\omega)u^p) - f = 0$$

(Taking $\theta = \frac{1}{k}$).

as the discretisation of :

$$\frac{du}{dt} + \mathcal{A}(u) - f = 0 \qquad u(0) = u^o$$

- take $\alpha = \beta$ and suppose that $\mathcal{A}$ is a matrix with diagonal elements all equals to 1 then we find the classical relaxation method with relaxation parameters $\frac{1}{\omega_j + \theta_j}$ and hypothesis (2.4) implies that we are here in the case of under relaxation and GAUSS-SEIDEL methods.

- The sequence $\{k(p)\}$ constitutes the "delays" : notion introduced by CHAZAN-

MIRANKER for multiprocessors relaxation algorithms which can also be mixted with "Theoretical algorithm" (1.7) in the following manner :

$\forall p \in N$ let $h(p) \subset \{1,\ldots,\alpha\}$; $k^1(p) = \{k_1^1(p),\ldots,k_i^1(p),\ldots k_\alpha(p)\} \in N^\alpha$, $k^2(p) \in N$ and we suppose that :

- $\forall i \in \{1,\ldots,\alpha\}$ $\{p \in N \mid i \in h(p)\}$ is indefinite
- $\exists s \in N$ such that $\forall i \in \{1,\ldots,\alpha\}$ $0 \leq k_i^1(p) \leq s(p)$ $0 \leq k^2(p) \leq s(p)$
- $\forall i \in \{1,\ldots,\alpha\}$ $p \to p-k_i^1(p)$ is isotone, and $p \to p-k^2(p)$ is also isotone.

Then we consider the algorithm :

$$\begin{cases} u_j^{p+1} = G_j(\ldots,u_i^{p-k_i^1(p)},\ldots, u^{p-k^2(p)}) & \text{if } j=h(p) \\ u_j^{p+1} = u_j^p & \text{if } j \neq h(p) \end{cases}$$

Proposition 2 extends without difficulties to this more general situation. It seems that a particular interest of block relaxation methods in quasi-variational inequation problems lies in the fact that, these problems being often of great size, it can be necessary to decompose the arrays, which contain iterates vectors, in sub arrays corresponding to block sub problems, which can be contained in central memory of the processor used. Block relaxation can be interpreted as an iterative coordination method between these subproblems, which needs relatively simple (sequential), exchanges between central memory, and peripheral memories (disks).

BIBLIOGRAPHIE

ALBRECHT, J.

Fehlerschranken und konvergenzbescheunigung Bei einer Monotonen oder Alterniendev iterationsfolge.

Numer. Math., 4, (1962), 196-208.

BAIOCCHI, C. ; COMINCIOLI, V. ; GUERRI, L. ; VOLPI, G.

Free boundary problems in the theory of fluid flow through pourous media : a numerical approach.

Calcolo 10, (1973), 1-86.

BENSOUSSAN, A. ; GOURSAT, M. ; LIONS, J.L.

C.R. Acad. Sci. PARIS Sér. A, (1973), 1279.

BIRKHOFF, G. ; KELLOGG, R.

Solution of equilibrium equations in thermal networks.

Proc. Symp. generalized Networks BROOKLYN, (1966), 443-452.

BOHL.

Nichtlineare aufgaben in halb geordneten raumen.

Numer. Math. 10, (1964), 220-231.

CHARNAY, M.

Itérations chaotiques sur un produit d'espaces.

C.R. Acad. Sci. PARIS, t. 279, (1964).

CHARNAY, M. ; MUSY, F. ; ROBERT, F.

Itérations chaotiques série-parallèle pour des équations non linéaires de point fixe. *A paraître dans Aplikace Mathematiky.*

CHARNAY, M.

Thèse de 3ème cycle, Université Claude Bernard, LYON (1975).

CHARNAY, M. ; MUSY, F.

Sur le théorème de Stein-Rosenberg.

R.A.I.R.O., R-2, (1974), 95-108.

CHAZAN, D. ; MIRANKER, W.

Chaotic relaxation.

Linear algebra and its appl., 2, (1969), 199-222.

COLLATZ, L.

Functional analysis and numerical mathematics.

Springer Verlag, Berlin, Tranl. by H. OSER, Acad. Press, New-York, (1966).

DONNELLY, J.D.P.

Periodic chaotic relaxation.

Linear algebra and its appl., 4, (1971), 117-128.

DURAND, J.F.

L'algorithme de Gauss-Seidel appliqué à un problème unilatéral non symétrique.

R.A.I.R.O., R-2, (1972), 23-30.

FIORIOT, J.-Ch. ; HUARD, P.

Relaxation chaotique en optimisation.

Publication du laboratoire de Calcul de l'Université de Lille, (1974).

GORENFLO, R. ; SCHAUM, H.J.

Monoton einschliessende Iterationsverfahren für invers-isotone diskretisierung nicht-linearer Zwei-Punkt-Randwertaufgaben zweiter Ordnung.

Lectures Notes in Math., 395, Springer-Verlag, Berlin, (1974), 177-198.

KANTOROVICH, L.

The method of successive approximations for functionnal equations.

Acta Math., 71, (1939), 63-97.

KRASNOSELSKII.

Approximate solution of operator equations.

Walters Nordhoff Publishing Groningen, (1972).

LEGAY.

Exposé au Colloque sur les méthodes numériques en calcul scientifique et technique.

AFCET, Chatenay-Malabry, Nov. 1974.

MAURIN.

Exposé à la 7th IFIP Conference on optimisation techniques.

NICE, Septembre 1975.

MIELLOU, J.C.

C.R. *Acad. Sci. PARIS, 278, série A, (1974), p. 957.*

MIELLOU, J.C.

Exposé au séminaire IMAG (Univ. de GRENOBLE, avril 1974), et au Colloque National d'Analyse Numérique de GOURETTE, Juin 1974.

MIELLOU, J.C.

C.R. Acad. Sci. Paris, 280, série A, (1975), 233-236.

MOSCO, U. ; SCARPINI, F.

Complementarity systems and approximation of variational inequalities.

R.A.I.R.O., R-1, 9ème année, (1975), 83-104.

ORTEGA, J.M. ; RHEINBOLDT, W.C.

Iterative solution of non linear equations in severale variables.

Academic Press, (1970).

OSTROWSKI, A.

Determination mit überwiegender Haupt diagonale und die absolute Konvergenz von linearen Iterationsprozessen.

Commun. Math. Helv., 30, (1955), 175-210.

PORSHING, T.

Jacobi and Gauss-Seidel Methods for non linear network problems.

SIAM J. Numer. Anal. 6, (1969), 437-449.

RHEINBOLDT, W.C.

J. Math. Anal. and Appl. 32, (1970), 274-307.

ROUX, J.

EDF, Bulletin de la direction des études et recherches. Série C, Math.-Inform., n° 2, (1972), 77-90 ; n° 1, (1973), 43-54.

SCHECHTER, S.

Relaxation methods for linear equations.

Commun. Pure and Appl. Math., 12, (1959), 313-335.

SCHECHTER, S.

Iteration methods for non linear problems.

Trans. AMS, 104, (1962), 179-189.

SCHRODER, J.

An wendung von fixpunksatzen bei der numerischen behandlung nichtlinearen gleichungen in halb geordneten raumen.

Arch. National Mech. Anal., 4, 177-192.

TARTAR, L.

C.R. Acad. Sci. PARIS, 278, série A, (1974), 1193.

ECLATEMENT DE CONTRAINTES EN PARALLELE POUR LA MINIMISATION D'UNE FORME QUADRATIQUE

Guy PIERRA

I.U.T. 1 INFORMATIQUE
Université C. BERNARD
43 boulevard du 11.11.1918
69621 VILLEURBANNE

Laboratoire de Mathématiques Appliquées
Université Scientifique et Médicale
de GRENOBLE

INTRODUCTION

On étudie ici le problème de la construction du projecteur sur une intersection de convexes d'un espace de Hilbert, c'est-à-dire, en changeant éventuellement de produit scalaire, celui de la minimisation de fonctionnelles quadratiques coercives.

Dans le cas où les convexes initiaux sont suffisamment simples, il peut être intéressant de décomposer le problème en sous-problèmes ne mettant chacun en jeu qu'un seul de ces convexes. De telles méthodes, consistant à "éclater les contraintes" [1], ont été proposées, en particulier, dans [2] et [4]. En associant comme dans [7], au problème posé sur une intersection de convexes, un problème équivalent posé sur le produit de ces convexes, on obtient ici, par décomposition, une méthode parallèle d'éclatement des contraintes.

Cette méthode est basée sur la réalisation, en parallèle, de projections sur chacun des convexes initiaux. Une extrapolation, [7] , qui s'introduit naturellement dans le cadre de l'espace produit, permet d'affranchir l'algorithme des questions d'"angles" qui influencent beaucoup les méthodes habituelles de projections, et d'accélérer, très considérablement, la convergence.

A titre d'applications, on présente, dans une deuxième partie, et en utilisant un formalisme dû à [3], quelques méthodes de décomposition d'opérateurs dans des problèmes elliptiques.

Ce travail utilise largement les résultats obtenus dans [7] ; il a été décrit de façon plus détaillé dans [8] ; certains développements sont présentés dans [8] et [9].

Le plan est le suivant :

§ 1 - PROJECTION SUR UNE INTERSECTION DE CONVEXES :

I - POSITION DU PROBLEME :

Soient :

V un espace de Hilbert réel, identifié à son dual, de produit scalaire, norme et distance respectivement notés $(.,.)$, $\|.\|$, $d(.,.)$

$C_i \subset V$, $i=1,\dots,n$, des ensembles convexes fermés

$$C = \bigcap_{i=1}^{n} C_i$$

$g \in V$ un élément donné.

On considère le problème :

(P) "Trouver $u \in V$ tel que $\|u-g\| = \min [\|v-g\| \; ; v \in C]$ " .

Nous supposerons que le problème n'est pas dégénéré et ferons les hypothèses :

(H1) $C = \bigcap_{i=1}^{n} C_i \neq \emptyset$

(H2) $g \notin C$

Le problème (P) consiste à chercher la projection de g sur une intersection de convexes et nous nous proposons d'obtenir cette projection en utilisant des projections sur chacun des convexes C_i , $i=1,\dots,n$.

II - PRINCIPE DE LA METHODE :

1. Définitions :

Etant donné un couple ordonné a, b de points d'un espace de Hilbert X, On convient de noter :

$$H(a,b) = \{v \in X \; , \; (v-b,b-a)_X \geq 0\} \; .$$

On remarque que $H(a,a) = X$; Si $a \neq b$, $H(a,b)$ est un demi-espace fermé sur lequel b est la projection de a.

Etant donné un triplet ordonné de points a, b, c, d'un espace de Hilbert X, on note Q(a,b,c) la projection de a sur $H(a,b) \cap H(b,c)$.

2. Théorème [2]

Etant donné un triplet ordonné de points a, b, c, d'un espace de Hilbert X, on note :

$$\lambda = (b-a,c-b)_X \; ; \; \mu = \|c-b\|_X^2 \; ; \; \nu = \|a-b\|_X^2 \; ; \; \eta = \mu\nu - \lambda^2 \; ;$$

les quatre éventualités suivantes constituent une partition des cas possibles :

1. $\eta = 0$, $\lambda < 0$ Alors $H(a,b) \cap H(b,c) = \emptyset$
2. $\eta = 0$, $\lambda \geq 0$ Alors $Q(a,b,c) = c$
3. $\eta \neq 0$, $\lambda\mu - \eta \geq 0$ Alors $Q(a,b,c) = a+(1+\frac{\lambda}{\mu})(c-b)$
4. $\eta \neq 0$, $\lambda\mu - \eta < 0$ Alors $Q(a,b,c) = b + \frac{\mu}{\eta}(\nu(c-b)-\lambda(b-a))$

Ce théorème est démontré dans [2] dont on a conservé les notations.

3. Principe de la méthode :

L'idée de la méthode pour chercher la projection d'un point $g \in V$ sur un convexe $C \subset V$ est alors la suivante :

Supposons qu'à l'étape p de l'algorithme, nous connaissions un élément $u^p \in V$ tel que $H(g,u^p) \supset C$; On construit alors un nouveau demi-espace $H(u^p,b^{p+1})$ contenant C et l'on définit l'itéré suivant $u^{p+1} \in V$ par $u^{p+1} = Q(g,u^p,b^{p+1})$; cette projection étant calculée à l'aide du théorème II.2.

La définition des demi-espaces successifs, $H(u^p,b^{p+1})$ contenant C, va être obtenue, comme dans [7] en associant au problème (P) un problème équivalent posé dans un espace produit.

III - APPROXIMATION EXTERIEURE D'UNE INTERSECTION DE CONVEXES

On introduit maintenant [7] :

$\mathcal{V} = \prod_{i=1}^{n} V$ doté de la topologie produit par le produit scalaire $((.,.)) = \sum_{i=1}^{n}(.,.)$. Sa norme et sa distance sont notées $|||\,.\,|||$ et $d((.,.))$

$\mathcal{C} = \prod_{i=1}^{n} C_i \subset \mathcal{V}$

et le convexe diagonal : $\mathcal{D} = \{\mathcal{b} = (b,b,\dots,b) \in \mathcal{V} \; ; \; b \in V\}$.

Le convexe diagonal, $\mathscr{D}$, est isomorphe à V par la bijection j : $V \to \mathscr{D}$; $x \in V \to j(x) = (x,x,\ldots,x) \in \mathscr{D} \subset \mathscr{V}$. Nous n'identifierons pas V et $\mathscr{D}$ et, pour alléger les notations, nous noterons l'image par j d'un élément de V par la lettre ronde correspondante ($j(b) = \mathscr{b}$; $j(x) = \mathscr{x}$ ...).

Rappelons ici quelques propriétés établies dans [7] :

1. Proposition :

L'image par j de l'intersection des n convexes C_i est l'intersection du convexe produit $\mathscr{C}$ et du convexe diagonal $\mathscr{D}$:

$$j(\bigcap_{i=1}^{n} C_i) = \mathscr{C} \cap \mathscr{D} .$$

2. Proposition :

(i) La projection d'un point $\mathscr{b} = (b,b,\ldots,b) \in \mathscr{D}$ sur $\mathscr{C}$ a pour composantes les projections de b sur chacun des convexes C_i :

$$\mathrm{Proj}_{\mathscr{C}}\, \mathscr{b} = (\mathrm{Proj}_{C_1} b, \mathrm{Proj}_{C_2} b, \ldots, \mathrm{Proj}_{C_n} b$$

(ii) La projection d'un point $v = (v_1,\ldots,v_n) \in \mathscr{V}$ sur $\mathscr{D}$ est l'image par j du barycentre des n points $v_1,v_2,\ldots,v_n$:

$$\mathrm{Proj}_{\mathscr{D}}\, v = j\,(\frac{1}{n} \sum_{i=1}^{n} v_i)$$

3. Nouvelle formulation du problème (P) :

Si l'on note donc $\mathscr{g} = j(g) = (g,g,\ldots,g)$, il résulte de la définition de la norme de $\mathscr{V}$ que le problème (P) est équivalent, dans $\mathscr{V}$, au problème :

(P') "Trouver $\mathscr{u} \in \mathscr{C} \cap \mathscr{D}$ tel que

$$|||\mathscr{u} - \mathscr{g}|||^2 = \min\, [|||\mathscr{v} - \mathscr{g}|||^2 \; ; \mathscr{v} \in \mathscr{C} \cap \mathscr{D}\,] " .$$

Nous nous proposons d'utiliser, pour résoudre le problème (P'), l'idée de l'algorithme II.3 et définissons donc une méthode de constructions successives de demi-espace contenant $\mathscr{C} \cap \mathscr{D}$. Donnons, tout d'abord,un résultat préliminaire :

4. Lemme :

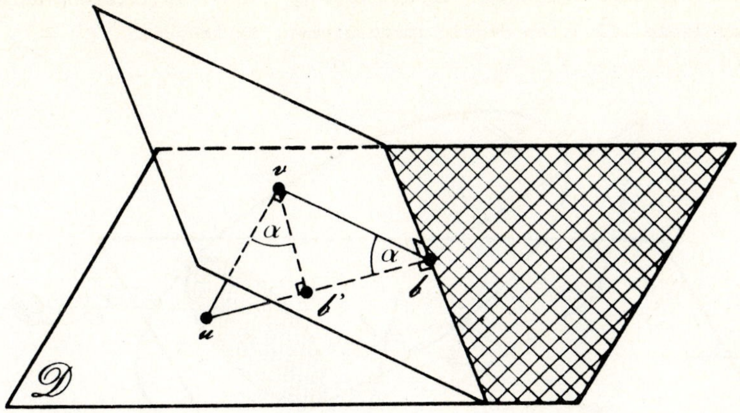

Figure 1

Etant donné un couple $(u, v) \in \mathcal{D} \times \mathcal{V}$ et en notant b' la projection de v sur $\mathcal{D}$:

1. Si $u = b' = v$, alors $\mathcal{D} \cap H(u,v) = \mathcal{D}$

2. Si $u = b' \neq v$, alors $\mathcal{D} \cap H(u,v) = \emptyset$

3. Si $u \neq b'$, alors on définit — cf. [7] — l'extrapolé b par:

$$b = b' + \frac{|||b'-v|||^2}{|||b'-u|||^2}(b'-u) = u + \frac{|||v-u|||^2}{|||b'-u|||^2}(b'-u)$$

et les deux propositions suivantes sont équivalentes :

i) $x \in \mathcal{D} \cap H(u,v)$

ii) $x \in \mathcal{D} \cap H(u,b)$

La démonstration, très simple, de ce lemme est basée sur l'égalité des deux angles notés α dans la figure. Elle est donnée dans [8].

Le lemme précédent nous donne un procédé constructif de demi-espaces successifs contenant l'ensemble, non vide d'après l'hypothèse (H1), $\mathscr{C} \cap \mathscr{D}$. Considérons, en effet, un élément $u \in \mathscr{D}$; $u \notin \mathscr{C}$. De l'application au couple $(u, \mathrm{Proj}_{\mathscr{C}}\, u)$ du lemme III.4, on déduit immédiatement le théorème :

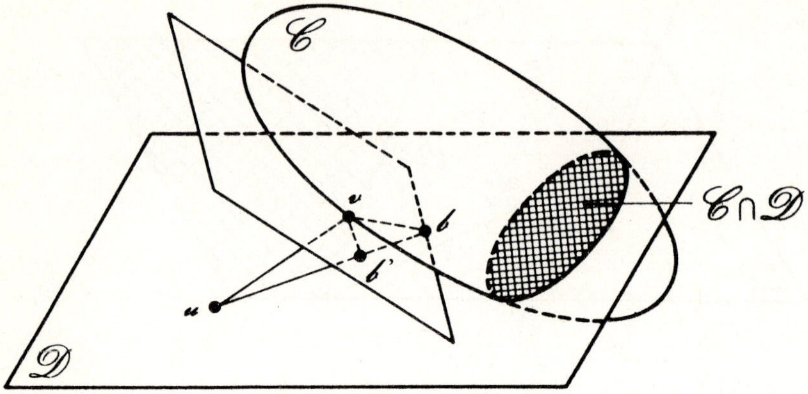

Figure 2

5. Théorème d'extrapolation dans $\mathscr{V}$:

On suppose maintenant que l'hypothèse (H1) est vérifiée, et l'on considère un élément $u \in \mathscr{D}$. Si $u \notin \mathscr{C}$, on définit :

$$v = \mathrm{Proj}_{\mathscr{C}}\, u$$

$$b' = \mathrm{Proj}_{\mathscr{D}}\, v$$

$$b = u + \frac{|||v-u|||^2}{|||b'-u|||^2}\,(b'-u) ,$$

et l'on a :

i) $\quad b' \neq u$

ii) $\quad \mathscr{C} \cap \mathscr{D} \subset \mathscr{D} \cap H(u,b)$.

Si nous exprimons dans l'espace V le théorème d'extrapolation, il vient, compte tenu de la proposition III.2 :

6. Théorème d'extrapolation dans V :

L'hypothèse (H1) étant vérifiée, on considère un élément

$u \in V \; ; \; u \notin \bigcap_{i=1}^{n} C_i$.

On définit :

$$v_i = \mathrm{Proj}_{C_i} u \qquad , \; i=1,\ldots,n$$

$$b' = \frac{1}{n} \sum_{i=1}^{n} v_i$$

$$b = u + \frac{\sum_{i=1}^{n} \| u-v_i \|^2}{n \; \| b'-u \|^2} \cdot (b'-u)$$

et l'on a :

i) $\qquad b' \neq u$

ii) $\qquad C = \bigcap_{i=1}^{n} C_i \subset H(u,b)$.

IV - DESCRIPTION DE LA METHODE :

Supposons donc qu'à l'étape p-1 de l'algorithme, nous connaissions $\mathcal{u}^{p-1} \in \mathcal{D}$ tel que $\mathcal{C} \cap \mathcal{D} \subset H(\mathcal{g}, \mathcal{u}^{p-1})$.

Si $\mathcal{u}^{p-1} \in \mathcal{C}$, il est clair que $\mathcal{u}^{p-1}$ est solution de (P')

Si $\mathcal{u}^{p-1} \notin \mathcal{C}$, on construit alors une nouvelle approximation extérieure $H(\mathcal{u}^{p-1}, \mathcal{b}^p) \supset \mathcal{C} \cap \mathcal{D}$ par application du théorème III.5 et l'on définit le nouvel itéré $\mathcal{u}^p$ par :

$$\mathcal{u}^p = Q(\mathcal{g}, \mathcal{u}^{p-1}, \mathcal{b}^p) .$$

Nous pouvons maintenant décrire, dans l'espace V, l'algorithme proposé :

1. Algorithme :

Partant de $u^\circ = g$, le passage de u^{p-1} à u^p s'effectue de la façon suivante :

1) On calcule pour i=1,...,n

$$v_i^p = \mathrm{Proj}_{C_i}\ u^{p-1}$$

si $v_i^p = u^{p-1}$, i=1,...,n , on pose $u = u^{p-1}$ et on stoppe l'algorithme sinon :

2) On définit :

$$b'^p = \frac{1}{n} \Sigma\ v_i^p$$

et l'extrapolé :

$$b^p = u^{p-1} + \frac{\sum_i \|v_i^p - u^{p-1}\|^2}{n\ \|b'^p - u^{p-1}\|^2}\ (b'^p - u^{p-1})\ . \qquad (A)$$

3) On calcule alors :

$$u^p = Q(g, u^{p-1}, b^p)\ .$$

Puisque $u° = g$, $H(g,u°) = V \supset C$, on en déduit donc , par récurrence sur p que $H(g,u^{p-1}) \supset C$. Il résulte alors du théorème d'extrapolation III.6 que si $u^{p-1} \notin C$, alors $u^{p-1} \neq b'^p$ et $H(g,u^p) \supset C$.

2. Remarques

1) Comme dans [7] (cf. [7], I §6) l'on peut, en modifiant la définition de la norme de $\mathcal{V}$, faire intervenir, au lieu de b'^p, un barycentre pondéré.

2) De même, il est possible de ne pas effectuer à chaque pas tout ou partie de l'extrapolation, c'est-à-dire de prendre pour certaines valeurs de p :

$$b^p = u^{p-1} + [1+\rho(\frac{\sum_i \|v_i^p - u^{p-1}\|^2}{n.\|b'^p - u^{p-1}\|^2} - 1)](b'^p - u^{p-1})\ ;\ 0 \leq \rho \leq 1$$

3) On vérifie — cf. [8] lemme II.3.5 — que lorsque u^{p-1} appartient à certains des convexes C_i, il suffit, pour calculer l'extrapolé b^p, d'appliquer la formule (A) au seuls n' convexes (n'< n) auxquels u^{p-1} n'appartient pas.

3. Remarque

Lorsque les projections sur chacun des convexes C_i ne peuvent pas être obtenues facilement, on peut, en associant à la méthode précédente un croisement d'algorithmes, linéariser, à chaque pas, les contraintes. Cette méthode est présentée dans [8] .

V - CONVERGENCE

On a le résultat suivant :

1. Théorème :

Les hypothèses (H1) et (H2) étant supposées vérifiées, si la suite $\{u^p\}$ définie par l'algorithme IV.1 est finie, son dernier élément est solution de (P) ; si la suite est infinie, elle converge fortement vers la solution de (P).

Démonstration :

Si la suite $\{u^p\}$ est finie, cela signifie que $u^p \in C$; et puisque $C \subset H(g,u^p)$ on en déduit que u^p est solution de (P). Supposons donc la suite $\{u^p\}$ infinie.

1) La suite $\{\|u^p-g\|\}$ est convergente.

En effet, par définition de u^p, $u^p \in H(g,u^{p-1})$. La suite $\{\|u^p-g\|\}$ est donc non décroissante. De plus, puisque $H(g,u^p) \supset C \neq \emptyset$, cette suite est bornée ; elle est donc convergente.

On en déduit en particulier :

$$\lim_{p\to\infty} [\|u^{p-1}-g\|^2 - \|u^p-g\|^2] = 0 ,$$

et, puisque $u^p \in H(g,u^{p-1})$, $(u^p-u^{p-1},u^{p-1}-g) \geq 0$, d'où :

$$\lim_{p\to\infty} \|u^p-u^{p-1}\|^2 = 0 ;$$

de même, puisque $u^p \in H(u^{p-1},b^p)$:

$$\lim_{p\to\infty} \|u^{p-1}-b^p\|^2 = \lim_{p\to\infty} |||\mathcal{u}^{p-1}-\mathcal{b}^p|||^2 = 0 . \tag{A}$$

2) $\lim_{p\to\infty} ||| \mathcal{u}^{p-1}-v^p |||^2 = 0$

Appliquant le lemme III.4 au couple $(\mathcal{u}^{p-1},v^p)$, on obtient :

$$\mathcal{b}^p \in H(\mathcal{u}^{p-1},v^p) .$$

Les propriétés des projections impliquent donc, compte tenu de (A) :

$$\lim_{p\to\infty} ||| \mathcal{u}^{p-1}-v^p |||^2 = 0$$

$$\lim_{p\to\infty} \|u^{p-1}-v_i^p\|^2 = 0 , \qquad i=1,\dots,n . \tag{B}$$

3) Convergence de la suite :

La suite $\{u^p\}$ étant bornée, il existe une sous-suite, notée $\{u^q\}$, convergeant faiblement vers un élément $u \in V$. De la faible s.c.i. de la distance à un convexe on déduit, compte tenu de (B), que $u \in C$.

Enfin, puisque $C \subset H(g,u^p)$:

$$\|u^{p-1}-g\| \leq \|u^p-g\| \leq \|x-g\| \quad , \quad \forall\, x \in C \;, \forall\, p \in \mathbb{N} \,,$$

et l'on déduit alors de la faible s.c.i. de la norme, selon un argument classique, et pour toute la suite $\{u^p\}$:

$$\begin{cases} u = \mathrm{Proj}_C\, g \\ \lim\limits_{p\to\infty} \|u^p-g\| = \|u-g\| \\ \lim\limits_{p\to\infty} \|u-u^p\| = 0 \,, \end{cases}$$

ce qui établit le théorème.

§ 2 - DECOMPOSITION D'OPERATEURS ELLIPTIQUES VIA UN ECLATEMENT PARALLELE DE CONTRAINTES :

La traduction de certains problèmes elliptiques sous la forme de problème (P) - cf. § 1 - est effectuée ici en utilisant un formalisme de Y. HAUGAZEAU [3] . Le problème posé, par exemple, sur $H^1(\Omega)$, est transformé en un problème dans un espace V qui est une puissance de $L^2(\Omega)$, et dans lequel $H^1(\Omega)$ est plongé. L'approximation est effectué à partir d'éléments de $[L^2(\Omega)]^r$, il s'agit donc d'une approximation externe, et l'appartenance de la solution à $H^1(\Omega)$ se traduit, dans le nouveau problème, par une introduction de contraintes. L'utilisation de l'algorithme précédent, avec traitement parallèle des contraintes, correspond, au niveau du problème initial, à une décomposition en parallèle de l'opérateur elliptique. Cette décomposition est de même type que celle utilisée dans la méthode des directions alternées [11] ou dans la méthode de décomposition de [1] et entraîne une convergence assez rapide.

Je remercie Monsieur Y. HAUGAZEAU pour d'utiles conversations concernant cette partie et Monsieur R. CHIFFLET pour sa participation aux essais numériques.

Nous présentons ici deux exemples d'applications de la méthode, l'un pour un problème linéaire l'autre pour un problème non linéaire, ainsi que quelques remarques sur des essais numériques. D'autres exemples ainsi que l'ensemble essais numériques effectués seront présentés par ailleurs.

I - EXEMPLES DE PROBLEMES

1. Problème de Neumann

Soit Ω un ouvert borné régulier de R^n, de frontière Γ . On cherche à minimiser, sur $H^1(\Omega)$, la fonctionnelle :

$$(M) \qquad J(v) = \lambda\int_\Omega |v(x)|^2\, dx + \sum_{i=1}^{n} \int_\Omega \left|\frac{\partial v}{\partial x_i}\right|^2 dx - 2\int_\Omega f(x).v(x)dx \quad , \lambda > 0$$

avec f donnée dans $L^2(\Omega)$.

Il s'agit — cf. [3],[10]— du classique problème de Neumann :

$$(N) \qquad \begin{cases} -\Delta u + \lambda u = f & \text{pp dans } \Omega \\ \gamma_1 u = 0 & \text{pp sur } \Gamma \ . \end{cases}$$

Nous noterons $(.,.)$ et $\|.\|$ le produit scalaire et la norme de $L^2(\Omega)$. On introduit alors l'espace :

$$V = [L^2(\Omega)]^{n+1} = \{(v_o,v_1,\dots,v_n)\ ;\ v_i \in L^2(\Omega)\ ;\ i=0,\dots,n\}$$

doté de la topologie produit, et, en notant A_i , $i=1,\dots,n$, l'opérateur :

$$v_o \in L^2(\Omega) \to \frac{\partial v_o}{\partial x_i} \in L^2(\Omega) ,$$

les sous-ensembles convexes fermés de V :

$$C_i = \{v = (v_o,v_1,\dots,v_n) \in V\ ;\ v_i = A_i\, v_o\} \qquad i=1,\dots,n .$$

L'injection de $H^1(\Omega)$ dans V :

$$y \in H_1(\Omega) \to (y, \frac{\partial y}{\partial x_i}, \dots, \frac{\partial y}{\partial x_n})$$

est une isométrie de $H^1(\Omega)$ et de $\bigcap_{i=1}^{n} C_i$ dans V doté du produit scalaire

$$\langle\langle .,. \rangle\rangle = \sum_{i=1}^{n} (.,.).$$

On vérifie facilement — cf. [10] — que la forme bilinéaire sur $V \times V$:

$$a(u,v) = \lambda(u_o,v_o) + \sum_{i=1}^{n} (u_i,v_i)$$

est, avec λ strictement positif, continue et coercive sur V dotée de la métrique précédente. Elle définit donc un nouveau produit scalaire que nous noterons $((.,.))$, et une norme notée $|||\ .\ |||$.

Le problème (M) devient :

"Trouver $u \in C = \bigcap_{i=1}^{n} C_i \subset V$ tel que $F(u) = \min\ [F(v)\ ;\ v \in C]$ "

avec : $F(v) = \lambda \|v_o\|^2 + \sum_{i=1}^{n} \|v_i\|^2 - 2(f,v_o)$;

ou encore, en introduisant :

$$g = (\frac{1}{\lambda} f,0,\dots,0) \in V :$$

"Trouver $u \in \bigcap_{i=1}^{n} C_i$ tel que $|||\ u-g\ |||^2 = \min\ [|||\ v-g\ |||^2\ ;\ v \in \bigcap_{i=1}^{n} C_i]$ " , c'est-à-dire la recherche de la projection de g sur l'intersection des convexes C_i, $i=1,\dots,n$. Nous appliquons donc à ce problème l'algorithme IV.1 du § 1.

On vérifie facilement — cf. [8] — les formules de calcul des projections sur les ensembles C_i, $i=1,\dots,n$. En prenant, pour fixer les idées, $n = 2$, l'algorithme est alors le suivant :

Algorithme :

Partant de $v^\circ = g$, le passage de v^{p-1} à v^p se fait de la façon suivante :

1) On calcule $x^p = \mathrm{Proj}_{C_1} v^{p-1}$ c'est-à-dire :

$$x_o^p = (\lambda I + A_1^* A_1)^{-1} (\lambda v_o^{p-1} + A_1^* v_1^{p-1})$$

$$x_1^p = A_1 x_o^p$$

$$x_2^p = v_2^{p-1}$$

et $y^p = \mathrm{Proj}_{C_2} v^{p-1}$:

$$y_o^p = (\lambda I + A_2^* A_2)^{-1} (\lambda v_o^{p-1} + A_2^* v_2^{p-1})$$

$$y_1^p = v_1^{p-1}$$

$$y_2^p = A_2 y_o^p .$$

2) On coordonne en calculant :

$$b'^p = \frac{1}{2} (x^p + y^p)$$

et l'on extrapole :

$$b^p = v^{p-1} + \rho (b'^p - v^{p-1})$$

avec
$$\rho = \frac{\lambda \| x_o^p - v_o^{p-1} \|^2 + \| x_1^p - v_1^{p-1} \|^2 + \lambda \| y_o^p - v_o^{p-1} \|^2 + \| y_2^p - v_2^{p-1} \|^2}{2(\lambda \| b_o'^p - v_o^{p-1} \|^2 + \sum_{i=1}^{2} \| b_i'^p - v_i^{p-1} \|^2)}$$

3) Enfin, on calcule v^p par :

$$v^p = Q(g, v^{p-1}, b^p) .$$

Du théorème V du § 1 on déduit :

<u>Proposition</u> :

$$v_o^p \to u \quad \text{solution de (N) dans } L^2(\Omega) \text{ Fort}$$

$$v_i^p \to \frac{\partial u}{\partial x_i}\ ,\ i=1,2\ ,\ \text{dans } L^2(\Omega) \text{ Fort}$$

<u>Remarques</u> :

1) Si la discrétisation est effectuée par différence finie en utilisant les méthodes classiques d'approximations — cf. [10] , [11] — le calcul de x_o^p — ainsi que celui de y_o^p — demande la résolution d'un système linéaire, découplé par lignes du domaine, et dont la matrice tri-diagonale est d'inversion tout à fait standard. Ces systèmes, qui doivent être résolus à chaque itération, sont les mêmes que ceux rencontrés dans la méthode des directions alternées [11] ou dans celle des pas fractionnaires [10] , [1] .

2) La projection $Q(g,v^{p-1},b^p)$ effectuée selon le théorème (§ 1, II.2) demande seulement le calcul d'un produit scalaire, les autres paramètres résultant de calculs précédents.

3) L'extension de l'algorithme au cas n = 3 est immédiate.

2. <u>Un problème non linéaire</u> — cf. [5] — :

On en donnera seulement les grandes lignes. Il s'agit du problème aux limites :

$$(M)\quad \begin{cases} -\ \Delta u + u + 2\,u^3 = f & \text{pp. dans } \Omega \\ \gamma_1\, u = 0 & \text{pp. sur } \Gamma\ , \end{cases}$$

qui se traduit, sur l'espace $W = L^4(\Omega) \cap H^1(\Omega)$, par le problème d'optimisation :

(Q) "Trouver $u \in W$ tel que $J(u) = \min\,[J(v)\ ;\ v \in W]$ "

avec $$J(v) = \|v\|^2 + \sum_{i=1}^{n} \left\|\frac{\partial x}{\partial x_i}\right\|^2 + \|v^2\|^2 - 2(f,v)$$

en notant, à nouveau, $\|.\|$ et $(.,.)$ la norme et le produit scalaire de $L^2(\Omega)$, et avec f donné dans $L^2(\Omega)$.

Si l'on introduit $V = [L^2(\Omega)]^{n+2}$ et l'injection de W dans V :

$$v \in W \to (v, \frac{\partial v}{\partial x_1}, \ldots, \frac{\partial v}{\partial x_n}, v^2),$$

on démontre — cf. [3] — que le problème (Q) est équivalent, en notant $|||\cdot|||$, $|||\cdot|||^2 = \sum_{i=0}^{n+1} ||\cdot||^2$, la norme de V et $g = (f,0,\ldots,0)$, au problème :

(Q') "Trouver $u \in C = \bigcap_{i=1}^{n+1} C_i$ tel que $|||u-g|||^2 = \min [|||v-g|||^2 \, ; \, v \in C]$ " ,

avec :

$$C_i = \{(v_o,v_1,\ldots,v_{n+1}) \in V \, ; \, v_i = \frac{\partial v_o}{\partial x_i}\} \qquad i=1,\ldots,n ,$$

$$C_{n+1} = \{(v_o,v_1,\ldots,v_{n+1}) \in V \, ; \, v_{n+1}(x) \geq v_o^2(x) \quad \text{pp. dans } \Omega\} \quad :$$

c'est-à-dire à un problème de projection sur une intersection de convexes auquel la méthode de la première partie s'applique.

II - ESSAIS NUMERIQUES

Nous avons considéré deux problèmes de Neumann très voisin qui avait déjà été traité, le premier, dans [10], par la méthode des pas fractionnaires, le second dans [5] par les méthodes de Cholewski, relaxation par blocs et coordination.

Il s'agit des deux problèmes :

$$(1) \quad \begin{cases} -\Delta u + 2u = f & \text{p.p. dans } \Omega =]0,1[\times]0,1[\\ \frac{\partial u}{\partial n}\Big|_\Gamma = 0 & \text{p.p. sur } \Gamma \\ f = (2\pi^2 + 2)\cos \pi x_1 \cos \pi x_2 \end{cases}$$

$$(2) \quad \begin{cases} -\Delta u + u = f & \text{p.p. dans } \Omega =]0,1[\times]0,1[\\ \frac{\partial u}{\partial n}\Big|_\Gamma = 0 & \text{p.p. sur } \Gamma \\ f = (2\pi^2 + 1)\cos \pi x_1 \cos \pi x_2 \end{cases}$$

La solution de ces deux problèmes est : $u = \cos \pi x_1 \cos \pi x_2$.

Quelques remarques :

1 - Le passage de la forme (1) à la forme (2) du problème se traduit , après discrétisation par différences finies, par une modifications de la "dominance" de la diagonale de la matrice du système. Cette modification n'a aucune influence sur le comportement de notre méthode : les éléments itérés sont les mêmes, seuls changent les coefficients d'extrapolation.

2 - Dans tous les exemples étudiés, la convergence semble extrêmement rapide pour les premières itérations. Ainsi pour les problèmes (1) et (2) la solution est atteinte en une itération. Par la suite, et au voisinage immédiat de la solution, des oscillations légères peuvent se produire, sur la valeur de l'erreur relative ; la décroissance du paramètre $\mu = \|b^p - u^{p-1}\|$ est par contre toujour monotone.

3 - Bien que non indispensable, le travail en double précision accroît significativement l'efficacité de la méthode.

4 - C'est l'extrapolation qui fait la rapidité de la méthode. Dans le cas où l'extrapolation n'est pas effectuée (méthode du barycentre) les résultats pour le problème (1) avec un pas de discrétisation de 1/20 ont été les suivants. On a calculé l'erreur relative :

$$e = \frac{\Sigma|\text{valeur calculée au point } (i,j) - \text{solution } (i,j)|}{\Sigma\,|\text{ solution } (i,j)|}$$

Itérations	avec extrapolation	sans extrapolation
N = 1	$e = 1.8665 \times 10^{-3}$ *	$e = 0.8346$
N = 40	$e = 1.8665 \times 10^{-3}$	$e = 9.2453 \times 10^{-3}$

* L'erreur 1.86×10^{-3} est l'erreur due à la discrétisation.

Résultats

1 - Nous donnons les résultats du problème (2) avec un pas de h = 1/40

Méthode	Coordination (α) (éléments finis)	Cholewski (α)	Relax-bloc (α)	Eclatement (β) parallèle
Erreur	$4,9 \times 10^{-4}$	$4,9 \times 10^{-4}$	$4,9 \times 10^{-4}$	$4,6 \times 10^{-4}$
temps de calcul	3 sec.	3,22 sec.	5 sec.	3,1 sec.
place mémoire	180 K	660 K	180 K	35 K

(α) donnés dans [5] , calculs effectués sur IBM 360-91.

(β) calculs effectués en D.P. sur IRIS 80 (plusieurs fois moins rapide).

2 - Pour le problème (1), le nombre d'itération nécessaire pour la méthode des pas fractionnaires dépend d'un paramètre τ. cf [10]. Le temps d'une itération de pas fractionnaire est environ 20 % inférieur à celui d'une itération d'éclatement en parallèle. Le nombre d'itération pour assurer la convergence des deux méthodes est de :

Méthode	h 1/10	h = 1/20
Pas fractionnaire cf. [10]	$N \gg 5$ (dépend de τ)	$N \gg 5$
Eclatement parallèle	N = 1	N = 1

REFERENCES

[1] A. BENSOUSSAN, J.L. LIONS, R. TEMAM : "Sur les méthodes de décomposition, de décentralisation et de coordination, et applications". Cahier IRIA, N° 11 (1972) 5-190.

[2] Y. HAUGAZEAU : Thèse, Paris (1968).

[3] Y. HAUGAZEAU : Communication au Colloque d'Anglet, (juin 1971).

[4] P.J. LAURENT, B. MARTINET : "Méthodes duales pour le calcul du minimum d'une fonction convexe sur une intersection de convexes". Symposium on Optimisation, Nice (1969) in Lect. Notes Math. 132, Springer Verlag, Berlin (1970), 159-180.

[5] P. LEMONNIER : "Résolution numérique d'équations aux dérivées partielles par décomposition et coordination". Cahier IRIA, n° 11 (1972), 191-239.

[6] J.L. LIONS, E. MAGENES : "Problèmes aux limites non homogènes et applications" Dunod, Vol.1, Paris (1968).

[7] G. PIERRA : "Méthodes de projections parallèles extrapolées relatives à une intersection de convexes". Rapport de Recherche N° 7, Mathématiques Appliquées et Informatique, Grenoble (1975). (A paraître).

[8] G. PIERRA : "Projection sur une intersection de convexes et applications". Rapport de Recherche N° 9, Mathématiques Appliquées et Informatique, Grenoble (1975).

[9] G. PIERRA : "Une nouvelle méthode de programmation quadratique". Rapport de Recherche N° 10, Mathématiques Appliquées et Informatique, Grenoble, (1975).

[10] R. TEMAM : "Analyse Numérique". P.U.F., Paris (1970)

[11] R.S. VARGA : "Matrix iterative analysis". Prentice Hall (1962).

APPLICATION DE LA METHODE DE PENALISATION AUX PROBLEMES DE CONTROLE EN NOMBRES ENTIERS

C. Saguez

IRIA-LABORIA

B.P. 5 Domaine de Voluceau

78150 Le Chesnay (France)

Les actions que nous pouvons avoir sur un système, présentent souvent un caractère discret (ouverture - fermeture d'un circuit, choix optimal de matériaux dans un catalogue,...). Dans cet article nous étudions des problèmes de contrôle optimal de ce type où le contrôle intervient sous la forme :

$\sum_{i=1}^{n} \alpha_i \chi_{A_i}$ avec $\{A_i\}_{i=1,\ldots,n}$ une famille donnée d'intervalles

χ_{A_i} la fonction caractéristique de A_i

$\alpha_i \in \{0,1\}$ $\forall$ $i=1,\ldots,n$ les variables de contrôle.

Après avoir présenté la méthode de pénalisation pour les problèmes d'optimisation en variables mixtes, nous appliquons celle-ci aux problèmes de contrôle optimal de systèmes gouvernés par une équation parabolique ([8] YVON). Nous donnons les résultats numériques obtenus. Enfin nous montrons que ce principe s'applique également à certaines équations non linéaires et aux équations aux valeurs propres du deuxième ordre.

I - Pénalisation en variables mixtes.

1.1. Position du problème.

Soient : V, Y deux espaces de Banach réflexifs.

I un ensemble discret fini $\{\alpha_1,\dots,\alpha_p\}$

f une application de $I \times Y$ dans $\mathbb{R}$, convexe continue sur Y

g une application de $I \times Y$ dans V, continue de $I \times Y$ faible dans V faible.

On suppose de plus que f et g vérifient une des deux hypothèses suivantes :

$$(1.1) \quad \begin{cases} f(\alpha,y) \to +\infty \\ \|y\|_Y \to +\infty \end{cases} \quad \text{et } G(\alpha) = \{y \mid y \in Y,\ g(\alpha,y) = 0\} \text{ convexe, fermé dans } Y$$

$$(1.1)' \quad \text{l'ensemble } \{y \mid \|g(\alpha,y)\|_V \leq C_1 \text{ , } f(\alpha,y) < C_2 \text{ , } \alpha \in I\} \text{ est borné dans } Y$$

On considère alors le problème suivant :

$$(1.2) \quad \begin{cases} \text{Min } f(\alpha,y) \\ g(\alpha,y) = 0 \\ \alpha \in I,\ y \in Y \end{cases}$$

Proposition 1 : Sous les hypothèses précédentes, le problème (1.2) admet au moins une solution.

Remarque 1 : en supposant $f(\alpha,y)$ deux fois continument différentiable en (α,y) et $I = \{0,1\}^k$, on peut se ramener à la minimisation d'une fonctionnelle convexe.

Remarque 2 : Si $f(\alpha,y)$ est strictement convexe et sous l'hypothèse (1.1), il existe un problème admettant comme unique solution, une solution du problème (1.2) Cette remarque nous permettra par la suite de supposer que le problème étudié admet une solution unique.

1.2. Le problème pénalisé.

On définit le problème pénalisé suivant :

$$(1.3) \quad \begin{cases} \text{Min } f(\alpha,y) + \frac{1}{\varepsilon} \|g(\alpha,y)\|_V^2 \\ \alpha \in I \text{ , } y \in Y \end{cases}$$

On a alors la proposition suivante :

Proposition 2 :

Si $(\alpha_\varepsilon^*, y_\varepsilon^*)$ est une solution de (1.3), tout point adhérent faible de la suite $(\alpha_\varepsilon^*, y_\varepsilon^*)$ est solution de (1.2).

De plus si (1.2) admet une solution unique, alors

$$\alpha_\varepsilon^* = \alpha^* \quad \text{pour } \varepsilon \text{ suffisamment petit.}$$

$$y_\varepsilon^* \xrightarrow[\varepsilon \to 0]{} 0 \text{ dans } Y \text{ faible}$$

avec (α^*, y^*) solution du problème (1.2).

Donc d'après la remarque 2, on voit qu'il existe ε_0 tel que pour $\varepsilon < \varepsilon_0$, $\alpha_\varepsilon^* = \alpha^*$ et alors on obtient y^* en résolvant le problème

$$\begin{cases} \text{Min } f(\alpha^*, y) \\ \quad g(\alpha^*, y) = 0 \\ \quad y \in Y \end{cases}$$

On constate que les difficultés numériques de convergence et de stabilité rencontrées souvent dans les méthodes de pénalisation n'existent plus ici, la convergence ayant lieu pour ε suffisamment petit.

Mais il faut remarquer que le problème pénalisé est en général un problème en variables mixtes difficile à résoudre. Cependant on verra que dans un certain nombre de cas, ce problème prend une forme simplifiée qui permet une résolution relativement facile.

II - Application au problème de contrôle optimal en nombres entiers d'un système gouverné par une équation parabolique.

2.1. Position du problème.

Soient : - V, H deux espaces de Hilbert avec $V \subset H \subset V'$ avec injection continue et densité.

- $a(t;\varphi,\Psi)$ une famille de formes linéaires continues sur V telles que

(2.1)
$$\begin{cases} \forall \varphi,\ \Psi \in V \quad t \to a(t;\varphi,\Psi) \text{ mesurable sur }]o,T[\\ \text{et } |a(t;\varphi,\Psi)| \leq c \,\|\varphi\|_V \|\Psi\|_V \end{cases}$$

(2.2)
$$\begin{cases} \text{il existe } \lambda \text{ tel que} \\ a(t;\varphi,\varphi) + \lambda\, |\varphi|_H^2 \geq \alpha \|\varphi\|_V^2 \qquad \forall \varphi \in V \\ \qquad\qquad\qquad\qquad\qquad\qquad \forall t \in]o,T[\end{cases}$$

On écrit

$$a(t;\varphi,\Psi) = (A(t)\varphi,\Psi)_{V' \to V}$$

avec $A(t) \in \mathcal{L}(L^2(o,T;V),\ L^2(o,T;V'))$

- $\mathcal{U}$ l'espace des fonctions en escalier sur $]o,T[$, muni du produit scalaire de $L^2(o,T)$.

- $B \in \mathcal{L}(\mathcal{U},\ L^2(o,T;V')$

On définit alors l'état du système par :

(2.3)
$$\begin{cases} \dfrac{\partial y}{\partial t} + Ay = f + Bu \\ y(x,o) = y_o(x) \end{cases} \qquad \text{avec } f \in L^2(o,T;V') \quad y_o \in H .$$

On prend u sous la forme

(2.4)
$$u = \sum_{i=1}^{n} \mu_i \alpha_i \chi_{[a_i,b_i]}$$
avec n un entier fixé

$[a_i,b_i]$ une famille d'intervalles de $]o,T[$

$\chi_{[a_i,b_i]}$ la fonction caractéristique de $[a_i,b_i]$

$\mu_i \in \mathbb{R}$, $\alpha_i \in \{o,1\}$ $\forall i=1,\dots,n$

On note $I = \{o,1\}^n$, I_{ad} un sous-ensemble de I.

On définit la fonctionnelle

(2.5)
$$J(\alpha) = \|y - z_d\|^2_{L^2(o,T;H)} + \nu\|u\|^2_{L^2(o,T)} \qquad \nu \geq 0$$

avec z_d donné dans $L^2(o,T;H)$.

On considère alors le problème de contrôle :

$$(2.6)\qquad \begin{cases} \text{Trouver } \alpha^* \in I_{ad} \text{ tel que} \\ J(\alpha^*) = \underset{\alpha \in I_{ad}}{\text{Min}}\, J(\alpha) \end{cases}$$

2.2. Le problème pénalisé.

Soit l'espace de Hilbert $Y = \{y \mid y \in L^2(o,T;V)\ ;\ \frac{\partial y}{\partial t} + Ay \in L^2(o,T;V')\}$.

muni de la norme

$$\|y\|_Y^2 = \|y\|^2_{L^2(o,T;V)} + \|\frac{\partial y}{\partial t} + Ay\|^2_{L^2(o,T;V')} + \|y(x,o)\|_H^2 .$$

On obtient la fonctionnelle pénalisée

$$(2.7)\qquad J_\varepsilon(\alpha,y) = J(\alpha) + \frac{1}{\varepsilon_1}\|\frac{\partial y}{\partial t} + Ay - f - Bu\|^2_{L^2(o,T;V')} + \frac{1}{\varepsilon_2}\|y(x,o) - y_o(x)\|_H^2$$

Le problème pénalisé s'écrit alors:

$$(2.8)\qquad \begin{cases} \text{Trouver } (\alpha_\varepsilon^*,y_\varepsilon^*) \in I_{ad} \times Y \text{ tel que} \\ J_\varepsilon(\alpha_\varepsilon^*,y_\varepsilon^*) = \underset{(\alpha,y) \in I_{ad} \times Y}{\text{Min}}\, J_\varepsilon(\alpha,y) \end{cases}$$

Alors les résultats du I) sont applicables et donc en particulier tout point adhérent faible de $(\alpha_\varepsilon^*,y_\varepsilon^*)$ est solution de (2.6).

Le problème (2.8) est un problème de minimisation quadratique - quadratique en variables mixtes. Dans le cas où les intervalles $]a_i,b_i[$ sont disjoints, le problème pénalisé est équivalent à un problème continu ($\alpha \in [o,1]^n$). C'est dans ce cas que la méthode de pénalisation s'avère particulièrement intéressante.

2.3. Méthodes numériques.

On notera par y les variables continues et α les variables entières.

a) Dans le cas où le problème se ramène à un problème continu, on peut utiliser une méthode de décomposition, en minimisant successivement par rapport à α, puis par rapport à y et en itérant jusqu'à convergence. On a alors le résultat suivant : si d désigne la distance sur Y et d_1 la distance euclidienne dans $\mathbb{R}^n$ on note

$$W_k^\varepsilon(\alpha,y) = \{(\beta,z) \in I \times Y \mid d(y,z) \leq \varepsilon\ ,\ d_1(\beta,\alpha) = k\}$$

alors la méthode précédente converge vers un point (α,y) tel que

$$J_\varepsilon(\alpha,y) \leq J_\varepsilon(\beta,z) \qquad \forall\ (\beta,z) \in W_1^\varepsilon(\alpha,y)$$

b) Dans le cas où le problème se ramène à un problème continu, on peut également utiliser une méthode du type Lagrangien augmenté développée par Rockafellar ([5] Rockafellar). On se trouve dans les conditions d'application de cette méthode et en particulier si P désigne le problème primal et D le problème dual alors:

$$\text{Inf}(P) = \text{Max}(D)$$

c) Dans le cas général, on a un problème quadratique en variables mixtes, alors en posant $\gamma_{ij} = \alpha_i \alpha_j$, on peut se ramener à un problème linéaire en (α,γ) avec contraintes linéaires sur α, γ. On peut alors développer l'algorithme suivant ([7] Saguez)

1. On résoud le problème continu associé par une méthode de gradient conjugué réduit.
2. Si la solution α est entière, c'est la solution du problème, sinon on génère une coupe de Gomory (ceci est possible car on se trouve alors à un sommet du simplexe).
3. On itère alors ce processus, jusqu'à convergence. Celle-ci est assurée par la théorie des coupes de Gomory.

2.4. Exemple Numérique.

On prend l'état du système solution de

$$\begin{cases} \dfrac{\partial y}{\partial t} - \Delta y = f + \sum_{i=1}^{n} \alpha_i \mu_i \chi_{[a_i b_i]} & \text{sur } \Omega \times]o,T[= Q \\ y\big|_{\Sigma} = 0 & \\ y(x,o) = y_o(x) & \text{avec } f \in L^2(Q) \;;\; y_o(x) \in L^2(Q) \end{cases}$$

On définit la fonctionnelle

$$J(\alpha) = \int_\Omega \int_o^T |y - z_d|^2 \, dx\, dt \quad ; \; z_d \in L^2(Q)$$

On pose

$$I_{ad} = I = \{o,1\}^n$$

Nous avons pris $f(x,t) = 0 \;;\; y_o(x) = 0 \;;\; \nu = 0$

Nous avons appliqué les méthodes a) et b) et nous avons obtenu des résultats identiques.

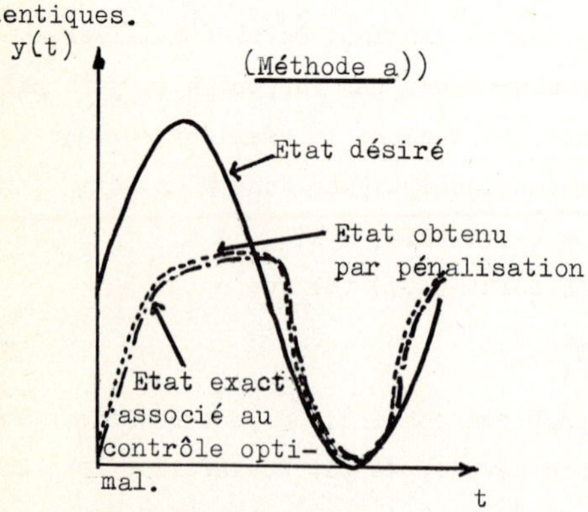

Etat du système pour x = 0.5

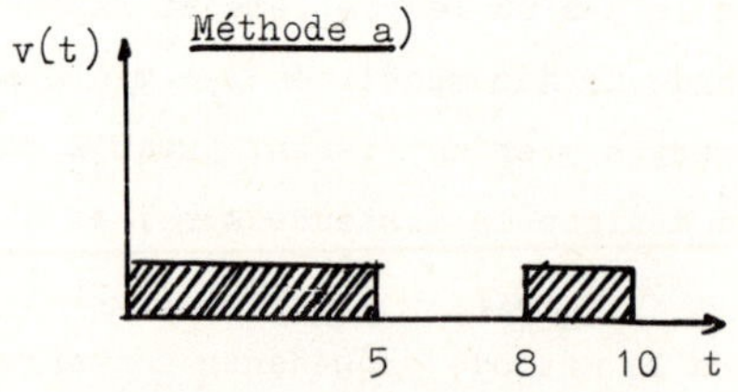

Contrôle optimal obtenu pour $\varepsilon = 1$

Méthode b)

0,9970 ; 1,0023 ; 0,9997 ; 0,9989 ; 1,0068 ; -0,0052 ; -0,0019 ; -0,0063 ; 1,013 ; 1,0111 .

Exemple 1 : $z_d(x,t) = 1 + \sin 2\Pi t \qquad \mu_i = 10 \quad \forall\, i$

<u>Exemple 2</u> : $z_d(x,t) = 5 \sin 2 \Pi t$

$\mu_i = 1 \quad \forall i$

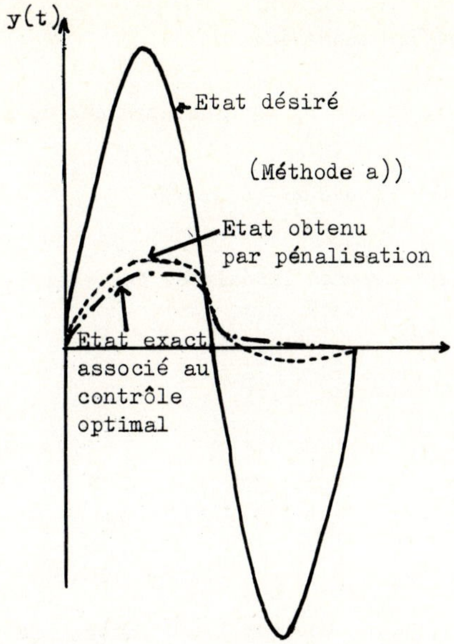

Etat du système pour x = 0,5

<u>Méthode a)</u>

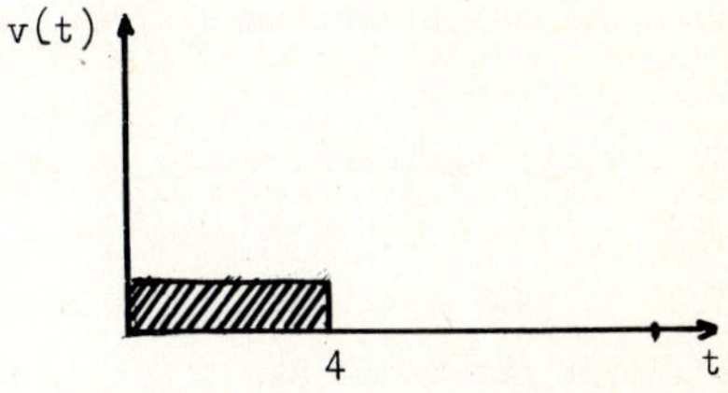

<u>Contrôle optimal obtenu pour</u> $\varepsilon = 1$

<u>Méthode b)</u> on obtient pour le contrôle :

1,0037 ; 0,9990 ; 1,0039 ; 1,0014 ; - 0,0007 ; -0,0039 ; -0,0035 ; -0,0043 ; -0.0026 ; -0,0005 .

<u>Exemple 3</u> : $z_d(x,t) = 4\,x(x-1) \sin 2 \Pi t$

$\mu_i = 3 \quad \forall i$

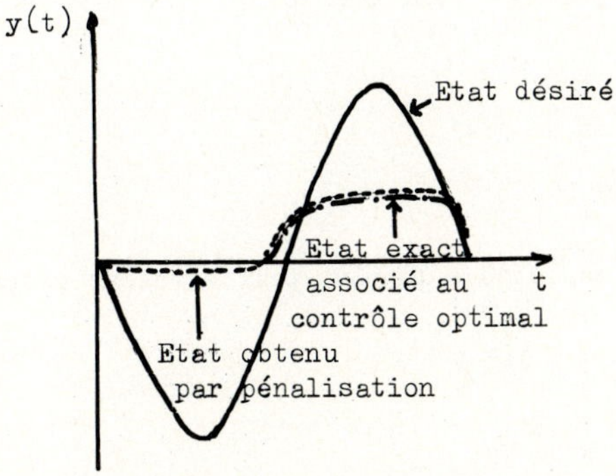

<u>Etat du système pour x</u> = 0,5

<u>Méthode a)</u>

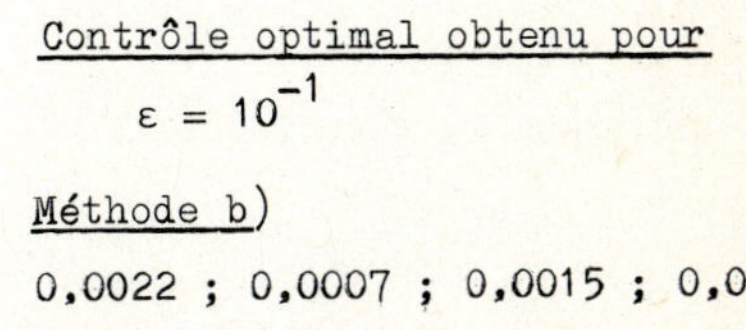

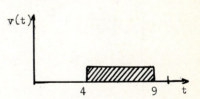

<u>Contrôle optimal obtenu pour</u>

$\varepsilon = 10^{-1}$

<u>Méthode b)</u>

0,0022 ; 0,0007 ; 0,0015 ; 0,0042 ; 1,0052 ; 0,9923 ; 0,9955 ; 0,9942 ; 0,9923 ; -0,0090 .

Les temps de calcul sont de l'ordre de 1 mn 30 s sur CII 10070 pour la méthode a) et de 2 mn 30 s sur CII 10070 pour la méthode b).

III - Autres applications pour des problèmes de contrôle optimal.

3.1. Tout d'abord nous pouvons remarquer que cette méthode est appréciable, chaque fois que le contrôle intervient sous la forme

$$u = \sum_i \alpha_i f_i \qquad \text{avec } f_i \text{ une fonction donnée dans } \mathcal{U}$$

en particulier on obtient un système linéaire en α, si les f_i sont orthogonaux deux à deux.

On obtient également cette propriété quand $u = \sum_{i=1}^{n} \alpha_i \delta(x - b_i)$.

3.2. On peut appliquer cette méthode à des sytèmes non linéaires. Par exemple lorsque l'état est donné par :

$$\begin{cases} \dfrac{\partial y}{\partial t} - \dfrac{\partial^2 y}{\partial x^2} + \left(\sum_{i=1}^{n} \alpha_i \mu_i \chi_{[a_i, b_i]}\right) y = f & \text{sur } \Omega \times]o,T[\\ y|_\Sigma = 0 \\ y(x,o) = y_o(x) \end{cases}$$

avec $f \in L^2(Q)$, $y_o(x) \in L^2(\Omega)$, $\{[a_i, b_i]\}$ $i = 1,\dots,n$ une famille d'intervalles de Ω.

Pour cette exemple on a des résultats de convergence analogue au II).

3.3. Cas d'un problème aux valeurs propres :

On se donne le système :

$$\begin{cases} - a \dfrac{d^2 y}{dx^2} + \left(\sum_{i=1}^{n} \alpha_i \mu_i \chi_{[a_i, b_i]}(x)\right) y = \lambda y & \text{sur } \Omega =]o,R[\\ \dfrac{dy}{dn}(o) = 0 \\ y(R) = 0 \end{cases}$$

alors l'état du système est la fonction propre associée à la plus petite valeur propre positive, telle que $y(o) = 1$

On prend la fonctionnelle

$$J(\alpha) = \int_\Omega |y - zd|^2 \, d\Omega + \nu(\lambda - \lambda_o)^2$$

avec $z_d \in L^2(\Omega)$, $\nu > 0$, λ_o donné dans $\mathbb{R}$.

On considère le problème de contrôle.

Trouver $\alpha^* \in \{o,1\}^n = I$ tel que

$$J(\alpha^*) = \min_{\alpha \in \{o,1\}^n} J(\alpha).$$

Sur cet exemple on peut montrer que la méthode de pénalisation s'applique et on obtient un résultat de convergence analogue au II). ([7] Saguez] .

Conclusion :

Cette technique de pénalisation permet donc de ramener le problème de contrôle à une minimisation en variables mixtes et dans un certain nombre de cas à un problème continu. On peut ainsi avoir dans un temps de calcul très inférieur aux méthodes d'énumération, la solution exacte ou au moins une bonne estimation de celle-ci.

BIBLIOGRAPHIE

[1] BOALE : On quadratic programming.
(Naval Research Logesties Quaterly Vol 6, 1969)

[2] GONDRAN : Une approche par pénalisation sur les variables des programmes quadratiques bivalents aux contraintes d'égalité.
(Bulletin E.D.F Etudes et Recherches N° 1 1972).

[3] HAMMER-RUBIN : Some Remarks on quadratic programming with o-1 variables.
(R.I.R.O Vol 3 1970).

[4] LIONS : Contrôle optimal de systèmes gouvernés par des équations aux dérivées partielles.
(Dunod - Gauthier-Villars 1968).

[5] ROCKAFELLAR : Augmented Lagrange multiplier function and duality in non convex programming.
(SIAM J. Control Vol 12 N° 2 May 1974).

[6] SAGUEZ : Application de la méthode de pénalisation aux problèmes de contrôle en nombres entiers.
(Rapport Laboria à paraître).

[7] YVON : Application de la pénalisation à la résolution d'un problème de contrôle optimal.
(Thèse 3ème cycle Paris 1968).

UNE NOUVELLE METHODE DE DECOMPOSITION DES GRANDS SYSTEMES OU LA PARTITION PRECEDE L'AFFECTATION

par A. Thuaire et J.P. Malengé

I.M.A.N. et LA 190

Université de Nice

06000-NICE FRANCE

I - INTRODUCTION

La modélisation des systèmes physiques, économiques, sociologiques conduit parfois à de grands systèmes de relations ($N > 100$) où le nombre de variables intervenant dans chaque relation est faible ($M \sim 10$). L'expérience montre que ces grands systèmes peuvent souvent être décomposés en une séquence ordonnée de sous-systèmes où seules les relations appartenant à un même sous-système doivent être résolues simultanément. Le but d'une méthode de décomposition est d'identifier les différents sous-systèmes et de déterminer leur ordre de résolution. Il est évident que la résolution du système décomposé est plus simple que celle du système initial.

La décomposition d'un grand système est liée à la structure de celui-ci et ne dépend pas de la forme des relations. On représente cette structure sous forme matricielle de la façon suivante :

- A chaque relation et à chaque variable sont respectivement associées une ligne et une colonne d'une matrice booléenne appelée matrice d'occurence ou matrice structurale.
- Un élément a_{ij} de cette matrice est tel que a_{ij} =1 si, et seulement si, la variable x_j intervient dans la relation f_i.

Soit par exemple le système de huit équations :

$f_1\ (x_1,x_5) = 0$

$f_2\ (x_2,x_7,x_8) = 0$

$f_3\ (x_1,x_3,x_4,x_6) = 0$

$f_4\ (x_1,x_2,x_3,x_4,x_6) = 0$

$f_5\ (x_2,x_5,x_7) = 0$

$f_6\ (x_1,x_5,x_8) = 0$

$f_7\ (x_5,x_8) = 0$

$f_8\ (x_2,x_3,x_6,x_7,x_8) = 0$

La matrice d'occurence associée à ce système est la matrice booléenne A de la fig.1

	1	2	3	4	5	6	7	8
1	1				1			
2		1					1	1
3	1		1	1		1		
4	1	1	1	1		1		
5		1			1		1	
6	1				1			1
7					1			1
8		1	1			1	1	1

fig.1. Matrice A

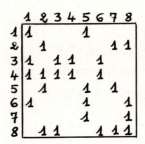

Il faut noter que la matrice d'occurence d'un système donné est toujours définie à une permutation près des lignes et des colonnes. En permutant les lignes et les colonnes de la matrice précédente on peut par exemple obtenir la matrice B de la fig. 2

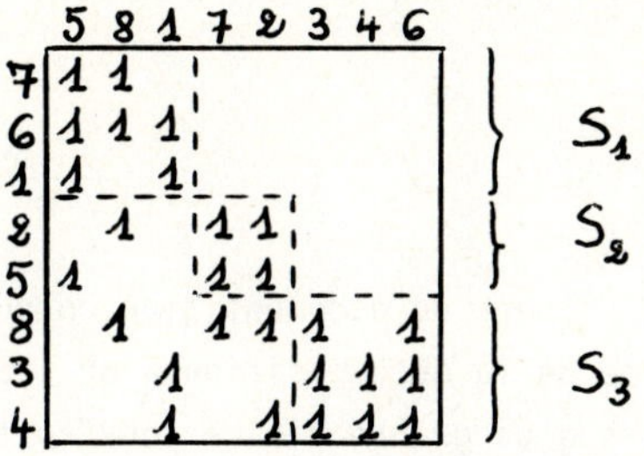

fig.2. Matrice B

Il apparaît clairement sur cette seconde matrice que le système donné peut être décomposé en trois sous systèmes S_1, S_2 et S_3. On commencera donc, pour résoudre ce système, par résoudre les équations 7, 6 et 1 pour obtenir la valeur des variables 5, 8 et 1. On peut alors résoudre les équations 2 et 5 puis les équations 8, 3 et 4 et déterminer ainsi la valeur de toutes les variables du système.

Le but d'une méthode de décomposition est donc de trouver, parmi toutes les combinaisons possibles des lignes et des colonnes d'une matrice d'occurence donnée au départ, une combinaison (il en existe plusieurs) qui représente cette matrice sous une forme triangulaire par blocs.

De nombreux algorithmes de décomposition ont déjà été publiés. La plupart d'entre eux font intervenir trois phases distinctes : affectation, partition et hiérarchisation, sur lesquelles nous reviendrons en détail au § II. La programmation de ces trois phases met en évidence les points suivants :

- La capacité en mémoire est proportionnelle à N^2
- Les temps d'exécution obtenus sont proportionnels au mieux à N^2

Ces algorithmes ne sont de ce fait utilisables que sur de gros ordinateurs.

Le but de cet article est de présenter une nouvelle méthode de décomposition diminuant notablement la place mémoire et le temps d'exécution et applicable sur de petits ordinateurs. On démontre comment, à partir d'une propriété remarquable des systèmes partitionnables, on peut ramener un problème de décomposition à un problème de minimisation. On donne également les caractéristiques d'un algorithme qui, programmé en FORTRAN et en BASIC, a permis d'obtenir :

- un encombrement mémoire proportionnel à N x M
- des temps d'exécution proportionnels à N

II - BIBLIOGRAPHIE

Les principaux algorithmes de décomposition connus sont basés sur le même principe : ramener le problème de la décomposition à un problème de recherche des composantes fortement connexes d'un graphe dirigé construit à partir du système donné. On y distingue pratiquement trois phases : affectation, partition et hiérarchisation.

II - 1. L'affectation

Cette première phase a pour but d'associer à chaque relation du système donné une variable intervenant dans cette relation. Deux variables différentes ne peuvent pas être associées à la même relation et vice-versa. On détermine ainsi un ensemble de couples variable-relation permettant de résoudre le système donné. Il est évident que la solution n'est en général pas unique. On démontre cependant que la décomposition finale ne dépend pas de la solution choisie. C'est dire en particulier que cette affectation pourra être modifiée au moment de la résolution proprement dite.

Parmi les algorithmes les plus efficaces pour ce type particulier de problème d'affectation on peut citer l'algorithme des chaînes alternées (K1) et l'algorithme de Steward (S1). Pour parvenir à une affectation totale ces deux algorithmes procèdent par itérations à partir d'une affectation partielle de base.

Un système affecté peut être représenté par une combinaison des lignes et des colonnes de la matrice d'occurence où tous les éléments de la diagonale principale sont égaux à un. La fig.2 montrait une matrice affectée et partitionnée, la fig.3 montre la même matrice uniquement affectée.

	1	7	6	4	2	8	5	3
1	1						1	
2		1			1	1		
3	1		1	1				1
4	1		1	1	1			1
5		1			1		1	
6	1					1	1	
7						1	1	
8		1	1		1	1		1

fig.3. Matrice d'occurence affectée

II - 2. Le graphe dirigé associé - le partitionnement

Une affectation étant trouvée, on peut définir sur l'ensemble des relations du système la relation binaire suivante :

$$f_i \,\mathcal{R}\, f_j \text{ si la variable associée à } f_i \text{ apparaît dans } f_j$$

On associe ainsi au système donné un graphe dirigé $(\mathcal{F},\mathcal{R})$, $\mathcal{F}$ désignant l'ensemble des relations. On démontre alors que l'ensemble des sous-systèmes irréductibles du système donné coïncide avec l'ensemble des composantes fortement connexes du graphe $(\mathcal{F},\mathcal{R})$.

Partitionner revient donc à rechercher l'ensemble des composantes fortement connexes d'un graphe. Parmi les algorithmes les plus efficaces on peut citer l'algorithme des puissances de matrices (H1), l'algorithme de Steward (S2) et l'algorithme de Billingsley (B1).

II - 3. La hiérarchisation

La hiérarchisation a pour but de déterminer l'ordre de résolution des différents sous-systèmes irréductibles de la partition. C'est une opération simple à réaliser :

- on construit un graphe $G' = (\mathcal{B},\mathcal{R}')$ où $\mathcal{B}$ désigne l'ensemble des sous-systèmes irréductibles et $\mathcal{R}'$ la relation binaire sur $\mathcal{B}$:

$$B1 \;\mathcal{R}'\; B2 \iff \left\{ \begin{array}{l} \text{La variable associée à une des relations du sous-} \\ \text{système B1 intervient dans une des relations du} \\ \text{sous-système B2.} \end{array} \right.$$

- on ordonne les sommets de ce graphe sans circuit sous forme d'arbre.

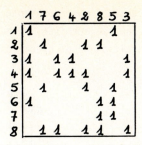

La matrice d'occurence de la fig.4 représente le système donné au § I sous une forme partitionnée mais non hiérarchisée. La matrice de la fig.5 représente le même système sous une forme partitionnée et hiérarchisée.

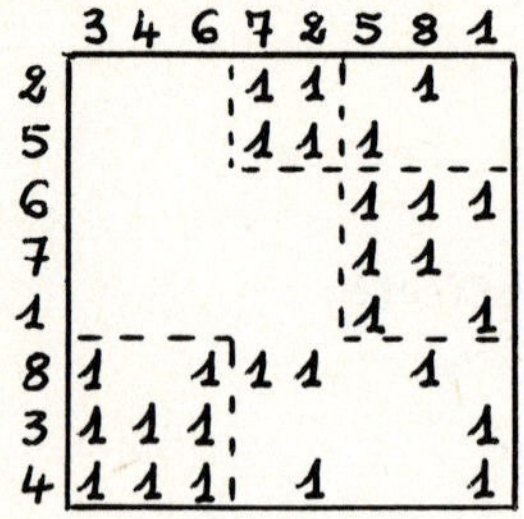

fig.4. Partition mais non hiérarchisation

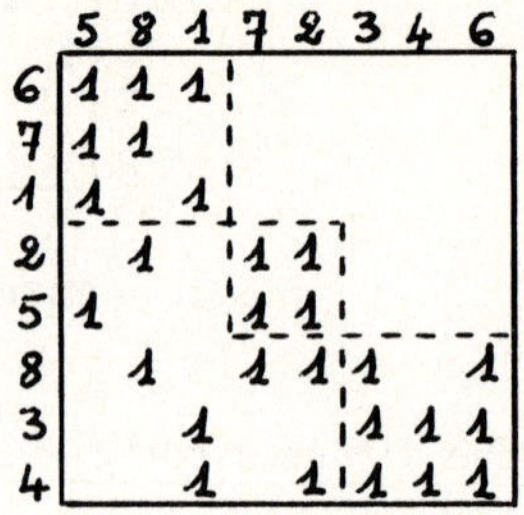

fig.5. Partition et hiérarchisation

Pour simplifier l'exposé on appellera par la suite matrice partitionnée une combinaison des lignes et des colonnes de la matrice d'occurence représentant le système sous une forme partitionnée et hiérarchisée.

II - 4. Critique des différentes méthodes

Les méthodes précédentes ont comme principal inconvénient de nécessiter beaucoup de place en mémoire : elles travaillent sur des matrices (matrice d'occurence puis matrice associée au graphe **(F,R)**) mémorisées sous forme de tableaux N x N . De plus ces méthodes sont relativement lentes et leur temps d'exécution est proportionnel à N^2 (A1), ce qui est normal : les algorithmes d'affectation et de partition effectuent des recherches dans des matrices N x N .

III - PRINCIPE DE NOTRE METHODE

Il nous a semblé que le défaut majeur des méthodes existantes était précisement de rechercher une affectation avant de partitionner, et cela pour trois raisons :

1- on a généralement intérêt à couper un problème en ses parties avant de le résoudre (D1)

2- parmi les $(N!)^2$ combinaisons possibles des lignes et des colonnes de la matrice d'occurence, les matrices partitionnées sont beaucoup plus nombreuses que les matrices affectées.

3- affecter un sous-système irréductible est une opération simple : il s'agit par définition d'une composante fortement connexe et les algorithmes d'affectation convergent alors très vite.

Le principe de notre méthode est très simple : chercher d'abord la partition, affecter ensuite chacun des sous-systèmes. En fait d'après la remarque 3 le seul problème est de partitionner, l'affectation d'un sous-système étant très peu coûteuse. Le paragraphe suivant montre comment on peut, grâce à une propriété remarquable des matrices partitionnées, ramener ce problème à un problème de minimisation d'une quantité numérique $\mu(A)$ associée à chaque combinaison des lignes et des colonnes de la matrice d'occurence.

IV - UNE PROPRIETE REMARQUABLE DES SYSTEMES PARTITIONNABLES

Sur la matrice partitionnée de la fig.6 (qui est la même que celle de la fig.5) on constate que la ligne brisée qui délimite la zone de zéros de la partie supérieure droite n'est pas monotone décroissante.

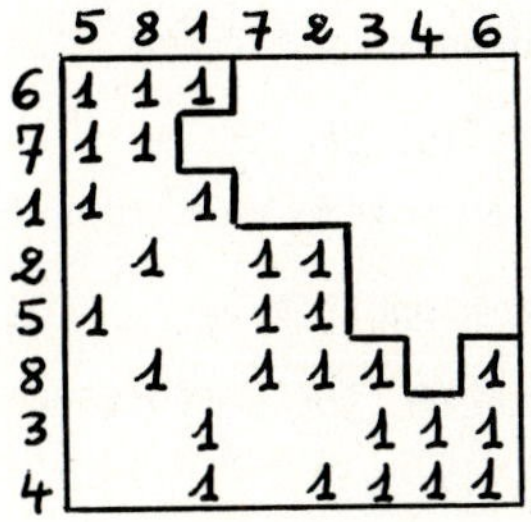

fig.6.

Mais, par des permutations n'intervenant que sur les lignes et les colonnes relatives à un même sous-système, on peut rendre cette séparatrice monotone décroissante (fig.7)

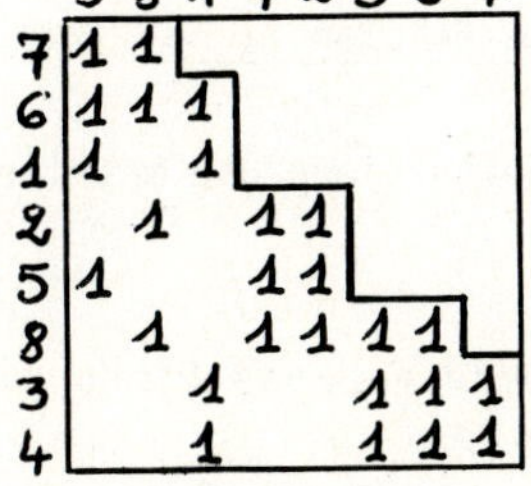

fig.7.

Soit μ le nombre de zéros du coin supérieur droit. Le but de ce paragraphe est de montrer qu'il suffit de maximiser μ, sur l'ensemble des matrices d'occurence ayant la séparatrice monotone décroissante, pour obtenir une matrice partitionnée. Après avoir précisé les notations, nous énoncerons plus rigoureusement cette propriété et la démontrerons.

IV - 1. Notations

Pour faciliter leur identification les relations et les variables seront numérotées de 1 à N et désignées par ce seul numéro.

F : ensemble des N relations du système donné

E : ensemble des N variables intervenant dans ces relations

A : matrice d'occurence du système. Les relations et les variables sont placées respectivement en ligne et en colonne dans l'ordre $1,2,\ldots,N$

S_A : ensemble des matrices X pouvant être obtenues par permutation des lignes et des colonnes de la matrice A

L_X : vecteur permutation ligne d'un élément X de S_A

C_X : vecteur permutation colonne d'un élément X de S_A

$\mathcal{D}_A$: sous-ensemble de S_A constitué par les matrices triangulaires inférieures par blocs représentant le système partitionné et hiérarchisé

$\overline{\mathcal{D}_A}$: complémentaire de $\mathcal{D}_A$ par rapport à S_A

$\mathcal{E}_A$: ensemble des matrices échelon (cf définition ci-dessous)

x_{ij} : terme général d'une matrice X

$\operatorname{Inf}(i/x_{ij} \neq o)$: valeur minimale de l'indice i pour laquelle le terme x_{ij} n'est pas nul

$\operatorname{Sup}(j/x_{ij} \neq o)$: valeur maximale de l'indice j pour laquelle le terme x_{ij} n'est pas nul

IV - 2. Définition d'une matrice échelon ligne

On appelle matrice échelon ligne une matrice appartenant à S_A et vérifiant la propriété

$$\operatorname{Sup}(j/x_{ij} \neq o) \leq \operatorname{Sup}(j/x_{i+1,j} \neq o) \qquad \forall i \in \{1,2,\ldots,N\}$$

En d'autres termes, si l'on considère chaque ligne de la matrice d'occurence comme la représentation binaire d'un nombre (poids faible à gauche), une matrice échelon ligne est une matrice dont les lignes sont classées par ordre croissant de leur bit de poids le plus fort. Si on considère la matrice d'occurence A de la fig.1 la matrice ci-dessous (fig.7) est une matrice échelon ligne de S_A:

	1	2	3	4	5	6	7	8
1	1				1			
3	1		1	1		1		
4	1	1	1	1		1		
5		1			1		1	
2		1					1	1
6	1				1			1
7					1			1
8		1	1			1	1	1

fig.7. Matrice échelon ligne

On notera $\mathcal{L}_A$ le sous-ensemble de S_A constitué par les matrices échelon ligne et φ_L la transformation qui associe à un élément de S_A un élément de $\mathcal{L}_A$

IV - 3. Définition d'une matrice échelon colonne

On appelle matrice échelon colonne une matrice appartenant à S_A et vérifiant la propriété :

$$\text{Inf } (i/x_{ij} \neq o) \leqslant \text{Inf } (i/x_{i,j+1} \neq o) \qquad \forall j \in \{1,2,\ldots,N\}$$

En d'autres termes, si l'on considère chaque colonne de la matrice d'occurence comme la représentation binaire d'un nombre (poids faible en bas), une matrice échelon colonne est une matrice dont les colonnes sont classées par ordre décroissant de leur bit de poids le plus fort. Reprenons la matrice d'occurence A de la fig.1, la matrice de la fig.8 est une matrice échelon colonne appartenant à S_A:

	1	5	2	7	8	3	4	6
1	1	1						
2			1	1	1			
3	1					1	1	1
4	1		1			1	1	1
5		1	1	1				
6	1	1			1			
7		1			1			
8			1	1	1	1		1

fig.8. Matrice échelon colonne

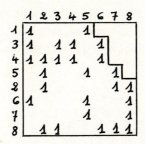

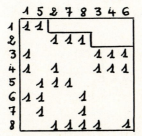

On notera $\mathcal{C}_A$ le sous-ensemble de S_A constitué des matrices échelon colonne et φ_c la transformation qui permet d'associer à un élément de S_A un élément de $\mathcal{C}_A$.

IV - 4. Définition d'une matrice échelon

On appelle matrice échelon une matrice appartenant à l'ensemble $\mathcal{L}_A \cap \mathcal{C}_A$ On notera $\mathcal{E}_A$ l'ensemble des matrices échelon. On peut remarquer que tout élément de $\mathcal{E}_A$ possède une séparatrice de la zone de zéros du coin supérieur droit monotone décroissante. Si l'on considère à nouveau le système donné au § I, la matrice de la fig.9 est une matrice échelon de $\mathcal{E}_A$

	3	6	1	4	5	8	2	7
3	1	1	1	1				
1			1		1			
6			1		1	1		
7					1	1		
4	1	1	1	1			1	
2						1	1	1
5					1		1	1
8	1	1				1	1	1

fig.9. Matrice échelon

IV - 5. Enoncé et démonstration de la propriété

Nous allons démontrer que la matrice échelon qui a le plus grand nombre de zéros dans le coin supérieur droit est partitionnée (si elle est partitionnable).

On pose par définition, pour tous les éléments X appartenant à S_A

$$\mu(X) = \sum_{i=1}^{N} \sup (j / x_{ij} \neq 0)$$

On note X^* l'élément de S_A tel que :

$$\mu(X^*) = \operatorname*{Inf}_{X \in \mathcal{E}_A} \mu(X)$$

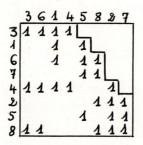

IV - 5. 1. Enoncé de la propriété P

Si A est la matrice d'occurence d'un système partitionnable $X^* \in \mathcal{E}_A \cap \mathcal{D}_A$

IV - 5. 2. Démonstration de la propriété P

Supposons vraie la proposition suivante Q que l'on démontrera au § V :

$$\forall X \in \mathcal{E}_A \cap \overline{\mathcal{D}}_A \quad \exists Z \in \mathcal{E}_A \cap \mathcal{D}_A \quad \text{tel que} \quad \mu(Z) < \mu(X)$$

La démonstration de la propriété P est alors très simple : il suffit de choisir $X_o \in \mathcal{E}_A \cap \overline{\mathcal{D}}_A$ tel que :

$$\mu(X_o) = \inf_{X \in \mathcal{E}_A \cap \overline{\mathcal{D}}_A} \mu(X)$$

Il nous reste donc, pour que la propriété P soit démontrée, à démontrer la proposition Q.

V - DEMONSTRATION DE LA PROPOSITION Q

Considérons un système F de N relations et de N variables partitionnable en k sous-systèmes irréductibles $F_1, F_2, \ldots, F_k$. Le sous-système F_1 est le premier de la hiérarchie, F_k est le dernier.

On pose $\mathcal{F} = \bigcup_{i=1}^{k-1} F_i$ et on note $\mathcal{V}$ et V les sous-systèmes de variables correspondant respectivement à $\mathcal{F}$ et F_k.

Le principe de la démonstration est le suivant :

- on se donne une matrice échelon quelconque X associée au système F
- on suppose que X n'est pas partitionnée mais que l'on connait les numéros des équations et des variables qui forment les différents sous-systèmes.
- on construit (§ V.1) une matrice X" partitionnée en deux sous-systèmes $\mathcal{F}$ et F_k et telle que (§ V.2) $\mu(X'') < \mu(X)$
- on itère sur le sous-système réductible $\mathcal{F}$ et on aboutit de proche en proche à une matrice Y complètement partitionnée (§ V.3) avec bien sûr, $\mu(Y) < \mu(X)$

- on montre pour terminer qu'une transformation ne comportant que des permutations à l'intérieur de chaque sous-système permet, à partir de Y, d'aboutir à une matrice échelon Z partitionnée et telle que $\mu(Z) \leq \mu(Y)$.

Il peut paraître choquant de supposer au début, que l'on connait les numéros des variables et des équations de chaque sous-système, donc la partition. En fait il ne faut pas oublier que ce paragraphe a pour but de démontrer une propriété des systèmes partitionnables et non pas de démontrer la convergence d'un algorithme.

V - 1. Construction de la matrice X"

La matrice X" est l'image de la matrice X par la composée $f_2 0 f_1$ de deux transformations décrites ci-dessous.

V - 1.1. Définition de la tranformation f_1

$$\mathcal{E}_A \cap \overline{\mathcal{D}}_A \xrightarrow{f_1} S_A$$

$$X \longmapsto f_1(X) = X'$$

La matrice image X' est définie à l'aide des vecteurs $L_{X'}$ et $C_{X'}$ de la façon suivante :

a) Le vecteur ligne $L_{X'}$ est obtenu en plaçant en tête les numéros des relations appartenant à $\mathcal{F}$, en respectant l'ordre qu'elles avaient les unes par rapport aux autres dans la matrice X. On place ensuite les relations appartenant à F_k.

b) On ne modifie pas le vecteur colonne : $C_{X'} = C_X$

Reprenons à titre d'exemple la matrice échelon de la fig.9. Son image par l'application f_1 est la matrice de la fig.10.

$L_{X'}$ (lignes), $C_{X'}$ (colonnes)

	3	6	1	4	5	8	2	7
1			1		1			
6			1		1	1		
7					1	1		
2						1	1	1
5					1		1	1
3	1	1	1	1				
4	1	1	1	1			1	
8	1	1				1	1	1

fig.10. Matrice X'

V - 1.2. Définition de la transformation f_2

$$f_1\,(\mathcal{E}_A \cap \overline{\mathcal{D}}_A) \xrightarrow{\;f_2\;} S_A$$

$$X' \longmapsto f_2(X') = X''$$

La matrice image X" représente le système sous une forme partitionnée et hiérarchisée en deux sous-systèmes $\mathcal{F}$ et F_k. Elle est définie à l'aide des vecteurs $L_{X''}$ et $C_{X''}$ de la façon suivante :

a) le vecteur colonne $C_{X''}$ est obtenu en plaçant en tête les numéros des variables appartenant à $\mathcal{V}$, en respectant l'ordre qu'elles avaient les unes par rapport aux autres dans la matrice X'. On place ensuite les variables appartenant à V.

b) on ne modifie pas le vecteur ligne : $L_{X''} = L_{X'}$

La matrice X" de la figure 11 est l'image par f_2 de la matrice X' de la figure 10.

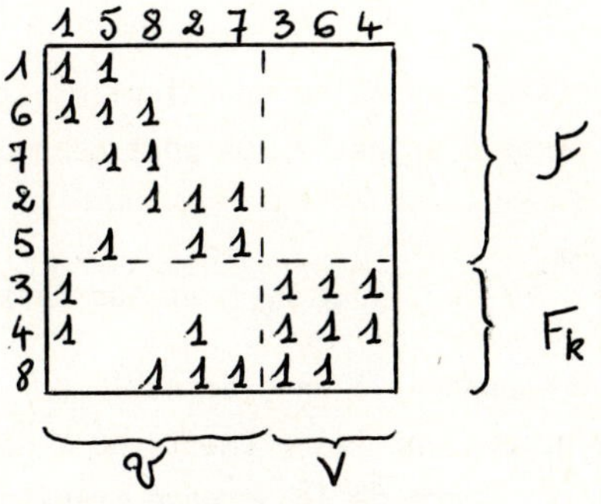

fig.11 Matrice X"

V - 2. Comparaison de μ(X") et de μ(X)

Le vecteur colonne de X n'est pas modifié par l'application f_1. On peut donc écrire que $\mu(X') = \mu(X)$. Par contre l'application f_2 fait subir deux types de variations au nombre μ :

- une diminution apportée par les relations appartenant à $\mathcal{F}$
- une augmentation apportée par les relations appartenant à F_k

Nous calculons ci-dessous ces deux variations.

V - 2.1. La diminution

On pose :

$p = \text{Card}\ \mathcal{F} = \text{Card}\ \mathcal{V}$

$q_{x'}(i)$: position de la variable numéro i dans $C_{x'}$

$q_{x''}(i)$: position de la variable numéro i dans $C_{x''}$

$$T = \left\{ i \in V \ / \ q_{x'}(i) \neq q_{x''}(i) \right\}$$

A l'issue de la première transformation, les p relations appartenant à $\mathcal{F}$ occupent les p premières lignes de la matrice X'. Par suite les p premiers éléments x'_{ij} des colonnes associées aux variables de V sont nuls. En d'autres termes :

(1) $$\text{Inf}\ (i\ /x'_{i\ell} \neq 0) \quad > \quad p \qquad \forall \ell \in \left\{ 1,2,\ldots,N \right\} \ / \ C_{x'}(\ell) \in T$$

Au cours de la deuxième transformation le déplacement de chaque élément de T produit, d'après (1), une diminution d'une unité du nombre μ pour chacune des p premières lignes. La diminution totale sera par conséquent de p x Card T.

V - 2.2. L'augmentation

Il n'est pas possible de faire un calcul exact de l'augmentation totale, celle-ci variant d'un système à un autre. On peut néanmoins en déterminer un majorant. Pour cela on pose :

$$T' = \left\{ i \in \left\{1,2,\ldots,N\right\} \ / \ L_{x'}(i) \in F_k \quad \text{et Sup}\ (j/x'_{ij} \neq 0) \ < \ \text{Sup}\ (j/x''_{ij} \neq 0) \right\}$$

Pour une ligne donnée correspondant à un élément de T', l'augmentation ne peut dépasser p. En effet, l'ordre qu'avaient les éléments de V dans $C_{x'}$ les uns par rapport aux autres a été respecté au cours de la seconde transformation et le contraire contredirait alors l'hypothèse Card $\mathcal{F}$ = p. On peut donc en déduire que, dans le cas le plus défavorable où tous les éléments de T' subissent une augmentation égale à p, l'augmentation totale est égale à p x Card T'.

V - 2.3. Démonstration de la relation $\mu(X'') < \mu(X)$

D'après V-2.1. et V-2.2. on a :

$$\mu(X'') \leqslant \mu(X') - p \times \text{Card}\ T + p \times \text{Card}\ T'$$

Nous allons maintenant démontrer que, dans l'hypothèse où chaque élément de T' subit une augmentation de μ égale à p, on a, quel que soit le système, Card T' < Card T. Cette démonstration tient dans les deux remarques suivantes :

Remarque 1 : Tout élément m appartenant à T' vérifie après la seconde transformation l'inégalité ci-dessous :

$$\text{Sup } (j \;/\; x''_{mj} \neq 0) \leqslant p + \text{Card } T$$

En d'autres termes aucune variable appartenant à $\complement_V T$ n'intervient dans l'ensemble des relations correspondant à T'

Remarque 2 : $\forall\, k \in T' \quad$ et $\quad \forall\, l \notin T' \;/\; l > p \quad$ on a $\quad k < l$

En d'autres termes les relations correspondant à T' sont placées, à l'issue de la première transformation, en tête du sous-système F_k. Cela provient du fait que l'ordre qu'avaient les éléments de F_k les uns par rapport aux autres dans L_x a été respecté dans $L_{x'}$.

On déduit de ces deux remarques et de l'inégalité (1) que, dans l'hypothèse où Card T = Card T', F_k est partitionnable en deux sous-systèmes $F_k^{(1)}$ et $F_k^{(2)}$, $F_k^{(1)}$ désignant le sous-ensemble de relations correspondant à T' et $F_k^{(2)}$ son complémentaire par rapport à F_k. Ceci contredit bien entendu l'hypothèse de départ : "F_k sous-système irréductible" et on a donc dans tous les cas Card T' ≠ Card T.

Il est évident par ailleurs que le cas Card T' > Card T conduit, pour des raisons analogues, à $F_k^{(1)}$ sous-système indéterminé. On a donc dans tous les cas :

$$\mu(X'') < \mu(X).$$

V - 3. Construction de la matrice Y

Les transformations f_1 et f_2 que nous venons d'appliquer sur le système F peuvent être appliquées successivement sur les sous-systèmes : $\mathcal{F} = \bigcup_{i=1}^{k-1} F_i$,

$\bigcup_{i=1}^{k-2} F_i \;,\; \bigcup_{i=1}^{k-3} F_i \;,\; \ldots \;,\; F_1 \cup F_2$.

On obtient ainsi de proche en proche une matrice Y telle que :

$$Y \in \mathcal{D}_A \quad \text{et} \quad \mu(Y) < \mu(X)$$

V - 4. Construction de la matrice Z

Dans certains cas la matrice Y, qui appartient à $\mathcal{D}_A$ et à $\mathcal{C}_A$, n'appartient pas à $\mathcal{L}_A$ (fig.11). Nous allons démontrer que, dans tous les cas, on peut associer à la matrice Y une matrice Z appartenant à $\mathcal{E}_A \cap \mathcal{D}_A$ et telle que $\mu(Z) \leq \mu(Y)$. Cette démonstration achèvera la démonstration de la proposition Q et par conséquent (§ IV-5.) celle de la propriété P.

L'ensemble $\mathcal{E}_A$ a été défini au(§ IV-4.) comme l'intersection des ensembles $\mathcal{L}_A$ et $\mathcal{C}_A$. Rappelons (§ IV-2.) que la transformation φ_L, qui permet d'associer à un élément de S_A un élément unique de $\mathcal{L}_A$, est obtenue en classant par ordre croissant les lignes de la matrice suivant leur bit de poids le plus fort (poids faible à gauche) et que la transformation φ_C, qui permet d'associer à un élément de S_A un élément unique de $\mathcal{C}_A$, est obtenue en classant par ordre décroissant les colonnes de cette matrice suivant leur bit de poids le plus fort (poids faible en bas).

La matrice Y appartient à $\mathcal{C}_A$ mais la matrice φ_L (Y) n'appartient pas forcément à $\mathcal{C}_A \cap \mathcal{L}_A$ c'est à dire à $\mathcal{E}_A$. En effet, dans certains cas, un classement des lignes provoque un déclassement des colonnes. Nous allons démontrer qu'une transformation φ_C suivant cette transformation φ_L sur Y permet d'obtenir une matrice Z élément de $\mathcal{E}_A$(§ V-4.1.). Nous démontrerons ensuite (§ V-4.2.) que $\mu(\varphi_C \circ \varphi_L(Y))$ est inférieur ou égal à $\mu(Y)$

V - 4.1. La matrice $\varphi_C \circ \varphi_L(Y)$

On pose par définition :

$$\mathcal{U} = \{k \in \{1,2,\ldots,N\} \ / \ \exists j \in \{1,2,\ldots,N\} \text{ tel que } \operatorname{Sup}(j/y_{ij} \neq 0) = k\}$$

La démonstration repose sur les deux remarques suivantes :

Remarque 1 : D'après la définition de φ_L, les colonnes appartenant à $\mathcal{U}$ restent classées les unes par rapport aux autres dans la matrice φ_L (Y)

Remarque 2 : L'ordre des colonnes de $\mathcal{U}$ les unes par rapport aux autres sera encore conservé par une transformation φ_C suivant φ_L puisque, de par sa définition, φ_C ne fait qu'intercaler entre les lignes classées les lignes non classées.

On peut déduire de ces deux remarques que la transformation φ_C ne modifie pas le classement des lignes. Comme d'autre part les transformations φ_C et φ_L peuvent être faites indépendemment sur chaque bloc de la matrice Y, et ne modifient donc pas la répartition des équations et des variables de chaque sous-système, la matrice $Z = \varphi_C \circ \varphi_L(Y)$ appartient à la fois à $\mathcal{E}_A$ et à $\mathcal{D}_A$ donc à $\mathcal{E}_A \cap \mathcal{D}_A$.

V - 4.2. Comparaison de $\mu(\varphi_C \circ \varphi_L(Y))$ et de $\mu(Y)$

La transformation φ_L ne modifie pas l'ordre des colonnes d'une matrice. Par conséquent, quelque soit la matrice Y, $\mu(Y) = \mu(\varphi_L(Y))$. Pour démontrer l'inégalité :

$$\mu(\varphi_C \circ \varphi_L(Y)) \leqslant \mu(Y)$$

il suffit donc de démontrer l'inégalité :

$$\mu(\varphi_C \circ \varphi_L(Y)) \leqslant \mu(\varphi_L(Y))$$

Nous allons voir que cette seconde inégalité découle de la définition de φ_C

Au cours de la transformation φ_C , toute colonne que l'on déplace fait subir au nombre μ deux types de variations :

a) une augmentation d'une unité de μ pour la ligne contenant le premier élément non nul de la colonne déplacée.

b) Une diminution d'une unité pour toutes les lignes comprises entre la position, dans $\varphi_L(Y)$, du premier élément non nul de la colonne déplacée et la position de cet élément après φ_C. Il est évident que, pour chaque colonne déplacée, il y a au moins une ligne concernée puisque, dans le cas contraire, la colonne n'aurait pas été déplacée.

On déduit de ce qui précède que l'augmentation totale du nombre μ sur $\varphi_C \circ \varphi_L(Y)$ est toujours inférieure ou égale à la diminution totale et donc que l'inégalité : $\mu(\varphi_C \circ \varphi_L(Y)) \leqslant \mu(\varphi_L(Y))$, est vérifiée quelque soit la matrice Y (obtenue après les transformations f_1 et f_2).

VI - MISE EN OEUVRE DE LA METHODE

L'utilisation de la propriété P pour le partitionnement d'un grand système consiste donc à trouver un algorithme permettant de construire une suite de matrices échelon dont la valeur de μ est décroissante. Si cet algorithme converge, la matrice finale sera partitionnée. En fait il n'est pas nécessaire d'atteindre le minimum de μ : il existe plusieurs matrices échelon partitionnées qui n'ont pas la même valeur de μ (fig.12).

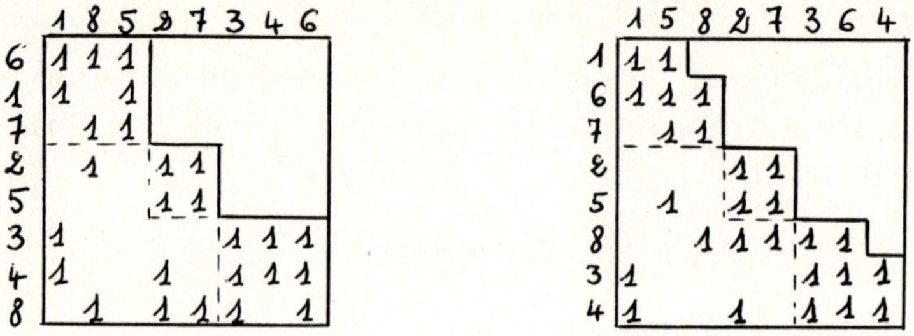

$\mu = 43$ $\mu = 41$

fig.12.

L'algorithme que nous avons écrit possède les trois caractéristiques suivantes :

- il fonctionne par dichotomie : il cherche à partitionner la matrice donnée en deux blocs puis il itère dans chacun des blocs
- il nécessite peu de place en mémoire (cf. VI-1.)
- sa convergence n'est pas prouvée théoriquement mais pratiquement

VI - 1. Mémorisation de la matrice d'occurence

Contrairement à certaines méthodes existantes (puissances de matrices par exemple), la méthode décrite ici ne tirerait aucun profit d'une matrice d'occurence mémorisée sous forme d'un tableau N x N. C'est pourquoi, afin d'occuper une place mémoire aussi petite que possible, la matrice d'occurence est mémorisée sous forme d'un tableau N x P construit de la façon suivante :

- chaque ligne représente une relation et contient les numéros des variables intervenant dans cette relation.
- le nombre P de colonnes est le nombre maximum de variables intervenant dans une relation du système donné.

VI - 2. Description sommaire de l'algorithme

L'algorithme sera décrit en détail dans un autre article. On se contente ici d'en donner les grandes lignes.

La partie principale de l'algorithme est celle qui, étant donnée une matrice échelon X_i, permet d'en obtenir une autre X_{i+1} telle que $\mu(X_{i+1}) \leq \mu(X_i)$, avec arrêt en cas d'égalité stricte. On utilise pour cela essentiellement cinq transformations φ_L, φ_c, φ_S, φ_F et γ_S. Les transformations φ_L et φ_c ont été respectivement décrites au IV - 2. et IV - 3. Les transformations φ_S, φ_F et γ_S sont décrites ci-dessous.

VI - 2.1. Définition de φ_S

$$\begin{aligned} S_A &\longrightarrow S_A \\ X &\longmapsto \varphi_S(X) \end{aligned}$$

La matrice image $\varphi_S(X)$ est telle que :

- $C_{\varphi_S(X)} = C_{\varphi_c(X)}$

- $L_{\varphi_S(X)} = L_{\varphi_L(X)}$

VI - 2.2. Définition de φ_F

$$\begin{aligned} S_A &\longrightarrow S_A \\ X &\longmapsto \varphi_F(X) \end{aligned}$$

La matrice image $\varphi_F(X)$ est telle que

- $C_{\varphi_F(X)} = C_{\varphi_S \circ \varphi_S}$

- $L_{\varphi_F(X)} = L_{\varphi_L(X)}$

VI - 2.3. Définition de γ_S

$$\begin{aligned} \mathcal{E}_A &\longrightarrow S_A \\ X &\longmapsto \gamma_S(X) \end{aligned}$$

La matrice γ_S (X) est définie par :

- $C\gamma_S(X)$: permutation colonne qui minimise la somme des écarts entre les deux derniers élements non nuls de chaque ligne de X

- $L\gamma_S(X)$: permutation ligne qui minimise la somme des écarts entre les deux premiers éléments non nuls de chaque colonne de X.

La figure 13 montre un exemple d'application de γ_S .

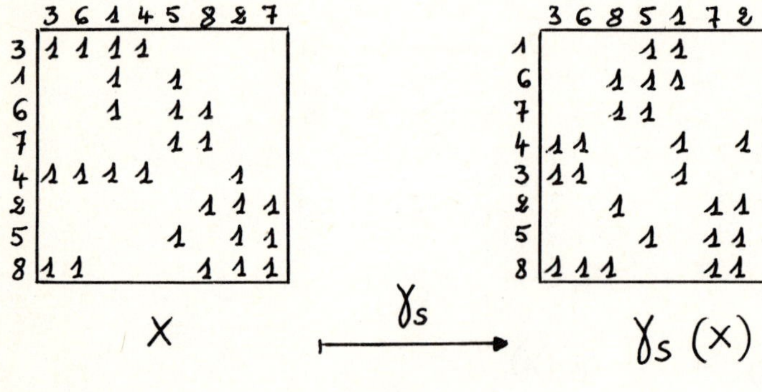

fig.13.

VI - 2.4. Construction de X_{i+1} à partir de X_i

Etant donné une matrice X_o appartenant à S_A, on pose :

$$Y_{X_o} = \begin{cases} \varphi_C \circ \varphi_L (X_o) \text{ si } & \mu\left(\varphi_C \circ \varphi_L (X_o)\right) < \mu\left(\varphi_L \circ \varphi_C (X_o)\right) \\ \varphi_L \circ \varphi_C (X_o) \text{ sinon} & \end{cases}$$

X_{i+1} est alors définie par les formules

(1) X_o donnée au départ

(2) $X_1 = Y_{\varphi_F}(X_o)$

(3) $X_{i+1} = Y_{\varphi_F}(\gamma_S(X_i))$

VI - 3. Les résultats

L'algorithme que nous venons de décrire très sommairement a été testé sur des systèmes partitionnables en deux sous-systèmes irréductibles, l'un des principes de la méthode étant de travailler par dichotomie. Les calculs ont été effectués sur un ordinateur T1600 Télémécanique de 16 k mots de 16 bits et ont porté sur plus de cent cinquante systèmes (N=10 à 50) partitionnables, générés aléatoirement. Dans tous les cas l'algorithme a partitionné rapidement le système donné. Les résultats détaillés seront publiés en même temps que l'algorithme.

VII - CONCLUSION

L'idée de base consistant à partitionner avant d'affecter nous a permis de trouver un algorithme qui s'est révélé efficace (rapidité et encombrement). Il est cependant probable que cet algorithme est loin d'être le plus performant et on peut penser que la poursuite des recherches dans cette voie sera fructueuse.

REFERENCES

B1 Billingsley, D.S., Chem. Eng. Sci. 22 (4), 719 (1967) (Eng)

H1 Himmelblau, D.M., Chem. Eng. Sci. 21 (5), 425-38 (1966) (Eng)

H2 Himmelblau, D.M.,and Bischoff, K.B., "Process Analysis and Simulation", Wiley, New York (1961)

D1 Descartes, "Discours de la méthode" (1637)

K1 Kaufmann, "Introduction à la combinatorique en vue de ses applications", DUNOD (1968)

K2 Kevorkian, A.K., and Snoek J., "Decomposition in large scale systems", NATO Advanced Study Institute (1972)

L1 Ledet, W.P., Ph. D. Thesis Univ. of Texas (1968)

S1 Steward, D.V., SIAM Rev. 4,321 (1962)

S2 Steward, D.V., J. SIAM Numer. Anal. Ser B, Vol 2, 2 (1965)

ON THE MULTIVARIABLE CONTROL OF NUCLEAR REACTORS USING THE STATE FEEDBACK APPROACH

S. TZAFESTAS and N. CHRYSOCHOIDES

Department of Reactors, N.R.C. "Demokritos"
Aghia Paraskevi, Attiki, Athens, Greece

ABSTRACT

Recently an effort was made to design nuclear reactor systems via state-variable feedback techniques. On the other hand, a great amount of research has been concentrated on the multivariable state feedback control methodology, for its own, and a variety of useful results have been derived. The purpose of the paper is to apply a general multivariable state feedback control technique to nuclear reactor systems (e.g. multiregion reactors, coupled core reactors, etc.). Two fundamental design tasks are considered, namely, noninteraction and realization of desired transfer functions. This technique requires the system under control to be given in its phase canonical form and provides explicit expressions for the feedback control law matrices required. Two nuclear reactor examples are considered and fully worked out.

1. INTRODUCTION

The application of the state variable feedback control technique to nuclear reactor systems seems to be a promising approach with many advantages over the classical and the optimal control techniques providing a kind of link between them. Some studies concerning the application of this technique to nuclear reactors are involved in [1]-[7]. In general, the objective of this technique is to realize exactly given dynamics by feeding back some or all of the state variables through appropriate gains. The desired system dynamics is usually described by a given transfer function which is completely specified by its zeroes, poles and d.c. gains.

The work described in [1]-[3] is constrained to single-input single-output reactor systems and derives the solution (i.e. the feedback gains) by a direct comparison of the transfer function of the closed-loop system with that of the desired model, and an equation of the equal-power terms. The full theory of this method may be found in [4]. These results were extended in [5] to reactor systems having m inputs and m outputs through Gilbert's technique of canonical decoupling

[9]. A similar technique is described in [6]. A further extension of this method to the case where additional compensation is required to meet the desired criteria was made in [7].

From the point of view of pure control theory this problem has received independent attention and a substantial amount of results are already available [8]-[19]. Particular attention was given to two subproblems of the general state variable feedback control design problem, namely the decoupling problem [8]-[10], and the eigenvalue control problem [11].

The purpose of the present paper is to investigate the applicability to multi-input multi-output nuclear reactors of a recent state variable feedback control technique [11], [15], [21] which is based on the assumption that the system under control is given in, or transformed into, its phase canonical form. The main problem is that of noninteracting system design, and the method provides simultaneous input-output and state variable decoupling. The pole and d.c. gain assignment can be accomplished simultaneously with the decoupling, but to control the zeros additional compensators are used as in [7]. Decoupling (noninteraction) in an actual coupled-core reactor is obtained by negating the neutron coupling between the cores. For comparison purposes the nuclear reactor examples studied in [5]-[7] are considered and completely worked out by the present technique.

2. STATE EQUATIONS OF MULTIPLE-CORE REACTORS

A coupled-core reactor is a critical reactor composed by two or more independently subcritical cores. The coupling effect is the result of mutual exchange of leakage neutrons between cores. In such reactors in order to apply successful control to the power levels of the cores independently, the effects of the neutron coupling must be balanced, i.e. a decoupled or noninteracting system is to be designed.

Consider a multiple-core reactor in which each core is coupled only with the neighbouring cores. Assume for simplicity one delayed neutron group for each core, and small neutron travel time between cores. Then, including the negative temperature feedback, the state equations (linearized) for a 3-core system are*

---

*Details of derivation together with an introduction of nuclear reactors in state space are given in [2].

$$\frac{dx_1}{dt} = \frac{D_{11}+\beta_1}{\tau_1} x_1 + \frac{D_{12}}{\tau_2} x_2 + \lambda_1 x_4 - \frac{\alpha_1 n_1^o}{\tau_1} x_7 + \frac{n_1^o}{\tau_1} x_{10}$$

$$\frac{dx_2}{dt} = \frac{D_{21}}{\tau_1} x_1 - \frac{2D_{22}+\beta_2}{\tau_2} x_2 + \frac{D_{23}}{\tau_3} x_3 + \lambda_2 x_5 - \frac{\alpha_2 n_1^o}{\tau_2} x_8 + \frac{n_2^o}{\tau_2} x_{11}$$

$$\frac{dx_3}{dt} = \frac{D_{32}}{\tau_2} x_2 - \frac{D_{33}+\beta_3}{\tau_3} x_3 + \lambda_3 x_6 - \frac{\alpha_3 n_3^o}{\tau_3} x_9 + \frac{n_3^o}{\tau_3} x_{12}$$

$$\frac{dx_4}{dt} = \frac{\beta_1}{\tau_1} x_1 - \lambda_1 x_4, \quad \frac{dx_7}{dt} = k_1 x_1 - m_1 x_7, \quad \frac{dx_{10}}{dt} = x_{13}$$

$$\frac{dx_5}{dt} = \frac{\beta_2}{\tau_2} x_2 - \lambda_2 x_5, \quad \frac{dx_8}{dt} = k_2 x_2 - m_2 x_8, \quad \frac{dx_{11}}{dt} = x_{14} \qquad (1)$$

$$\frac{dx_6}{dt} = \frac{\beta_3}{\tau_3} x_3 - \lambda_3 x_6, \quad \frac{dx_9}{dt} = k_3 x_3 - m_3 x_9, \quad \frac{dx_{12}}{dt} = x_{15}$$

$$\frac{dx_{13}}{dt} = -\Theta_1 x_{13} + \mu_1 u_1, \quad \frac{dx_{14}}{dt} = -\Theta_2 x_{14} + \mu_2 u_2, \quad \frac{dx_{15}}{dt} = -\Theta_3 x_{15} + \mu_3 u_3$$

$$y_1 = x_1, \quad y_2 = x_2, \quad y_3 = x_3$$

Here x_1, x_2, x_3 are power levels in cores 1,2,3 correspondingly, x_4, x_5, x_6 are concentrations of delayed neutrons x_7, x_8, x_9 are control rod rates, and u_1, u_2, u_3 are the control inputs for cores 1,2,3 respectively. The parameters involved have the following interpretation with i=1,2,3.

τ_i	prompt neutron generation time in core i
β_i	delayed neutron fraction in core i
n_i^o	steady state neutron power level in core i
D_{ij}	neutron coupling coefficient from core j to core i
λ_i	decay constant of the delayed neutron emitter in core i
α_i	reactivity-temperature proportionality constant of core i
k_i	temperature-power proportionality constant of core i
m_i	heat removal coefficient of core i
Θ_i	inverse time constant of rod controller in core i

In this model the control rods are assumed to be driven by electric motors and hence each error signal produces a proportional motor speed. Since the reactivity is proportional to control rod position (not to control rod speed) one must integrate the output of the control rod driver to obtain the reactivity. A signal flow graph of this model indicating the coupling among the cores as well as the control channels is given in Fig. 1.

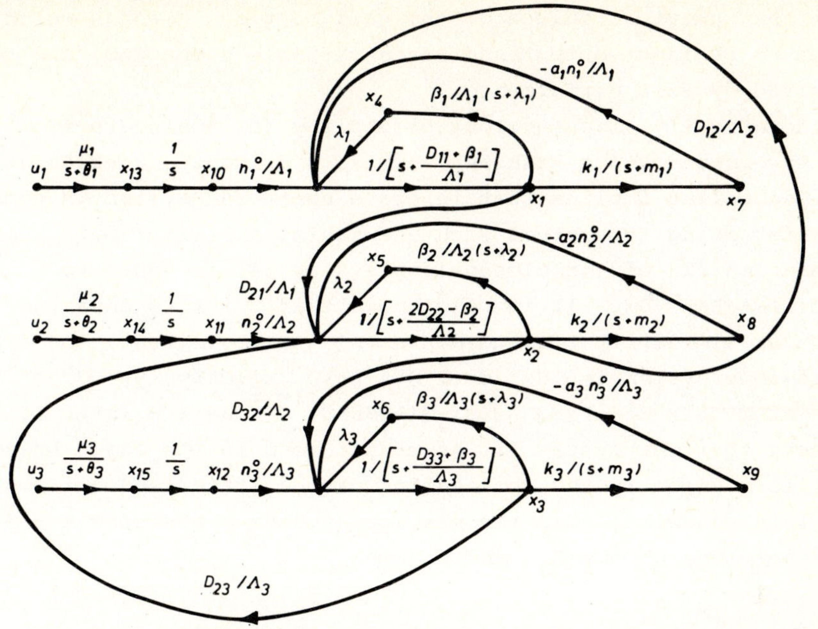

Fig. 1. Signal flow graph of the 3-coupled-core reactor

In matrix form this 3-core reactor system (as any multiple core reactor system) can be written as:

$$\frac{dx(t)}{dt} = Ax(t)+Bu(t), \quad y(t) = Cx(t) \tag{2}$$

where $x(t) = [x_1, x_2, \ldots, x_{15}]^T$ is the state vector, $u(t) = [u_1, u_2, u_3]^T$ is the input vector, $y(t) = [y_1, y_2, y_3]^T$ is the output vector, and the matrices A,B,C have obvious definitions. It is the state-space model (2) which will be utilized in the present paper.

3. THE STATE-VARIABLE FEEDBACK CONTROL PROBLEM

3.1 Statement of the problem

In general u(t) is an m-dimensional and y(t) is a p-dimensional vector. Introducing the linear state feedback control law

$$u(t) = Kx(t) + Nw(t) \tag{3}$$

where w(t) is a new input vector of dimensionality m', and K,N are matrix-valued gains of appropriate dimensions, yields the closed-loop system

$$dx_c/dt = (A+BK)x_c + BNw, \quad y = Cx_c \tag{4}$$

Here it is assumed that $p=m'=m$. The problem under consideration is to calculate the numerical values of the gain matrices K and N which cause the system to be input-output and state variable decoupled, and to possess required dynamic performance.

Mathematically, input-output decoupling (or noninteraction) implies that the input-output transfer matrix is diagonal, whereas state variable decoupling implies that in state space the system is composed by m noninteracting subsystems each one having one input-output pair. The transfer matrix of the closed-loop system (4) is equal to $H_c(s) = C(sI-A-BK)^{-1}BN$. It is well known that $H_c(s)$ is invariant under a nonsingular similarity transformation $x_c' = Qx$. In fact, the transformed closed-loop system is $dx_c'/dt = Q(A+BK)Q^{-1}x_c'+QBNw$, $y= CQ^{-1}x_c'$ and has the transfer matrix $H_c'(s)=CQ^{-1}\left[(sI-Q(A+BK)Q^{-1}\right]^{-1}QBN = H_c(s)$. It is assumed here that the system (2) is transformed in its input-Luenberger canonical form prior to the application of the state feedback control law. In this canonical form the matrices A,B have the form $A = \left[A_{ij}\right]$, $B = B_i$, where the blocks A_{ij} and B_i are

$$A_{ii} = \left[\begin{array}{c|c} 0 & \overbrace{\quad I \quad}^{\sigma_i-1} \\ \hline \multicolumn{2}{c}{\underbrace{\alpha_{ii}^{1}\cdots\alpha_{ii}^{\sigma_i}}_{\sigma_i}} \end{array}\right]\Big\}{\scriptstyle\sigma_i-1} = \left[\begin{array}{c} A_{ii}^{*} \\ \hline A_{ii}^{**} \end{array}\right] \qquad B_i = \left[\begin{array}{c} \overbrace{\qquad 0 \qquad}^{m} \\ \hline 0\ldots\underset{\substack{\uparrow\\ \text{ith position}}}{1}\ldots 0 \end{array}\right]\Big\}{\scriptstyle\sigma_i-1} = \left[\begin{array}{c} B_i^{*} \\ \hline B_i^{**} \end{array}\right] \tag{5}$$

$$A_{ij} = \left[\begin{array}{c} 0 \\ \hline \alpha_{ij}^{1}\cdots\alpha_{ij}^{\sigma_i} \end{array}\right]\Big\}{\scriptstyle\sigma_i-1} = \left[\begin{array}{c} A_{ij}^{*} \\ \hline A_{ij}^{**} \end{array}\right]$$

and the matrix C is not required to have any special form.

By using a new similarity transformation $x_c=M\hat{x}$ to the closed-loop system (4) yields the system

$$d\hat{x} = \hat{A}\hat{x} + \hat{B}w, \quad y=\hat{C}\hat{x} \tag{6}$$

where

$$\hat{A} = M^{-1}(A+BK)M, \quad \hat{B} = M^{-1}BN, \quad \hat{C} = CM \tag{7}$$

Clearly, a Luenberger canonical system with matrices $\hat{A} = \mathrm{diag}\left[\hat{A}_1,\ldots,\hat{A}_m\right]$, $\hat{B}^T = \left[\hat{B}_1^T,\ldots,\hat{B}_m^T\right]$, $\hat{C} = \left[\hat{C}_1,\ldots,\hat{C}_m\right]$, where σ_i are defined in [16],

and

$$\hat{A}_i = \begin{bmatrix} 0 & \vdots & I \\ \hline \hat{\alpha}_{i1} & \cdots & \hat{\alpha}_{i\sigma_i} \end{bmatrix} \sigma_i - 1 , \quad \hat{B}_i = \begin{bmatrix} 0 \\ \hline 0 \ldots 1 \ldots 0 \end{bmatrix} , \quad \hat{C}_j = \begin{bmatrix} 0 \\ \hline \hat{c}_{j1} \cdots \hat{c}_{j\sigma_i} \\ \hline 0 \end{bmatrix} \tag{8}$$

ith position

is input-output decoupled and consists of m decoupled (noninteracting) single input single-output subsystems.

Hence, the combined input-output and state variable decoupling problem under consideration here is reduced to that of selecting K,N, and M so as to satisfy the conditions in (7), with $\hat{A},\hat{B},\hat{C}$ having the form (8). The control of the system poles is accomplished by suitably choosing the parameters $\hat{\alpha}_{ij}$, $j=1,2,\ldots,\sigma_i$, $i=1,2,\ldots,m$, whereas the d.c. gains are controlled by suitably selecting $\hat{c}_{jk}$. Of course it must be noted here that not all of $\hat{c}_{jk}$ are free to be selected arbitrarily, since they are constrained by the zeros of the system under control, i.e. by the structure of system (2).

3.2 Solution of the problem

The pure input-output decoupling problem has been studied by Falb and Wolovich [8], and the canonical decoupling problem has been considered by Gilbert [9]. They show that the necessary and sufficient condition for a matrix pair $\{K,N\}$ to exist such that the state feedback control (3) yields an input-output decoupled closed-loop system is the nonsingularity of the matrix

$$D = \begin{bmatrix} c_1 A^{d_1} B \\ \vdots \\ c_m A^{d_m} B \end{bmatrix} \tag{9}$$

where the indexes $d_i (i=1,2,\ldots,m)$ are defined as

$$d_i = \begin{cases} \min\{ j : c_i A^j B \neq 0, j = 0,1,\ldots,n-1\} \\ n-1, \text{ if } c_i A^j B = 0 \text{ for all } j \end{cases} \tag{10}$$

The present method is based on the fact that the decoupleability of system (2) by the control law (3), as well as the indexes d_i are invariant under a non-singular similarity transformation M. We shall consider two cases: (i) the system (2) has no inherent coupling in the sense of Gilbert 9 i.e. $|D| \neq 0$, and (ii) the system has weak inherent coupling, i.e. $|D| \neq 0$ but $|H(s)| \neq 0$.

No inherent coupling

Decompose the matrix M in blocks M_{ij}, M^*_{ij}, and M^{**}_{ij}, equidimensional with the corresponding blocks A_{ij}, A^*_{ij}, and A^{**}_{ij} in the Luenberger form of (2), and write $M = [M_{ij}]$ and $M^T_{ij} = [M^{*T}_{ij} \mid M^{**T}_{ij}]$. Introducing the matrices

$$\underline{A}^* = [A^*_{ij}],\quad \underline{B}^* = [B^*_i],\quad \underline{M}^* = [M^*_{ij}],\quad \underline{A}^{**} = [A^{**}_{ij}],\quad \underline{B}^{**} = [B^{**}_i],\quad \underline{M}^{**} = [M^{**}_{ij}]$$

the first two conditions in (7) can be grouped as

$$\underline{A}^* + \underline{B}^*K = \underline{M}^*\hat{A}M^{-1},\quad \underline{A}^{**} + \underline{B}^{**}K = \underline{M}^{**}\hat{A}M^{-1},\quad \underline{B}^*N = \underline{M}^*\hat{B},\quad \underline{B}^{**}N = \underline{M}^{**}\hat{B} \qquad (11)$$

Clearly, $\underline{B}^{**} = I$ (unity matrix), and hence the two relations in (11) involving $\underline{B}^{**}$ give

$$K = \underline{M}^{**}\hat{A}M^{-1} - \underline{A}^{**},\quad N = \underline{M}^{**}\hat{B} \qquad (12)$$

Taking into account the fact that $\underline{B}^* = 0$ the other two relations in (11) yield

$$\underline{A}^*M = \underline{M}^*\hat{A},\quad \underline{M}^*\hat{B} = 0 \qquad (13)$$

The third condition (7) together with conditions (13) constitute the set of equations which determine M. If this set of linear algebraic equations has a solution matrix M with $|M| \neq 0$ then (12) provides the desired state feedback matrix pair required.

Weak inherent coupling

In this case prior to applying the feedback law (3) with K,N being given by (12), one uses a $\bar{n}$-dimensional precompensator of the type

$$d\bar{x}/dt = \bar{A}\bar{x} + \bar{B}\bar{u} \quad \text{with} \quad u = F_1\bar{u} + F_2\bar{x} \qquad (14)$$

and obtains an overall precompensated system with state vector x and matrices A,B,C, where

$$\tilde{x} = \begin{bmatrix} x \\ \hline \bar{x} \end{bmatrix},\quad \tilde{A} = \left[\begin{array}{c|c} A & BF_2 \\ \hline 0 & \bar{A} \end{array}\right],\quad \tilde{B} = \begin{bmatrix} BF_1 \\ \hline \bar{B} \end{bmatrix},\quad \tilde{C} = [\, C \mid 0 \,]$$

Of course care must be taken here to transform the matrices A and B in the Luenberger canonical form. Usually, one starts by using a 1-dimensional precompensator (i.e. with $\bar{n}=1$). If there still exists weak inherent coupling one uses a 2-dimensional precompensator, and so on until the resulting system has no inherent coupling.

3.3 Control of zeros

The state variable feedback is adequate if one desires, simultaneously with the decoupling, to control the poles and the d.c. gains only. To control one or more zeros one must use suitable precompensators which implies that the state dimensionality of the overall system is increased. However, when introducing precompensators prior to decoupling special care is required, since even if the uncompensated system has not inherent coupling, the compensated one may have. A first method of overcoming this difficulty was proposed in [7] and is summarized in the following theorem. "Given a system of the type (2) having no inherent coupling, the series compensator $\dot{\bar{x}} = \bar{A}\bar{x}+\bar{B}\bar{u}$, $u=\bar{x}+\bar{E}\bar{u}$, where $\bar{A},\bar{B}$ and $\bar{E}$ are diagonal matrices of dimensions mxm and $\bar{B}$ is nonsingular, does not introduce inherent coupling if (i) $\bar{E}=0$ or (ii) $\bar{E}$ is nonsingular". This theorem implies that to preserve decoupling when adding precompensators every compensator must involve both a pole and a zero or only a pole. The drawback of this method is that there is the possibility of loosing zeros. This is avoided if the system is decoupled prior to the introduction of the precompensators, since the addition of series precompensators to a system that is already decoupled does not influence decoupling.

The second method of introducing precompensators is based on exactly this observation and was also proposed in [7] in the form of the following theorem. "Suppose that the kth decoupled subsystem of a system decoupled by state feedback, or by other means, has z_k zeros and n_k poles, whereas the kth precompensator has z_k^c zeros and n_k^c poles. Then the $n_k+n_k^c$ poles of each augmented subsystem can be controlled by state feedback, but the $z_k+z_k^c$ zeros are not affected by state feedback". In fact, by reordering the state variables, the kth precompensated subsystem has the transfer function $F_k(s)\ F_k^c(s)$, where $F_k(s)$ is the transfer function of the kth subsystem of the original decoupled uncompensated system, and $F_k^c(s)$ is the transfer function of the kth cascade compensator. Now, from single-input single-output state variable control theory it is known that all $n_k+n_k^c$ poles are controlled, but the zeros are fixed to be the zeros of $F_k(s)F_k^c(s)$. Thus, combining the results of section 3.2 with the second method outlined here, the following control design procedure is proposed.

Step 1: Transform the system under control into its Luenberger input canonical form.

Step 2: Specify the Luenberger canonical decoupled model by using the desired poles and d.c. gains.

Step 3: Compute the required state feedback pair $\{K,N\}$ and the similarity transformation M.

Step 4: Introduce cascade precompensators to the decoupled single output subsystems in order to control the zeros as desired.

4. APPLICATION EXAMPLES

Example 1

As a first example consider a special case of the coupled-core reactor model (1), namely one with identically-coupled identical cores, in which the delayed neutrons and the control rod dynamics are neglected. By a convenient relabelling of the state variables the state equations of this system take the form

$$
\begin{aligned}
\dot{x}_1 &= -\frac{2D}{\tau}x_1 - \frac{\alpha n_o}{\tau}x_2 + \frac{D}{\tau}x_3 + \frac{D}{\tau}x_5 + \frac{n_o}{\tau}u_1, \quad \dot{x}_2 = kx_1 - mx_2 \\
\dot{x}_3 &= \frac{D}{\tau}x_1 - \frac{2D}{\tau}x_3 - \frac{\alpha n_o}{\tau}x_4 + \frac{D}{\tau}x_5 + \frac{n_o}{\tau}u_2, \quad \dot{x}_4 = kx_3 - mx_4 \qquad (15) \\
\dot{x}_5 &= \frac{D}{\tau}x_1 + \frac{D}{\tau}x_3 - \frac{2D}{\tau}x_5 \frac{-\alpha n_o}{\tau}x_6 + \frac{n_o}{\tau}u_3, \quad \dot{x}_6 = kx_5 - mx_6
\end{aligned}
$$

where D, τ, m, k, and n_o are the common neutron coupling coefficient, effective generation time, heat removal coefficient, power-temperature proportionality constant, and steady state power level, respectively. Here x_1, x_3, x_5 are the neutron power levels, and x_2, x_4, x_6 the temperatures in cores 1,2,3 respectively. The parameter values are $D=0.1$, $\tau=0.1$ sec, $k=10^{-5}$, $n_o=10^5$W, $\alpha=10^{-3}$/deg and $m=10^{-2}\text{sec}^{-1}$. Hence, the state equations in (15) take the form

$$
\begin{aligned}
\dot{x}_1 &= 2x_1 - 10^3x_2 + x_3 + x_5 + 10^6u_1, \quad \dot{x}_2 = 10^{-5}x_1 - 10^{-2}x_2 \\
\dot{x}_3 &= x_1 - 2x_3 - 10^3x_4 + x_5 + 10^6u_2, \quad \dot{x}_4 = 10^{-5}x_3 - 10^{-2}x_4 \\
\dot{x}_5 &= x_1 + x_3 - 2x_5 - 10^3x_6 + 10^6u_3, \quad \dot{x}_6 = 10^{-5}x_5 - 10^{-2}x_6
\end{aligned}
$$

The measured outputs are $y_1=x_1, y_2=x_3$ and $y_3=x_5$. These equations can be written in the form (2) with matrices

$$A_o = \left[\begin{array}{cc|cc|cc} -2 & -10^3 & 1 & 0 & 1 & 0 \\ 10^{-5} & -10^{-2} & 0 & 0 & 0 & 0 \\ \hline 1 & 0 & -2 & -10^3 & 1 & 0 \\ 0 & 0 & 10^{-5} & -10^{-2} & 0 & 0 \\ \hline 1 & 0 & 1 & 0 & -2 & -10^{-3} \\ 0 & 0 & 0 & 0 & 10^{-5} & -10^{-2} \end{array}\right] \quad B_o^T = \left[\begin{array}{cc|cc|cc} 10^6 & 0 & 0 & 0 & 0 & 0 \\ 0 & 0 & 10^6 & 0 & 0 & 0 \\ 0 & 0 & 0 & 0 & 10^6 & 0 \end{array}\right] \quad C_o = \left[\begin{array}{cc|cc|cc} 1 & 0 & 0 & 0 & 0 & 0 \\ 0 & 0 & 1 & 0 & 0 & 0 \\ 0 & 0 & 0 & 0 & 1 & 0 \end{array}\right] \tag{16}$$

Prior to applying the control techniques the decoupleability of the system is checked. Since

$$D = \begin{bmatrix} c_1 B \\ c_2 B \\ c_3 B \end{bmatrix}_0 = 10^6 \begin{bmatrix} 1 & 0 & 0 \\ 0 & 1 & 0 \\ 0 & 0 & 1 \end{bmatrix}$$

for which $|D| \neq 0$, there is no inherent coupling, and so the system is directly decoupleable without the need to introduce any precompensator. To find the similarity matrix Q which transforms the system in the Luenberger form (5), we construct a matrix

$$L = \left[b_{o1},\ A_o b_{o1}, \ldots, A_o^{\sigma_1 - 1} b_{o1}, b_{02}, \ldots, A_o^{\sigma_2 - 1} b_{02}, \ldots, A_o^{\sigma_m - 1} b_{om} \right]$$

consisting of n linearity independent columns of the controllability matrix $P = \left[B_o, A_o B_o, \ldots, A_o^{n-1} B_o \right]$. Here $\sigma_1 = \sigma_2 = \sigma_3 = 2$, and so

$$L = \begin{bmatrix} 10^6 & -2\text{x}10^6 & 0 & 10^6 & 0 & 10^6 \\ 0 & 10 & 0 & 0 & 0 & 0 \\ 0 & 10^6 & 10^6 & -2\text{x}10^6 & 0 & 10^6 \\ 0 & 0 & 0 & 10 & 0 & 0 \\ 0 & 10^6 & 0 & 10^6 & 10^6 & -2\text{x}10^6 \\ 0 & 0 & 0 & 0 & 0 & 10 \end{bmatrix}$$

Setting $\varepsilon_o = 0$, $\varepsilon_k = \sum_{i=1}^{k} \sigma_i$ $(k=1,2,\ldots,m)$, and letting γ_k be the ε_kth row of L^{-1}, the matrix Q is given by

$$Q = \begin{bmatrix} \gamma_1 \\ \vdots \\ \gamma_1 A_o^{\sigma_1 - 1} \\ \gamma_2 \\ \vdots \\ \gamma_m A_o^{\sigma_m - 1} \end{bmatrix}$$

In the present case $\varepsilon_o=0, \varepsilon_1=\sigma_1=2$, $\varepsilon_2=\sigma_1,+\sigma_2=4$, $\varepsilon_3=\sigma_1+\sigma_2+\sigma_3=6$, and

$$L^{-1} = \begin{bmatrix} 10^6 & 2x10^{-1} & 0 & 10^{-1} & 0 & -10^{-1} \\ 0 & 10^{-1} & 0 & 0 & 0 & 0 \\ 0 & -10^{-1} & 10^6 & 2x10^{-1} & 0 & -10^{-1} \\ 0 & 0 & 0 & 10^{-1} & 0 & 0 \\ 0 & -10^{-1} & 0 & -10^{-1} & 10^{-6} & 2x10^{-1} \\ 0 & 0 & 0 & 0 & 0 & 10^{-1} \end{bmatrix} \begin{matrix} \\ \leftarrow \varepsilon_1\text{th row} \\ \\ \leftarrow \varepsilon_2\text{th row} \\ \\ \leftarrow \varepsilon_3\text{th row} \end{matrix}$$

Hence

$$Q = \begin{bmatrix} \gamma_1 \\ \gamma_1 A_o \\ \gamma_2 \\ \gamma_2 A_o \\ \gamma_3 \\ \gamma_3 A_o \end{bmatrix} = 10^{-1} \begin{bmatrix} 0 & 1 & 0 & 0 & 0 & 0 \\ 10^{-5} & -10^{-2} & 0 & 0 & 0 & 0 \\ 0 & 0 & 0 & 1 & 0 & 0 \\ 0 & 0 & 10^{-5} & -10^{-2} & 0 & 0 \\ 0 & 0 & 0 & 0 & 0 & 1 \\ 0 & 0 & 0 & 0 & 10^{-5} & -10^{-2} \end{bmatrix}$$

$$Q^{-1} = 10 \begin{bmatrix} 10^3 & 10^5 & 0 & 0 & 0 & 0 \\ 1 & 0 & 0 & 0 & 0 & 0 \\ 0 & 0 & 10^3 & 10^5 & 0 & 0 \\ 0 & 0 & 1 & 0 & 0 & 0 \\ 0 & 0 & 0 & 0 & 10^3 & 10^5 \\ 0 & 0 & 0 & 0 & 1 & 0 \end{bmatrix}$$

The matrices of the input Luenberger canonical form are found to be

$$A=QA_oQ^{-1} = \left[\begin{array}{cc|cc|cc} 0 & 1 & 0 & 0 & 0 & 0 \\ -3x10^{-2} & -2.001 & 10^{-2} & 1 & 10^{-2} & 1 \\ \hline 0 & 0 & 0 & 1 & 0 & 0 \\ 10^{-2} & 1 & -3x10^{-2} & -2.001 & 10^{-2} & 1 \\ \hline 0 & 0 & 0 & 0 & 0 & 1 \\ 10^{-2} & 1 & 10^{-2} & 1 & -3x10^{-2} & -2.001 \end{array}\right] \tag{17}$$

$$B=QB_o= \left[\begin{array}{c|c|c} 0 & 0 & 0 \\ 1 & 0 & 0 \\ \hline 0 & 0 & 0 \\ 0 & 1 & 0 \\ \hline 0 & 0 & 0 \\ 0 & 0 & 1 \end{array}\right], \quad C=C_oQ^{-1} = \left[\begin{array}{cc|cc|cc} 10^4 & 10^6 & 0 & 0 & 0 & 0 \\ \hline 0 & 0 & 10^4 & 10^6 & 0 & 0 \\ \hline 0 & 0 & 0 & 0 & 10^4 & 10^6 \end{array}\right]$$

The above suggest that the canonical decoupled model appropriate in the present case has the matrices

$$\hat{A}=\left[\begin{array}{cc|cc|cc} 0 & 1 & & & & \\ \hat{a}_{11} & \hat{a}_{12} & \multicolumn{2}{c|}{O} & \multicolumn{2}{c}{O} \\ \hline & & 0 & 1 & & \\ \multicolumn{2}{c|}{O} & \hat{a}_{21} & \hat{a}_{22} & \multicolumn{2}{c}{O} \\ \hline & & & & 0 & 1 \\ \multicolumn{2}{c|}{O} & \multicolumn{2}{c|}{O} & \hat{a}_{31} & \hat{a}_{32} \end{array}\right], \hat{C}=\left[\begin{array}{cc|cc|cc} \hat{c}_{11} & \hat{c}_{12} & \multicolumn{2}{c|}{O} & \multicolumn{2}{c}{O} \\ \hline \multicolumn{2}{c|}{O} & \hat{c}_{21} & \hat{c}_{22} & \multicolumn{2}{c}{O} \\ \hline \multicolumn{2}{c|}{O} & \multicolumn{2}{c|}{O} & \hat{c}_{31} & \hat{c}_{32} \end{array}\right] \quad (18)$$

$$\hat{B}=B$$

The canonical decoupling conditions are $\underline{A}^*M=\underline{M}^*\hat{A}$, $M^*\hat{B}=0$ and $CM=\hat{C}$ where $M=\left[\mu_{ij}\right]$ is a 6x6 matrix, and

$$\underline{A}^* = \begin{bmatrix} 0 & 1 & 0 & 0 & 0 & 0 \\ 0 & 0 & 0 & 1 & 0 & 0 \\ 0 & 0 & 0 & 0 & 0 & 1 \end{bmatrix}, \quad M^* = \begin{bmatrix} \mu_{11}\mu_{12}\cdots\mu_{16} \\ \mu_{31}\mu_{32}\cdots\mu_{36} \\ \mu_{51}\mu_{52}\cdots\mu_{56} \end{bmatrix}$$

Solving these conditions for $M(|M|)\neq 0)$ one obtains $M=10^{-6}\text{diag}\left[\hat{c}_{12},\hat{c}_{12},\hat{c}_{22},\hat{c}_{22},\hat{c}_{32},\hat{c}_{32}\right]$ subject to the constrains $\hat{c}_{i2}=10^2\hat{c}_{i1}$ (i=1,2,3). The inverse matrix M^{-1} equal to $M^{-1}=10^6\text{diag}\left[\hat{c}_{12}^{-1}, \hat{c}_{12}^{-1}, \hat{c}_{22}^{-1}, \hat{c}_{22}^{-1}, \hat{c}_{32}^{-1}, \hat{c}_{32}^{-1}\right]$. The required feedback gain matrices are given by $K=\underline{M}^{**}\hat{A}M^{-1}-\underline{A}^{**}$ and $N=\underline{M}^{**}\hat{B}$, where

$$\underline{A}^{**} = \begin{bmatrix} 3\times10^{-2} & -2.001 & 10^{-2} & 1 & 10^{-2} & 1 \\ 10^{-2} & 1 & -3\times10^{-2} & -2.001 & 10^{-2} & 1 \\ 10^{-2} & 1 & 10^{-2} & 1 & -3\times10^{-2} & -2.001 \end{bmatrix}$$

$$\underline{M}^{**} = \begin{bmatrix} \mu_{21}\cdots\mu_{26} \\ \mu_{41}\cdots\mu_{46} \\ \mu_{61}\cdots\mu_{66} \end{bmatrix} = 10^{-6}\begin{bmatrix} 0 & \hat{c}_{12} & 0 & 0 & 0 & 0 \\ 0 & 0 & 0 & \hat{c}_{22} & 0 & 0 \\ 0 & 0 & 0 & 0 & 0 & \hat{c}_{32} \end{bmatrix}$$

Hence

$$K = \left[\begin{array}{cc|cc|cc} \hat{a}_{11} & \hat{a}_{12} & \multicolumn{2}{c|}{O} & \multicolumn{2}{c}{O} \\ \hline \multicolumn{2}{c|}{O} & \hat{a}_{21} & \hat{a}_{22} & \multicolumn{2}{c}{O} \\ \hline \multicolumn{2}{c|}{O} & \multicolumn{2}{c|}{O} & \hat{a}_{31} & \hat{a}_{32} \end{array}\right] - A^{**}, \quad N=10^{-6}\begin{bmatrix} \hat{c}_{12} & & O \\ & \hat{c}_{22} & \\ O & & \hat{c}_{32} \end{bmatrix} \quad (19)$$

The feedback matrix gains for the original system (16) are $N_o=N$, and

$$K_o = KQ = 10^{-1}\left[\begin{array}{c|c|c|c|c|c} 10^{-5}\alpha_{12} & \alpha_{11}-10^{-2}\alpha_{12} & 0 & 0 & 0 & 0 \\ \hline 0 & 0 & 10^{-5}\alpha_{22} & \alpha_{21}-10^{-2}\alpha_{22} & 0 & 0 \\ \hline 0 & 0 & 0 & 0 & 10^{-5}\alpha_{32} & \alpha_{31}-10^{-2}\alpha_{32} \end{array}\right]$$

$$+\left[\begin{array}{c|c|c|c|c|c} 2.001\text{x}10^{-6} & 0.99\text{x}10^{-3} & -10^{-6} & 0 & -10^{-6} & 0 \\ \hline -10^{-6} & 0 & 2.001\text{x}10^{-6} & 0.99\text{x}10^{-3} & -10^{-6} & 0 \\ \hline -10^{-6} & 0 & -10^{-6} & 0 & 2.001\ \text{x}10^{-6} & 0.99\text{x}10^{-3} \end{array}\right]$$

The closed-loop transfer function is found to be

$$H_c(s) = C(sI-A-BK)^{-1}BN = C_o(sI-A_o-B_oK_o)^{-1}B_oN_o$$

$$= \text{diag}\left[\frac{\hat{c}_{12}(s+10^{-2})}{s^2-\hat{\alpha}_{12}s-\hat{\alpha}_{11}},\ \frac{\hat{c}_{22}(s+10^{-2})}{s^2-\hat{\alpha}_{22}s-\hat{\alpha}_{21}},\ \frac{\hat{c}_{32}(s+10^{-2})}{s^2-\hat{\alpha}_{32}s-\hat{\alpha}_{31}}\right] \quad (20)$$

We observe that the resulting closed-loop system is composed by three noninteracting second-order systems. Clearly, we can control the poles and the d.c. gains of each subsystem by selecting the parameters $\hat{\alpha}_{ij}$, $i=1,2,3$, $j=1,2$, and $\hat{c}_{12}$, $i=1,2,3$ of the canonically decoupled model (18). The zeros however are fixed at $s=-10^{-2}$ and cannot be controlled by the state feedback (19). The most important advantage of this canonically decoupled system achieved is that besides the input-output decoupling one also has decoupling between states belonging to different subsystems (cores). This implies that the three subsystems composing the whole system are completely noninteracting and so one can control the output (power level) as well as the state responses of each subsystem independently.

Now, let us examine the problem of controlling the zeros. Suppose for example that each subsystem in the canonically decoupled model must have two poles at $s_{1,2}=(-3+j\sqrt{3})/2$, a zero at $s=0$, and a d.c. gain equal to $10^6/3$. This means that the desired transfer function of each subsystem is $10^6/(s^2+3s+3)$. The denominator implies that each subsystem has an undamped natural frequency $\omega_n=\sqrt{3}$ and a damping ratio $\zeta=3/2\omega_n = \sqrt{3/2}$. Now, since the zero at $s=-10^{-2}$ of each subsystem cannot be controlled by state feedback, a cascade precompensator must be added to each subsystem with a pole at $s=-10^{-2}$ and a zero at the desired position $s=0$. Hence, one obtains the equality

$$\frac{1}{s+10^{-2}}\left[\frac{\hat{c}_{i2}(s+10^{-2})}{s^2-\hat{\alpha}_{i2}s-\hat{\alpha}_{i1}}\right] = \frac{10^6}{s^2+3s+3} \qquad (i=1,2,3)$$

from which it follows that $\hat{c}_{i2}=10^6$ and a $\hat{\alpha}_{i1}=\hat{\alpha}_{i2}=-3$ (i=1,2,3). In state space form each subsystem has the equations

$$\frac{d}{dt}\begin{bmatrix}x_1\\x_2\end{bmatrix}_i=\begin{bmatrix}0&1\\\hat{\alpha}_{i1}&\hat{\alpha}_{i2}\end{bmatrix}\begin{bmatrix}x_1\\x_2\end{bmatrix}_i+\begin{bmatrix}0\\1\end{bmatrix}u_i,\ y_i=\begin{bmatrix}\hat{c}_{i1},10^2\hat{c}_{i1}\end{bmatrix}\begin{bmatrix}x_1\\x_2\end{bmatrix}_i$$

and each cascade precompensator the equations $\dot{\bar{x}}_i=-10^{-2}\bar{x}_i+\bar{u}_i,\ u_i=\bar{x}_i$. Hence the state equations of each precompensated canonically decoupled subsystem are

$$\frac{d\tilde{x}_i}{dt}=\left[\begin{array}{cc|c}0&1&0\\\hat{\alpha}_{i1}&\hat{\alpha}_{i2}&1\\\hline 0&0&-10^{-2}\end{array}\right]\tilde{x}_i+\left[\begin{array}{c}0\\0\\\hline 1\end{array}\right]\bar{u}_i,\ \tilde{x}_i=\left[\begin{array}{c}\begin{bmatrix}x_1\\x_2\end{bmatrix}_i\\\hline \bar{x}_i\end{array}\right]$$

$$y_i=\left[\begin{array}{cc|c}\hat{c}_{i1}, & 10^2\hat{c}_{i1} & 0\end{array}\right]\tilde{x}_i$$

It is easy to verify that the transfer function of the preceding subsystem is $y_i(s)/\bar{u}_i(s)=\hat{c}_{i2}/(s^2-\hat{\alpha}_{i2}s-\hat{\alpha}_{i1})$ as desired.

Example 2

The second example is a two coupled-core reactor system with control rod dynamics:

$$\dot{x}_1=-\frac{D}{\tau}x_1-\frac{\alpha n_o}{\tau}x_2+\frac{D}{\tau}x_3+\frac{n_o}{\tau}x_5,\ \dot{x}_2=kx_1-mx_2,\ \dot{x}_5=-\Theta x_5+\mu u_1$$

$$\dot{x}_3=\frac{D}{\tau}x_1-\frac{D}{\tau}x_3-\frac{\alpha n_o}{\tau}x_4+\frac{n_o}{\tau}x_6,\ \dot{x}_4=kx_3-mx_4,\ \dot{x}_6=-\Theta x_6+\mu u_2 \tag{21}$$

$$y_1=x_1,\quad y_2=x_3$$

where $\Theta=10\ \text{sec}^{-1}$, $\mu=1$ and the remaining parameters have the same values as in example 1. Introducing the parameter values one finds that the matrices A_o,B_o and C_o of the state space description are

$$A_o=\begin{bmatrix}1&-10^3&1&0&10^6&0\\10^{-5}&-10^{-2}&0&0&0&0\\1&0&-1&-10^3&0&10^6\\0&0&10^{-5}&-10^{-2}&0&0\\0&0&0&0&-10&0\\0&0&0&0&0&-10\end{bmatrix},\quad B_o^T=\begin{bmatrix}0&0&0&0&1&0\\0&0&0&0&0&1\end{bmatrix},\quad C_o=\begin{bmatrix}1&0&0&0&0&0\\0&0&1&0&0&0\end{bmatrix} \tag{22}$$

The system has no inherent coupling since

$$D = \begin{bmatrix} c_1 AB \\ c_2 AB \end{bmatrix}_o = 10^6 \begin{bmatrix} 1 & 0 \\ 0 & 1 \end{bmatrix}, \quad |D| \neq 0$$

The similarity matrix Q and its inverse are found to be

$$Q = \begin{bmatrix} 0 & 10^{-1} & 0 & 0 & 0 & 0 \\ 10^{-6} & -10^{-3} & 0 & 0 & 0 & 0 \\ -1.01\text{x}10^{-6} & -0.99\text{x}10^{-3} & 10^{-6} & 0 & 1 & 0 \\ 0 & 0 & 0 & 10^{-1} & 0 & 0 \\ 0 & 0 & 10^{-6} & -10^{-3} & 0 & 0 \\ 10^{-6} & 0 & -1.01\text{x}10^{-6} & -0.99\text{x}10^{-3} & 0 & 1 \end{bmatrix}$$

$$Q^{-1} = \begin{bmatrix} 10^4 & 10^6 & 0 & 0 & 0 & 0 \\ 10 & 0 & 0 & 0 & 0 & 0 \\ 0 & 0 & 0 & 10^4 & 10^6 & 0 \\ 0 & 0 & 0 & 10 & 0 & 0 \\ 2\text{x}10^{-2} & 1.01 & 1 & -10^{-2} & -1 & 0 \\ -10^{-2} & -1 & 0 & 2\text{x}10^{-2} & 1.01 & 1 \end{bmatrix}$$

and the input-Luenberger form of (22) is found to be

$$A = QA_o^{-1} = \left[\begin{array}{ccc|ccc} 0 & 1 & 0 & 0 & 0 & 0 \\ 0 & 0 & 1 & 0 & 0 & 0 \\ -2\text{x}10^{-1} & -10.12 & -11.01 & 10^{-1} & 10.01 & 1 \\ \hline 0 & 0 & 0 & 0 & 1 & 0 \\ 0 & 0 & 0 & 0 & 0 & 1 \\ 10^{-1} & 10.01 & 1 & -2\text{x}10^{-1} & -10.12 & -11.01 \end{array}\right]$$

$$B^T = (QB_o)^T = \left[\begin{array}{ccc|ccc} 0 & 0 & 1 & 0 & 0 & 0 \\ 0 & 0 & 0 & 0 & 0 & 1 \end{array}\right], \quad C = C_oQ^{-1} = \left[\begin{array}{ccc|ccc} 10^4 & 10^6 & 0 & 0 & 0 & 0 \\ \hline 0 & 0 & 0 & 10^4 & 10^6 & 0 \end{array}\right] \tag{23}$$

The matrices (23) suggest the following canonically decoupled Luenberger model

$$A=\left[\begin{array}{ccc|ccc} 0 & 1 & 0 & & & \\ 0 & 0 & 1 & & 0 & \\ \hat{a}_{11} & \hat{a}_{12} & \hat{a}_{13} & & & \\ \hline & & & 0 & 1 & 0 \\ & 0 & & 0 & 0 & 1 \\ & & & \hat{a}_{21} & \hat{a}_{22} & \hat{a}_{23} \end{array}\right] \qquad \hat{C}=\left[\begin{array}{ccc|ccc} \hat{c}_{11} & 10^2\hat{c}_{11} & 0 & 0 & 0 & 0 \\ \hline 0 & 0 & 0 & \hat{c}_{21} & 10^2\hat{c}_{21} & 0 \end{array}\right]$$

$$\hat{B}=B,\ \hat{c}_{12}=10^2\hat{c}_{11},\ \hat{c}_{22}=10^2\hat{c}_{21}$$

In the present case the decoupling conditions give $M=10^{-4}\mathrm{diag}\left[\hat{c}_{11},\hat{c}_{11},\hat{c}_{11},\hat{c}_{21},\hat{c}_{21},\hat{c}_{21}\right]$ and the decoupling matrix gain pair $\{K,N\}$ is found to be

$$K=\left[\begin{array}{ccc|ccc} \hat{a}_{11} & \hat{a}_{12} & \hat{a}_{13} & 0 & 0 & 0 \\ \hline 0 & 0 & 0 & \hat{a}_{21} & \hat{a}_{22} & \hat{a}_{23} \end{array}\right]$$

$$-\left[\begin{array}{ccc|ccc} -2\times10^{-1}, & -10.12, & -11.01 & 10^{-1}, & 10.01, & 1 \\ \hline 10^{-1}, & 10.01, & 1 & -2\times10^{-1} & -10.12, & -11.01 \end{array}\right]$$

$$N=10^{-4}\begin{bmatrix} \hat{c}_{11} & 0 \\ 0 & \hat{c}_{21} \end{bmatrix}$$

Finally, the closed-loop transfer matrix function is

$$H_c(s)=C(sI-A-BK)^{-1}BN=\mathrm{diag}\left[\frac{\hat{c}_{12}(s+10^{-2})}{s^3-\hat{a}_{13}s^2-\hat{a}_{12}s-\hat{a}_{11}},\ \frac{\hat{c}_{22}(s+10^{-2})}{s^3-\hat{a}_{23}s^2-\hat{a}_{22}s-\hat{a}_{21}}\right]$$

which can be treated for pole, d.c. gain, and zero control as in Example 1.

5. CONCLUSIONS

The technique presented in this paper is applicable to coupled nuclear reactor systems which, owing to the large number of states of each core and the large number of cores possible, belong to the class of large multivariable systems. Spatially-distributed-core reactors can be treated using this technique by subdividing the core into a number of coupled subcores. Actually, the results of this paper constitute the first part of a work aiming to apply the state feedback approach to complete power reactor systems [20]. This will first require an extension of the method to systems with time delays in the state and/or the control variables. An extension of the method for treating the inputs and outputs in groups is also under investigation. When the system involves some states which are not accessible to direct measurements one must use output feedback or generate the unknown state from the measured states upon which they are dependent. This problem is essentially open. A general FORTRAN program which will provide the solution for systems with large matrices is being developed. The method is general and can be used not only in reactor systems but in all cases where simultaneous input-output and state variable decoupling, pole control, d.c. gain control, and zero control is desired.

It is noted that one may reverse the order of steps 3 and 4 in the algorithm of sec. 3.3, i.e. introduce the precompensator prior to decoupling, by defining the matrices of the precompensated input-Luenberger canonical system under control as

$$A' = \left[\begin{array}{c|c} A & 0 \\ \hline 0 & 0 \end{array}\right], \quad B' = \left[\begin{array}{c|c} B & 0 \\ \hline 0 & 1 \end{array}\right], \quad C' = \left[\begin{array}{c|c} C & C^{11} \\ \hline C^{21} & C^{22} \end{array}\right]$$

where the elements of C^{11}, C^{21}, C^{22}, except of the requirement to be selected such that the system has not inherent coupling, are otherwise arbitrary. Of course in this case there is again the possibility of loosing or introducing undesired zeros, and so the algorithm must be preferred in the order step 1, step 2, step 3, step 4.

REFERENCES

1. WEAVER, L. and VANASSE, R. "State Variable Feedback Control of Multiregion Reactors", Nucl.Sci.Eng., 29, 264-271 (1967).
2. WEAVER,L., "Reactor Dynamics and Control: State-Space Techniques", Chapters 8 and 9 (American Elsevier, New York, 1968).
3. HERRING, J., SCHULTZ, D., WEAVER, L. and VANASSE R., "Design of Linear and Nonlinear Control Systems via State Variable Feedback with Application in Nuclear Reactors Control", Engineering Experiment Station Report, University of Arizona, Tucson (Feb. 1967).
4. MELSA J. and SCHULTZ D., "Linear Control Systems" Chapters 3 and 9 (McGraw-Hill, New York, 1969).
5. SLIVINSKY, C. and WEAVER, L., "Reactor Control Using a New Multivariable Design Technique", Nucl.Sci.Eng., 37, 163-166 (1969).
6. RAJU G., and STELZER M., "Noninteracting Control System Design for a Coupled Core Nuclear Reactor", IEEE Trans.Nucl.Sci.Eng., 541-548 (1970).
7. SLIVINSKY, C., SCHULTZ,D. and WEAVER, L., "State Variable Feedback and Series Compensation of Multivariable Systems", Nucl.Sci.Eng., 38, 125-129 (1970).
8. FALB, P. and WOLOVICH, W., "Decoupling in the Design and Synthesis of Multivariable Control Systems", IEEE Trans.Auto.Control, AC-12, 651-659 (1967).
9. GILBERT, E. "The Decoupling of Multivariable Systems by State Feedback", SIAM J. Control, 7, 50-63 (1969).
10. TZAFESTAS, S. and PARASKEVOPOULOS, P., "On the Decoupling of Multivariable Control Systems with Time Delays", Int.J. Control, 17, 405-415 (1973).
11. PARASKEVOPOULOS, P. and TZAFESTAS, S., "New Results in Feedback Modal-Controller Design", Int.J.Control, 21, 911-928 (1975).

12. BROCKETT, R. "Poles, Zeros and Feedback: State Space Interpretation", IEEE Trans.Auto. Control, AC-10, 129-135 (1965).
13. TZAFESTAS, S. and PARASKEVOPOULOS P., "On the Exact Model-Matching Controller Design", Proc. 1974 IEEE Conf. Decision and Control, Phoenix, Arizona (1974), Also to appear in IEEE Trans.Auto Control (1975).
14. WOLOVICH,W. and FALB, P., "On the Structure of Multivariable Systems", SIAM J. Control, 7, 437-451 (1969).
15. PARASKEVOPOULOS, P., "On the Model Matching of Multivariable Systems", Doctoral Thesis, Faculty of Engineering, University of Patras (1975).
16. LUENBERGER, D., "Canonical Forms for Linear Multivariable Systems", IEEE Trans. Auto. Control, AC-12, 290-293 (1967).
17. WANG, S. and DESOER, C., "The Exact Model Matching of Linear Multivariable Systems", IEEE Trans.Auto Control, AC-17, 491-497 (1972).
18. WOLOVICH, W., "The Use of State Feedback for Exact Model-Matching", SIAM J. Control, 10, 512-523 (1972).
19. LANDAU, I., "A Survey of Model Reference Adaptive Techniques-Theory and Applications", Automatica, 10, 353-379 (1974).
20. DUNCOMBE, E. and RATHBONE, D., "Optimization of the Response of a Nuclear Reactor Plant to Changes in Demand", IEEE Trans. Auto Control, AC-14, 277-283 (1969).
21. TZAFESTAS, S., "Indirect model matching technique for multicontroller systems, Electronics Letters, 11, 353-354 (1975).

UN ALGORITHME DE MINIMISATION DE FONCTIONS CONVEXES AVEC OU SANS CONTRAINTES " L'ALGORITHME D'ÉCHANGES"

C.CARASSO

Université de Saint-Etienne

23, rue du Dr Paul Michelon

42100 SAINT-ETIENNE FRANCE

Résumé.

On étudie un algorithme permettant de trouver le minimum d'une forme linéaire sur une intersection finie ou infinie de demi-espaces. Cet algorithme s'applique à une large classe de problèmes d'optimisation convexe (programmation mathématique, théorie de l'approximation,...)

I. INTRODUCTION

On s'interesse à la résolution d'un problème d'optimisation dans $\mathbb{R}^n$ du type :

(1-1) (P) $\alpha = \mathrm{Inf}[(x|z) \mid x \in C \cap V_t]$

avec $C = \{x \in \mathbb{R}^n \mid c(s) \leqq (x|a(s)) \quad \forall s \in S\}$, dans lequel z est un élément fixe de $\mathbb{R}^n$, S est un compact quelconque, c et a deux applications continues de S dans respectivement $\mathbb{R}$ et $\mathbb{R}^n$. L'ensemble V_t désigne le translaté d'un sous espace vectoriel V de dimension p de $\mathbb{R}^n$. On supposera que V_t est de la forme :

$$V_t = \{x \in \mathbb{R}^n \mid (x|u_i) = c_i \quad i = p+1,\dots,n\}.$$

La plupart des problèmes d'optimisation convexe peuvent se mettre sous la forme (P). Nous verrons au paragraphe 3 des exemples de problème mis sous cette forme. La forme (P) n'est pas la plus générale; on peut, en utilisant systèmatiquement la dualité et l'analyse convexe, étudier le problème (P) en se plaçant dans un espace vectoriel topologique localement convexe X en dualité avec X'. Pour une étude de ce type nous renvoyons à [3] et [7].

Nous étudions un algorithme appelé "algorithme d'échange" qui permet, moyennant des hypothèses sur (P) et sur le déroulement de l'algorithme, de construire une suite d'éléments x^k de V_t tels que :

$$\lim_{k\to\infty}(x^k|z) = \alpha \text{ et } \max_{s\in S}[c(s)-(x^k|a(s))] \leqq \eta^k \qquad \eta^k>0$$

et $\lim_{k\to\infty} \eta^k = 0$.

Lorsqu'on s'interesse à la recherche du meilleur approximant dans un espace vectoriel normé, l' "algorithme d'échange" est l'algorithme de REMES généralisé ([6]). La convergence est alors démontrée sans faire l'hypothèse classique (et forte !) de "condition de HAAR vérifiée" (voir [1]).

II. DESCRIPTION DE L'ALGORITHME D'ECHANGE

On suppose que α est fini et que z n'appartient pas à $\mathcal{L}(u_{p+1},\ldots,u_n) = V$.

On fera les hypothèses suivantes :

H1 Il existe $x^* \in V_t$ tel que $c(s) < (x^* | a(s))$ pour tout $s \in S$

H2 Il existe $u_1,\ldots,u_p \in a(S)$ et $\rho_1,\ldots,\rho_n \in \mathbb{R}$ avec $\rho_i > o$ $i=1,\ldots,p$ tels que :

i) $\mathcal{L}(u_1,\ldots,u_n) = \mathbb{R}^n$

ii) $\sum_{i=1}^{n} \rho_i u_i = z$

En notant $\mathcal{L}(u_1,\ldots,u_n)$ le sous-espace vectoriel engendré par $u_1,\ldots,u_n$.

Passage de l'itération k à l'itération k+1

Soient β et ε des scalaires positifs donnés et "petits".

On suppose qu'à l'itération k on dispose de p éléments $u_1,\ldots,u_p \in a(S)$, de p éléments $c_1,\ldots,c_p \in \mathbb{R}$ tels que $u_i = a(s_i)$ et $c_i = c(s_i)$ $(i=1,\ldots,p)$ et de n coefficients $\rho_1,\ldots,\rho_n \in \mathbb{R}$ tels que :

$$\sum_{i=1}^{n} \rho_i u_i = z \qquad \text{avec } \rho_i > o \quad i=1,\ldots,p \tag{2-1}$$

et d'une précision η $(o \leq \eta)$.

Pour passer à l'itération k+1 on franchit les étapes suivantes :

étape 1 : Déterminer $x \in \mathbb{R}^n$ tel que :

$$(x | u_i) = c_i \qquad i=1,\ldots,n \tag{2-2}$$

étape 2 : Faire $\eta := \eta/2$

étape 3 : Calculer $\text{APPUI}(x,\eta) = (d,\bar{u},\bar{c}) \in \mathbb{R} \times \mathbb{R}^n \times \mathbb{R}$

étape 4 : si $d < \beta$ aller à l'étape 10 sinon aller à l'étape 5.

étape 5 : Déterminer $\rho' = (\rho'_1,\ldots,\rho'_n) \in \mathbb{R}^n$ tel que

$$\sum_{i=1}^{n} \rho'_i u_i = -\bar{u} \tag{2-3}$$

étape 6 : Déterminer $\alpha_o \in \mathbb{R}$ et $\hat{i} \in \{1,\ldots,p\}$ tels que :

$$\alpha_o = \frac{\rho'_{\hat{i}}}{\rho_{\hat{i}}} = \underset{i=1,\ldots,p}{\text{Min}} \frac{\rho'_i}{\rho_i} \tag{2-4}$$

étape 7 : Echanger $u_{\hat{i}}$ avec $\bar{u}$ $(u_{\hat{i}} := \bar{u})$ et $c_{\hat{i}}$ avec $\bar{c}$ $(c_{\hat{i}} := \bar{c})$

étape 8 : Poser

(2-5) $\rho_i := \rho_i - \frac{\rho_i'}{\alpha_o}$ $\quad i = 1,\ldots,n \quad i \neq \hat{\imath}$

$\rho_{\hat{\imath}} := -\frac{1}{\alpha_o}$.

Si $\rho_i > o$ pour $i=1,\ldots,p$ aller à l'étape 1 sinon aller à l'étape 9.

étape 9 : L'algorithme n'est pas "itératif".

étape 10: Si $\beta+\eta > \varepsilon$ aller à l'étape 2 sinon x est solution du problème $\text{Inf}[(x|z) \mid c(s) \leqq (x|a(s)) + \varepsilon \quad \forall s \in S \quad x \in V_t]$

L'application APPUI qui intervient à l'étape 3 est une application qui, à $x \in \mathbb{R}^n$ et $\eta \in \mathbb{R}$ associe l'élément $(d,\bar{u},\bar{c})$ $\mathbb{R} \times \mathbb{R}^n \times \mathbb{R}$ tel que si l'on pose :

$\bar{d} = \underset{s \in S}{\text{Max}}[c(s)-(x|a(s))]$ on ait :

$o \leqq d = \bar{c}-(x|\bar{u})$ et

$\bar{d}-\eta \leqq d \leqq \bar{d}$

REMARQUES.

1°/Si on note k le numéro de l'itération on a :

$u_i = a(s_i^k)$ $i=p+1,\ldots,n$; $c_i=c(s_i^k)$; $\bar{u}=c(\hat{s}^k)$

$\bar{c} = c(\hat{s}^k)$ et $\rho_i=\rho_i^k$ $(s_i^k \in S$ et $\hat{s}^k \in S)$.

2°/ La matrice du système linéaire (2-3) est la transposée de la matrice du système linéaire (2-2). On peut donc numériquement utiliser (2-2) pour résoudre (2-3). De même, à chaque passage à l'étape 1, le système linéaire à résoudre ne diffère du précédent que par une ligne; il n'est donc en général pas utile de recommencer intégralement sa résolution.

3°/ L'élément η qui intervient dans APPUI à l'étape 3 représente la précision avec laquelle est approché $\underset{s \in S}{\text{Max}}[c(s)-(x|a(s))]$. Dans de nombreux problèmes on peut prendre $\eta = o$.

4°/ Après passage à l'étape 8 on a :

(2-6) $$\sum_{\substack{i=1 \\ i\neq\hat{\imath}}}^{n} \rho_i u_i + \rho_{\hat{\imath}}\bar{u} = z \quad \text{ou} \quad \sum_{\substack{i=1 \\ i\neq\hat{\imath}}}^{p} \rho_i^{k+1} a(s_i^k) + \sum_{i=p+1}^{n} \rho_i^{k+1} u_i + \rho_{\hat{\imath}}^{k+1} a(\hat{s}^k) = z$$

5°/ Les éléments a,c et S du problème (P) n'interviennent qu'à l'étape 3 dans APPUI.

6°/ Pour permettre de calculer aisément les éléments de départ (itération o) vérifiant (2-1) on peut résoudre le problème :

(P') $\alpha = \text{Min}[(x|z) \mid c(s) \leqq (x|a(s)) \quad s \in S, \ \|x\| \leqq r \text{ et } x \in V_t]$

où r est suffisamment grand pour que (P) et (P') aient la même solution.

En remarquant que $\|x\| \leqq r$ peut s'écrire sous la forme :

$$-r \leq (x|t) \quad \forall t \in S' = \{x \in \mathbb{R}^n \mid \|x\| \leq 1\}$$

le problème (P') s'écrit :

(P') $\alpha = \text{Min}[(x|z) \mid c'(s) \leq (x|a'(s)) \quad \forall s \in S \cup S' \quad \text{et} \quad x \in V_t]$

avec

$$a'(s) = \begin{cases} a(s) & \text{si } s \in S \\ s & \text{si } s \in S' \end{cases} \qquad c'(s) = \begin{cases} c(s) & \text{si } s \in S \\ r & \text{si } s \in S \end{cases}$$

L'application APPUI du problème (P') étant l'application qui, à $x \in \mathbb{R}^n$, associe $(d',\bar{u}',\bar{c}') \in \mathbb{R} \times \mathbb{R}^n \times \mathbb{R}$ définis par :

$$d' = \text{Max}\,[d, \|x\|-r] \qquad (\bar{u}',c') = \begin{cases} (\bar{u},\bar{c}) & \text{si } d'=d \\ (-\frac{x}{\|x\|},-r) & \text{si } d' = \|x\|-r \end{cases}$$

les éléments d, $\bar{u}$ et $\bar{c}$ étant définis par l'application APPUI du problème (P).
Les éléments de départ pour (P') sont pris sous la forme $u_i = \varepsilon_i s_i$ $(i=1,\dots,p)$ avec $s_i \in S'$ et ε_i=signe ρ_i $(i=1,\dots,p)$

III. EXEMPLES D'APPLICATIONS

Nous donnons deux exemples classiques d'applications; pour d'autres exemples voir [4].

3-1 MINIMISATION D'UNE FONCTION CONVEXE DERIVABLE AVEC CONTRAINTES DERIVABLES.

On considère le problème :

(P) $\alpha = \text{Inf}[g(t^*,x) \mid g(t,x) \leq o \quad \forall t \in K \quad \text{et} \quad x \in W_t]$

où les applications $x \longmapsto g(t,x)$ de $\mathbb{R}^{n-1}$ dans $\mathbb{R}$ sont, pour tout $t \in K \cup \{t^*\}$ convexes et dérivables; K est un compact et on a :

$$W_t = \{x \in \mathbb{R}^{n-1} \mid (x|u_i') = c_i \quad i=p+1,\dots,n\}$$

On suppose qu'il existe $x_o \in \mathbb{R}^{n-1}$ tel que l'ensemble :

$$L = \{x \in \mathbb{R}^{n-1} \mid g(t^*,x)<g(t^*,x_o) \quad g(t,x) \leq o\}$$ soit non vide et compact

Pour tout $x \in L$ on a ([8] p.242) :

$$g(t,x) = \underset{y \in L}{\text{Max}}[\, g(t,y)+(g_x'(t,y)|x-y)]$$

où $g_x'(t,y)$ désigne le gradient en y de $x \mapsto g(t,x)$. On suppose $(t,y) \longmapsto g_x'(t,y)$ continue.

Le problème (P) peut donc s'écrire :

$$\alpha = \text{Inf}[((x,x_n)|(o,1)) \mid g(t,y)-(g_x'(t,y)|y) \leq ((x,x_n)|(-g_x'(t,y),\delta_t))$$
$$\forall (t,y) \in (K \cup \{t^*\}) \times L \quad (x,x_n) \in V_t]$$

avec $\quad V_t = \{(x,x_n) \in \mathbb{R}^{n-1} \times \mathbb{R} \mid (x,x_n)|u_i) = c_i \quad i=p+1,\dots,n\}$

où $u_i = (u_i',o)$ et $\delta_t = \begin{cases} 1 & \text{si } t = t^* \\ o & \text{sinon} \end{cases}$

On retrouve la forme (1-1). L'application APPUI de ce problème fait correspondre à $(x,x_n) \in \mathbb{R}^{n-1} \times \mathbb{R}$ et $o \leqq \eta$ l'élément $(d,\bar{u},\bar{c}) \in \mathbb{R} \times \mathbb{R}^n \times \mathbb{R}$ défini par :

$$\bar{u} = (-g_x'(\tilde{t},x),\ \delta_{\tilde{t}})$$

$$\bar{c} = g(\tilde{t},x) - (g_x'(\tilde{t},x)|x)$$

$$o \leqq d = g(\tilde{t},x) - \delta_{\tilde{t}} x_n$$

avec $\quad \bar{d} - \eta \leqq d \leqq \bar{d} \quad$ où $\quad \bar{d} = \underset{t \in K \cup \{t^*\}}{\text{Max}} [g(t,x) - \delta_t x_n]$

REMARQUE :

Si K est fini, on a aisément $d=\bar{d}$; on peut donc prendre $\eta=o$. S'il n'y a pas de contrainte il suffit de faire $K = \phi$. Si les fonctions $g(t,.)$ sont linéaires et l'ensemble K fini l'algorithme d'échange est l'algorithme du simplexe appliqué au problème dual de (P).

Les hypothèses H1 et H2 s'écrivent :

H1 Il existe $x^* \in W_t$ tel que $g(t,x^*) < o$ pour tout t de K.

H2 Il existe $(t_1,y_1),\dots,(t_p,y_p) \in (K \cup \{t^*\}) \times L$ et $\rho_1,\dots,\rho_n \in \mathbb{R}$ avec $\rho_i > o \quad i=1,\dots,p$ tels que :

i) $\sum_{i=1}^{p} \rho_i g_x'(t_i,y_i) + \sum_{i=p+1}^{n} \rho_i u_i' = o \quad$ et $\quad \sum_{i=1}^{p} \rho_i \delta_{t_i} = 1$

ii) $\mathbb{R}^n = \mathcal{L}((-g_x'(t_i,y_i),\ \delta_{t_i})\ i=1,\dots,p;\ (u_i',o)\ i=p+1,\dots,n)$

3-2 MEILLEUR APPROXIMANT DANS UN CONVEXE D'UN ESPACE NORME.

On considère le problème :

$$\text{(P)} \qquad \alpha = \text{Inf} \left[\| f_o - \sum_{i=1}^{n-1} x_i f_i \| \mid g(t,x) \leqq o \quad \forall t \in K \text{ et } x \in W_t \right]$$

où $\|.\|$ représente la norme d'un espace vectoriel E de dual topologique E'; $f_i \in E$ $i=o,\dots,n-1$; K est compact, $x \mapsto g(t,x)$ est dérivable et on a :

$$W_t = \{x \in \mathbb{R}^{n-1} \mid (x|u_i') = c_i \quad i=p+1,\dots,n\}$$

En remarquant que ([9]) : $\quad \|g\| = \underset{\ell \in S'}{\text{Max}} <g,\ell> \quad \forall g \in E$

où $S' = \{\ell \in E' \mid \|\ell\| \leqq 1\}$ $\quad <g,\ell> = \ell(g)$, le problème (P) s'écrit :

$$\alpha = \text{Inf}[((x,x_n)|(o,1)) \mid <f_o,\ell> \leqq ((x,x_n)|(<f_1,\ell>,\dots,<f_{n-1},\ell>,1)) \ \forall \ell \in S'$$

$$g(t,y)-(g'_x(t,y)|y) \leq ((x,x_n)|(-g'_x(t,y),o)) \ \forall(t,y) \in K\times L \quad \text{et} \quad (x,x_n) \in V_t]$$

où L est un ensemble supposé non vide et compact de la forme :

$$L = \{x \in \mathbb{R}^{n-1} \mid \|f_o - \sum_{i=1}^{n-1} x_i f_i\| < \|f_o - \sum_{i=1}^{n-1} x_i^\circ f_i\| \quad \text{et} \quad g(t,x) \leq o \quad \forall t \in K\}$$

L'application APPUI,de ce problème qui a la forme (1-1), fait correspondre à $(x,x_n) \in \mathbb{R}^{n-1} \times \mathbb{R}$ et $o \leq \eta$ (à partir des éléments

$$d_1 = \|f - \sum_{i=1}^{n-1} x_i f_i\| - x_n \quad , \quad d_2 = \underset{t \in K}{\text{Max}}\, g(t,x),$$

$$\bar{d} = \text{Max}(d_1,d_2),$$

$$\bar{\ell} \in S' \quad \text{tel que :} \quad d_1+x_n-\eta \leq \langle f- \sum_{i=1}^{n-1} x_i f_i, \bar{\ell}\rangle \leq d_1+x_n \quad \text{et} \quad \tilde{t} \in K$$

tel que $d_2-\eta \leq g(\tilde{t},x) \leq d_2$) l'élément $(d,\bar{u},\bar{c}) \in \mathbb{R} \times \mathbb{R}^n \times \mathbb{R}$ tel que :

$$\bar{d}-\eta \leq d \leq \bar{d}$$

$$\bar{u} = \begin{cases} (\langle f_1,\bar{\ell}\rangle,\ldots,\langle f_{n-1},\bar{\ell}\rangle,1) \\ (-g'_x(\tilde{t},x),o) \end{cases} \quad \text{et} \quad \bar{c} = \begin{cases} \langle f_o,\bar{\ell}\rangle & \text{si } \bar{d}=d_1 \\ g(\tilde{t},x)-(g'_x(\tilde{t},x)|x) & \text{si } \bar{d}=d_2 \end{cases}$$

REMARQUE :

Dans C(K) muni de la norme du max on retrouve l'algorithme de REMES, algorithme généralisé par LAURENT dans le cas d'un espace norme ([6]).

IV. CONVERGENCE DE L'ALGORITHME D'ECHANGE

THEOREME DE CARACTERISATION.

Si H1 est vérifiée alors $\bar{x} \in C \cap V_t$ est solution de (P) si et seulement si, il existe k ($1 \leq k \leq p$) éléments $s_i \in S$, **k** coefficients $\rho_i > o$ tels que :

i) $c(s_i) = (\bar{x}|a(s_i)) \quad i=1,\ldots,k$

ii) $\bar{\ell} = \sum_{i=1}^{p} \rho_i a(s_i) \in V^\perp + z$

DEMONSTRATION:

On a $\alpha = \underset{x \in \mathbb{R}^n}{\text{Inf}}\, f(x)$ avec $f(x)=(x|z)+\chi_C(x)+\chi_{V_t}(x)$. L'élément $\bar{x}$ est solution de (P) si et seulement si $o \in \partial f(\bar{x})$. Les fonctionnelles χ_C et χ_{V_t} étant d'après H1 finies et continues en x^* on a :

$$\partial f(\bar{x}) = \{z\} + \partial\chi_C(\bar{x}) + \partial\chi_{V_t}(\bar{x})$$

On a $C = \{x \in \mathbb{R}^n \mid r(x) \leq o\}$ avec $r(x) = \underset{s \in S}{\mathrm{Max}}[c(s)-(x|a(s))]$;
donc ([6] p.386 et p.355) :

$\partial\chi_C(\bar{x}) = cc\partial r(\bar{x})$ et $\partial r(\bar{x}) = -\overline{\mathbf{co}}\ a(F(\bar{x}))$ avec

$F(\bar{x}) = \{s \in S | c(s) = (\bar{x}|a(s))\}$ donc, $F(\bar{x})$ étant compact :

$\partial\chi_C(\bar{x}) = -cc\ a(F(\bar{x}))$. On a aussi $\partial\chi_V(\bar{x}) = V^\perp$. La relation $o \in \partial f(\bar{x})$ est donc équivalente à l'existence d'un élément $\bar{\ell} \in cc\ a(F(\bar{x})) \cap (V^\perp+z)$ qui s'écrit aussi sous la forme ii) en appliquant le théorème de CARATHEODORY ([6] p.74).

Q.E.D.

THEOREME DE DUALITE (**voir** [4] pour la démonstration).

Si H1 et H2 sont vérifiées alors :

i) Il existe $\bar{x} \in C \cap V_t$ tel que $(\bar{x}|z) = \alpha$

ii) $\alpha = (\tilde{x}|z) + \mathrm{Max}\left[\sum_{i=1}^{k} \rho_i c(s_i) - (\tilde{x}| \sum_{i=1}^{k} \rho_i a(s_i)) \mid 1 \leq k \leq p\right.$

$$\left.\rho_i > o;\ s \in S\ \sum_{i=1}^{k} \rho_i a(s_i) \in V^\perp + z\right]$$

où $\tilde{x}$ est un élément fixe de V_t et $V^\perp = \mathcal{L}(u_{p+1},\ldots,u_n)$.

On suppose vérifiées les hypothèses H1 et H2.

LEMME 4-1 :

A chaque itération k on a $\rho_i \geq o$ $i=1,\ldots,p$ et $\mathbb{R}^n = \mathcal{L}(u_i\ \ i=1,\ldots,n)$.

DEMONSTRATION :

1°/ Montrons que dans (2-4) on a $\alpha_o < o$. Si on avait $\alpha_o \geq o$ on aurait $\rho_i' \geq o$ $i=1,\ldots,p$ et en multipliant scalairement (2-3) par $x^k \in V_t$ on aurait :

$$\sum_{i=1}^{p} \rho_i' c(s_i^k) + (x^k | a(\hat{s}^k)) = o$$

(en posant $x=x^k$, $u_i=a(s_i^k)$, $c_i=c(s_i^k)$ $\bar{u}=a(\hat{s}^k)$, $\bar{c}=c(\hat{s}^k)$, $\bar{d}=\bar{d}^k$ les éléments obtenus à l'étape k).

Par définition on a :

$$o < c(\hat{s}^k) - (x^k | a(\hat{s}^k)) = \bar{d}^k - \beta^k$$

d'où

$$\sum_{j=1}^{p} \rho_j' c(\hat{s}_j^k) + c(\hat{s}^k) = \bar{d}^k - \beta^k$$

Mais d'après H1 :

$$o = \sum_{j=1}^{p} \rho_j^{!}(x^*|a(s_j^k))+(x^*|a(\hat{s}^k)) > \sum_{j=1}^{p} \rho_j^{!}c(s_j^k)+c(\hat{s}^k) = \bar{d}^k-\beta^k$$

ce qui est impossible, donc $\alpha_o < o$.

2°/ Supposons qu'à l'instant k on ait :

(4-1) $\quad \mathbb{R}^n = \mathcal{L}(u_i;\ i=1,\dots,n)$ et $\sum_{i=1}^{n} \rho_i u_i = z$

Montrons que ces relations sont encore vérifiées à l'itération k+1. Raisonnons par l'absurde. Soit $x \neq o$ tel que $x \in V^{\perp} = \mathcal{L}(u_{p+1},\dots,u_n)$ et $x = \sum_{i=1}^{p} \mu_i a(s_i^{k+1})$

on a aussi : $x = \sum_{\substack{i=1 \\ i\neq \hat{i}}}^{p} \mu_i a(s_i^k) + \mu_{\hat{i}^k} a(\hat{s}^k) \in V^{\perp}$ d'où :

$$a(\hat{s}^k) = v + \sum_{\substack{i=1 \\ i\neq \hat{i}^k}}^{p} \alpha_i a(s_i^k)$$ avec $v \in V^{\perp}$. En portant cette valeur dans (2-6)

il vient : $\sum_{i=1}^{p} (\rho_i^{k+1} + \alpha_i \rho_{\hat{i}^k}^{k+1}) u_i + \sum_{i=p+1}^{n} \rho_i^{k+1} u_i = z$ ce qui est contradictoire avec (4-1).

Q.E.D.

On supposera que dans le déroulement de l'algorithme d'échange décrit au paragraphe 2 on ne passe jamais par l'étape 9 soit :

H3 Quelle que soit l'itération k on a $\rho_i > o \quad i=1,\dots,p$

On dit alors que l'algorithme est itératif.

Remarquons que, dans le problème (1-1), on peut théoriquement se passer de la contrainte $x \in V_t$. On a en effet le lemme suivant dont la démonstration est évidente :

LEMME 4-2

Soit dans $\mathbb{R}^p$ le problème suivant :

(4-2) $\quad \alpha - (\tilde{x}|z) = \mathrm{Inf}[(x'|z') \mid c'(s) \leq (x'|a'(s)) \quad \forall s \in S]$

où, si $u_1,\dots,u_p$ désigne une base de $V = \{x \in \mathbb{R}^n \mid (x|u_i) = o \quad i=p+1,\dots,n\}$
et $\tilde{x}$ un élément fixe de V_t, on a :

$$z' = ((u_1|z),\dots,(u_p|z)) \in \mathbb{R}^p$$
$$a'(s) = ((u_1|a(s)),\dots,(u_p|a(s))) \in \mathbb{R}^p$$
$$c'(s) = c(s) - (\tilde{x}|a(s))$$

L'algorithme d'échange et les hypothèses H1,H2 et H3 des problèmes (1-1) et (4-2) sont équivalents.

On a plus précisément $x^k = \sum_{i=1}^{p} x_i'^k u_i + x$ où x^k est l'élément de V_t obtenu à la $k^{\text{ième}}$-itération sur (1-1) et $x'^k = (x_1'^k,\dots,x_p'^k)$ l'élément de $\mathbb{R}^p$ obtenu à la $k^{\text{ième}}$-

itération sur (4-2).

On supposera dans ce qui suit que l'algorithme d'échanges avec les hypothèses H1, H2 et H3 es appliqué au problème (1-1) avec $V_t = \mathbb{R}^n$. (on a donc p=n).

Comme dans la démonstration du lemme 4-1 on utilisera un indice k pour indiquer le numéro de l'itération.

On pose :

$$(4\text{-}3) \qquad \alpha^k = \sum_{i=1}^{n} \rho_i^k c(s_i^k) = (x^k | z)$$

LEMME 4-3 :

La suite $\{\alpha^k\}$ est croissante et bornée supérieurement par $(x^* | z)$.

On a : $\alpha^{k+1} - \alpha^k = \rho_{\hat{i}^k}^{k+1} \ [c(\hat{s}^k) - (x^k | a(\hat{s}^k))]$

DEMONSTRATION :

On a par construction :

$c(s_j^k) = (x^k | a(s_j^k)) \quad j=1,\ldots,n \quad$ et $\quad c(s_j^{k+1}) = (x^k | a(s_j^{k+1})) \quad j=1,\ldots,n \quad j \neq \hat{i}^k$

On a donc :

$$\alpha^{k+1} = (x^k \mid \sum_{i=1}^{n} \rho_i^{k+1} a(s_i^{k+1})) - (x^k | \rho_{\hat{i}^k}^{k+1} a(s_{\hat{i}^k}^{k+1})) + \rho_{\hat{i}^k}^{k+1} c(s_{\hat{i}^k}^{k+1})$$

De (2-6), (4-3) et $\hat{s}^k = s_{\hat{i}^k}^{k+1}$ on tire :

$$\alpha^{k+1} = \alpha^k + \rho_{\hat{i}^k}^{k+1} [c(\hat{s}^k) - (x^k | a(\hat{s}^k))].$$

L'hypothèse H3 entraîne $\rho_{\hat{i}^k}^{k+1} > 0$ donc $\{\alpha^k\}$ est croissante. De l'hypothèse H1 on tire :

$$\alpha^k = \sum_{j=1}^{n} \rho_j^k \ c(s_j^k) < (x^* | \sum_{j=1}^{n} \rho_j^k \ a(s_j^k)) = (x^* | z)$$

Q.E.D.

LEMME 4-4 :

S'il existe s>o tel que $0 < s \leqq \rho_{\hat{i}^k}^{k+1}$ alors :

$$\lim_{t \to \infty} d^k = 0 \quad \text{et} \quad \lim_{k \to \infty} \alpha^k = \alpha$$

DEMONSTRATION :

1°/ Du lemme 4-3 et de la définition de d^k on **tire** :

$$\alpha^{k+1} - \alpha^k \geqq \rho_{\hat{i}^k}^{k+1} (\bar{d}^k - \eta^k)$$

d'où :

$$0 \leqq \bar{d}^k \leqq \frac{1}{s}(\alpha^{k+1} - \alpha^k) + \eta^k$$

D'après 4-3 et l'étape 2 de l'algorithme on a $\{\alpha^k\}$ qui converge et $\lim_{k\to\infty} \eta^k = o$ d'où $\lim_{k\to\infty} \bar{d}^k = \lim_{k\to\infty} d^k = o$

2°/ On pose $\tilde{\alpha} = \lim_{k\to\infty} \alpha^k$; de $\alpha^k \leqq \alpha$ on tire $\tilde{\alpha} \leqq \alpha$.

Supposons que l'on ait $\tilde{\alpha} < \alpha$, d'après le théorème de caractérisation on a:

$$\alpha = \sum_{i=1}^{n} \rho_i c(s_i) \qquad \text{avec} \quad \sum_{i=1}^{n} \rho_i a(s_i) \in V^{\perp} + z \qquad \rho_i \geqq o \quad i=1,\dots,n$$

Posons $\alpha - \tilde{\alpha} = \mu > o$. Quel que soit $k \in \mathbb{N}$ il existe $\bar{s} \in S$ tel que

$$c(\bar{s}) - (x^k | a(\bar{s})) \geqq \mu\left(\sum_{i=1}^{n} \rho_i\right)^{-1} = \beta$$

(Sinon pour un indice $\bar{k}$ on aurait $c(s) - (x^{\bar{k}} | a(s))$ $\forall s \in S$ et par suite $\sum_{i=1}^{n} \rho_i c(s_i) - (x^{\bar{k}}\ z) = \alpha - \alpha^{\bar{k}} < \mu$)

On a alors : :

$$\bar{d}^k \geqq c(\bar{s}) - (x^k | a(\bar{s})) \geqq \mu\left(\sum_{j=1}^{n} \rho_j\right)^{-1}$$ on ne pourrait donc avoir $\lim_{k\to\infty} \bar{d}^k = o$

Q.E.D.

LEMME 4-5 :

Si on note A l'enveloppe convexe de $z \cup -a(S)$, il existe une constante $w > o$ telle que :

$$A \cup -A \subset wA$$

DEMONSTRATION :

D'après H2 il existe $\rho_j > o$ et $s_j \in S$ $(j=1,\dots,n)$ tel que

$$z = \sum_{i=1}^{n} \rho_i a(s_i)$$

Par suite :

$$o = \frac{1}{K} z + \sum_{i=1}^{n} \frac{\rho_i}{K}(-a(s_i)) \qquad \text{avec} \quad K = 1 + \sum_{i=1}^{n} \rho_i > o$$

et

$$o = \frac{1}{K}(-z) + \sum_{i=1}^{n} \frac{\rho_i}{K} a(s_i)$$

donc o appartient à l'intérieur de $A \cup -A$. On note $B(o,\varepsilon)$ une boule de centre o et de rayon ε contenue dans $A \cup -A$; S étant compact, $A \cup -A$ est borné, il existe donc $\mu > o$ tel que $A \cup -A \subset B(o,\mu)$.

On a donc $B(o,\mu) = \frac{\mu}{\varepsilon} B(o,\varepsilon) \subset \frac{\mu}{\varepsilon} A$. En posant $w = \frac{\mu}{\varepsilon}$ on a le résultat.

Q.E.D.

THEOREME 4-1 :

Les hypothèses H1,H2 et H3 étant vérifiées on a :

i) $\lim_{k\to\infty} (x^k|z) = \alpha$

ii) il existe une sous-suite $\{x^{\psi(k)}\}$ de $\{x^k\}$ telle que :

$$\lim_{k\to\infty} \operatorname{Max}_{s\in S} [c(s)-(x^{\psi(k)}|a(s))] = o$$

DEMONSTRATION :

Si pour tout i la suite $\{\rho_i^k\}_{k\in\mathbb{N}}$ est bornée inférieurement par un scalaire strictement positif on a d'après le lemme 4-4 :

$\lim_{k\to\infty} \alpha^k = \lim(x^k|z) = \alpha$ et $\lim_{k\to\infty} d^k = o$ ce qui par définition de d^k entraîne $\lim_{k\to\infty} \bar{d}^k = o$ et démontre le théorème.

Supposons maintenant qu'il existe $i_1 \in \{1,\dots,n\}$ tel que $\{\rho_{i_1}^k\}$ ne soit pas bornée inférieurement par un scalaire strictement positif. On peut alors extraire une sous-suite $\phi_1 : \mathbb{N} \to \mathbb{N}$ telle que : $\lim_{k\to\infty} \rho_{i_1}^{\phi_1(k)} = o$. S'il existe $i_2 \in \{1,\dots,n\} - i_1$ tel que $\lim \rho_{i_2}^{\phi_1(k)} = o$ on extrait de $\{\rho_{i_2}^{\phi_1(k)}\}$ une sous-suite définie par $\phi_2 : \mathbb{N}\to\mathbb{N}$ telle que $\lim_{k\to\infty} \rho_{i_2}^{\phi_2(k)} = o$. On continue l'opération jusqu'à ce qu'on ait une sous-suite définie par $\phi_q : \mathbb{N} \to \mathbb{N}$ telle que :

1°/ $\lim_{k\to\infty} \rho_i^{\phi_q(k)} = o$ pour tout $i \in I_o = \{i_1,\dots,i_q\}$

2°/ il existe $\mu>o$ tel que pour tout $i\in I_1 = \{1,\dots,n\} - I_o$ on ait $\mu<\rho_i^{\phi_q(k)}$.

Remarquons que I_1 est non vide car, de la relation $z = \sum_{i=1}^n \rho_i^k a(s_i^k)$ on tire :

$$o < \frac{||z||}{\operatorname{Max}_{s\in S}|a(s)|} \leq \sum_{i=1}^n \rho_i^k$$

Démontrons le lemme technique :

LEMME 4-6 :

Il est impossible que, pour k fixé, l'on ait pour tout k'> k

$$\{s_i^{\phi_q(k')} \mid i\in I_1\} \subset \{s_i^{\phi_q(k)} \mid i=1,\dots,n\}$$

DEMONSTRATION :

D'après l'hypothèse (H1) on a : $\operatorname{Max}_{s\in S}[c(s)-(x^*|a(s))] = d<o$ d'où :

$$\sum_{i=1}^{n}\rho_i^k\ c(s_i^k)-(x^*|\sum_{i=1}^{n}\rho_i^k\ a(s_i^k)) \leqq d\sum_{i=1}^{n}\rho_i^k < o$$

Par suite, en tenant compte de (2-6) :

$$\sum_{i=1}^{n}\rho_i^k \leqq \frac{1}{|d|}\ (|\tilde{\alpha}| + \|x^*\| . \|z\|) \qquad \text{avec } \tilde{\alpha} = \lim_{k\to\infty} \alpha^k$$

La suite $\{\rho_i^k\}_{k\in\mathbb{N}}$ est donc bornée pour tout $i\in\{1,\dots,n\}$. On peut donc extraire de $\{\rho_{i_{q+1}}^{\phi_q(k)}\}_{k\in\mathbb{N}}$ pour $i_{q+1}\in I_1$ une sous-suite définie par $\{\rho_{i_{q+1}}^{\phi_{q+1}(k)}\}_{k\in\mathbb{N}}$ convergeant vers $\rho_{i_{q+1}} > o$; on peut de même extraire de $\{\rho_{i_{q+2}}^{\phi_{q+1}(k)}\}_{k\in\mathbb{N}}$ $(i_{q+2}\in I_1)$ une sous-suite définie par $\phi_{q+2} : \mathbb{N} \longrightarrow \mathbb{N}$ telle que :

$$\lim_{k\to\infty} \rho_{i_{q+2}}^{\phi_{q+2}(k)} = \rho_{i_{q+2}} > o.$$

En opérant ainsi sur tous les indices de I_1, on dispose d'une sous suite définie par $\phi\ : \mathbb{N} \longrightarrow \mathbb{N}$ telle que :

$$\lim_{k\to\infty} \rho^{\phi(k)} = o \quad \text{pour } i\in I_o$$

et

$$\lim_{k\to\infty} \rho^{\phi(k)} = \rho_i \quad \text{pour } i\in I_1 = \{1,\dots,n\} - I_o \quad \text{avec } o<\mu\leqq\rho_i$$

Si le lemme n'est pas vérifié on a :

$$z = \sum_{i\in I_o}\rho_i^{\phi(k')}\ a(s_i^{\phi(k')}) + \sum_{i\in I_1}\rho_i^{\phi(k')}\ a(s_i^{\phi(k)}) \qquad \text{pour tout } k'>k.$$

En faisant tendre k' vers l'infini on aurait $z = \sum_{i\in I_1}\rho_i\ a(s_i^{\phi(k)})$ ce qui est impossible car d'après H3 on a $z = \sum_{i=1}^{n}\rho_i^{\phi(k)}a(s_i^{\phi(k)})$ avec $\rho_i^{\phi(k)} > o$. <u>Le lemme est donc démontré.</u>

Quitte à extraire de $\{\rho_i^{\phi_q(k)}\}_{k\in\mathbb{N}}$ une nouvelle sous suite définie par $\phi : \mathbb{N} \longrightarrow \mathbb{N}$, on peut supposer d'après le lemme que l'ensemble $\{s_i^{\phi(k+1)} \mid i\in I_1\}$ contient au moins un point nouveau par rapport à l'ensemble $\{s_i^{\phi(k)} \mid i=1,\dots,n\}$. Notons $s_{e(k)}^{\phi(k+1)}$ le dernier élément nouveau introduit et supposons qu'il ait été introduit à l'itération $\psi(k)$ $(\phi(k) \leqq \psi(k) < \phi(k+1))$. On a donc :

$$s_i^{\phi(k+1)} = \begin{cases} s_i^{\psi(k)} & \text{si } i\neq e(k) \\ \bar{s}^{\psi(k)} & \text{si } i=e(k) \end{cases} \qquad i\in I_1$$

De la relation :

$$\alpha^{\phi(k+1)} = \sum_{i\in I_0} \rho_i^{\phi(k+1)} c(s_i^{\phi(k+1)}) + \sum_{i\in I_1} \rho_i^{\phi(k+1)} c(s_i^{\phi(k+1)})$$

on déduit en retranchant la quantité :

$$\alpha^{\psi(k)} = \sum_{i=1}^{n} \rho_i^{\phi(k+1)} (x^{\psi(k)} \mid a(s_i^{\phi(k+1)}))$$

$$\alpha^{\phi(k+1)} - \alpha^{\psi(k)} = \sum_{i\in I_0} \rho_i^{\phi(k+1)} [c(s_i^{\phi(k+1)}) - (x^{\psi(k)} | a(s_i^{\phi(k+1)}))]$$

$$+ \sum_{\substack{i\in I_1 \\ i\neq e(k)}} \rho_i^{\phi(k+1)} [c(s_i^{\psi(k)}) - (x^{\psi(k)} | a(s_i^{\psi(k)}))]$$

$$+ \rho_{e(k)}^{\phi(k+1)} (\bar{d}^{\psi(k)} - \xi^{\psi(k)})$$

avec $d^k = \bar{d}^k - \xi^k \qquad (o \leq \xi^k \leq \eta^k)$.

En tenant compte de la définition de $x^{\psi(k)}$ et en ajoutant $\sum_{i\in I_0} \rho_i^{\phi(k+1)} (x^{\psi(k)} | z)$ aux deux membres il vient :

(4-4) $$\alpha^{\phi(k+1)} + (\varepsilon(k)-1)\alpha^{\psi(k)} = B(k) + \rho_{e(k)}^{\phi(k+1)} (d^{\psi(k)} - \xi^{\psi(k)})$$

avec : $\varepsilon(k) = \sum_{i\in I_0} \rho_i^{\phi(k+1)}$

et

$$B(k) = \sum_{i\in I_0} \rho_i^{\phi(k+1)} [c(s_i^{\phi(k+1)}) - (x^{\psi(k)} | a(s_i^{\phi(k+1)}) - z)]$$

En posant $\omega^* = \underset{s\in S}{\mathrm{Max}} |c(s)|$ on a:

$$|B(k)| \leq \varepsilon(k)(\omega^* + 2\,\mathrm{Max}\{|(x^{\psi(k)}|t)| \mid t \in (-a(S)) \cup \{z\}\}$$

D'après le lemme 4-5 il existe $w > o$ telle que : $-A \cup A \subset wA$ avec A enveloppe convexe de $(-a(S)) \cup \{z\}$. On a alors :

$$\underset{t\in A}{\mathrm{Max}} |(x^{\psi(k)}|t)| = \underset{t\in(-A)\cup A}{\mathrm{Max}} (x^{\psi(k)}| t) \leq \underset{t\in wA}{\mathrm{Max}} (x^{\psi(k)}|t)$$

d'où :

$$\mathrm{Max}\{|(x^{\psi(k)}|t)| \mid t\in -a(S) \cup \{z\}\} \leq w \underset{t\in -a(s)\cup\{z\}}{\mathrm{Max}} (x^{\psi(k)}| t).$$

On a par ailleurs :

$$\underset{t\in -a(S)}{\text{Max}} \; (x^{\psi(k)}|t) \leqq \underset{s\in S}{\text{Max}} \; [c(s)-(x^{\psi(k)}|a(s))] + \omega^* = \bar{d}^k + \omega^*$$

et $\quad (x^{\psi(k)}|z) = \alpha^{\psi(k)} \leqq \bar{d}^{\psi(k)} + d^* \qquad$ avec $\alpha^* = (x^*|z) \qquad$ (lemme 4-3)

On a donc : $\quad \underset{t\in A}{\text{Max}} \; |(x^{\psi(k)}|t)| \leqq w\bar{d}^{\psi(k)} + w \; \text{Max}(\omega^*,\alpha^*)$

et par suite :

$$|B(k)| \leqq \varepsilon(k)\left[\omega^* + w\text{Max}(\omega^*,\alpha^*) + w\bar{d}^{\psi(k)}\right]$$

En posant

$d' = \omega^* + w\text{Max}(\omega^*,\alpha^*) > o$, (4-4) s'écrit :

$$\alpha^{\phi(k+1)} + (\varepsilon(k)-1) \, \alpha^{\psi(k)} + \varepsilon(k)d' + \rho^{\phi(k+1)}_{e(k)} \quad \xi^{\psi(k)} \geqq \rho^{\phi(k+1)}_{e(k)} - w \, \varepsilon(k)\bar{d}^{\psi(k)}$$

L'indice e(k) appartenant à I_1, on a :

$$o < \mu \leqq \rho^{\phi(k+1)}_{e(k)}$$

La suite $\bar{d}^{\psi(k)}$ est donc majorée, pour k suffisamment grand, par la suite $\{\beta_k\}$ avec :

$$\beta_k = \frac{1}{\mu - w\varepsilon(k)} \left(\alpha^{\phi(k+1)} + (\varepsilon(k)-1)\alpha^{\psi(k)} + \varepsilon(k)d' + K\xi^{\psi(k)}\right)$$

On a par construction $\lim\limits_{k\to\infty} \xi^{\psi(k)} = o$; la convergence de $\{\alpha^k\}$ et celle de $\varepsilon(k)$ vers zéro entraîne $\lim\limits_{k\to\infty} \beta_k = o$ et par suite $\lim\limits_{k\to\infty} \bar{d}^{\psi(k)} = o$ ce qui démontre ii) du théorème.

On a toujours $\alpha^k \leqq \alpha$; en faisant un raisonnement analogue à celui du lemme 4-3 on a $\lim\limits_{k\to\infty} \alpha^{\psi(k)} = \alpha$ et $\{\alpha^k\}$ étant croissante on a $\lim\limits_{k\to\infty} \alpha^k = \alpha$ ce qui achève la démonstration du théorème.

Q.E.D.

REMARQUE :

L'hypothèse H3 est, dans le cas de l'algorithme de REMES, beaucoup moins forte que la "condition de HAAR". Le théorème précédent assure la convergence de l'algorithme dès qu'il se "déroule bien"(Hypothèse H3).

REFERENCES :

[1] CARASSO C. : Etude de l'algorithme de REMES en l'absence de conditions de HAAR,Numer.Math., (1972), 165-178.

[2] CARASSO C. : Densité des hypothèses assurant la convergence de l'algorithme de REMES, R.A.I.R.O.,(1972), R.3, 69-84.

[3] CARASSO C. : Thèse, Grenoble (1973).

[4] CARASSO C. : Pré-Publication n°3, Université de Saint-Etienne (1975)

[5] GOLDSTEIN A.A. : Constructive real analysis, Haaper and Row (1967).

[6] LAURENT P.J. : Approximation et Optimisation. Hermann, Paris (1972).

[7] LAURENT P.J. : Conference on Approximation Theory, Austin (Texas)(1973).

[8] ROCKAFELLAR R.T. : Convex analysis, Princ.Univ.Press, (1970).

[9] SINGER I. : Best approximation in normed linear spaces by elements of linear subspaces. Springer-Verlag (1970).

A REMARK ON MULTIPLIER METHODS FOR NONLINEAR PROGRAMMING

M. Cirinà

Istituto Matematico del Politecnico

10129 Torino, Italy

ABSTRACT. This paper is concerned with certain aspects of multiplier methods where the solution of a constrained minimization problem is obtained by means of a sequence of unconstrained minimizations of an augmented Lagrangian $L(x,y,r)$, followed each by an iteration on the Lagrange multiplier vector y. In spite of the growing recognition that multiplier methods are among the most effective constrained minimization methods, the value to be given to the penalty parameter r does not seem yet to have received enough attention. This paper - related to work done recently by Bertsekas and Polyak - contains a result in such direction, namely the following one: if G and Q are given matrices and Q is positive definite on the kernel of G, then it is produced r^* such that for all $r > r^*$, $Q + r\, G^T G$ is positive definite on the whole space. Also we prove a lemma - related to a known one - about Hilbert space operators with uniformly bounded inverses, that may be useful in extending the result above to more general situations. To test the computational value of the estimate r^* arrived at here, a computer program is being tested and some numerical results are reported.

1. INTRODUCTION

Let $f, g_1, \ldots, g_m$ be real valued functions defined on R^n, $n \geqslant m > 0$, and $g = (g_1, \ldots, g_m)$. Consider the problem

(P) $\qquad$ minimize $f(x)$, subject to $g(x) = 0$

and the sequence of unconstrained minimizations

$$L(x_k, y_k, r_k) = \min \{L(\cdot, y_k, r_k) \mid x \in R^n\} \tag{1.1}$$

where $\{y_k\},\{r_k\}$ are to be chosen appropriately and the augmented Lagrangian L -cf. Arrow and Solow [1], Fletcher [5], Hestenes [6] and the work of Rockafellar, see [10,11] and their references - is defined by

(1.2) $L(x,y,r) = f(x) + (y,g(x)) + \frac{r}{2} \|g\|^2$, $(x,y,r) \in R^n \times R^m \times]o, \infty[$.

The penalty method for solving (P) - see for instance Fiacco and McCormick [4] - can be thought of as the minimizations (1.1) with the choice $y_k = 0$ and $r_k \nearrow \infty$. The numerical difficulties (instability) that may arise in such procedure for "r_k large" can in general be avoided or reduced by using the iteration

(1.3) $$y_{k+1} = y_k + r_k\, y(x_k)$$

for which the condition $r_k \nearrow \infty$ is not necessary: this choice of y_k characterizes the multiplier methods - see Hestenes [6] and Powell [9] - that are receiving growing recognition as one of the most effective class of methods for constrained minimization.

In the actual process of solving a specific problem (P) with such a scheme one has of course to decide which values the penalty parameter r_k is to be given; it may be easily found out that haphazard guesses may lead to a substantial waste of time and effort: indeed, roughly speaking, when the violation of the constraints is penalized too little (1.1) may end up generating points away from the feasible set,and when it is penalized too much the points generated by (1.1) may have tendency to jump around.

In a recent paper, Bertsekas [3] has succeded in proving convergence and rate of convergence results for the multiplier method ; actually, under the standard 2nd order sufficiency conditions for $\bar{x}$ to be an isolated local minimum (i.e. when - see Luenberger [7] , p.226 - f and g are C^2 in a ball $B(\bar{x}, \varepsilon)$, at $\bar{x}$ the differential dg of g has rank m and $df + \bar{y}^T dg = 0$ for some $\bar{y} \in R^m$, at $(\bar{x}, \bar{y}, o)$ the hessian of $L(\cdot, \bar{y}, 0)$ is positive definite on the kernel of $dg(\bar{x})$) and under a Lipschitz condition on $d^2f, d^2g_1, \ldots, d^2g_m$ in $B(\bar{x}, \varepsilon)$ he proves that given a bounded subset $\mathcal{Y} \subseteq R^m$ $\exists\, r_1 \geqslant 0$ and $M_1 \geqslant 0$ such that for all $r \geqslant r_1$ and $y \in \mathcal{Y}$

(i) $\exists$ one minimizer $x(y,r)$ of $L(\cdot,y,r)$ in $B(\bar{x}, \varepsilon)$, $\varepsilon > 0$

(ii) $$|x(y,r)-\bar{x}| \text{ and } |y+rg(x(y,r))-\bar{y}| \leqslant \frac{M_1\, |y-\bar{y}|}{r} .$$

In the proof, such conclusions are obtained by requiring - in particular - r to be so large as to make

(1.4) $$d^2L(\bar{x},\bar{y},o) + r\, dg^T(\bar{x})dg(\bar{x}) \quad \text{positive definite.}$$

In section 2 it is given a computable estimate of r^* such that (1.4) holds for all $r > r^*$; actually, we feel that such result (local estimate theorem) is a special case of a global one.

The main difficulty for obtaining such more general result is that a certain set of matrices should possess uniformly bounded inverses. However, we shall not investigate the global problem here, except for setting it (in section 2), congecturing its solution and extending (in section 3) a lemma in Polyak [8] about Hilbert space operators possessing uniformly bounded inverses, to a more general situation. Actually we feel that such more general lemma or some variant of it, besides having some interest in themselves, may be useful for establishing our congecture about the global result.

Finally some numerical results, obtained by making use of r^* in a simple way within the standard multiplier method, are reported in section 4 and compared with analogous ones available in the letterature.

2. A LOCAL ESTIMATE OF r^*

We shall first state the general problem we are concerned with. Let the norm $\|\cdot\|$ be defined on the space of finite dimensional linear operators $L(R^n, R^m)$ by $\|A\| = \sup\{(Ax, Ax)^{1/2}/(x,x)^{1/2} : x \in R^n\}$, for $n \geq m > 0$ consider a map $G : X \subseteq R^n \to L(R^n, R^m)$ and let $N(G(x))$ be the kernel of $G(x)$, for each $x \in X$; also, consider a map $Q : Z = X\times Y \subseteq R^n\times R^m \to L(R^n, R^n)$ and for r real let Q_r be defined by

$$Q_r(x,y) = Q(x,y) + r\, G^T(x)G(x) \quad , \ (x,y) \in X\times Y$$

The problem we are concerned with is the following:

> if it is known that for $(x,y) \in Z$, $Q(x,y)$ has a coerciveness property on the subspace $N(G(x))$; when and how it possible to estimate r^* such that for every $r > r^*$ $Q_r(x,y)$ has such coerciveness property on all of R^n ?

Here we shall give an answer to such problem in the simpler case where the sets X and Y contain just one point each.

Theorem 2.1 (Local estimate)

Let n, m be positive integers $(n \geq m)$, and suppose that
(i) $G \in L(R^n, R^m)$ has rank m;
(ii) $Q \in L(R^n, R^n)$ is symmetric and $(Qs,s) > 0$, all $s \in N(G)$, $s \neq 0$.

For r real consider $Q_r = Q + r\, G^T G$ and let r^* be defined by (2.3) be low. Then

$$r > r^* \Rightarrow (Q_r s, s) > 0 \quad ,\text{all } s \in R^n,\ s \neq 0.$$

Remark 2.2. Since Q_r is a matrix of numbers, to say that it is coercive it is the same as saying that it is positive definite.

Proof of theorem 2.1.

Let us observe that there are congruent transformations $B(\cdot)B^T$ that reduce $G^T G$ to a diagonal matrix; such transformations - by a classical result of Sylvester - leave invariant the number of positive and the number of zero eigenvalues. A natural choice for one such transformation is given by $B \in L(R^n, R^n)$ defined as

$$(2.1) \qquad B = [G^T | e_{j_1}, \ldots, e_{j_{n-m}}]^{-1}$$

where e_{j_p} is the j_p-th column of the identity matrix and j_p is the row index of the p-th row of G^T that is linearly dependent on the previous ones. Indeed it can be easily seen that such B acts as left inverse of G and

$$BG^T GB^T = \begin{bmatrix} I & 0 \\ 0 & 0 \end{bmatrix} \quad , \ I \quad m\times m .$$

Let us now partition $BQ_r B^T$ symmetrically

$$(2.2) \qquad BQ_r B^T = BQB^T + \begin{bmatrix} rI & 0 \\ 0 & 0 \end{bmatrix} = \begin{bmatrix} P_r & \Pi^T \\ \Pi & p \end{bmatrix} \quad , \ P_r \quad m\times m \ ;$$

put

$$M(p;\Pi) = \begin{cases} 2\|p^{-1}\|\,\|\Pi\|^2 - 2\|\Pi\| + \|p\| & \text{if } m < n \\ 0 & \text{else} \end{cases}$$

and define r^* by

$$(2.3) \qquad r^* = \max\{0, \max_{1\leq i\leq m} [-(BQB^T)_{ii} + \sum_{j=1,\neq i}^{m} |(BQB^T)_{ij}| + M]\} .$$

We claim that such r^* will do; indeed let $r > r^*$ be given. It has to be shown that

$$(2.4) \qquad (s, Q_r s) > 0 \qquad \text{all } s \in R^n,\ s \neq 0.$$

Since G^TG is positive semidefinite, from the definition of Q_r it is seen that (2.4) with $r \geq 0$ holds if s is restricted to the kernel of G i.e.

$$(2.5) \qquad (s, Q_r s) > 0 \ , \ \text{all } s \in N(G) \ , \quad s \neq 0 \ ;$$

for $s \in R^n$ let $w \in R^n$ be defined by $s = B^T w$, call w' (respectively w") the vector formed by the first m (last n-m) components of w and observe that

$$(s, Q_r s) = (w, BQ_r B^T w) = (w', P_r w') + 2(w", \Pi w') + (w", p\, w")$$

implies in view of the consequence of (2.5)

$$s \in N(G) \quad \Leftrightarrow \quad w' = 0 \quad ,$$

that the submatrix p is positive definite; thus M is a well defined real number.
We now claim that for $w \in R^n$ with $(w,w) = 1 = \|w'\|^2 + \|w"\|^2$ one has

$$(2.6) \quad (w, BQ_r B^T w) \geqslant (\lambda(P_r) - \|p\| + 2\|\Pi\|)\|w'\|^2 - 2\sqrt{2}\, \|\Pi\|\ \|w'\|^2 + \|p^{-1}\|^{-1}$$

where $\lambda(\cdot)$ denotes the least eigenvalue of the argument; indeed using the Cauchy-Schwarz inequality, the fact that $(1-\sqrt{2}t)^2 \geqslant 0$ for t real and that $(w", p\, w") \geqslant \lambda(p)(1-\|w'\|^2)$ one obtains

$$(w, BQ_r B^T w) \geqslant (\lambda(P_r) - \lambda(p) + 2\|\Pi\|)\ \|w'\|^2 - 2\sqrt{2}\ \|\Pi\|\ \|w'\| + \lambda(p)$$

and hence (2.6), because $\|p^{-1}\|^{-1} = \lambda(p) \leqslant \|p\|$.
From (2.6) one has

$$(2.7) \quad 2\|\Pi\|^2 - (\lambda(P_r) - \|p\| + 2\|\Pi\|)\|p^{-1}\|^{-1} < 0 \ \Rightarrow\ (w, BQ_r B^T w) > 0, \text{ all } w,\ \|w\| = 1$$

from which it is seen that the proof of the theorem is completed as soon as it is shown that

$$(2.8) \qquad \lambda(P_r) > M(p; \Pi) ;$$

this inequality does indeed hold because it follows from Gershgorin's theorem and the choice of r*: in fact, there is an integer $i(1 \leqslant i \leqslant m)$ such that

$$-\lambda(P_r) + [\ (BQB^T)_{ii} + r] \leqslant \sum_{j=1, \neq i}^{m} |(BQB^T)_{ij}| \quad ,$$

and this inequality implies (2.8) because $r > r^*$.

If the condition "X and Y contain just one point each" is dropped and the hypotheses of theorem 2.1 are somewhat strengthened so as to make the matrices $B(x)$ (i.e. the left inverses of $G^T(x)$, $x \in X$) uniformly bounded in x, we congecture that a result similar to the one above should hold. In the following section we shall examine some conditions insuring that linear operators depending on parameters possess uniformly bounded inverses .

3. UNIFORMLY BOUNDED INVERSES

In this section we generalize a result of Polyak, [8] lemma 1, about Hilbert space operators with uniformly bounded inverses and then give a variant of it.

Let H, h be real Hilbert spaces; their inner product as well as the induced norms will be denoted by the same symbols $(\cdot\ ,\ \cdot)$ and $\|\cdot\|$, when there is no risk of misunderstanding; let the product space $H\times h$ be normed by $\|\cdot\|^2 = \|\cdot\|^2_H + \|.\|^2_h$, the space $L(H,h)$ of bounded linear operators from H to h be normed by $\|C\| = \sup\{\|Cs\|_h/\|s\|_H : s \neq 0\}$, and let X, Z, R be subsets of the Euclidean spaces R^ν, R^μ and R^ρ respectively. Suppose

(3.1) $\{C_x : x \in X\}$ is a bounded family of operators $C_x \in L(H,h)$ satisfying $\|C_x^* s\| \geq \sqrt{k_1}\ \|s\|$, all $s \in h$ and $x \in X$, some $k_1 > 0$, where C_x^* denote the adjoint of C_x ;

(3.2) $\{U_z : z \in Z\}$ is a bounded family of symmetric operators $U_z \in L(H,H)$ satisfying $(U_z s,s) \geq k_2\ \|s\|^2$, all $s \in H$, $z \in Z$, some $k_2 > 0$; let k_3 be a positive number such that $\|U_z\| \leq k_3$, all $z \in Z$;

(3.3) $\{u_r: r \in R\}$ is a family of operators $u_r \in L(h,h)$ satisfying $(u_r s,s) \leq k_4\|s\|^2$, all $s \in h$ and $r \in R$, with k_4 real.

For each $(x,z,r) \in P = X\times Z\times R$, consider the operator $A=A_{x,z,r}: H\times h \to H\times h$ defined by

$$A = \begin{bmatrix} U_z & C_x^* \\ C_x & u_r \end{bmatrix}$$

If the families (3.1), (3.2) reduce to a single operator and if the u_r's are the specific negative operators $-I/r$ where I is the identity and $r \in]o, \infty[$, Polyak proves in [8] that the inverse of A exists and is bounded uniformly in r. Here we obtain a similar result for the more general situation (3.1), (3.2) and (3.3), where in particular (3.3) does not esclude u_r to be the null operator, a case important for some nonlinear programming methods ,and indeed allows the operators u_r to be somewhat positive; the simplicity of the proof is enhanced by the explicit exhibition of the inverse of A.

Lemma 3.1 (Formula for the inverse operator)

Suppose that (3.1), (3.2), and (3.3) with $k_4 < k_1/k_3$ hold; for each $(x,z,r) \in P$ define the operator $F = F_{x,z,r}$ by

$$(3.4) \qquad F_{x,z,r} = \begin{bmatrix} U^{-1}+Sv^{-1}S^* & -Sv^{-1} \\ -v^{-1}S^* & v^{-1} \end{bmatrix}$$

where $U^{-1} = U_z^{-1}$ and the operators $S = S_{x,z}$, $v = v_{x,z,r}$ are defined by $S = U_z^{-1}C_x^*$, $v = u_r - C_x U_z^{-1} C_x^*$. Then

(i) $\exists$ K real: $\|F_{x,z,r}\| \leq K \ \forall \ (x,z,r) \in P$;

(ii) $\forall \ (x,z,r) \in P$: $F_{x,z,r}$ is the inverse of $A_{x,z,r}$.

Proof

In view of (3.1), (3.3) and the choice of k_4, the family of operators on h $\{-v_{x,z,r}\}$ is uniformly coercive because

$$(3.5) \qquad (-v_{x,z,r}s,s) = -(u_r s,s) + (C_x U_z^{-1}C_x^* s,s) \geq -k_4\|s\|^2 + \frac{1}{\|U_z\|}\|C_x^*s\|^2$$
$$\geq (\frac{k_1}{k_3} - k_4)\|s\|^2, \quad \text{all } s \in h \text{ and } (x,z,r) \in P$$

and hence $\{v_{x,z,r}^{-1}\}$ is a bounded family because (3.5) implies, in addition to the existence of the inverses, that

$$(3.6) \qquad \|v_{x,z,r}^{-1}\| \leq \frac{1}{(k_1/k_3)-k_4}, \quad \text{all } (x,z,r) \in P ;$$

thus the operator $F_{x,z,r}$ is well defined. A direct check of its product with $A_{x,z,r}$ shows that indeed (ii) holds. Also $\{U_z^{-1}\}$ is a bounded subset of L(H,H) because the uniform coerciveness condition in (3.2) implies

$$(3.8) \qquad \|U_z^{-1}\| \leq \frac{1}{k_2}, \quad \text{all } z \in Z ;$$

in view of (3.6), (3.8) and (3.1), all the operators appearing in the definition (3.4) belong to bounded families, hence the uniform boundedness conclusion (i) is a simple consequence of the fact that

$$\|F_{x,z,r}\| \leq (\|U^{-1}+Sv^{-1}S^*\|+\|Sv^{-1}\|+\|v^{-1}S^*\|+\|v^{-1}\|)|_{x,z,r} .$$

The lemma is thus established.

We shall now give a variant of lemma 1: roughly speaking, if the set $\{u_r\}$ is sufficienty coercive, the hypotheses on $\{C_x\}$ can be weakened. More precisely, let us assume that

(3.1)' $\{C_x : x \in X\}$ is a bounded subset of $L(H,h)$; let k_5 be a number such that $\|C_x^*\| \leqslant k_5$, all $x \in X$;

(3.3)' $\{u_r : r \in R\}$ is a subset of $L(h,h)$ satisfying: $(u_r s,s) \geqslant k_6\|s\|^2$, all $s \in h$ and $r \in R$, $k_6 > 0$.

Lemma 3.2.

Under the hypotheses (3.1)', (3.2), and (3.3)' with $k_6 > k_5^2/k_2$ the conclusions of lemma 3.1 hold.

Proof

Observe that

$$(v_{x,z,r}\, s,s) = (u_r s,s)-(C_x U_z^{-1} C_x^* s,s) \geqslant k_6\|s\|^2 - \|U_z^{-1}\|\ \|C_x^*\|^2\|s\|^2$$
$$\geqslant (k_6-k_5^2/k_2)\ \|s\|^2$$

holds for all $s \in h$ and $(x,z,r) \in P$; hence

$$\|v_{x,z,r}^{-1}\| \leqslant (k_6-k_5^2/k_2)^{-1} .$$

From now on the proof continues as in lemma 3.1 and is omitted.

Let us now return to the general problem formulated at the beginning of section 2 and make more precise our congecture about its solution.

In view of the lemmas above, the rank condition as well as the positive definiteness condition of theorem 2.1 should be strengthened so as to hold uniformly in the parameters. More precisely we congecture that under the following type of hypotheses

- the map G is bounded and satisfies: $\|G^T(x)\ s\| \geqslant k_1\ \|s\|$, for all $s \in R^m$, all $x \in X$, and some $k_1 > 0$, where $G^T(x)$ denotes the transpose of $G(x)$;
- the map Q is bounded, its values are symmetric (matrices), and satisfies $(Q(x;y)\ s,\ s) \geqslant k_2\ \|s\|^2$, for all $s \in N(G(x))$, all $(x;y) \in X\times Y$, and some $k_2 > 0$;

it is likely to hold a global estimate of the following type

$$r^* = \max\{0,\ \sup_{x,y}\ \max_{1\leqslant i\leqslant m}\ [-(BQB^T|_{x,y})_{ii} + \sum_{j=1,\neq i}^{m} |(BQB^T|_{x,y})_{ij}| + \hat{M}]\} .$$

4. SOME NUMERICAL RESULTS

The estimate of the penalty parameter arrived at here has been used - in a simple way - within the standard multiplier method scheme, and implemented in a computer program. This latter has been tested in particular on the following example of constrained minimization problem (P):

$$(4.1) \quad \text{minimize } f(x) = x_1 x_2 x_3 x_4 x_5 \text{ , subject to}$$
$$g_1(x) = x_1^2+x_2^2+x_3^2+x_4^2+x_5^2 - 10 = 0$$
$$g_2(x) = x_2 x_3 - 5x_4 x_1 = 0$$
$$g_3(x) = x_5^3+x_2^3 + 1 = 0$$

whose solution (to 4 figures) is

$$f(\bar{x}) = -2.920, \quad \bar{x} = (-0.7636 , 1.596 , 1.827 , -0.7636 , -1.717).$$

Numerical results for (4.1) were given originally by Powell [9] and subsequently by Asaadi [2] where, incidentally, there is a misprint in $g_2(x_2 - x_3$ instead of $x_2x_3)$. Actually in [2], the initial point x_o used in solving (4.1) is declared

$$x_o = (-1, 1.5, 2, -1, -2) \quad ,$$

and (4.1) is among a number of examples solved by means of a few different nonlinear programs with the purpose of comparing their efficiency. Let Ni stand for number of iterations (outer iterations if there are as in the multiplier method inner iterations i.e. within each unconstrained minimization) and Ne stand for number of function evaluations. From [2] one sees that to solve (4.1) to 4 figures (Ni; Ne) is

(4.2)
- (7;148) with the method of Kowalik, Osborne and Ryan
- (6, 90) with Powell's version of the multiplier method
- (6,126) with Fiacco and McCormick penalty method (SUMT).

By using the minimizations (1.2) with the iterations (1.3) initiated at $y_o = 0$ and with the penalty parameter r_k equal for all k to the constant r^* computed in correspondance of the initial point, we have obtained the solution with (Ni; Ne) = (3;78). The usefulness of inserting the estimate r^*, is not however in such possible slight improvement of the efficiency in terms of "number of function evaluations" as used in [2]; actually in [2] it is not said how has been fixed r_k nor are tabulated initial guesses if any - i.e. minimizations (1.2) and iteration (1.3) - that might have been necessary in order to find "good values" of the penalty parameter, for instance in the computations (4.2); such possible guess work, that need not indeed be negligible, is reduced or avoided by using an estimate such as r*.

REFERENCES

[1] Arrow K. J. and Solow, R. M., "Gradient Methods for Constrained Maxima with Weakened Assumptions", in Studies in Linear and Nonlinear Programming, Arrow, K., Hurwicz, L., and Uzawa, H. (eds.), Stanford University Press, Stanford, 1958.

[2] Asaadi J., "A Computational Comparison of some Non-linear Programs", Math. Prog., Vol. 4, 1973, pp. 144-154.

[3] Bertsekas D. P., "On Penalty and Multiplier Methods for Constrained Minimization", SIAM J. on Control, to appear.

[4] Fiacco A.V. and McCormick G. P., "Nonlinear Programming: Sequential Unconstrained Minimization Techniques", J. Wiley, New York, 1968.

[5] Fletcher R., "A class of methods for nonlinear programming with termination and convergence properties", in Integer and Nonlinear Programming, J. Abadie (editor), North-Holland, 1970.

[6] Hestenes M.R., "Multiplier and Gradient Methods", Journal of Optimization Theory and Applications, Vol. 4, No. 5, 1969; pp. 303-320.

[7] Luenberger D.G., "Introduction to Linear and Nonlinear Programming", Addison-Wesley, 1973.

[8] Polyak B.T., "The Convergence Rate of the Penalty Function Method", Zh. Vychisl. Mat. Mat. Fiz., Vol. 11, No. 1, 1971, pp. 3-11.

[9] Powell M.J.D., "A Method for Nonlinear Constraints in Minimization Problems", in Optimization, R. Fletcher (ed.), Academic Press, pp. 283-298.

[10] Rockafellar R.T., "Augmented Lagrange Multiplier Functions and Duality in Nonconvex Programming", SIAM J. Control , Vol. 12, No.2, 1974.

[11] Rockafellar R.T., "New applications of duality in convex programming", written version of talk at 7th International Symposium on Math. Programming (the Hague, 1970) and elsewhere, published in the Proc. of the 4th Conference on Probability (Brasov, Romania, 1971).

OPTIMISATION SANS CONTRAINTES :

Construction d'une famille d'algorithmes à convergence quadratique par la linéarisation.

J. DENEL

Dept Informatique, Université de Lille I
B.P. 36 59650 VILLENEUVE D'ASCQ

Abstract :

An algorithm using first order approximation of a real-valued function is described for the solution of the unconstrained maximization of a continuous differentiable function ; convergence is proved and a rate of convergence is determined under the usual second order assumptions. Identity is shown between the proposed algorithm and the more general conjugate directions algorithm.

0. INTRODUCTION - NOTATIONS

Dans le problème de l'optimisation sans contraintes d'une fonction différentiable, on distingue deux types d'approche : les méthodes directes et les méthodes de prédiction selon la terminologie de Powell. Les premières requièrent une optimisation unidimensionnelle à chaque étape tandis que la maximisation de la fonction sur une droite n'est plus utilisée dans les secondes.

Les méthodes du premier type qui nous intéressent ici sont caractérisées par l'utilisation de directions conjuguées successives. On trouve par exemple dans cette classe de méthodes, d'une part les méthodes de gradients conjugués (dues à Hestenes et Stiefel [15] et appliquées à l'optimisation par Fletcher et Reeves [13] tout d'abord puis par exemple par Polak et Ribière [20]) et d'autre part les méthodes quasi-newton (Broyden [2,3], Davidon [7], Fletcher et Powell [12] et de nombreuses autres).

Ces algorithmes, encore appelés méthodes de "métrique-variable" ont suscité la formulation de schémas généraux (Adachi [1], Greenstadt [14], Huang [16]) pour le cas quadratique.

Huang montre que le déroulement itératif de ces méthodes est identique (pour les mêmes conditions initiales) dans le cas quadratique ; Dixon [9] donne une condition nécessaire et suffisante du déroulement identique des algorithmes de Huang pour une fonction différentiable quelconque. Les problèmes de convergence sont abordés par Powell [21] pour la méthode D.F.P., Broyden [3] pour ses méthodes de rang 1 et 2, Daniel [6] et Polak [20] pour leur méthode. De même certains auteurs prouvent la convergence superlinéaire de quelques algorithmes (Powell [21], Broyden-Dennis-Moré [4]) tandis que Cohen montre la convergence quadratique de trois algorithmes particuliers de gradients conjugués soumis à une réinitialisation tous les r pas ($r \geq n$).

En fait, le trait essentiel de ces méthodes est qu'elles rentrent, pour le cas quadratique, dans un schéma très général qui consiste à choisir, à chaque pas, une direction de recherche conjuguée de toutes les précédentes. Cependant toutes les méthodes restreignent, à chaque pas, le choix d'une nouvelle direction par des formules de récurrence sur les matrices de "métrique-variable".

Nous proposons dans ce papier un algorithme qui permet de conserver, à chaque pas, la généralité du schéma cité ci-dessus. Dans le cas quadratique, l'ensemble des directions conjuguées de toutes les précédentes s'identifie avec l'ensemble des solutions optimales d'un programme linéaire construit à partir de linéarisation de la fonction ; les programmes linéaires successifs ne diffèrent que par une seule contrainte supplémentaire ; ils ont au plus n contraintes. L'application au cas non quadratique suit le même schéma.

A une étape donnée, la détermination d'une direction conjuguée des précédentes revient, en définitive, à résoudre un système d'équations linéaires. Nous montrons que le schéma proposé permet d'atteindre les solutions de ce système sous-déterminé par l'utilisation de la méthode duale-simplex (enchaînement de programmes linéaires). De plus il est montré en I.4 l'analogie entre la recherche de ces solutions et la recherche de "centres" au sens de Huard [17] du polyèdre défini par des linéarisations du domaine non linéaire. En fait, en utilisant la notion de F-distance [17], l'ensemble des directions admissibles (i.e conjuguées des précédentes) est identique à un sous espace d'"égale distance" des linéarisations de la fonction.

L'algorithme présenté ici, est l'application à l'optimisation sans contraintes d'une méthode introduite dans l'algorithme partiel de la méthode des centres linéarisée de Huard [18] pour accélérer sa convergence pratique lors de la résolution du problème général d'optimisation non linéaire sous contraintes [8].

Dans la section I, après la description de l'algorithme, il est prouvé, sous l'unique hypothèse de continue-différentiabilité, que tout point d'accumulation de la suite générée est un point stationnaire. La section II montre l'identité avec le schéma de directions conjuguées rappelé ci-dessus. La vitesse de convergence de l'algorithme proposé est déterminée dans la section III. Sous les hypothèses usuelles relatives aux dérivées du second ordre, il est prouvé un taux de convergence d'ordre 2.

NOTATIONS :

- sur $\mathbb{R}^n$

  x^t le vecteur ligne transposé de x

  (x,y) le produit scalaire usuel sur $\mathbb{R}^n$

  $|x|$ norme induite sur $\mathbb{R}^n$ de x

  $||A||$ la norme de matrices carrées engendrée par la norme de vecteurs $|.|$

  $[a_1,\dots,a_k]$ le sous espace vectoriel engendré par les vecteurs a_j $j = 1,\dots,k$

  $[a_1,\dots,a_k]^\perp$ l'orthogonal de $[a_1,\dots,a_k]$

- pour une fonction numérique $f : \mathbb{R}^n \to \mathbb{R}$ 2 fois différentiable

$\nabla f(x)$ le gradient de f en x

$H(x)$ le hessien de f en x (matrice des dérivées secondes)

$\hat{f}$ l'approximation quadratique de f en x_o

I. Description de l'algorithme. Convergence

On considère le problème de maximisation sans contraintes :

$$(P)\left|\begin{array}{l}\text{maximiser } f(x)\\ \quad x \in \mathbb{R}^n\end{array}\right.$$

où f est une fonction continûment différentiable telle que, pour un point quelconque x_o de $\mathbb{R}^n$, l'ensemble $E_o = \{x \in \mathbb{R}^n \mid f(x) \geq f(x_o)\}$ est compact. On se donne $\delta \in]0,1]$.

I.1) Description de l'algorithme A

à l'itération i on dispose d'un point x_i de départ, d'une direction arbitraire c_o vérifiant $|c_o| = 1$ et $(\nabla f(x_i),c_o) \geq \delta|\nabla f(x_i)|$.

On pose $y_o = x_i$.

1) définir $y_1 = y_o + \Theta_o c_o$ tq $(\nabla f(y_1),c_o) = o$ et $\Theta_o > 0$

$\omega_1 = y_o + \gamma_o c_o$ tq $f(\omega_1) = f(y_o)$

2) étape k ($k<n$)

on dispose de $y_o,\dots,y_k$ $\quad y_j = y_{j-1} + \Theta_{j-1}c_{j-1}$ et $(\nabla f(y_j),c_{j-1}) = 0$

$y_o,\omega_1,\dots,\omega_k$ $\quad \omega_j = y_{j-1} + \gamma_{j-1}c_{j-1}$ et $f(\omega_j) = f(y_o)$

* choisir c_k de norme 1 tq

$$(\nabla f(y_o),c_k) = (\nabla f(\omega_1), c_k) = \dots = (\nabla f(\omega_k),c_k)$$

** définir $y_{k+1} = y_k + \Theta_k c_k$ avec $(\nabla f(y_{k+1}),c_k) = 0$

$\omega_{k+1} = y_k + \gamma_k c_k$ avec $f(\omega_{k+1}) = f(y_o)$

stop si $\nabla f(y_{k+1}) = 0$.

*** si $k+1 = n$ aller en 3 sinon retour en 2 avec $k+1$ au lieu de k.

3) poser $x_{i+1} = y_n$ et fin de l'itération i.

I.2) CONVERGENCE

Théorème 1 : Sous l'hypothèse précédente sur f *(continue-différentiabilité) tout point d'accumulation* x^* *de la suite* (x_i) *vérifie* $\nabla f(x^*) = 0$.

Démonstration : Nous utiliserons le théorème de Zangwill que l'on trouve dans Dubois [10] ou FIOROT-HUARD [11]. Ce résultat est rappelé ci-dessous :

$E \subset R^n$ est compact, $P \subset E$ fermé, γ une application multivoque $E \to \mathcal{P}(E)$

$h : E \to R$ continue sur E

si $\forall\, x \in E-P$ a) $\gamma(x) \neq \emptyset$

b) $x' \in \gamma(x) \Longrightarrow h(x') > h(x)$

alors l'algorithme :

$$\left\| \begin{array}{l} x_0 \in E,\ x_{i+1} \in E \text{ tq } h(x_{i+1}) \geq h(y_i) \text{ avec } y_i \in \gamma(x_i) \\ \qquad x_{i+1} = x_i \text{ sinon} \end{array} \right.$$

génère une suite $(x_i)_{i \in N}$ telle que tout point d'accululation x^* de $(x_i)_{i \in N}$ appartient à P si l'application γ est sup-continue en x^*.

De manière évidente, pour l'application de ce résultat, nous définirons $P = \{x \in \mathbb{R}^n \mid \nabla f(x) = 0\} \subset E_0$ et l'application multivoque : $E_0 \to \mathcal{P}(E_0)$ par :

$$\gamma : x \to \gamma(x) = \{y \in E_0 \mid y = x + \lambda h \ ;\ h \in \mathbb{R}^n,\ |h| = 1,$$
$$(\nabla f(x),h) \geq \delta\, |\nabla f(x)|,\ (\nabla f(y),h) = 0\}$$

Prouvons que γ est sup-continue sur E_0 :

si $(x_k)_{k \in N}$ est une suite de E_0 convergeant vers x

$(y_k)_{k \in N}$ une suite de E_0 convergeant vers y telle que $y_k \in \gamma(x_k)$ $\forall k$

alors par définition de $\gamma(x_k)$, $\forall k$ on a

$$- \ y_k = x_k + \lambda_k h_k \quad \lambda_k = |y_k - x_k|,\ |h_k| = 1$$

$$- \ (\nabla f(x_k),h_k) \geq \delta |\nabla f(x_k)|$$

$$- \ (\nabla f(y_k),h_k) = 0$$

comme $|h_k| = 1$ $\forall k$, h_k appartient à un compact de $\mathbb{R}^n$, il existe donc une sous-suite $k' \in N' \subset N$ telle que $h_{k'} \to \bar{h}$ et le triplet $(x_{k'}, y_{k'}, h_{k'})_{k' \in N'}$ converge vers $(x,y,\bar{h})$.

Dans ces conditions, f étant continûment différentiable, le passage à la limite dans les trois relations ci-dessus entraîne $y \in \gamma(x)$. D'autre part en prenant pour fonction h la fonction f, il est évident que le résultat de Zangwill s'applique et tout point d'accumulation x^* de la suite générée par l'algorithme A vérifie $\nabla f(x^*) = 0$.

I.4) CONSTRUCTION DES DIRECTIONS c_k ADMISSIBLES A L'ETAPE k DE A

Par définition c est admissible à l'étape k si et seulement si

$$c \in [a_1,\ldots,a_k]^\perp \text{ avec } a_1 = \nabla f(\omega_1) - \nabla f(y_o)$$

Une méthode simple pour construire le sous espace des directions admissibles consiste à envisager à l'étape k le programme linéaire à n+1 variables $x \in \mathbb{R}^n$, $\mu \in \mathbb{R}$

$$\text{PL}(k) \left| \begin{array}{ll} \max \mu & \\ (\nabla f(y_o),\ x-y_o) = \mu & \\ (\nabla f(\omega_j),\ x-\omega_j) = \mu & j = 1,\ldots,k \\ (x,\mu) \in \mathbb{R}^{n+1} & \end{array} \right.$$

toute direction d'infinitude de PL(k) est une direction admissible pour l'étape k de l'itération i de A.

Il est à noter que le passage de l'étape k à l'étape k+1 introduit une contrainte supplémentaire au programme PL(k) ; la résolution de PL(k+1) peut donc s'effectuer par la technique duale-simplex. Notons, par ailleurs, que la résolution de PL(k) revient à chercher l'ensemble des points à "égale distance" au sens de Huard [17] de toutes les contraintes d'un polyèdre linéaire défini par

$$\{x \mid (\nabla f(y_o),x-y_o) \geq 0,\ (\nabla f(\omega_j),x-\omega_j) \geq 0\ j=1,\ldots,k.$$

Les contraintes de ce cône polyèdrique d'appui au domaine $E_i = \{x \mid f(x) \geq f(x_i)\}$ sont les linéarisations de la fonction f aux points $y_o,\omega_1,\ldots,\omega_k$.

II. APPLICATION DE L'ALGORITHME $\mathcal{A}$ A L'OPTIMISATION D'UNE FONCTION QUADRATIQUE

Considérons le problème (P.Q)

$$\left| \begin{array}{l} \max f(x) = \frac{1}{2} x^t Qx + qx \\ x \in \mathbb{R}^n \end{array} \right. \qquad \text{Q matrice symétrique définie négative.}$$

Le principal résultat de cette section est la preuve de la convergence finie de A pour résoudre P.Q.

Proposition 2.1.

Pour toute étape k, *on suppose que* $c_o, c_1, \ldots, c_{k-1}$ *sont conjuguées entre elles alors*

a) *si* $c \in \mathbb{R}^n$ *vérifie*

$$(\nabla f(y_o), c) = (\nabla f(\omega_1), c) = \ldots = (\nabla f(\omega_k), c)$$

c *est conjuguée des directions* $c_o, c_1, \ldots, c_{k-1}$

b) *réciproquement, si* c *est conjuguée de* $c_o, c_1, \ldots, c_{k-1}$ *alors (2.1) est vérifiée.*

Démonstration :

a) prouvons tout d'abord $(c, Qc_o) = 0$

partant de $(\nabla f(\omega_1), c) = (\nabla f(y_o), c)$ et de $\nabla f(\omega_1) - \nabla f(y_o) = \gamma_o Q c_o$ le résultat est évident. Supposons donc $(c, Qc_\ell) = 0$ $\ell = 0, \ldots, j-1$ et prouvons $(c, Qc_j) = 0$. En utilisant $(\nabla f(\omega_{j+1}, c) = (\nabla f(y_o), c)$ il vient

$$0 = (\nabla f(\omega_{j+1}) - \nabla f(y_o), c) = \gamma_j (c, Qc_j) + \sum_{r=1}^{j} \Theta_{r-1}(c, Qc_{r-1})$$

d'où, avec l'hypothèse de récurrence, la conclusion

b) réciproquement, si c est conjuguée de $c_o, c_1, \ldots, c_{k-1}$, on écrit comme précédemment

$$\omega_k - \omega_k = \gamma_{k-1} c_{k-1} - \gamma_{j-1} + \sum_{r=j+1}^{k-1} \Theta_{r-1}\, c_{r-1}$$

avec $j \leq k$

d'où $(c, Q(\omega_k - \omega_j)) = 0$ et (2.1) est vérifiée.

Théorème 2 : Pour une fonction quadratique strictement concave, l'algorithme A *avec un point de départ* x_o *quelconque et une direction quelconque* c_o, *non orthogonale au gradient de* f *en* x_o, *fournit à la première itération en au plus* n *étapes, la solution de (P.Q).*

C'est un résultat classique sur les algorithmes de directions conjuguées.

En conclusion l'algorithme A appliqué à une fonction quadratique s'identifie donc à l'algorithme le plus général de directions conjuguées.

III. Vitesse de convergence de l'algorithme A dans le cas général

III.0) INTRODUCTION

Hypothèses sur f

H1) pour $x_0 \in \mathbb{R}^n$, $E_0 = \{x \mid f(x) \geq f(x_0)\}$ compact

H2) f est 2 fois continûment différentiable et vérifie une condition de Lipschitz au voisinage de l'optimum :
$\exists B > 0 \quad ||H(x) - H(x')|| \leq B|x-x'|$

H3) $\exists \ell > 0 \ \forall x \in E_0 \ \forall y \in \mathbb{R}^n \quad \ell|y|^2 \leq |(y,H(x).y)|$

$\exists L > 0 \ \forall x \in E_0 \ ||H(x)|| \leq L$

La convergence de A vers l'unique point α solution optimale du problème (P) est assurée par le théorème 1. Le résultat de cette section porte sur la vitesse de convergence de A qui est, à partir d'un certain rang i_0 d'ordre 2 ie
$\exists c < +\infty \ \forall i \geq i_0 \ |x_{i+1} - \alpha| \leq c|x_i - \alpha|^2$.
Signalons enfin que notre résultat est semblable à celui de Cohen [5] qui sous les mêmes hypothèses sur f montre la convergence quadratique des 3 algorithmes : Daniel, Fletcher et Reeves, Polak et Ribière.

III.1) VITESSE DE CONVERGENCE

Théorème 3 : L'algorithme A, sous les hypothèses H1, 2, 3 a un taux de convergence d'ordre 2.

Démonstration en III.2.

L'approche utilisée dans ce papier pour mesurer le taux de convergence de A est la même que celle de Cohen ou originellement celle de Daniel [6]. Donnons-nous, pour l'itération i, les points engendrés par A

$$y_0 = x_i, y_1, \ldots, y_n = x_{i+1}$$

et les directions $c_0, c_1, \ldots, c_{n-1}$ associées, et l'approximation quadratique $\hat{f}$ de f en x_i.

Nous allons montrer qu'il est possible à l'étape j (j=0,...,n-1) de définir, à partir de la direction c_j, une direction $\hat{c}_j$ conjuguée, pour la matrice des dérivées secondes de $\hat{f}$, par rapport aux directions précédentes $\hat{c}_o, \hat{c}_1, \ldots, \hat{c}_{j-1}$ et telle que

- $|c_j - \hat{c}_j| = \sigma(|x_i - \alpha|)$

- l'étape j de A appliqué à $\hat{f}$ avec la direction $\hat{c}_j$ fournisse un point $\hat{y}_{j+1}$ vérifiant $|\hat{y}_{j+1} - y_{j+1}| = \sigma(|x_i - \alpha|^2)$.

Si pour j=0,...,n-1 ce qui précède est possible (voir preuve en III.2), le théorème II implique que pour un rang q($\leq$n), le point $\hat{y}_q$ sera l'optimum de $\hat{f}$, que nous notons $\hat{\alpha}$, et on aura la relation $|y_q - \hat{\alpha}| = \sigma(|x_i - \alpha|^2)$.

D'autre part, il est connu (voir par exemple Ortega et Rheinboldt [19 page 312] que $\hat{\alpha} = \hat{y}_q$ est obtenu par une itération de la méthode de Newton (ordre 2) avec x_i comme point de départ ; ceci, avec ce qui précède, prouve que

$$\exists i_o \quad i \geq i_o \quad |y_q - \alpha| = \sigma(|x_i - \alpha|^2)$$

De plus, un développement de Taylor à l'ordre 2 en α permet de montrer

$$|x_{i+1} - \alpha| \leq \sqrt{\frac{L}{\ell}}\, |y_q - \alpha|$$

Cette dernière remarque prouvera le résultat énoncé.

III.2) DEMONSTRATION DU THEOREME 3

La démonstration est constructive pour chaque j et utilise une série de lemmes présentés en annexe.

a) Définissons $\hat{c}_o = c_o$ et $\hat{y}_1 = y_o + \hat{\Theta}_o \hat{c}_o$, $\hat{\omega}_1 = y_o + \hat{\gamma} \hat{c}_o$ avec

$$(\nabla \hat{f}(\hat{y}_1), \hat{c}_o) = 0 \text{ et } \hat{f}(\hat{\omega}_1) = f(y_o)$$

on a $|\hat{y}_1 - y_1| = \sigma(x_i - \alpha|^2)$

en effet $|\hat{y}_1 - y_1| = |\Theta_o c_o - \hat{\Theta}_o \hat{c}_o| = |\hat{\Theta}_o - \Theta_o|$ puisque $\hat{c}_o = c_o$ et $|\hat{c}_o| = 1$

or par construction, y_1 est le maximum de f sur la direction c_o, il existe donc

$$\eta_o \in (y_o, y_1) \text{ tq } \Theta_o = \frac{(\nabla f(y_o), c_o)}{(c_o, H(\eta_o) . c_o)} \quad \text{de même } \hat{\Theta}_o = -\frac{(\nabla f(y_o), c_o)}{(c_o, H(x_i) . c_o)}$$

les hypothèses $H_{2,3}$ et le lemme A1 entraînent le résultat.

b) supposons que pour ℓ=0,...,j-1 des directions $\hat{c}_\ell$ de norme 1 admissibles pour l'algorithme A appliqué à $\hat{f}$ ont été construites ; on dispose donc des points :

$y_o, \hat{y}_1, \ldots, \hat{y}_j$: $\hat{y}_{\ell+1} = \hat{y}_\ell + \hat{\Theta}_\ell \hat{c}_\ell$ $(\nabla\hat{f}(\hat{y}_{\ell+1}), \hat{c}_\ell) = 0$ $\hat{\Theta}_\ell$ peut être négatif

$y_o, \hat{\omega}_1, \ldots, \hat{\omega}_j$: $\hat{\omega}_{\ell+1} = \hat{y}_\ell + \hat{\gamma}_\ell \hat{c}_\ell$ $\hat{f}(\hat{\omega}_{\ell+1}) = f(y_o)$ avec $\hat{\gamma}_\ell > 0$

d'après la remarque de I.4 si $b_k = \nabla\hat{f}(\hat{\omega}_k) - \nabla f(y_o)$ $k=1,\ldots,j$ on a $\hat{c}_\ell \in [b_1,\ldots,b_\ell]^\perp$ $\ell=1,\ldots,j-1$.

Faisons l'hypothèse de récurrence :

(3.1) 1) $|y_{\ell+1} - \hat{y}_{\ell+1}| = \sigma(|x_i-\alpha|^2)$

$\ell = 0,\ldots,j-1$

(3.2) 2) $|c_\ell - \hat{c}_\ell| = \sigma(|x_i-\alpha|)$

d'après le a) ci-dessus, (3.1) et (3.2) sont vraies pour $\ell=0$.

D'après les remarques de III.1) nous devons associer à la direction c_j (connue par le déroulement fixé de A à l'itération i), une direction $\hat{c}_j$ conjuguée de $\hat{c}_o,\ldots,\hat{c}_{j-1}$.

Définissons donc $\hat{c}_j = \dfrac{\tilde{c}_j}{|\tilde{c}_j|}$ avec

(3.3) $\tilde{c}_j = c_j + \sum_{k=0}^{j-1} \lambda_k \hat{c}_k$ les λ_k déterminés pour que $(\tilde{c}_j, b_\ell) = 0$ $\ell=1,\ldots,j$, ceci étant une condition nécessaire et suffisante pour que c_j soit conjuguée des précédentes.

Prouvons a) que (3.2) est vérifiée pour $\ell=j$ avec cette définition

b) que (3.1) est vérifiée si $\hat{y}_{j+1}$ est le maximum de $\hat{f}$ sur la direction $\hat{c}_j$ passant par $\hat{y}_j$.

a) Montrons que $|\hat{c}_j - c_j| = \sigma(|x_i-\alpha|)$

Proposition 3.1. $\forall k$, $k=0,\ldots,j-1$, $|\lambda_k| = \sigma(|x_i-\alpha|)$

Démonstration : remarquons que $(c_j, a_k) = 0$ $k=1,\ldots,j$ par définition de l'algorithme A (cf. I.4) ; en conséquence, le second membre du système linéaire d'inconnues (λ_k) a pour composantes les produits scalaires $d_r = (c_j, a_r - b_r)$ $r=1,\ldots,j$.

Le lemme A4 entraîne

(3.4) $\quad |d_k| = \sigma(|x_i-\alpha|^2)$

D'autre part, puisque $b_k = \nabla f(\hat{\omega}_k)-\nabla f(y_o) = H(x_i).(\hat{\omega}_k-y_o)$ on peut écrire la relation de récurrence sur les vecteurs b_k

(3.5)
$$\begin{cases} b_1 = \hat{\gamma}_o H(x_i).c_o \\ b_k = b_{k-1} - (\hat{\gamma}_{k-2}-\hat{\Theta}_{k-2})\, H(x_i).\hat{c}_{k-2} + \hat{\gamma}_{k-1} H(x_i).\hat{c}_{k-1} \quad k=2,\ldots,j \end{cases}$$

le système linéaire donnant les α_k est triangulaire inférieur et du fait de la conjugaison des directions $\hat{c}_\ell$ ($\ell=0,\ldots,j-1$) les éléments diagonaux sont

$$(\hat{c}_{k-1},b_k) = \hat{\gamma}_{k-1}\ (c_{k-1},H(x_i).\hat{c}_{k-1}) \qquad k=1,\ldots,j$$

et en vertu du lemme A2 on a

(3.6) $\quad |(\hat{c}_{k-1},b_k)| \geq m_2\ |x_i-\alpha|$

ceci prouve que $|\lambda_o| = \sigma(|x_i-\alpha|)$ et que si

(3.7) $\quad |\lambda_o|,\ldots,|\lambda_{k-1}|$ $k=1,\ldots,j-1$ sont des $\sigma(|x_i-\alpha|)$, alors la k+1^e équation entraîne $|\lambda_k| = \sigma(|x_i-\alpha|)$; en effet elle s'écrit :

$$\lambda_k(\hat{c}_k,b_{k+1}) + \sum_{r=0}^{k-1} \lambda_r(\hat{c}_r,b_{k+1}) = (c_j,a_{k+1}-b_{k+1})$$

par majoration

$$|\lambda_k|\ |(\hat{c}_k,b_{k+1})| \leq |(\hat{c}_j,a_{k+1}-b_{k+1})| + \sum_{r=0}^{k-1} |\lambda_r|\ |(\hat{c}_r,b_{k+1})|$$

la relation (3.4), le lemme 1, l'hypothèse (3.7) entraînent, avec la relation (3.6)

$$|\lambda_k| \leq M|x_i-\alpha|$$

<u>cqfd</u>.

Maintenant si nous notons $u \in \mathbb{R}^n$, $u = \tilde{c}_j-c_j$ la proposition 3.1 avec $|\hat{c}_r| = 1$ $r=0,\ldots,j-1$ entraîne

(3.8) $\quad |u| = \sigma(|x_i-\alpha|)$ et donc le résultat

(3.9) | la relation (3.3) définit une direction conjuguée des directions $\hat{c}_\ell$ $\ell=0,\ldots,j-1$ telle que $|\hat{c}_j-c_j| = \sigma(|x_i-\alpha|)$

b) Montrons, qu'avec cette définition, $|\hat{y}_{j+1}-y_{j+1}| = \sigma(|x_i-\alpha|^2)$

Proposition 3.2 $|\Theta_j-\hat{\Theta}_j| = \sigma(|x_i-\alpha|^2)$.

Démonstration : Par construction de y_{j+1} et $\hat{y}_{j+1}$ (maxima exacts) on a

$$|\Theta_j-\hat{\Theta}_j| = \frac{|(\hat{c}_j,H(x_i).\hat{c}_j)\ (\nabla f(y_j),c_j) - (c_j,H(\eta_j).c_j)\ (\hat{\nabla f}(\hat{y}_j),\hat{c}_j)|}{(\hat{c}_j,H(x_i).\hat{c}_j)(c_j,H(\eta_j).c_j)}$$

expression que nous pouvons majorer avec l'hypothèse H3 par

$$\begin{aligned}
\ell^2|\Theta_j-\hat{\Theta}_j| \leq\ & |(\hat{c}_j-c_j,\ H(x_i).(\hat{c}_j+c_j))\ (\nabla f(y_j),c_j)| \\
& + |(c_j,[H(x_i)-H(\eta_j)].c_j)\ (\nabla f(y_j),c_j)| \\
& + |(c_j,H(\eta_j).c_j)\ (\nabla f(y_j),c_j-\hat{c}_j)| \qquad (3.10) \\
& + |(c_j,H(\eta_j).c_j)(\nabla f(y_j)-\ f(\hat{y}_j),\hat{c}_j)| \\
& + |(c_j,H(\eta_j).c_j)(\nabla f(\hat{y}_j)-\hat{\nabla f}(\hat{y}_j),\hat{c}_j)|
\end{aligned}$$

le lemme A1, les relations (3.1) et (3.9) fournissent la preuve de la proposition 3.2 en majorant chaque ligne de (3.10).

Proposition 3.3 $|y_{j+1}-\hat{y}_{j+1}| = \sigma(|x_i-\alpha|^2)$

Démonstration :

$$y_{j+1} = y_j + \Theta_j c_j \text{ et } \hat{y}_{j+1} = \hat{y}_j + \hat{\Theta}_j\hat{c}_j \text{ d'où}$$

$$|y_{j+1}-\hat{y}_{j+1}| \leq |y_j-\hat{y}_j| + |\hat{\Theta}_j||c_j-\hat{c}_j| + |\Theta_j-\tilde{\Theta}_j|$$

or l'hypothèse H_3 entraîne $|\hat{\Theta}_j| \leq \frac{1}{\ell}\ |(\hat{\nabla f}(\hat{y}_j),\hat{c}_j)|$ mais

$\hat{\nabla f}(\hat{y}_j) = \nabla f(y_o) + \sum_{k=1}^{j} \hat{\Theta}_{k-1}\ H(x_i)\ \hat{c}_{k-1}$ et par la conjuguaison des directions $\hat{c}_r$ il vient $(\hat{\nabla f}(\hat{y}_j),\hat{c}_j) = (\nabla f(y_o),\hat{c}_j)$. Le lemme A2 prouve alors $|\hat{\Theta}_j| = \sigma(|x_i-\alpha|)$.

La démonstration s'achève par (3.1) pour ℓ=j-1, (3.9) et la proposition (3.2).

En conclusion, ce qui précède, prouve qu'il existe un algorithme de directions conjuguées, associé au déroulement de l'itération i de A appliqué à f, convergeant vers $\hat{\alpha}$ en au plus n itérations. Les explications de III.1 apportent la preuve du théorème 3.

ANNEXE

Lemme A1.

* $\forall t\ f(t) \geq f(x_i) \Longrightarrow |t-x_i| \leq (1+\sqrt{\frac{L}{\ell}})\ |x_i-\alpha|$

* $\forall t\ \hat{f}(t) \geq f(x_i) \Longrightarrow |t-x_i| \leq \frac{2L}{\ell}\ |x_i-\alpha|$

* $|\nabla f(x_i)| < L|x_i-\alpha|$

* $|\nabla f(x_i)| \geq \ell|x_i-\alpha|$

* *sous l'hypothèse* $(\nabla f(x_i),c_0) \geq \delta|\nabla f(x_i)|$, *on a* $\frac{1}{\gamma_0} \geq m|x_i-\alpha|$

* $\hat{f}(\hat{y}_1) - f(x_i) \geq m_1|x_i-\alpha|^2$

* $\forall x \in E_0\ |f(x)-\hat{f}(x)| = \delta(|x_i-\alpha|^3)$

* $|\nabla f(y_j)| = \delta(|x_i-\alpha|)$

* $|\nabla f(y_j) - \nabla f(\hat{y}_j)| \leq L|y_j-\hat{y}_j|$

* $|\nabla f(\hat{y}_j) - \nabla\hat{f}(\hat{y}_j)| = \sigma(|x_i-\alpha|^2)$

* $|a_k| = |\nabla f(\omega_k) - \nabla f(x_i)| = \sigma(|x_i-\alpha|)\ \forall k < n$

* $|b_k| = |\nabla\hat{f}(\hat{\omega}_k) - \nabla f(x_i)| = \sigma(|x_i-\alpha|)$

Les démonstrations des points précédents utilisés dans la démonstration du théorème 3 sont évidentes par l'utilisation de développements de Taylor.

Lemme A2. $\forall k,\ k=0,\dots,q-1,\ \hat{\gamma}_k \geq m_2|x_i-\alpha|$

$\hat{\gamma}_k$ est défini par $\hat{\omega}_{k+1} = \hat{y}_k + \hat{\gamma}_k\hat{c}_k\ \hat{f}(\hat{\omega}_{k+1}) = f(x_i)$ et $\hat{\gamma}_k > 0$

1) si $(\nabla\hat{f}(\hat{y}_k),\hat{c}_k) \geq 0$, $\hat{\gamma}_k$ est défini par

(A1) $\quad 0 = -\hat{f}(\hat{\omega}_{k+1})+\hat{f}(\hat{y}_k) + \hat{\gamma}_k(\nabla\hat{f}(\hat{y}_k),\hat{c}_k) + \frac{\hat{\gamma}_k^2}{2}(\hat{c}_k,H(x_i).\hat{c}_k)$

d'où $\quad \frac{\hat{\gamma}_k^2}{2} \mid (\hat{c}_k,H(x_i).\hat{c}_k)\mid \geq \hat{f}(\hat{y}_k)-f(x_i) \geq \hat{f}(\hat{y}_1) - f(x_i) \geq m_1|x_i-\alpha|^2$

2) si $(\nabla\hat{f}(\hat{y}_k),\hat{c}_k) < 0$, $\hat{\gamma}_k$ est racine positive de (A.1)

Considérons le point $\hat{\omega}_{k+1}$ et sa projection u sur le convexe $\{x \mid \hat{f}(x) \geq \hat{f}(\hat{y}_k)\}$ on a $\hat{f}(u) = \hat{f}(\hat{y}_k)$ et $u-\hat{\omega}_{k+1} = rc \quad r > 0$ en notant

$$c = \frac{\nabla\hat{f}(u)}{|\nabla\hat{f}(u)|}.$$

Il est évident que $\hat{\gamma}_k > r$ (projection) ; d'autre part r est la racine positive de l'équation $\frac{r^2}{2}\beta - r(\nabla f(u),c) + \beta' = 0$

avec $\left|\begin{array}{l}\beta = (c,H(x_i).c) < 0\\ \beta' = \hat{f}(u) - f(x_i) > 0\end{array}\right.$

$$\text{on a } r = \frac{2\beta'}{(\nabla f(u),c) + \sqrt{(f(u),c)^2 - 2\beta\beta'}}$$

or $(\nabla f(u),c) \leq k_1 \, |x_i-\alpha|$; de plus $-\beta \leq L$ et $\beta' \leq k_2|x_i-\alpha|^2$; il vient

$$r \geq \frac{2\beta'}{k|x_i-\alpha|} \geq m_2|x_i-\alpha|$$

<u>*Lemme A3*</u> : $\forall j \leq q, \forall k = 1,\ldots,j \; |\omega_k-\hat{\omega}_k| = \sigma(|x_i-\alpha|^2)$

<u>Démonstration</u> : On utilise l'hypothèse de récurrence (3.1) et (3.2) par construction

$$\omega_k = y_k + (\gamma_{k-1}-\Theta_{k-1})\, c_{k-1} = y_k + \varepsilon_{k-1}\, c_{k-1}$$

$$\hat{\omega}_k = \hat{y}_k + (\hat{\gamma}_{k-1}-\hat{\Theta}_{k-1})\, \hat{c}_{k-1} = \hat{y}_k + \hat{\varepsilon}_{k-1}\, \hat{c}_{k-1}$$

par choix (remarque du I.2) on a $\varepsilon_{k-1} > 0$ et par construction du point $\hat{\omega}_k$ ($\hat{\gamma}_{k-1} > 0$ quel que soit le signe de Θ_{k-1}), d'où,

$$|\hat{\omega}_k-\omega_k| \leq |y_k-\hat{y}_k| + |\varepsilon_{k-1}\, c_{k-1} - \hat{\varepsilon}_{k-1}\, \hat{c}_{k-1}|$$

(A.2) $$\leq |y_k-\hat{y}_k| + |\hat{\varepsilon}_{k-1}||\hat{c}_{k-1} - c_{k-1}| + |\hat{\varepsilon}_{k-1} - \varepsilon_{k-1}|$$

(A.3) $\quad - |y_k - \hat{y}_k| = \sigma(|x_i-\alpha|^2)$ pour $k \leq j$ $\quad$ cf. hypothèse (3.1)

(A.4) $\quad - |\hat{c}_{k-1} - c_{k-1}| = \sigma(|x_i-\alpha|)$ pour $k \leq j$ $\quad$ cf. hypothèse (3.2)

Maintenant, ε_{k-1} et $\hat{\varepsilon}_{k-1}$ sont racines positives des équations

$$\text{(A.5)} \quad \left| \begin{array}{l} \frac{1}{2}(\varepsilon_{k-1})^2 (c_{k-1}, H(\eta_k).c_{k-1}) + f(y_k) - f(x_i) = 0 \qquad \eta_k \in (y_k, \omega_k) \\ \\ \frac{1}{2}(\hat{\varepsilon}_{k-1})^2 (\hat{c}_{k-1}, H(x_i).\hat{c}_{k-1}) + \hat{f}(\hat{y}_k) - f(x_i) = 0 \end{array} \right.$$

comme le montre un développement de Taylor de $f(\omega_k)$ et $\hat{f}(\hat{\omega}_k)$ en y_k et $\hat{y}_k$ en conséquence

$$\text{(A.6)} \qquad |\varepsilon_{k-1} - \hat{\varepsilon}_{k-1}| = \frac{|\hat{\varepsilon}^2_{k-1} - \varepsilon^2_{k-1}|}{\varepsilon_{k-1} + \hat{\varepsilon}_{k-1}} \leq \frac{1}{\hat{\varepsilon}_{k-1}} |\varepsilon^2_{k-1} - \hat{\varepsilon}^2_{k-1}|$$

$$\text{on a } \hat{\varepsilon}^2_{k-1} = 2 \frac{\hat{f}(\hat{y}_k) - f(x_i)}{|(\hat{c}_{k-1}, H(x_i).\hat{c}_{k-1})|} \geq 2 \frac{\hat{f}(\hat{y}_1) - f(x_i)}{|(\hat{c}_{k-1}, H(x_i)\hat{c}_{k-1})|}$$

$$\geq \frac{2}{L} m_1 |x_1 - \alpha|^2$$

par l'hypothèse H_3 et le lemme A1.

En utilisant les équations (A5) on écrit

$$\frac{\varepsilon^2_{k-1} - \hat{\varepsilon}^2_{k-1}}{2} = \frac{(c_{k-1}, H(\eta_k).c_{k-1})\ (\hat{f}(\hat{y}_k) - f(x_i)) - (\hat{c}_{k-1}, H(x_i).\hat{c}_{k-1})\ (f(y_k) - f(x_i))}{(c_{k-1}, H(\eta_k).c_{k-1})\ (\hat{c}_{k-1}, H(x_i).\hat{c}_{k-1})}$$

le dénominateur est minoré par ℓ^2 (cf. H_3) et le numérateur, en valeur absolue, est majoré par

$$|(c_{k-1} - \hat{c}_{k-1}, H(\eta_k).(\hat{c}_{k-1} + c_{k-1}))|\,|\hat{f}(\hat{y}_k) - f(x_i)|$$

$$+ |(\hat{c}_{k-1}, [H(\eta_k) - H(x_i)]\ .\ \hat{c}_{k-1})|\,|\hat{f}(\hat{y}_k) - f(x_i)|$$

$$+ |(\hat{c}_{k-1}, H(x_i).\hat{c}_{k-1})|\,|\hat{f}(\hat{y}_k) - \hat{f}(y_k)|$$

$$+ |(\hat{c}_{k-1}, H(x_i).\hat{c}_{k-1})|\,|\hat{f}(y_k) - f(y_k)|$$

l'hypothèse H_2, les hypothèses de récurrence (3.1) et (3.2), le lemme A1 montrent que le numérateur est un $\sigma(|x_i-\alpha|^3)$.

En utilisant (A.6) et (A.7) il vient :

$$|\varepsilon_{k-1} - \hat{\varepsilon}_{k-1}| = \sigma(|x_i-\alpha|^2)$$

Il est , par ailleurs, facile de prouver en utilisant (A.5) et l'hypothèse H_3 que

$$|\hat{\varepsilon}_{k-1}| = \sigma(|x_i-\alpha|)$$

d'où avec $(A_{2,3,4})$, $|\omega_k-\hat{\omega}_k| = \sigma(|x_i-\alpha|^2)$

cqfd.

Lemme A4 :

$$\forall j,\ j \leqslant q,\ \forall k = 1,\dots,j\ |a_k-b_k| = \sigma(|x_i-\alpha|^2).$$

Démonstration :

$$a_k = \nabla f(\omega_k) - \nabla f(x_i) = H(\eta_k).(\omega_k-x_i)\quad \eta_k \in (x_i,\omega_k)$$

$$b_k = \hat{\nabla f}(\hat{\omega}_k) - \nabla f(x_i) = H(x_i).(\hat{\omega}_k-x_i)$$

d'où

$$a_k - b_k = [H(\eta_k) - H(x_i)].(\omega_k-x_i) + H(x_i)\ .\ (\omega_k-\hat{\omega}_k)$$

en utilisant H_2, H_3 et les lemmes A1 et A3 il vient :

$$|a_k - b_k| = \sigma(|x_i-\alpha|^2)$$

cqfd.

B I B L I O G R A P H I E

[1] ADACHI (N)
On variable-metric algorithms.
Journal of Optimization theory, vol.7 (6) 1971, p.391-410.

[2] BROYDEN (C.G.)
Quasi-Newton methods and their application to function minimization.
Mathematics of Computation, vol. 21 (99) 1967 p.368-381.

[3] BROYDEN (C.G.)
The convergence of a class of double rank minimization algorithms.
J. Inst. Math. and App. vol.6 (1) 1970, p.76-90.

[4] BROYDEN-DENNIS-MORE
On the local and superlinear convergence of quasi-newton methods.
J. Inst. Math. and App. vol.12 (3) 1973, p.223-245.

[5] COHEN (A.I.)
Rate of convergence of several conjugate gradient algorithms.
SIAM Num. Analysis, vol.9 (2) 1972 p.248-259.

[6] DANIEL (J.W.)
The conjugate gradient method for linear and non linear operator equations.
SIAM Num. Analysis, vol.4 (1) 1967, p.10-26.

[7] DADIDON (W.C.)
Variance algorithm for minimization.
Computer Journal vol.10 (4) 1968, p.406-410.

[8] DENEL (J.)
- *Résolution de problèmes d'optimisation non linéaires par la méthode des centres linéarisée.*
  Bul. D.E.R. de E.D.F. série C n°1 1973.
- *Adaptation and performance of the linearized method of centers.*
  cahiers du CERO, vol.16 (4) 1974, p.447-457.

[9] DIXON (L.C.W.)
Variable metric algorithms : Necessary and sufficient conditions for identical behavior of non quadratic function.
J. of Optimization vol.10 (1) 1972, p.34-40.

[10] DUBOIS (J.)
Theorems of convergence for improved non linear programming algorithms.
Operations Res. vol.21 1973, p.328-332.

[11] FIOROT-HUARD
Composition et réunion d'algorithmes généraux.
Publication 43, mai 1974, Laboratoire de Calcul Université de Lille I France.

[12] FLETCHER-POWELL
A rapidly convergent descent method for minimization.
Computer Journal vol.6 (2) 1963, p.163-168.

[13] FLETCHER-REEVES
Function minimization by conjugate gradients.
Computer Journal vol.7 (2) 1964, p.149-154.

[14] GREENSTADT (J.)
Variations on the variable metric methods.
Math. of Computation vol.24 (109) 1970, p.1-30

[15] HESTENES-STIEFEL
Methods of conjugate gradients for solving linear systems.
Journal of Res. of NBS vol.49 (6) 1952, p.409.

[16] HUANG (H.Y)
Unified approach to quadratically convergent algorithms for function minimization.
Journal of Optimization vol.5 (6) 1970, p.405-423.

[17] HUARD (P.)
Resolution of mathematical programming problems with non linear constraints by the method of centers.
Non linear programming (Ed. Abadie) North Holl. Publishing C° 1967.

[18] HUARD (P.)
Programmation mathématique convexe.
RIRO (7) 1968 p.43-59.

[19] ORTEGA-RHEINBOLDT
Iteration solution of non linear equations in several variables.
Academic Press 1970.

[20] POLAK-RIBIERE
Note sur la convergence de méthodes de directions conjuguées.
RIRO n°16 1969, p.35-43.

[21] POWELL (M.J.D.)
. *Rank one methods for unconstrained optimization.*
Integer and non linear programming (Ed. Abadie) 1970, North Holland.
. *On the convergence of the variable metric algorithm.*
Journal of Inst. Maths. Appl. vol.7 (1) 1971, p.21-36.

OPTIMIZATION IN LARGE PARTLY NONLINEAR SYSTEMS

Arne Drud
IMSOR
The Institute of Mathematical Statistics
and Operations Research
The Technical University of Denmark
DK 2800 Lyngby - Denmark

0. Abstract.

This paper describes a method for optimizing large partly nonlinear systems. The method is based on the GRG-algorithm, that solves problems with nonlinear objective function and nonlinear equality constraints. The original GRG-algorithm is described and its relations with LP are stressed. Some storage problems in large problems are discussed, and a special inversion procedure for the GRG-algorithm is presented. Some special kinds of constraints, inequalities and linear constraints, are considered, and it is shown, how their special features can be utilized. Finally some computational results with the method are given.

1. Introduction.

The purpose of the research project, that is described in this paper, was to develop an optimization procedure, that could be used in connection with large econometric models.

Econometric models have some features, that are important for the way the optimization procedure is designed. First of all, most constraints in the model are equality constraints. Many equations are linear, but there are usually also some nonlinear equations. There can be lower and/ or upper bounds on many variables. And finally there are only a few active variables in each constraint.

The GRG-algorithm by J. Abadie, [1], [2], and [3], is made for problems of this kind, and a first approach was therefore to use the GRG69 computer program, that ranges very high on the list of Colville, [5]. But we very soon ran into problems, because the program needed too much core storage for arrays.

This paper describes, how a GRG-type program can be designed, taking into account the sparseness of the Jacobian of the constraints. The program uses less core storage than the GRG69 program, and due to a fast inversion procedure, it is also faster.

In section 2 the ideas in the GRG-algorithm are described, and some important subproblems are described in section 3 and 4. Section 5 explains how the Jacobian can be stored, and section 6 contains an inversion procedure specially designed for the GRG-algorithm. Section 7 is on the special treatment of inequalities and linear equations, and section 8 shows some computational results. The conclusion is in section 9.

2. The Generalized Reduced Gradient Method.

In this section the main ideas in the GRG-method will be explained for easy reference in the rest of the paper.

Consider the problem:

$$\max \quad z = f(\underline{x}) \tag{1}$$

$$\text{subject to} \quad \underline{g}(\underline{x}) = \underline{0} \tag{2}$$

$$\text{and} \quad \underline{\alpha} \leq \underline{x} \leq \underline{\beta} \tag{3}$$

where $\underline{x}$, $\underline{\alpha}$, and $\underline{\beta}$ are m-vectors and $\underline{g}$ is an n-dimensional vectorfunction, $(n < m)$. The functions f and $\underline{g}$ are assumed to be at least one time differentiable with known continuous partial derivatives.

A set of n equalities like (2) can be used to eliminate n variables. Partition the $\underline{x}$-vector into $\underline{x}_b$ and $\underline{x}_n$, where $\underline{x}_b$ has n elements and $\underline{x}_n$ has m-n elements, and transform (2) into $\underline{x}_b = \underline{g}_1(\underline{x}_n)$. Maybe it is not possible to find an analytic expression for $\underline{g}_1$, but from the theory of implicit functions we have the following theorem: If $(\underline{x}_b^o, \underline{x}_n^o)$ satiesfies $\underline{g}(\underline{x}_b^o, \underline{x}_n^o) = \underline{0}$ and the Jacobian $\frac{\partial \underline{g}}{\partial \underline{x}_b}$ has rank n, i.e. is nonsingular, then in a neighbourhood of $(\underline{x}_b^o, \underline{x}_n^o)$ it is possible to transform $\underline{g}(\underline{x}_b, \underline{x}_n) = \underline{0}$ into $\underline{x}_b = \underline{g}_1(\underline{x}_n)$. The function $\underline{g}_1$ is differentiable, and the Jacobian is found by implicit differentiation:

$$\frac{\partial \underline{g}}{\partial \underline{x}_b} \cdot \frac{\partial \underline{x}_b}{\partial \underline{x}_n} + \frac{\partial \underline{g}}{\partial \underline{x}_n} = \underline{0}$$

$$\frac{\partial \underline{g}_1}{\partial \underline{x}_n} = \frac{\partial \underline{x}_b}{\partial \underline{x}_n} = -\left(\frac{\partial \underline{g}}{\partial \underline{x}_b}\right)^{-1} \cdot \frac{\partial \underline{g}}{\partial \underline{x}_n} \tag{4}$$

The equation $\underline{x}_b = \underline{g}_1(\underline{x}_n)$ is now introduced in the objective function:

$$z = f(\underline{x}_b,\underline{x}_n) = f(\underline{g}_1(\underline{x}_n),\underline{x}_n) = F(\underline{x}_n)$$

where $F(\underline{x}_n)$ is differentiable with derivative:

$$\frac{\partial F}{\partial \underline{x}_n} = \frac{\partial f}{\partial \underline{x}_b}\cdot\frac{\partial \underline{x}_b}{\partial \underline{x}_n} + \frac{\partial f}{\partial \underline{x}_n} = \frac{\partial f}{\partial \underline{x}_n} - \left(\frac{\partial f}{\partial \underline{x}_b}\cdot\left(\frac{\partial \underline{g}}{\partial \underline{x}_b}\right)^{-1}\right)\cdot\frac{\partial \underline{g}}{\partial \underline{x}_n} \qquad (5)$$

The problem (1), (2), and (3) can now be reformulated:

$$\max \quad z = F(\underline{x}_n) \qquad (6)$$

$$\text{subject to} \quad \underline{\alpha}_n \leq \underline{x}_n \leq \underline{\beta}_n \qquad (7)$$

$$\text{and} \quad \underline{\alpha}_b \leq \underline{x}_b \leq \underline{\beta}_b \qquad (8)$$

$$\text{where} \quad \underline{x}_b = \underline{g}_1(\underline{x}_n) \qquad (9)$$

The transformations done until now are very similar to those done in an LP-problem. $\underline{x}_b$ is the set of basic variables. They are introduced to compensate for changes in the non-basic $\underline{x}_n$-variables, so that (2) will still hold. The derivatives $\frac{\partial F}{\partial \underline{x}_n}$ are similar to the reduced costs in an LP-problem. They measure the influence on the objective function of changes in the non-basic variables, taking into account the correspon-ding changes in the basic variables. $\frac{\partial F}{\partial \underline{x}_n}$ is called the reduced gradient. The matrix $\frac{\partial \underline{g}}{\partial \underline{x}_b}$ plays the same role as the basic-matrix does in LP.

A major difference between this problem and an LP-problem is, that in the optimal solution this problem can have more than n variables between bounds. Thus, it is not possible to use an optimization procedure like the simplex, that only works with basic solutions.

The major steps in the GRG-algorithm are now as shown in fig. 1. In 2) the basic variables are chosen strictly between the bounds, so that they can actually compensate for changes in $\underline{x}_n$ without exceeding a bound at once. The vector $\underline{u}$ in 4) is similar to the simplex multipliers in LP, and $\underline{rg}$ is the reduced gradient found in (5). $\underline{h}$ is the reduced gradient projected on the simple inequality constraints, and if it is zero, the Kuhn-Tucker conditions are satiesfied. While choosing the optimal θ in 8) it can be valuable to notice, that $\frac{\partial z}{\partial \theta}\big|_{\theta=0} = \underline{h}^T\cdot\underline{rg}$. Step 8b) will be described in next section.

The algorithm in fig. 1 uses the steepest ascend method. If the nonbasic variables are the same from iteration to iteration, we are optimizing $F(\underline{x}_n)$ with the same $\underline{x}_n$-variables, and therefore it is possible to use

1) Find a first feasible solution, $\underline{x}^o$.

2) Calculate $\frac{\partial g}{\partial \underline{x}}$ and choose n variables among the $\underline{x}$-variables, $\underline{x}_b$, so that $\underline{\alpha}_b < \underline{x}_b^o < \underline{\beta}_b$ with a strict inequality, and so that $\frac{\partial g}{\partial \underline{x}_b}$ is nonsingular. Name the rest of the variables $\underline{x}_n$.

3) Find the inverse of $\frac{\partial g}{\partial \underline{x}_b}$, $(\frac{\partial g}{\partial \underline{x}_b})^{-1}$.

4) Compute $\underline{u}^T = -(\frac{\partial f}{\partial \underline{x}_b})^T \cdot (\frac{\partial g}{\partial \underline{x}_b})^{-1}$.

5) Compute $\underline{rg} = \frac{\partial f}{\partial \underline{x}_n} + \underline{u}^T \cdot \frac{\partial g}{\partial \underline{x}_n}$

6) Find $\underline{h}$ as: $h_i = \begin{cases} 0 \text{ if } rg_i < 0 \text{ and } x_{ni}^o = \alpha_{ni} \\ 0 \text{ if } rg_i > 0 \text{ and } x_{ni}^o = \beta_{ni} \\ rg_i \text{ else} \end{cases}$

7) If $\underline{h} = \underline{0}$, then stop.

8) Choose θ to maximize $F(\underline{x}_n^o + \theta \cdot \underline{h})$ by solving a)-c) for different θ-values.

a) $x_{ni} = \begin{cases} \alpha_{ni} \text{ if } x_{ni}^o + \theta \cdot h_i < \alpha_{ni} \\ \beta_{ni} \text{ if } x_{ni}^o + \theta \cdot h_i > \beta_{ni} \\ x_{ni}^o + \theta \cdot h_i \text{ else} \end{cases}$

b) Find $\underline{x}_b = \underline{g}_1(\underline{x}_n)$

c) $z = f(\underline{x}_b, \underline{x}_n)$

9) Store the best solution until now in $\underline{x}^o$ and go to 2.

Figur 1. The GRG-algorithm.

a conjugate gradient method. Fletcher-Reeves method [6] is probably the best method, because it requires a very limited amount of core storage.

3. The implicit function $\underline{x}_b = \underline{g}_1(\underline{x}_n)$.

A very important problem in optimization problems with equality constraints is to stay on the surface defined by $\underline{g}(\underline{x}) = \underline{0}$, or in the words of the GRG-algorithm: Find $\underline{x}_b = \underline{g}_1(\underline{x}_n)$.

The function $\underline{g}_1$ is not known explicitly, so $\underline{x}_b$ must be found as a solution to the set of equations $\underline{g}(\underline{x}_b, \underline{x}_n) = \underline{0}$, where $\underline{x}_n$ is known, i.e. n nonlinear equations must be solved for n unknown.

The classical method is the Newton-Raphson method:

$$g(\underline{x}_b+\Delta\underline{x}_b,\underline{x}_n) \simeq g(\underline{x}_b,\underline{x}_n) + \frac{\partial g}{\partial \underline{x}_b}\cdot\Delta\underline{x}_b = \underline{0}$$

$$\Delta\underline{x}_b = -\left(\frac{\partial g}{\partial \underline{x}_b}\right)^{-1}\cdot g(\underline{x}_b,\underline{x}_n)$$

$$\underline{x}_b := \underline{x}_b + \Delta\underline{x}_b$$

It is an iterative method, and with a good starting point it is very fast. As a stop criterion we can use

$$\sum_{i=1}^{n}(g_i(\underline{x}_b,\underline{x}_n))^2 < \varepsilon_{new}$$

or any other convenient measure of precision.

It can be very timeconsuming to calculate $\frac{\partial g}{\partial \underline{x}_b}$ and invert it in each new point. But $\frac{\partial g}{\partial \underline{x}_b}$ is continuous, so for small steplengths the matrix calculated in step 2 of the GRG-algorithm and the inverse from step 3 can be used. When a constant Jacobian is used, the method is sometimes called the pseudo Newton-Raphson method. The use of the same Jacobi-matrix and its inverse both in step 4 of the GRG-algorithm and to compute the implicit function g_1 is one of the main advantages of the GRG-algorithm.

If the steplength θ in step 8 is too large, $g(\underline{x}_b,\underline{x}_n)$ can be far from $\underline{0}$, and the inverse Jacobian from step 2 may be far from the inverse Jacobian in the actual point, so the pseudo Newton-Raphson method does not converge. The best treatment of this problem is probably to stop after a fixed number of iterations and then decrease the value of θ.

4. Lower and upper bounds on the basic variables.

Until now only the bounds on the nonbasic variables have been treated. When the set of nonlinear equations $g(\underline{x}_b,\underline{x}_n) = \underline{0}$ is solved with respect to $\underline{x}_b$, the guess for $\underline{x}_b$ is changed in each pseudo Newton-Raphson iteration. What can be done, if one or more basic variables exceed a bound? The problem is basically the same that occure in LP, when a basic variable becomes too small or too large.

The situation is: $\underline{x}_b$ lies within the bounds and $\underline{x}_b+\Delta\underline{x}_b$ lies outside. Choose the largest α so that $\underline{x}_b+\alpha\Delta\underline{x}_b$ lies inside or on the bounds, and replace $\underline{x}_b$ by $\underline{x}_b + \alpha \cdot \Delta\underline{x}_b$. This operation should decrease the error in $\underline{g}(\underline{x}_b, \underline{x}_n) = \underline{0}$, but not as much as for $\alpha = 1$. Now one of the basic variables is at a bound, and it cannot compensate for changes in $\underline{x}_n$. Therefore perform a change of basis: Choose an $\underline{x}_n$-variable, that lies strictly between the bounds and introduce it as a basic variable, while the one at bound is transfered to $\underline{x}_n$. The only thing we need in order to proceed with the pseudo Newton-Raphson method is the inverse Jacobian $\left(\frac{\partial \underline{g}}{\partial \underline{x}_b}\right)^{-1}$ with the new basic variables $\underline{x}_b$.

Only one column in $\frac{\partial \underline{g}}{\partial \underline{x}_b}$ is exchanged with another. This is exactly what happens in a change of basis in the simplex procedure, and the procedure for updating $\left(\frac{\partial \underline{g}}{\partial \underline{x}_b}\right)^{-1}$ is the same. The actual calculations depend on the way $\left(\frac{\partial \underline{g}}{\partial \underline{x}_b}\right)^{-1}$ is represented and will be described after the section on inversion procedures.

5. Storing the Jacobian.

Large problems means in this paper problems with more than 250 constraints and more than 300 variables. Therefore the Jacobian will be a matrix with more than 75,000 elements, and even for a large modern computer, this is a large matrix.

However in most reallife problems only a few variables are active in each equation. This means, that may be only 1000 - 2000 out of the possible 75,000 elements in the Jacobian are nonzero. So if only the nonzero elements are stored with a reference to the place in the matrix, a lot of core storage can be saved.

In the different steps of the GRG-algorithm we always use one column of the Jacobian at a time. So it is very efficient to store the elements column by column in a long vector with another integer-valued vector telling the row-number of each element, and a shorter vector telling where in the long vectors the different columns start.

Example:

$$\frac{\partial \underline{g}}{\partial \underline{x}} = \begin{Bmatrix} 1. & 0. & 0. & 0. & 9. \\ 0. & 3. & -5. & 7. & 0. \\ 0. & 4. & 6. & 0. & -10. \\ 2. & 0. & 0. & -8. & 11. \end{Bmatrix}$$

Storage pattern:

Start column vector: 1 , 3 , 5 , 7 , 9 , 12
Element value vector: 1., 2., 3., 4., -5., 6., 7., -8., 9., -10., 11.
Row number vector: 1 , 4 , 2 , 3 , 2 , 3 , 2 , 4 , 1 , 3 , 4

6. Inversion procedures.

Step 3 of the GRG-algorithm in fig. 1 read: "find the inverse of $\frac{\partial g}{\partial \underline{x}_b}$, $\left(\frac{\partial g}{\partial \underline{x}_b}\right)^{-1}$."

We do not need the inverse, which can easily be a 62,500 element matrix, in explicit form. Some representation of the inverse, that makes it possible to multiply a vector from the right or from the left with the inverse, is sufficient.

Fortunately these problems have been worked on for a long time in connection with linear programming, where the two matrix multiplications are used to find the simplex multipliers and to find a transformed column Some of the techniques have been described in [4], [8], [9], [10], and [11].

Section 6.1 will describe the idea of representing the inverse as a string of eta-vectors, and section 6.2 describes an inversion procedure, that performs the inversion and the selection of the basic variables simultaneously.

6.1 The inverse on product form.

According to the definition of an inverse matrix, the product $\underline{x} = \underline{\underline{A}}^{-1}\cdot\underline{y}$ is the solution to the set of equations $\underline{\underline{A}}\cdot\underline{x} = \underline{y}$. A set of linear equations is usually solved by performing a set of row operations on $\underline{\underline{A}}$ and $\underline{y}$. When $\underline{\underline{A}}$ has been transformed into the unit matrix $\underline{\underline{I}}$, then $\underline{y}$ has been transformed into $\underline{\underline{A}}^{-1}\cdot\underline{y} = \underline{x}$.

Consider the following row operations: multiply row i by $1/a_{ii}$, multiply row i by $-a_{ji}/a_{ii}$ and add it to row j, j = 1,...,n, j $\neq$ i. The rowoperations are equivalent to a multiplication from the left with the matrix $\underline{\underline{E}}_i$:

$$
\underline{\underline{E}}_i = \begin{Bmatrix} 1 & 0 & \dots & -a_{1i}/a_{ii} & \dots & 0 \\ \dots & \dots & \dots & \dots & \dots & \dots \\ 0 & 0 & \dots & 1/a_{ii} & \dots & 0 \\ \dots & \dots & \dots & \dots & \dots & \dots \\ 0 & 0 & \dots & -a_{ni}/a_{ii} & \dots & 0 \end{Bmatrix}
$$

Now the inverse can be represented as a sequence of $\underline{\underline{E}}$-matrices, each specifying some of the row operations. The $\underline{\underline{E}}$-matrices are usually called eta-vectors - vectors, because they only differ from the unit matrix in one vector.

The matrix $\underline{\underline{E}}_i$ can be stored very compact. First of all only the column $a_{.i}$ should be stored, and in this column it is enough to store the nonzero elements. As shown in section 6.2 it is possible to keep the number of nonzero elements low by choosing the pivot elements a_{ii} carefully. And with fewer eta-elements less core storage should be used, and fewer row operations must be performed during the use of the inverse.

6.2 The GRG-inversion procedure.

There is a main difference between the linear programming inversion procedure and the procedure, that is used in the GRG-problem. In the later there is an extra degree of freedom, because there are usually more variables, that can be choosen as basic variables, than needed.

The procedure described here is heavily based on the procedure for LP-problems described by Hellerman and Rarick, [8] and [9].

First a basic observation must be made. The first eta-vector is created very easily - it is created directly from the column in which we pivot first. The second eta-vector must be chosen from the matrix after the row operations symbolized by $\underline{\underline{E}}_1$ have been performed. But if the column to be used for pivot next is chosen, so it has no element in the first pivotrow, the row operations does not effect this column.

It is seen, that if the matrix can be transformed into lower-triangular form by rearranging rows and columns, and if the pivotelements are chosen along the diagonal from the upper left corner, no columns will have to be transformed before the eta-vectors are created.

Usually it is not possible to transform the matrix into lower-triangular form. The Hellerman-Rarick procedure tries systematically to create a lower-triangular matrix, and if it is not possible, the procedure selects the columns to be transformed (the so-called spikes) in such a way, that the rest of the matrix becomes almost triangular.

The procedure in its form for the GRG-problem will be described with a small illustrative example. In the following, only the columns strictly between bounds are used. Fig. 2 shows the zero-nonzero pattern of the matrix.

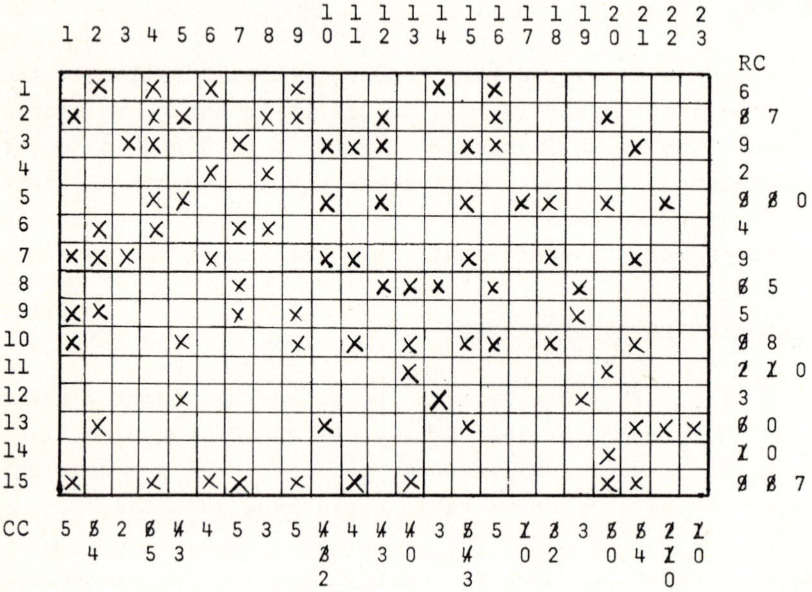

Figur 2: The zero-nonzero pattern of the matrix.

The first steps identifies the part of the matrix, that directly is lower-triangular:

1. Find the rowcounts (RC) and columncounts(CC), i.e. the number of nonzero elements in each row and column.
2. Is there an $RC_i = 1$? If not, go to 3.
   Find the corresponding column, choose the element as pivot-element, delete the column and revise RC and CC, and go to 2.
3. If all rows have a pivotelement, then go to 17.
4. Is there an $CC_j = 1$? If not, go to 5.
   a. If there are more columns with $CC_j = 1$, choose one with the smallest number of nonzero elements.

b. Find the corresponding row, choose the element as pivotelement, delete the row and revise RC and CC and go to 4.

5. If all rows have a pivotelement, then go to 17.

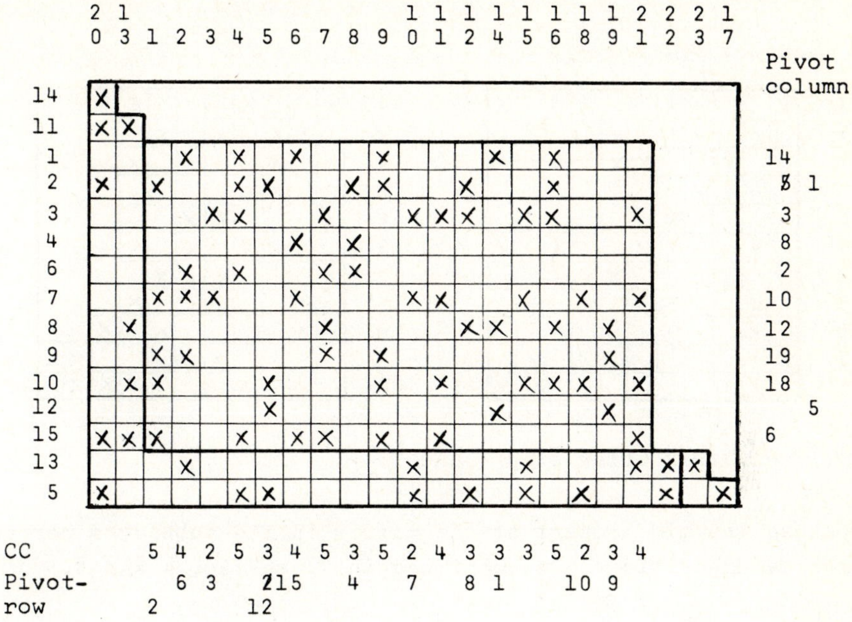

Figur 3: The matrix after step 1 to 5.

After rearranging the matrix, it is now divided into three parts: two lower-triangular parts and a part with either no elements or at least two elements pr row and pr column. See fig. 3. From now on only the part in the middle is considered.

6. Perform a temporary pivotassignment:
   a. Consider all columns starting with the columns with the smallest CC_j. Use the first nonzero element found in a row, that has no pivotelement yet, as pivot element. If a column cannot be assigned a pivot element after this procedure, the next column is considered.
   b. If all rows have a pivot element, then go to 7.
   c. Use the Ford and Fulkerson labelling method to reassign pivot elements, until we have a maximum matrix transversal, see [7].
7. Now consider the matrix of those columns, that were assigned pivot elements in step 6. Decompose this matrix into minimal nondecomposable submatrices, f.ex. with the predecessor-successor method in Hellerman and Rarick [9], or with the method of Steward [12].

Step 6 and 7 used on the example results in the situation in fig. 4. After the assignment in step 6a, row 12 is still without a pivot element, which is found by the labeling procedure.

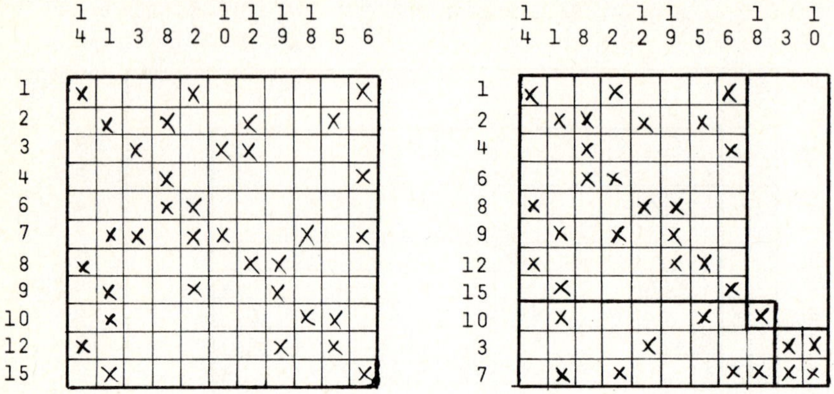

Figur 4: The matrix after step 6 and 7.

8. Decompose the middle part of the matrix in the subblocks corresponding to the decomposition in submatrices in 7, se fig. 4 and 5.

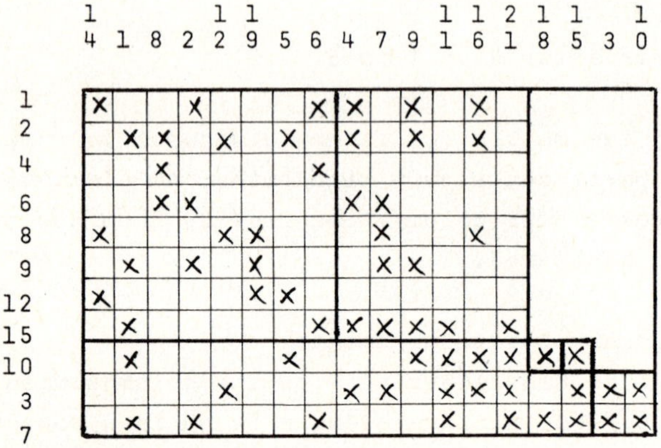

Figur 5: The subblocks formed in step 8.

The following steps 9 to 16 are now done for each subblock, starting with the upper left corner. The temporary pivotassignments are not used any more.

9. Find RC and CC for this subblock.
10. Perform the operations in step 4 on the subblock.

The result is now a matrix with at least two elements per row and per column, so it is not possible to get a lower triangular matrix directly. The procedure is now to select the columns, that should be transformed or that should be left out completely. These columns are called spikes. For the selection of spikes, Hellerman and Rarick defined a tallyfunction in the following way:

$t_k(j)$ = number of nonzero elements in column j, appearing in rows with RC $\leq$ k.

The selection of a spike is now done with the following procedure, where k starts as the smallest RC:

a. S = The set of columns for which $t_k(j)$ is maximum.
b. If S has only one element, then return.
c. If $t_{max} > 1$, then go to g.
d. k:= the smallest RC > k.
e. S:= the subset of S for which $t_k(j)$ is maximum.
f. Go to b
g. S:= the subset of S for which CC is maximum.
h. If S has only one element, then return.
i. S:= the subset of S for which the total column count is maximum.
j. Return with any element from S as the spike.

The ideas behind the selection procedure are:

a. Make the smallest rowcount smaller, so we can get a row with $RC_i = 1$.
b. If there are more possibilities, choose the one that will bring most small rowcounts down.
c. If there are still more possibilities, choose the column with the largest number of elements.

Now the pivot selection procedure can be continued:

11. k:= the smallest RC in rows without a pivot element.
12. If k = 0, go to 16, if k = 1, go to 14, else go to 13.
13. Select a spike, save its number, delete the column, revise RC and CC, and go to 11.
14. If there are more $RC_i = 1$, then go to 15, else select row no i and the corresponding column for final pivot, delete the column, revise RC and CC, and go to 11.
15. Select the pivot column. The procedure for selecting spikes is used with point i exchanged by
    i1. S:= the subset of S for which the total columncount is minimum.
    Choose the pivot row as one with $RC_i = 1$, delete the column, revise

RC and CC, and go to 11.

16. The last column in the list of spikes is transformed with the eta-vectors, and the resulting column is transformed into an eta-vector with pivot in a row with $RC_i = 0$ and without a pivot element. If all rows in the block have a pivot element now, this block has been finished, else go to 11.

Fig. 6 shows how the procedure works on the largest subblock. Fig. 7 shows the matrix after all rearrangements have been done, and fig. 8 shows the corresponding eta-vectors. Some columns has been divided into two eta-vectors, so that the transformation of the spike will only create new elements in the upper part of the spike. This idea is described by Beale in [4].

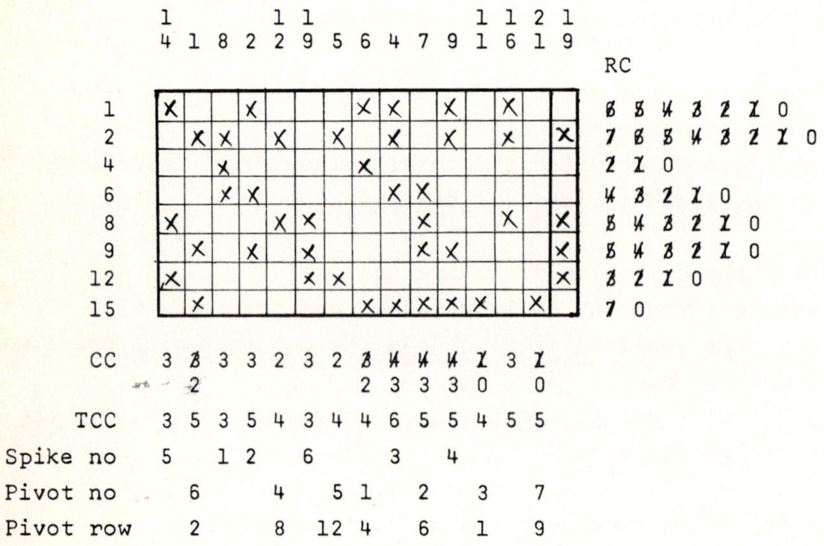

Figur 6: The final pivot selection in a subblock. RC and CC are row- and columncounts, TCC is columncounts in the original matrix, Spike no and Pivot no indicates, when a column has been chosen for either spike or pivot, Pivot row tells, in which row the pivot of the column can be found.

Sometimes a pivotelement will be found to be too small. In this case it is relatively easy to delete the corresponding column, because it is linearly dependent on earlier columns, and use the last spike-column instead.

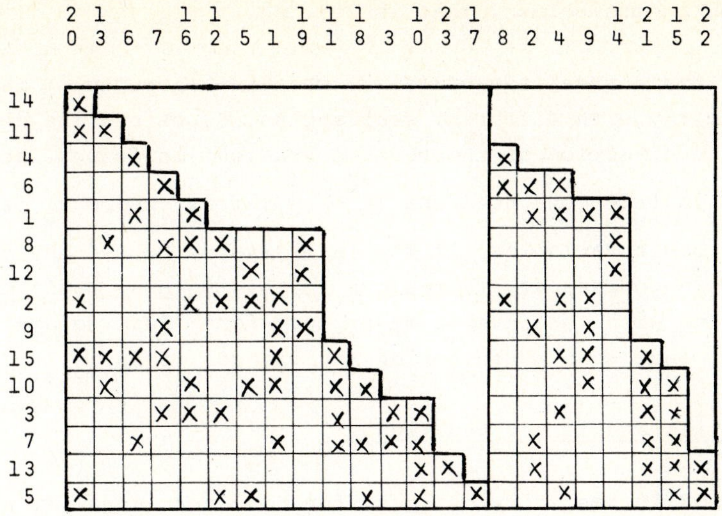

Figur 7: The matrix after the final rearrangements.

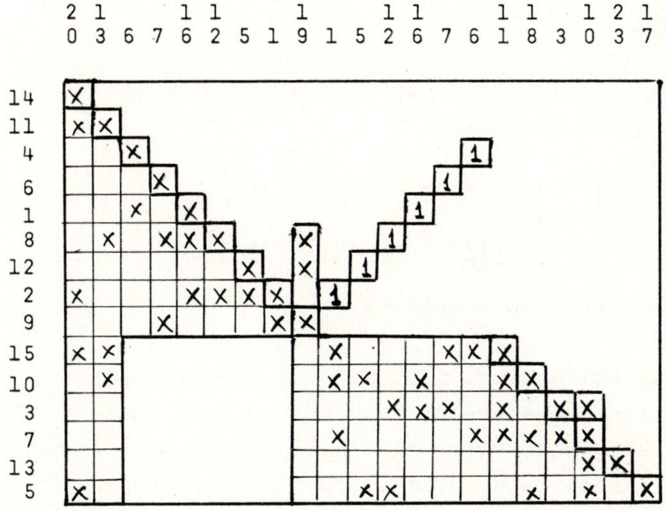

Figur 8: The resulting eta-vectors.

6.3 Updating the inverse after changes of basis.

In section 4 we saw, that a change of basis can occure in a Newton-Raphson iteration, and that the operations needed to update the inverse are exactly the same as those of the simplex method, i.e. transform the incomming column and use this column as an extra eta-vector with pivot in the same row as the column, that is leaving.

This method has a great advantage. In the GRG-algorithm it is often nessecary to try with different steplengths, θ, before the optimal steplength in the direction of the reduced gradient is found. Each time a new θ-value is tried, we need the $(\frac{\partial g}{\partial \underline{x}_b})^{-1}$ matrix, where $\underline{x}_b$ is the vector of original basic variables. If the last θ-value caused any changes of basis, it is now nessecary to restore the old inverse basis matrix. This is easily done with the inverse on product form: just delete the extra eta-vectors, that were added during the changes of basis.

7. Inequalities and linear equations.

The GRG-method is basically designed for nonlinear equality constraints. This section shows, what happens with inequalities and with linear equations.

7.1 Inequalities.

It is always possible to change an inequality into an equality constraint by adding a slack variable:

$$g_i(\underline{x}) \geq 0$$

$$g_i(\underline{x}) - s_i = 0, \quad 0 \leq s_i < \infty$$

There are now two cases to consider:

The constraint is not active, i.e. $s_i > 0$. Now the combined inversion and basis-selection procedure will select s_i as a basic variable during step 4, and the eta-vector will only have one nonzero element, the pivot element -1. It is easy to show, that the $\underline{u}$-vector in the GRG-algorithm will have a zero in position i, which means, that row no i has no influence on the reduced gradient. And in the Newton-Raphson iterations, equation no i will converge very fast, because s_i will just be assigned the value $g_i(\underline{x})$.

The constraint is active, i.e. $s_i = 0$. Now s_i cannot be used as a basic variable, and almost all calculations will be as if the equation had been $g_i(\underline{x}) = 0$.

It can be seen, that the practical calculations are very similar to those of a relaxation procedure, where only the active constraints are used when deciding where to go in next iteration.

7.2 Linear equations.

Although a model is nonlinear, it will usually have many equality constraints, such as continuity constraints, equations found by linear regression etc. Linear equations are mathematically much simpler than nonlinear ones, and it is indeed possible to take advantage of this simplicity.

Linear equations will have zero error after the first Newton-Raphson iteration, because the linear approximation used in the Newton-Raphson procedure is identical to the linear equation itself. Therefore it is not nessecary to compute the value of the $\underline{g}$-vector for the linear equations after the first Newton-Raphson iteration. This again will make the calculation of $\left(\frac{\partial \underline{g}}{\partial \underline{x}_b}\right)^{-1} \cdot \underline{g}$ faster, since it is not nessecary to perform the multiplications, where one of the factors is zero.

8. Computational experience.

8.1 User routines.

The computer program, that has been developed, uses four problem defining subroutines.

The first subroutine calculates the value of the objective function, and the second finds the gradient of the objective function. It is only nessecary to recompute the variable terms in the gradient, because the vector is not destroyed by the optimization routine.

The third subroutine calculates $\underline{g}$, the value of the constraints, and the fourth calculates the Jacobian. Again it is only nessecary to recompute the variable elements in the Jacobian.

8.2 Computer experiments.

The program, that is written in FORTRAN IV, has been tested at different problems, and fig. 9 shows some results. The inversion procedure used is an earlier version, that creates slightly more eta-elements than the one described in section 6.2. The idea in section 7.2 has not yet been

	Problem 1	Problem 2
Number of variables	65	136
Number of constraints, linear	44	75
nonlinear	10	41
total	54	116
Number of Jacobi elements, constant	132	263
variable	23	93
total	155	356
Avarage number of elements in the inverse on product form	222	615
Core storage in K bytes:		
User supplied subroutines	4	6
Arrays	13	27
Optimization routine and buffers	69	69
Total	86	102
Execution time in sec CPU:		
Until first feasible solution	0.41	1.70
Pr. iteration after first feasible solution	0.14	0.56
Total time incl input/output, ca.	6	20

Figur 9: Caracteristics of some test problems run on an IBM 370/165 with a FORTRAN H compiler.

implemented.

9. Conclusion.

This paper shows, how it is possible to use the ideas of the GRG-algorithm on large nonlinear models, if special care is taken to reduce the requirement for core storage.

Many methods and ideas from large scale linear programming can be used, directly or in a slightly changed form on the GRG-algorithm, because of the relations between the GRG-algorithm and linear programming.

Although the largest testproblem solved until now has had only 116 constraints, it is reasonable to expect, that problems with 500 constraints and 600 variables can be solved within a region of 250 K bytes.

10. References.

1. Abadie, J. and J. Carpentier: "Generalization of the Wolfe Reduced Gradient Method to the Case of Nonlinear Constraints", in Optimization, R. Fletcher (ed.), Academic Press, 1969.

2. Abadie, J.: "Application of the GRG Algorithm to optimal control problems", in Integer and Nonlinear Programming, J.Abadie (ed.), North-Holland, 1970.

3. Abadie, J.: "Optimization Problems with Coupled Blocks", in Economic Computation and Economic Cypernetics Studies and Research, p 5-26, vol 1970.4.

4. Beale, E.M.L.: "Sparseness in Linear Programming", in Large Sparse Sets of Linear Equations, J.K. Reid (ed.), Academic Press, 1971.

5. Colville, A.R.: "A comparative study of nonlinear programming codes", in Procedings of the Princeton Symposium on Mathematical Programming, H.W. Kuhn (ed.), Princeton University Press, 1970.

6. Fletcher, R. and C.M. Reeves:"Function Minimization by Conjugate Gradients", in British Computer Journal, vol 7, 1964.

7. Ford, L.R. and D.R. Fulkerson: "Flows in Networks", Princeton University Press, 1962.

8. Hellerman, E. and D. Rarick: "Reinversion with the Preassigned Pivot Procedure", in Mathematical Programming, vol 1, 1971.

9. Hellerman, E. and D. Rarick: "The partitioned Preassigned Pivot Procedure", in Sparse Matrices and their Applications, D.J. Rose and R.A. Willoughby (eds.), Plenum Press, 1972.

10. Markowitz, H.M.: "The elimination form of the inverse and its application to Linear Programming", in Management Science, vol 3, 1957.

11. Orchard-Hays, W.: "Advanced Linear Programming Computing Techniques", McGraw-Hill, 1968.

12. Steward, D.V.:"On an approach to techniques for the analysis of the structure of large systems of equations", in SIAM Review, vol 4, 1962.

A NEW BRANCH AND BOUND APPROACH FOR CONCAVE MINIMIZATION PROBLEMS

Reiner Horst

Fachbereich Mathematik der Technischen Hochschule D 61 Darmstadt, Schloßgartenstr. 1

Introduction

In this paper we shall consider the following programming problem: Find a vector $\bar{x} \in D$ satisfying

$$f(\bar{x}) = \min_{x \in D} f(x). \qquad \text{(Problem P)}$$

It is assumed throughout this paper that the set $D \subset \mathbb{R}^n$ is convex, compact and $f: A \rightarrow \mathbb{R}^n$ is a concave function on an open convex set A, $D \subset A \subset \mathbb{R}^n$.

For solving problem P a branch and bound algorithm is proposed, which solves a sequence of problems in each of which the objective function is linear. The main difference between this approach and previous approaches ([1],[4]) is the use of simplex-partitions instead of rectangular ones and a different refining rule such that the convergence proof does not rely on the concept of convex envelopes and the algorithm handles non-separable functions.

In the next section we shall give some useful preparatory results. The subsequent section contains a description of the algorithm and the last one an associated convergence theorem.
A generalization of this algorithm will appear in [3]. Proofs of statements given in [3] will be omitted here.

Simplex-partitions and affine subfunctions

Starting with a n-dimensional enclosing simplex S^1, $D \subset S^1 \subset A$, we shall get a sequence of subproblems by partitioning S^1 in from step to step refined subsimplices S^{ki} and by constructing affine underestimating functions $L^{ki}: S^{ki} \rightarrow \mathbb{R}$. This will be done in such a way that each L^{ki} will be the best convex subfunction of f over S^{ki}, that is for all convex functions $g^{ki}: S^{ki} \rightarrow \mathbb{R}$ satisfying $g^{ki}(x) \leq f(x)\ \forall\, x \in S^{ki}$ we shall have $g^{ki}(x) \leq L^{ki}(x) \leq f(x)\ \forall\, x \in S^{ki}$. The chosen refining rule than will yield a convergence theorem.

Definition 1:

Let S be a n-dimensional simplex in the euclidian space $\mathbb{R}^n$. A set $\{S^1,\dots,S^r\}_S$ of finitely many n-dimensional simplices S^i $(i=1,\dots,r)$ is said to be a simplex-partition of S if

$$S = \bigcup_{i=1}^{r} S^i,$$

$$S^i \cap S^j = S^{(ij)}, \qquad i,j \in \{1,\dots,r\},\ i\neq j,$$

where $S^{(ij)}$ denotes a (n-1)-dimensional simplex-face of both S^i and S^j.

Lemma 1:

Let $S \subset \mathbb{R}^n$ be a n-dimensional simplex (n-simplex), characterized by its n+1 affinely independent vertices $x_1,\dots,x_{n+1} \in \mathbb{R}^n$. We write $S = \{x_1,\dots,x_{n+1}\}$. Moreover let x be a point in S but no vertex:

$$x = \sum_{i=1}^{n+1} \alpha_i x_i,\ \alpha_i \geq 0 (i=1,\dots,n+1),\ \sum_{i=1}^{n+1} \alpha_i = 1;$$

$x \neq x_i$ $(i=1,\dots,n+1)$. Replacing one vertex x_i of S with $\alpha_i > 0$ by x we obtain a n-subsimplex $S^i = \{x_1,\dots,x_{i-1},x,x_{i+1},\dots,x_{n+1}\}$ of S.

The set of all by means of x in this way constructible subsimplices form a simplex-partition of S.

Proof:

It is well known from linear algebra, that given a set of affinely independent vectors $S = \{x_1,\dots,x_{n+1}\}$ and a vector $x = \sum_{i=1}^{n+1} \alpha_i x_i, \alpha_i \geq 0$ $(i=1,\dots,n+1)$, $\sum_{i=1}^{n+1} \alpha_i = 1$; $x \neq x_i$ $(i=1,\dots,n+1)$, then if any vector $x_i \in S$ for which $\alpha_i > 0$ is removed from S and x is added to S, the new collection S^i of (n+1) vectors is also a set of affinely independent vectors. It follows that all S^i constructed according to Lemma 1 are n-dimensional subsimplices of S and obviously we have $\bigcup S_i \subset S$.

To prove $S \subset \bigcup S^i$ take any element $\bar{x} \in S$, $\bar{x} \neq x$, and consider the ray emanating from x through $\bar{x}$ up to the boundary point $x^* \in S$ where it leaves S ($x^* = x$ is possible):

$$x^* = \sum_{j=i_1}^{i_m} \alpha_j^* x_j,\ \alpha_j^* \geq 0\ (j=i_1,\dots,i_m),\ \sum_{j=i_1}^{i_m} \alpha_j^* = 1;$$

$m<n+1$ since x^* is element of a simplex-face of dimension less than n. We have

$$\bar{x} = \beta x + (1-\beta)x^*, \quad 0 \leq \beta < 1$$

or

$$\bar{x} = \beta x + \sum_{j=i_1}^{i_m} \beta_j x_j, \quad \beta_j = (1-\beta)\kappa_j^*,$$

$$\beta_j \geq 0 \, (j=i_1,\ldots,i_m), \quad \beta + \sum_{j=i_1}^{i_m} \beta_j = 1$$

proving $x \in \bigcup S^i$.

$S^i \cap S^j = S^{(ij)} \, (i \neq j)$, where $S^{(ij)}$ is a (n-1)-dimensional simplex-face of both S^i and S^j, results immediately from the construction of S^i and S^j.

Lemma 2:

Let $S = \{x_1,\ldots,x_{m+1}\} \subset \mathbb{R}^n$ be a n-simplex and $f\colon S \to \mathbb{R}$ be a concave function. Then there is exactly one affine function $L\colon \mathbb{R}^n \to \mathbb{R}$ satisfying $L(x_i) = f(x_i)$ $(i=1,\ldots,n+1)$. Moreover $L(x)$ is the convex envelope of f over S.

Proof:

Every affine function L may be represented by $L(x) = \lambda^T x + \lambda_o$, $\lambda \in \mathbb{R}^n$, $\lambda_o \in \mathbb{R}$. The determinant of the system of linear equations

$$\lambda^T x_i + \lambda_o = f(x_i) \qquad (i=1,\ldots,n+1)$$

is different from zero, since the n+1 vertices x_i $(i=1,\ldots,n+1)$ are affinely independent. $L(x) \leq f(x)$ for all $x \in S$ and moreover that $L(x)$ actually is the convex envelope of f over S (the best convex subfunction in the sense mentionned above) follows easily by concavity of f. A rigorous proof can be found in [2].

The minimum values of f and L over S are attained in a vertex of S. It follows

$$\min_{x \in S} f(x) = \min_{x \in S} L(x) \quad .$$

Minimizing L over S instead of f yields the minimum value of f. Unfortunately this is not true for minimizing points:

$$\{\bar{x} \mid f(\bar{x}) = \min_{x \in S} f(x)\} \subset \{\bar{x} \mid L(\bar{x}) = \min_{x \in S} L(x)\}$$

and in general there are minimizing points of L not minimizing f over S.

In our approach for solving Problem P we shall solve a sequence of subproblems in each of which we minimize L over $S \cap D$ instead of f. An associated convergence theorem stated in the last section will show that by suitable refining of simplex-partitions it is possible to generate a sequence of points $\{x^k\}$ having accumulation points, each of which yields the desired solution value. But the above discussion explains why we will get weaker results for the solutions points.

The next two Lemmata state how to choose a suitable refining rule. We omit the proofs, here because they are given in [3].

Lemma 3:

Let $S^1 \subset \mathbb{R}^n$ be a n-simplex and construct a sequence of n-simplices in the following way: $S^k \rightarrow S^{k+1}$ $(k=1,2,\ldots)$: Choose the midpoint x^k of one of the longest edges of S^k and let S^{k+1} be one of the two subsimplices constructed according to Lemma 1 using x^k instead of x. Then there is a $\tilde{x} \in \mathbb{R}^n$ satisfying

$$\lim_{k\to\infty} S^k := \bigcap_{k=1}^{\infty} S^k = \{\tilde{x}\}.$$

Lemma 4:

Let $S^1 \subset \mathbb{R}^n$ be a n-simplex and $\{S^k\}_{k=1}^{\infty}$ be a sequence of n-simplices constructed from S^1 according to Lemma 3. Furthermore let $f: A \rightarrow \mathbb{R}$ be a concave function on an open set A, $S^1 \subset A \subset \mathbb{R}^n$, and $\{L^k\}_{k=1}^{\infty}$ the sequence of affine functions each L^k defined over S^k according to Lemma 2. Then we have

a) $L^{k-1}(x) \leq L^k(x) \leq f(x)$ for all $x \in S^k$, $k=1,2,\ldots$;

b) the sequence $\{L^k\}$ is equicontinuous in the following sense: For every $\varepsilon > 0$ there exists a $\delta = \delta(x^*,\varepsilon)$ not dependent on k such that

$$|L^k(x) - L^k(x^*)| \leq \varepsilon$$

whenever $x^*, x \in S^k$ and $|x-x^*| \leq \delta$.

Description of the Algorithm

Step 0:

0.1.: Choose a n-simplex $S^1 \subset \mathbb{R}^n$ such that $D \subset S^1 \subset A$.
(A is supposed to be "large" enough of course such that S^1 exists.)
(If in problem (P) $D = D_1 \cap D_2$, D_1 convex, compact and $D_2 := \{x = (x_1,\dots,x_n) \in \mathbb{R}^n \mid x_i \geq 0,\ i=1,\dots,n\}$, then one method to construct S^1 is solving the optimization problem

$$z_o := \max_{x \in D} \sum_{i=1}^{n} x_i$$

and taking $S^1 = \{x_1^1,\dots,x_{n+1}^1\}$, $x_{n+1}^1 = \Theta$, the origin, $z_i^1 = z_o e_i$ $(i=1,\dots,n)$, where e_i denotes the i-th unit-coordinate vector.)

0.2.: Determine the affine function $L^o(x)$ over S^1 according to Lemma 2.

0.3.: Compute x^o from

$$L^o(x^o) := \min_{x \in D} L^o(x) \ .$$

0.4.: Compute $f(x^o)$. If $L^o(x^o) = f(x^o)$, then stop the algorithm. If $L^o(x^o) < f(x^o)$, then go to step 1.

Step 1:

1.1.: Construct the simplex-partition $\{S^{11}, S^{12}\}_{S^1}$ according to Lemma 3.

1.2.: Determine the affine functions $L^{11}: S^{11} \to \mathbb{R}$, $L^{12}: S^{12} \to \mathbb{R}$ according to Lemma 2.

1.3.: Compute x^{1i} from

$$L^{1i}(x^{1i}) = \min_{x \in D \cap S^{1i}} L^{1i}(x)$$

for those $i \in \{1,2\}$ with $D \cap S^{1i} \neq \emptyset$.

1.4.: Compute x^1 from

$$L^1(x^1) := \min_i L^{1i}(x^{1i}).$$

1.5.: Compute $f(x^1)$, If $L^1(x^1) = f(x^1)$, then stop the algorithm. If $L^1(x^1) < f(x^1)$, then go to step 2.

<u>Step k (k=2,3,...):</u>

Assume x^{k-1} $S^{k-1,k}$.

k.1.: Construct the simplex-partition $\{S^{k1}, S^{k2}\}_{S^{k-1,k}}$ of $S^{k-1,k}$ according to Lemma 3.

k.2.: Determine the affine functions $L^{ki}: S^{ki} \rightarrow \mathbb{R}$ $(i=1,2)$ according to Lemma 2.

k.3.: Set $S^{k,2+i} = S^{k-1,i}$ $(i=1,\ldots,k-1)$,
$L^{k,2+i} = L^{k-1,i}$ $(i=1,\ldots,k-1)$.

k.4.: Compute x^{ki} from

$$L^{ki}(x^{ki}) = \min_{x \in D \cap S^{ki}} L^{ki}(x), \quad D \cap S^{ki} \neq \emptyset \quad (i=1,\ldots,k+1)$$

(we have to take into account all subsimplices of S constructed up to now and look for the minimum over the collection of all affine subfunctional at hand. x^{ki} is already known for $i > 2$).

k.5.: Compute x^k from

$$L^k(x^k) := \min_i L^{ki}(x^{ki}).$$

k.6.: Compute $f(x^k)$. If $L^k(x^k) = f(x^k)$, then stop the algorithm. If $L^k(x^k) < f(x^k)$, then go to step k+1.

<u>Main properties of the algorithm</u>

The main properties of the algorithm are stated in the following theorem.

<u>Theorem</u>

a) Let $\bar{x} \in D$ be a solution of problem P. Then
$$L^k(x^k) \leq L^{k+1}(x^{k+1}) \leq f(\bar{x}) \leq \min_{j=1,\ldots,k+1} f(x^j) \leq \min_{j=1,\ldots,k} f(x^j) \quad k=1,2,\ldots;$$

b) If the algorithm stops after step j, then x^j is a solution of problem P;

c) If the algorithm is not terminated after finitely many steps, then it generates a sequence $\{x^k\}$ of feasible points one accumulation point of which is a solution. Moreover the limit of the numbers $\{L^k(y^k)\}$ is $f(\bar{x})$ where $\{y^k\}$ is any convergent subsequence of $\{x^k\}$.

The theorem is a consequence of Lemmata 1 to 4, a more general version is proved in [3] . The simple illustrative example discussed in the course of the presentation at the conference is also given in [3].

First feq tests had been rather successful finding good upper and lower bounds for the solution after a few steps.

References

[1] J.E. Falk and R.M. Soland, An algorithm for separable nonconvex programming problems, Management Sci. 15(1969), 550-569.

[2] R. Horst, Zur Charakterisierung affin-linearer Hüllfunktionale, (to appear in Zeitschrift für Angewandte Mathematik und Mechanik)

[3] R. Horst, An algorithm for nonconvex programming problems, (to appear in Mathematical Programming)

[4] R.M. Soland, An algorithm for separable nonconvex programming II: nonconvex constraints, Management Sci. 17(1971), 759-773.

MATHEMATICAL PROGRAMMING AND THE COMPUTATION OF OPTIMAL TAXES FOR ENVIRONMENTAL POLLUTION CONTROL †

Stephen E. Jacobsen
Engineering Systems Department
School of Engineering and Applied Science
University of California, Los Angeles, Ca. 90024/USA

ABSTRACT

This paper considers some theoretical and computational problems that arise when trying to find optimal taxes for environmental pollution control. The paper takes cognizance of the reality of mixed-economy difficulties (and, therefore, Lagrangian decomposition is not appropriate), and also demonstrates that a "property-rights" approach to environmental quality control may not be appropriate. The paper presents a water quality control problem which highlights the difficulties. In addition, the resulting mathematical program is nonconvex and a solution algorithm is presented.

† This research was supported by the National Science Foundation, Grant ENG 74-02629

1. TAXATION AND DECENTRALIZABLE PROGRAMS

Consider the separable mathematical program, denoted by (P),

$$\begin{aligned} \max \quad & \sum_{j=1}^{k} f_j(x^j) \\ \text{s.t.} \quad & \sum_{j=1}^{k} g_j(x^j) \leq 0 \\ & x^j \in X^j, \quad j = 1,\ldots,k . \end{aligned} \qquad \text{(P)}$$

Here, x^j represents the decision vector of the $j^{\underline{th}}$ subsector and X^j represents the set from which decisions must be drawn. In addition, there is a set of constraints upon resources which link the subsectors. That is, each g_j is a vector-valued function in R^m and $g_j(x^j)$ represents the amounts of the m scarce resources consumed when the $j^{\underline{th}}$ subsector utilizes the decision vector x^j. The function

$$\sum_{j=1}^{k} f_j(x^j)$$

is understood to be the "benefit measure" of a central authority (C.A.) as a function of the decisions of the k subsectors.

Let $\lambda = (\lambda_1,\ldots,\lambda_m) \geq 0$ be an m-vector of per unit taxes upon the m scarce resources. Let

$$L_j(x^j, \lambda) = f_j(x^j) - \lambda g_j(x^j) .$$

<u>Definition 1</u>: The system (P) is <u>centralized,</u> with respect to λ, if the decision maker of the $j^{\underline{th}}$ subsector, $j = 1,\ldots,k$, seeks to solve

$$\begin{aligned} \max \quad & L_j(x^j, \lambda) \\ \text{s.t.} \quad & x^j \in X^j \end{aligned} \qquad (P_j(\lambda))$$

when confronted with the nonnegative tax vector λ. The system (P) is said to be <u>mixed</u> if this is not the case.

<u>Definition 2</u>: We say a centralized system is <u>capable of decentralization</u> if the Lagrangian of (P) possesses a saddle-point.

We now see the meaning of <u>decentralization.</u> Since $(\bar{x}, \bar{\lambda})$ is a saddle-point for (P), it must be the case that $\bar{x}$ is optimal for (P). Therefore, if the C.A. knows an appropriate per unit tax vector, $\bar{\lambda}$, the centralized system assumption implies that the $j^{\underline{th}}$ coordinate vector, $\bar{x}^j$, of $\bar{x}$ is an optimizer of the $j^{\underline{th}}$ subsector's problem $(P_j(\bar{\lambda}))$.

This implies that if $\bar{\lambda}$ presented to each of the subsectors then, if $P_j(\bar{\lambda})$ has a unique optimal solution, the optimal vector $\bar{x} = (\bar{x}^1,..,\bar{x}^k)$ will be recovered by individual subsystem optimization. If some of the subsectors have more than one optimal vector for $P_j(\bar{\lambda})$ it may occur that the vector presented to the C.A., by subsector individual optimization, will not produce an optimal vector for the problem facing the C.A. (i.e., problem P).

Centralized systems capable of decentralization have nice implications for optimization by economic iteration without complete information by the C.A. (e.g., see Dantzig [3]). This Walrasian tatonnement interpretation is the basis of dual algorithms for solving problems such as (P) (e.g., see Uzawa [13], Huard [6], Falk [4]).

2. THE BENEFIT MEASURE

Effective intervention in any sphere of economic activity must be based upon a comparison of some notion of benefits and losses associated with various policies. Such, of course, must also be the basis of efforts for environmental pollution control. A C.A.'s policies to achieve a better environment will generally be translated, in the form of higher prices, to the consumers of products produced by "polluting" firms. What constitutes an appropriate measure of the reduction in consumer welfare is a difficult question that is yet to be completely resolved. Traditionally, economists have relied on the consumer's surplus notion. Recently, Willig [14] has given a rather definitive treatment of the notion of consumer's surplus and has shown that under fairly general conditions the consumer's surplus closely approximates the so-called equivalent and compensating variations.

Assume there are n consumer goods under consideration. Let $D_i(p_1,\ldots,p_i,\ldots,p_n)$ be the demand function for the $i^{\underline{th}}$ good as a function of the prices of the n consumer goods. Assume $p^o = (p_1^o,\ldots,p_n^o)$ is the initial price vector and the C.A. has implemented a policy which changes the consumer price vector to p^1. The loss in consumer welfare is defined to be

$$\sum_{i=1}^{n} \int_{p_i^o}^{p_i^1} D_i(\bar{p}_i)\, du_i$$

where

$$\bar{p}_i = (p_1^1,\ldots,p_{i-1}^1,u_i,p_{i+1}^o,\ldots,p_n^o) \quad .$$

To this loss we added producers' loss (i.e., change in profits) and C.A.

loss (i.e., change in tax take) to arrive at the expression

$$S(p^o,p^1) = \sum_{i=1}^{n} \int_{p_i^o}^{p_i^1} D_i(\bar{p}_i)du_i + \pi^o - \pi^1 + T^o - T^1$$

as a measure of loss due to the C.A.'s policy which causes a consumer good price vector change from p^o to p^1 .

Under independence of demands (i.e., the quantity of the $i^{\underline{th}}$ good demanded depends only upon its own price) and some mild additional assumptions, it can be shown that

$$S(p^o, p^1) = - \sum_{i=1}^{n} \int_{o}^{q_i^1} [D_i^{-1}(u_i) - S_i^{-1}(u_i)]du_i$$

$$+ \sum_{i=1}^{n} \int_{o}^{q_i^o} [D_i^{-1}(u_i) - S_i^{-1}(u_i)]du_i \quad ,$$

where $q_i = D_i(p_i)$ and S_i^{-1} is the marginal cost of production function for the $i^{\underline{th}}$ firm. Therefore, under these assumptions, we see that minimizing total loss is equivalent to maximizing total surplus (i.e., the sum of consumer and producer surplus).

3. A WATER POLLUTION EXAMPLE

In this section we present a simple example which highlights the theoretical and computational difficulties of computing optimal taxes. We assume there are k firms located on a one-directional stream and that each firm must dispose of a certain amount of waste which in turn, depends upon the factor usage vector of the firm. For simplicity of notation we assume independence of demands. Let D_j^{-1} denote the inverse demand function for the $j^{\underline{th}}$ firm's product; let y_j denote the product output level of the $j^{\underline{th}}$ firm and let $C_j(y_j, p^j)$ denote the production cost, to firm j, of producing output level y_j when the per unit factor price is p^j. Let x^j denote the factor usage vector of firm j and let ϕ_j denote the $j^{\underline{th}}$ firm's production function. Then, assuming each firm is a price taker on factor markets, we have that

$$C_j(y_j, p^j) = \min_{x^j \geq 0} \{p^j x^j \mid \phi_j(x^j) \geq y_j\} .$$

Let $W_j(x^j)$ denote the amount of waste which the $j^{\underline{th}}$ firm must dispose of when using factor vector x^j. Let $T_j(\alpha_j, W_j(x^j))$ denote the $j^{\underline{th}}$

firm's treatment cost function as a function of the total amount of waste and the percentage-of-treatment variable, α_j. For simplicity, we assume waste discharged to the stream consumes only dissolved oxygen and that the C.A. has set maximum allowable dissolved oxygen deficits. Therefore, the mathematical program is

$$\text{maximize} \quad \sum_{j=1}^{k} \left[\int_0^{y_j} D_j^{-1}(z_j)\,dz_j - C_j(y_j,p^j) - T_j(\alpha_j, W_j(x^j)) \right]$$

$$\begin{aligned} \text{subject to } & D_{j\ell}(W_1(x^1),\ldots,W_j(x^j), \alpha_1,\ldots,\alpha_j) \le \bar{D}_{j\ell}, && \ell=0,\ldots,m_j \quad \text{(PcA)} \\ & && j=1,\ldots,k \\ & \phi_j(x^j) - y_j \ge 0, && j=1,\ldots,k \\ & 0 \le \alpha_j \le 1,\ x^j \ge 0,\ y_j \ge 0 && j=1,\ldots,k \end{aligned}$$

where $\bar{D}_{j\ell}$ is the greatest allowable dissolved oxygen deficit in reach $j\ell$ between the $j^{\underline{th}}$ and $(j+1)^{\underline{st}}$ firms, m_j is the number of reaches between the $j^{\underline{th}}$ and $(j+1)^{\underline{st}}$ firms, and the set of constraint functions, representing dissolved oxygen deficit, are derived from the Streeter-Phelps equations (e.g., see Revelle, Loucks, and Lynn [11]). To be specific, we are assuming

$$D_{j\ell}(W_1(x^1),\ldots,W_j(x^j), \alpha_1,\ldots,\alpha_1)= \sum_{i=1}^{j} a_{j\ell,i}W_i(x^i)(1-\alpha_i) \qquad (1)$$

where

$a_{j\ell,i}$ = number of pounds of oxygen demanded, in reach $j\ell$ between the $j^{\underline{th}}$ and $(j+1)^{\underline{st}}$ firms, due to one pound of BOD discharged to the stream at location io (the location of the $i^{\underline{th}}$ firm upstream of the $j^{\underline{th}}$ firm),

and

$W_i(x^i)$ = number of pounds of BOD that the $i^{\underline{th}}$ firm must dispose of (by treatment or discharge to stream).

It is clear that these functions crucially depend upon the waste load and treatment decision variables of the firms along the stream. The dissolved oxygen deficit constraint functions are generally <u>nonconvex</u> functions of the firms' decision variables x^j and α_j, $j=1,\ldots,k$. In fact, even if W_j is a linear function of x^j, then $W_j(x^j)(1-\alpha_j)$ is a pseudo-concave and nonconvex function of (x^j, α_j). Therefore, $D_{j\ell}$ given by (1) is not even, generally, a pseudo-concave function of $(x^1,\ldots,x^j, \alpha_1,\ldots,\alpha_j)$ since sums of pseudo-concave functions are not generally pseudo-concave. In fact, in many applications it is also the case that the treatment capital investment functions $T_j(\alpha_j,W_j(x^j))$

exhibit economies-of-scale in the argument W_j. That is, $-T_j$ is convex or has convex segments in the argument W_j. This implies that the objective function facing the C.A. (the objective function of P_{cA}) is not generally a concave function of the decision variables of the problem.

Note also that a significant amount of information appears to be required by the C.A. In particular, knowledge of D_j^{-1}, C_j, ϕ_j, and T_j is required. However, under our assumptions, knowledge of ϕ_j is equivalent to knowledge of C_j(e.g., see Jacobsen [7,8]).

Recall that the major advantage of the decentralization of a centralized system is that system optimization can be carried out without complete information on the part of the C.A. Since the information requirements for the C.A. appear to be quite great for P_{cA}, we now investigate whether or not P_{cA} is a centralized system capable of decentralization. Because P_{cA} generally possesses a nonconcave objective subject to numerous nonconvex constraints, no claim can generally be made regarding the existence of a saddle-point. That is, the system generally is not capable of decentralization.

Moreover, if the $j^{\underline{th}}$ firm is confronted with a tax vector λ^* it is not the case that the $j^{\underline{th}}$ firm will seek to solve

$$\begin{aligned} &\text{maximize} \quad L_j(y_j, x^j, \alpha_j; \lambda^*) \\ &\text{subject to} \quad \phi_j(x^j) - y_j \geq 0 \qquad\qquad P_j(\lambda^*) \\ &\qquad\qquad 0 \leq \alpha_j \leq 1,\ x^j \geq 0,\ y_j \geq 0 \end{aligned}$$

where

$$L_j(y_j, x^j, \alpha_j, \lambda^*) =$$

$$= \int_0^{y_j} D_j^{-1}(z_j)dz_j - C_j(y_j, p^j) - T_j(\alpha_j, W_j(x^j)) - \lambda^* a^j (1-\alpha_j) W_j(x^j) ,$$

and where a^j is the $j^{\underline{th}}$ column of the set of constraints (1). That is, a profit maximizing firm, say, will attempt to maximize the difference between revenue and cost (production cost plus treatment cost plus the tax cost of waste discharge to the waterway). Therefore, a profit mazimizing firm will, rather, attempt to optimize a function $\bar{L}_j$, subject to the same constraints as in $P_j(\lambda^*)$, where

$$\bar{L}_j(y_j, x^j, \alpha_j, \lambda^*) =$$

$$= D_j^{-1}(y_j)y_j - C_j(y_j, p^j) - T_j(\alpha_j, W_j(x^j)) - \lambda^* a^j (1-\alpha_j) W_j(x^j).$$

Therefore,

$$L_j(y_j,\ x^j,\ \alpha j;\ \lambda^*) - \bar{L}_j(y_j,\ x^j,\ \alpha_j;\ \lambda^*) =$$

$$= \int_o^{y_j} D_j^{-1}(z_j)dz_j - D_j^{-1}(y_j)y_j,$$

the consumers' surplus associated with output level y_j. That is, P_{cA} is not a centralized system.

The fact that such systems are neither centralized nor capable of decentralization has, unfortunately, serious consequences for both the Walrasian tatonnement interpretation and algorithmic effectiveness of the various dual algorithms. That is, the rather elegant connection between decentralization and the amount of information needed (by the C.A.) to optimize is lost for mixed economies. Secondly, the dual price procedures, viewed as just computer algorithms for optimization, require the existence of saddle-points. Note that, in principle, the C.A. can centralize the system P_{cA} by offering, to the firms, schedules of the consumers' surplus

$$S_j(y_j) = \int_o^{y_j} D_j^{-1}(u_j)du_j - D_j^{-1}(y_j)y_j$$

However, this raises other distributional problems beyond the scope of the present paper.

4. PROPERTY RIGHTS

In an attempt to deal with information requirements, economists have often suggested a "property rights" approach. Briefly, the idea is that a C.A. will provide a fixed number of "pollution permits," each of which allows the owner to dump a fixed number of units (e.g., one) of polluting material. The firms are then free to buy and sell these permits on an open market and, therefore, an equilibrium price will be arrived at. That is, the market mechanism itself will provide the appropriate environmental charge so that environmental services (for instance, the assimilative capacity of a stream) will not be under-valued. This strategy is, perhaps, best articulated by Dales [2]. In this section we demonstrate that such a strategy is generally not valid.

We begin by answering the following question:

If a market for disposal rights provides an equilibrium price, then what measure (if any) of benefits is being maximized by the creation of such a market and price?

To answer this question, we proceed as follows. Assume there are n firms indexed $i = 1,..,n$. Let x^i be the decision vector for the ith firm and let $f_i(x^i)$ be the profit accrued to the ith firm when it decides upon the input vector x^i. Let $W_i(x^i)$ be the solid waste produced for disposal by the ith firm when the ith firm is using the input vector x^i. Let w be the regional limit (per unit time) upon the quantity of waste that the authority will accept for disposal services.

If there is a price λ per unit of waste, then the ith firm will seek to solve its own profit maximization problem. That is, the firm seeks to compute its modified optimal profits.

$$\mathscr{L}_i(\lambda) = \text{maximize } f_i(x^i) - \lambda W_i(x^i)$$
$$\text{subject to } x^i \varepsilon X^i ,$$

where X^i is the set of feasible decisions available to the ith firm. Let $x^i(\lambda)$ be an optimizing vector for the above optimization problem of the ith firm and let

$$w_i(\lambda) = W_i(x^i(\lambda))$$

denote the resulting level of waste offered for disposal. Then,

$$w(\lambda) = \sum_{i=1}^{n} w_i(\lambda)$$

denotes the total quantity demanded for waste disposal services when λ is the per unit price for such services. It is well known that the function $\mathscr{L}_i(\lambda)$ is convex in λ.

Also, by non-negativity of the function $W_i(x^i)$, it can easily be shown that $\mathscr{L}_i(\lambda) \geq \mathscr{L}_i(\gamma)$ and $w_i(\lambda) \geq w_i(\gamma)$ if $\lambda < \gamma$. This implies that $w(\lambda)$ is a non-increasing function of the waste disposal price λ.

Now assume $\lambda^* \geq 0$ is an equilibrium price. That is, λ^* is such that

$$w(\lambda^*) = w$$

or, equivalently, λ^* is the price which equates the fixed supply w with demand $w(\lambda)$. Now, when the ith decision unit is faced with waste disposal price λ^*, it will solve its own optimization problem and decide upon vector $x^i(\lambda^*)$. Therefore, for each $i = 1,..,n$, the following condition holds:

$$x^i(\lambda^*) \text{ solves } \max f_i(x^i) - \lambda^* W_i(x^i) \text{ subject to } x^i \in X^i \quad .$$

Also, by assumption, the following two conditions hold:

$$w = w(\lambda^*) = \sum_{i=1}^{n} W_i(x^i(\lambda^*)) \quad ,$$

and

$$0 = \lambda^* \left[\sum_{i=1}^{n} W_i(x^i(\lambda^*)) - w \right] \quad .$$

Of course, these conditions comprise the statement that $(x^*, \lambda^*) = (x^1(\lambda^*),...,x^n(\lambda^*), \lambda^*)$ is a saddle-point for the optimization problem

$$\max \sum_{i=1}^{n} f_i(x^i)$$

$$\text{subject to} \sum_{i=1}^{n} W_i(x^i) \leq w \qquad (2)$$

$$x^i \in X^i \quad , \quad i = 1,..,n \quad .$$

We therefore have an answer to the above question.

> *If a waste disposal authority sets a quantity limit w and if an efficient private market acts to equilibrate supply and demand for the limited disposal services, then the resulting actions of the n decision makers are such that the sum of the individual objective functions of the n decision makers is maximized subject to the authority's quantity constraint.*

The attractiveness of using property rights to generate a market for environmental services is somewhat mitigated in certain situations. For instance, if there are several firms discharging wastes to a waterway and a market is created for which these firms are to purchase limited quantities of the stream's assimilative capacity, our result implies that the sum of the firms' individual objectives will be maximized. However, if some of the firms have considerable monopoly or oligopoly power in their respective product markets, the result will then be that the product output levels of each of the firms will diverge from the optimal levels which would occur if we were to use the benefit measure developed above. Broadly speaking, there seems to be no justification for maximizing the sum of the polluting firms' profits subject to a stream assimilative capacity constraint. This methodological problem of welfare economics seems to have gone unnoticed by Dales.

While the above argument demonstrates that a "property rights" approach may be methodologically incorrect, it is also the case that such an approach may also be technically incorrect. In particular, suppose we approve of maximizing the sum of firms' profits. Then the above argument demonstrates that an equilibrium price implies the existence of a saddle-point for optimization problem (2). But suppose such a problem does not have a saddle-point as is likely to be the case for problems such as P_{cA}(i.e., there are economies-of-scale or nonconvex constraints). Then it is clear that an equilibrium price cannot be found by any ordinary market mechanism. For instance, consider the following simple example

$$\begin{aligned} &\max \quad x^2 + y^2 \\ &\text{s.t.} \quad \tfrac{1}{2}x + 2y \leq 3 \\ &0 \leq x \leq 2 \ , \ 0 \leq y \leq 2 \ . \end{aligned}$$

Then

$$x(\lambda) = \begin{cases} 2 \ , & \lambda \in [0,4] \\ 0 \ , & \lambda \geq 4 \end{cases} \ .$$

Therefore, $w_1(\lambda) = \frac{1}{2}x(\lambda)$ (note that $x(\lambda)$ and, therefore, $w_1(\lambda)$ are not single-valued). Also,

$$y(\lambda) = \begin{cases} 2 \ , & \lambda \in [0,1] \\ 0 \ , & \lambda \geq 1 \end{cases}$$

and $w_2(\lambda) = 2y(\lambda)$. Therefore

$$w(\lambda) = \begin{cases} 5 \ , & \lambda \in [0,1] \\ 1 \ , & \lambda \in [1,4] \\ 0 \ , & \lambda \geq 4 \end{cases}$$

and hence there is no intersection of the demand curve $w(\lambda)$ with the fixed supply of 3 units. What market behavior would occur, when an equilibrium price does not exist, is not clear.

5. A SOLUTION PROCEDURE

The above sections demonstrate that a Walrasian price adjustment process, for the purpose of arriving at prices which will induce firms to act optimally, is virtually useless. In particular, many environmental problems, such as P_{cA}, do not correspond to centralized systems capable of decentralization. This implies that a great deal of economic information is required by the C.A. in order to solve P_{cA} and to then set some policy (i.e., taxes) to encourage firms to behave optimally. Nevertheless, it behooves the C.A. to do the best it can as far as a solution of P_{cA} is concerned. The following discussion is based upon that of Hillestad and Jacobsen [5] .

Under some fairly general additional assumptions we sketch an algorithm which is based upon the fact that the constraints of P_{cA} can be converted to a set of convex constraints plus a set of reverse convex constraints (a single constraint $g(x) \leq 0$ is said to be a reverse convex constraint if g is a concave function). The term "reverse convex" is taken from Meyer [10]. Once this conversion is accomplished, a combination of Kelley [9] cuts for the convex constraints and Tui [12] cuts for the reverse convex constraints can be used to develop the algorithm.

The constraints of P_{cA} can be converted to the above mentioned form as follows. Let $z_j = W_j(x^j)$, $d_j = 1 - \alpha_j$, $j=1,\ldots,k$. Let $D_{j\ell}$ denote the j x j diagonal matrix whose $i^{\underline{th}}$ diagonal term is $a_{j\ell,i}$. Let

$$Q_{j\ell} = \begin{bmatrix} 0 & D_{j\ell}/2 \\ D_{j\ell}/2 & 0 \end{bmatrix}$$

and let $z^j = (z_1,\ldots,z_j)$, $d^j = (d_1,\ldots,d_j)$. Then the constraint (1) can be rewritten as

$$(z^j,d^j)Q_{j\ell}(z^j,d^j)' \leq \bar{D}_{j\ell} \quad . \qquad (3)$$

Now, by the principle of diagonal dominance, we can express (3) as the difference of convex quadratics. That is, (3) is equivalent to

$$(z^j,d^j)Q^1_{j\ell}(z^j,d^j)' - (z^j,d^j)Q^2_{j\ell}(z^j,d^j)' \leq \bar{D}_{j\ell} \tag{4}$$

where

$$Q^1_{j\ell} = Q_{j\ell} + \varepsilon_{j\ell} I \quad ,$$

$$Q^2_{j\ell} = \varepsilon_{j\ell} I \quad ,$$

and $\varepsilon_{j\ell}$ is a number chosen so that

$$\varepsilon_{j\ell} > \max_{i=1,\ldots,j} \left\{ a_{j\ell,i}/2 \right\} \quad .$$

Now, by the introduction of another variable, we can express (4) as

$$\begin{aligned} (z^j,d^j)Q^1_{j\ell}(z^j,d^j)' - u_{j\ell} &\leq \bar{D}_{j\ell} \\ (z^j,d^j)Q^2_{j\ell}(z^j,d^j)' - u_{j\ell} &\leq 0 \\ -(z^j,d^j)Q^2_{j\ell}(z^j,d^j)' + u_{j\ell} &\leq 0 \end{aligned} \tag{5}$$

and where the first and second constraints of (5) are convex and the third constraint of (5) is of the reverse convex type.

Now, under the assumptions that each production function ϕ_j is concave and each waste load function W_j is either convex or concave, the constraints of P_{cA} can be equivalently rewritten as a set of convex and a set of reverse convex constraints as follows:

$$\begin{aligned} (z^j,d^j)Q^1_{j\ell}(z^j,d^j)' - u_{j\ell} &\leq \bar{D}_{j\ell} \\ (z^j,d^j)Q^2_{j\ell}(z^j,d^j)' - u_{j\ell} &\leq 0 \qquad \ell=0,\ldots,m_j \\ & \qquad\qquad j=1,\ldots,k \\ -(z^j,d^j)Q^2_{j\ell}(z^j,d^j)' + u_{j\ell} &\leq 0 \\ z_j - W_j(x^j) &\leq 0 \qquad j=1,\ldots,k \\ -z_j + W_j(x^j) &\leq 0 \\ \phi_j(x^j) - y_j &\geq 0 \qquad j=1,\ldots,k \\ 0 \leq d_j \leq 1, \; x^j \geq 0, \; y_j &\geq 0 \quad . \end{aligned} \tag{6}$$

Similarly, it is easy to show that if $T_j(\alpha_j,z_j)$ is quadratic or is of the form

$$T_j(\alpha_j,z_j) = T_{j1}(\alpha_j)T_{j2}(z_j)$$

then also the treatment cost function terms of the objective of P_{cA} can be incorporated into the constraints in the form of convex and reverse convex functions. Then, by introducing a new variable, the objective can be placed in the constraints and by then bounding the feasible region by a bounded convex polyhedron our problem P_{cA} is converted to the form

$$\begin{aligned} &\max\ cx \\ \text{subject to}\ & h(x) \le 0 \\ & g(x) \le 0 \\ & A_o x \le b_o \\ & x \ge 0 \end{aligned}$$

where h is a vector of convex functions and g is a vector of concave functions.

We let

$$\begin{aligned} Ax &\le b \\ x &\ge 0 \end{aligned}$$

denote the set of linear constraints at a generic iteration of the algorithm. The algorithm is composed of the following steps:

1) Let x^o be a basic optimal solution for

$$\begin{aligned} &\max\ cx \\ \text{subject to}\ & Ax \le b \\ & x \ge 0\ . \end{aligned}$$

2) If $h(x^o) \le 0$ and $g(x^o) \le 0$, then x^o is optimal. Otherwise, let $h_k(x^o) = \max_i \{h_i(x^o)\}$, $g_\ell(x^o) = \max_j \{g_j(x^o)\}$ and generate a cut for the function h_k or g_ℓ which solves

$$\max \{h_k(x^o), g_\ell(x^o)\}\ .$$

3) Append the cut to the linear constraints and return to Step 1. When the cut is to be made for h_k, the usual Kelley cut

$$h_k(x^o) + \nabla h_k(x^o)(x-x^o) \le 0$$

is used. To describe the Tui cut, when the cut is to be made for g_ℓ, we proceed as follows. Assume A is m x n and that the basic optimal solution x^o (including slacks) is nondegenerate. This implies, since the feasible region for the linear constraints is assumed to be bounded, that the vertex x^o has n neighboring vertices, denoted by x^{oi}, i=1,...,n. Let $d^{oi} = x^{oi} - x^o$, i=1,...,n. Let $\bar{\alpha}_j$, i=1,...,n, denote optimal solutions for the n one-dimensional problems

$$\max \ \alpha_i$$
$$\text{subject to} \quad g_\ell(x^o + \alpha_i d^{oi}) \geq 0 \ .$$

The Tui cut is defined by the half-space (not containing x^o) whose bounding hyperplane passes through the n linearly independent points

$$z^{oi} = x^o + \bar{\alpha}_i d^{oi}, \quad i=1,\ldots,n \ .$$

That is, compute the hyperplane passing through the points z^{oi} and denote this hyperplane by $ax = \gamma$. If $ax^o < \gamma$, the Tui cut is $ax \geq \gamma$. If $ax^o > \gamma$, the Tui cut is $ax \leq \gamma$. Of course, it is well known that the direction vectors d^{oi} are proportional to vectors of the form

$$(-\bar{a}_{1,m+i},\ldots,-\bar{a}_{m,m+i}, 0,\ldots,0,\underset{\uparrow\, i^{\text{th}} \text{ nonbasic coordinate}}{1}, 0,\ldots,0),\ i=1,\ldots,n,$$

and the latter are readily available in the final simplex tableau which produced x^o, the basic optimal solution. Therefore, these latter vectors are to be used as direction vectors rather than pivoting to actually compute the neighboring vertices x^{oi}, $i=1,\ldots,n$.

Also, the sequence of cuts does not cut away any portion of the feasible region. In particular, it is well known that a Kelley cut (for a convex constraint) cannot delete any portion of the feasible region. To see that the same is true for a Tui cut we proceed as follows. Let g_j be any concave constraint such that $g_j(x^o) > 0$. Observe that if

$$x \in \text{conv}\,\{x^o, z^{o1},\ldots,z^{on}\}$$

then there exist nonnegative weights whose sum is one and

$$g_j(x) \geq \lambda_o\, g_j(x^o) + \lambda_{o1} g_j(z^{o1}) + \ldots + \lambda_{on} g_j(x^{on}) = \lambda_o g_j(x^o) \quad .$$

Assume the Tui cut is given by $ax \leq \gamma$ (i.e., $ax^o > \gamma$) and let x be any feasible solution. Therefore, $g_j(x) \leq 0$. If $g_j(x) < 0$, then

$$x \notin \text{conv}\,\{x^o, z^{o1},\ldots,z^{on}\}$$

and therefore $ax \leq \gamma$. If $g_j(x) = 0$ and

$$x \in \text{conv}\,\{x^o, z^{o1},\ldots,z^{on}\},$$

it must be the case that x can be written as a convex combination of only $z^{o1},\ldots,z^{on}$ and, therefore, $ax = \gamma$.

Therefore, the cutting plane algorithm above produces a sequence of nonincreasing upper bounds for the optimal value.

It may occur that x^o is degenerate and there will generally be more than n neighboring vertices and, therefore, the Tui cut cannot be unambiguously executed. In fact, it can be shown that if one arbitrarily chooses n directions to form the cut, some part of the feasible region may be deleted. However, in such situations we may use the cut proposed by Carvajal-Moreno [1]. Carvajal's cut is motivated by the Tui cut and it can be shown that both are the same when x^o is nondegenerate.

Even though we have not established a priori convergence conditions, the algorithm does produce a sequence of nonincreasing upper bounds. Moreover, the algorithm has been found to solve complex nonconvex programs (of the type of this paper) and, in the absence of other methods for such problems, appears to be a very useful tool.

Moreover, often lower bounds for P_{cA} can be obtained. For instance, if T_j is linear in W_j and W_j is convex in x^j then P_{cA} is a concave program for a fixed set of α's and is, similarly, a concave program for a fixed set of x's. Therefore, to perhaps obtain a good lower bound, we can fix the α's and optimize over the y's and x's. Then, fixing the y's and x's at the previous optimal values, we optimize over the α's, etc. Such a procedure, in conjunction with the cutting plane algorithm, produces upper and lower bounds for P_{cA}, an otherwise relatively intractable problem.

Of course, when the C.A. has obtained optimal or near optimal solutions for P_{cA} it is still faced with the problem of how to set taxes so as to encourage firms to act optimally. About all that can be said at this point is that it may be possible to perform this "coordination" in the following manner. Given all the information needed to solve P_{cA} the C.A. can then find each firm's optimal factor mix and treatment response as a function of tax parameters in the firms' profit functions. One would then hope that there are tax parameters, for each firm, which would force the firms to choose the corresponding optimal values of P_{cA}. However, it can be shown that this is generally not the case when considering environmental taxes and therefore the "coordination" problem for such systems is still generally unsolved.

REFERENCES

1. R. Carvajal-Moreno, "Minimization of Concave Functions Subject to Linear Constraints," Operations Research Center ORC 72-3, University of California, Berkeley, (1972).

2. J. H. Dales, Pollution, Property and Prices, University of Toronto Press, 1968.

3. G. B. Dantzig, Linear Programming and Extensions, Princeton University Press, (1963).

4. J. E. Falk, "Lagrange Multipliers and Nonlinear Programming," Journal of Math Analysis and Applications, 19, 141-159, 1967.

5. R. J. Hillestad and S.E. Jacobsen, "A Cutting Plane Algorithm for Water Quality Control," submitted to J. ORSA.

6. P. Huard, "Convex Programming-Dual Algorithm," Operations Research Center ORC 63-20, University of California, Berkeley, 1963.

7. S. E. Jacobsen, "Production Correspondences," Econometrica, 38 (5), 754-771, 1970.

8. S. E. Jacobsen, "On Shephard's Duality Theorem," Journal of Economic Theory, 4 (3), 458-464, 1972.

9. J. E. Kelley, Jr., "The Cutting-Plane Method for Solving Convex Programs," J. Soc. Indust. Appl. Math., 8, (4), 703-712, (1960).

10. R. Meyer, "The Validity of a Family of Optimization Methods," SIAM J. Control, 8, 41-54, (1970).

11. C. S. Revelle, D.P. Loucks, and W.R. Lynn, "Linear Programming Applied to Water Quality Management," Water Resour. Res., 4, 1-9, (1968).

12. H. Tui, "Concave Programming Under Linear Constraints," Soviet Mathematics, 5, (6), (1964).

13. H. Uzawa, "Iterative Methods for Concave Programming,"Chapter 10 of Studies in Linear and Nonlinear Programming, K. Arrow, L. Hurwicz H. Uzawa (eds.), Stanford University Press, 1958.

14. R. Willig, "Consumer's Surplus: A Rigorous Cookbook," Institute for Mathematical Studies in the Social Sciences, Tech. Report No. 98, Stanford University, (1973).

ON LARGE SCALE LINEAR FRACTIONAL PROGRAMS

Á. Kovács - J. Stahl, Budapest

1. In the present profit regulating system the objective of a Hungarian firm can be expressed [4] as the maximization of the ratio

$$H = \frac{P}{\sigma W + K} \qquad /1/$$

where

P = profit

σ = coefficient given by the planning authorities

W = wage cost

K = capital (sum of the fixed and the average working capital)

In the case of a large multidivision company this suggests the consideration of the following programming problem:x)

$$\sum_j A_j x_j = b \qquad /j=1, 2, \ldots/$$

$$B_j x_j = b_j$$

$$x_j \geq 0$$

$$\max \frac{\sum_j c_j x_j + \gamma}{\sum_j d_j x_j + \delta} \qquad /2/$$

Several decomposition procedures can be developed for solving this problem. Two such procedures, obtainable through transforming the above problem into a linear program to which the Dantzig-Wolfe principle is applied, are considered. The economic interpretations of the procedures are also discussed.

---

x) Matrices are denoted by capital letters, vectors by small letters and scalars by greek letters.

2. In this section it is shown, how the Dantzig-Wolfe principle can be applied for solving /2/ linear fractional programming problem.

It is assumed that for each j the set $S_j = \{x_j \mid Bx_j = b_j , x_j \geq 0\}$ is non-empty.

This is a necessary condition for having a feasible solution to problem /2/. Checking wether these necessary conditions are satisfied easily can be built in the algorithm suggested below. Furtherly it is assumed that $\sum_j d_j x_j + \delta > 0$ for each $x = /\ldots x_j \ldots/$ satisfying the constraints in /2/. In the above economic problem this does not mean a restriction since the value of the capital used always can be regarded positive.

Denote $\bar{x}_{j1}, \bar{x}_{j2}, \ldots, \bar{\bar{x}}_{j1}, \bar{\bar{x}}_{j2}, \ldots,$ the elements of the canonical decomposition of the polyhedron S_j, where one and two bars represent the bounded and unbounded components, respectively [3].

As it is in the linear programming case, problem /2/ is equivalent to the problem (with variables λ_{jk} and μ_{jk}) [2]

$$\sum_j \Big(\sum_k \bar{a}_{jk} \lambda_{jk} + \sum_k \bar{\bar{a}}_{jk} \mu_{jk}\Big) = b$$

$$\sum_k \lambda_{jk} = 1 \qquad /j=1,2,\ldots/$$

$$\lambda_{jk}, \mu_{jk} \geq 0$$

$$\max \frac{\sum_j \Big(\sum_k \bar{\gamma}_{jk} \lambda_{jk} + \sum_k \bar{\bar{\gamma}}_{jk} \mu_{jk}\Big) + \gamma}{\sum_j \Big(\sum_k \bar{\delta}_{jk} \lambda_{jk} + \sum_k \bar{\bar{\delta}}_{jk} \mu_{jk}\Big) + \delta} \qquad /3/$$

where

$$\bar{a}_{jk} = A_j \bar{x}_{jk} \qquad \bar{\bar{a}}_{jk} = A_j \bar{\bar{x}}_{jk}$$

$$\bar{\gamma}_{jk} = c_j \bar{x}_{jk} \qquad \bar{\bar{\gamma}}_{jk} = c_j \bar{\bar{x}}_{jk}$$

$$\bar{\delta}_{jk} = d_j \bar{x}_{jk} \qquad \bar{\bar{\delta}}_{jk} = d_j \bar{\bar{x}}_{jk} .$$

Following the algorithm suggested by Charnes and Cooper [1] for solving a linear fractional programming problem one can introduce the variable

$$\tau = \frac{1}{\sum_j (\sum_k \bar{\delta}_{jk} \lambda_{jk} + \sum_k \bar{\bar{\delta}}_{jk} \mu_{jk}) + \delta}$$

$$\alpha_{jk} = \frac{\lambda_{jk}}{\sum_j (\sum_k \bar{\delta}_{jk} \lambda_{jk} + \sum_k \bar{\bar{\delta}}_{jk} \mu_{jk}) + \delta}$$

$$\beta_{jk} = \frac{\mu_{jk}}{\sum_j (\sum_k \bar{\delta}_{jk} \lambda_{jk} + \sum_k \bar{\bar{\delta}}_{jk} \mu_{jk}) + \delta}$$

and obtain the linear programming programming problem

$$\sum_j (\sum_k \bar{a}_{jk} \alpha_{jk} + \sum_k \bar{\bar{a}}_{jk} \beta_{jk}) - b\tau = 0$$

$$\sum_k \alpha_{jk} - \tau = 0 \qquad /j=1,2,\dots/$$

$$\sum_j (\sum_k \bar{\delta}_{jk} \alpha_{jk} + \sum_k \bar{\bar{\delta}}_{jk} \beta_{jk}) + b\tau = 1$$

$$\alpha_{jk}, \beta_{jk} \geq 0 \qquad /4/$$

$$\max \left[\sum_j (\sum_k \bar{\gamma}_{jk} \alpha_{jk} + \sum_k \bar{\bar{\gamma}}_{jk} \beta_{jk}) + \gamma\tau\right].$$

This problem can be solved by using the Dantzig-Wolfe principle and the following algorithm is obtained:

a) Determine an optimal basic feasible solution to /4/. Denote $(f \dots f_j \dots f)$ a system of dual variables corresponding to this basic solution.

b) Consider the LP problems

$$B_j x_j = b_j$$

$$x_j \geq 0 \qquad /j=1,2,\dots/ \qquad /5/$$

$$\max (c_j - fA_j - f d_j) x_j$$

c) If for each j

$$(c_j - fA_j - \varphi d_j)\bar{x}_j \leq \varphi_j \qquad /6/$$

where $\bar{x}_j$ is an optimal solution of /5/, the procedure is terminated.

If $\tau \neq O$ in the regarded basic feasible solution of /4/, $x = \frac{1}{\tau}$ $(\ldots \sum_k \alpha_{jk} \bar{x}_k + \sum_k \beta_{jk} \bar{\bar{x}}_k \ldots)$ is an optimal solution of /1/.

If $\tau = O$, the actual objective function value in /4/ - which is the optimal value of /4/ - is an upper bound for the objective function values in /2/ and moving along the direction $\bar{\bar{x}} = (\ldots \sum_k \beta_{jk} \bar{\bar{x}}_k \ldots)$ one can get a solution for /2/ for which the objective is arbitrarily close to this value.

d) If for some j the optimal solution of /5/ does not satisfy the condition /6/, or /5/ does not have an optimal solution, new columns are introduced into the coefficient matrix of problem /4/, the same way as in the LP case. That is to say problem /4/ is augmented by the column $(A_j \bar{x}_j, O, \ldots 1, \ldots \ldots O, d_j \bar{x}_j, c_j \bar{x}_j)$, where $\bar{x}_j$ is an extremal solution of /5/ not satisfying condition /6/ and/or by the column $(A_j \bar{\bar{x}}_j, O, \ldots, O, O, d_j \bar{\bar{x}}_j, c_j \bar{\bar{x}}_j)$, where $\bar{\bar{x}}_j$ is an extremal direction obtained in the case where /5/ was found unbounded, for which

$$(c_j - fA_j - \varphi d_j)\bar{\bar{x}}_j > O.$$

Now the procedure is continued from a) with the augmented central program. If for this new matrix problem /4/ is unbounded, /2/ is also unbounded and the procedure is completed. Otherwise the procedure is continued from b).

If for problem /2/ it is also assumed that for each solution of /5/ $d_j x_j > O$, then the sector problem can be formulated as

$$B_j x_j = b$$

$$x_j \geq O$$

$$\max \frac{(c_j - fA_j)x_j - \varphi_j}{d_j x_j} \qquad /7/$$

and the optimality condition is that for each j the maximum value of /7/ is at most φ.

For the solution of problem /3/ the algorithm proposed by Martos [5] also can be applied. As it was shown by Wagner [6] for each solution obtained by /6/ corresponds a solution obtained by /2/ and vice versa, and starting from the same basic feasible solution generally in each step the same basic feasible solutions are obtained.

If we apply the Martos procedure the solution must be kept as long as possible in the bounded part of the convex polyhedron defined by the constraints. This can be obviously fulfilled if the constraints are given explicitly which is not the case for /3/. Naturally this problem does not arise if the convex polyhedron defined by the constraints of /3/ is bounded, which often can be assumed in a real-life economic model.

For describing the application of the Martos-algorithm consider the two LP problems, both with constraints /3/ and one by taking the numerator and the other the denominator of the objective function in /3/ to be objective function. Denote $(p \ldots \pi_j \ldots)$ and $(r \ldots \varrho_j \ldots)$ the vectors of the provisional dual prices corresponding to a basic feasible solution of the LP problems above, respectively, and denote Δ and Γ the objective function values in this basic solution. (i. e. Δ is the value of the numerator and Γ is the value of the denominator of the objective function in /3/.) A solution is optimal if for each λ_{jk},

$$\Delta(\bar{\gamma}_{jk} - p\bar{a}_{jk} - \pi_j) - \Gamma(\bar{\delta}_{jk} - r\bar{a}_{jk} - \varrho_j) \leq 0$$

and for each μ_{jk}

$$\Delta(\bar{\bar{\gamma}}_{jk} - p\bar{\bar{a}}_{jk}) - \Gamma(\bar{\bar{\delta}}_{jk} - r\bar{\bar{a}}_{jk}) \leq 0 .$$

It follows that a sector problem of the form

$$B_{xj} = b \; ;$$

$$x_j \geq O \qquad /8/$$

$$\max \left\{ \Delta (c_j - pA_j)\, x_j - \Gamma \left[(d_j - rA_j)\, x_j - \delta_j \right] \right\}$$

can be defined, while the other steps of the decomposition procedure remain the same.

If problem /8/ is bounded and the optimum value is positive, the vector submitted to the central program always will have a positive element, for example the element corresponding to the $\sum_k \lambda_{jk} = 1$ constraint and so the new solutions will be kept in the bounded part of the convex polyhedral defined by the constraints of problem /3/.

If one of the sector problems in unbounded, an appropriate $\bar{\bar{x}}_j$ can be chosen by solving the LP problem

$$B_j x_j = O$$

$$e x_j = 1 \qquad /9/$$

$$\Delta (c_j - pA_j) - \Gamma (d_j - rA_j)\, x_j \geq \varepsilon$$

$$\max e\, B^{-1} A_j x_j$$

where B^{-1} is the appropriate part of the actual basic solution of /3/, e is a vector all elements of which are 1 and ε is a sufficiently small positive number.

3. Now we turn to the economic interpretations of the above procedures. The iterations in a Dantzig-Wolfe type decomposition can be described as steps in a two-way planning process:

1) The center gives the initial values of the centrally determined parameters for the divisions optimization problems

2) Each subdivision submits a feasible subdivisional program based on the centrally given parameters

3) The center examines weather the optimum of the whole problem is

reached as a combination of the subdivisional programs, and if this is not the case preassings a revised set of parameters and the process is repeated.

The procedures always have this general structure but differences are found in the objective functions of the subdivisional programs and in the centrally given parameters.

In the first case discussed above a subdivision maximizes the objective function

$$\left[c_j - fA_j - \varphi(\sigma w_j + k_j)\right] x_j \ .$$

The following interpretation can be given to this objective function. The subdivision maximizes the profit, but the profit of each activity is reduced by the sum of the "taxes" paid for the use of the central resources $(f A_j x_j)$ and for the use of the factors of production $(\varphi \sigma w_j + k_j) x_j$.

The elements of f corresponding not to a central resource but to a central task will be negative (or zero) so beside the taxes there can be subsidies too.

This decomposition procedure defines a decentralized decision making process in which the center determines the taxes and subsidies after comparing the centrally computed φ_j values and the optimal profit values of the subdivisions and solving a new central problem. The optimal profit values in the sectors generally will be positive, which appears to be preferable to a system in which the optimal values of the sectors' objective function would be zero.

In the second formulation - when it can be assumed that for each possible sector program $\sigma w_j + k_j$ will be positive - the objective function of the sector can be stated as

$$\max \frac{(c_j - fA_j)\, x_j - \varphi_j}{(\sigma w_j + k_j)\, x_j} \ .$$

This objective function requires the maximization of the ratio of the sectors profit to the weighted value of the additional (new) resource and capital requirements. The profit of the sector in this case is reduced by the taxes paid after the use of the central resources and by an individual profit tax. Accordingly in the corresponding decentralized decision making system the center determines the resource taxes and the individual sector taxes, after comparing the centrally computed φ value and the values of the objective functions of the sectors and solving a new master program. The objective function values of the sectors in this system will always be positive.

The third possibility discussed above defines the following objective function in the sectors

$$c_j x_j - pA_j x_j - \frac{\Gamma}{\Delta}\left[/\sigma w_j + k_j / x_j - r_j A_j x_j - \varsigma_j\right] .$$

It also can be regarded as a special profit function, in which the profit is reduced by the taxes paid for the use of the central resources and for the use of the factors of production. However now this latter tax is determined in several steps. First the factor use is modified and this modification is also made in two steps: there is an individual modification element (ς_j) and a modification element determined by the use of the central resources ($r_j A x_j$).

The sector pay a factor tax after this modified value of factor use and the tax rate is determined by the actual value of the objective function ($\frac{\Gamma}{\Delta}$). This objective function appears to be highly complicated, although there can be cases where this system can be preferred to a simpler one because it offers the possibility of a dual valuation of the scarce resources and takes into account the differences in the factor needs of the subdivisions.

In the corresponding decentralized decision making system the center determines two sets for resource taxes (p and r), the factor tax rate ($\frac{\Gamma}{\Delta}$) and the subdivisional factor use tax (ς_j).

It does not seem appropriate to rank the above decentralized decision making systems, the best suitable one can be chosen only after taking into account the local circumstances.

REFERENCES

[1] CHARNES, A. - COOPER, W. W.: Programming with linear fractional functionals. Naval Research Logistics Quarterly, 9/2 (1962).

[2] DANTZIG, G. B. - WOLFE, Ph.: The decomposition algorithm for linear programs. Econometrica, 29/4 (1961).

[3] GOLDMANN, A. I.: Resolution and separation theorems for polyhedral convex sets. In: Linear inequalities and related systems. Edited by KUHN, H. W. and TUCKER, A. W, Princeton, 1956. University Press.

[4] KOVÁCS, Á.: On the normative behavious of the firm in the Hungarian probit system, in Colloquia Mathematica Societatis János Bolyai: 12. Progress in Operations Research, Eger 1974.

[5] MARTOS, B.: Hyperbolic programming. Naval Research Logistics Quarterly, June-Sept. (1964).

[6] WAGNER, H. M. - YUAN, J. S. C.: Algorithmic equivalence in linear fractional programming. Management Science, 14/5 (1968).

SOME REMARKS ON GENERALIZED LAGRANGIANS

S. Kurcyusz

Technical University of Warsaw
Institute of Automatic Control
ul. Nowowiejska 15/19
00-665 WARSAW, POLAND

Abstract

In the paper a definition is given of a class of generalized Lagrangians, and some simple properties of them are discussed, especially those related to the topology in the set of constraints. A general formulation of the method of multipliers is presented and a theorem characterising convergence of this method in case of linear-quadratic problems in Hilbert space. Numerical examples of computing the optimal control of time lag systems to terminal functions are presented. The results indicate that the effectiveness of the method of multipliers depends on the choice of the norm in the set of constraints.

0. INTRODUCTION

The paper presents a definition and selected simple properties of a class of generalized Lagrangians. Many authors have recently furnished definitions and discussed the properties of generalized Lagrange functionals - see e.g. [1] [6] [7] [8] [22] [26].

However, their research has been primarily limited to nonlinear or convex programming in R^n or to convex problems in more general spaces [14] . A more detailed study of so called shifted penalty functionals in Hilbert space has been presented in [30] [31] . The theoretical investigations are intimately related to practical problems of computational techniques and become especiallny fruitful when applied to the mentioned shifted penalty method (often called the method of multipliers) [10] [16]. A number of papers appeared discussing the convergence, the rate of convergence and the relation to the more abstract duality framework [2] [15] [29] [19] [20] and others. Except [23] [25] [30] [31] all consideration have been carried out in R^n as the space of constraints values.

The goal of this paper is to extend several notions to a general class of optimization problems including those with infinite dimensional or integer constraints. Only the main facts are given in order to rather indicate the possibility of generalization than describe it in detail. The duality theory for these generalized Lagrangians, leads in a natural way to φ-conjugates of functions, [27] [28] which resemble Fenchel conjugates with the scalar product (or duality) substituted by an arbitrary function of two variables. The presentation below has much in common with the thesis of Seidler [26]. However, some simple properties are shown connected with the topology in the set of constraints values: in finite dimensional case essentially one topology is used, while generally it plays non-negligible role. In the last two paragraphs the application of the method of multipliers to a problem with linear operator constraints is discussed and numerical examples are described which show that the proper choice of the penalizing norm, consistent with the topological features of the constraining operator, influences strongly the computational effectiveness of the algorithm. The examples presented are optimal control problems for time-lag systems.

1. PRELIMINARIES

Suppose that two sets are given, Y and P, a family $\{Y_p\}_{p \in P}$ of subsets of Y indexed by $p \in P$ and a functional $Q : Y \longrightarrow R$. Then a family of optimization problems can be defined:

(1.1) minimize $Q(y)$ over $y \in Y_{p_0}$

where p_0 can change over P.

Such a problem statement allows the discussion of linear, nonlinear as well as integer programming problems and extremal problems with infinitely many constraints, for instance those of optimal control theory. Besides, the description is simple and clear.

The family $\{Y_p\}_{p \in P}$ determines a family $\{P_y\}_{y \in Y}$ of "inverse images" in P :

(1.2) $p \in P_y \iff y \in Y_p$.

Problem (1.1) can be equivalently stated as

(1.3) minimize $Q(y)$ over y: $p_0 \in P_y$.

The family $\{Y_p\}_{p \in P}$ introduces also in P a natural partial ordering:

$$(1.4) \qquad p \leq p' \iff Y_p \subset Y_{p'}.$$

If we suppose for a while that this partial order is such that for every $y \in Y$ with $P_y \neq \emptyset$ there exists $\inf_{p \in P_y} p \in P_y$ then (1.4) is equivalent to

$$\text{minimize } Q(y) \quad \text{over } y: \inf_{p \in P_y} p \leq p_o .$$

This reformulated problem is in no way easier to treat then (1.1). However, it suggests a relaxation of (1.1) into what is called a "surrogate problem" [5] [9] . Given a functional $f : P \to \mathbf{R}$, solve

$$(1.5) \quad \text{minimize } Q(y) \quad \text{over } y: \inf_{p \in P_y} f(p) \leq f\,(p_o).$$

Any solution to (1.1) clearly solves (1.5) and it is always possible to find a function f such that (1.5) is equivalentt to (1.1).

Suppose now that instead of one function f we have a family of such functions, namely, a function $\varphi : P \times W \to R$ where W is another set. Any $w \in W$ defines a function $\varphi(\cdot,w): P \to R$ which can be viewed as the "distribution of prices" on perturbations $p \in P$.

Then the value

$$(1.6) \qquad K(y,w) = \inf_{p \in P_y} \varphi(p,w) - \varphi(p_o,w)$$

can be used to measure the distance from a given y to the set of admissible solutions Y_{p_o} in terms of prices corresponding to the "distribution of prices" $w \in W$. Assume further that

$$(1.7) \qquad \sup_{w \in W} K(y,w) = \begin{cases} +\infty & , \ p_o \notin P_y \\ 0 & , \ p_o \in P_y \end{cases} .$$

Then solving (1.1) is clearly equivalent to solving the following problem without constraints:

(1.8) minimize $\sup_{w \in W} L(y,w)$ over $y \in Y$

where $L(y,w)$ we define to be the <u>generalized Lagrangian</u> associated with (1.1):

(1.9) $L(y,w) = Q(y) + K(y,w) =$

$$= Q(y) + \inf_{p \in P_y} \varphi(p,w) - \varphi(p_o,w) .$$

For similar definitions see e.g. [6] [22] and especially [26]. All have been given for nonlinear programs with constraints in R^n.

We give now the examples.

Suppose P is a topological vector space, $S : Y \to P$ - an operator and $D \subset P$ a (closed) convex cone with vertex at zero. Consider the following nonlinear program:

(1.10) minimize $Q(y)$ subject to $p_o - S(y) \in D$.

Then for $p \in P$, $Y_p = \{ y \in Y: \ p - S(y) \in D \}$ and the partial order (1.4) coincides with the partial ordering introduced by the cone D.

(i) Let $W = D^*$, the dual cone, and $\varphi(p,w) = \langle w, p \rangle$. Then

$$K(y, w) = \langle w, S(y) - p_o \rangle$$

and L is the classical Lagrange functional. (1.7) clearly holds.

(ii) Suppose P is a lattice with respect to the order introduced by D, i.e. for each $p \in P$ $\max(0,p) = p^+$ exists (all function spaces are lattices with respect to the cone of nonnegative functions). Suppose $\Psi : P \to R$ is a functional satisfying the following conditions ($p \leq p'$ means $p' - p \in D$):

(a) $0 \leq p \leq p' \Rightarrow \Psi(p) \leq \Psi(p')$

(b) $0 \leq p < p' \Rightarrow \Psi(p) < \Psi(p')$ where $p < p' \Leftrightarrow p \leq p',\ p \neq p'$.

(c) $\Psi(p) = \Psi(p^+ + (-p)^+)$.

Let $W = R_+ \times P$, so that $w \in W$ is a pair $w = (\varrho, v)$, $\varrho \geqslant 0$, $v \in P$. Define

(1.11) $\varphi(p,w) = \varrho\Psi(p - v)$.

According to the definition of Y_p,

(1.12) $P_y = \{p \in P:\ p - S(y) \in D\} = \{p \in P:\ p \geqslant S(y)\}$.

Hence

$$\inf_{p \in P_y} \varphi(p,w) = \varrho \inf_{p \geqslant S(y)} \Psi(p-v) = \varrho \inf_{p' \geqslant S(y) - v} \Psi(p') = \varrho\Psi((S(y) - v)^+) \;,$$

(1.13) $K(y,w) = \varrho\Psi((S(y) - v)^+) - \varrho\Psi(p_o - v)$.

It remains to verify (1.7). If $p_o \notin P_y$ then

$(S(y) - p_o)^+ > 0$ and by (b)

$$K(y,(\varrho, p_o)) = \varrho[\Psi((S(y) - p_o)^+) - \Psi(0)] \xrightarrow[\varrho \to \infty]{} +\infty$$

If $p_o \in P_y$ then $S(y) \leq p_o$, $(S(y) - v)^+ \leq (p_o - v)^+$ and in virtue of (a) and (c)

$$K(y,w) = \varrho[\Psi((S(y) - v)^+) - \Psi(p_o - v)] \leqslant$$
$$\leqslant \varrho[\Psi((S(y) - v)^+) - \Psi((p_o - v)^+)] \leqslant 0 \;,$$

while

$$K(y, (\varrho, p_o)) = \varrho[\Psi(0) - \Psi(0)] = 0.$$

(iii) Suppose P is a Hilbert space. Set $W = R_+ \times P$, as before and take $\alpha > 0$. Define

$$\varphi(p,w) = \varrho \,\| p - v \|^\alpha \;.$$

We have from (1.12):

$$\inf_{p \in P_y} \varphi(p,w) = \varrho \inf_{p' \in S(y) - v + D} \| p' \|^\alpha = \varrho \,\| (S(y) - v)^{D^*} \|^\alpha$$

where p^{D^*} denotes the projection of p onto D^* - see [31].

Then

(1.14) $K(y,w) = \varrho \| (S(y) - v)^{D^*} \|^{\alpha} - \varrho \| p_o - v \|^{\alpha}$.

Property (1.7) is verified similarly as above. For $\alpha = 2$, $L(y,w)$ with this K is the augmented Lagrangian of Wierzbicki [30] [31].

(iv) We now specialize the two preceding examples to the case of $P = R^n$, $D = \{p = (p^1,\ldots,p^n): p^i \geqslant 0,\ i = 1,\ldots,m,\ p^i = 0,\ i = m+1,\ldots,n\}$.

Suppose $\zeta : R \to R$ is a monotone (strictly increasing) in R_+, nonnegative function with $\zeta(0) = 0$.

Define ψ under (ii) by

$$\psi(p) = \sum_{i=1}^{n} \zeta(p^i) .$$

Then the generalized Lagrangian (1.9) with K as in (1.13) is practically the generalized Lagrangian employed by Mangasarian [15].

Assume now for simplicity that $m = n = 1$ and $p_o = 0$ so that we have only one inequality constraint and that $\zeta(a) = a^2$, $a \in R$. Then the above $K(y,w)$ becomes a special case of (1.14) with $P = R$, $D = D^* = R_+$, $\alpha = 2$:

$$K(y,w) = \varrho ((S(y) - v)^+)^2 - \varrho v^2 .$$

Substitute $\lambda = 2\varrho v$ and note that

$$K(y,w) = \varrho\, \gamma(S(y), \lambda/\varrho)$$

where

$$\gamma(a,b) = ((a - \tfrac{1}{2} b)^+)^2 - \tfrac{1}{4} b^2 = \begin{cases} a^2 - ab, & a \geqslant \tfrac{1}{2} b \\ -\tfrac{1}{4} b^2, & a \leqslant \tfrac{1}{2} b . \end{cases}$$

One can therefore obtain also the augmented Lagrangian of Rockafellar [19].

For further use we shall need the following definition. The primal functional $\hat{Q} : P \to R$ is defined by

$$(1.15) \qquad \hat{Q}(p) = \inf_{y \in Y_p} Q(y).$$

In virtue of (1.7) the optimal value for (1.1) is

$$(1.16) \qquad \hat{Q}(p_o) = \inf_{y \in Y} \sup_{w \in W} L(y,w).$$

2. DUALITY

The theory of Lagrange multipliers for convex problems is strongly related to the theory of Fenchel conjugate functions in convex analysis [11] [14] [18] . During last several years some attempts have been made to extend the tools of convex analysis to nonlinear problems (e.g. [11]). In particular, the notion of φ-conjugate functions has been introduced by Weiss [27] and Vogel [28] and applied to the study of augmented Lagrangians similar to (1.9) by Seidler [26]. Given the primal problem (1.16) its dual may be formulated:

$$(2.1) \qquad \text{Find} \quad \sup_{w \in W}(D) = \sup_{w \in W} \inf_{y \in Y} L(y,w)$$

Always

$$(2.2) \qquad \sup(D) \leq \hat{Q}(p_o).$$

Define

$$\hat{L}(w) = \inf_{y \in Y} L(y,w)$$

and note that

$$\hat{L}(w) = \inf_{y \in Y} \inf_{p \in P_y} \{ Q(y) + \varphi(p,w) - \varphi(p_o,w) \} =$$

$$= \inf_{p \in P} \inf_{y \in Y_p} \{ Q(y) + \varphi(p,w) - \varphi(p_o,w) \} =$$

$$= - \varphi(p_o,w) + \inf_{p \in P} \{ \varphi(p,w) + \hat{Q}(p) \}.$$

For a given function $F : P \to R$ Weiss [27] defines its $-\varphi$ - conjugate by

(2.3) $$F^*(w) = \sup_{p \in P} \{-\varphi(p,w) - F(p)\} = -\inf_{p \in P} \{\varphi(p,w) + F(p)\}$$

and the second $-\varphi$-conjugate of F by

(2.4) $$F^{**}(p) = \sup_{w \in W} \{-\varphi(p,w) - F^*(w)\}.$$

Therefore we may write:

(2.5) $$\hat{L}(w) = -\varphi(p_o,w) - \hat{Q}^*(w)$$

(2.6) $$\sup(D) = \hat{Q}^{**}(p_o).$$

Therefore the problem when inequality (2.2) becomes an equality, so that we have the weak duality

(2.7) $$\inf_y \sup_w L(y,w) = \sup_w \inf_y L(y,w),$$

is equivalent to asking when

(2.8) $$\hat{Q}(p_o) = \hat{Q}^{**}(p_o).$$

The well known theorem of Moreau-Fenchel [11] states that (2.8) holds for a convex function $\hat{Q}$ defined on a topological vector space P at any p_o if and only if $\hat{Q}$ is l.s.c. Somehow similar requirements are needed in our general case.

Theorem 2.1

(compare [7])

(3.7) or, equivalently, (2.8) holds iff there is a sequence $\{w_n\}_{n=1}^{\infty} \subset W$ satisfying

(2.9) $$\hat{Q}(p) - \hat{Q}(p_o) \geq \varphi(p_o,w_n) - \varphi(p,w_n) - \frac{1}{n} \qquad \forall p \in P.$$

Proof.

$\hat{Q}(p_o) = \hat{Q}^{**}(p_o) = \sup_w \inf_p \{\varphi(p,w) - \varphi(p_o,w) + \hat{Q}(p)\}$ iff for each integer $n \geq 0$ an $w_n \in W$ exists such that

$$\hat{Q}(p_o) \leq \inf_p \{\varphi(p,w) - \varphi(p_o,w) + \hat{Q}(p)\} + \frac{1}{n}$$

or, which is the same

$$\hat{Q}(p_o) \leq \varphi(p,w) - \varphi(p_o,w) + \hat{Q}(p) + \frac{1}{n} \qquad \forall p \in P$$

as stated in (2.9).

In the theorem of Moreau-Fenchel the two properties are distinguished: convexity and lower semicontinuity. Lying aside the question of what may be here called " φ-convexity" (some definitions exist [28] but they are not very constructive) we observe some simple facts concerning the lower semicontinuity of Q and its conjugates.

In the sequel let τ_w (resp. τ_p) denote any topology in W (resp. P) such that all functions $\varphi(p,\cdot)$, $p \in P$ (resp. $\varphi(\cdot,w)$, $w \in W$) are u.s.c. in τ_w (τ_p). Observe that functions $\varphi(\cdot,w)$, $w \in W$ generate at least one such topology in P, for instance the weakest possessing this property. It is interesting to note that the convergence in P in this weakest topology is characterized as follows:

$$p_n \xrightarrow[n]{} p_o \Longleftrightarrow \overline{\lim_n}\, \varphi(p_n,w) \leq \varphi(p_o,w) \qquad w \in W .$$

so that $\{p_n\}$ converges to p_o iff it converges in terms of all "distributions of prices" $\varphi(\cdot,w)$, $w \in W$. In the examples described in the preceding section this topology is equivalent to weak topology in P (example (i)) and to norm topology (example (iii)).

<u>Proposition 2.1</u>

For every function $F: P \to R$, F^* is l.s.c. in τ_w and F^{**} in τ_p.

<u>Proof.</u>

It is sufficient to show [11] that for any $\alpha \in R$ the level set $\{ w \in W : F^*(w) \leq \alpha \}$ is closed. It is, as a product of closed sets:

$$\{w : F^*(w) \leq \alpha\} = \{w: -\varphi(p,w) - F(p) \leq \alpha \quad \forall p \in P\} =$$

$$= \bigcap_{p \in P} \{w: \varphi(p,w) \geq -F(p) - \alpha\} .$$

The proof for F^{**} is analogous.

Proposition 2.2

If $\hat{Q}(p_o) = \hat{Q}^{**}(p_o)$ then $\hat{Q}$ is l.s.c. in τ_p at p_o.

Proof.

For any integer $n \geq 1$ take w_{2n} as in (2.9). Since $\varphi(\cdot, w_{2n})$ is u.s.c. in τ_p, for some neighborhooed U of p_o

$$\varphi(p,w_{2n}) \leq \varphi(p_o,w_{2n}) + \frac{1}{2n} \qquad \forall p \in U.$$

Combining this with (2.9) we have

$$\hat{Q}(p) \geq \hat{Q}(p_o) - \frac{1}{n} \qquad \forall p \in U .$$

Lower semicontinuity of the primal functional $\hat{Q}(p)$ over Y is characterized in the following way (compare Dolecki [3] for the case of linear operator equality constraints).

Theorem 2.2

For $\alpha \in R$ denote $A_\alpha = \{p: \hat{Q}(p) \leq \alpha\}$, $PB_\alpha = \{p: p \in P_y, Q(y) \leq \alpha\}$.

Then $\hat{Q}$ is l.s.c. on P (in any topology τ) iff

(2.10) $$A_\alpha = \bigcap_{\varepsilon > 0} \overline{PB}_{\alpha+\varepsilon} \qquad \forall \alpha \in R.$$

Proof.

Sufficiency of (2.10) is clear, since all level sets of $\hat{Q}$ are closed.

Necessity. Observe that for $\alpha \in R$

$$\hat{Q}(p) \leq \alpha \iff \forall \varepsilon > 0 \quad \exists y \in Y_p \quad Q(y) \leq \alpha + \varepsilon$$

so that

(2.11) $$A_\alpha = \bigcap_{\varepsilon > 0} PB_{\alpha+\varepsilon}$$

and

$$A_\alpha \subset \bigcap_{\varepsilon > 0} \overline{P B}_{\alpha + \varepsilon}$$

To prove the converse inclusion, take $p \in \bigcap_{\varepsilon > 0} \overline{P B}_{\alpha + \varepsilon}$. Since $\hat{Q}$ is l.s.c., for any $\delta > 0$ there is a neighborhood U of p such that

(2.12) $$\hat{Q}(p) \leq \hat{Q}(p') + \delta \qquad \forall p \in U$$

Now let $\varepsilon > 0$; there is $p' \in U \cap P B_{\alpha + \varepsilon}$, i.e.

$$Q(y) \leq \alpha + \varepsilon \qquad \text{for some} \quad y \in Y_{p'}.$$

Together with (2.12) this yields

$$\hat{Q}(p) \leq \hat{Q}(p') + \delta \leq Q(y) + \delta \leq \alpha + \delta + \varepsilon .$$

Since $\delta, \varepsilon > 0$ were arbitrary, $\hat{Q}(p) \leq \alpha$ and $p \in A_\alpha$. Therefore (2.10) holds.

Example

Suppose Y is a topological, P a topological Hausdorff vector space and we consider the nonlinear program (1.10) described previously

$$\text{minimize} \quad Q(y) \qquad \text{subject to} \quad p - S(y) \in D$$

where $S : Y \longrightarrow P$ is continuous and $D \subset P$ is a closed set. If the level sets B_α, $\alpha \in R$ of Q are compact then the sets

$$P B_\alpha = \{p : p = S(y) + d, \ d \in D, \ y \in B_\alpha\} = S(B_\alpha) + D$$

are closed for any α, being a sum of a closed set and a compact. In virtue of (2.11) this means that (2.10) holds and $\hat{Q}$ is l.s.c.

In the context of generalized Lagrange functionals (1.9) also other questions can be discussed, for instance the problem of strong duality instead of weak duality (2.7):

(2.13) $$\inf_y \sup_w L(y,w) = \max_w \inf_y L(y,w).$$

Supposing a maximizing $\hat{w} \in W$ exists (and is known!), it would be possible to compute the optimal value $\hat{Q}(p_o)$ in one unconstrained minimization of $L(\cdot, w)$, which is a very desirable property. It turns out that (2.13) is equivalent to supporting $\hat{Q}$ at p_o by $\varphi(p_o, \hat{w}) - \varphi(\cdot, \hat{w})$ for some (maximizing) $\hat{w} \in W$:

$$\hat{Q}(p) - \hat{Q}(p_o) \geqslant \varphi(p_o, \hat{w}) - \varphi(p, \hat{w}) \qquad \forall p \in P.$$

(Compare [19] [31]).

One may further proceed with deriving different problems from (1.1) with help of functions $\varphi(\cdot, w)$ (constrained Lagrangian minimization, surrogate problems etc.) and discussing their properties. See [1] [6] [7] [8] [9].

3. EVERETT THEOREM AND THE METHOD OF MULTIPLIERS

<u>Theorem</u>

Suppose $\bar{y}$ satisfies

$$(3.1) \qquad L(\bar{y}, w) \leqslant L(y, w) + \varepsilon \qquad \forall y \in Y$$

and $\bar{p} \in P_{\bar{y}}$ is such that

$$(3.2) \qquad K(\bar{y}, w) = \varphi(\bar{p}, w) - \varphi(p_o, w)$$

Then $\bar{y}$ is an ε -solution to the problem (1.1) with p_o changed to $\bar{p}$, i.e.

$$Q(\bar{y}) \leqslant \hat{Q}(\bar{p}) + \varepsilon \quad , \qquad \bar{y} \in Y_{\bar{p}}.$$

<u>Proof</u>. is classical [4] [1]

From the algorithmic point of view it would be most desirable to establish the existence of a $w \in W$ such that $\bar{y}$ determined by (3.1) would satisfy the constraints, i.e. $p_o \in P_{\bar{y}}$. Then the whole constrained problem (1.1) would reduce to the single minimization of $L(\cdot, w)$ without constraints. Since such a $w = \hat{w}$ would be generally unknown, one should apply an iterative scheme to find it.

Assume from now on that for any $w \in W$ and $y \in Y$ with $P_y \neq \emptyset$ the set $P(y, w)$ is nonempty, where

$$P(y, w) = \{ p \in P_y : K(y, w) = \varphi(p, w) - \varphi(p_0, w) \} .$$

Given $\varepsilon \geq 0$ and $w \in W$ denote by $Y(\varepsilon, w)$ the set of all $\bar{y}$'s satisfying (3.1), i.e. being ε-minimal points for $L(\cdot, w)$ over Y. Finally, denote

$$P_\varepsilon(w) = \bigcup_{\bar{y} \in Y(\varepsilon, w)} P(\bar{y}, w) .$$

Then one can find an ε-solution to (1.1) minimizing ε-approximately $L(\cdot, w)$ over Y iff a $\hat{w} \in W$ exists solving the inclusion

(3.3) $$p_0 \in P_\varepsilon(w) .$$

Suppose now that:

(i) P is a Banach space, $W = V \subset P$ and for $w = v \in V$

$$\varphi(p, v) = \varrho \psi(p - v)$$

where $\varrho > 0$ is treated as a parameter.

(ii) For some $\varrho = \varrho_0 > 0$ and each $v \in V$ an element y_v exists minimizing $L(\cdot, v)$ over Y.

(iii) (for simplicity) For this $\varrho = \varrho_0$ and all $v \in V$ the set $P_0(v)$ contains precisely one element, say p_v.

Under these assumptions (3.3) becomes:

$$p_0 = p_v$$

or

(3.4) $$v = v + p_0 - p_v = T(v).$$

This is a fixed point problem and one may try to solve it iteratively using the method of successive approximations: given initial $v_1 \in V$, one takes

$$(3.5) \qquad v_n = T(v_{n-1}), \qquad n = 2,3,\dots .$$

The algorithm therefore works as follows:

1. Select $\varrho = \varrho_o$ and initial v_1. Set $n: = 1$.
2. Minimize $L(\cdot, v_n)$ over Y obtaining $y_n = y_{vn}$ and $p_n = p_{v_n}$.
3. Update v_n by setting

$$(3.6) \qquad v_{n+1} := v_n + \eta(p_o - p_n), \quad \eta > 0 \text{ fixed.}$$

4. Set $n: = n+1$ and go to 2.

It should be noted that the formula (3.6) is independent of the function Ψ and ϱ_o chosen, although y_n and p_n clearly depend on them.

This is the algorithm (shifted penalty method, method of multipliers) of Hestenes [10] and Powell [16] for equality constraints in R^n, generalized to inequality constraints in R^n and in Hilbert space by Wierzbicki [29] [30]. The rate of convergence of this algorithm for $P = R^n$ has been investigated in 1971 by Wierzbicki [29] and later by several authors, most completely by Bertsekas [2]. All of them used quadratic functionals. For the theoretical discussion of this algorithm see also [10] [19] [20] for $P = R^n$ [30] [31] for $P =$ a Hilbert space, [23] [24] [25] for variational and optimal control problems and others.

It is clear that one may attempt to solve (3.4) by other methods than successive approximations, e.g. Newton or variable metric method, provided the operation $v \longmapsto v - T(v)$ possesses desirable properties.

4. THE SHIFTED PENALTY METHOD FOR LINEAR OPERATOR CONSTRAINTS

In this section we shall investigate briefly the application of the shifted penalty method to an important special case of the optimization problem (1.1) and present a theorem precising the conditions under which this method is convergent. The next section contains numerical examples.

Suppose Y, P are Hilbert spaces and $S \in \mathcal{L}(Y, P)$. The problem is

$$(4.1) \qquad \text{minimize } \|y\|^2 \quad \text{subject to} \quad Sy = p_o .$$

This is a special case of (1.10) with the cone D reduced to $\{0\}$. For every $p_o \in \text{im}S$ (4.1) has a unique solution $\hat{y}(p_o)$. We set $W = V = \text{im}S$ and $\varphi(p,v) = \varrho \|p - v\|^2$, as in the preceding seation, treating here ϱ as a parameter. The shifted penalty method has been formulated above and the Lagrangian is here

$$(4.2) \qquad L(y, v) = \|y\|^2 + \varrho \|Sy - v\|^2 - \varrho \|p_o - v\|^2 .$$

<u>Theorem 4.1</u>

The shifted penalty method with $\eta = 1$ converges for each $p_o \in \text{im}S$, initial $v_1 \in \text{im}S$ and $\varrho > 0$ if, and only if $\text{im}S$ is a closed subspace of P. In this case the rate of convergence is estimated by:

$$(4.3) \qquad \frac{\|v_{n+1} - v_n\|}{\|v_n - v_{n-1}\|} \leq \frac{1}{1 + \varrho\varkappa} \qquad \text{where } \varkappa > 0.$$

<u>Proof.</u>

Sufficiency. The mapping T defined by (3.4) can be here explicity expressed. For any $v \in P$, unique y_v exists minimizing $L(\cdot, v)$ over P. By (4.2), this y_v satisfies

$$L'(y_v, v) = y_v + \varrho S^*(S y_v - v) = 0.$$

The corresponding p_v is equal to $p_v = S\, y_v$ and hence from the above equation

$$y_v = -\, \varrho\, S^{*}(p_v - v),$$

$$p_v = S\, y_v = -\, \varrho\, SS^{*}(p_v - v)$$

and

$$(I + \varrho\, S\, S^{*})\, p_v = \varrho\, SS^{*} v\,. \tag{4.4}$$

Since imS is closed, by Banach closed range theorem [32] $\mathrm{im}S^{*}$ also is. Operator S^{*} is an injection from imS onto $\mathrm{im}S^{*}$; both these subspaces being complete, S possesses a continuous inverse S^{*-1} : $\mathrm{im}S^{*} \longrightarrow \mathrm{im}S$ in virtue of the open mapping theorem [32]. Denote $\varkappa = \| S^{*-1} \|^{-2}$ (norm computed with respect to $\mathrm{im}S^{*}$) and $A = I + \varrho\, SS^{*}$. Clearly, $A^{*} = A$ and $A : \mathrm{im}S \longrightarrow \mathrm{im}S$. Moreover, for each $p \in \mathrm{im}S$

$$\langle p, A\, p\rangle = \langle p, (I + \varrho\, SS^{*})p\rangle = \| p \|^{2} + \varrho \| S^{*} p \|^{2} \geqslant$$

$$\geqslant \| p \|^{2} + \varrho\varkappa \| p \|^{2} = (1 + \varrho\varkappa)\, \| p \|^{2}.$$

In virtue of Lax-Milgram theorem [32], $A \in \mathcal{L}$ (imS, imS) admits the continuous inverse on imS and

$$\| A^{-1} \| \leqslant \frac{1}{1 + \varrho\varkappa}\,. \tag{4.5}$$

Now, for $v \in \mathrm{im}S$, p_v may be calculated from (4.4):

$$p_v = A^{-1}\, \varrho\, SS^{*}\, v = A^{-1}\,(Av - v) = v - A^{-1}\, v.$$

The mapping T: imS $\longrightarrow$ P is defined by (3.4):

$$T(v) = p_o - p_v + v = p_o + A^{-1} v$$

Hence for every p_o, $v \in \mathrm{im}S$, $T(v) \in \mathrm{im}S$ and for $v', v'' \in \mathrm{im}S$

$$\| T(v') - T(v'') \| \leqslant \| A^{-1} \|\, \| v' - v'' \| \leqslant \frac{1}{1 + \varrho\varkappa} \| v' - v'' \|\,.$$

Since $\varkappa > 0$, T is a contraction mapping in a complete metric space imS. Hence the algorithm converges to a point $v = T(v)$

with the rate of convergence (4.3).

Necessity. If the algorithm is convergent, then the equation

$$v = T(v)$$

admits a solution for each $p_o \in \text{im}S$. According to the definition of T (see the preceding section), for each $p_o \in \text{im}S$ an element $v \in P$ exists such that the unique optimal solution $\hat{y}(p_o)$ to (4.1) minimizes $L(\cdot, v)$ over Y. By (4.2), $\hat{y}(p_o)$ must satisfy

$$L'(\hat{y}(p_o), v) = \hat{y}(p_o) + \varrho S^*(S\hat{y}(p_o) - v) = 0$$

so that

(4.5) $$\hat{y}(p_o) \in \text{Im}S^*. \qquad \forall p_o \in \text{im}S .$$

Suppose imS is not closed, so that by closed range theorem [32] $(\ker S)^{\perp} \setminus \text{im}S^*$ contains an element y_o. Select $p_o = Sy_o$. For any y such that $Sy = p_o$ obviously $\langle y_o, y - y_o\rangle = 0$, so that

$$\|y\|^2 - \|y_o\|^2 = \langle y + y_o, y - y_o\rangle = \langle y - y_o, y - y_o\rangle = \|y - y_o\|^2 .$$

Hence $y_o = \hat{y}(p_o)$, the unique solution to (4.1). Then (4.5) yields a contradiction, since $y_o \notin \text{im}S^*$. Therefore imS must be closed.

The "if" part of theorem 4.1 can be also obtained from a more general theorem [31]. The condition that imS be closed in P is intimately related to the existence of Lagrange multipliers for the problem (4.1.) with any $p_o \in \text{im}S$ [13] [21]. In the convex case the existence of Lagrange multipliers is sufficient for the convergence of the method of multipliers, as has been shown by Rockafellar for $P = R^n$ [20].

In the course of sufficiency proof the constant in (4.3) was taken to be $\|S^{*-1}\|^{-2}$. Thus, the smaller is the norm of

$\| S^{*-1} \|$, the quicker the convergence. Therefore the rate of convergence depends on the norm chosen; this is confirmed also by the numerical results.

It is also possible to apply the shifted penalty method with $\eta \neq 1$; but then η and ϱ must satisfy

$$0 < \eta < 1 + \frac{\varkappa \varrho}{2 + \varkappa \varrho} ,$$

the rate of convergence is expressed by

$$\frac{\| v_{n+1} - v_n \|}{\| v_n - v_{n-1} \|} \leq |1 - \eta| + \frac{\varrho}{1 + \varkappa \varrho}$$

and is the best at $\eta = 1$.

5. NUMERICAL EXAMPLES

Two optimal control problems for linear time-lag systems with fixed final function were solved numerically.

Example 1

minimize $Q(y) = \int_0^2 y^2(t)dt$

for the control $y \in L^2(0,t)$

subject to constraints

$$(5.1) \qquad \begin{cases} \dot{x}(t) = - x(t-1) + y(t) & \text{a.e. in } [0,2] \\ x(t) = 0 & \forall t \in [-1,0] \end{cases}$$

$$(5.2) \qquad x(t) = - \frac{1}{2} (t-1)^2 \qquad \forall t \in [1,2]$$

This problem was taken from [12]. This is the special case of (4.1). The operator S is defined as follows: given control $y(\cdot)$, solve (5.1) and put $Sy = x \,|_{[1,2]}$. The element p_o is defined by (5.2). For each $y \in L^2(0,2)$, $Sy \in W_1^2(1,2)$, the Sobolev space of absolutely continuous functions with square integrable desirative. Since $W_1^2(1,2) \subset L^2(1,2)$, one may use here at least two spaces of constraints: $P_1 = L^2(1,2)$ and $P_2 = W_1^2(1,2)$. imS is closed in P_2 and not closed in P_1. Consequently, two Lagrangians can be used (we neglect the term $\varrho \| p_0 - v \|^2$ since it does not influence the computations):

$$L_1(y,v) = \int_0^2 y^2(t)dt + \varrho \int_1^2 (x(t) - v(t))^2 dt$$

$$L_2(y,v) = \int_0^2 y^2(t)dt + \varrho (x(2) - v(2))^2 + \varrho \int_1^2 (\dot{x}(t)-\dot{v}(t))^2 dt$$

The method used was (after the problem was discretized) the original algorithm of Powell [16] ,which increases ϱ if the improvement in the constraints violation was too small. Computations have been carried out for several values of initial ϱ_0. Both algorithms were the same and the computational effort per one evaluation of L_1 or L_2 was practically the same. The results are:

ϱ_0	Number of evaluation of L		Final constraint violation	
	$P_1 = L^2(1,2)$	$P_2 = W_1^2(1,2)$	$P_1 = L^2(1,2)$	$P_2 = W_1^2(1,2)$
0.01	742	152	$0.921 \cdot 10^{-3}$	$0.44 \cdot 10^{-3}$
1.0	816	157	$0.983 \cdot 10^{-3}$	$0.92 \cdot 10^{-3}$
10	1153	161	$0.452 \cdot 10^{-3}$	$0.64 \cdot 10^{-3}$

Computations have been performed on an IBM-360 in Fortran. The constraint violation was in both cases measured as

$$\max_{t \in [1,2]} \left| x(t) + \frac{1}{2}(t-1)^2 \right| .$$

<u>Example 2</u>. (K.M. Przyłuski [17]).

minimize $Q(y) = \int_0^3 (y(t) - z(t))^2\, dt$ [1]

over $y \in L^2(0,3)$ subject to the constraints:

$$(5.3)\quad \begin{cases} \dot{x}_1(t) = u(t) & \text{a.e. in } [0,3] \\ \dot{x}_2(t) = x_1(t-1) & \\ x_1(t) = x_2(t) = 0 & \forall\, t \in [-1,0] \end{cases}$$

$$(5.4)\quad \begin{cases} x_1(t) = t - 1 \\ x_2(t) = \frac{1}{2}(t-2)^2 \end{cases} \qquad \forall\, t \in [2,3]$$

where $z(t) = \begin{cases} 0 & , \; t \in [0, 3/2] \\ 1 & , \; t \in (3/2, 3]. \end{cases}$

For given control $y \in L^2(0,3)$ the operator Sy is defined by $Sy = (x_1, x_2)\big|_{[2,3]}$ where (x_1, x_2) solve (5.3). The element p_o is determined by (5.4). Since $x_2(\cdot)$ is absolutely continuous, one can put either $P_1 = W_1^2(2,3\,;\,R^2)$ or $P_2 = W_1^2(2,3) \times W_2^2(2,3)$. imS is not closed in P_1 and closed in P_2 [13]. The corresponding Lagrangians are (also without the term $\varrho\|p_o - v\|^2$):

$$L_1(y,w) = \int_0^3 (y(t) - z(t))^2 dt + \varrho|x(3) - v(3)|^2 + \int_2 |\dot{x}(3) - \dot{v}(3)|^2 dt$$

$$L_2(y,w) = \int_0^3 (y(t)-z(t))^2 dt + \varrho|x(3)-v(3)|^2 + \varrho|\dot{x}_2(3) - \dot{v}_2(3)|^2 +$$

$$+ \varrho \int_2^3 \left(|\dot{x}_1(t) - \dot{v}_1(t)|^2 + |\ddot{x}_2(t) - \ddot{v}_2(t)|^2 \right) dt.$$

The computational effort per one evaluation of L_1 or L_2 was more less the same. The results are:

---

1) This is not exactly the problem of (4.1) type but can be transformed to (4.1) by a simple shift of zero in Y.

Discretization of [0,3]	ρ_0	Number of evaluation of L		Final constraint violation	
		$P_1 = W_1^2$	$P_2=W_1^2 x W_2^2$	$P_1 = W_1^2$	$P_2 = W_1^2 x W_2^2$
12	1	200	88	$0.649\cdot 10^{-3}$	$0.960\cdot 10^{-3}$
	10	184	60	$0.624\cdot 10^{-3}$	$0.912\cdot 10^{-3}$
	100	146	50	$0.682\cdot 10^{-3}$	$0.466\cdot 10^{-3}$
90	1	1090	44	> 0.012	$0.246\cdot 10^{-2}$
	10		40		$0.177\cdot 10^{-2}$
	100		30		$0.177\cdot 10^{-2}$

The constraint violation was here measured by

$$c = \max(|x_1(3)-2|, |x_2(3)-\tfrac{1}{2}|, \max_{t\in[2,3]} (|\dot{x}_1(t)-1|, |\dot{x}_2(t)- t + 2|))$$

Summarising, in these two cases the computational effort for solving the problem with similar accuracy was 2-3 times smaller for the Lagrangian L_2, employing the proper norm. For more detailed description and discussion of these results see [13] [17].

6. CONCLUSIONS

A class of generalized Lagrangians has been defined associated with a family of extremal problems with general constraining set. The relation of these Lagrangians to others found in literature was indicated; it occurs that many facts proved for generalized Lagrangians of nonlinear programs with R^n constraints remain true in much more general setting. Some simple properties of these Lagrangians have been shown, especially those related to the topology in the set of constraints values; also the Everett theorem and an abstract formulation of the method of multipliers (shifted penalty technique) were given. A theorem has been provided characterizing the convergence of this method in the case of linear-quadratic problems in Hilbert space. The last section contains numerical examples of the application of the algorithm to time-delay optimal control problems.

The results show that the behavior of the finitedimensional algorithm applied to the discretized version of inifinite-dimensional problem may depend on topological properties of the original problem.

REFERENCES

[1] M. Bellmore, H.J. Greenberg, J.J. Jarvis - Generalized penalty - function concepts in mathematical optimization - Opns. Res., 18 (1970), No 2.

[2] D.P. Bertsekas - Combined primal-dual and penalty method for constrained minimization - SIAM J. Control 13 (1975), No 3.

[3] S. Dolecki - Bounded controlling sequences, inf - stability and certain penalty procedures, to appear.

[4] H. Everett III - Generalized Lagrange multipliers method for solving problems of optimum allocation of resources - Opns. Res. 11 (1963), No 3.

[5] F. Glover - Surrogate constraints - Opns. Res. 16 (1968), 741-769

[6] F.J. Gould - Extensions of Lagrange multipliers in nonlinear programming - SIAM J. Appl. Math. 17 (1969) No 6.

[7] F.J. Gould - Nonlinear pricing: applications to concave programming - Opns. Res. 19 (1971), No 4.

[8] H.J. Greenberg - The generalized penalty function/surrogate model - Opns. Res. 21 (1973), No 1.

[9] H.J. Greenberg, W.P. Pierskalla - Surrogate mathematical programming - Opns. Res. 18 (1970), 924-939.

[10] M.R. Hestenes - Multiplier and gradient methods. In: Computing methods in optimization problems - 2, ed. by: L.A. Zadeh, L.W.: Neustadt, A.V. Balakrishnan, Academic Press 1969, 143-164.

[11] A.D. Ioffe, W.M. Tikhomirov - Theory of extremal problems (in Russian) - Nauka, Moscow 1974.

[12] M.Q. Jacobs, T.J. Kao - An optimum setting problem for time-lag systems - J. Math. Anal. Appl, 40 (1973), 687-707.

[13] S. Kurcyusz - Necessary optimality conditions for problems with function space constraints (in Polish) - Ph. D. Thesis, Instytut Automatyki, Politechnika Warszawska, Warsaw 1974.

[14] P.J. Laurent - Approximation et optimization - Herrman, Paris 1972

[15] O.L. Mangasarian - Unconstrained Lagrangians in nonlinear programming - SIAM J. Control 13 (1975), No 4.

[16] M.J.D. Powell - A method for nonlinear constraints in minimization problems, in: Optimization, ed. by R. Fletcher, Academic Press 1969, 283-298.

[17] K.M. Przyłuski - Application of the shifted penalty method to dynamic optimization of delay processes (in Polish) - M. Sc. Thesis, Instytut Automatyki, Politechnika Warszawska, Warszawa 1974.

[18] R.T. Rockafellar - Convex analysis - Princeton University Press, Princeton 1970.

[19] R.T. Rockafellar - Augmented Lagrange multiplier functions and duality in nonconvex programming - SIAM J. Control, 12 (1974), No 2.

[20] R.T. Rockafellar - The multiplier method of Hestenes and Powell applied to convex programming - J. Opt. Theory Appl. 10 (1973).

[21] S. Rolewicz - Functional analysis and control theory (in Polish) - PWN, Warsaw 1974.

[22] J.D. Roode - Generalized Lagrangian functions and mathematical programming - in: Optimization, ed. by R. Fletcher, Academic Press 1969.

[23] R.D. Rupp - A method for solving a quadratic optimal control problem - J. Opt. Theory Appl. 9 (1972), No 4.

[24] R.D. Rupp - Approximation of the Classical Isoperimetric Problem - J.Opt.Theory Appl. 9 /1972/,pp.251-264

[25] R.D. Rupp - A nonlinear optimal control minimization technique - Trans. AMS, 178 (1973), 357-381.

[26] K.H. Seidler - Zur Dualisierung in der nichtlinearen optimierung - Ph. D. Thesis, Technische Hochschule Ilmenau, Ilmenau 1972.

[27] W. Vogel - Duale Optimierungsaufgaben und Sattelpunktsatze - Unternehmensforschung 13 (1969), 1-28.

[28] E.A. Weiss - Konjugierte Funktionen - Arch. Math. 20 (1969), 538-545.

[29] A.P. Wierzbicki - A penalty function shiffting method in constrained static optimization and its convergence properties - Archiwum Automatyki i Telemechaniki 16 (1971), 395-416.

[30] A.P. Wierzbicki, A. Hatko - Computational methods in Hilbert space for optimal control problems with delays - Proc. of 5-th IFIP Conference on Optimization Techniques, Rome 1973.

[31] A.P. Wierzbicki, S. Kurcyusz - Projection on a cone, generalized penalty functionals and duality theory - Institute of Automatic Control, Technical University of Warsaw Technical Report No 1/1974.

[32] K. Yosida - Functional analysis - Springer Verlag, Berlin 1966.

SUBGRADIENT OPTIMIZATION, MATROID PROBLEMS AND HEURISTIC EVALUATION

F. Maffioli
Istituto di Elettrotecnica ed Elettronica
Politecnico di Milano, 20133 Milano (Italy)

Abstract

Many polynomial complete problems can be reduced efficiently to three matroids intersection problems. Subgradient methods are shown to yield very good algorithms for computing tight lower bounds to the solution of these problems. The bounds may be used either to construct heuristically guided (branch-and-bound) methods for solving the problems, or to obtain an upper bound to the difference between exact and approximate solutions by heuristic methods. The existing experience tend to indicate that such bounds would be quite precise.

1. FOREWORD

Consider the following three apparently unrelated problems.

3-dimensional assignment . Given n men for n jobs and n different time slots and a weighting c_{ijk} for assigning the i-th man to the j-th job in the k-th time slot, find an assignment of maximum total weight.

Traveling salesman problem. A salesman has to go from city 1 to n-1 other different cities. Given the distance matrix of the n cities find a path of minimum total length going through each city at most once.

A Sequencing problem. Let there be n jobs to be processed on a single machine and let job i be requiring T_i units of time, having a deadline D_i after which a penalty P_i has to be paid. Find the sequencing of the jobs which minimizes the overall penalty to be paid.

As it will appear in the following these are three instances of a general class of problems, namely those reducible to 3-matroid intersection problems. The purpose of this work is to present a general method to obtain tight bounds to the optimal solution to these problems in order to be able to estimate fairly accurately the error by which any heuristically obtained solution would be affected, In section 3 a few notions from matroid theory will be reviewed. Section 4 will be devoted to matroid problem reduction. Section 5 will develop the main algorithm for calculating the bounds. Section 6 will show how the three above mentioned problems can be approached as matroid problems. Some conclusions and areas for further research will be outlined in section 7.

This work was completed while the author was on leave as Research Associate at the Electronics Research Laboratory of the University of California at Berkeley with a NATO Senior Fellowship.

2. COMPLEXITY OF ALGORITHMS AND PROBLEMS REDUCTION

A problem is said to be in P if an algorithm for its solution exists whose computing time is a polynomial function in the size of the problem. Karp [1] has shown that many problems which are (probably) not in P can be reduced one to the other so that one of them being in P would imply that all of them are. We say that a problem "reduces" to another if there exists an algorithm in P which would yield the solution to the second once the first is solved. The three problems mentioned above are (probably) not in P. Even the best algorithms known are of exponential complexity and for large problems only heuristic methods can be used successfully.

3. MATROID AXIOMATICS [2]

Let $E = \{e_1, e_2, \ldots, e_n\}$ be a finite set of elements and $\mathfrak{I}$ a non-empty family of subsets of E such that:

1) if $I \subset J \in \mathfrak{I}$ then $I \in \mathfrak{I}$;

2) if $I, J \in \mathfrak{I}$ and $|I| = |J|+1$, then there exists an element $e \in I-J$ such that $J+e \in \mathfrak{I}$.

Then $M = (E,\mathfrak{I})$ is a matroid and the members of $\mathfrak{I}$ are called its independent sets. A maximal independent set is called a base. A minimal dependent set is called a circuit. All bases of a matroid have the same cardinality. As an example let E be the set of edges of a linear graph and $\mathfrak{I}$ the set of forests of the graph: this is the graphic matroid of the graph. Else let E be the set of columns of a matrix and $\mathfrak{I}$ the family of sets of columns which are linearly independent (over any field): this is a matric matroid of the matrix. As a third example let E be any finite set and let π be a partition of E into r disjoint subsets $S_1, S_2, \ldots, S_r$. Let $d = (d_1, d_2, \ldots, d_r)$ be a r-dimensional vector and

$$\mathfrak{I} = \{I: I \subseteq E \ \& \ |I \cap S_i| \leq d_i \ , \ i = 1,2,\ldots,r\} .$$

Then $M = (E,\mathfrak{I})$ is a partition matroid.

Let there be a weighting function $w : E \rightarrow R^+$. The problem of finding an independent set $\bar{I}$ of M having maximum total weight (or equivalently a base of minimum total weight) is solved by the "greedy" algorithm: "include in $\bar{I}$ the element of maximum weight among those not yet included, disregarding an element only if it would destroy independence once included into $\bar{I}$" [3].

Let $m = |E|$ and $c(m)$ be the complexity order of the method for testing independence (TI) in M. Then the greedy algorithm has a complexity at worst of order $mc(m)$ and if TI $\in$ P the greedy algorithm also belongs to P.

Let M_1 and M_2 be two given matroids. Then $M = (E_1 \cup E_2, \mathfrak{I})$ where

$\mathfrak{I} = \{I : I = I_1 \cup I_2 \ \& \ I_1 \in \mathfrak{I}_1, I_2 \in \mathfrak{I}_2\}$

is a matroid called the sum of M_1 and M_2.

4. MATROID PROBLEMS REDUCTION

Let there be k matroids $M_1, M_2, \ldots, M_k$ over the same set E. A subset I of E which is independent in all of them is called an intersection. Consider also a partition of E into p disjoint subsets $P_1, P_2, \ldots, P_p$ and let h be the maximum cardinality of them. A very general matroid problem is the following. Let $\mathfrak{I}_i$ be the family of independent sets of the i-th matroid. Find the subset I of E of maximum weight such that

$$I \in \bigcap_{i=1}^{k} \mathfrak{I}_i \qquad (3$$

and such that $|I \cap P_j| = \{|P_j| \text{ else } 0\}$ for $j = 1,2,\ldots,p$. (4

Any set obeying (4) is called a h-parity set, where

$$h = \max_j |P_j| .$$

It can be shown that due to the results of Lawler [4] this problem can be reduced to a 3-matroid intersection problem on a set E' containing 2km elements where two of the matroids are partition matroids. The reader is referred to [4] for the corresponding reductions.

5. A SUBGRADIENT ALGORITHM

The method for obtaining bounds will be formulated for the special instance of the 3-matroid intersection problem which is yielded by the reductions mentioned in the previous section.

Let $M_i = (E, \mathcal{J}_i)$, $i = 1,2,3$ be three matroids defined over the same set of elements E and having respectively ranks n,n and n+r where $2n = |E|$. Let $w : E \to R^+$ be a given weighting function. The first two matroids, M_1 and M_2, are particular partition matroids corresponding to the following partitions of E.

$$E = \{A_1, A_2, \ldots, A_r\} = \{B_1, B_2, \ldots, B_r\}$$

$$|A_i| = |B_i| = 2 \ , \ i = 1,2, \ldots, r.$$

So that

$$\mathcal{J}_1 = \{I : I \subseteq E \ \& \ |I \cap A_i| \leq 1, \quad i=1,2,\ldots,r \}$$

$$\mathcal{J}_2 = \{I : I \subseteq E \ \& \ |I \cap B_i| \leq 1, \quad i=1,2,\ldots,r \}$$

Let $\pi = (\pi_1, \pi_2, \ldots, \pi_r)$ be a real r-dimensional vector. Then a new weighting function w' may be defined for each $e \in E$ as

$$w'(e) = w(e) + \sum_{i=1}^{r} \pi_i \left[|\{e\} \cap A_i| + |\{e\} \cap B_i| \right]$$

Then for any $X \subseteq E$,

$$w'(X) = w(X) + \pi \cdot \mu(X)$$

where

$$\mu(X) = (|A_1 \cap X| + |B_1 \cap X|, \ \ldots, \ |A_r \cap X| + |B_r \cap X|)$$

Let now $\bar{I}$ be such that

$$w'(\bar{I}) = \max \{w'(X) : X \in \mathcal{J}_3 \text{ and } |X| \leq r\}$$

It is always impossible to find $\bar{I}$ by the greedy algorithm. Obviously $|\bar{I}| = r$. Let now I_1 and I_2 be two intersections of maximum cardinality of the three given matroids. Then

$$|I_i \cap A_j| \leq 1 \qquad\qquad i = 1,2$$

$$|I_i \cap B_j| \leq 1 \qquad\qquad j = 1,2,\ldots,r.$$

On the other hand, since $|I_1| = |I_2| = r$,

$$\mu(I_1) = \mu(I_2) = \varepsilon$$

where ε is a vector whose entries are all equal to 2. Let now I^o be a maximum weight intersection. Then by definition

$$w'(I^o) \leq w'(\bar{I})$$

so that

$$w(I^o) + \pi \cdot \mu(I^o) \leq w'(\bar{I})$$

i.e.

$$w(I^o) \leq w'(\bar{I}) - \pi \cdot \varepsilon = f(\pi)$$

and $f(\pi)$ is a valid upper bound to the value of the optimum solution for any π. The tightest of such bounds will be obtained by

$$f(\pi^*) = \min_{\pi} f(\pi)$$

Subgradient methods have been extensively studied and used in recent literature [5,6,7,8]. In our case the gradient of $f(\pi)$ is

$$\nabla f(\pi) = \mu(\bar{I}) - \varepsilon$$

and can easily be obtained applying the greedy algorithm to M_3 under the weighting w'. A simple iterative scheme for computing π^* is the following.

<u>Step 1</u> $\pi = 0$, $i = 0$.

<u>Step 2</u> $i+1 \rightarrow i$, $w(e) + \sum_{j=1}^{r} \pi_j \left[|A_j \cap \{e\}| + |B_j \cap \{e\}| \right] \rightarrow w(e)$ for all e E.

<u>Step 3</u> Find a maximum weight independent set $\bar{I}$ of M_3 by the greedy algorithm. Then $f(\pi) = w(\bar{I}) - \pi \cdot \varepsilon$ and $g = \mu(\bar{I}) - \varepsilon$.

<u>Step 4</u> If $g = 0$ stop: $I^o = \bar{I}$. If i = maximum number of iterations stop: $f(\pi)$ is the best obtainable upper bound.

<u>Step 5</u> $\pi + tg \rightarrow \pi$, t being a suitable scalar. Go to step 2.

A maximum number of iterations has to be chosen since the method could fail to converge to a case for which $g = 0$. The reader is referred to [5,6] for further informations about this behaviour. For choosing t one can adopt a rough criterion as f.i. $t = 1$ as in [5], or more refined criteria as suggested in [5] and used in [7]. Also a modified subgradient direction can be used to speed up the search [9].

6. EXAMPLES

The three dimensional assignment and the traveling salesman problems are already in the form of 3-matroid intersection problems. For the first one has to consider three partition matroids one for each index of the cost coefficients. For the second problem two of the matroids are partition matroids generated by out and in-degrees at every node of the graph, while the third matroid is the graphic matroid.

As far as the sequencing problem is concerned consider the graph of figure 1 together with the following theorem.

Theorem (Edmonds & Fulkerson [10]). Let $G = (N,A)$ be a graph and let $E \subset N$ be any subset of N. Consider the family $\mathfrak{I}$ of all the subsets I of E such that there exists a matching of G covering all the nodes of I. Then $M = (E, \mathfrak{I})$ is a matroid called the matching matroid of G.

Apply this theorem to the graph of figure 1, E being the set of nodes on the left corresponding to the various jobs identified by as many nodes as the required units of time, the penalty of each job being divided in any arbitrary way among its nodes. If each job corresponds at maximum to h nodes, one has a h-parity matroid problem on a matching matroid, since a subset of E of maximum weight which can be covered by a matching corresponds to an optimum scheduling.

7. CONCLUSIONS

An immediate development of this work would be the implementation of a computer code in order to test the algorithm for various problem instances. The present experience although quite promising has been almost entirely confined to the traveling salesman problem: there are no reasons however against the hope of obtaining as good results on other problems. The elementary iteration of section (5) could be improved following [7,9] and [11]. Further research is needed to explore the full class of problems to which this method applys as well as to see which modifications if any would be needed to approach problems apparently not belonging to this class such as the Steiner network problem and the quadratic assignment.

i	D_i	P_i	T_i
1	2	9	2
2	4	7	3
3	2	5	1
4	6	3	2

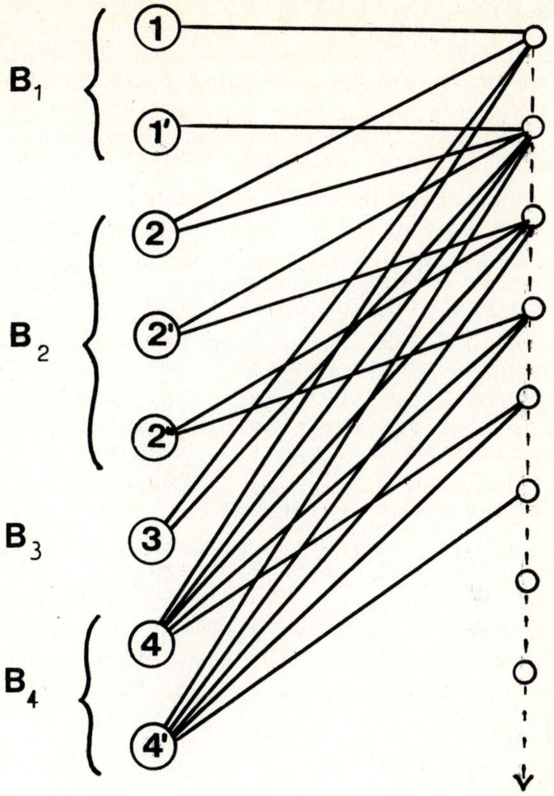

Units of jobs Units of time

Fig. 1

REFERENCES

[1] R.M. Karp, "Reducibility among combinatorial problems", Symp. on Complexity of Computations (Miller & Tatcher eds.) Plenum Press 1972

[2] H. Whitney, "On the abstract properties of linear dependence", Amer. J. Math. 57 (1935) 509-533.

[3] J. Edmonds, "Matroids and the greedy algorithm", Math. Programm. 1 (1971) 127-136.

[4] E.L. Lawler , "Polynomial bounded and (apparently) non-polynomial bounded matroid computations", Combinatorial Algorithms (Rustin ed.) Algorithmics Press 1972.

[5] M. Held & R.M. Karp, "The traveling salesman problem and minimum spanning trees: part II", Math. Programm. 1 (1971) 6-25.

[6] P.M. Camerini, L. Fratta & F. Maffioli, "A heuristically guided algorithm for the traveling salesman problem", J. of the Institution of Computer Sciences 4 (1973) 31-35.

[7] K.H. Hansen & J. Krarup, "Improvements of the Held-Karp algorithm for the symmetric traveling salesman problem" Math.Progr.7(1974)87-96.

[8] M. Held, P. Wolfe & H.P. Crowder, "Validation of subgradient optimization", Math. Programm. 6 (1974) 62-88.

[9] P.M. Camerini, L. Fratta & F. Maffioli, "Relaxation methods improved by modified gradient techniques", Conference on Opns. Res., Eger (Hungary) August 1974.

[10] J. Edmonds & D.R. Fulkerson, "Transversals and matroid partition", J. Res. NBS 69B (1965) 147-153.

[11] H. Crowder, "Computational improvements for subgradient optimization", IBM Res. Rep. RC 4907 (21841) 1974.

ACKNOWLEDGMENT

The author is pleased to acknowledge the many fruitful discussions with E.L. Lawler, R.M. Karp and P.M. Camerini on this subject.

THEORETICAL AND PRACTICAL ASPECTS OF COORDINATION BY PRIMAL METHOD

K.B. Malinowski, J. Szymanowski

Technical University of Warsaw
Institute of Automatic Control
ul. Nowowiejska 15/19
00-665 Warsaw, Poland

Introduction

In this paper a multilevel method of primal type for solving large scale optimization problems is considered. The multilevel methods of optimization are now in phase of fast development/[4] , [9]/, since they form a basis for large system control/[3]/. The Primal Method/[2]/ has many advantages when compared with for example, the methods of balance type developed by Lasdon and other scientists/[13]/. However considerable difficulties still arise when we want to make from the Primal Method an efficient numerical tool for solving large mathematical programming problems. There are also many theoretical aspects of this method which are not fully examined yet. These problems and difficulties are discussed in the sequel.

1. Problem formulation and general properties

1.1. Decomposition

Let us suppose that we are given the following mathematical programming problem.

$$\min_{x} f(x)$$

subject to

$$F(x) \in Y_o \qquad (1)$$
$$x \in X_o$$

where $X_o \subset X$, $Y_o \subset Y$; X , Y are linear-topological Hausdorff spaces l.t. spaces and

$f : X \rightarrow R$ is the performance index

$F : X \rightarrow Y$ is the constraint operator

Problems of this kind represent a very general class of optimization problems in operations research, optimal control theory etc.etc. If problem (1) is a very complex one then there may be profitable or even-necessary to apply some multilevel method of optimization for solving it. Methods of multilevel type cannot be used directly to problem (1) and their application is possible only after problem (1) has been transformed /at least formally/ into another form by the decomposition process. Since in this paper we focuse our investigations on the primal methods of multilevel programming, we will only describe here shortly the decomposition used in this particular case.

Generally speaking the process of decomposition consists of three steps /which are of course very strongly connected together/:

1. choose appropriate l.t. Hausdorff spaces M and V
2. choose and define map $\varphi: M \times V \rightarrow X$
3. form the new mathematical programming problem

$$\min_{m,v} \left[Q(m, v) = \Psi(Q^1(m^1, v), \ldots, Q^N(m^N, v)) \right] \qquad (2)$$

subject to

$$G^i(m^1,v) \in Y^i_o \subset Y^i, \quad i= 1,\dots,N$$

$$m^i \in M^i_o \subset M^i$$

$$T(v) \in Z_o \subset Z$$

$$v \in V_1 \subset V$$

where

$$m = (m^1,\dots,m^N) \in M^1 \quad x\dots xM^N = M$$

$T : V \rightarrow Z$ (Z - l.t. space)

$G^i: M^i \times V \rightarrow Y^i$ (Y^i - l.t. space, $i = 1,\dots,N$)

$G^i: M^i \times V \rightarrow R$, $i = 1,\dots,N$ and

$\Psi: R^N \rightarrow R$ is continuous function atrictly preserving partial ordering is $\Omega \subset R^N$ x/

The set of points (m,v) satisfying all constraints in (2) we denote by W.

We say that the decomposition is consistent with problem (1) iff

i/ solution of (2) exists if the solution of (1) exists

ii/ for every solution $(\hat{m}, \hat{v})$ of (2) the point $\hat{x}$ such that

$$\hat{x} = \varphi(\hat{m}, \hat{v}) \tag{3}$$

is the solution of (1).

To fulfill the above demands spaces M, V and mapping φ should satisfy some general assumptions which are summarized in [16] for the case, when

$$Q = f \circ \varphi$$

$$G = (G,\dots, G^N, T) = F \circ \varphi$$

---

x/ it is: $\forall\, a^1 = (a^1_1,\dots,a^1_N)$, $a^2 = (a^2_1,\dots, a^2_N) \in \Omega \subset R^N$ such that $a^1_i \geqslant a^2_i$ $(i = 1,\dots,N)$ and $a^1_j > a^2_j$ for at least one $j \in \{1,\dots, N\}$ the following relation holds: $\Psi(a^1) > \Psi(a^2)$.

Apart of unquestionably essential demands i/, ii/ it would be also desireable to preserve through the decomposition process such features as, for example, the convexity properties of initial optimization problem.

1.2. Description of Primal Method

The optimization problem in form (2) can be solved by two-stage minimization which is the basic idea for the concept of Primal Method. To introduce this method we define first the Infimal Problems of the following kind:

Infimal Problem (I P) $i = 1,\dots,N$

$$\min_{m^i} Q^i(m^i,v) \tag{4}$$

subject to

$$m^i \in M^i_v = \left\{ m^i \in M^i : m^i \in M^i_o ,\ G^i(m^i,v) \in Y^i_o \right\}$$

where $v \in V$ is fixed.

We denote by $\hat{Q}^i(v)$ the value of $Q^i(\hat{m}^i_v, v)$ where $\hat{m}^i_v$ is the solution of IP for given v /if such solution exists/.

Remark

Sometimes it may be desireable /ref. [12]/ to use, when performing the computations, some functional $J^i_v : M^i \rightarrow R$ instead of $Q^i(\cdot, v)$ in (4).

Functional J^i_v have of course to satisfy the following property: $\hat{m}^i_v$ minimizes J^i_v on M^i_v iff $\hat{m}^i_v$ minimizes $Q^i(\cdot,v)$ on M^i_v.

Neverthless in any case the value $\hat{Q}^i(v)$ has to be computed in every Infimal Problem after solving it.

Since the IP's make sense only in the case when $M^i_v \neq \emptyset$ we define set $V_o \subset V$ as follows:

$$v \in V_o \text{ iff } M^i_v \neq \emptyset \quad \text{for } i = 1,\dots,N \tag{5}$$

The Supremal second-level problem has the form:

Supremal Problem (SP) :

$$\min_v \left[\hat{Q}(v) = \Psi(\hat{Q}^1(v), \dots, \hat{Q}^N(v)) \right] \tag{6}$$

subject to

$$v \in V_s = \left\{ v \in V : \quad v \in V_o , \quad v \in V_1 , \quad T(v) \in Z_o \right\} \tag{7}$$

It should be noted here, that however v have to belong to V_s , the coordinator / it is Supremal Problem decision maker / can send to infimal problems v such that $v \notin V_1$ or /and/ $T(v) \notin Z_o$ but necessarily $v \in V_o$.

In the general case the analytical expression for $\hat{Q}(\cdot)$ /see (6)/ is unknown and so the SP is solved according to the following iterative scheme /see fig. 1 / :

1) for given v^k the Infimal Problems IP are solved and the value $\hat{Q}(v^k)$ is sent to Supremal Problem (SP)

2) on the base of the knowledge of $\hat{Q}^i(v^k)$, /$i = 1,\dots,N$ /and eventually some other informations from IP/ the coordinator chooses new value of coordination variable v^{k+1} , sends it to IP and step 1) is repeated unless $\hat{Q}(v^k) = \min\limits_{v \in V_s} \hat{Q}(v)$, in which case the coordination is finished.

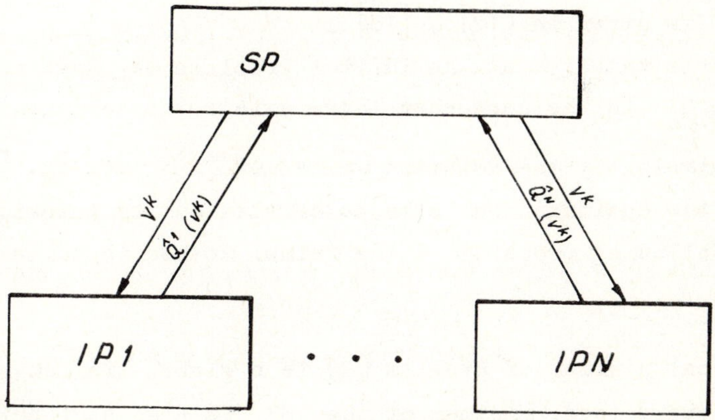

Fig. 1

The general scheme of information flow in the Primal Method

1.3. General properties

In this section we will summarize the main properties of the Primal Method.

Definition 1

The problem (1) is coordinable by the Primal Method if there exists $\hat{v} \in V_s$ such that $\hat{Q}(\hat{v}) = \min_{v \in V_s} \hat{Q}(v)$ and $\hat{x}_o = \varphi(\hat{m}_{\hat{v}}, \hat{v})$ is the solution of (1), where $\hat{m}_{\hat{v}} = (\hat{m}^1_{\hat{v}}, \ldots, \hat{m}^N_{\hat{v}})$ and m^i_v, $i = 1, \ldots, N$ are the solutions of IP with given and fixed v.

Remark

It should be noted that if problem (1) is coordinable by the Primal Method then for some solution $(\hat{m}, \hat{v})$ of (2) we have $\hat{m}_{\hat{v}} = \hat{m}$.

Theorem 1

Let us suppose that the solution $\hat{x}$ of problem (1) exists and that for every $v \in V_o$ there exist the solutions $\hat{m}^i_v$ of IP.

Then problem (1) is coordinable by the Primal Method.

The proof is given in [12], [16].

It should be noted, that the infimal problems may have not solution for every $v \in V_o$ in the case when there exists the solution $\hat{x}$ of (1) or equivalently the solution $(\hat{m}, \hat{v})$ of (2) /see eg. [12]/.

Since the optimization is to be carried out by numerical methods the following property of the Primal Method is quite important.

Theorem 2

If any local minimum of problem (2) is a global minimum of this problem then every local minimum of the SP is also a global minimum of SP.

Moreover if problem (2) has a global minimum then the SP has also a global minimum.

The proof is given in [1], [12].

In [12] it has been shown that the infimal problems IP may have local minima even when problem (2) has no them and so the more detailed conditions have to be examined.

Definition 2

We say that mapping $R : A \supset A_o \longrightarrow B$ /A,B - linear spaces / is concave on A_o with respect to the convex cone $S \subset B$ if $\forall x_1, x_2 \in A_o$ and $\forall \varrho \in [0,1]$

$$-\varrho R(x_1) - (1-\varrho) R(x_2) + R(\varrho x_1 + (1-\varrho) x_2) \in S$$

Theorem 3

Assume that

(i) M_o^i , V_1 are convex, closed sets,

(ii) Y_o^i , Z_o are convex, closed cones,

(iii) the sets M_v^i are compact for every $v \in V_o$, $i = 1,\dots,N$,

(iv) the functionals Q^i are lower semi-continuous and convex on $M_o^i \times V$,

(v) the mappings G^i are continuous and concave on $M_o^i \times V$ with respect to Y_o^i , $i = 1,\dots,N$ and the mapping T is continuous and concave on V with respect to Z_o.

Then sets M_v^i are convex for every $v \in V_o$, set V_o is convex and closed, functional $\hat{Q}(\cdot)$ is convex and finite on V_o.
If, moreover, the set W has the nonempty interior then V_o has also the nonempty interior and $\forall v \in \text{int } V_o$ sets M_v^i have the nonempty interiors.

The proof is given in [11] , [12] .

In the following sections we will investigate the more detailed properties of the Primal Method, especially important for numerical applications.

2. Theoretical properties of the Supremal Problem

Some of the SP properties have been just summarized in Theorem 3. In this section we will discuss two questions. One is about the representation of sets V_o, V_s and the problems connected with it; the other is about differentiability properties of the SP performance index $\hat{Q}$.

2.1. Representation of the feasible sets V_o, V_s.

Usually, when numerical methods of optimization are applied to a given problem, they require analytical expressions or at least some subroutines for the feasible set representation. When solving the SP we meet this problem because the SP feasible set V_s has been up to now specified only by (7). We may of course assume that the set V_1 is known explicitly as well as the constraint mapping T. So in further considerations we will for simplicity take $V_1 = V$. But there is the serious problem with set V_o. It is so far defined implicitly by (5); this definition does not give us any analytical representation of V_o and the knowledge of such representation is even more essential than in standard mathematical programming problems just because for $v \notin V_o$ the infimal problems are not well defined.

In many cases it is possible to obtain the analytical expression for V_o /see [3], [10], [11] /. This means that we can find / when the decomposition is being made /the mapping $h: V \rightarrow R^{\ell}$, $\ell \in \mathcal{N}$ such that

$$V_o = \{v \in V : h(v) \leqslant 0\} \tag{8}$$

To illustrate this idea we give a simple example: let $V = R^2$ and suppose that we have two infimal problems in which there are the local constraints:

$$1)\ \|m^1\| \leqslant v_1, \qquad v = (v_1, v_2) \in R^2,$$

$$2)\ \|m^2\| \leqslant v_2$$

where m^1, m^2 belong to Banach spaces M^1, M^2.

Suppose also that there is one supremal problem constraint

$$v_1 + v_2 \leqslant a, \qquad a > 0.$$

In this case the mapping h has the following form: $h(v) = v$ and $V_o = \{ v \in R^2 : v_1 \geqslant 0, v_2 \geqslant 0 \} = R^2_+$

while

$$V_s = \{ v \in R^2 : v_1 \geqslant 0, v_2 \geqslant 0, v_1 + v_2 \leqslant a \}.$$

In a general case it may be not possible to decribe set V_o as in (8). We may only to prove the following theorem:

Theorem 4

Assume that:

(i) Y^i /$i = 1,\dots,N$/ are Banach spaces and Y^i_o are convex cones, $Y^i_o \subset Y^i$,

(ii) M^i_o are convex sets,

(iii) mappings G^i are concave on $M^i_o \times V$ with respect to convex cones Y^i_o,

(iv) for every $v \in V$ the sets $G^i(M^i_o, v)$ are compact /e.g. when M^i_o are compact and $G^i(\cdot, v)$ are continuous /.

Then $v \in V_o$ if and only if

$$\max_{m^i \in M^i_o} \langle \lambda^i , G^i(m^i, v) \rangle \geqslant 0^{x/} \qquad (9)$$

for every $\lambda^i \in \Lambda^i$, where

$$\Lambda^i = \{ \lambda^i \in Y^{i*} ; \lambda^i \in Y^{i*}_o , \| \lambda^i \| = 1 \}.$$

The proof of the above theorem is given in [12].

Theorem 4 shows that set V_o is, in a general case, determined by infinite number of inequalities (9). If Y^i are separable Banach spaces then it is sufficient to take only countable number of inequalities (9). However it is still troublesome to consider all of them. There is another problem connected with maximum operation performed in (9) $\max_{m^i \in M^i_o}$ - since that maximization has to be carried on for

x/ By $\langle a,b \rangle$ we denote the value of linear continuous functional $a \in X^*$ on $b \subset X$ /X-Banach space, X^*-dual space to X/.
If $X_o \subset X$ and X_o is a cone then by $X^*_o \subset X^*$ we denote cone conjugated to X_o.

every v of interest. There is one important case when this is not required, namely when $G^i(m^i,v) = G_1^i(m^i) + G_2^i(v)$. In some cases / [12] / it is possible to prove that we can take only finite numer of inequalities (9).

Summarizing then there are essentiatly two main problems connected with set V_o. First of them is that we do not know in general the analytical representation of this set. The second is, that for $v \notin V_o$ the infimal problems are not well defined /their feasible sets are empty/ and so it is not possible to apply for solving the SP for example exterior penalty function numerical methods for constrained problems. In section 3 we will describe some methods which may prove capable of handling in some cases the difficulties mentioned above.

2.3. Differentiability properties of the SP performance index $\hat{Q}$.

It is very well known that differentiability properties of performance index are of extreme importance for application of numerical methods of mathematical programming.

In this section we assume that M^i, V, Y^i, Z are reflexive Banach spaces, Y_o^i, Z_o are closed convex cones, mappings G^i, T are Fréchet differentiable and concave with respect to Y_o^i, Z_o. Q is assumed to be convex on $M \times V$ and bounded on every bounded set in $M \times V$. M^i are convex, closed sets and W is bounded, satisfying the Slater's regularity condition / [5] /.

Then we can prove the following important theorem.

Theorem 5

If $M_o^i = M^i$ for $i = 1,\dots,N$ and $v_o \in V_o$ then the SP performance index $\hat{Q}$ has in v_o subdifferential $\partial\hat{Q}(v_o)$ on V_o 1/ of the form:

$$\partial \hat{Q}(v_o) = \left\{ p \in V^* : \; p = q_{v_o} + \sum_{i=1}^{N} H_{v_o}^{i*} \lambda^i \right\} \tag{10}$$

1/ $p \in \partial\hat{Q}(v_o) \subset V^*$ iff $\forall\, v \in V_o \quad \hat{Q}(v) - \hat{Q}(v_o) \geq \langle p, v - v_o \rangle$.

The elements $p \in \partial\hat{Q}(v_o)$ are called subgradients of $\hat{Q}(\cdot)$ at $v = v_o$.

where

λ^i , $i = 1,\dots,N$ and the /possible/ solution of infimal problems $\hat{m}_{v_o} = (\hat{m}_{v_o}, \dots, \hat{m}^N_{v_o})$ satisfy the conditions:

$$\left.\begin{array}{l} q_{\hat{m}^i_{v_o}} + H^{i*}_{\hat{m}^i_{v_o}} \lambda^i = 0 \\ \langle \lambda^i , G^i(\hat{m}^i_{v_o} , v_o) \rangle = 0 \\ \lambda^i \in - Y^{i*}_o , \quad i = 1,\dots,N \end{array}\right\} \qquad (11)$$

and

$H^i_{\hat{m}^i_{v_o}}$ - Fréchet derivate with respect to m^i of G^i in the point $(\hat{m}^i_{v_o} , v_o)$,

$H^i_{v_o}$ - Fréchet derivative with respect to v_o of G^i in the point $(\hat{m}^i_{v_o} , v_o)$,

$q_{\hat{m}^i_{v_o}} \in \partial_{m^i} Q(\hat{m}_{v_o} , v_o)$ / ∂_{m^i} - subdifferential of Q with respect to m^i /.

$q_{v_o} \in \partial_v Q(\hat{m}_{v_o} , v_o)$ / ∂_v - subdifferential of Q with respect to v /.

The proof is given in [12] .

<u>Lemma 1</u>

If $v_o \in \text{int } V_o$ then the subdifferential set (10) of the above theorem is nonempty, convex and weakly compact.

The lemma can be proved directly / [12] /, but the result can be also obtained from theorem 3 since convex and finite functional has the nonempty, convex and weakly compact subdifferential / in Banach reflexive space/ in every interier point of its domain / e.g. [6] /.

<u>Remark</u>

It can be seen from Theorem 5 that the sufficient conditions for $\hat{Q}$ to be Gâteaux differentiable at $v_o \in V_o$ are: uniqueness of the infimal problems solutions $\hat{m}^i_{v_o}$ / due for example to the strict convexity of Q /, existence and uniqueness of λ^i satisfying (11) and Gâteaux differentiability of Q with respect to v at the point $(\hat{m}^i_{v_o}, v_o)$.

Theorem 5 can be generalized / [12] / to the case when M^i_o are convex closed cones with nonempty interiors. The differentiability conditions for G^i may be rejected / [12] / if Y^i are finite dimensional Euclidean spaces and $Y^i_o = R^{n_{y^i}}_+$.
In this case the assumption about concavity of G^i may be weakened to the quasi-concavity assumption / [11] /.

It may be there a considerable difficulty if $v_o \in \partial V_o$ /∂V_o - boundary of V_o / since in this case it may happen that $\partial \hat{Q}(v_o)$ is empty / [12] /.

The above theoretical results form a basis for the SP algorithm developed in the next section. They also show us the complexity of problems connected with applications of the Primal Method.

3. The SP coordination algorithms

The Primal Method requires in a natural way the coordination strategy for solving the SP being some numerical minimization procedure /see [2] , [12] /. Theorem 1 gives the conditions under which the solution of the SP exists and Theorem 2 quarantee that there are not local minima in the SP when there are not local minima in the decomposed optimization problem. This fact is very important for the application of any minimization procedure as the coordination strategy. However there still is a question what minimization procedures could be applied. The problems connected with set V_o and differentiability properties of $\hat{Q}$ which have been discussed before make it clear that we cannot just use standard procedures of mathematical programming. So some special algorithms should be developed. There is moreover one additional and a very important demand which ought be satisfied by these algorithms. They should need the least possible number of the goal function evaluations to reach the solution since every evaluation of $\hat{Q}(v)$ requires the solving of all infimal problems which may be a very time consuming task.

In this section we will describe the algorithms for finding the best direction of improvement for $\hat{Q}$ at the points belonging to V_o. We will also outline three algorithms for solving the SP. The first of them is based upon the idea of subdifferential of $\hat{Q}$, the two others are heuristic algorithms capable to handle the most serious problems connected with coordination in the Primal Method.

We assume everywhere in this section that $V_1 = V$, Y^i, V are finite dimensional spaces and all the assumptions made at the beginning of section 2.3 hold. We assume also that mapping T is Gâteaux differentiable.

3.1. The best direction of improvement

In every point $v \in V_o$ in which the subdifferential of $\hat{Q}$ is nonempty and compact / and so by lemma 1 in every interior point of V / we can compute the best direction of improvement for $\hat{Q}$. This direction is defined as follows:

$$D\hat{Q}(v;\ s_v) = \min_s D\hat{Q}(v;\ s)^{x/} \tag{12}$$

subject to

$\| s \| \leq 1$ $^{xx/}$ $\qquad v + s \in V_o$

and

$\langle \nabla t_k(v), s \rangle \geq 0$ for $k \in K(v) = \{k:\ t_k(v) = 0\}$.

where t_k is k-th component of $T = (t_1 \dots t_{n_z})$

$t_k : V \longrightarrow R$ and $Z_o = R_+^{n_z}$.

$D\hat{Q}(v;\ s)$ is connected with $\partial\hat{Q}(v)$ in the following way /see [6]/.

$$D\hat{Q}(v:\ s) = \max_{p \in \partial\hat{Q}(v)} \langle p, s \rangle \tag{13}$$

Since theorem 5 gives us the representation of $\partial\hat{Q}(v)$, we can write:

$$D\hat{Q}(v:\ s) = \max_{q_v, \hat{m}_v, \lambda^i} \left[\langle q_v, s \rangle + \sum_{i=1}^{N} \langle \lambda^i, H_v^i s \rangle \right] \tag{14}$$

subject to constraints (11) and $q_v \in \partial_v Q(\hat{m}_v, v)$.
Eq. (14) shows that in general case it is very complicated to compute even $D\hat{Q}(v:\ s)$ and so far more complicated to compute s_v from (12).

The computation of $D\hat{Q}(v:\ s)$ is simplified in the case when the infimal problems solutions $\hat{m}_v^i$ are unique. If moreover Q is Gâteaux differentiable with respect to v /and so q_v is unique / and m then (14) reduces to <u>linear programming problem</u> /λ_j^i corresponding to nonactive constraints are set to zero /. As it can be easily seen the problem (14) in such a case can be decomposed into N independent problems:

$$\max_{\lambda^i} \left\{ \langle \nabla_v Q(\hat{m}_v, v), s \rangle + \langle \lambda^i, H_v^i s \rangle \right\}$$

---

x/ $D\hat{Q}(v;\ s)$ is defined as follows:

$$D\hat{Q}(v;\ s) = \lim_{\alpha \to 0^+} \frac{\hat{Q}(v + \alpha s) - \hat{Q}(v)}{\alpha}$$

xx/So the direction s_v is the best in the sense of the norm in $V = R^{n_v}$, which has not to be the euclidean norm.

where

$$\lambda^{i} = (\lambda_{1}^{i}, \ldots, \lambda_{n_{y^i}}^{i}) \tag{14a}$$

λ_{j}^{i} corresponding to nonactive constraints are set to zero and

$$\nabla_{m^i} Q(\hat{m}_v, v) + H^{i*}_{\hat{m}^i_v} \lambda^{i} = 0$$

$$\lambda^{i} \in - Y_{o}^{i*}$$

/we assume here that $Y_o^i = R_+^{n_{y^i}}$ /.

So if some procedure for solving the SP is carrying on the search in any given direction s then it may be checked by solving N linear programming problems if s is still the direction of improv - ement. This may save some computational effort for solving the /non-linear/ infimal problems in the case when s is no longer the direction of improvement for $\hat{Q}$.

To compute s_v we have to solve in general the min max problem (12), (14). However if the assumptions mentioned above are satisfied and (14) is the linear programming problem we can transform /see [8], [11] /the min max problem (12), (14) to the following linear programming problem /ref. [11]/if in space $V = R^{n_v}$ we define the norm as

$$\|v\| = \max_{i = 1,\ldots,n_v} |v_i|$$

$$\min_{s,z^1,\ldots,z^N} \left[\langle \nabla_v Q(\hat{m}_v, v), s\rangle + \sum_{i=1}^{N} \langle -\nabla_{m^i} Q(\hat{m}^i_v, v), z^i\rangle\right] \tag{15}$$

subject to

$$z^{i} \in M^{i}, \quad i = 1,\ldots,N$$

$H^{oi}_{m^i} z^i \leqslant H^{oi}_{v} s$ / $H^{oi}_{m^i}$, H^{oi}_{v} - Jacobians of active constraints only/

$$-1 \leqslant s_j \leqslant 1, \quad j = 1,\ldots,n_v$$

and

$$\langle \nabla t_k(v), s\rangle \geqslant 0 \quad \text{for } k \in K(v).$$

Obviously, we are interested only in finding s_v solving the above problem.

There are however some difficulties which can be encountered when solving problem (15) . First, it may happen, if the infimal problems are not solved absolutely correctly, then the problem (15) has an unbounded solution. Second, the application of (15) will require considerable exchange of information between the SP and the IP-s which may be sometimes nondesired. Besides it the problem (15) has to be solved only in such points of V_o in which $\hat{Q}$ is not differentiable /the measure of such interior points of V_o is zero- see [15] /. If $\hat{Q}$ is differentiable at $v \in V_o$ we can find the gradient of $\hat{Q}$ /$\nabla\hat{Q}(v)$ / in another ways/ which may be simpler/, eq. by solving the eqs. (11) to find λ^i and substituting to (10) . In the sequel we will describe the method for finding $\nabla\hat{Q}(v)$ based on the shifting penalty function method used for solving the infimal problems.

Summarizing then is seems that the following scheme could be preferable in some cases /e.g. in the coordination for improvement of real processes [3] , [12] /:

(i) check by solving (14) if the direction s of search used by the SP procedure is indeed the direction of improvement of $\hat{Q}$; if it is, then solve the infimal problems for $v^{k+1} = v^k + \tau_k s$ /where τ_k is some specified positive number/ and repeat (i) ; if not, then go to (ii),

(ii) choose the new direction s or /if necessary/ perform the computations required to find s_v / e.g. by solving (15) /. Then go to (i) .

It seems that the above scheme should be further investigated.

3.2. The subgradient algorithm

We will discribe now shortly the algorithm, which can be used for solving the SP. We assume here that the assumptions of theorem 5 are fulfilled and V is Euclidean space.

Suppose at first that $V_s = V_o = V$ /this assumption will be partially relaxed later/, and that some sequence $\{\rho_k\}$ of real positive numbers is given having the following properties:

$$\varrho_k \to 0^+ \quad \text{and} \quad \sum_{k=1}^{\infty} \varrho_k = +\infty$$

/eq. the sequence $\left\{\frac{1}{k}\right\}$ /.

Starting from any point $v^o \in V_o$ the subgradient algorithm generates the sequence v^k according to:

$$v^{k+1} = v^k - \varrho_{1k}\, p^k , \quad k = 0,1,\ldots \tag{16}$$

where

$$p^k \in \partial \hat{Q}\,(v^k),$$

$$\varrho_{1k} = \begin{cases} 0 & \text{if } p^k = 0 \\ \dfrac{\varrho_k}{\|p^k\|} & \text{if } p^k \neq 0 \end{cases}$$

Convergence of the above algorithm has been proved in [1] .

It should be noted that after solving the infimal problems we can rather easily compute some $p^k \in \partial \hat{Q}\,(v^k)$, but algorithm (16) is supposed to be very slow because of the proporties of sequence $\{\varrho_k\}$.

Remark

If $V_s \subset V_o = V$ then the above algorithm has to be modified. After computing $v^{(k+1)'}$ from (16) it is necessary to project $v^{(k+1)'}$ onto V_s to generate the new point v^{k+1} of $\{v^k\}$.

3.3. The local penalty function algorithm.

As we have mentioned above there is a need to develop the practical method for finding $\nabla \hat{Q}\,(v)$ whenever it exists. It has been shown in [8] that it can be done by application of the shifting penalty function method to solve the infimal problems. We assume here that $M_o^i = M^i$, $G^i = (-g_1^i ,\ldots, -g_{n_{yi}}^i)$, $\Psi(Q^1(m^1, v) ,\ldots, Q^N(m^N , v)) = \sum_{i=1}^{N} Q^i(m^i , v)$ and Q^i , g_j^i are differentiable functions.

The penalty /shifting/ function method means the repetive minimization of the following function /we consider the i-th infimal problem/:

$$K^i(m^i,v,w^i,\Theta^i) = Q^i\ m^i, v + \frac{1}{2}\sum_{j=1}^{n_g^i} w^i_j\,(g^i_j(m^i,v) + \Theta^i_j)\ \cdot \qquad (17)$$

$$\cdot \max\left(0, g^i_j(m^i, v) + \Theta^i_j\right)]$$

where $j = 1,\ldots, n_g i$, $w^i_j \geqslant 0$ are the penalty coefficients and Θ^i_j are penalty shifts.

Suppose that at point $\hat{m}^i_v$ - the solution of i-th infimal problem - the following conditions are satisfied:

$$\nabla_{m^i}\ K^i(\hat{m}^i_v, v, \hat{w}^i, \hat{\Theta}^i) = 0$$

$$g^i_j(\hat{m}^i_v, v) \leqslant 0\ ,\quad j = 1,\ldots, n_g i \qquad (18)$$

$$\hat{\Theta}^i_j \begin{cases} \geqslant 0 & \text{for } j \in J^i(v) = \left\{j:\ g^i_j(\hat{m}^i_v, v) = 0\right\} \\ = 0 & \text{for } j \notin J^i(v) \end{cases}$$

Then it may be shown, that

$\lambda^i_j = \hat{w}^i_j \cdot \hat{\Theta}^i_j$ / $j = 1,\ldots, n_g i$ / is the solution of egs. (11).

So if the Lagrange multipliers are determined uniquely then they can be directly computed from the penalty function.

From the above result and Theorem 5 it follows that

$$\nabla \hat{Q}^i(v) = \nabla_v Q^i(\hat{m}^i_v, v) + \sum_{j=1}^{n_{gi}} \lambda^i_j\ \nabla_v\ g^i_j(\hat{m}^i_v, v) =$$

$$= \nabla_v Q^i(\hat{m}^i_v, v) + \sum_{j=1}^{n_{gi}} \hat{w}^i_j \cdot \hat{\Theta}^i_j\ \nabla_v\ g^i_j(\hat{m}^i_v, v) = \nabla_v K^i(\hat{m}^i_v, v, \hat{w}^i, \Theta^i) \qquad (19)$$

If in the point v exists $\hat{Q}(v)$ then it can be computed according to (19) as:

$$\nabla \hat{Q}(v) = \sum_{i=1}^{N} \nabla \hat{Q}^i(v) \qquad (20)$$

The algorithm presented in this paragraph is designed in the way to avoid the difficulties connected with set V_o /ref. [8] /.

The shifted penalty function method has the property that in most cases the penalty coefficients w^i_j do not have to be increased to infinity to find the solution $\hat{m}^i_v$. They will of course increase to infinity if set M_v^i is empty.

In our algorithm we use the modified two-level optimization problem. The aim of this modification is to obtain the SP being the unconstraint minimization problem. The modification is achieved in the following way:

if during the solving of the infimal problems for given $v \in V$ by the shifted penalty method

(1) the routine termination criteria are satisfied with $w_j^i \leq w_{max}$

or

(2) some of the penalty coefficients w_j^i become greater than w_{max}

then the SP functional value is set as

$$\hat{Q}(v) = \sum_{i=1}^{N} K^i(\hat{m}_v^i, v, \hat{w}^i, \hat{\theta}^i)$$

with $\hat{m}_v^i$, $\hat{w}^i$, $\hat{\theta}^i$ being the actual values obtained when solving the infimal problems.

$\nabla \hat{Q}(v)$ is then computed according to (19), (20) and used by some coordination strategy /e.g. by the conjugate gradient method/.

The local penalty function algorithm /Fig.2/ is as follows:

(i) Set the initial values /with reasonaby great w_o/,

(ii) Solve the modified two-level optimization problem. After this problem is solved we have $\hat{v}^k$, $\hat{m}^k$.

(iii) If $\hat{v}^k \in V_o$ in the sense that for some specified $\varepsilon > 0$

$$g_j^i(\hat{m}^k, \hat{v}^k) \leq \varepsilon, \ j = 1,\dots, n_y^i \ ; \ i = 1,\dots,N$$

then we assume that $\hat{v}^k$ is the solution of our SP.

If $\hat{v}_k \notin V_o$ /it is some infimal problem is stopped because w_j^i gets greater than w_{max} /then set $w_{max} \Leftarrow w_{max} + \Delta w$ and go to (ii) with $\hat{m}^k$, $\hat{v}^k$ treated as the initial point there.

This algorithm have been tested and it have appeared that it can behave quite well and /as usual in numerical applications/the success depends heavily on the skill with choice of the routine parameters for the shifted penalty function method and the SP procedure /see point 4.1. of this paper/.

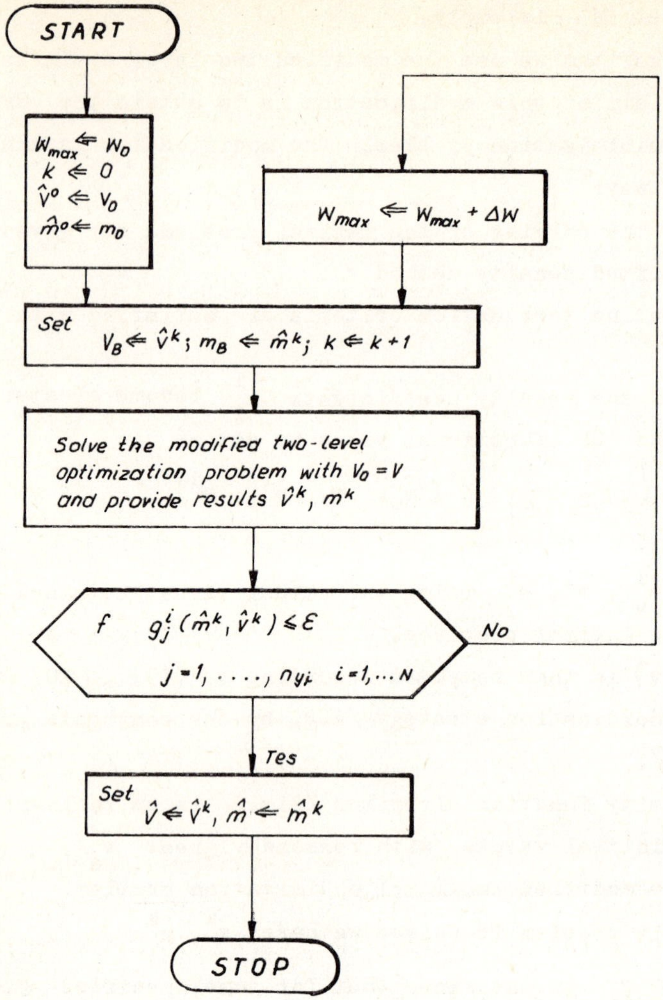

Fig. 2

The block diagram of the local penalty function algorithm

3.4. The methods of feasible directions

The algorithm described above may be applied even in the situation when we do not know the representation of set V_o. However if we know this representation in the form (8) then we may try to avoid the repetive two-level optimization which is the essential drawback of the algorithm from section 3.3. The same trouble is encountered when using the interior penalty function method to solve the SP if the representation (8) of V_o is known. The repetitions of two-level optimization may be avoided when we apply for solving the SP the methods of feasible directions. We do not describe these methods here since they are rather well known and mention only some modifications that have to be made before applying the methods of feasible directions for solving the SP. There are essentialy two such modifications:

i) Any method of feasible directions should need the values of $\hat{Q}(v)$ /or $\nabla\hat{Q}(v)$/ in the points of V_o only.
So if we want to apply gradient projection methods /such as Rosen's, Wolfe's or Polak's methods/ then they will have to be modified /ref. [7]/.

ii) The existence of gradient /or subdifferential/ of $\hat{Q}$ and so the existence of the best direction for improvement/ which is essential for the methods of feasible directions/ is by lemma 1 quaranted in the interior points of V_o only. There are two, in certain sence dual, ways to deal with that trouble:

1/ we may, for the SP procedure, define the "new" set $V_{o\varepsilon}$ in the place of the set V_o, it is

$$V_{o\varepsilon} = \left\{ v \in V : \ h(v) + \varepsilon \leq 0 \right\} \tag{21a}$$

where ε is specified vector of positive real numbers.
If h is continuous mapping and $V_{o\varepsilon} \neq \emptyset$ then $V_{o\varepsilon} \subset \operatorname{int} V_o$.

2/ the infimal problems feasible sets may be expanded by defining the sets $M^i_{v\alpha}$

$$M^i_{v\alpha} = \left\{ m^i \in M^i : \ m^i \in M^i_o, \ g^i_j(m^i, v) \leq \alpha^i_j, \ j = 1,\dots n_y i \right\} \tag{21b}$$

where $\alpha^i = (\alpha_1^i, \dots, \alpha_{n_y^i}^i)$ is some vector of positive real numbers, $i = 1,\dots,N$.
After such modification V_o is itself in the interior of "new" V_o - set, say $V_{o\alpha}$ and it is V_o which is used when solving the SP.

Remark

The approaches 1/, 2/ are dual in the sense, that for any positive α, ε the following relation holds:

$$\hat{q}_\varepsilon \geq \hat{q} \geq \hat{q}_\alpha \tag{22}$$

where $\hat{q}$ is the optimal value of original problem performance idex and $\hat{q}_\varepsilon$, $\hat{q}_\alpha$ are optimal values of modified problems. The relation (22) gives the lower and upper bounds for $\hat{q}$ and thus may be valuable for numerical applications.

4. Computational results and conclussions

Some of the algorithms described above has been tested on the following decomposed nonlinear programming problem:

i/ the infimal problems:

1.

$$\min_{m^1} \left[Q^1(m^1, v) = 100(m^1 - v_1)^2 + (1 - v_1)^2 \right]$$

subject to

$$m^1 \in M_v^1 = \left\{ m^1 \in R : \begin{array}{l} 40\ m^1 + v_1 \leqslant 0 \\ (m^1)^2 - v_1 - 0.04 \leqslant 0 \end{array} \right\}$$

2.

$$\min_{m^2} \left[Q^2(m^2, v) = (10 v_1 + v_2)^2 + 5(m_1^2 - m_2^2)^2 + \right.$$

$$\left. + (v_1 - 2 m_1^2)^4 + 10(v_2 - m_2^2)^4 \right]$$

subject to

$$m^2 \in M_v^2 = \left\{ m^2 \in R^2 : (m_1^2)^2 + (m_2^2)^2 - 0.008 \leqslant 0 \right\}$$

3.

$$\min_{m^3} \left[Q^3(m^3, v) = 100(m^3 - (v_2)^3)^2 + (1 - v_2)^2 \right]$$

subject to

$$m^3 \in M_v^3 = \left\{ m^3 \in R : m^3 + v_2 - 0.53 \leqslant 0,\ - m^3 \leqslant 0 \right\}$$

ii/ the supremal problem

$$\min_{v} \left[\hat{Q}(v) = \hat{Q}^1(v) + \hat{Q}^2(v) + \hat{Q}^3(v) \right]$$

subject to

$$v \in V_s = \left\{ v \in R^2 : v \in V_o,\ v_2 \geqslant 0 \right\}$$

where

$$V_o = \left\{ v \in R^2 : - v_1 - 0{,}04 \leqslant 0,\ v_1 - 40 \leqslant 0,\ v_2 - 0{,}53 \leqslant 0 \right\}.$$

The constraints has been selected in such a way that some of them were active at the solution. For every $v \in \text{int } V_o$ there exists $\nabla \hat{Q}(v)$ and $\hat{v}$ /the solution of SP/ belongs to int V_o. It should be noted here, that multilevel methods of optimization and so the Primal Method are not designed for the solution of such simple problems as the one described above. This problem has been used to test the coordinations algorithms only and its relative simplicity has made it possible to solve the problem many times.

4.1. Computational results

Two types of algorithms described in section 3 have been used to solve the problem formulated above;

The first of them was the local penalty function algorithm /LSPF/. Next three algorithms belonging to class of feasible directions methods were applied:

1/ Rosen's gradient projection method for linear constraints /FDRL/,

2/ modified Rosen's gradient projection method for nonlinear constraints / FDRN/,

3/ modified Polak's hybrid method /FDPH/.

These algorithms have been chosen from the methods described in section 3 because the fast convergence of them had been expected when applied to test problem. Let us note that $\hat{Q}(\cdot)$ is differentiable in this problem. Up to now there are not computational results available for the other methods /the best direction of improvement method and the subgradient algorithm/ designed mainly for the case when $V_o = V$ but $\hat{Q}(\circ)$ is not differentiable.

The results of computations with LSPF method are presented in Table 1.

Table 1

LFG	LFD	EPS	CEE	CE	NF
26	1396	10^{-5}	10^{-7}	10^{-6}	0.006%
44	1523	10^{-8}	10^{-7}	10^{-6}	0.08%
x/ 36	1560	10^{-10}	10^{-9}	10^{-8}	optimal value xx/
36	1712	10^{-10}	10^{-10}	10^{-9}	optimal value

x/ the algorithm has been stopped because the terminal criterion could not have been fulfilled.

xx/ $\hat{q} = \hat{Q}(\hat{v}) = 1.446028$

where

LFG - number of the SP performance index evaluations,

KFD - total numer of the IP performance indices evaluations /in all infimal problems/.

EPS - parameter of the termination criterion for the SP /termination criterion - $\|\nabla \hat{Q}(v)\| \leqslant EPS$/.

CEE - parameter of the termination criterion for the IP unconstrained minimization /performed by the conjugate gradient method/.

CE - parameter of the termination criterion for the IP /in the case, when $w_j^i \leqslant w_{max}$ - see section 3.3/.

NF - value of the global performance index.

In LSPF method w_o has been taken as 10^5. In every case this value has appeared to be sufficient to satisfy the termination criterion after one two-level optimization just because $\hat{v} \in \text{int } V_o$. The initial point has been always assumed to be $v_{10} = v_{20} = m_o^1 = m_{10}^2 = m_o^3 = 1$. It can be easily seen that for $v_1 = v_2 = 1$ the constraints of the second and the third infimal problem cannot be satisfied by any values of local variables. So in the first step of the SP procedure $\hat{Q}$ is increased by large $/ \gg \hat{Q}(v)/$ penalty term. In few steps the SP procedure has brought v to set V_o. In the next iterations the constraints of V_o have been practically satisfied because of $\hat{v} \in \text{int } V_o$ and the value of w_o. It should be noted that the computational effort has been approximately the same as when we have solved the decomposed problem by one-level method. This, because of simplicity of the tested problem, is rather encouraging. The conjugate gradient method has been used to perform all unconstrained minimizations.

The results of computations with three methods of feasible directions /EDRL, FDRN, FDPH/ are presented in Table 2.

Table 2

Name of algorithm	LFG	LFD	EPS	CEE	CE	NF
FDRL	47	3519	10^{-4}	10^{-10}	10^{-6}	1.446030
	50	4014	10^{-5}	10^{-12}	10^{-6}	1.446029
FDRN	69	4604	10^{-2}	10^{-8}	10^{-6}	1.446050
	99	8750	10^{-2}	10^{-10}	10^{-6}	1.446029
	106	9671	10^{-2}	10^{-12}	10^{-6}	1.446028
FDPH	23	1507	10^{-2}	10^{-8}	10^{-6}	1.446050
	20	1521	10^{-3}	10^{-9}	10^{-6}	1.446028
	25	2095	10^{-3}	10^{-11}	10^{-6}	1.446029

The notation in Table 2 is the same as in Table 1.

In every case the initial point has been taken as $v_o = (0.1, 0.1)$, $m_o^1 = m_{10}^2 = m_{20}^2 = m_o^3 = 1$.

The IP were solved by the shifting penalty algorithm with the conjugate gradient method. The application of this method has made it possible to compute $\nabla\hat{Q}(v)$ according to (19), (20).

It can be seen that the modified hybrid method has appeared to be much faster than the two Rosen algorithms. Also the number of the evaluations of the infimal performance indices when applying the hybrid method was similiar as for the algorithm LSPF. However the example has undoubtely appeared to be in favor of this last algorithm since $\hat{v} \in \operatorname{int} V_o$ and so there have no been any repetitions of the two-level optimization. One may expect that for $\hat{v} \in \partial V_o$ the hybrid method of feasible directions may appear to be the most effective.

4.2. Conclusions

In this paper we have tried to show that some problems connected with the Primal Method can be theoretically and practically solved. It seems that the most important results are theorems 2,3,5 and, from the practical point of view, the algorithm described in section

3.3. as well as the modified methods of feasible directions. There are still however the numerous questions to be investigated and answered at; e.g. the application of the best direction of improvement method /section 3.1/, the methods for finding some representation of set V_o in the presence of nonlinear constraints and the vast family of nonconvex problems. The most interesting applications of the Primal Method can be seen in the control of large real processes/ref. [3] /; same problems connected with these applications have been just investigated / [12] , [17] /, yet many of them are still opened. We hope that this paper makes same contribution to the studies on the Primal Method and its applications.

The authors would like to express their gratitude to Professor W. Findeisen and other members of the academic staff of the Institute of Automatic Control for many helpful discussions.

References

[1] A. Auslender - Problémes de Minimax via L'Analyse Convexe et les Inégalites variationnelles: Théorie et Algorithmes. Springer -Verlag, Berlin, Heidelberg, New York 1972.

[2] Findeisen W. - Parametric Optimization by Primal Method in Multilevel Systems. IEEE Trans. on Syst.Sci.and Cybernetics, vol. SSC - 4, no. 2, 1968.

[3] Findeisen W. - Multilevel Control Systems, WNT, Warsaw, 1974 /in Polish/.

[4] Findeisen W., Szymanowski J., Wierzbicki A.- Computational Methods of Optimization, to appear in WNT, Warsaw /in Polish/.

[5] Golshtein E.G. - Teoriya dvoistvennosti v matematicheskom programmirovani i ee prilozheniya, Nauka, Moscow 1972/in Russian/

[6] Ioffe A., Tikhomirov W.- Teoriya ekstremalnykh zadach. Nauka Moskow 1974.

[7] Kiczko R.- Analysis of the application of the methods of feasible directions to multilevel problems of optimization. MSc.thesis, Institute of Automatic Control,Tech.University of Warsaw, Warsaw 1974 /in Polish/.

[8] Kręglewski T., Malinowski K. -Application of the best direction of improvement method on the supremal level of parametric optimization. Proceedings of the VI KKA,Poznań 1974,Poland /in Polish/.

[9] Lasdon L.S. - Optimization Theory for Large Systems. The MacMillan Comp.,New York 1970.

[10] Malinowski K.- Applicability conditions for multilevel methods of static optimization. Podstawy sterowania. Vol.2,z.4. Kraków 1972 /in Polish/.

[11] Malinowski K. - The Properties of Parametric Optimization by Primal Methods in Multilevel Optimization Problems. Systems Science. Vol.1, Wrocław 1975.

[12] Malinowski K. - Multilevel methods of mathematical programming and complex system optimization. PhD Dissertation. Institute of Automatic Control.Techn.University of Warsaw,Warsaw 1974 /in Polish/.

[13] Mesarowic M., Macko D., Takahara Y.- Theory of Hierarchical, Multilevel, Systems. Academic Press, New York 1970.

[14] Polak E.-Computational Methods in Optimization. Academic Press, New York 1971.

[15] Rockafellar P.T. - Convex Analysis. Princeton University Press, 1970.

[16] Woźniak A.- Conditions of applicability for two-level parametric optimization. Podstawy sterowania.Vol.3.z.1. Kraków 1973 / in Polish/

[17] Woźniak A. - Two-level system optimization by parametric method. Proceedings of the VI KKA, Wrocław 1974, Poland / in Polish/.

[18] Pshenichnyi B.N. - Convex multi-valued mappings and their conjugates. In Mathematical Models in Economics (ed. J.Łoś and M.W.Łoś). NORTH-HOLLAND Pub. Comp.- Amsterdam and PWN - Warszawa, 1974.

ON THE IMPLEMENTATION OF REDUCED GRADIENT METHODS

H. Mukai and E. Polak

Department of Electrical Engineering and Computer Sciences
and the Electronics Research Laboratory
University of California, Berkeley, California 94720

ABSTRACT

Until now, the implementation of reduced gradient methods had to be improvised empirically, since procedures for the truncation of the inner iterations, in the feasibility restoration stage, have not been analyzed with respect to convergence of the overall algorithm. This paper presents an implementation of one reduced gradient method. While retaining all the attractive features of the classical reduced gradient methods, this implementation incorporates, explicitly, efficient procedures for truncating the inner iterations to a finite number. In the paper, we present the properties of the restoration subalgorithm and we prove the convergence of the new algorithm under fairly general assumptions.

Research sponsored by the Joint Services Electronics Program Contract F44620-71-C-0087, the U.S. Army Research Office--Durham Contract DAHC04-73-C-0025 and the National Science Foundation Grant GK-37672.

1. Introduction

There are both practical and theoretical difficulties associated with reduced gradient methods as stated in [1], [2], [5], [10]. Both of these difficulties stem from the fact that implementation of these methods has to be improvised empirically, since procedures for the truncation of the infinite inner iteration, in the feasibility restoration stage, have not been analzed with respect to convergence of the overall algorithm. This paper presents an extension of the reduced gradient method in [6,11]: it takes a gradient projection step whenever it is not possible to take a reduced gradient step, and it incorporates explicitly the required procedures for truncating the inner iterations to a finite number. We state the properties of the restoration subalgorithms and we prove the convergence of the implementable algorithm under fairly general assumptions.

2. Algorithm Model

In this section we present a slightly generalized version of the algorithm model for the implementation of approximations presented in [7]. For this purpose, we consider the following abstract

Problem 1: Let T be a closed subset of $\mathbb{R}^n$. Given a nonempty subset $\Delta \subset T$ of desirable points, find a point in Δ. ⌑

We shall denote the Euclidean norm on $\mathbb{R}^n$ by $\|\cdot\|$; and we shall use the notation $B(z,\rho) = \{x \in \mathbb{R}^n \mid \|x - z\| \leq \rho\}$; $\mathbb{R}_+ = [0,\infty)$, $\mathbb{R}_{++} = (0,\infty)$. We shall say that a sequence $\{z_i\}_{i=0}^{\infty}$ is compact if it is contained in a compact set.

Our algorithm model makes use of a cost function $f : \mathbb{R}^n \to \mathbb{R}$, an approximating iteration map $A : \mathbb{R}^n \times \mathbb{R}_{++} \to 2^{\mathbb{R}^n}$, a proximity function $p : \mathbb{R}^n \to \mathbb{R}_+$ which is used to provide a measure of closeness of a point from the set T, and a restoration map $Q : \mathbb{R}^n \times \mathbb{R}_{++} \to 2^{\mathbb{R}^n}$, which is used to "drive" points into T. Finally, let $P : \mathbb{R}_+ \to 2^{\mathbb{R}^n}$ be defined by

$$P(\varepsilon) = \{x \in \mathbb{R}^n \mid p(x) \leq \varepsilon\} \tag{1}$$

Assumption 1

(i) $f : \mathbb{R}^n \to \mathbb{R}$ is continuously differentiable.

(ii) $p : \mathbb{R}^n \to \mathbb{R}_+$ is continuous, and $P(0) = T$.

(iii) For any $x \in \mathbb{R}^n$, $\varepsilon > 0$, $Q(x,\varepsilon) \subset P(\varepsilon)$.

(iv) For any $\varepsilon > 0$ and $x \in P(\varepsilon)$, $Q(x,\varepsilon) = \{x\}$.

(v) For any $z \in \mathbb{R}^n$, if the infinite sequences $\{x_i\}$ and $\{\varepsilon_i\}$ satisfy $x_i \in Q(z,\varepsilon_i)$, for $i = 0,1,2,\ldots$, and $\varepsilon_i \to 0$ as $i \to \infty$, then $\{x_i\}$ has a finite number of accumulation points, all of which are in T.

(vi) For any compact subset C of $\mathbb{R}^n$, there exist constants $M,e \in \mathbb{R}_{++}$, such that

$$\forall\varepsilon \in (0,e],\ \forall x \in C \cap P(\varepsilon),\ \forall\varepsilon' \in (0,\varepsilon],\ Q(x,\varepsilon') \subset B(x,M\varepsilon). \tag{2}$$

(vii) For any $\varepsilon > 0$ and $x \in P(\varepsilon)$, $A(x,\varepsilon) \subset P(\varepsilon)$.

(viii) For any $z \in T \sim \Delta$, there exist $\rho > 0$, $\delta < 0$, and $e > 0$ such that $\forall \varepsilon \in (0,e]$, $\forall x \in B(z,\varepsilon)$, $\forall y \in A(x,\varepsilon)$,

$$f(y) - f(x) \leq \delta < 0 \qquad (3)$$ ¤

Algorithm Model

Parameters: $\gamma > 0$, $\beta \in (0,1)$.

Data: $z_0 \in \mathbb{R}^n$, $\varepsilon_0 > 0$.

Step 0: Set $i = 0$, $j = 0$, $\varepsilon = \varepsilon_0$.

Step 1: Compute an $x \in Q(z_i,\varepsilon)$.

Step 2: Compute a $y \in A(x,\varepsilon)$.

Step 3: If $f(y) - f(x) \leq -\gamma\varepsilon$, go to step 4; else set $x_j = x$, $\varepsilon = \varepsilon\beta$, $j = j + 1$ and go to step 1.

Step 4: Set $z_{i+1} = y$, $i = i + 1$ and go to step 1. ¤

The above Model is an implementation of the limiting case obtained by setting $\varepsilon_0 = 0$ in the data, above. The following result can be deduced from Theorem 1 in [7].

Theorem 1: (i) If the Algorithm Model stops at a particular z_i and constructs an infinite sequence $\{x_j\}$, then $\{x_j\}$ has a finite number of accumulation points, all of which are in Δ. (ii) If the Algorithm Model constructs a compact infinite sequence $\{z_i\}$, then any accumulation point of $\{z_i\}$ is in Δ. ¤

3. A Structure for the Iteration Map A

In the application we shall discuss in this paper, as well as in some others, the map A introduced in the preceding section has a very complex description and it is quite cumbersome to show directly that it satisfies parts (vii) and (viii) of Assumption 1. Because of this, we introduce an intermediate stage, viz., a structure for A, which guarantees that parts (vii) and (viii) of Assumption 1 are satisfied.

Thus, let Y be an open subset of $\mathbb{R}^n$, and let $\mathcal{J}_1, \mathcal{J}_2$ be two finite sets of indexing parameters, let $\tilde{\mathcal{J}}_1 : Y \to 2^{\mathcal{J}_1}$ and $\tilde{\mathcal{J}}_2 : Y \to 2^{\mathcal{J}_2}$ be two direction selection functions. Finally, let $D_1 : Y \to 2^{\mathbb{R}^n}$, $D_2 : \mathbb{R}^n \to 2^{\mathbb{R}^n}$ be two direction sets defined by

$$D_k(x) = \{d_k(x,I) \mid I \in \tilde{\mathcal{J}}_k(x)\}$$

where for each $x \in \mathbb{R}^n$ the functions $d_k(\cdot,I)$ are well defined for all $I \in \tilde{\mathcal{J}}_k(x)$ on some neighborhood N_x^k of x.

In terms of these, we now define $\phi_1 : Y \to \mathbb{R}$, $\phi_2 : \mathbb{R}^n \to \mathbb{R}$ by

$$\phi_k(x) = \max_{I \in \tilde{\mathcal{J}}_k(x)} \langle \nabla f(x), d_k(x,I) \rangle \quad , \ k = 1,2 \qquad (4)$$

and $\mathcal{J}_1^* : Y \to 2^{\mathcal{J}_1}$, $\mathcal{J}_2^* : \mathbb{R}^n \to 2^{\mathcal{J}_2}$ by

$$\mathcal{J}_k^*(x) = \{I \in \tilde{\mathcal{J}}_k(x) \mid \langle \nabla f(x), d_k(x,I) \rangle = \min_{J \in \tilde{\mathcal{J}}_k(x)} \langle \nabla f(x), d_k(x,J) \rangle\} \quad , k = 1,2 \qquad (5)$$

Assumption 2: For k = 1,2,

(i) For any $z \in T \cap Y$ ($z \in T$ for k = 2) and any $I \in \tilde{\mathcal{J}}_k(z)$, $p(z+\lambda d_k(z,I))/\lambda] \to 0$ as $\lambda \to 0$.

(ii) For any $x \in Y$ ($x \in \mathbb{R}^n$ for k = 2), $\phi_k(x) \leq 0$.

(iii) $\phi_k^{-1}(0) \cap T \subset \Delta$.

(iv) For any $x \in Y$ ($x \in \mathbb{R}^n$ for k = 2) and $I \in \tilde{\mathcal{J}}_k(x)$, there exists a neighborhood N of x such that $d_k(\cdot,I)$ is continuous on N.

(v) For any $x \in Y$ ($x \in \mathbb{R}^n$ for k = 2) there exists a neighborhood N of x such that for all $y \in N$, $\tilde{\mathcal{J}}_k(y) \subset \tilde{\mathcal{J}}_k(x)$. ⌑

Finally, let $\tilde{Y} \subset Y$ be such that the closure of $\tilde{Y}$ is also contained in Y. We now define the approximating iteration map A in terms of the functions introduced above, which we state in the form of a subalgorithm, as is the case in practice.

Iteration Map A

Parameters: $\gamma > 0$, $\lambda_s > 0$, $\lambda_e \in (0,\lambda_s/\varepsilon_0)$,† $\alpha \in (0,1)$, $\beta \in (0,1)$.

Data: $x \in \mathbb{R}^n$, $\varepsilon > 0$.

Step 1: If $x \in \tilde{Y}$, set k = 1; else set k = 2.

Step 2: Compute $\phi_k(x)$, $\mathcal{J}_k^*(x)$, and select any $I \in \mathcal{J}_k^*(x)$.

Step 3: If $\phi_k(x) \leq -\gamma\varepsilon$, go to step 4; else, set $A(x,\varepsilon) = \{x\}$, and stop.

Step 4: Set $\lambda = \lambda_s$.

Step 5: If $\lambda \geq \lambda_e \varepsilon$, go to step 6; else, set $A(x,\varepsilon) = \{x\}$ and stop.

Step 6: If $f(x+\lambda d_k(x,I)) - f(x) \leq \alpha\lambda\phi_k(x)$, go to step 7; else, set $\lambda = \lambda\beta$ and go to step 5.

Step 7: Compute a $y \in Q(x+\lambda d_k(x,I),\varepsilon)$.

Step 8: If $f(y) - f(x) \leq \alpha\lambda\phi_k(x)$, set $A(x,\varepsilon) = \{y\}$, and stop; else set $\lambda = \lambda\beta$ and go to step 5. ⌑

We now proceed to show that, defined as above, under Assumption 2, A satisfies parts (vii) and (viii) of Assumption 1.

Proposition 1: The functions ϕ_k, k = 1,2 are upper semicontinuous. ⌑

This result follows directly from Assumptions 1(i) and 2(iv), 2(v). Next, from the structure of A and Assumption 1(iii) we obtain that A satisfies Assumption 1(vii).

Proposition 2: The map A, defined in this section satisfies part (viii) of Assumption 1.

Proof: Let $z \in T \cap Y \sim \Delta$, and let $C \subset Y$ be a compact neighborhood of z. Then by Assumption 1(iv) and (vi), there exist M > 0 and e > 0 such that

$$\forall x \in C \cap P(e),\ \forall \varepsilon \in (0,e],\ Q(x,\varepsilon) \subset B(x,Mp(x)). \qquad (6)$$

---

† ε_0 is as defined in the Algorithm Model.

Next, because $P(e) \cap C$ is a neighborhood of z, it follows from Assumption 2(iv) and (v) that there exist $r_0 > 0$, $r_1 > 0$ and $\bar{\lambda} \in (0,\lambda_s]$ such that $B(z,r_1) \subset Y$, $B(z,r_1) \subset P(e) \cap C$, and $\forall x \in B(z,r_0)$, $\tilde{\mathcal{J}}_1(x) \subset \tilde{\mathcal{J}}_1(z)$. Furthermore, $\forall x \in B(z,r_0)$, $\forall I \in \tilde{\mathcal{J}}_1(x)$

$$x + \lambda d_1(x,I) \in B(z,r_1) \subset P(e) \cap C \tag{7}$$

It now follows from (6) and (7) that there exists a compact set U such that

$$\forall x \in B(z,r_0),\ \forall \lambda \in [0,\bar{\lambda}],\ \forall I \in \tilde{\mathcal{J}}_1(x),\ \forall \varepsilon \in (0,e], Q(x+\lambda d_1(x,I),\varepsilon)$$

$$\subset B(x+\lambda d_1(x,I),\ Mp(x+\lambda d_1(x,I))) \subset U \tag{8}$$

Since $f(\cdot)$ is Lipschitz continuous on U because of Assumption 1(i), there exists an $L > 0$ such that

$$\forall x \in B(x,r_0),\ \forall \lambda \in [0,\bar{\lambda}],\ \forall I \in \tilde{\mathcal{J}}_1(x),\ \forall \varepsilon \in (0,e],\ \forall y \in Q(x+\lambda d_1(x,I),\varepsilon),$$

$$f(y) - f(x+\lambda d_1(x,I)) \le L\|y - x - \lambda d_1(x,I)\| \le LMp(x+\lambda d_1(x,I)) \tag{9}$$

Next, making use of the first order Taylor expansion, we get, $\forall \in [0,\lambda_s]$, $\forall I \in \tilde{\mathcal{J}}_1(x)$, $\forall x \in B(z,r_0)$,

$$f(x+\lambda d_1(x,I)) - f(x) = \lambda\langle\nabla f(x),d_1(x,I)\rangle + \lambda\int_0^1 \langle\nabla f(z+t\lambda d_1(z,I))$$

$$- \nabla f(z),d_1(z,I)\rangle\, dt \le \lambda\alpha\phi_1(x) + \lambda(1-\alpha)\phi_1(x)$$

$$- \sup_{t\in[0,1]} \|\nabla f(x+\lambda d_1(x,I)) - \nabla f(x)\|\cdot\|d_1(x,I)\| \tag{10}$$

Now, since $\phi_1(\cdot)$ is upper semi-continuous and $\phi_1(z) < 0$, there exists a $\rho_z^* \in (0,r_0]$ such that $\phi_1(x) \le \frac{1}{2}\phi(z)$ for all $x \in B(x,\rho_z^*)$, and there exists an integer $\ell_1 \ge 0$ such that $\forall x \in B(z,\rho_z^*)$, $\forall I \in \tilde{\mathcal{J}}_1(x)$, $\forall \lambda \in [0,\lambda_s\beta^{\ell_1}]$

$$\frac{1}{2}(1-\alpha)\phi_1(z) + \sup_{t\in[0,1]} \|\nabla f(x+t\lambda d_1(x,I)) - \nabla f(x)\| \le \frac{1}{4}(1-\alpha)\phi_1(z) \le 0 \tag{11}$$

so that $\forall x \in B(z,\rho_z^*)$, $\forall I \in \tilde{\mathcal{J}}_1(x)$, $\forall \lambda \in [0,\lambda_s\beta^{\ell_1}]$

$$f(x+\lambda d_1(x,I)) - f(x) \le \lambda\alpha\phi_1(x) \tag{12}$$

Because of Assumption 2(i) there exists an integer $\ell_{1z} \geqq \ell_1$ such that $LMp(z+\lambda_{1z}d_1(z,I)) \leqq -\frac{1}{8}\lambda(1-\alpha)\phi_1(z)$, where $\lambda_{1z} = \lambda_s\beta^{\ell_{1z}}$. It now follows from Assumptions 2(iv), (v) that there exists $\rho_z \in (0,\rho_z^*]$ such that

$$\forall x \in B(z,\rho_z),\ LMp(x+\lambda_{1z}d_1(x,I)) \leqq -\frac{1}{4}\lambda_{1z}(1-\alpha)\phi_1(z) \tag{13}$$

It follows from (9) through (11) and (13) that $\forall x \in B(z,\rho_{1z})$, $\forall I \in \tilde{\mathcal{J}}_1(x)$, $\forall y \in Q(z+\lambda_{1z}d_1(x,I),\varepsilon)$, $\forall \varepsilon \in (0,e]$,

$$f(y) - f(x) = f(y) - f(x+\lambda_{1z}d_1(x,I)) + f(x+\lambda_{1z}d_1(x,I)) - f(x)$$

$$\leqq \lambda_{1z}\alpha\phi_1(x) + \frac{1}{4}\lambda_{1z}(1-\alpha)\phi_1(z) + LMp(x+\lambda_{1z}d_1(x,I))$$

$$\leqq \lambda_{1z}\alpha\phi_1(x) \tag{14}$$

Now let $e_{1z} \in (0,e]$ be such that $\lambda_{e_{1z}} \geq \lambda_{1z}$, and $\frac{1}{2}\phi_1(z) \leq -\gamma e_{1z}$, let $\varepsilon \in (0,e_{1z}]$, and for each $x \in B(z,\rho_{1z})$ and $I \in \tilde{\mathcal{J}}_1(x)$, let $\tilde{\lambda}_1(x,\varepsilon,I)$ be such that the tests in steps 6 and 8 of the Iteration Map A are satisfied. Then because of (12) and (14), for all $x \in B(x,\rho_{1z})$, for all $\varepsilon \in (0,e_{1z}]$, for all $I \in \tilde{\mathcal{J}}_1(x)$ and for any $y \in Q(x+\lambda_1(x,\varepsilon,I)d_1(x,I),\varepsilon)$, we get $\tilde{\lambda}_1(x,\varepsilon,I) \geqq \lambda_{1z}$ so that

$$f(y) - f(x) \leq \alpha\tilde{\lambda}_1(x,\varepsilon,I)\phi_1(x) \leq \alpha\lambda_{1z}\phi_1(z)/2 < 0 \tag{15}$$

which is of the form of (3).

Now suppose that $z \in T \sim \Delta$. Then by an identical argument to the one used above, we can show that there exist $\rho_{2z} > 0$, $e_{2z} > 0$ and a $\lambda_{2z} > 0$ such that for any $x \in B(z,\rho_{2z})$, for any $\varepsilon \in (0,e_{2z}]$ for any $I \in \tilde{\mathcal{J}}_2(x)$, and for any $y \in Q(x+\tilde{\lambda}_2(x,\varepsilon,I)d_2(x,I),\varepsilon)$, we get

$$f(y) - f(x) \leq \alpha\lambda_{2z}\phi_2(z)/2 < 0 \tag{16}$$

Consequently, since the closure of $\tilde{Y}$ is contained in Y, we now get the following conclusion. Suppose $z \in T \sim \Delta$ satisfies $z \notin Y$, then, from the above, there exists a ball $B(z,\rho_{2z})$, a $\lambda_{2z} > 0$ and an $e_{2z} > 0$ such that $B(z,\rho_{2z}) \cap \tilde{Y} = \phi$, and $\forall x \in B(z,\rho_{2z})$, $\forall \varepsilon \in (0,e_{2z}]$, the map A sets $y = x + \tilde{\lambda}_2(x,\varepsilon,I)d_2(x,I)$, with $I \in \tilde{\mathcal{J}}_2^*(x)$, and (16) holds for all such y. Next, suppose that $z \in T \sim \Delta$ and $z \in Y$. Then, from the above, there exists a $\rho_z = \min\{\rho_{1z},\rho_{2z}\}$ and an $e_z = \min\{e_{1z},e_{2z}\}$, such that for all $x \in B(z,\rho_z)$, and all $\varepsilon \in (0,e_z]$, the map A sets $y = x + \tilde{\lambda}_1(x,\varepsilon,I)d_1(x,I)$, with $I \in \mathcal{J}_1^*(x)$, or $y = x + \tilde{\lambda}_2(x,\varepsilon,I)d_2(x,I)$, with $I \in \mathcal{J}_2^*(x)$, depending on whether $x \in \tilde{Y}$ or not, but in either event, we get

$$f(y) - f(x) \leq \frac{1}{2}\alpha \max\{\lambda_{1z}\phi_1(z), \lambda_{2z}\phi_2(z)\} < 0 \tag{17}$$

$\forall x \in B(z,\rho_z)$, $\forall \varepsilon \in (0,e_z]$, $\forall y \in A(x,\varepsilon)$. Thus, the iteration map A, defined in this section satisfies Assumption 1(viii), which completes our proof. ¤

4. An Implementable Reduced Gradient Method

We shall now present an implementable generalization of the reduced gradient method analyzed by Luenberger [6] (Ex. 18, p. 275) for the case of linear equality constraints only. Thus, we shall consider the problem

$$\min\{f(x) \mid g(x) = 0,\ x \geqq 0\} \tag{18}$$

where $f : \mathbb{R}^n \to \mathbb{R}^1$, $g : \mathbb{R}^n \to \mathbb{R}^m$ $(m \leq n)$.

We shall denote the components of a vector by superscripts; given $x \in \mathbb{R}^n$, $x^+ \in \mathbb{R}^n_+$, $x^- \in \mathbb{R}^n_-$ are defined componentwise by $x^{+i} = \max\{0,x^i\}$, $x^{-i} = \min\{0,x^i\}$, $i = 1,2,\dots,n$, respectively; $T = \{x \in \mathbb{R}^n \mid g(x) = 0, x \geqq 0\}$.

Assumption 3:

(i) The function f is continuously differentiable.

(ii) The function g is twice continuously differentiable.

(iii) For all $x \in \mathbb{R}^n$, the pair of matrices$(\frac{\partial g(x)}{\partial x}, I_n)$, where I_n is the n×n identity matrix, satisfies the LI condition [9], viz., $\frac{\partial g(x)}{\partial x}^T \psi + I_n \nu = 0$ and $\nu \geqq 0$, implies that $\psi = 0$ and $\nu = 0$. ¤

Proposition 3: The set T satisfies the Kuhn-Tucker constraint qualification (see Theorem 3.3.17 in [4]). ¤

Proposition 4: The system $\frac{\partial g(x)}{\partial x} v = a$, $v \geqq b$, has a solution for any $x \in \mathbb{R}^n$, $a \in \mathbb{R}^m$, $b \in \mathbb{R}^n$ if and only if Assumption 3(iii) is satisfied (see [9]). ¤

The algorithm in this section seeks Kuhn-Tucker points, which we define as being points $x \in T$ such that there exist multipliers $\nu \geqq 0$ and ψ satisfying $-\nabla f(x) + \frac{\partial g(x)}{\partial x}^T \psi + \nu = 0$, and $\langle \nu, x \rangle = 0$. We identify Δ with the set of Kuhn-Tucker points.

Now, given $I = \{i_1,i_2,\dots,i_\ell\} \subset \{1,2,\dots,n\}$ $(i_1<i_2<\dots<i_\ell)$ and $x \in \mathbb{R}^n$, we define $x_I \in \mathbb{R}^\ell$, the m×ℓ matrix $\frac{\partial g(x)}{\partial x_1}$ and $\nabla_I f(x) \in \mathbb{R}^\ell$ by

$$x_I = (x^{i_1}, x^{i_2}, \dots, x^{i_\ell})^T \tag{19}$$

$$\frac{\partial g(x)}{\partial x_1} = \begin{bmatrix} \frac{\partial g^1(x)}{\partial x^{i_1}} & \dots, & \frac{\partial g^1(x)}{\partial x^{i_\ell}} \\ \vdots & & \vdots \\ \frac{\partial g^m(x)}{\partial x^{i_1}} & \dots, & \frac{\partial g^m(x)}{\partial x^{i_\ell}} \end{bmatrix} \tag{20}$$

$$\nabla_I f(x) = \left(\frac{\partial f(x)}{\partial x^{i_1}}, \frac{\partial f(x)}{\partial x^{i_2}}, \dots, \frac{\partial f(x)}{\partial x^{i_\ell}}\right)^T \tag{21}$$

Next, we define $\mathcal{J}_2$ to be a singleton and $\mathcal{J}_1$ to be the set

$$\mathcal{J}_1 = \{I \subset \{1,2,\dots,n\} \,\big|\, |I| = n - m\} \tag{22}$$

where $|I|$ denotes the cardinality of I. For any $I \in \mathcal{J}_1$ we shall denote its complement in $\{1,2,\dots,n\}$ by $\bar{I}$. And we define $\tilde{\mathcal{J}}_1 : \mathbb{R}^n \to 2^{\mathcal{J}_1}$, $\tilde{\mathcal{J}}_2 : \mathbb{R}^n \to 2^{\mathcal{J}_2}$ and $p : \mathbb{R}^n \to \mathbb{R}_+$ by

$$\tilde{\mathcal{J}}_1(x) = \{I \in \mathcal{J}_1 \mid x^i \leq x^j \ \forall i \in I, \forall j \in \bar{I}\}^\dagger \tag{23a}$$

$$\tilde{\mathcal{J}}_2(x) \equiv \mathcal{J}_2 \tag{23b}$$

† This is the variable selection scheme used by Zangwill [11] and Luenberger [6].

$$p(x) = (\|g(x)\|^2 + \|x^-\|^2)^{1/2} \tag{24}$$

Note that p, as defined above, satisfies Assumption 1(ii).

We shall also need the following quantities:

$$p_1(x) \triangleq \frac{1}{2}\|g(x)\|^2 \tag{25a}$$

$$p_2(x) \triangleq \frac{1}{2}p(x)^2 \tag{25b}$$

and for any $I \in \mathcal{J}_1$, E(I) will denote the n×n permutation matrix satisfying

$$x = E(I)\begin{pmatrix} x_{\bar{I}} \\ x_I \end{pmatrix} \tag{26}$$

and $t : \mathbb{R}^n \times \mathcal{J}_1 \to \mathbb{R}_+$ is defined by

$$t(x,I) = \left|\det \frac{\partial g(x)}{\partial x_{\bar{I}}}\right| \tag{27}$$

Referring to Sec. 3, we see that we must define the restoration map Q before we can define the iteration on map A. Let $v_1 : \mathbb{R}^n \times \mathcal{J}_1 \to \mathbb{R}^n$ and $v_2 : \mathbb{R}^n \to \mathbb{R}^n$ be defined by

$$v_1(x,I) = -E(I)\begin{bmatrix} \dfrac{\partial g(x)}{\partial x_{\bar{I}}}^{-1} g(x) \\ \text{-----} \\ 0 \end{bmatrix} \quad \text{whenever } t(x,I) \neq 0 \tag{28a}$$

$$v_2(x,I) = \arg\min_{v \in \mathbb{R}^n} \{\|v\|^2 \mid g(x) + \frac{\partial g(x)}{\partial x} v = 0,\ x + v \geqq 0\} \tag{28b}$$

We need not define v_1 at (x,I) such that t(x,I) = 0.

Restoration Subalgorithm $q_j : \mathbb{R}^n \times \mathbb{R}_{++} \times \mathcal{J}_1 \to \mathbb{R}^n$, j = 1,2

Parameters: $a > 0$, $\alpha \in (0,1)$, $\beta \in (0,1)$.

Data: $z \in \mathbb{R}^n$, $\varepsilon > 0$, $I \in \mathcal{J}_1$ (when j = 1), j.

Step 0: If j = 2, go to step 6; else go to step 1.

Step 1: If $p(z) \leq \varepsilon$, set $q_1(z,\varepsilon,I) = z$ and stop; else, go to step 2.

Step 2: Set $x_0 = z$ and k = 0.

Step 3: If $t(x_k,I) > a$ and $x_k \geqq 0$, go to step 4; else, go to step 6.

Step 4: If $p(x_k) \leq \varepsilon$, set $q_1(z,\varepsilon,I) = x_k$ and stop; else go to step 5.

Step 5: Set

$$x_{k+1} = x_k + \beta^{\ell_k} v_1(x_k,I) \tag{29}$$

where $\ell_k \geq 0$ is the smallest integer satisfying

$$p_1(x_k + \beta^{\ell_k} v_1(x_\ell,I)) \leq (1-\alpha\beta^{\ell_k})p_1(x_k), \tag{30}$$

set k = k + 1 and go to step 3.

Step 6: Set $x_0 = z$ and k = 0.

Step 7: If $p(x_k) \leq \varepsilon$, set $q_j(z,\varepsilon,I) = x_k$ and stop; else, go to step 8.

Step 8: Set

$$x_{k+1} = x_k + \beta^{\ell_k} v_2(x_k), \tag{31}$$

where $\ell_k \geq 0$ is the smallest integer satisfying

$$p_2(x_k + \beta^{\ell_k} v_2)) \leq (1 - \alpha\beta^{\ell_k}) p_2(x_k), \tag{32}$$

set k = k + 1 and go to step 7. ⌑

We now define $Q : \mathbb{R} \times \mathbb{R}_{++} \to \mathbb{R}^n$ by

$$Q(x,\varepsilon) = \{q_j(x,\varepsilon,I) \mid I \in \tilde{\mathcal{I}}_1(x),\ j = 1,2\} \tag{33}$$

Since q is either a globalized Newton's method in $\mathbb{R}^m$ or a globalized Newton's method in $\mathbb{R}^n$, we see that Assumption 3(ii), (iii) and the following Assumption 4 ensures that Q, as defined in (33), satisfies Assumption 1(iii), (iv), (v) and (vi) (for a proof, see [7]).

Assumption 4: If the restoration subalgorithm q_j constructs an infinite sequence $\{x_k\}_{k=0}^{\infty}$ for some $z \in \mathbb{R}^n$, $\varepsilon > 0$ and $I \in \mathcal{I}_1$, then the constructed sequence $\{x_k\}$ has an accumulation point. ⌑

Next, we define the sets Y and $\tilde{Y}$, as follows. Let $\omega_1 >> 0$, $\omega_2 > 0$ be two parameters, then

$$Y = \{x \in \mathbb{R}^n \mid t(x,I) > 0,\ \forall I \in \tilde{\mathcal{I}}_1(x)\} \tag{34}$$

$$\tilde{Y} = \{x \in \mathbb{R}^n \mid t(x,I) > \omega_1,\ \forall I \in \mathcal{K}(x)\} \tag{35}$$

where $\mathcal{K} : \mathbb{R}^n \to 2^{\mathcal{I}_1}$ is defined by

$$\mathcal{K}(x) = \{I \in \mathcal{I}_1 \mid x^i \leq x^j + \omega_2,\ \forall i \in I,\ \forall j \in \bar{I}\} \tag{36}$$

Clearly, as defined in (34), Y is an open set and since $\omega_2 > 0$, $\mathcal{K}(x) \supset \tilde{\mathcal{I}}_1(x)$, so that $\tilde{Y} \subset Y$.

Proposition 5: The closure of the set $\tilde{Y}$ (as defined in (35)) is contained in Y (as defined in (34)).

Proof: Let $\{x_i\}_{i=0}^{\infty}$ be a sequence in $\tilde{\mathbf{Y}}$ converging to a point x^*. Then, for some i_0, $\mathcal{K}(x_i) \supset \tilde{\mathcal{I}}_1(x^*)$ for all $i \geq i_0$, and hence we must have $t(x^*,I) \geq \omega_1 > 0$ for all $I \in \tilde{\mathcal{I}}_1(x^*)$. Hence $x^* \in Y$. ⌑

We now proceed to define the function $d_1(\cdot,\cdot)$. For every $x \in Y$ and $I \in \tilde{\mathcal{I}}_1(x)$, we define the reduced gradient by

$$r(x,I) = \nabla_I f(x) - \frac{\partial g(x)}{\partial x_I}^T \frac{\partial g(x)}{\partial x_{\bar{I}}}^{-1^T} \nabla_{\bar{I}} f(x) \tag{37}$$

Next, we define the vector $d_1(x,I)_I$ componentwise by

$$d_1(x,I)_I^i = \begin{cases} -r(x,I) & \text{if } -r^i(x,I) > 0 \\ -\max\{x_I^i, 0\} r^i(x,I) & \text{otherwise} \end{cases} \tag{38a}$$

and the vector $d_1(x,I)_{\bar I}$ by

$$d_1(x,I)_{\bar I} = -\frac{\partial g(x)}{\partial x_{\bar I}}^{-1}\frac{\partial g(x)}{\partial x_1} d_1(x,I)_I \tag{38b}$$

Finally, we define $d_1(x,I)$ by

$$d_1(x,I) = E(I)\begin{pmatrix} d_1(x,I)_{\bar I} \\ d_1(x,I)_I \end{pmatrix} \tag{39}$$

The following hypothesis corresponds to the nondegeneracy assumption of linear programming.

<u>Assumption 5</u>: For any $x \in T$ and $I \in \tilde{\mathcal{I}}_1(x)$, $x_{\bar I} > 0$. ¤

<u>Lemma 2</u>: Assumption 2 is satisfied for k = 1.

<u>Proof</u>: (i) Let $z \in T \cap Y$, and let $I \in \tilde{\mathcal{I}}_1(z)$. Then $d_1(z,I)$ is well defined. Next, because of (38b),

$$\frac{\partial g(z)}{\partial x} d_1(z,I) = \frac{\partial g(z)}{\partial x_I} d_1(z,I)_I + \frac{\partial g(z)}{\partial x_{\bar I}} d_1(x,I)_{\bar I} = 0 \tag{40}$$

Hence, since $g(z) = 0$, we get

$$g(z+\lambda d_1(z,I)) = \lambda \int_0^1 \left[\frac{\partial g}{\partial x}(z+s\, d_1(z,I)) - \frac{\partial g(z)}{\partial x}\right] d_1(z,I)ds \tag{41}$$

and hence, since g is twice continuously differentiable, there exists an $L \in \mathbb{R}_{++}$ such that $\forall\lambda \in [0,1]$,

$$\|g(z+\lambda d_1(z,I))\| \leq \lambda^2 L\|d_1(z,I)\|^2 \tag{42}$$

Next, since by Assumption 5, $z_{\bar I} > 0$, there exists a $\lambda_0 > 0$ such that $z_{\bar I} + \lambda d_1(z,I)_{\bar I} \geqq 0$ for all $\lambda \in [0,\lambda_0]$. Next since $d_1(z,I)_I^i \geqq 0$ for all i such that $z_I^i = 0$, it follows that there exists a $\bar\lambda_0 \in (0,\lambda_0]$ such that $z + \lambda d_1(z,I) \geq 0$ (i.e., $(z+\lambda d_1(z,I))^- = 0$, for all $\lambda \in [0,\bar\lambda_0]$). Hence, for all $\lambda \in [0,\bar\lambda_0]$, $p(z+\lambda d_1(z,I)) = \|g(z+\lambda d_1(z,I)\|$, and the desired result now follows from (42).

(ii) By direct calculation, for any $x \in Y$,

$$\phi_1(x) = \max_{I\in\tilde{\mathcal{I}}_1(x)} \langle \nabla f(x), d_1(x,I)\rangle \tag{43}$$

Now, for any $I \in \tilde{\mathcal{I}}_1(x)$,

$$\begin{aligned}\langle \nabla f(x), d_1(x,I)\rangle &= \langle \nabla_I f(x), d_1(x,I)_I\rangle + \langle \nabla_{\bar I} f(x), d_1(x,I)_{\bar I}\rangle \\ &= \langle r(x,I), d_1(x,I)_I\rangle \leq 0 \end{aligned} \tag{44}$$

and we are done.

(iii) Suppose $z \in T \cap Y$ is such that $\phi_1(z) = 0$. Then there exists an $I \in \tilde{\mathcal{I}}_1(z)$ such that $\langle \nabla f(z), d_1(z,I)\rangle = \langle r(z,I), d_1(z,I)_I\rangle = 0$. Let $\psi = \frac{\partial g(z)}{\partial x_{\bar I}}^{-1T} \nabla f_{\bar I}(z)$, $\nu_I = r(z,I)$

and $\nu_{\bar{I}} = 0$. Then we get $\nu = E(I)\begin{pmatrix}\nu_{\bar{I}}\\ \nu_I\end{pmatrix} \geq 0$.

$$-\nabla_{\bar{I}} f(z) + \frac{\partial g(z)}{\partial x_{\bar{I}}}^T \psi + \nu_{\bar{I}} = 0 \tag{45a}$$

$$-\nabla_I f(z) + \frac{\partial g(z)}{\partial x_I}^T \psi + \nu_I = 0 \tag{45b}$$

and $\langle \nu, z \rangle = 0$, so that z is a Kuhn-Tucker point, i.e., $z \in \Delta$.

(iv) This follows from the continuity of t.

(v) Let $x \in Y$ be arbitrary and let $i_1 \neq i_2 \neq \ldots \neq i_n$ in $\{1,2,\ldots,n\}$ be such that $x^{i_1} \leq x^{i_2} \leq \ldots \leq x^{i_n}$. There must exist two integers α, β, $\alpha \leq n - m \leq \beta$ such that $x^{i_1} \leq x^{i_2} \leq x^{i_3} \ldots \leq x^{i_{\alpha-1}}$ $x^{i_\alpha} = x^{i_{\alpha+1}} = \ldots = x^{i_\beta} < x^{i_{\beta+1}} \leq \ldots \leq x^{i_n}$, and hence $\tilde{\mathcal{J}}_1(x) = \{I \in \{i_1, i_2, \ldots, i\} \,|\, |I| = n - m \text{ and } i_1, i_2, \ldots, i_{\alpha-1} \in I\}$. By inspection, there exists a neighborhood N of x such for all $y \in N$, $y^{i_1}, y^{i_2}, \ldots, y^{i_{\alpha-1}} < y^{i_\alpha}, \ldots, y^{i_\beta} < y^{i_{\beta+1}}, \ldots, y^{i_n}$, and hence $\tilde{\mathcal{J}}_1(y) \subset \tilde{\mathcal{J}}_1(x)$ for all $y \in N$. This complete our proof. ¤

Next, for any $x \in \mathbb{R}^n$, and any I in $\tilde{\mathcal{J}}_2(x)$ we define $d_2(x,I)$ by

$$d_2(x,I) = \arg\min_{d \in \mathbb{R}^n} \|\{\nabla f(x) + d\|^2 \;\Big|\; \frac{\partial g(x)}{\partial x} d = 0,\ x^+ + d \geqq 0\} \tag{46}$$

Thus, d_2 is defined to be independent of I, i.e. I here is an inoperative variable. The formula (46) is that of gradient projection, as discussed in [7] and the fact that Assumption 2 is satisfied for k = 2 follows directly from the results in [7]. Consequently, we can summarize our results as follows.

Theorem 2: The iteration map A defined for problem (18) by (22), (23a,b), (24) (34) and (46) satisfies the conditions (vii) and (viii) of Assumption 1, the function p defined in (24) satisfies the conditions (ii) of Assumption 1 and the map Q, defined by (33) satisfies the conditions (iii) through (vi) of Assumption 1. ¤

Corollary: Consider problem (18) and the implementable reduced gradient algorithm defined by the Algorithm Model, the Iteration map A subprocedure of Sec. 3, with all quantities defined as in this section for problem (18) and Q defined as in (33). Then the conclusions of Theorem 1 apply to any sequences $\{z_i\}$ and $\{x_j\}$ that this algorithm has constructed. ¤

REFERENCES

[1] J. Abadie and J. Carpentier, Generalization of the Wolfe Reduced Gradient Method for the Case of Nonlinear Constraints," in Optimization, ed. by R. Fletcher, Academic Press, 1969.

[2] J. Abadie and J. Guigou, "Numerical Experiments with the GRG Method," in Integer and Nonlinear Programming, ed. by J. Abadie, North-Holland Pub. Co., Amsterdam, 1970.

[3] L. Armijo, "Minimization of Functions having Continuous Partial Derivatives," Pacific J. Math., vol. 16, pp. 1-3, 1966.

[4] M. D. Canon, C. D. Cullum, Jr. and E. Polak, Theory of Optimal Control and Mathematical Programming, McGraw-Hill, 1970.

[5] D. Babay and D. G. Luenberger, "Efficiently Converging Minimization Methods Based on the Reduced Gradient," SIAM Journal on Control, in press.

[6] D. G. Luenberger, Introduction to Linear and Nonlinear Programming, Addison-Wesley Pub. Co., 1973.

[7] H. Mukai and E. Polak, "On the Use of Approximations in Algorithms for Optimization Problems with Equality and Inequality Constraints," Electronics Research Laboratory Memorandum No. 489, University of California, Berkeley, November 1974.

[8] E. Polak, Computational Methods in Optimization, Academic Press, 1971.

[9] S. M. Robinson, "Extension of Newton's Method to Mixed Systems of Nonlinear Equations and Inequalities," Tech. Sum. Rept. no. 1161, Mathematical Research Center, University of Wisconsin, 1971.

[10] P. Wolfe, "Methods for Nonlinear Constraints," in Nonlinear Programming, ed. by J. Abadie, North-Holland Pub. Co., Amsterdam, 1967.

[11] W. I. Zangwill, Nonlinear Programming: A Unified Approach, Prentice-Hall, Englewood Cliffs, New Jersey, 1969.

CONTRIBUTION TO DUBOVITSKIY AND MILYUTIN'S OPTIMIZATION FORMALISM

Ludmila Rigby
Department of Computing and Control
Imperial College of Science and Technology
London SW7 2BZ
United Kingdom

ABSTRACT

This paper is a contribution to the unified approach of Halkin, Neustadt, Gamkrelidze and others to the theory of necessary conditions for general optimization problems.

The basic problem is formulated in terms of real linear topological spaces, mappings between them and a partial ordering determined by a proper convex cone. It includes, therefore, problems with both scalar- and vector-valued optimality criteria.

Optimality conditions are developed in terms of Gâteaux and Fréchet differentials of given mappings and linear continuous functionals on the spaces concerned, making use of the Dubovitskiy and Milyutin's formalism.

INTRODUCTION

We develop necessary and sufficient optimality conditions for a Mathematical Programming Problem, employing the Dubovitskiy and Milyutin's formalism [1,2]. This is built round a form of the separation principle for finite families of convex sets with empty intersection, each set corresponding to an approximation to a constraint in the problem.

The Mathematical Programming Problem is of considerable generality and covers a wide range of applications. We take the underlying set to be a real linear topological space and allow for vector-valued objective function as well as for infinitely many equality and inequality constraints.

Our findings differ in minor respects from known results in the literature, principally over weakening of differentiability requirements on the constraint and objective function in obtaining the necessary conditions. We feel though that the main contribution is in presenting a simple, complete proof of the results.

1. FORMULATION OF THE BASIC PROBLEM

Take X, Y_1, Y_2, Y_3 real linear topological spaces, mappings $F : X \to Y_1$, $G : X \to Y_2$, $H : X \to Y_3$, set $Q \subseteq X$, proper convex cones $C \subseteq Y_1$, $S \subseteq Y_2$ with non-empty interiors, and a fixed element $e \in Y_3$. The set Ω of *feasible points* is defined as follows:

$$\Omega = \{x \in X; x \in Q, G(x) \in S, H(x) = e\}.$$

We introduce the Basic Problem:

Basic Problem

Find $x \in \Omega$ such that

$$\{z \in \Omega; F(z) - F(x) \in C \setminus \{\theta\}\} = \phi.$$

Every such element is called *optimal*. (Here, and subsequently, θ denotes the zero element in the space concerned.)

This will be recognised as a mathematical programming problem over a real linear topological space with multivalued objective function in the presence of constraints, of a form similar to that studied in [3].

The set Q comprises the implicit constraints, while the set $\{z \in X; H(z) = e, G(z) \in S\}$ defines the explicit (equality and inequality) constraints. We make no assumptions concerning finite dimensionality of the ranges of the constraint functions.

2. BASIC THEORY

Let X be a real linear topological space. (Throughout this paper we shall suppose all topologies to be Hausdorff.)

The real linear space of all linear continuous functionals on X is denoted by X^*.

2.1 We begin with definitions of "conical approximation" and "polar cone", as these are the two basic concepts in formulating and deriving the necessary conditions for optimality.

A list of useful properties is given below. Verification of these properties is routine and so no proofs are given.

Let Q be a subset of X, $x \in X$ and N a base of neighbourhoods of zero. The following sets

$$K(x,Q) = \{z\in X;\ \exists\, \varepsilon>0\ \forall\, \alpha\in(0,\varepsilon)\ \ x+\alpha z\in Q\}$$

$$C(x,Q) = \{z\in X;\ \forall\, \varepsilon>0\ \exists\, \alpha\in(0,\varepsilon)\ \ x+\alpha z\in Q\}$$

$$N(x,Q) = \{z\in X;\ \exists\, \varepsilon>0\ \ \exists\, U\in \mathcal{N}\ \forall\, \alpha\in(0,\varepsilon)\ \forall\, \omega\in z+U\ \ x+\alpha\omega\in Q\}$$

$$M(x,Q) = \{z\in X;\ \forall\, \varepsilon>0\ \forall\, U\in \mathcal{N}\ \exists\, \alpha\in(0,\varepsilon)\ \ \exists\, \omega\in z+U\ \ x+\alpha\omega\in Q\}$$

are called *conical approximations* of Q with respect to x.

The following terminology is usually attached:

K,C - cones of approximations

N - cone of internal directions

M - cone of tangents

Precise definitions of these cones and notation vary in the literature; our notation is as used in [4].

<u>Proposition 1</u>

Let P and Q be subsets of X, $x \in X$. Then

(a) $N(x,Q) \subseteq K(x,Q) \subseteq C(x,Q) \subseteq M(x,Q)$;

(b) $K(x,Q)$ and $C(x,Q)$ are cones;
$N(x,Q)$ is an open cone;
$M(x,Q)$ is a closed cone;

(c) $N(x,P) \cap N(x,Q) = N(x,P\cap Q)$;
$K(x,P) \cap K(x,Q) = K(x,P\cap Q)$;
$N(x,P) \cap M(x,Q) \subseteq M(x,P\cap Q)$;
$K(x,P) \cap C(x,Q) \subseteq C(x,P\cap Q)$;

(d) $N(x,X) = X$; $M(x,\phi) = \phi$;

(e) $N(x,Q) = N(x, \text{int}\ Q)$;
$M(x,Q) = M(x,\bar{Q})$;

(f) if $P \subseteq Q$ then
$N(x,P) \subseteq N(x,Q)$, $K(x,P) \subseteq K(x,Q)$,
$C(x,P) \subseteq C(x,Q)$, $M(x,P) \subseteq M(x,Q)$;

(g) if Q is cone then
$K(\theta,Q) = Q$, $N(\theta,Q) = \text{int}\ Q$;

(h) if Q is convex then
$K(x,Q) = C(x,Q)$, $N(x,Q) = K(x, \text{int}\ Q)$

$N(x,Q)$, $K(x,Q)$, $C(x,Q)$ and $M(x,Q)$ are convex

(i) if Q is convex, int $Q \neq \phi$ then

$N(x,Q)^- = M(x,Q)$, $N(x,Q) = \text{int } M(x,Q)$.

Let Q be a subset of X. The set

$$Q^+ = \{f \in X^*;\ f(x) \geq 0 \text{ for all } x \in Q\}$$

is called the *polar cone* of Q.

Proposition 2

Let P and Q be subsets of X. Then

(a) Q^+ is a convex cone in X^*;

(b) if $P \subseteq Q$ then $Q^+ \subseteq P^+$;

(c) $X^+ = \{\theta\}$, $\{\theta\}^+ = X^*$;

(d) $Q^+ = (\bar{Q})^+$;

(e) if Q is convex, int $Q \neq \phi$, then $Q^+ = (\text{int } Q)^+$;

(f) if Q is a subspace in X then
$Q^+ = \{f \in X^*;\ f(x)=0 \text{ for all } x \in Q\}$.

Proposition 3

Take Q a convex subset of X, $f \in X^*$. Consider the assertions:

(a) $f(x) \leq f(y)$ for all $y \in Q$;

(b) $f \in K(x,Q)^+$;

(c) $f \in N(x,Q)^+$;

Then if $x \in Q$, (a) $\Leftrightarrow$ (b), and if $x \in \bar{Q}$, int $Q \neq \phi$, (a) $\Leftrightarrow$ (b) $\Leftrightarrow$ (c).

Proposition 4

Take Q a cone in X, $x \in \bar{Q}$ and $f \in X^*$. If $f(x) \leq f(y)$ for all $y \in Q$, then $f(x) = 0$ and $f \in Q^+$.

2.2 We make essential use of the following consequence of the Hahn-Banach Theorem:

Separation Theorem

Let $Q_0, Q_1, \ldots, Q_n$ $(n \geq 1)$ be non-empty convex sets in X, $Q_1, \ldots, Q_n$ open and $\theta \in \bar{Q}_i$, $i = 0, 1, \ldots, n$. Then $\bigcap_{i=0}^{n} Q_i = \phi$ if and only if there exist linear continuous functionals $f_0, f_1, \ldots, f_n$ on X, not all zero, such that

(a) $f_i \in Q_i^+$, $i = 0, 1, \ldots, n$;

(b) $\sum_{i=0}^{n} f_i = \theta$.

If $\bigcap_{i=0}^{n} Q_i = Q \neq \phi$, then $Q^+ = \sum_{i=0}^{n} Q_i^+$.

This result was originally stated in [1]. The proof given in [2] for X normed space, carries over to this more general setting unaltered.

In the rest of this section let X and Y be real linear topological spaces and $C \subseteq Y$ a convex cone with non-empty interior.

We say that mapping $T : X \to Y$ is *C-convex*, if

$$T(\lambda x+\mu y) - \lambda T(x) - \mu T(y) \in C$$

for all $x, y \in X$ and $\lambda, \mu \in (0,1)$, $\lambda + \mu = 1$.

Proposition 5

Suppose that $T : X \to Y$ is C-convex and $T(\theta) = \theta$.

(a) If $T^{-1}(\text{int } C) \neq \phi$ and $g \in [T^{-1}(C)]^+$, then there exists $f \in C^+$ such that $f(T(x)) \leq g(x)$ for all $x \in X$.

(b) If $T^{-1}(\text{int } C) = \phi$, then there exists non-zero functional $f \in C^+$ such that $f(T(x)) \leq 0$ for all $x \in X$.

Proof

Let us denote

$$Q_0 = \{(T(x)-\omega, -g(x));\ x\in X,\ \omega\in\bar{C}\},$$

$$Q_1 = \{(y,z);\ y \in \text{int } C,\ z>0\}.$$

Then, evidently, Q_0 is a convex set in $Y \times R$, $(\theta,0) \in Q_0$, and Q_1 is an open convex cone in $Y \times R$. It is easily verified that $Q_0 \cap Q_1 = \phi$, for if it was not so, then there would exist $x \in X$ and $\omega \in \bar{C}$ such that $T(x) - \omega \in \text{int } C$ and $g(x) < 0$. But $T(x) \in C$ because $\omega \in \bar{C}$, and so $g(x) \geq 0$, which is a contradiction.

From the Separation Theorem we conclude, that there exists a non-zero functional $(h,\alpha) \in Y^* \times R$, where $(h,\alpha) \in Q_1^+$ and $-(h,\alpha) \in Q_0^+$. Since $T^{-1}(\text{int } C) \neq \phi$ it follows that $\alpha > 0$ and $h(T(x)-\omega) - \alpha g(x) \leq 0$ for all $x \in X$ and $\omega \in \bar{C}$. We set $f = h/\alpha$. Putting $x = \theta$ and $\omega = \theta$, respectively, we in turn get that $f \in C^+$ and $f(T(x)) \leq g(x)$ for all $x \in X$. To conclude the proof we notice, that if $T^{-1}(\text{int } C) = \phi$, $\text{int } C \cap \{T(x)-\omega;\ x\in X,\ \omega\in\bar{C}\} = \phi$. Applying the Separation Theorem again we obtain a non-zero functional $f \in Y^*$ such that $f(T(x)-\omega) \leq 0$ for all $x \in X$ and $\omega \in \bar{C}$. Therefore $f \in C^+$ and $f(T(x)) \leq 0$ for all $x \in X$.

If $T : X \to Y$ is linear and continuous, then T^* denotes the dual mapping $T^* : Y^* \to X^*$ defined by

$$T^*(f) = f \circ T \text{ for all } f \in Y^*.$$

Corollary

Suppose that $T : X \to Y$ is linear and continuous.

(a) If $T^{-1}(\text{int } C) \neq \phi$, then $[T^{-1}(C)]^+ = T^*(C^+)$.

(b) If $T^{-1}(\text{int } C) = \phi$, then there exists a non-zero functional $f \in C^+$ such that $f \circ T = \theta$.

We take note also of the following standard result.

Proposition 6

Let X, Y be Banach spaces and $T : X \to Y$ linear and continuous projection. Then $[T^{-1}(\{\theta\})]^+ = T^*(Y^*)$.

2.3 To state the results of this sub-section, we need to recall ([5]) the definitions of Fréchet and Gâteaux differentiability.

The mapping $T : X \to Y$ is said to be *Gâteaux differentiable* (briefly G-differentiable) at $x \in X$, if the limit

$$\lim_{t \to 0^+} \frac{T(x+th)-T(x)}{t}$$

written $DT(x,h)$, exists for all $h \in X$. Then $DT(x)$ denotes the mapping $DT(x,\cdot) : X \to Y$.

We say that T is *strongly G-differentiable* at $x \in X$, if for all $h \in X$

$$DT(x,h) = \lim_{\substack{t \to 0^+ \\ k \to h}} \frac{T(x+tk)-T(x)}{t}$$

The mapping $T : X \to Y$, where X and Y are normed spaces, is said to be *Fréchet differentiable* (briefly F-differentiable) at $x \in X$ with F-derivative $DT(x) : X \to Y$, if $DT(x)$ is linear and continuous, and for all $h \in X$

$$\lim_{h \to \theta} \frac{||T(x+h)-T(x)-DT(x)(h)||}{||h||} = 0$$

If moreover, for all $h, k \in X$

$$\lim_{\substack{(h,k)\to(\theta,\theta)\\ h\neq k}} \frac{||T(x+h)-T(x+k)-DT(x)(h-k)||}{||h-k||} = 0$$

then T is said to be *strongly F-differentiable* at x.

The next three propositions follow directly from the above definitions.

Proposition 7

Suppose that T is G-differentiable at $x \in X$. Then

(a) $DT(x)^{-1}(K(T(x), \text{int } C)) \subseteq K(x, T^{-1}(\text{int } C))$;

(b) $DT(x)^{-1}(\text{int } C) \subseteq K(x, T^{-1}(T(x)+ \text{int } C))$.

Proposition 8

Let T be linear mapping and $x \in X$. Then

$$\{h\in X;\ DT(x,h)=\theta\} = K(x,T^{-1}(\{T(x)\})).$$

Proposition 9

Suppose that T is strongly G-differentiable at $x \in X$. Then

(a) $DT(x)^{-1}(N(T(x), \text{int } C)) \subseteq N(x, T^{-1}(\text{int } C))$;

(b) $DT(x)^{-1}(\text{int } C) \subseteq N(x, T^{-1}(T(x)+ \text{int } C))$.

Proposition 10

Let X, Y be Banach spaces and suppose that T is strongly F-differentiable at $x \in X$, $DT(x) : X \to Y$ is a projection. Then

$$\{h\in X;\ DT(x)(h)=\theta\} = M(x,T^{-1}(\{T(x)\})).$$

Proof

Let us denote $K = \{h\in X;\ DT(x)(h)=\theta\}$. Since every strongly F-differentiable mapping is strongly G-differentiable, it follows immediately that $M(x,T^{-1}(\{T(x)\})) \subseteq K$.

Conversely, let $k \in K$ and $\varepsilon > 0$ be given. Suppose that $Y \neq \{\theta\}$, for if $Y = \{\theta\}$, then $K = X = M(x,X)$. K is a closed subspace in X, so X/K is also a Banach space and there exists a linear continuous bijection $A : X/K \to Y$, $A \neq \theta$, such that $DT(x) = A \circ f$, where $f : X \to X/K$ is the canonical projection $f(x) = x + K$. By the open-mapping theorem A^{-1} is continuous, hence A is an isomorphism.

Mapping T is strongly F-differentiable, therefore there exists $\delta > 0$ such that

(1) $$||h_1|| < \delta, ||h_2|| < \delta \Rightarrow ||T(x+h_1)-T(x+h_2)-DT(x)(h_1-h_2)|| < c||h_1-h_2||,$$

where $c = \dfrac{1}{4||A^{-1}||}$

Further there exists $\alpha \in (0,\varepsilon)$ such that

(2) $$||h-k|| < \varepsilon \Rightarrow \alpha||h|| < \delta \text{ and}$$

(3) $$||T(x+\alpha k)-T(x)|| < \alpha\varepsilon c/2.$$

We now define sequences $\{k_n\}$ in X and $\{t_n\}$ in X/K in the following manner:

$$k_0 = k, \quad t_0 = K \quad \text{and}$$

(4) $$t_{n+1} = t_n - 1/\alpha\ A^{-1}[T(x+\alpha k_n)-T(x)],$$

(5) $$k_{n+1} \in t_{n+1} \text{ such that } ||k_{n+1}-k_n|| < 2||t_{n+1}-t_n||, \ n \geq 0.$$

Using (3) we note that $||t_1|| < \varepsilon/8$, and $||k_1-k|| < \varepsilon/4$. We show by induction that the sequence $\{k_n\}$ has the following properties:

(6) $$||k_{n+1}-k_n|| < 2^{-n}||k_1-k||,$$

(7) $$||k_{n+1} - k|| < \varepsilon/2$$

This would imply that $\{k_n\}$ and $\{t_n\}$ are Cauchy sequences in X and X/K, respectively, and therefore convergent. If $h = \lim k_n$, $s = \lim t_n$, then $h \in s$ and using (4) we conclude that

$$s = s - 1/\alpha\ A^{-1}[T(x+ h)-T(x)], \text{ i.e.}$$

$$T(x+\alpha h) = T(x) \text{ and } ||h-k|| < \varepsilon, \ \alpha \in (0,\varepsilon)$$

in other words $k \in M(x,T^{-1}(\{T(x)\}))$.

We saw that (6) and (7) are valid for $n = 0$. Supposing their validity for $m \leq n-1$, we show that they hold for n. Since $k_n \in t_n$, we have $A(t_n) = DT(x)(k_n)$, and

$$t_{n+1} = -A^{-1}\left[\frac{T(x+\alpha k_n)-T(x)}{\alpha} - DT(x)(k_n)\right].$$

Then

$$||t_{n+1}-t_n|| = \frac{||A^{-1}||}{\alpha}.||T(x+\alpha k_n)-T(x+\alpha k_{n-1})-DT(x)(k_n-k_{n-1})||$$

$$< \frac{||A^{-1}||}{\alpha}.c\alpha||k_n-k_{n-1}|| = 1/4||k_n-k_{n-1}||.$$

Hence

$$||k_{n+1}-k_n|| < 2||t_{n+1}-t_n|| < 1/2||k_n-k_{n-1}|| < 2^{-n}||k_1-k||,$$

and

$$||k_{n+1}-k\ || \leq \sum_{i=0}^{n} ||k_{i+1}-k_i|| \leq 2||k_1-k|| < \varepsilon/2$$

which completes the proof.

3. THE MAIN RESULTS

In this section we consider the Basic Problem. First, under the assumption X and Y_3 are Banach spaces, we give necessary conditions for optimality, in the presence of equality and inequality constraints:

Theorem 1

Let us suppose that either

(a) there exists a non-empty open convex cone $K \subseteq N(x,Q)$,

(b) mappings F and G are strongly G-differentiable at x, mapping H is strongly F-differentiable at x;

or

(a') there exists a non-empty open convex cone $K \subseteq C(x,Q)$,

(b') mappings F anf G are G-differentiable at x, mapping H is linear and continuous,

and

(c) DF(x) and DG(x) are C-convex and S-convex, respectively, and continuous, DH(x) has a closed range.

If x is optimal, then there exist functionals $\alpha \in X^*$, $\lambda \in Y_1^*$, $\mu \in Y_2^*$ and $\nu \in Y_3^*$ such that

(I) $[\alpha+\lambda\circ DF(x)+\mu\circ DG(x)+\nu\circ DH(x)](h) \leq 0$ for all $h \in X$;

(II) $\lambda \neq \theta$ or $\mu \neq \theta$ or $\nu \neq \theta$;

(III) $\alpha \in K^+$, $\lambda \in C^+$, $\mu \in S^+$ and $\mu(G(x)) = 0$.

In the absence of equality constraints, we may dispense with the assumption that X be a Banach space:

Theorem 2

Let us suppose that either

(a) there exists a non-empty convex cone $K \subseteq M(x,Q)$,

(b) mappings F and G are strongly G-differentiable at x;

or

(a') there exists a non-empty convex cone $K \subseteq C(x,Q)$,

(b') mappings F and G are G-differentiable at x;

and

(c) DF(x) and DG(x) are C-convex and S-convex, respectively, and continuous.

If x is optimal, then there exist functionals $\alpha \in X^*$, $\lambda \in Y_1^*$ and $\mu \in Y_2^*$ such that

(I) $[\alpha+\lambda\circ DF(x)+\mu\circ DG(x)](h) \leq 0$ for all $h \in X$;

(II) $\lambda \neq \theta$ or $\mu \neq \theta$;

(III) $\alpha \in K^+$, $\lambda \in C^+$, $\mu \in S^+$ and $\mu(G(x)) = 0$.

The following results, Theorems 3 and 4, consider the Basic Problem when the objective function is single-valued. Under certain convexity assumptions, the necessary conditions in the above theorems become sufficient for optimality, if the multiplier associated with the objective function is non-zero (in particular, if the appropriate Slater's condition holds). As above, we may develop our results in a more general framework where equality constraints are absent.

Theorem 3

Let us suppose that

(a) x is a feasible point;

(b) Q is a convex set with non-empty interior;

(c) mappings F and G are C-convex and S-convex, respectively, and G-differentiable at x, where DF(x) and DG(x) are continuous, mapping H is linear and continuous;

(d) there exist functionals $\alpha \in X^*$, $\mu \in Y_2^*$, $\nu \in Y_3^*$ and a real number λ such that

  (I) $[\alpha+\lambda DF(x)+\mu\circ DG(x)+\nu\circ H](h) \leq 0$ for all $h \in X$;

  (II) $\lambda \neq 0$ or $\mu \neq \theta$ or $\nu \neq \theta$,

  (III) $\alpha(x) \leq \alpha(y)$ for all $y \in Q$,

  (IV) $\lambda \in C^+$, $\mu \in S^+$ and $\mu(G(x)) = 0$.

Then

(1) If $\lambda \neq 0$, x is optimal.

(2) If H is a projection and there exists $z \in \text{int } Q$ such that $G(z) \in \text{int } S$, $H(z) = e$, then $\lambda \neq 0$.

Theorem 4

Let us suppose that

(a) x is a feasible point;

(b) Q is convex;

(c) mappings F and G are C-convex and S-convex respectively, and G-differentiable, where DF(x) and DG(x) are continuous;

(d) there exist functionals $\alpha \in X^*$, $\mu \in Y_2^*$ and a real number λ such that

(I) $[\alpha+\lambda DF(x)+\mu \circ DG(x)](h) \leq 0$ for all $h \in X$;

(II) $\lambda \neq 0$ or $\mu \neq \theta$,

(III) $\alpha(x) \leq \alpha(y)$ for all $y \in Q$,

(IV) $\lambda \in C^+$, $\mu \in S^+$ and $\mu(G(x)) = 0$.

Then

(1) If $\lambda \neq 0$, x is optimal.

(2) If there exists $z \in Q$ such that $G(z) \in \text{int } S$, then $\lambda \neq 0$.

4. PROOF OF THE MAIN RESULTS

Here we prove the theorems of Section 3. We shall see that these results follow simply from the Separation Theorem using the properties of conical approximations and polar cones developed in Section 2.

Proof of Theorem 1

Let us suppose that x is optimal and assumptions (a), (b) and (c) hold.

We define

$$P = \{z \in X;\ F(z)-F(x) \in C\backslash\{\theta\}\},$$
$$K_1 = \{h \in X;\ DF(x,h) \in \text{int } C\},$$
$$K_2 = \{h \in X;\ DG(x,h) \in N(G(x), \text{int } S)\}, \text{ and}$$
$$K_3 = \{h \in X;\ DH(x,h) = \theta\}.$$

Note that K_1 and K_2 are open convex sets, $\theta \in \bar{K}_1$, $\theta \in \bar{K}_2$ and K_3 is a subspace in X.

We first show that the conclusion of the theorem follows trivially if either $K_1 = \phi$ or $K_2 = \phi$ or DH(x) is not a projection.

Suppose that $K_1 = \phi$. Then using Propositions 2 and 5 we conclude that there is a non-zero functional $\lambda \in (\text{int } C)^+ = C^+$ such that

$$\lambda \circ DF(x,h) \leq 0 \text{ for all } h \in X.$$

Now suppose that $K_2 = \phi$. Using Proposition 5 again it follows that there is a non-zero functional $\mu \in N(G(x), \text{int } S)^+$ such that

$$\mu \circ DG(x,h) \leq 0 \text{ for all } h \in X.$$

Propositions 2, 3 and 4 imply that $\mu \in S^+$ and $\mu(G(x)) = 0$.

If $DH(x)$ is not a projection onto Y_3, then the range of $DH(x)$ is a proper closed subspace of Y_3, so by Hahn-Banach theorem there exists a non-zero functional $\nu \in Y_3^*$ such that $\nu \circ DH(x) = \theta$.

Now suppose that both K_1 and K_2 are non-empty and $DH(x)$ is a projection. It follows immediately from Propositions 8 and 10 that

$$K_1 \subseteq N(x, F^{-1}(F(x) + \text{int } C)) \subseteq N(x,P),$$
$$K_2 \subseteq N(x, G^{-1}(\text{int } S)) \subseteq N(x, G^{-1}(S)), \text{ and}$$
$$K_3 = M(x, H^{-1}(\{e\})).$$

We point out that

$$Q \cap P \cap G^{-1}(S) \cap H^{-1}(\{e\}) = \phi \tag{1}$$

by virtue of x being optimal, and consequently using Proposition 1, we conclude that

$$N(x,Q) \cap N(x,P) \cap N(x,G^{-1}(S)) \cap M(x,H^{-1}(\{e\})) = \phi \tag{2}$$

and so

$$K \cap K_1 \cap K_2 \cap K_3 = \phi \tag{3}$$

By the Separation Theorem there exist functionals $f \in K^+$, $f_1 \in K_1^+$, $f_2 \in K_2^+$, $f_3 \in K_3^+$, not all zero, such that

$$f + f_1 + f_2 + f_3 = \theta.$$

From Propositions 5 and 6 it follows that there are functionals $\lambda \in (\text{int } C)^+$, $\mu \in N(G(x), \text{int } S)^+$ and $\nu \in Y_3^*$, not all zero, such that for all $h \in X$

$$\lambda \circ DF(x,h) \leq f_1(h),$$
$$\mu \circ DG(x,h) \leq f_2(h), \text{ and}$$
$$\nu \circ DH(x) = f_3.$$

We set $\alpha = f$. Consequently for all $h \in X$

$$[\alpha + \lambda \circ DF(x) + \mu \circ DG(x) + \nu \circ DH(x)](h) \leq 0.$$

As before we observe that $\lambda \in (\text{int } C)^+$ implies $\lambda \in C^+$ and $\mu \in N(G(x), \text{int } S)^+$ implies that $\mu \in S^+$ and $\mu(G(x)) = 0$.

Now let us suppose that (a'), (b') and (c) hold. Using Propositions 7 and 9 we get the following inclusions:

$$K_1 \subseteq K(x,F^{-1}(F(x)+\text{int } C)) \subseteq K(x,P)$$
$$K_2 \subseteq K(x,G^{-1}(\text{int } S)) \subseteq K(x,G^{-1}(S)), \text{ and}$$
$$K_3 = K(x,H^{-1}(\{e\})).$$

Finally we observe that, reasoning almost exactly as before, but with

$$C(x,Q) \cap K(x,P) \cap K(x,G^{-1}(S)) \cap K(x,H^{-1}(\{e\})) = \phi \tag{4}$$

instead of (2) and with

$$K_2 = \{h \in X, DG(x,h) \in K(G(x), \text{int } S)\}$$

possibly replacing the earlier definition of K_2, we can draw the same conclusions.

We confine ourselves to proving Theorems 1 and 3 only, in as much as virtually the same arguments are used to prove Theorems 2 and 4.

In proving Theorem 2, however, we do not make use of Propositions 6 and 10, and so we need not require X to be a Banach space.

Proof of Theorem 3

Firstly suppose that $\lambda \neq 0$ and let $z \in X$, $z \neq x$, be another feasible point, i.e.

$$z \in Q,\ G(z) \in S \text{ and } H(z) = e.$$

We write $h = z - x$. In view of the convexity of F and G

$$DF(x,h) \in F(z) - F(x) + \bar{C},$$
$$DG(x,h) \in G(z) - G(x) + \bar{S},$$

from which it follows that

$$\lambda DF(x,h) \geq \lambda[F(z)-F(x)],$$
$$\mu \circ DG(x,h) \geq 0.$$

Moreover we have

$$\nu \circ H(h) = \nu(H(z)-H(x)) = 0, \text{ and}$$
$$\alpha(h) = \alpha(z) - \alpha(x) \geq 0.$$

Therefore

$$\lambda[F(z)-F(x)] \leq \alpha(h) + \lambda DF(x,h) + \mu \circ DG(x,h) + \nu \circ H(h) \leq 0.$$

If $C = [0,\infty)$, then $\lambda > 0$ and $F(z) - F(x) \leq 0$, i.e. $F(z) - F(x) \notin C\backslash\{\theta\}$. If $C = (-\infty,0]$, then $\lambda < 0$ and $F(z) - F(x) \geq 0$, so again $F(z) - F(x) \notin C\backslash\{\theta\}$. Hence $\{z\in\Omega,\ F(z)-F(x)\in C\backslash\{0\}\} = \phi$, which proves that x is optimal.

Now let H be a projection and suppose that there exists some $z \in \text{int } Q$ such that $G(z) \in \text{int } S$ and $H(z) = e$. Writing $h = z - x$ and $K = K(x,Q)$, we use

$$K_2 = \{h\in X;\ DG(x,h)\in K(G(x),\ \text{int } S)\}, \text{ and}$$
$$K_3 = \{h\in X;\ H(h)=\theta\}$$

as before.

By hypothesis $h \in K(x,Q)$, since $x,\ z \in Q$ and Q is convex, moreover $G(x) + \alpha DG(x,h) \in (1-\alpha)G(x) + \alpha G(z) + \bar{S} \subseteq \text{int } S$ for all $\alpha \in (0,1]$, so $h \in K_2$, and $H(h) = H(z) - H(x) = \theta$, i.e. $h \in K_3$. Therefore $K \cap K_2 \cap K_3 \neq \phi$.

Now suppose that $\lambda = 0$. We write

$$f = \alpha$$
$$f_2 = -\alpha - \nu \circ H$$
$$f_3 = \nu \circ H.$$

Then

$$f + f_2 + f_3 = \theta.$$

Proposition 3 implies that $f \in K^+$. Since H is a projection, $f_3 \neq \theta$ if $\nu \neq \theta$, and $f_3 \in K_3^+$. Further we note that $\mu \in S^+$ and $\mu(G(x)) = 0$ imply that $\mu \in K(G(x),\ \text{int } S)^+$; in view of the inequality $\mu \circ DG(x,k) \leq f_2(k)$ for all $k \in X$ we see that $f_2 \in K_2^+$. Moreover $f_2 \neq \theta$ if $\mu \neq \theta$, for in that case $0 < \mu \circ DG(x,h) \leq f_2(h)$. Because $f + f_2 + f_3 = \theta$, and either $f_2 \neq \theta$ or $f_3 \neq \theta$, we conclude from the Separation Theorem that

$$K \cap K_2 \cap K_3 = \phi$$

which is a contradiction.

5. CONCLUDING REMARKS

As emphasised in the introduction, our main concern in this paper has been to give a simple, complete derivation of the optimality conditions. We conclude, however, by drawing attention to some minor differences with available results. Research into necessary conditions for optimality has been limited largely to the case when the range of the equality constraint function is finite dimensional [1,2,3,6,7]. Only recently has attention been given to the more general situation studied here, where this finite dimensionality requirement is disposed with [8,9,10]. Theorem 1 is in a sense complementary to [3, theorem 6.1]. The Theorem gives necessary conditions for optimality under different differentiability assumptions on F and G (Gâteaux differentiability to a continuous C-convex, resp. S-convex function) as compared with [3] (strong Gâteaux differentiability to a C-convex, resp. S-convex function). This is achieved at the cost of expressing the necessary conditions with respect to a smaller "convex approximation" to the underlying set Q.

It does not appear possible, retaining the present level of generality, to remove the hypothesis in Theorems 1 and 2, that the G-differentials are continuous; that it can be dispensed with in [3] leans heavily on the finite dimensionality of the range of H.

Finally we mention that we slightly generalise results in [8] to the extent that the development here is in real linear topological spaces and also in that their strongest result [8, theorem 2.3] is stated for H continuously Fréchet differentiable, while only strong Fréchet differentiability is here required.

REFERENCES

1. Dubovitskiy A.Ya., and Milyutin A.A., Extremum Problems in the Presence of Restrictions, (USSR) Journal of Computational Mathematics and Mathematical Physics, Vol. 5, No. 1, 1965.

2. Halkin H., A Satisfactory Treatment of Equality and Operator Constraints in the Dubovitskii-Milyutin Optimizational Formalism, Journal of Optimization Theory and Applications, Vol. 6, No. 2, 1970.

3. Neustadt L.W., A General Theory of Extremals, Journal of Computer and System Sciences, Vol. 3, No. 1, 1969.

4. Vlach M., On Necessary Conditions of Optimality in Linear Spaces, Comment. Math. Univ. Carolinae, Vol. 11, No. 3, 1970.

5. Nonlinear Functional Analysis, Proceedings of an Advanced Seminar, Ed. L.B. Rall, Academic Press, 1971.

6. Neustadt L.W., An Abstract Variational Theory with Applications to a Broad Class of Optimization Problems, I General Theory, SIAM Journal on Control, Vol. 4, No. 3, 1966.

7. Neustadt L.W., An Abstract Variational Theory with Applications to a Broad Class of Optimization Problems, II Applications, SIAM Journal on Control, Vol. 5, No. 1, 1967.

8. Bazaraa M.S. and Goode J.J., Necessary Optimality Criteria in Mathematical Programming in Normed Linear Spaces, Journal of Optimization Theory and Applications, Vol. 11, No. 3, 1973.

9. Craven B.D., Nonlinear Programming in Locally Convex Spaces, Journal of Optimization Theory and Applications, Vol. 10, No. 4, 1972.

10. Virsan C., Necessary Conditions for Optimization Problems with Operational Constraints, SIAM Journal on Control, Vol. 8, No. 4, 1970.

A PERTURBATION THEORY APPROACH TO NON-LINEAR PROGRAMMING

James Thurber
Andrew Whinston
Purdue University
West Lafayette, IN 47907/USA

I. Perturbation Theory and Its Applications

The majority of problems which occur in mathematical physics are stated in the form of differential equations with some prescribed boundary and/or initial conditions. Generally, it is impossible to obtain closed form analytical solutions to such problems. A long standing method for obtaining approximate solutions is perturbation theory. One or more parameters enter the problem and for some fixed values of the parameters (usually taken to be zero), the problem is exactly solvable analytically. The equations and their solutions are expanded in terms of these parameters and one thereby obtains approximate solutions of the original problem.

We apply this method to non-linear mathematical programming, in order to obtain a sequence of approximate solutions which converges to solutions of the original problem. As a corollary to this approach, we are able to derive sufficient and/or necessary conditions for local optimum solutions in a very general setting. In particular, when the normal first order conditions fail due to the presence of cusps, we derive a new set of sufficient conditions.

The perturbation series approach leads to a new algorithmic development which is discussed in this paper and will be further amplified in a sequel.

II. Direct Expansion

The problem to be considered in this paper is:

Minimize $F = F(Z)$ (1)

Subject to the constraints

$G_\ell(Z) = 0, \quad 1 \leq \ell \leq L$ (2)

where $Z = (Z_1, \ldots, Z_R)$ is a vector with real components, and $F(Z)$ and the $G_\ell(Z)$ are real holomorphic functions in the components of Z.

We study the problem in a neighborhood of a particular point

$$Z_o = (Z_1^o, Z_2^o, \ldots, Z_R^o) \qquad (3)$$

and on feasible paths in this neighborhood passing through this point. These paths are constructed by means of a perturbation series to be described. In the neighborhood of Z_o, F and G_ℓ, $1 \leq \ell \leq L$ have power series expansions with non-zero radii of of convergence. For a justification of the form of the perturbation expansion which we assume in this section, see reference 6.

In order to give a geometric interpretation of the terms in the perturbation series, it will be convenient to express the Taylor expansions of the constraint functions in terms of tensor notation. For our purposes, we will regard the tensor notation as an extension of matrix notation. In matrix notation, we have two indices i and j to denote the element m_{ij} in the i-th row and the j-th column of the matrix M. A matrix is a special case of a tensor. For tensors, we allow an arbitrary number of indices. A tensor of valence V will have V indices. Thus, a component of a tensor T of valence V will be denoted by $t_{j_1, j_2, \ldots, j_V}$. Tensors of the same valence where the corresponding indices have the same range of values form a linear vector space, and they can be linearly combined to give tensors of the same type. We will consider such tensor spaces for complex as well as real valued components. Temporarily, we will only consider tensors where the indices j all have the same range. Below, it is assumed that this is always the case.

A tensor with exactly one index is a vector. A tensor with no indices is referred to as a scalar or simply a number. As is the case with matrices and vectors, tensors can also be multiplied. Two types of multiplication will concern us, namely tensor multiplication and contraction. The product P of two tensors S and T with components $s_{j_1}, \ldots, {}_{j_r}$ and $t_{i_1}, \ldots, {}_{i_u}$ respectively is the tensor with components

$$P_{j_1, \ldots j_r i_1 \ldots i_u} = s_{j_1 \ldots j_r} t_{i_1 \ldots i_u} . \tag{4}$$

Multiplication of tensors is generally non-commutative; for example,

$$P_{12} = S_1 T_2 \neq S_2 T_1 = P_{21} . \tag{5}$$

Tensor multiplication of tensors is, however, associative

$$R(ST) = (RS)T \tag{6}$$

and distributive

$$R(S + T) = RS + RT \tag{7}$$

$$(S + T)R = SR + TR . \tag{8}$$

Contraction consists of tensor multiplication followed by a summation over pairs of indices which are equal; one member of each pair is taken from the two terms in the product. For example, suppose that the indices have the range $1 \leq j \leq 3$. T is of the form t_{ij}, and S is of the form s_{kl}; then the tensor product ST of the form $t_{ij}s_{kl}$ can be contracted on the pair of indices j and k to give a tensor with the components

$$\sum_{j=1}^{3} t_{ij}s_{j\ell}$$

where here, k is equated to j, and then j is summed. This is of course just equivalent to ordinary matrix multiplicated of S and T, where tensors of valence two are treated as ordinary matrices. On the other hand, we could consider S to be of the form

$$S_{k\ell m}$$

and T to be of the form

$$t_{ijnq} ;$$

then ST is of the form $S_{k\ell m}t_{ijnq}$, and we could contract the indices ℓ and q to give a tensor of the form

$$U_{kmijn} = \sum_{\ell=1}^{3} S_{k\ell m}t_{ijn\ell} . \tag{9}$$

We could also have contracted S and T on the two pairs of indices ℓ and q as well as k and n to give

$$V_{mij} = \sum_{\ell=1,\ k=1}^{3\ 3} S_{k\ell m}t_{ijk\ell} . \tag{10}$$

Tensor multiplication of S and T followed by contraction on n pairs of indices will be denoted by

$$S\ n^{\bullet}\ T .$$

In the above examples, we could write

$$U = S\ 1^{\bullet}\ T \quad \text{and} \tag{11}$$

$$V = S\ 2^{\bullet}\ T . \tag{12}$$

In tensor analysis, one usually has two kinds of indices - superscripts as well as subscripts, and contraction occurs between indices of different type (i.e., super and sub-scripted indices); however, we will deal only with the case of so-called Cartesian tensors and the distinction will not be necessary for our purposes. We will find it useful also to regard contraction as represented by ordinary matrix multiplication. For example, in the tensor $S_{k\ell m}$, we can regard the pair of indices $k\ell$ as corresponding to the 3^2 components of a nine dimensional vector space and similarly for the indices nq of t_{ijnq}. From this point of view, we can regard S as a 9x3 matrix and T as a 9x9 matrix, and the 3x9 matrix V is formed by ordinary matrix multiplication from S and T. Thus, $V = S^t . T$ where the superscript t denotes the transpose and the dot, ordinary matrix multiplication.

In the following, we will find both the tensor and matrix interpretations of contraction useful. We will now relax the restriction that the indices of tensors all have the same range; however, we can only contract on pairs of indices which have the same range of values.

One can extend the idea of tensors to tensor operators; thus the gradient operator $\nabla = \left(\frac{\partial}{\partial Z_1}, \frac{\partial}{\partial Z_2}, \ldots, \frac{\partial}{\partial Z_R}\right)$ can be regarded as a vector (tensor of valence one) operator with components $\frac{\partial}{\partial Z_j}$ for $1 \leq j \leq R$; and ∇^2, the tensor product of ∇ with itself is a partial differential operator with the components

$$\frac{\partial^2}{\partial Z_i \partial Z_j} \quad \text{for } 1 \leq i, j \leq R ; \tag{13}$$

and similarly for ∇^3, and so on.

If in ∇^2, we equate the indices i and j and contract, we have

$$\nabla 1^{\bullet} \nabla = \sum_{i=1}^{R} \frac{\partial^2}{\partial Z_i^2} \tag{14}$$

which is the well known Laplace operator in R variables.

In our use of tensors, mostly symmetric tensors will occur. A tensor T of valence u is said to be symmetric if for every component $t_{i_1, \ldots, i_n}$ of T, we have

$$t_{i_1, \ldots, i_u} = t_{i_1', \ldots, i_u'} \tag{15}$$

where the indices $i_1', \ldots, i_u'$ are an arbitrary permutation of the indices $i_1, \ldots, i_u$. Basically the reason that mainly symmetric tensors are of interest to us is that in the Taylor Expansions of analytic functions to which the tensors will be applied, we know the order in which partial derivatives or more generally, directional derivatives are applied, is immaterial. We rearrange the order, and the result is the same.

Clearly sums and linear combinations of symmetric tensors are symmetric; however, the product of two symmetric tensors is in general, not symmetric. Thus we need the concept of a permutation operator acting on a tensor. A permutation operator Ω acts on a tensor T by permuting the indices of T to give another tensor $\tilde{T}$.

For example, let T be the tensor of valence four having components $t_{ijk\ell}$, and let Ω be the permutation operator

$$\Omega = \begin{pmatrix} 2143 \\ 1234 \end{pmatrix} . \tag{16}$$

That is, Ω takes the first index i and puts it in the place of the second index j. The second index j is put in the place of the first index i. The third index k is put in the place of the fourth index ℓ, and the fourth index ℓ is put in the place of the third index k.

Clearly there are $24 = 4!$ different such permutation operators Ω for tensors of valence four.

The symmetrized tensor $\hat{T}$ corresponding to a tensor T of valence n is defined to be

$$\hat{T} = \frac{1}{n!}\left(\sum_{k=1}^{n!} \Omega_k T\right), \tag{17}$$

where the Ω_k are the n! different permutation operators defined on tensors of valence n.

If T is a symmetric tensor, then clearly

$$\hat{T} \equiv T . \tag{18}$$

In the application to the Taylor expansions of the constraint functions, the tensors which are tensor products of the gradient operator are clearly symmetric tensor operators. These tensors ∇^n are applied to the vector of constraint functions G at a particular point $Z = Z_0$. The resultant tensor $\nabla_0^n G$ will be symmetric in all the indices arising from ∇^n but not from the index labeling the components of G. We will be interested in the left null spaces of the operators $\nabla_0^n G$ and their orthogonal complementary spaces, since it is from these considerations that we will be able to scale the variables properly in terms of the perturbation parameter ϵ. The left indices are those corresponding to ∇^n, and the right indices are those belonging to G. From the symmetry in the left indices, we see that the projection operators π_n defining the left null spaces and the operators C_n giving the projection operators for the corresponding orthogonal complementary spaces must be symmetric in these same indices which are in fact the only indices occurring in the definition of the operators.

π_n and C_n which satisfy the relations

$$\pi_n\pi_n = \pi_n \tag{19}$$

$$C_nC_n = C_n \tag{20}$$

$$\pi_nC_n = C_n\pi_n = 0 \text{(the 0-operator)}. \tag{21}$$

From the symmetry in the indices of π_n and C_n, we have the following useful property, namely, if $T_1T_2 \cdot \ldots T_k$ is a product of tensors lying in the domains of π_n and C_n, then a total contraction of all the indices of the product $T_1 \cdot \ldots\cdot T_k$ $\pi_nT_1T_2 \cdot \ldots\cdot T_k$ with the corresponding indices of π_n gives a symmetric tensor, and similarly for $C_nT_1T_2 \cdots T_k$, and consequently

$$\pi_nT_1T_2 \cdot \ldots\cdot T_k = \pi_nT_{1'} T_{2'} \ldots\cdot T_{k'} \, . \tag{22}$$

where $1', 2', \ldots, k'$ is an arbitrary permutation of 1, 2, ..., k, and similarly,

$$C_n T_1 T_2, \ldots, T_k = C_n T_1, T_2, \ldots T_k, \tag{23}$$

The functions defining the constraints are written as a row vector

$$G(Z) = (G_1(Z), G_2(Z), \ldots, G_L(Z)), \tag{24}$$

and the Taylor Expansions take the form

$$G(Z_o + \Delta Z) = \sum_{n=0}^{\infty} \frac{(\Delta Z)^n \, n^{\cdot} \nabla_o^n G}{n!} \tag{25}$$

where the following notational conventions are used:

$$\text{a)} \quad \Delta Z = (\Delta Z_1, \Delta Z_2, \ldots, \Delta Z_R) \tag{26}$$

and by $(\Delta Z)^n$, one means the n-fold tensor product; thus $(\Delta Z)^2$ is an R x R matrix with i,j entry equal to $\Delta Z_i \Delta Z_j$. We set $(\Delta Z)^o = 1$ by definition. The term $(\Delta Z)^3$ would be a triply indexed array with i, j, kth entry equal to ΔZ_i, ΔZ_j, ΔZ_k, etc.

b) ∇ is the gradient operator written as a column vector. By ∇^n, again one means the tensor product, and by $\nabla_o^n G$, one means the n-fold gradient operator applied to G at the value $Z = Z_o$.

c) $n^{\cdot}$ refers to an n-fold contraction of the indices of $(\Delta Z)^n$ and ∇^n. Thus

$$\frac{(\Delta Z)^n n^{\cdot} \nabla^n}{n!} \tag{27}$$

is the usual scalar operator giving the nth order terms in the Taylor Expansions. This representation of the Taylor Expansions enables us to describe in a simple fashion the perturbation expansion.

In the perturbation expansion, a perturbation parameter ϵ occurs, and the form of the expansion is a Puiseaux series in ϵ. That is, the series consists of terms which are fractional powers of ϵ, namely terms of the form ϵ^α where α is a non-negative rational number. In fact, for our consideration, α will always be of the form $\frac{P}{N}$, where N is some fixed positive integer, and P is some non-negative integer. For practical calculations where in general, only a finite number of terms of the perturbation expansion are considered, ϵ is effectively a scaling parameter which determines the percent error we are willing to tolerate in satisfying the constraints. It will be convenient to regard $\nabla_o^n G$ as an R^n by L matrix, and of particular interest are the left null spaces of $\nabla_o^n G$ and their orthogonal complements. Our problem now is to determine the form of the solution to the constraints (2) locally. By assumption, the expansion converges for

$$|\Delta Z_i| \leq r > 0 \tag{28}$$

for some radius of convergence r. We are interested in the ΔZ such that $G(Z_o) = 0$, and $G(Z_o + \Delta Z) = 0$. We will justify later the expansion

$$\Delta Z = \sum_{j=1}^{\infty} \epsilon^{\frac{P_j}{N}} \tilde{Z}^j \quad \text{(Note: the j in } \tilde{Z}^j \text{ is a superscript or index and not an exponent)} \tag{29}$$

which we now assume to be valid.

The first operator or matrix of interest is the $\nabla_o^1 G$ and the corresponding term in the Taylor Expansion is $\Delta Z\ 1 \cdot \nabla_o^1 G$. For ΔZ having a direction belonging to the left null space of $\nabla_o^1 G$, there is no contribution to the linear term in the Taylor Expansion. Thus, for an arbitrary ΔZ, the only contribution to the linear term is the orthogonal projection of ΔZ onto the orthogonal complement of the left null space of $\nabla_o^1 G$.

Let π_1 denote the projection operator for the left null space of $\nabla_o^1 G$. Then, $\pi_1 \Delta Z$ does not occur in the linear approximations to the constraints. Consider next the operator

$$\frac{\nabla_o^2 G}{2!} \tag{30}$$

and the tensor product $(\Delta Z)^2$ written in the form:

$$(\Delta Z)^2 = (C_1 \Delta Z)^2 + (C_1 \Delta Z)(\pi_1 \Delta Z) + (\pi_1 \Delta Z)(C_1 \Delta Z) + (\pi_1 \Delta Z)^2 \tag{31}$$

where $C_1 = 1 - \pi_1$. (32)

If $\pi_2(\pi_1 \Delta Z)^2 = (\pi_1 \Delta Z)^2$, then $\pi_1 \Delta Z$ as well as belonging to the left null space of $\nabla_o^1 G$ also is such that its tensor product with itself belongs to the left null space of $\frac{\nabla_o^2 G}{2!}$ and clearly does not contribute to leading order to the quadratic term in the Taylor Expansion.

For $\Delta Z \neq 0$, more generally, let us assume for $k \geq 2$, that

$$\begin{aligned} &C_1 \Delta Z = 0, \text{ hence } \pi_1 \Delta Z \neq 0 \\ &C_2 (\Delta Z)^2 = 0 \\ &\vdots \\ &C_{k-1} (\Delta Z)^{k-1} = 0; \\ \text{but} \quad &C_k (\Delta Z)^k \neq 0; \end{aligned} \tag{33}$$

then we will regard ΔZ as a term of order $\epsilon^{1/k}$ where ϵ is a small perturbation parameter in terms of which we scale our expansion which will give us an expansion for solving the constraints at regular points or in the neighborhood of cusps.

To see how the expansion works, we proceed as follows. We begin with the linear term in the Taylor expansion, namely

$$(\Delta Z)\ 1 \cdot \nabla_o^1 G = (C_1 \Delta Z + \pi_1 \Delta Z)\ 1 \cdot \nabla_o^1 G \tag{34}$$

$$= C_1 \Delta Z \; 1^{\bullet} \; V_o^1 G$$

$C_1 \Delta Z$ we take to be of order ϵ.

Since $\pi_1 \Delta Z \; 1^{\bullet} \; V_o^1 G = 0$, $\pi_1 \Delta Z$ will occur in the Taylor expansion no earlier than the tensor product of ΔZ with itself, we have

$$(\Delta Z)^2 = (C_1 \Delta Z)^2 + (C_1 \Delta Z)(\pi_1 \Delta Z) + (\pi_1 \Delta Z)(C_1 \Delta Z) + (\pi_1 \Delta Z)^2 \; . \tag{35}$$

The only part of $(\Delta Z)^2$ which contributes to the second degree terms in the Taylor expansion, namely

$$\frac{(\Delta Z)^2 \; 2^{\bullet} \; V_o^2 G}{2!}$$

is $C_2(\Delta Z)^2$. Since C_2 is symmetric in its indices,

$$C_2(C_1 \Delta Z)(\pi_1 \Delta Z) = C_2(\pi_1 \Delta Z)(C_1 \Delta Z); \tag{36}$$

thus

$$C_2(\Delta Z)^2 = C_2(C_1 \Delta Z)^2 + 2C_2(C_1 \Delta Z)(\pi_1 \Delta Z) + C_2(\pi_1 \Delta Z)^2 \; . \tag{37}$$

We wish to determine those terms of $C_2(\Delta Z)^2$ which are of order ϵ. Clearly $C_2(C_1 \Delta Z)^2$ is of order ϵ^2, and to a first approximation, we can for small ϵ, neglect this contribution to $C_2(\Delta Z)^2$.

It is reasonable to choose a vector of the form $\pi_1 \Delta Z$ to be of order $\epsilon^{\frac{1}{2}}$ for then $C_2(\pi_1 \Delta Z)^2$ will be of order ϵ. If $C_2(\pi_1 \Delta Z)^2 \neq 0$, we make this choice. The following difficulty arises if two vectors ΔZ_1 and ΔZ_2 are such that

$$C_2(\pi_1 \Delta Z_1)^2 \neq 0 \text{ and } C_2(\pi_1 \Delta Z_2)^2 \neq 0 \; ; \tag{38}$$

we cannot in general assert that $C_2\left\{\pi_1\left(\alpha \Delta Z_1 + \beta \Delta Z_1\right)\right\}^2 \neq 0$ for arbitrary α and β, so we proceed as follows. For ΔZ_1 such that

$$\pi_1 \Delta Z_1 = \Delta Z_1 \tag{39}$$

and

$$C_2(\pi_1 \Delta Z_1)^2 \neq 0 \; , \tag{40}$$

choose ΔZ_1 to be of order $\epsilon^{\frac{1}{2}}$.

Next choose ΔZ_2 to be linearly independent of ΔZ_1 and such that

$$C_2(\pi_1 \Delta Z_2)^2 \neq 0 \tag{41}$$

if possible. Then choose ΔZ_3 to be linearly independent of ΔZ_1 and ΔZ_2 such that

$$\pi_1 \Delta Z_3 = \Delta Z_3 \tag{42}$$

and such that

$$C_2(\pi_1 \Delta Z_3)^2 \neq 0 \tag{43}$$

if possible and continue as far as possible obtaining linearly independent

$$\Delta Z_1, \Delta Z_2, \ldots, \Delta Z_n$$

such that

$$\pi_1 \Delta Z_k = \Delta Z_k \tag{44}$$

and

$$C_2(\pi_1 \Delta Z_k)^2 \neq 0 \text{ for } 1 \leq k \leq n . \tag{45}$$

Then ΔZ_k for $1 \leq k \leq n$ are taken to be of order $\epsilon^{\frac{1}{2}}$.

The vectors in the subspace determined by C_1 and $\Delta Z_1, \ldots, \Delta Z_n$ may not generate the full vector space of possible ΔZ. If they do, then we are in a position to assign a unique order ϵ^{α} to each vector; if not, then there must exist vectors $\Delta Z \neq 0$ such that

$$\pi_1 \Delta Z = \Delta Z, \tag{46}$$

$$C_2(\pi_1 \Delta Z)^2 = 0, \tag{47}$$

or equivalently

$$\pi_2(\pi_1 \Delta Z)^2 = (\pi_1 \Delta Z)^2 . \tag{48}$$

We must determine the order of such a ΔZ. Such a ΔZ might satisfy the relation

$$C_2(\Delta Z)W \neq 0 \tag{49}$$

where W is in the subspace generated by $\Delta Z_1, \Delta Z_2, \ldots, \Delta Z_n$. If this is so, let us consider vectors H of the form

$$H = \lambda \Delta Z + W \tag{50}$$

for arbitrary W and scalars λ.

By assumption,

$$\pi_1 H = \lambda \pi_1 \Delta Z + \pi_1 W = \lambda \Delta Z + W = H, \tag{51}$$

and therefore,

$$C_2(\pi_1 H)^2 = \lambda^2 C_2(\Delta Z)^2 + 2\lambda C_2(\Delta Z)W + C_2 W^2 \tag{52}$$

using the symmetry in the indices of C_2.

Since $C_2(\Delta Z)^2 = 0$, and $C_2 \Delta Z W \neq 0$, we have

$$C_2(\pi_1 H)^2 = 2\lambda C_2(\Delta Z)W + C_2 W^2 \neq 0 \tag{53}$$

for λ sufficiently large.

But $\pi_1 H = H$ is linearly independent of $\Delta Z_1, \Delta Z_2, \ldots, \Delta Z_n$, and thus we could have chosen $H = \Delta Z_{n+1}$, a contradiction. Another possibility is that ΔZ might satisfy a relation of the form

$$C_2(\Delta Z)u \neq 0 \tag{54}$$

where

$$\pi_1 u = u. \tag{55}$$

Such a u is of order ϵ, and unless the order of ΔZ is chosen to be $\epsilon^0 = 1$, we cannot have $C_2(\Delta Z)u$ of order ϵ. Thus if ΔZ is not of order $\epsilon^0 = 1$, we have

$$C_2(\Delta Z)u = 0. \tag{56}$$

Thus we must look to cubic or higher degree terms to find the order of ΔZ. Therefore, we look for ΔZ such that ΔZ is linearly independent of $\Delta Z_1, \ldots, \Delta Z_n$, and the vectors V such that $C_1 V = C$ and such that

$$\pi_1 \Delta Z = \Delta Z, \tag{57}$$

$$C_2(\pi_1 \Delta Z)^2 = 0, \tag{58}$$

and

$$C_3(\pi_1 \Delta Z)^3 \neq 0 \tag{59}$$

If we find such a ΔZ, we take it to be of order $\epsilon^{1/3}$ as before we look for another such ΔZ linearly independent of the first, choose it to be of order $\epsilon^{1/3}$, and so on until we obtain a maximal such linearly independent set of vectors, all of order $\epsilon^{1/3}$.

As before, we can show that no vector satisfying

$$\pi_1 \Delta Z = \Delta Z \tag{60}$$

$$C_2(\pi_1 \Delta Z)^2 = 0 \tag{61}$$

linearly independent of the above vectors of order $\epsilon^{1/3}$ and occurring as a factor of T such that

$$C_3 T \neq 0 \tag{62}$$

can occur. Similarly we continue and obtain a maximal linearly independent set of vectors ΔZ of order $\epsilon^{1/k}$ satisfying

$$\begin{aligned} &\pi_1 \Delta Z = \Delta Z \\ &C_2(\pi_1 \Delta Z)^2 = 0 \\ &\vdots \\ &C_{k-1}(\pi_1 \Delta Z)^{k-1} = 0 \\ &C_k(\pi_1 \Delta Z)^k \neq 0 \,. \end{aligned} \tag{63}$$

Since our original vector space is of finite dimension, there must be a maximum finite such value of k, and all other vector ΔZ must satisfy the conditions

$$\pi_1 \Delta Z = \Delta Z \tag{64}$$

and

$$C_n(\pi_1 \Delta Z)^n = 0 \tag{65}$$

for all positive integers n. In addition, for all T in the domain of C_n for which ΔZ occurs as a factor and all other factors are of order ϵ or $\epsilon^{\frac{1}{2}}$ or ... $\epsilon^{1/k}$, we have

$$C_n T = 0$$

By an easy extension of the argument used in the construction of H (see (50) with as many λ as ΔZ, we can similarly show that a T in which more than one such different ΔZ occur as factors, we have also

$$C_n T = 0 \tag{67}$$

for T in the domain of C_n.

Let ΔZ_i denote the above factors of T, and now define

$$H_n = \sum_{i=1}^{\ell} \lambda_i \Delta Z_i + R \tag{68}$$

where R consists of terms corresponding to C_{n-1}, C_{n-2}, ..., C_1. By assumption,

$$C_n H_n^{\,n} = 0 \,. \tag{69}$$

On the other hand, expanding $H_n^{\,n}$, we have a term having some of the ΔZ_i as factors which is different from 0. There is a non zero such term with maximum index i, $1 \leq i \leq \ell$. We can then choose this λ_j so large in ratio to the remaining λ_i that if the sum of the terms with λ_j is $\neq 0$, then $C_n H_n^{\,n} \neq 0$. If there is only one such term by assumption, it is $\neq 0$, and we have the desired contradiction. If not, then among these terms there must be a non zero term with maximum index $i < j$. Denote this index by k. We then choose λ_j and λ_k sufficiently large but with λ_k small in ratio to λ_j and repeat this process until after a finite number of steps, we arrive at the desired contradiction. If there are any $\Delta Z \neq 0$ for which

$$C_n T = 0 \tag{70}$$

whenever ΔZ occurs as a factor T for all $n \geq 1$, then ΔZ corresponds to a variable a linear combination of the original variables, which does not occur in the Taylor expansions of the constraint function. That is, the constraint functions of G are locally independent of ΔZ. By the analyticity of the G, this means that the constraint functions are globally independent of ΔZ. By holding the remaining variables fixed, we optimize with respect to ΔZ. This gives a new criterion function which is independent of such ΔZ. We can then proceed to optimize with respect to the remaining variables ignoring such ΔZ, providing the criterion function is such that the interchange of limits is valid. For rational criterion functions with denominator bounded away from zero, this is certainly the case. If there is any problem of order of limits, we simply keep such ΔZ, treating them as parameters in the criterion function. We are now prepared to give the proper scaling of each component ΔZ in terms of a perturbation parameter ϵ where

$$\epsilon \geq 0 , \tag{71}$$

and ρ_k is the projection operator for terms of index k. Define

$$\begin{aligned} \rho_o(\Delta 2) &= C_1 \Delta Z = \epsilon \tilde{W}_o \\ \rho_{k_1^*} \Delta Z &= \epsilon^{\frac{1}{k_1^*}} \tilde{W}_1 \\ &\vdots \\ \rho_{k_n^*} \Delta Z &= \epsilon^{\frac{1}{k_n^*}} \tilde{W}_1 . \end{aligned} \tag{72}$$

The vectors $\tilde{W}_o$, $\tilde{W}_1$, ..., and $\tilde{W}_n$ are in general, functions of ϵ but are of order

$$\tilde{W}_j = O(1) = O(\epsilon^o) \text{ for } o \leq j \leq n. \tag{73}$$

and

$$\Delta Z = \epsilon \tilde{W}_o + \epsilon^{\frac{1}{k_1^*}} \tilde{W}_1 + \dots + \epsilon^{\frac{1}{k_n^*}} \tilde{W}_n . \tag{74}$$

The leading or dominant terms of the Taylor expansion are

$$C_1 \Delta Z \; 1^\circ \; V_o^1 G + \sum_{j=1}^{n} \frac{\left(\rho_{k_j^*} \; \Delta Z\right)^{k_j^*} \dot{k_j^*} \; V_o^{k_j^*} G}{(k_j^*)\,!}$$

$$= \epsilon\left(\tilde{W}_o\, 1^{\cdot}\, V_o^1 G + \sum_{j=1}^{n} \frac{(\tilde{W}_j)^{k_j^*}\, k_j^{*\cdot}\, V_o^{k_j^*} G}{(k_j^*)!}\right) \tag{75}$$

With the above scaling in terms of ϵ, the dominant terms are all of the same order in ϵ. The first approximation to the non-linear constraints (1) becomes

$$\tilde{W}_o^*\, 1^{\cdot}\, V_o^1 G + \sum_{j=1}^{n} \frac{(\tilde{W}_j^*)^{k_j^*}\, k_j^{*\cdot}\, V_o^{k_j^*} G}{(k_j^*)!} = 0 \tag{76}$$

and if $\tilde{W}_o$ can be chosen as the dependent variable

$$\tilde{W}_o = -\left\{ \sum_{j=1}^{n} \frac{(\tilde{W}_j^*)^{k_j^*}\, k_j^{*\cdot}\, V_o^{k_j^*} G}{(k_j^*)!}\right\} (V_o^1 G)^+ \tag{77}$$

where $(V_o^1 G)^+$ denotes the Penrose pseudo inverse of $V_o^1 G$ where $\tilde{W}_j^*$ denotes the zeroth order terms in the Puiseaux Expansions of

$$\tilde{W}_j,\ 0 \le j \le n. \tag{78}$$

To obtain (77), simply expand (2) in a Taylor series as given by (74), then substitute relations (72) into (2), expand each of the $\tilde{W}_j$ as Puiseaux series and collect together all terms having the same power of ϵ. Solve for $\tilde{W}_o^*$ using the Penrose pseudo inverse $(V_o^1 G)^+$, and since by definition $\tilde{W}_o^*$ is independent of ϵ, we have (77).

To obtain higher order approximations to the constraints, if we can take the

$$\tilde{W}_j \text{ for } 1 \le j \le n \tag{79}$$

as the independent variables and write

$$\tilde{W}_o = -\left(\sum_{m=1}^{\infty} \frac{(\Delta Z)^m\, m^{\cdot}\, V_o^m G}{m!}\right) \frac{(V_o^1 G)^+}{\epsilon}\,. \tag{80}$$

Higher approximations to $\tilde{W}_o$ than given by (77) can be given by iteration as follows

$$\tilde{W}_o^{(\ell+1)} = -\left(\sum_{m=1}^{\infty} \frac{\left[(\Delta Z)^{(\ell)}\right]^m\, m^{\cdot}\, V_o^m G}{m!}\right) \frac{(V_o^1 G)^+}{\epsilon}\,. \tag{81}$$

By $(\Delta Z)^{(\ell)}$, we mean the ℓth approximation to ΔZ, the 0-th approximation to $\tilde{W}_o$ is given by (77).

We take

$$\tilde{W}_j^{(\ell+1)} = \tilde{W}_j^{(\ell)} = \tilde{W}_j \text{ for } 1 \le j \le n \tag{82}$$

since the $\tilde{W}_j$ $(1 \le j \le n)$ are the independent variables, and the limit as $\ell \to \infty$ of

$\tilde{W}_o^{(\ell)}$ is $\tilde{W}_o$.

When

$$\rho_{k^*_\infty} = 0 \text{ but } C_1 = 0, \tag{83}$$

we have a situation where in general the Kuhn-Tucker constraint qualifications are violated and there is no linear approximation which represents the non-linear constraints in a reasonable manner. The example of a cusp that we treat in the example section is just such a situation. Nonetheless our prescription for scaling the variables in terms of powers of ϵ remains valid. The terms corresponding to the lowest powers of ϵ could be treated similarly to the way we treated the independent variables $\tilde{W}_o$ above. However, in general coupled non-linear algebraic equations of degree k_1^* would have to be solved.

The situation where $\rho_{k^*_\infty} \neq 0$ is more complicated. A vector ΔZ such that $\Delta Z = \rho_{k^*_\infty} \Delta Z$ corresponds to a direction for which there is no effective constraint. In this case, all derivatives mixed or otherwise in the direction ΔZ are equal to 0, and the direction is unconstrained. The value of ΔZ places no restriction on variables in directions orthogonal to ΔZ.

The directions corresponding to $\rho_{k_1^*}, \rho_{k_2^*}, \ldots, \rho_{k_n^*}$ are scaled as before. For each ΔZ such that $\Delta Z = \rho_{k^*_\infty} \Delta Z$ it may happen that it occurs in a term of the Taylor expansion of the criterion function with non zero coefficient where only derivatives in the directions ΔZ and/or directions corresponding to the projection operators $\rho_{k_1^*} \ldots, \rho_{k_n^*}$ occur. In this case, $\epsilon^0 = 1$ gives the correct scaling in such a way that ΔZ occurs in as low an order term as possible. Note that all exponents α of ϵ^α must be of the form $\alpha = \frac{P}{N}$ for some fixed integer N and some integer $P \geq 0$ (see reference 6). It may or may not happen that the term where ΔZ gives the dominant contribution is of order ϵ. Examples can be given to show that ΔZ may have its dominant contribution in a term of order higher than ϵ. We can now give a general sufficient condition for local optimality. At

$$Z = Z_o, \tag{84}$$

write

$$\Delta Z = C_1 \Delta Z + \rho_{k_1^*} \Delta Z + \cdots + \rho_{k_n^*} \Delta Z + \rho_{k^*_\infty} \Delta Z \tag{85}$$

where C_1 may or may not equal zero.

The dependent variables of ΔZ are now replaced by their expansion in terms of the independent variables and substituted into the expression

$$F(Z_o + \Delta Z) = F(Z_o) + \sum_{m=1}^{\infty} \frac{(\Delta Z)^m{}_m \circ \nabla_o^m F}{m!} \tag{86}$$

Ignoring $F(Z_o)$ we keep the dominant term in ϵ. If this term is of a definite sign and never zero unless all the independent variables are equal to zero, then we must be at a local optimum. We are at a maximum if the sign is negative and a minimum if it is positive. A necessary condition for local optimality is that the leading non-constant term be semi-definite. That is, it should be of a definite sign or possibly equal zero for non-zero values of the independent variables.

III. Example

As an example, we consider one that is often quoted in the literature to illustrate what can happen when the Kuhn-Tucker constraint qualification is violated.

The criterion function to be maximized is

$$F = X , \tag{87}$$

and the constraints are

$$\begin{aligned} &Y - (1-X)^3 \leq 0 \\ &X,Y \geq 0 . \end{aligned} \tag{88}$$

At the point $Y = 0$, $X = 1$, we have on writing $Y = \Delta Y$, $X = 1 + \Delta X$

$$\begin{aligned} &\Delta Y - (\Delta X)^3 \leq 0 \\ &1 + \Delta X \geq 0, \ \Delta Y \geq 0. \end{aligned} \tag{89}$$

Keeping only the linear terms in ΔX and ΔY and regarding ΔX as small compared to one, we have

$$\Delta Y \leq 0 \tag{90}$$

and

$$\Delta Y \geq 0 \tag{91}$$

or simply

$$\Delta Y = 0 \tag{92}$$

as our linear approximation to relation (88).

Since ΔX is taken to be small compared to one, there is effectively no sign constraint on ΔX in the linear approximations. Thus, according to the linear approximations, we seek to maximize

$$F = 1 + \Delta X \tag{93}$$

subject to $\Delta Y = 0$. To the linear approximations, we ought to be able to increase ΔX by some small positive amount, however clearly $X > 1$ is feasible, and it is obvious that $\Delta X = 0$ or $X = 1$ is in fact already the optimum value. Our method is designed, in fact, to deal with just such situations, where in addition, the result may no longer be obvious.

In order to prepare this problem for our technique, we rewrite it in the equivalent form

$$F = U^2 = X \qquad Y = V^2 \tag{94}$$

and

$$V^2 - (1 - U^2)^3 + W^2 = 0, \tag{95}$$

where U, V, and W are real valued variables. Setting $U_0 = 1$, $V_0 = 0$, and $W_0 = 0$, we write

$$U = 1 + \Delta U \qquad V = \Delta V \qquad W = \Delta W. \tag{96}$$

The Taylor expansion for $g = V^2 - (1 - U^2)^3 + W^2$ is simply

$$G = (\Delta V)^2 + (\Delta W)^2 + 8(\Delta U)^3 + 12(\Delta U)^4 + 6(\Delta U)^5 + (\Delta U)^6. \tag{97}$$

According to our prescription for scaling, we write

$$\Delta V = \epsilon^{1/2}\tilde{V} \qquad \Delta W = \epsilon^{1/2}\tilde{W}$$

and

$$\Delta U = \epsilon^{1/3}\tilde{U}. \tag{98}$$

Then

$$G = \epsilon(\tilde{V}^2 + \tilde{W}^2 + 8\tilde{U}^3) + 12\epsilon^{4/3}\tilde{U}^4 + 6\epsilon^{5/3}\tilde{U}^5 + \epsilon^2\tilde{U}^6. \tag{99}$$

To leading order, the constraints are solved by the relation

$$\tilde{U}^3 = -\frac{1}{8}(\tilde{V}^2 + \tilde{W}^2). \tag{100}$$

Clearly $\tilde{U} \leq 0$ and

$$F = 1 + 2\Delta U + (\Delta U)^2 = 1 + 2\epsilon^{1/3}\tilde{U} + \epsilon^{2/3}\tilde{U}^2. \tag{101}$$

Neglecting the trivial constant term 1 to leading order F is described by

$$2\epsilon^{1/3}\tilde{U} = \frac{-2\epsilon^{1/3}}{8^{1/3}}(\tilde{U}^2 + \tilde{W}^2)^{1/3}. \tag{102}$$

But $\epsilon > 0$, and the increment in F is negative definite. It is 0 only for $\tilde{U} = \tilde{W} = 0$. Thus, we know we must be at a local maximum.

REFERENCES

1. R. Bellman, Perturbation Techniques in Mathematics, Physics, and Engineering, Holt, Rinehart, and Winston, 1964.

2. T. L. Boullion and P. L. Odell, Generalized Inverse Matrices, Wiley-Interscience, 1971.

3. M. Van Dyke, Perturbation Methods in Fluid Mechanics, Academic Press, 1964.

4. R. C. Gunning and H. Bossi, Analytic Functions of Several Complex Variables, Prentice-Hall, 1965.

5. O. L. Magasarian, Nonlinear Programming, McGraw-Hill, 1969.

6. Bliss, Algebraic Functions, Dover Publishing Company, 1966.

AN INTRODUCTION TO
BOUNDED RATE SYSTEMS

C.Bruni - G.Koch
Istituto di Automatica
Università di Roma
Via Eudossiana, 18 - 00184 Roma

A. Germani
CSSCCA - CNR

SUMMARY

In this work a new class of nonlinear systems is introduced, for which the denomination "bounded rate systems" is proposed. This class appears to be quite relevant for its capability of modeling important physical phenomena in different field such as biology, ecology, engineering.

Bounded rate systems situate between bilinear and linear-in-control systems, so that a bounded rate system theory may be usefully investigated and developed exploiting already available results.

1. A NEW CLASS OF NONLINEAR SYSTEMS

The need of introducing nonlinear systems in modeling the physical world is presently an unquestioned fact especially in a number of important nontechnical fields like biology, ecology, socio-economics.

It is equally well know that the main obstacle in this direction is the difficulties presented by the analytical study of a general nonlinear system.

This situation motivates the actual trend, to spot out specific classes of nonlinear systems which possibly couple the advantages of not too difficult analytical study to the ability of modeling relevant classes of phenomena.

A noteworth step in this direction was the introduction of the class of bilinear systems, which in fact appears quite valid from an applicative point of view and in the mean time allowed to achieve a number of important theoretical results [1,2].

Another significant example of this trend was the study of a more general class of nonlinear systems, that is the systems linear in control for which relevant results are available expecially on the controllability and optimal control [1,3,4,5,6].

This paper is intended to give a futher contribution along the same line by introducing a new class of nonlinear systems. Denoting as usual by x the n-dimension state vector (1) and by u the p-dimension input vector, this class is defined as follows:

$$\dot{x}(t) = \phi(x) + N x u + Bu \qquad (1.1)$$

where the operator $\phi:\Omega \to R^n$, Ω open subset of R^n, is assumed to be locally Lipschitzian (2) with at most a linear growth with x, i.e.

$$\|\phi(x)\| \leq c_1 \|x\| + c_2 \ , \qquad x \in \Omega \qquad (1.2)$$

and $N: R^n \times R^p \to R^n$ and $B: R^p \to R^n$ are respectively bilinear and linear operators.

It is important to note that as it will be proved in Th.3.1, the assumptions on ϕ (local Lipschitziness and at most linear growth in Ω)

---

(1) The state vector is taken finite dimensional only for sake of simplicity. The extension to the infinite dimensional case is immediate.

(2) i.e. Lipschitzian within any closed bounded set in Ω.

guarantee existence and uniqueness of solutions of eq. (1.1) in Ω. They furthermore appear to be large enough to include all cases of practical interest.

This class situates between the ones of linear in control and bilinear systems. Indeed (1.1) is clearly linear in control with the restrictions that the free response $(u = 0)$ has a bounded growth rate and the u dependent term is at most linear in x. On the other hand, it is shown in the Appendix that the assumptions on ϕ are equivalent to assume:

$$\phi(x) = F(x)x + f(x) \tag{1.3}$$

where the operators $F:\Omega \to R^{n\times n}$ and $f:\Omega \to R^n$ are locally Lipschitzian with uniformly bounded range. This enlightens the structure of ϕ and at the same time shows that (1.1) may be also considered as a bilinear system with an instantaneous, locally Lipschitzian and uniformly bounded state feedback h (Fig.1), as soon as one defines

$$F(x) = A + N' h(x) \tag{1.4}$$

$$f(x) = B\, h(x) \tag{1.5}$$

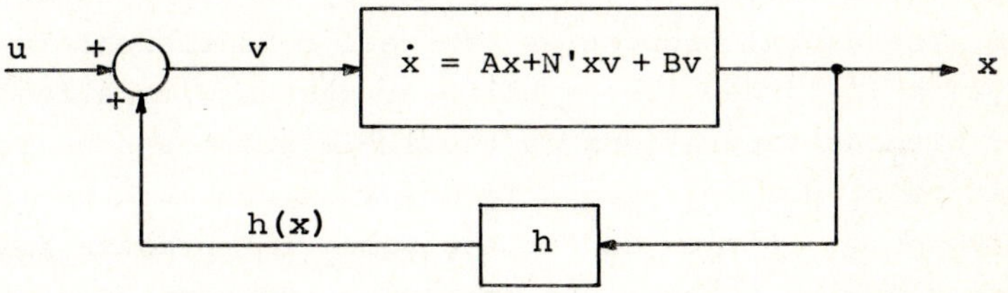

Fig.1

Due to the result (1.3), the considered class of systems may also be represented by the general form:

$$\dot{x}(t) = F(x)x + f(x) + Nxu + Bu \tag{1.6}$$

It is interesting to point out that the u-dependent part of the dynamics in (1.6) has essentially the same behaviour as in linear in control and bilinear systems, while the free response dynamic is indeed characteristic of the new class of systems. In fact it is neither linear as in bilinear systems nor a general nonlinear function as in linear in control systems but it is allowed to have at most a linear growth with x. For this reason in the following we shall denote the system (1.6) *bounded rate system*.

The motivation for the study of bounded rate systems is twofold. First of all it is important to underline the relevance of such a class of systems from an *applicative point of view*. Indeed, as is also shown by the few examples of sect.2, the bounded rate model (1.6) play a fundamental role in a number of important fields like in biology, ecology, engineering, socioeconomics.

Second, because of the structure of bounded rate systems at middle way between linear-in-control and bilinear systems, it is expected that important *theoretical results* may be achieved for such a class of systems both exploiting known results for linear-in-control systems and suitably extending bilinear systems properties.

As far as the case of bilinear systems in a feedback loop is concerned, it is important to note that if the feedback operator h in Fig. 1 is taken to be linear a *quadratic system* is arrived at.

Quadratic systems may be considered bounded rate within any bounded set in the state space, so that the local properties of the two classes must agree. In particular, if one proves that all the possible solutions for a given quadratic system must be confined within a bounded set in the state space,that system may be regarded as a bounded rate system. The global properties may well be different in that a quadratic system in large (i.e. a bilinear system with a feedback which is linear without saturations on the whole state space) is not obviously bounded rate. However such a feedback is not much realistic from an applicative point of view and at the same time creates some difficulty in proving global existence and uniqueness properties of the solution.

2. SOME EXAMPLES OF BOUNDED RATE SYSTEMS

In this section we intend to show how several natural and technical processes may be conveniently described by means of bounded rate models. We shall consider some important examples in the fields of biochemistry, population dynamics, and engineering.

2.1. Immune response

The immune response is the sequence of those phenomena which in a mammal are triggered by the injection of a foreign substance (antigen) and lead to the production of specific proteins (antibodies) able to bind the antigen and to neutralize it.

As shown in [8,9], under suitable assumptions, the immune response may be described by the following set of equations:

$$\frac{\partial C(K,t)}{\partial t} = \alpha_c \frac{1-KH}{1+KH} p_s(KH)C(K,t) - \frac{1}{\tau_c} C(K,t) + \beta \bar{p}_c(K) \tag{2.1}$$

$$\frac{\partial C_p(K,t)}{\partial t} = 2\alpha_c \frac{KH}{1+KH} p_s(KH)C(K,t) - \frac{1}{\tau_p} C_p(K,t) \tag{2.2}$$

$$\frac{\partial S(K,t)}{\partial t} = \alpha_s C_p(K,t) + \alpha_s' C(K,t) + 2\alpha_c \int_{-\infty}^{t} i(t-\theta) \frac{p_s(KH)}{1+KH} C(K,\theta)\,d\theta +$$

$$- Kc(K)S(K,t)\;H(t) + c(K)B(K,t) - \frac{1}{\tau_s} S(K,t) \tag{2.3}$$

$$\frac{\partial B(K,t)}{\partial t} = Kc(K)S(K,t)H(t) - c(K)B(K,t) - \frac{1}{\tau_B} B(K,t) \tag{2.4}$$

$$\frac{dH(t)}{dt} = -H(t) \int_{K_1}^{K_2} Kc(K)S(K,t)\,dK + \int_{K_1}^{K_2} c(K)B(K,t)\,dK - \frac{1}{\tau_H} H(t) +$$

$$+ \dot{H}_i(t) \tag{2.5}$$

where:

$C(K,t)$, $C_p(K,t)$, $S(K,t)$, $B(K,t)$ are respectively the concentration densities of immunocompetent and memory cells, plasma cells, antibody sites, immune complex, at time t, for a given affinity constant K (K ranges from K_1 to K_2);

$H(t)$, $H_i(t)$ are respectively the antigen concentration at time t and the amount of antigen introduced per unit volume of circulating fluids up to time t;

$p_s(KH)$ is the probability that a cell with affinity K be stimulated;

$\bar{p}_c(K)$ is the original distribution of immunocompetent cells with respect to K;

α_c is the proliferation rate constant of stimulated cells;

β is the production rate of immunocompetent cells from stem cells;

α_s, α'_s are respectively the rate constant of antibody production by plasma cells and the basal rate constant of antibody production by memory cells;

$i(t)$ is the additional antibody production intensity by a memory cell generated t instants ahead;

$c(K)$ is the dissociation reaction rate constant of the immune complex;

$\tau_c, \tau_p, \tau_s, \tau_B, \tau_H$ are the time constants for death or removal of the various species.

The set of equations (2.1)-(2.5) defines a nonlinear dynamical model for the immune response which is distributed with respect to K. In it $\bar{p}_c(K)$ and $\dot{H}_i(t)$ are two independent variables, which may be considered as inputs:

$$u(t) = \begin{bmatrix} \bar{p}_c(\cdot) \\ \dot{H}_i(t) \end{bmatrix} \tag{2.6}$$

Let us define the state vector $x(t)$:

$$x(t) = [C(\cdot,t)\, C_p(\cdot,t)\, S(\cdot,t)\, B(\cdot,t)\, \tilde{x}^T(\cdot,t)\, H(t)]^T \tag{2.7}$$

where $\tilde{x}(\cdot,t)$ is the state vector for the linear time invariant system with impulse response $i(t)$. From (2.1)-(2.5) we see that the u-dependent part is linear in u and independent of x, while the free response part includes products of state variables by suitable functions of x. Thus the set of equations (2.1)-(2.7) may be given the form:

$$\dot{x}(t) = F(x)\, x + Bu \tag{2.8}$$

where B is a linear time invariant operator. As far as the operator F is concerned, it may be proved [9] that all above mentioned functions of x are locally Lipschitzian and with uniformly bounded range, as soon as we restrict them to a suitably defined open subset Ω of the state space X. Furthermore, it was also proved that, choosing as initial state $x(t_o)$ any point in a given $S \subset \Omega$, possible solutions of (2.8) must evolve within S itself.

Therefore, as soon as $x(t_o)$ is chosen in S (which is the only possible choice in agreement with the physical meaning of x components), eq. (2.8) may be regarded as a particular case of (1.6).

2.2. Enzyme Kinetics

Let us denote by S,E,P,C respectively the substrate, the enzyme, the final reaction product and the substrate-enzyme complex. Once we keep the total amount of enzyme e_o and substrate s_o positive and constant, the general stoichiometric equation for the enzyme kinetics is [10]:

$$S + E \underset{K_{-1}}{\overset{K_1}{\rightleftarrows}} C \overset{K_2}{\rightarrow} P + E \qquad (2.9)$$

where K_1, K_{-1}, K_2 are positive reaction rate constants. Denoting the concentrations by low case letters eq. (2.9) leads to the dynamical equation of Michaelis-Menten:

$$\begin{aligned} \dot{s} &= K_{-1}\, c - K_1\, s\, e \\ \dot{p} &= K_2\, c \\ e + c &= e_o \\ s + c + p &= s_o \end{aligned} \qquad (2.10)$$

Defined:

$$x = \begin{bmatrix} x_1 \\ x_2 \end{bmatrix} = \begin{bmatrix} s \\ p \end{bmatrix} \qquad (2.11)$$

$$u = \begin{bmatrix} u_1 \\ u_2 \end{bmatrix} = \begin{bmatrix} s_o \\ e_o \end{bmatrix} \qquad (2.12)$$

eq. (2.10) take the form:

$$\dot{x} = F(x)x + N\,x\,u + Bu \tag{2.13}$$

where:

$$F(x) = \begin{bmatrix} -K_{-1} - K_1(x_1+x_2) & -K_{-1} \\ -K_2 & -K_2 \end{bmatrix} \tag{2.14}$$

$$N_1 = \begin{bmatrix} K_1 & 0 \\ 0 & 0 \end{bmatrix} \quad ; \quad N_2 = \begin{bmatrix} -K_1 & 0 \\ 0 & 0 \end{bmatrix} \tag{2.15}$$

$$B = \begin{bmatrix} K_{-1} & 0 \\ K_2 & 0 \end{bmatrix} \tag{2.16}$$

The system (2.13) is a quadratic one and may be looked at as bounded rate as soon as we show that for any initial condition $x(t_o)$ in a bounded subset S of R^2, any possible solution stays in S. Indeed, defined:

$$S = \{x \in R^2 : x_1 \in [0, s_o],\ x_2 \in [0, s_o]\} \tag{2.17}$$

for $x(t_o) \in S$ let $\bar{t} \geq t_o$ the first time in which any of the two components x_1, x_2 changes its sign. Due to the continuity of solutions, we have the following cases:

a) $$x_1(\bar{t}) = 0 \quad , \quad \dot{x}_1(\bar{t}) < 0 \tag{2.18}$$

Recalling (2.13), this means:

$$K_{-1}(u_1(\bar{t}) - x_2(\bar{t})) < 0 \tag{2.19}$$

and therefore $x_2(\bar{t}) > s_o$. This implies the existence of a $0 < t' < \bar{t}$ such that:

$$x_2(t') = s_o \quad , \quad \dot{x}_2(t') \geq 0 \tag{2.20}$$

From (2.13) we then deduce:

$$-K_1 x_1(t') \geq 0 \qquad (2.21)$$

which is in contrast with (2.18).

By a similar proof, we can also reject the possibility that x_1 changes its sign in $\bar{t}$ through a horizontal tangent flex point.

b) $$x_2(\bar{t}) = 0 \quad , \quad \dot{x}_2(\bar{t}) \leq 0 \qquad (2.22)$$

Recalling (2.13) this means:

$$x_1(\bar{t}) \geq u_1(\bar{t}) = s_o \qquad (2.23)$$

This implies the existence of a $0 < t' < \bar{t}$ such that:

$$x_1(t') = s_o \quad , \quad \dot{x}_1(t') \geq 0 \qquad (2.24)$$

From (2.13) we then deduce:

$$-x_2(t')(K_1 + K_{-1}) - K_1 s_o e_o \geq 0 \qquad (2.25)$$

which is in contrast with (2.22).

c) $$x_1(\bar{t}) = x_2(\bar{t}) = 0, \ \dot{x}_1(\bar{t}) \leq 0 \ , \ \dot{x}_2(\bar{t}) \leq 0 \qquad (2.26)$$

Recalling (2.13), this means:

$$K_2 u_1(\bar{t}) = K_2 s_o \leq 0 \qquad (2.27)$$

which is in contrast with the positiveness of s_o.

By a similar analysis, it can also be rejected the possibility that $x_1(t) > s_o$, or $x_2(t) > s_o$, for $t \geq t_o$.

2.3. Bacterial growth

The continuous bacterial colture problem is a typical problem in biochemical engineering [11]. A dynamical model for that process which was proposed in [12] in connection with an optimal control problem is the following:

$$\begin{aligned} \dot{b} &= \left(\frac{s}{K_2 s^2 + s + K_1} - q \right) b \\ \dot{s} &= (1 - s) q - \frac{sb}{K_2 s^2 + s + K_1} \end{aligned} \qquad (2.28)$$

where b, s are respectively the bacterial and nutritive substrate concentration in the growth vessel, q is the input and output flow rate and K_1, K_2 are suitable positive constants which determine the bacterial growth rate as a function of s. Defining:

$$x = \begin{bmatrix} x_1 \\ x_2 \end{bmatrix} = \begin{bmatrix} b \\ s \end{bmatrix} \tag{2.29}$$

$$u = q \tag{2.30}$$

eq. (2.28) take the form:

$$\dot{x} = F(x)x + Nxu + Bu \tag{2.31}$$

where:

$$F(x) = \begin{bmatrix} h(x_2) & 0 \\ -h(x_2) & 0 \end{bmatrix} \qquad h(x_2) = \frac{x_2}{K_2 x_2^2 + x_2 + K_1} \tag{2.32}$$

$$N = \begin{bmatrix} -1 & 0 \\ 0 & -1 \end{bmatrix}, \quad B = \begin{bmatrix} 0 \\ 1 \end{bmatrix} \tag{2.33}$$

As in previous cases, defined

$$S = \{x \in R^2 \; : \; x_1 \geq 0 \; , \; x_2 \geq 0\} \tag{2.34}$$

we first show that for $x(t_o) \in S$ and for non negative u, any possible solution of (2.31) stay in S. Indeed by integrating the first equation in (2.31) we obtain:

$$x_1(t) = x_1(t_o)\, e^{-\int_{t_o}^{t} [u(\tau) - h(x_2(\tau))]\, d\tau} \tag{2.35}$$

which for $x_1(t_o) \geq 0$ is always nonnegative. As far as x_2 is concerned, assume there exists a $\bar{t} \geq t_o$ such that:

$$x_2(\bar{t}) = 0 \quad , \quad \dot{x}_2(\bar{t}) < 0 \tag{2.36}$$

From (2.31) it then follows $\dot{x}_2(\bar{t}) = u(\bar{t}) < 0$ which is in contrast with the assumed nonnegativeness of u. With a similar reasoning, we can also reject the possibility of x_2 changing its sign through a horizontal tangent flex point.

Finally, h and therefore F is easily seen to be uniformly Lipschitzian and with uniformly bounded range on an open set $\Omega \subset S$. In fact, the derivative of $h(x_2)$ vanishes for $x_2 = \pm\sqrt{K_1/K_2} = \pm\hat{x}_2$.

In the case $4K_1K_2 > 1$, h is uniformly Lipschitzian on $\Omega = R^2$

$$\sup_{x\in\Omega}|h(x_2)| = |h(-\hat{x}_2)| < \infty \tag{2.37}$$

On the contrary, in the case $4K_1K_2 \leq 1$, we may take

$$\Omega = \{x \in R^2 \ : \ x_2 > \frac{\bar{x}_2}{2}\} \tag{2.38}$$

where $\bar{x}_2$ is the greatest singular point of $h(x_2)$; again $\Omega \subset S$, h is uniformly Lipschitzian on Ω and

$$\sup_{x\in\Omega}|h(x_2)| = \max(|h(\frac{\bar{x}_2}{2})|, \ |h(+\hat{x}_2)|) < \infty \tag{2.39}$$

In conclusion, eq. (2.31) may be looked at as a bounded rate system, for any $x(t_o) \in S$ and nonnegative u.

2.4. *Interacting Populations*

A well-known and fairly general model to describe the dynamics of n interacting species is given by the Volterra-Lotka equations [13,14]

$$\dot{x}_i = K_i x_i + \sum_{1}^{n}{}_j \alpha_{ij} x_j x_i , \quad i = 1,2,\ldots,n \tag{2.40}$$

where the x_i's are the population numbers of the i-th species, the K_i's are real numbers, not all positive, representing the intrinsic increasing rates and the α_{ij} are antisimmetric real numbers ($\alpha_{ij} = -\alpha_{ji}$; $\alpha_{ii} = 0$) representing the "predation efficiency" of the i-th species on the j-th one.

The model (2.40) is clearly a quadratic one and may be rewritten as:

$$\dot{x} = F(x)x \tag{2.41}$$

where:

$$F(x) = \begin{bmatrix} K_1 + \alpha_1^T x & 0 & \cdots & 0 \\ 0 & K_2 + \alpha_2^T x & \cdots & 0 \\ \cdots & \cdots & \cdots & \cdots \\ 0 & 0 & \cdots & K_n + \alpha_n^T x \end{bmatrix} \tag{2.42}$$

$$\alpha_i^T = [\alpha_{i1} \; \alpha_{i2} \; \cdots \; \alpha_{in}] \tag{2.43}$$

The model (2.41) is therefore bounded rate as soon as we prove that all possible solutions with initial state within a given bounded set S, stay in S itself.

Indeed, it was proved that if $\det[\alpha_{ij}] \neq 0$ (which may happen only for n even) then, the system (2.40) admits a unique nontrivial equilibrium point $\bar{x}$, which is in $\bar{R}_n^+ = \{x : x_i > 0, \quad i = 1,2,\ldots,n\}$ and is stable.

Furthermore, a scalar function $\phi(x)$ may be defined on $\bar{R}_n^+$:

$$\phi(x) = \sum_{1}^{n}{}_i \bar{x}_i \left(\frac{x_i}{\bar{x}_i} - \ln \frac{x_i}{\bar{x}_i} \right) \tag{2.44}$$

which on each trajectory takes a constant value, greater than or equal to $\sum_1^n{}_i \bar{x}_i$. As a consequence, any trajectory starting in $\bar{R}_n^+$ stays in $\bar{R}_n^+$.

For each constant $c \geq \sum_1^n{}_i \bar{x}_i$, we now define the set:

$$S_c = \{x \in \bar{R}_n^+ : \phi(x) \leq c\} \tag{2.45}$$

Clearly, being ϕ constant along the trajectory, S_c contains all the trajectories starting from points belonging to S_c itself. Finally, S_c is bounded; indeed, if $x \in S_c$, it easily follows that:

$$\frac{x_i}{\bar{x}_i} - \ln \frac{x_i}{\bar{x}_i} \leq \frac{c}{\bar{x}_i} \quad , \quad i = 1,2,\ldots,n \tag{2.46}$$

which implies a finite upper bound for each x_i.

2.5. Chemical reactor

The system under consideration is a continuous-flow stirred tank reactor, in which a single irreversible chemical reaction takes place. The dynamics of this system are easily described by material and heat balance equations [15]:

$$\dot{c} = \frac{F}{V}(c_o - c) - K_o \exp\left(-\frac{E}{RT}\right) c$$

$$\dot{T} = \frac{F}{V}(T_o - T) - \frac{U}{V\rho c_p}(T - T_k) + \frac{(-\Delta H)}{\rho c_p} K_o \exp\left(-\frac{E}{RT}\right) c \qquad (2.47)$$

where: c,T are respectively reactant concentration and reactor internal absolute temperature; c_o, T_o are the reactant concentration and temperature in the input flow; T_k is the coolant temperature. We also used the following notation:

F, input and output flow rate, assumed constant
V, reactor volume
$K = K_o \exp\left(-\frac{E}{RT}\right)$, specific reaction velocity constant at temperature T
U, product of area of the coolant surface in the coil and its heat transfer coefficient
$(-\Delta H) > 0$, enthalpy variation in the reaction, assumed exothermic
ρ, c_p, density and specific heat of the input flow.

We now define the constants:

$$K_1 = \frac{F}{V} \; ; \; K_2 = \frac{U}{V\rho c_p} \; ; \; K_3 = \frac{(-\Delta H)}{\rho c_p} \; ; \; K_4 = \frac{E}{R} \qquad (2.48)$$

and the state and input variables;

$$x = \begin{bmatrix} x_1 \\ x_2 \end{bmatrix} = \begin{bmatrix} c \\ T \end{bmatrix} \qquad (2.49)$$

$$u = \begin{bmatrix} u_1 \\ u_2 \\ u_3 \end{bmatrix} = \begin{bmatrix} c_o \\ T_o \\ T_k \end{bmatrix} \qquad (2.50)$$

Then eq. (2.47) take the form:

$$\dot{x} = F(x)x + Bu \qquad (2.51)$$

where:

$$F(x) = \begin{bmatrix} -\left(K_1 + K_o e^{-\frac{K_4}{x_2}}\right) & 0 \\ K_3 K_o e^{-\frac{K_4}{x_2}} & -(K_1 + K_2) \end{bmatrix} \qquad (2.52)$$

$$B = \begin{bmatrix} K_1 & 0 & 0 \\ & & \\ 0 & K_1 & K_2 \end{bmatrix} \tag{2.53}$$

The function F is locally Lipschitzian with uniformly bounded range in the open set $\Omega = \{x \in R^2 : x_2 > 0\}$. To prove that (2.54) is actually a bounded rate system, we now show that, defined:

$$S = \{x \in R^2 ;\ x_1 \geq 0,\ x_2 > 0\} \tag{2.55}$$

$$U = \{u : u_1(t) \geq 0,\ u_2(t) > 0,\ u_3(t) > 0\} \tag{2.56}$$

for any choice of $x(t_o)$ in S and of u in U (which are the only physically meaningful choices), any possible solution of (2.51) stays in S itself. Indeed, the first equation in (2.51) may be integrated out:

$$x_1(t) = x_1(t_o)\, e^{-\int_{t_o}^{t} \left(K_1 + K_o e^{-\frac{K_4}{x_2(\tau)}}\right) d\tau} +$$

$$+ \int_{t_o}^{t} e^{-\int_{\theta}^{t} \left(K_1 + K_o e^{-\frac{K_4}{x_2(\tau)}}\right) d\tau} K_1 u_1(\theta)\, d\theta \tag{2.57}$$

and therefore $x_1(t_o) \geq 0$ and $u_1(\theta) \geq 0$ imply $x_1(t) \geq 0$.

As far as x_2 is concerned, if we assume that there exists a time $\bar{t} \geq t_o$ such that:

$$x_2(\bar{t}) = 0 \quad , \quad \dot{x}_2(\bar{t}) \leq 0 \tag{2.58}$$

from the second equation in (2.51) we get:

$$K_1 u_2(\bar{t}) + K_2 u_3(\bar{t}) \leq 0 \tag{2.59}$$

which is in contrast with the assumed positiveness of $u_2(t)$, $u_3(t)$.

2.6. Nuclear reactor Kinetics

Other important examples of quadratic systems which are also bounded rate systems are the usually adopted models for the nuclear reactor kinetics.

As is extensively reported [16,17] a model for the free response of a point reactor with one group of delayed neutrons and one feedback region with Newton cooling is:

$$\dot{\nu} = - \frac{\alpha(T-T_o) + \beta}{l} \nu + \lambda c$$

$$\dot{c} = \frac{\beta}{l} \nu - \lambda c \tag{2.60}$$

$$\dot{T} = K(\nu - \nu_o) - \gamma(T - T_o)$$

where ν is the neutron density, c is the precursor density, T is an average reactor temperature, with equilibrium values respectively ν_o, $c_o = \frac{\beta}{l \cdot \lambda} \nu_o$, T_o. Furthermore:

α = temperature coefficient of reactivity

β = delayed neutron fraction

λ = precursor decay constant

l = neutron generation time

$1/K$ = reactor heat capacity

$1/\gamma$ = mean time for heat transfer to the coolant

Defining:

$$x = \begin{bmatrix} x_1 \\ x_2 \\ x_3 \end{bmatrix} = \begin{bmatrix} \nu - \nu \\ c - c_o \\ T - T_o \end{bmatrix} \tag{2.61}$$

equations (2.60) take the form:

$$\dot{x} = F(x)x \tag{2.62}$$

where:

$$F(x) = \begin{bmatrix} -\frac{\beta}{l} & \lambda & -\frac{\alpha}{l}(\nu_o + x_1) \\ \frac{\beta}{l} & -\lambda & 0 \\ K & 0 & -\gamma \end{bmatrix} \tag{2.63}$$

The system (2.62) is clearly a quadratic one so that all we have to prove is the uniform boundedness of its possible trajectories. For that purpose let us define the set S:

$$S = \{x \in R^3 : x_1 > -\nu_o,\ x_2 > -c_o\} \tag{2.64}$$

and the scalar function V on S:

$$V(x) = \nu_o \left[\frac{x_1}{\nu_o} - \ln\left(1 + \frac{x_1}{\nu_o}\right)\right] + c_o\left[\frac{x_2}{c_o} - \ln\left(1+\frac{x_2}{c_o}\right)\right] + \frac{\alpha}{2K1}x_3^2 \tag{2.65}$$

V is a Lyapunov function on S since V is positive definite on S and $\dot{V}$ is negative semidefinite along the trajectories of (2.62) which belong to S(*). Let us now define the subset $S_L \subset S$:

$$S_L = \{x \in R^3 : V(x) \leq L\} \tag{2.66}$$

Recalling (2.65) we see that each of the three terms at the RHS is non negative in S, and therefore each of them is bounded above by L in S_L. This means that in S_L each of the three components of x belongs to a closed bounded interval.

Thus for any L, S_L is closed and bounded. Moreover, for any initial state $x(t_o) \in S_L$, all the possible trajectories stay in S_L. Indeed, would this not be the case, there should exist a $\bar{t}$ such that $x(t) \in S$, $t \leq \bar{t}$, and $x(\bar{t}) \in S - S_L$. But this is impossible since $x(t) \in S$ for $t \leq \bar{t}$ implies that $V(x(\bar{t}))$ cannot be increased with respect to the initial value; so that $V(x(\bar{t})) \leq L$ and therefore $x(\bar{t}) \in S_L$.

As a conclusion, the model (2.62) is bounded rate in S_L, for any $L < \infty$.

2.7. Other examples

Other important examples of bounded rate systems may be found in virtually all the applicative fields when a multiplicative control is indeed a bounded function of the state itself. This for instance may happen in macro-economics. In [18,19] a model is proposed for the growth of a national economy, which in [10] is interpreted as a bilinear system: the state is the vector of total national output and the multiplicative control is the matrix of coefficients of material inputs. As a matter of fact, this control may be taken as a bounded function of the state, thus generating a bounded rate system.

Another interesting case in ecology is the insect-pest control model [20] where the state vector is the insect population at different ages. The multiplicative control is build up by two terms. The first

---

(*) Despite of the semidefiniteness of $\dot{V}$, it has been shown [17] that the origin, which is the only equilibrium point of (2.62) in S, is actually globally asymptotically stable.

one is an intrinsic control which describes the effects of overcrowding and is a bounded function of the state; the second one is the external control implemented by releasing a sterilized male population.

A final example in engineering is the heat exchanger: the coolant flow rate which acts as a multiplicative control on the heat exchange process [2] may well be thought as a bounded function of the temperature (state) by means of an external feedback loop.

3. EXISTENCE AND UNIQUENESS OF SOLUTIONS

A first important theoretical result that may be established for a bounded rate system defined by (1.6) :

$$\dot{x} = F(x)x + f(x) + Nxu + Bu \tag{1.6}$$

is related to the problem of existence and uniqueness of its solution. We have the following:

THEOREM 3.1. *Let us consider a bounded rate system (1.6) with the assumption: there exists a subset* $S \subset \Omega$ *and class* U *of continuous input functions such that for any initial condition* $x_o = x(t_o) \in S$ *and any* $u \in U$ *any possible solution of (1.6) takes value* $x(t) \in S$, $\forall t \geq t_o$.
Then for any $x_o \in S$, $u \in U$, *eq. (1.6) admite a unique solution for all* $t \geq t_o$.

PROOF. First of all, note that, being F, f locally Lipschitzian and uniformly bounded in $\Omega \supset S$, and being u continuous for any $(t_o, x_o) \in R^1 xS$, there exists a $R_o = \{(t,x): |t-t_o| \leq \alpha, \|x-x_o\| \leq \beta\}$ such that:

a) $R_o \subset \Omega$

b) the RHS of (1.6), in R_o, is uniformly bounded, continuous in t for each fixed x and uniformly Lipschitzian in x.

Therefore (1.6) admits a local unique solution through (t_o, x_o).

Furthermore the RHS of (1.6) in $Rx\Omega$ is continuous and recalling (1.2) is such that:

$$\begin{aligned} \|F(x(t))x(t) + f(x(t)) + Nx(t)u(t) + Bu(t)\| &\leq C_1 \|x(t)\| + C_2 + \\ &+ \|N\| \; \|u(t)\| \; \|x(t)\| + \|B\| \; \|u(t)\| = \\ &= K_1(t)\|x(t)\| + K_2(t) \end{aligned} \tag{3.1}$$

where $K_1(t)$, $K_2(t)$ are suitable scalar positive continuous functions of t.

The scalar equation:

$$\dot{z} = K_1(t)z + K_2(t) \tag{3.2}$$

admits a unique solution for any initial value $z(t_o) \in R$ and for all $t \geq t_o$. Consequently, due to Thm. 5.6.1 in [7] any local solution of (1.6) may be uniquely extended for all $t \geq t_o$.

REMARK 1. All the previously mentioned examples of bounded rate systems satisfy the assumptions of Thm. 3.1. and therefore admit a unique solution in large.

REMARK 2. A similar theorem, for systems evolving on group, is proved in [21]. The set of assumptions seems to be similar to that one of Thm. 3.1. and somehow more restrictive than the latter, in those cases in which both are applicable.

4. CONCLUSIONS

In the previous section we enlightened the relevance of the class of bounded rate systems. We showed how a number of important physical processes may be included in that class.

From the theoretical point of view a first general result for bounded rate systems was given in Sect. 3. We must however point out that the theory for such a class of systems is still to be developed and in our mind this should be an important task to be pursued.

In particular results on stability, controllability and optimal control would be useful. As far as stability is concerned, a first contribution was given in [9] where asymptotic stability for the immune response model was proved exploiting the general structure of a bounded rate system as a bilinear system in a uniformly bounded feedback loop.

This is a good example of how to achieve theoretical results for bounded rate systems by suitably exploiting already available results for bilinear systems.

As far as controllability is concerned, a number of important results are already available for the class of linear in control systems [3,4,5,6,21,24] which hopefully will be a useful starting point to build a controllability theory for bounded rate systems.

We also mention some papers [25,26] in which it is shown how to derive controllability properties by looking only at the u-dependent part of the state equation. Under this aspect bounded rate systems behave as bilinear systems, for which some results on controllability theory are already available [1].

APPENDIX

In order to prove eq. (1.3) we may state the following

THEOREM. *Given an open subset* Ω *of* R^n *a necessary and sufficient condition for a function* $\phi : \Omega \to R^n$ *to be locally Lipschitzian with the growth property:*

$$\|\phi(x)\| \leq C_1 \|x\| + C_2 \quad , \quad \forall x \in \Omega \tag{A.1}$$

where C_1, C_2 *are nonnegative constants, is that there exist two functions* $F : \Omega \to R^{n\times n}$, $f : \Omega \to R^n$ *which are locally Lipschitzian and with uniformly bounded range such that:*

$$\phi(x) = F(x)x + f(x) \quad , \quad \forall x \in \Omega \tag{A.2}$$

PROOF. *Sufficiency*. If (A.2) holds, then for x_1, x_2 in any closed bounded subset $M \subset \Omega$ we have:

$$\|\phi(x_1) - \phi(x_2)\| = \|F(x_1)x_1 + f(x_1) - F(x_2)x_2 - f(x_2)\|$$

$$\leq \|F(x_1)\| \, \|x_1 - x_2\| + \|F(x_1) - F(x_2)\| \, \|x_2\| + \|f(x_1) - f(x_2)\|$$

$$\leq \max_{x_1 \in M} \|F(x_1)\| \, \|x_1 - x_2\| + \max_{x_2 \in M} \|x_2\| \, L_F \, \|x_1 - x_2\| + L_f \|x_1 - x_2\|$$

$$\leq L_\phi \|x_1 - x_2\| \tag{A.3}$$

where L_F, L_f are the Lipschitz constants of F, f in M. Consequently ϕ is locally Lipschitzian in Ω. As far as (A.1) is concerned, we have:

$$\|\phi(x)\| = \|F(x)x + f(x)\| \leq \|F(x)\| \, \|x\| + \|f(x)\| \leq C_1 \|x\| + C_2$$

$$\forall x \in \Omega \tag{A.4}$$

where C_1, C_2 are the finite upper bounds for $\|F(x)\|$, $\|f(x)\|$ in Ω.

Necessity. Denoting by ϕ_i the i-th component of ϕ, it obviously is locally Lipschitzian and inequality (A.1) implies:

$$|\phi_i(x)| \leq C_1 \|x\| + C_2 \quad , \quad i = 1,2,\ldots,n \quad , \quad \forall x \in \Omega \tag{A.5}$$

so that necessity of (A.2) is proved as soon as we prove for each i the existence of two functions $F_i : \Omega \to R^n$, $f_i : \Omega \to R^1$, locally Lipschitzian and with uniformly bounded range, such that

$$\phi_i(x) = F_i^T(x)x + f_i(x) \qquad \forall x \in \Omega \tag{A.6}$$

Eq. (A.6) will now be proved by means of a constructive procedure.

For a fixed $\rho > 0$ we construct the function $f_i : \Omega \to R^1$ in the following way:

$$f_i(x) = \begin{cases} \phi_i(x) & , \quad x \in \Omega_i' = \{x \in \Omega : -C_1\rho - C_2 \le \phi_i(x) \le C_1\rho + C_2\} \\ C_1\rho + C_2 & , \quad x \in \Omega_i'' = \{x \in \Omega : \phi_i(x) > C_1\rho + C_2\} \\ -C_1\rho - C_2 & , \quad x \in \Omega_i''' = \{x \in \Omega : \phi_i(x) \le -C_1\rho - C_2\} \end{cases} \tag{A.7}$$

By construction f_i has a uniformly bounded range. Moreover, it is also locally Lipschitzian in Ω since for any choice of $x_1, x_2 \in \Omega$:

$$|f_i(x_1) - f_i(x_2)| \le |\phi_i(x_1) - \phi_i(x_2)| \tag{A.8}$$

We now observe that the distance of $\Omega_i'' \cup \Omega_i'''$ from the origin cannot be less than ρ. In fact, recalling (A.5), (A.7), if $x \in \Omega_i'' \cup \Omega_i'''$, than

$$C_1\|x\| + C_2 \ge |\phi_i(x)| > C_1\rho + C_2 \to \|x\| > \rho \tag{A.9}$$

Therefore the function $F_i : \Omega \to R^n$:

$$F_i(x) = \begin{cases} 0 & , \quad x \in \Omega_i' \\ \dfrac{\phi_i(x) - f_i(x)}{x^T x}\, x & , \quad x \in \Omega_i'' \cup \Omega_i''' \end{cases} \tag{A.10}$$

is well defined.

The functions f_i, F_i defined in (A.7), (A.10) already satisfy (A.6) in all Ω. Due to the growth property (A.5) F_i turns out to have a uniformly bounded range; indeed in Ω_i' we have $F_i(x) = 0$ and in $\Omega_i'' \cup \Omega_i'''$ we have:

$$\|F_i(x)\| = \frac{|\phi_i(x) - f_i(x)|}{\|x\|^2}\|x\| \le \frac{|\phi_i(x)|+|f_i(x)|}{\|x\|} \le$$

$$\le \frac{C_1\|x\| + C_2 + C_1\rho + C_2}{\|x\|} \le 2C_1 + \frac{2C_2}{\rho} \tag{A.11}$$

Finally because of the definition (A.10), F_i is a continuous function in Ω which in particular is a constant on Ω_i', while in $\Omega_i'' \cup \Omega_i'''$ is the product of locally Lipschitzian functions ($\phi_i(x)-f_i(x)$ and x) by a continuous uniformly bounded function $\left(\frac{1}{x^T x}\right)$. Therefore F_i is easily seen to be locally Lipschitzian on Ω.

REFERENCES

[1] C. BRUNI, G. DI PILLO, G. KOCH: *Bilinear systems: an appealing class of "nearly linear" systems in theory and applications.* IEEE Trans. on Automatic Control, vol. AC-19, n.4 August 1974.

[2] R.R.MOHLER: *Bilinear control processes with application to engineering, ecology and medicine.* Academic Press 1974.

[3] C. LOBRY: *Controllabilité des systèmes non lineares.* SIAM J. on Control, v. 8, n. 4, November 1970.

[4] G. HAYNES, H. HERMES: *Nonlinear controllability via Lie theory.* SIAM J. on Control, v. 8, n. 4, November 1970.

[5] R.W. BROCKETT: *System theory on group manifolds and coset space.* SIAM J. on Control, v. 10, n. 2, May 1972.

[6] V. JURDJEVIC: *Certain controllability properties of analytic control systems.* SIAM J. on Control, v. 10, n. 2, May 1972.

[7] G.E. LADAS, V. LAKSHMIKANTHAM: *Differential equations in abstract spaces.* Academic Press 1972.

[8] C. BRUNI, M.A. GIOVENCO, G. KOCH, R. STROM: *A dynamical model of humoral immune response.* To appear in mathematical biosciences. Rapp. Ist. Autom.Univ. di Roma, R. 74-22, Luglio 1974.

[9] C. BRUNI, M.A. GIOVENCO, G. KOCH, R. STROM: *The immune response as a variable structure system.* Variable structure systems in Biology and Socieconomics, Springer Verlag 1975.

[10] R.A. ALBERTY: *The enzymes.* Academic Press, 1959.

[11] I. MÁLEK: *Present state and perspectives of biochemical engineering.* Advances in Biochemical Engineering, vol. 3, Ed. T.K. Ghose, A. Fiechter, N. Blakebrough. Springer Verlag,1974.

[12] G. D'ANS, D. GOTTLIEB, P. KOKOTOVIC: *Optimal control of bacterial growth.* Automatica, vol. 8, Pergamon Press, 1972.

[13] R. ROSEN: *Dynamical system theory in biology.* Vol. 1, Wiley Interscience, 1970.

[14] GOEL, MAITRA, MONTROLL: *Models of interacting populations.* Reviews of modern physics. April 1971, Part. 1.

[15] O. BILANS, N.R. AMUNDSON: *Chemical reactor stability and sensitivity.* A.I.Ch.E.J., vol. 1, n. 4, 1955.

[16] E.P. GYFTOPULOS: *General reactor dynamics in the technology of nuclear reactor safety*. Vol. 1, Ed. T.J. Thompson, J.G. BECKERLEY, M.I.T. Press, 1964.

[17] D.L. HETRICK: *Dynamics of nuclear reactors*. The University of Chicago Press, 1971.

[18] W. LEONTIEFF: *The dynamics inverse*. Contribution to Input Output Analysis. Ed. Carter-Brody, North Holland, 1970.

[19] A.D. SMIRNOV: *Problems of constructing n optimal interbranch model of socialist reproduction*. Contribution to Input Output Analysis. Ed. Carter-Brody, North Holland, 1970.

[20] G.S. SAUER, J.L. MELSA: *Stochastic insect-pest control with variable observation policy*. IV San Diego Symposium in Nonlinear Estimation and Its Application. 10-12 Sept. 1973.

[21] V. JURDJEVIC, H.J. SUSSMANN: *Control systems on Lie groups*. Division of engineering and applied physics. Harvard University. Technical Report N. 628. Cambridge Massachusetts. November 1971.

[22] H. HERMES, G. HAYNES: *On the nonlinear control problem with control appearing linearly*. SIAM J. on Control, vol. 1, n. 2, 1963.

[23] H. HERMES: *Controllability and the singular problem*. SIAM J. on Control, vol. 2, n. 2, 1964.

[24] E.J. DAVISON, E.G. KUNZE: *Some sufficient conditions for the global and local controllability of nonlinear time-varying systems*. SIAM J. on Control, vol. 8, n. 4, 1970.

[25] H. TOKUMARU, N. ADACHI: *On the controllability of nonlinear systems*. Automatica, vol. 6, 1970.

[26] S.B. GERSHWIN, D.H. JACOBSON: *A controllability theory for nonlinear systems*. IEEE Trans. on A.C. Vol. AC-16, n.1, 1971.

Un calcul symbolique non commutatif pour les asservissements non linéaires et non stationnaires

Michel FLIESS[+]
Université Paris VIII
et
Centre d'Automatique de l'Ecole des Mines de Paris

Sommaire

(+) Travail sous contrat de recherche n° 75.133 de l'IRIA.

Introduction

Le but de cette Note est de promouvoir l'utilisation des polynômes et des séries formels en indéterminées non commutatives comme outil de calcul privilégié pour les asservissements non linéaires et (ou) non stationnaires.

Après divers rappels, nous montrons comment ces êtres algébriques peuvent représenter des systèmes donnés souvent, jusqu'alors, sous forme d'une série de Volterra. Aux séries rationnelles, introduites en théorie des automates et langages par M.P. Schützenberger (cf. [9]), correspondent les systèmes réguliers (ou bilinéaires) réalisables, étudiés selon d'autres méthodes par Brockett [4 , 6] , Bruni, DiPillo, Koch [8] , d'Alessandro, Isidori, Ruberti [1] . Polynômes et séries rationnelles permettent d'approcher uniformément, dans tout domaine compact donné, tout asservissement continu, ce tant dans la topologie des fonctions continues que de carré sommable. Ces faits ayant déjà été démontrés par l'auteur [15, 18, 19] , nous ne ferons que les passer en revue.

Le résultat original de cette Note concerne la résolution d'équations intégrales ou différentielles non linéaires en régime forcé. L'emploi des indéterminées non commutatives permet un calcul algorithmique bien plus facile que ceux proposés avec les séries de Volterra, notamment dans le cas non stationnaire (cf. Flake [14]). Ces méthodes sont applicables dans le cas stochastique à condition, évidemment, d'utiliser l'intégrale de Stratonovich [26] .

I.-Rappels sur les séries formelles non commutatives

a) Définitions et propriétés générales

Soient X un ensemble fini, non vide, l'alphabet, X^* le monoïde libre qu'il engendre. Un élément de X^* est un mot, l'élément neutre, ou mot vide, est noté "1". K étant un corps commutatif, soient $K\langle X\rangle$ et $K\langle\langle X\rangle\rangle$ les anneaux des polynômes et des séries formels, à coefficients dans K, en les indéterminées associatives $x \in X$ (non commutatives si card $X \geqslant 2$). Un élément $s \in K\langle\langle X\rangle\rangle$ est noté

$$s = \sum \left\{ (s,w)w \mid w \in X^* \right\} .$$

L'addition et le produit sont définis par:

$$s_1 + s_2 = \sum \left\{ ((s_1,w)+(s_2,w))w \mid w \in X^* \right\} ,$$

$$s_1 s_2 = \sum \left\{ \left(\sum_{w_1 w_2 = w} (s_1,w_1)(s_2,w_2) \right) w \mid w \in X^* \right\} .$$

$s \in K\langle\langle X\rangle\rangle$ est inversible ss'il existe une série s^{-1} telle que $ss^{-1}=1=s^{-1}s$. Comme dans le cas commutatif, s est inversible ssi son terme constant (s,1) est non nul. Un sous-anneau R de $K\langle\langle X\rangle\rangle$ est dit rationnellement clos ssi l'inverse de toute série inversible de R appartient encore à R. L'anneau $K\langle(X)\rangle$ des séries <u>rationnelles</u>, introduit par Schützenberger [25], est le plus petit sous-anneau rationnellement clos de $K\langle\langle X\rangle\rangle$ qui contienne $K\langle X\rangle$.

<u>Remarque</u>.-Lorsque X est réduit à une seule lettre x, une série rationnelle n'est autre que le développement de Taylor à l'origine d'une fraction rationnelle de la forme P/Q, où $P,Q \in K[x]$, $Q(0)\neq 0$.

Soit $K^{N\times M}$ l'ensemble des matrices à coefficients dans K, à N lignes et M colonnes. Une représentation (linéaire) $\mu : X^* \to K^{N\times N}$ est un homomorphisme du monoïde X^* dans le monoïde multiplicatif des matrices carrées d'ordre N. Le résultat fondamental suivant est connu sous le nom de théorème de Kleene-Schützenberger (cf. [25, 17]):

<u>Théorème 1.1</u>.-Une série $r \in K\langle\langle X\rangle\rangle$ est rationnelle si et seulement s'il existe un entier $N \geq 1$, une représentation $\mu : X^* \to K^{N\times N}$, des matrices ligne $\lambda \in K^{1\times N}$ et colonne $\gamma \in K^{N\times 1}$ tels que

$$r = \sum \{(\lambda \mu w \gamma) w \mid w \in X^*\} .$$

<u>Remarque</u>.-Un <u>langage</u> (<u>formel</u>) est une partie d'un monoïde libre X^*. Il peut être considéré comme série formelle en les indéterminées $x \in X$, à condition de prendre les coefficients dans le semi-anneau de Boole $\mathcal{B}=\{0,1\}$, avec 1+1=1, ou le semi-anneau <u>N</u> des entiers non négatifs, si l'on tient compte de l'ambiguïté. Sur de tels semi-anneaux, le concept de séries rationnelles se définit sans grand changement. Le théorème 1.1, dû à Schützenberger [25], y est encore valable. Il généralise le résultat original de Kleene d'après lequel un langage est rationnel ss'il est accepté par un automate fini. Pour plus de détails sur les liens des séries avec l'informatique et la linguistique mathématiques, nous renvoyons à Chomsky, Schützenberger [9], Eilenberg [12], Cohn [10].

La <u>matrice de Hankel</u> d'une série $s \in K\langle\langle X\rangle\rangle$ est le tableau infini $\mathcal{H}(s)$ dont lignes et colonnes sont indexées par X^*, de sorte que le coefficient d'indice $(u,v) \in X^* \times X^*$ est (s,uv). Le rang est immédiat à définir. Le théorème suivant généralise une propriété classique dans le cas d'une seule indéterminée. La démonstration peut se faire (cf. [17]) à l'aide de modules gauches ou droits sur $K\langle X\rangle$ qui sont l'analogue des modules introduits pour les systèmes linéaires par R.E. Kalman.

<u>Théorème 1.2</u>.-K étant un corps, une condition nécessaire et suffi-

sante pour qu'une série $r \in K\langle\langle X\rangle\rangle$ soit rationnelle est que la matrice de Hankel soit de rang fini $\bar{N}$. Il existe alors une représentation $\bar{\mu}: X^* \rightarrow K^{\bar{N}\times\bar{N}}$, des matrices ligne $\bar{\lambda} \in K^{1\times\bar{N}}$ et colonne $\bar{\gamma} \in K^{\bar{N}\times 1}$, telles que

$$r = \sum \{(\bar{\lambda}\,\bar{\mu}w\,\bar{\gamma})w \mid w \in X^*\} .$$

Soient un entier $N \geqslant 1$, une représentation $\mu: X^* \longrightarrow K^{N\times N}$, des matrices ligne $\lambda \in K^{1\times N}$ et colonne $\gamma \in K^{N\times 1}$ tels que

$$r = \sum \{(\lambda\,\mu w\,\gamma)w \mid w \in X^*\} .$$

Alors $N \geqslant \bar{N}$. Si $N = \bar{N}$, il existe une matrice inversible $P \in K^{\bar{N}\times\bar{N}}$ telle que $P\,\mu w P^{-1} = \bar{\mu}w$, $\bar{\lambda}P = \lambda$, $P\gamma = \bar{\gamma}$ (les représentations $\bar{\mu}$ et μ sont donc semblables). $\bar{\mu}$ est dite <u>réduite</u> (ou <u>minimale</u>).

<u>Application</u>.-Une représentation $\mu: X^* \longrightarrow K^{N\times N}$ est dite <u>nilpotente</u> ssi, pour tout mot non vide w, la matrice μw l'est. D'après un résultat dû à Levitzki, la représentation peut être triangularisée, en ce sens que toutes les matrices μw peuvent l'être simutanément.

<u>Proposition 1.3</u> (cf. [19]).-Une série rationnelle de $K\langle\langle X\rangle\rangle$ est un polynôme si et seulement si elle peut être produite par une représentation nilpotente. Alors, la représentation réduite est nilpotente.

b) Produit de Hurwitz

Le produit de séries formelles défini au début, dit parfois <u>produit de Cauchy</u>, n'est pas le seul possible. Le <u>produit de Hurwitz</u> ou <u>d'intercalement</u> (en anglais <u>shuffle product</u>) se définit par récurrence sur la longueur des mots:

$1 \sqcup\!\sqcup 1 = 1$, $\forall_X\ x\ 1 \sqcup\!\sqcup x = x \sqcup\!\sqcup 1 = x$, $\forall_X\ x,y\ \forall_{X^*}\ u,v\ ux \sqcup\!\sqcup vy = (u \sqcup\!\sqcup vy)x + (ux \sqcup\!\sqcup v)y$.

Il se prolonge par linéarité à $K\langle\langle X\rangle\rangle$:

$$s_1 \sqcup\!\sqcup s_2 = \sum \{(s_1,u_1)(s_2,u_2)u_1 \sqcup\!\sqcup u_2 \mid u_1,u_2 \in X^*\} .$$

Ce produit est associatif, commutatif et distributif par rapport à l'addition; il a pour élément neutre 1. Le produit de Hurwitz de deux polynômes est un polynôme. Par ailleurs, on peut démontrer (cf. [16]):

<u>Proposition 1.4</u>.-Le produit de Hurwitz de deux séries rationnelles est une série rationnelle.

c) Systèmes d'équations

Soit $\Xi = \{\xi_1, \ldots, \xi_M\}$ un ensemble de M lettres, les inconnues. Soit le système (E) de M équations $\xi_i = A_i$ $(i=1,\ldots,M)$, où A_i est une expression formée par la somme, possiblement infinie, de monômes où figurent la concaténation et le produit de Hurwitz des lettres de X et Ξ. Exemple: $x_1 \sqcup\!\sqcup (x_0 \xi_1)$. De plus, on impose aux monômes ne contenant que des ξ_j d'être nuls. A_i peut comprendre un terme constant non nul $A_i(0)$

Proposition 1.5.-Le système d'équations (E) admet un M-uple solution et un seul de séries de $K\langle\langle X\rangle\rangle$ que l'on peut obtenir par itération. Les coefficients des mots de longueur inférieure ou égale à k de l'itéré d'ordre k coïncident avec ceux de la solution.

Preuve.-La méthode est en tout point identique à celle utilisée pour les systèmes d'équations algébriques, étudiés en [9, 10], et qui généralisent les grammaires de Chomsky pour les langages algébriques ou context-free. Elle repose sur le fait que, $K\langle\langle X\rangle\rangle$ étant complet pour la topologie (X)-adique, on définit par substitution aux ξ_i d'éléments de $K\langle\langle X\rangle\rangle$ une contraction dans l'ensemble des M-uples de $K\langle\langle X\rangle\rangle$, où le terme constant de la i^e série est $A_i(0)$. Le M-uple solution est le point fixe.

II.-Asservissements non linéaires et séries formelles

K est désormais le corps $\underline{R}$ des réels ou celui $\underline{C}$ des complexes.

a) Asservissements réguliers (ou bilinéaires)

Un asservissement régulier (ou bilinéaire) réalisable a une description par espace d'état de la forme

$$(2.1)\qquad \begin{cases} \dot q(t)\ (=dq/dt) = (A_0 + \sum_{i=1}^{n} u_i(t)A_i)q(t) \\ y(t) = \lambda q(t) \end{cases},$$

où $q(t)$ [$q(0)$ donné] appartient au K-espace vectoriel d'état Q de dimension ici finie (d'où le terme réalisable), $A_0, A_1, \ldots, A_n : Q \to Q$, $\lambda : Q \to K$ sont des applications K-linéaires, $u_1, \ldots, u_n : \underline{R}_+ \to K$ ($\underline{R}_+ = [0, \infty[$) sont les entrées (ou commandes, ou gouvernes, ou contrôles) que l'on suppose localement intégrables par rapport à la mesure de Lebesgue.

Deux asservissements sont dits indiscernables ssi, pour les mêmes entrées, on obtient les mêmes sorties. On montre (cf. [3]) que tout asservissement linéaire

$$\begin{cases} \dot\eta(t) = F\eta(t) + \sum_{i=1}^{n} u_i(t)g_i \\ y(t) = H\eta(t) \end{cases},$$

d'espace d'état N de dimension finie, où $g_i \in N$ ($i=1,\ldots,n$) et $F: N \to N$, $H: N \to K$ sont des applications K-linéaires, est indiscernable d'un asservissement régulier réalisable.

Posons $\xi_0(t) = t$, $\xi_i(t) = \int_0^t u_i(\tau)d\tau$ ($i=1,\ldots,n$). La formule de Peano-Baker conduit à écrire:

$$(2.2)\quad y(t) = \lambda\left[1 + \sum_{k=1}^{\infty} \sum_{j_k,\ldots,j_1=1}^{n} A_{j_k}\ldots A_{j_1}\int_0^t d\xi_{j_k}\ldots d\xi_{j_1}\right] q(0) \quad ,$$

où l'intégrale est définie par récurrence sur la longueur:

$\int_0^t d\xi_j = \xi_j(t)$ $(j=0,1,\ldots,n)$, $\int_0^t d\xi_{j_k}\ldots d\xi_{j_1} = \int_0^t d\xi_{j_k}(\tau)\int_0^\tau d\xi_{j_{k-1}}\ldots d\xi_{j_1}$.
Soit l'alphabet $X=\{x_o,x_1,\ldots,x_n\}$ où x_j correspond à ξ_j.Guidés par (2.2),introduisons la série rationnelle $\underline{G}\in K\langle(X)\rangle$ donnée par la représentation $\mu : X^* \longrightarrow K^{N\times N}$ définie par $\mu x_j = A_j$,de sorte que

$$\underline{G} = \sum \{(\lambda \mu wq(0))w \mid w \in X^*\} \quad .$$

$\underline{G}$ est dite série <u>génératrice</u> de (2.1),qu'elle définit à une indiscernabilité près (cf. [15 ,19]).

Le théorème 1.2 donne la nature de la réalisation réduite d'un asservissement régulier (cf. [15 , 19]),résultat obtenu aussi par Brockett [3],Bruni,DiPillo,Koch [8],et d'Alessandro,Isidori, Ruberti [1] .

<u>Théorème 2.1</u>.-Il y a bijection canonique entre les asservissements réguliers réalisables,définis à une indiscernabilité près,et les séries rationnelles non commutatives.Un asservissement régulier réalisable,dont la matrice de Hankel de la série génératrice est de rang fini $\bar{N}$,est indiscernable de $(\bar{\Sigma})$

$$\begin{cases} \dot{\bar{q}}(t)=(\bar{A}_o+\displaystyle\sum_{i=1}^{n} u_i(t)\bar{A}_i)\bar{q}(t) \\ y(t)=\bar{\lambda}\bar{q}(t) \end{cases},$$

d'espace d'état $\bar{Q}$ de dimension $\bar{N}$.Tout asservissement régulier

$$\begin{cases} \dot{q}'(t)=(A'_o+\displaystyle\sum_{i=1}^{n} u_i(t)A'_i)q'(t) \\ y(t)=\lambda' q'(t) \end{cases},$$

indiscernable des précédents,a un espace d'état Q' de dimension $N' \geqslant \bar{N}$.Si $N'=\bar{N}$,il existe un isomorphisme $P:Q'\longrightarrow\bar{Q}$,tel que:

$$PA'_jP^{-1}=\bar{A}_j \quad (j=0,1,\ldots,n),\ \bar{\lambda}P=\lambda',\ Pq'(0)=\bar{q}(0) \quad .$$

$(\bar{\Sigma})$ est dit réduit.

<u>Application</u>.-Lorsque la série génératrice est un polynôme,on obtient un asservissement régulier,dit <u>nilpotent</u>,que la proposition 1.3 permet de caractériser ainsi:

<u>Proposition 2.2</u>.-Un asservissement régulier est nilpotent si et seulement s'il est indiscernable de l'asservissement

$$\begin{cases} \dot{q}(t)=(A_o+\displaystyle\sum_{i=1}^{n} u_i(t)A_i)q(t) \\ y(t)=\lambda q(t) \end{cases},$$

où le semi-groupe engendré par les matrices $A_o,A_1,\ldots,A_n$ est nilpotent.

D'après le théorème de Levitzki,$A_o,A_1,\ldots,A_n$ peuvent être simultanément triangularisées.

<u>Remarque</u>.-Marcus [23] utilise le terme nilpotent dans un sens légèrement plus général.

b) Séries de Volterra

Wiener [28], dès 1942, a tenté de mettre les asservissements non linéaires sous forme d'une série de Volterra (cf. [2]):

$$(2.3)\quad y(t)=h_0(t)+\int_0^t h_1(t,\tau_1)u(\tau_1)d\tau_1 + \int_0^t\int_0^{t} h_2(t,\tau_2,\tau_1)u(\tau_2)u(\tau_1)d\tau_2 d\tau_1 + \dots ,$$

où, pour simplifier, l'entrée et la sortie $u,y:\underline{R}_+ \rightarrow K$ sont supposées scalaires.

Les nombreux travaux, tant théoriques que pratiques, qui leur ont été consacrés, ont achoppé sur la détermination effective des noyaux $h_0, h_1, h_2, \dots$.

A une série $\underline{G} \in K\langle\langle x_0,x_1\rangle\rangle$, associons, par comparaison à (2.2), le système entrée-sortie donné par

$$y(t)=(\underline{G},1)+\sum_{k=0}^{\infty}\ \sum_{j_k,\dots,j_1=0,1}(\underline{G},x_{j_k}\dots x_{j_1})\int_0^t d\xi_{j_k}\dots d\xi_{j_1} ,$$

où $\xi_0(t)=t$, $\xi_1(t)=\int_0^t u(\tau)d\tau$.

Un asservissement, ainsi décrit par une série non commutative, est dit analytique. La série, dite encore génératrice, le caractérise à une indiscernabilité près.

Théorème 2.3 (cf. [18,19]).- Un asservissement est analytique si et seulement s'il peut être défini par une série de Volterra, dont les noyaux sont des fonctions analytiques.

Remarques.- (i) Pour ne pas alourdir l'exposé, nous n'étudierons pas les questions de convergence liées au développement en série non commutative.

(ii) Il y a stationnarité, c'est-à-dire invariance par translation temporelle, ssi le support de la série génératrice $\underline{G}$, défini par

$$\text{supp}\ \underline{G}=\{w \mid w\in X^*, (\underline{G},w)\neq 0\} ,$$

est de la forme

$$\text{supp}\ \underline{G} \subseteq \{x_0,x_1\}^* x_1 .$$

La multiplication de deux asservissements d'entrée u et de sorties y_1, y_2 est l'asservissement d'entrée u et de sortie, à l'instant t, $y_1(t)y_2(t)$. Une simple intégration par parties conduit à énoncer:

Proposition 2.4 (cf. [18,19]).- L'asservissement multiple de deux asservissements analytiques est analytique, de série génératrice le produit de Hurwitz des deux séries génératrices.

En vertu de la proposition 1.4, il vient si l'on appelle rationnel, polynômial un asservissement analytique dont la série génératrice l'est:

Corollaire 2.5 (cf. [18,19]).- L'asservissement multiple de deux asservissements analytiques rationnels (ou plynômiaux) est de même nature.

c) Approximations des asservissements

C'est la généralisation fonctionnelle, due à Fréchet (cf. [21], p. 78), du théorème d'approximation de Weierstrass qui justifie l'introduction par Wiener des séries de Volterra (cf. [2]). En effet, d'après ce résultat, les séries de Volterra tronquées, c'est-à-dire avec un nombre fini de noyaux non nuls, permettent d'approcher tout asservissement continu. On a un résultat analogue avec les asservissements anlytiques par application du théorème d'approximation de Weierstrass-Stone.

Soit $\mathfrak{F}$ l'espace des fonctions de $\underline{R}_+$ dans K que l'on suppose être l'ensemble des fonctions:

-continues muni de la topologie de la convergence compacte,

-localement de carré sommable par rapport à la mesure de Lebesgue, muni de la topologie suivante: la suite u_n tend vers zéro ssi, pour tout $t>0$, $\lim \int_0^t u_n^2(\tau)d\tau = 0$.

Un asservissement peut être considéré comme une application de $\underline{R}_+ \times \mathfrak{F}$ dans K obeissant au principe de causalité: la sortie $y(t)$ ne dépend de l'entrée $u(\tau)$ que pour $0 \leq \tau \leq t$. Un asservissement est dit continu ssi c'est une application continue de $\underline{R}_+ \times \mathfrak{F}$ dans K.

Théorème 2.6.-Dans tout domaine compact de $\underline{R}_+ \times \mathfrak{F}$, tout asservissement continu peut être uniformément approché par des asservissements analytiques que l'on peut choisir rationnels ou polynômiaux.

Corollaire 2.7.-Dans tout domaine compact de $\underline{R}_+ \times \mathfrak{F}$, tout asservissement continu peut être uniformément approché par des asservissements réguliers réalisables, que l'on peut choisir nilpotents.

Remarque.-Ces résultats d'approximations, démontrés en [19], avaient été énoncés en [18] pour la topologie de la convergence compacte des fonctions continues. Indépendamment et peu après, ils ont été retrouvés par Sussmann [27] pour la topologie de la convergence vague, fait qui implique le résultat de [18]. Par ailleurs, Krener [20] a prouvé que tout système de la forme

$$\begin{cases} \dot{q}(t) = f(q(t)) + u(t)g(q(t)) \\ y(t) = h(q(t)) \end{cases},$$

où f, g, h sont des fonctions C^∞, peut être approché par un asservissement régulier réalisable.

III.-Résolution d'équations différentielles non linéaires

a) Théorie

Soit l'asservissement, où l'entrée figure linéairement, décrit par:

$$(3.1)\quad \begin{cases} q(t)=q(0)+ta(t)+\displaystyle\int_0^t\Big[f(t,q(t))+\sum_{i=1}^{n}u_i(t)g_i(t,q(t))\Big]\,dt \\ y(t)=h(t,q(t)) \end{cases},$$

où le vecteur d'état q appartient à un K-espace vectoriel Q de dimension finie N, et où $a, f, g_1, \dots, g_n, h$ sont des fonctions analytiques dans un voisinage de l'origine contenant $(0, q(0))$ ((3.1) est un système intégrale à la Volterra généralisant (2.1)).

Proposition 3.1.-(3.1) définit un asservissement analytique dont la série génératrice peut être calculé par itération. On obtient ainsi une suite d'approximations par asservissements polynômiaux.

Preuve.-Soient $\Xi = \{\xi_1, \dots, \xi_N, \xi_{N+1}\}$ où $\xi_1, \dots, \xi_N$ correspondent aux N coordonnées de q et ξ_{N+1} à y. Rappelons que la multiplication à gauche par x_0 correspond à intégrer, celle par x_i à multiplier par u_i puis intégrer; de plus le produit des valeurs donne le produit de Hurwitz (cf. proposition 2.4). (3.1) conduit alors à écrire, sous forme matricielle:

$$(3.2)\quad \begin{cases} \begin{pmatrix}\xi_1\\ \vdots \\ \xi_N\end{pmatrix} = q(0)+x_0 ш A(x_0)+x_0F(x_0,\xi_1,\dots,\xi_N) \\ \qquad\qquad + \displaystyle\sum_{i=1}^{n} x_iG_i(x_0,\xi_1,\dots,\xi_N) \\ \xi_{N+1}=H(x_0,\xi_1,\dots,\xi_N) \end{cases},$$

où A,B,F,G,H sont des séries en les variables $x_0, \xi_1, \dots, \xi_N$ où le produit qui intervient est celui de Hurwitz.

La résolution par itération selon le § I.c de (3.2) est possible et correspond, de façon claire, à la méthode itérative de Picard pour les équations intégrales (ou différentielles).

D'après la proposition 1.5, on obtient un algorithme fournissant une suite de polynômes de degré inférieur ou égal au pas d'itération, dont les coefficients coïncident avec ceux de la série génératrice solution. Ces polynômes définissent des asservissements analytiques polynômiaux, ou réguliers nilpotents, qui convergent vers (3.1) dans l'une des topologies du § II.c.

Remarques.-(i) Brockett [5] a étudié les équations différentielles de la forme

$$\begin{cases} \dot q(t)=f(t,q(t))+\sum u_i(t)g_i(t,q(t)) \\ y(t)=h(t,q(t)) \end{cases}.$$

Il a prouvé que la solution pouvait se mettre sous la forme d'une série de Volterra lorsque f,g,h sont analytiques en q et seulement continues en t. Toutes les considérations qu'il développe sur le rayon de convergence et la prolongation analytique sont immédiatement transposables ici.

(ii)Une limitation importante de notre approche provient du fait que l'entrée est linéaire en (3.1).S'il n'en n'est pas ainsi,on peut, avec Brockett [7],prendre pour nouvelles entrées $v_1,\ldots,v_n$ où v_i est la dérivée de u_i et rajouter n dimensions q_{N+i} à l'espace d'état de sorte que

$$q_{N+i}(t)=u_i(t)=u_i(0)+\int_0^t v_i(\tau)d\tau.$$

(iii)Des équations assez semblables à (3.2) ont été rencontrées par Cori [11] pour définir des séries non commutatives associées à certains graphes planaires.

b)Exemples pratiques

Il importe parfois d'obtenir de manière explicite la solution d'un système de la forme (3.1).Lorsque le système est autonome,la transformation de Laplace multidimensionnelle offre un outil d'une certaine efficacité pour le calcul des noyaux de la série de Volterra.Si l'on n'utilise pas cet outil,et,dans le cas non autonome, on ne peut le faire,on aboutit à des équations aux dérivées partielles souvent inextricables liant les noyaux (cf. Flake [13,14]). Dans tous les cas,le calcul itératif de la série génératrice nous semble bien plus simple.

Exemple 1 (cf. Barrett [2]).-Soit l'équation dite de Duffing

$$\ddot{y} + a\dot{y} + by + ey^3=u(t) \quad .$$

Elle peut être réécrite sous la forme:

$$y(t)+a\int_0^t y(\tau)d\tau +b\int_0^t d\tau\int_0^\tau y(\tau')d\tau' \quad +e\int_0^t d\tau\int_0^\tau y^3(\tau')d\tau'=\int_0^t d\tau\int_0^\tau u(\tau')d\tau'+\alpha +\beta t,$$

où $y(0)=\alpha$, $\dot{y}(0)=\beta-a\alpha$ .La série génératrice $\underline{G}\in K\langle\langle x_0,x_1\rangle\rangle$ satisfait à l'équation:

$$\underline{G}+ax_0\underline{G}+bx_0^2\underline{G}+ex_0^2(\underline{G}\sqcup\!\sqcup\underline{G}\sqcup\!\sqcup\underline{G})=x_0x_1+\beta x_0+\alpha \quad .$$

On approxime par la suite de polynômes de $K\langle x_0,x_1\rangle$,qui définissent des asservissements polynômiaux ou réguliers nilpotents:

$-\alpha$;

$-\alpha+(\beta-a\alpha)x_0$;

$-\alpha+(\beta-a\alpha)x_0+(a^2\alpha-a\beta-b\alpha-e\alpha^3)x_0^2+x_0x_1$;

$-\alpha+(\beta-a\alpha)x_0+(a^2\alpha-a\beta-b\alpha-e\alpha^3)x_0^2+x_0x_1-ax_0^2x_1+$terme en x_0^3 ;etc....

Remarque.-Par d'autres méthodes,on a cherché à développer un calcul itératif des noyaux de séries de Volterra correspondant à des équations différentielles de forme identique (cf. Reddy,Reddy [24]).

Exemple 2 (cf. Brockett [5]).-Soit l'équation

$$\dot{y} = \sin y + u(t) \quad ,$$

où,pour simplifier les calculs,on suppose y(0)=0.Elle peut être réécrite sous la forme:

$$y(t)=\int_0^t \sin y(\tau)d\tau + \int_0^t u(\tau)d\tau$$

$$= \int_0^t \sum_{n=0}^{\infty} (-1)^n y^{2n+1}(\tau)/(2n+1)!\,d\tau + \int_0^t u(\tau)d\tau \quad .$$

La série génératrice $\underline{G} \in K\langle\langle x_0,x_1\rangle\rangle$ satisfait à:

$$\underline{G}=x_0\Big(\sum_{n=0}^{\infty} (-1)^n \underbrace{\underline{G}\sqcup\!\sqcup\ldots\sqcup\!\sqcup\underline{G}}_{2n+1 \text{ fois}}\Big) + x_1 \quad .$$

On approxime par la suite de polynômes de $K\langle x_0,x_1\rangle$:

- 0 ;
- x_1 ;
- $x_1+x_0x_1$;
- $x_1+x_0x_1-6x_0x_1^3-18x_0^2x_1^3-6x_0x_1^2x_0x_1$;etc... .

Exemple 3 (cf. Flake [14]).-Soit l'équation

$$\ddot{y} + t\dot{y} + y + y^2 = u(t) \qquad (y(0)=\dot{y}(0)=0) \quad .$$

Elle peut être réécrite sous la forme:

$$\dot{y}(t)+\int_0^t \tau\dot{y}(\tau)d\tau + \int_0^t d\tau\int_0^{\tau} \dot{y}(\tau')d\tau' + \int_0^t \Big(\int_0^{\tau} \dot{y}(\tau')d\tau'\Big)^2 d\tau = \int_0^t u(\tau)d\tau \quad .$$

La série génératrice $\underline{G} \in K\langle\langle x_0,x_1\rangle\rangle$ de y satisfait à:

$$\dot{\underline{G}}+x_0(x_0\sqcup\!\sqcup\,\dot{\underline{G}})+x_0^2\dot{\underline{G}}+x_0\left[(x_0\dot{\underline{G}})\sqcup\!\sqcup(x_0\dot{\underline{G}}) \right] = x_1 \quad .$$

La série génératrice $\underline{G}$ de y est donnée par:

$$\underline{G}=x_0\dot{\underline{G}} \quad .$$

On approxime par la suite de polynômes de $K\langle x_0,x_1\rangle$:

- 0 ;
- x_0x_1 ;
- $x_0x_1-2x_0^3x_1-x_0^2x_1x_0$;
- $x_0x_1-2x_0^3x_1-x_0^2x_1x_0+8x_0^4x_1+5x_0^3x_1x_0+2x_0^2x_1x_0^2-4x_0^4x_1^2$;etc... .

c)Cas stochastique

Lorsque l'entréee est une diffusion markovienne,on peut appliquer à (3.1) les méthodes itératives précédentes,cela sans changement à condition d'utliser l'intégrale stochastique de Stratonovich [26], qui,à l'inverse de celle d'Itô,obeit aux règles de calcul habituelles.Par une démonstration analogue,la proposition 3.1 reste valable et l'on obtient une suite d'asservissements plynômiaux,ou réguliers nilpotents,convergeant vers la solution en moyenne quadratique sur tout intervalle de temps borné.

Si,pour l'intégrale de Stratonovich,à un asservissement analytique rationnel correspond une système régulier donné par (2.1), rappelons que ce n'est pas aussi simple avec le calcul d'Itô.La première ligne de (2.1) est alors remplacée par:

$$\dot{q}(t)=\Big[\Big(A_0+1/2\sum_{i,j} R_{ij}(t)A_iA_j\Big)+\sum_i A_iu_i(t)\Big]\,q(t)\ ,$$

si $u={}^t(u_1,\dots,u_n)$ (^{t} indique la transposition matricielle) est un bruit blanc centré vérifiant $E(u(t)\,{}^tu(s))=R(t)\,\delta(t-s)$ (cf. Lo [22]).

Application.-Dans de nombreux problèmes,nos méthodes d'approximations devraient permettre un filtrage statistique non linéaire sous-optimal par utilisation des asservissements polynômiaux ou réguliers nilpotents.Illustrons nos dires par le résultat suivant de Marcus [23] :

Soit $\dot{u}(t)=F(t)u(t)+G(t)b(t)$,où b est un bruit blanc indépendant de la variable aléatoire gaussienne $u(0)$.La meilleure estimée de la sortie d'un asservissement polynômial,connaissant $z^t=\{z(\tau)\mid 0\leq\tau\leq t\}$, défini par $z(t)=H(t)u(t)+R^{1/2}(t)b'(t)$ ($R>0$,b' bruit blanc indépendant de b et $u(0)$,(F,G,H) complètement commandable et observable), peut être calculée par un système d'équations différentielles de dimension finie,conduit par l'innovation $d\nu(t)=dz(t)-H(t)\hat{u}(t|t)$ (la meilleure estimée $\hat{u}(t|t)$ de u est,bien sûr,fournie par un filtre de Kalman-Bucy).

Bibliographie

1.-d'Alessandro (P.),Isidori (A.) et Ruberti (A.).-Realization and structure theory of bilinear systems.SIAM J. Control,12,1974,p. 567-615.

2.-Barrett (J.F.).-The use of functionals in the analysis of non-linear physical models.J. Electronics Control,15,1963,p. 567-615.

3.-Brockett (R.W.).-On the algebraic structure of bilinear systems, in "Theory and Applications of Variable Structure Systems"(R.R. Mohler et A. Ruberti,éd.),p. 153-168.Academic Press,New York,1972.

4.-Brockett (R.W.).-Lie algebras and Lie groups in control theory, in "Geometric Methods in System Theory"(D.Q. Mayne et R.W. Brockett, éd.),p. 43-82.D. Reidel,Dordrecht,1973.

5.-Brockett (R.W.).-Volterra series and geometric control theory. Proc. 6th IFAC Congr.,Cambridge(Mass.),1975.

6.-Brockett (R.W.).-Nonlinear systems and differential geometry. Proc. IEEE,63,1975.

7.-Brockett (R.W.).-Functional expansions and higher order necessary conditions in optimal control,in "Algebraic System Theory" (G. Marchesini,éd.).Lect. Notes Math.,Springer-Verlag,Berlin,à paraître.

8.-Bruni (C.),DiPillo (G.) et Koch (G.).-Bilinear systems:an appealing class of "nearly linear" systems in theory and applicatio applications.IEEE Trans. Autom. Contr.,19,1974,p. 334-348.

9.-Chomsky (N.) et Schützenberger (M.P.).-The algebraic theory of context-free languages,in "Computer Programming and Formal Systems" (P. Braffort et D. Hirschberg,éd.),p. 118-161.North-Holland, Amsterdam,1963.

10.-Cohn (P.M.).-Algebra and language theory.Bull. London Math. Soc., 7,1975,p. 1-29.

11.-Cori (R.).-Un code pour les graphes planaires.Astérisque n° 27, 1975.

12.-Eilenberg (S.).-Automata,languages and machines,vol. A.Academic Press,New York,1974.

13.-Flake (R.H.).-Volterra series representation of nonlinear systems.AIEE Trans.,81,1963,p. 330-335.

14.-Flake (R.H.).-Volterra series representation of time varying nonlinear systems,Proc. 2nd IFAC Congr. (Bâle,1963),p. 91-99. Butterworths,London et R. Oldenbourg,Munich,1964.

15.-Fliess (M.).-Sur la réalisation des systèmes dynamiques bilinéaires.C.R. Acad. Sc. Paris,A-277,1973,p. 923-926.

16.-Fliess (M.).-Sur divers produits de séries formelles.Bull. Soc. Math. France,102,1974,p. 181-191.

17.-Fliess (M.).-Matrices de Hankel.J. Math. Pures Appl.,53,1974, p. 197-222.

18.-Fliess (M.).-Séries de Volterra et séries formelles non commutatives.C.R. Acad. Sc. Paris,A-280,1975,p. 965-967.

19.-Fliess (M.).-Un outil algébrique:les séries formelles non commutatives,in "Algebraic System Theory"(G. Marchesini,éd.).Lect. Notes Math.,Springer-Verlag,Berlin,à paraître.

20.-Krener (A.J.).-Bilinear and nonlinear realizations of input-output maps.SIAM J. Control,13,1975,p. 827-834.

21.-Lévy (P.).-Problèmes concrets d'analyse fonctionnelle.Gauthier-Villars,Paris,1951.

22.-Lo (J. T.-H.).-Signal detection on Lie groups,in "Geometric Methods in System Theory"(D.Q. Mayne et R.W. Brockett,éd.),p. 295-304.D. Reidel,Dordrecht,1973.

23.-Marcus (S.I.).-Estimation and analysis of nonlinear stochastic systems.Ph. D. Thesis,Dept. Electrical Engin. Comput. Sci.,M.I.T., Cambridge (Mass.),1975.

24.-Reddy (P.K.) et Reddy (D.C.).-Volterra kernels in nonlinear systems.Electronics Lett.,9,1973,p. 426-427.

25.-Schützenberger (M.P.).-On the definition of a family of automata. Inform. Control,4,1961,p. 245-270.

26.-Stratonovich (R.L.).-A new representation for stochastic integrals and equations (traduit du russe).SIAM J. control,4,1966,p. 362-371.

27.-Sussmann (H.J.).-Semigroup reprsentation,bilinear approximation of input-output maps,and generalized inputs,in "Algebraic System Theory"(G. Marchesini,éd.).Lect. Notes Math.,Springer-Verlag, Berlin,à paraître.

28.-Wiener (N.).-Nonlinear problems in random theory.M.I.T. Press, Cambridge(Mass.),1958.

THE NUMERICAL DESIGN OF FEEDBACK CONTROL SYSTEMS CONTAINING A SATURATION ELEMENT BY THE METHOD OF INEQUALITIES

Gray, J. O., Al-Janabi, T. H. Department of Electrical Engineering,
U.M.I.S.T.,
P.O. Box 88,
Sackville Street,
Manchester, M60 1QD, England

This paper is concerned with the numerical synthesis of compensators for single loop feedback control systems containing a single non-linear element of the saturation type. The approach adopted is that of the method of inequalities[1] where the problem is formulated by a set of inequalities which represent the closed loop performance required from the system as well as constraints of a physical engineering or financial kind. The inequalities define a set of points called the admissible space in the space of possible controller parameters and any point in this set represents an acceptable design. The approach differs from that of optimisation in that system performance is specified by a set of inequalities rather than a cost function, so that every aspect of the desired performance can be independently specified.

In previous work[1] the method was applied to the design of linear systems and here the method is extended to non-linear systems by the addition of a new inequality which ensures that every point in the admissible space represents a closed loop system which does not exhibit autonomous limit cycle oscillation. This inequlity is defined by using an enhanced sinusoidal describing function algorithm recently proposed by Mees[2] which takes account of the effects of higher harmonic signal components as well as the fundamental sinusoid in the search for regions of possible limit cycle operation. To simplify the numerical calculation of indicial responses, the non-linear element is replaced by an equivalent linear gain using the technique of the exponential describing function.[3] An example of use is given and the method evaluated.

2. The design method

The method of inequalities represents a completely general design philosophy which can be used to encompass constraints in both the time and frequency domain. In the extension to non linear systems it is simply required to find suitable representations for the non linear elements, a choice which will be determined by the way in which the closed loop specifications are to be formulated. For convenience the method of inequalities is now briefly reviewed and the algorithms used in this work outlined.

2.1 The method of inequalities

A design problem can be formulated in terms of a set of inequalities.

$$\phi_i \ (p) \leqslant C_i, \quad i = 1,2 \cdot \cdot \cdot \cdot \cdot \cdot m \quad \text{—— (1}$$

where the C_i are real numbers and p denotes the real vector $[p_1, \cdot \cdot \cdot \cdot \cdot \cdot \cdot \cdot \cdot p_n]$ and ϕ_i are real functions of p.

The inequalities (1) can represent performance specifications and system constraints which can be of a physical, engineering or financial kind and the components of p represent system parameters. Each inequality

$\phi\ (p) \leq C_i$ defines a set of S_i of points in the n dimensional space R^n ; the coordinates of this space are $p_1, p_2 \ldots\ldots p_n$. We can write

$$S_i = \{ p : \phi_i\ (p) \quad C_i \qquad \text{--(2}$$

where the boundary of S_i is defined by $\phi_i\ (p) = C_i$.

If there is a point p in R^n that satisfies simultaneously all the inequalities $\phi_i\ (p) \leq C_i \quad i = 1,2. \ldots\ldots m$ then p is inside every set S_i. Let S denote the intersection of all sets S_i; in the usual notation

$$S = \bigcap_{i=1}^{m} S_i$$

Thus p satisfies all the inequalities if and only if p is in S. We say that S is an admissable set and any p in S is an admissable point. This concept is illustrated graphically in Figure 1.

Computer programming techniques have been developed for the determination of controllers for single variable and multivariable linear systems using this design philosophy where the $\phi\ (p)$ relate to desired constraints on the loop indicial responses and loop interaction effects and also to practical constraints on such factors as controller parameter values and the magnitude of forcing functions. The direct extension of these techniques to non linear systems required the use of an algorithm to represent the non linear element. Ideally such an algorithm should be computationally simple yet sufficiently accurate to yield results which approximate closely to actual system responses.

2.2 The exponential input describing function

If for a step input sing al to the non linear feedback system of figure 2 the output increases monotonically to a final steady state value, then the input to the non linear element will be of a monotonically decreasing form as shown. This leads to the consideration of a model input signal to the non linearity which is an exponential time function (3). The exponential input describing function representation of the non linearity is determined by minimizing the integral-squared error in a linear approximation to the actual non linear output. The magnitude of this equivalent linear gain is thus determined as

$$N_E = \frac{\int_0^{\infty} x(t) \quad y\ \{x(t)\}\ dt}{\int_0^{\infty} x^2(t) \quad dt} \quad ; \quad \text{For } x\ (t) = Ee^{-t}/T$$

This relationship generally results in a simple analytical expression for most common static non linearities and is thus easy to compute. In practice, of course, the input to the non linearity will be of the more general form

$$E\ e^{-t/T} \cdot \ \text{sine}\{\omega t + \Phi\} + \text{bias}$$

where the harmonic content is determined by the systems dominant complex poles. However, if the percentage overshoot is constrained to 10% or less this harmonic component has little effect on the computed value of N_E. The effect of any bias level can be considered by a suitable adjustment to the value of E used in the derivation of N_E (4).

In this work the application of the exponential input describing function is limited to the study of single valued saturation type non linearities and all system linear elements are assumed to have essentially low pass frequency characteristics.

2.3 Limit Cycle Prediction

Sinusoidal input describing function techniques have been used widely in the study of limit cycle prediction in non linear single loop feedback systems where the application of complex frequency domain analysis has represented an intuitive extension of linear systems theory. Although the limitations of this approach are well documented, it has been shown recently (2) that the accuracy of limit cycle prediction can be improved by the addition of an error band to the $\hat{G}(j\omega)$ locus in the complex frequency comain. This error band is defined by a set of error circles centred on the $\hat{G}(j\omega)$ locus which represent the contribution of higher harmonic signal components and this artifice allows the possibility of a more rigorous study of limit cycle behaviour than any treatment based only on a fundamental sinusoidal analysis.

For the autonomous non linear feedback system of figure 3 where the single valued non linear element is constrained to a sector defined by lines with slopes α and β where $\beta > \alpha$, the radius of an error circle on $\hat{G}(j\omega)$ corresponding to the k harmonic is determined as

$$\sigma(\omega) = \left\{\frac{\beta-\alpha}{2}\right\}^2 \Big/ \left\{\rho(\omega) - \frac{\beta-\alpha}{2}\right\}$$

where

$$\rho(\omega) = \min_{\substack{k \neq 1 \\ k \text{ odd}}} \left| \frac{\alpha+\beta}{2} + \hat{G}(jk\omega) \right|$$

and $\rho(\omega)$ is defined only for values of ω on the set

$$\Gamma \triangleq \left\{ \omega : \rho(\omega) > \frac{\beta-\alpha}{2} \right\}$$

The envelope of all such circles over a subset of Γ' of Γ is the uncertainty band. The reason for choosing a subset is that as ω decreases below a certain value, the error circles become increasingly large and cease to give useful information. In the case of a complete intersection of the uncertainty band with the describing function locus as in figure 4 it can be shown that limit cycle operations can be rigorously predicted with a frequency in the range given by the span ω_1 to ω_2 and an amplitude lying within the range between X_1 and X_2. In addition the closed loop system can have no periodic oscillation of least frequency $\omega \in \Gamma'$ if there is no intersection of any part of the uncertainty band with the describing function locus. It should be noted that although this algorithm adds a degree of mathematical rigour to the normal intuitive sinusoidal describing function approach, the results obtained will only be particularly useful if the harmonic content of the signal is restrained within reasonable bounds and the requirement of essentially low pass linear system elements still remains.

3. Computational Procedures

The non linear system is assumed to have the structure shown in figure 5 where K(s) represents a precompensator with linear elements. A general structure for the elements of K(s) must first be chosen. This can be as complex as desired but it is found useful to start with very simple forms such as a proportional plus integral element or an elementary phase lead structure. The magnitude of the input step signal is now chosen and the equivalent gain derived for the non linear element using the exponential input describing function approximation. The linearised model is now examined for closed loop stability and a set of coefficients automatically computed for the elements of K to ensure this condition. For the chosen input step signal magnitude, the loop time response is now specified as a set of constraints which fix limits to such factors as rise time, percentage maximum overshoot and steady state error. Physical constraints can also be placed on the coefficients of the elements of K and on the magnitude of the input forcing function from the compensator.

Starting from the initial values of the coefficients of the elements of K which ensured closed loop stability for the linearised system, the design program now iterates to find suitable values for these elements which will meet the imposed constraints. At every stage in the iteration the stability of the linearised system is first checked and the derived values of the coefficients then used to seek for solutions of the harmonic balance equation in the complex frequency domain where the non linear element is now represented by its corresponding sinusoidal input describing function. Limit cycle operation is deemed to exist if, over a chosen frequency set Γ', error circles associated with either the third or fifth harmonic of the fundamental intersect the describing function locus. By incorporating this second computational procedure as a subroutine of the iteration sequence it is thus possible to ensure that each point in the admissable space represents a closed loop system which is theoretically free of limit cycle behaviour.

The computer program gives the result of each iteration so that the user is aware not only of the progress of the design but also of those particular specifications which are most difficult to meet within the limitations imposed by physical constraints and the nature of the compensator forms chosen for the design. The addition of a computer graphical facility allows system response to be studied in both the time and frequency domain at any stage in the design procedure if required.

4. An example of use

A precompensator is to be designed for the system shown in Figure 6 to ensure closed loop stability and a time response to a unit step signal which is defined by the following set of inequalities

Rise time $<$ 6 seconds

settling time $<$ 10 seconds

percentage overshoot $<$ 10%

maximum compensat or output $<$ 10

A simple compensator structure was chosen of the form

$$K(s) = \frac{K_1 (1 + K_2 s)}{1 + K_3 s}$$

where each K was restricted to a value 0.1

An initial investigation for limit cycle operation in the uncompensated system produced the computer graphical display shown in figure 7 which predicts a limit cycle with a magnitude in the range 1.3 → 1.6 and frequency in the range 1.9 → 2.05 Radians/sec. The limit cycle subsequently obtained from an analogue simulation of the uncompensated system is shown in trace d of figure 8 and has a measured magnitude of 1.43 and frequency 1.6 Radians/sec.

After five iterations of the design program a compensator of the form

$$K(s) = \frac{0.08 (1 + .035s)}{(1 + 0.0455s)}$$

was obtained

and the computed closed loop time response was as shown in trace b^1 of figure 8. The corresponding analogue computer simulation result is shown in trace b of figure 8. Traces a and c give the output response of the same simulated system with input step signal magnitudes of 0.5 and 2.0 respectively. The behaviour of the compensated system in the frequency domain is shown in the computer graphical display of figure 9 which indicates as expected, that there is now no intersection between the error circles on $\hat{G}(j\omega)$ and the describing function locus. The required design specifications have been achieved. Computation required approximately 30 seconds on the PDP 10 computer at U.M.I.S.T.

5. Evaluation

The successful extension of the method of inequalities to non linear feedback system design depends on the validity of the approximations used for representing the non linear element. Although simple in form, the exponential input describing function approximation has proved both remarkably accurate and robust in its prediction of indicial loop responses. Part of the reason for this must lie in the fact that while at the beginning of the iteration sequence the form of signal input to the non linear element may diverge appreciably from that assumed, as the computation proceeds the compensator structure enforces the required shape on the output time function which, in turn, increases the accuracy of the approximation and allows rapid convergence of the design. The convergence process can be further improved at little cost in computation time by calculating the magnitude of the input signals to the non linear element at each iteration and adjusting the values of the derived linear gains appropriately.

In theory, any derived compensator will meet the required performance in the time domain for the one input signal magnitude chosen at the beginning of the design. It is not difficult to recalculate the coefficients of the compensator elements over a range of magnitudes of input signal and derive a compromise set of coefficient values. In practice, over a very wide range of worked examples it has been found that provided the original input signal magnitude is chosen sensibly, the result produced by the exponential input describing function approximation is sufficiently robust to render this exercise generally unnecessary. This is illustrated in the example given.

Initially the non linear function was restricted to a simple saturation element as it was found that in the open loop condition the exponential input describing function approximation gave particularly good results with this type of non linearity. Within the context of the design program, however, equally good results have now been achieved with single valued non linear elements of different types and the range of application has thus been extended. The assumption of, essentially, low linear system elements still, of course, remains a vital prerequisite in any study involving the sinusoidal describing function philosophy. It has been recently shown (4) that the method can be extended to multi input, multi output, low pass linear systems containing a set of single valued, bounded non linear elements.

Within the method of inequalities the use of different algorithms than those discussed above is of course possible and these will be introduced as experience dictates to deal with particularly difficult system configurations which are not amenable to the simple approximations used here. It should be stressed that at every stage in the iteration process the user is faced with design decisions and the progress of the design and even its ultimate success will depend on a sensible evaluation of both initial constraints and possible controller structures. The user must also have some initial insight into those factors which influence the system under consideration and be aware at all times of the limitations inherent in the algorithms used in the computational procedure.

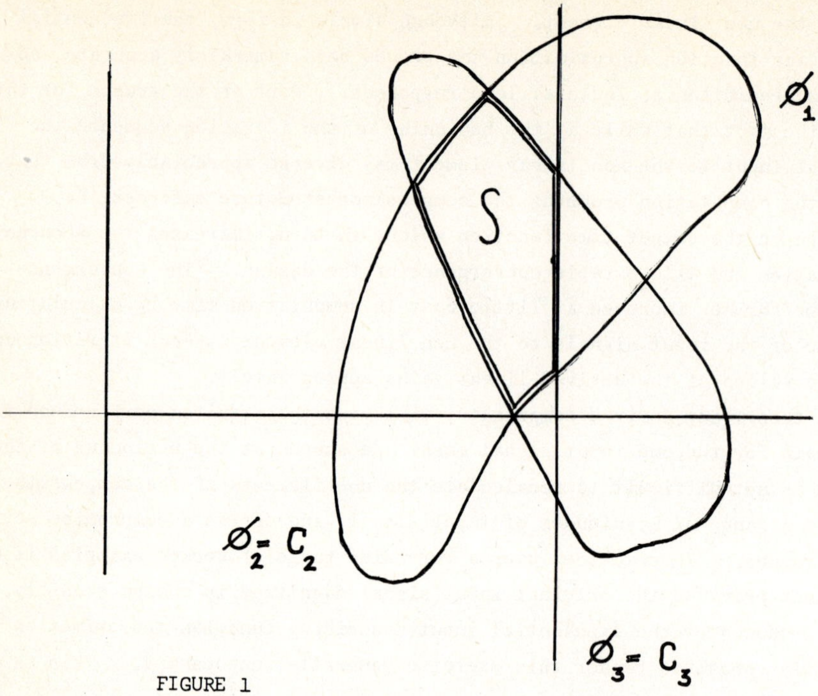

FIGURE 1

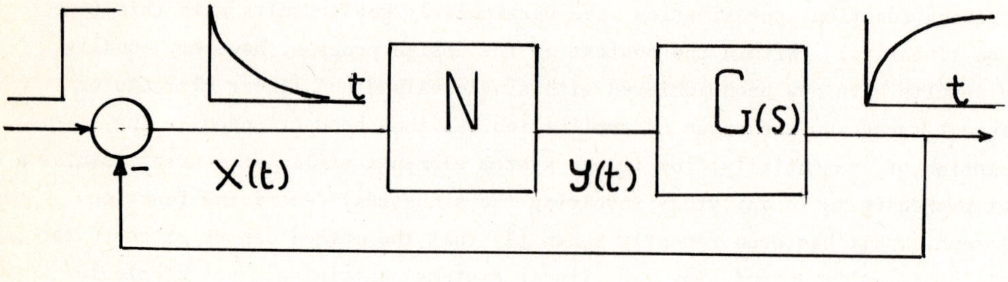

FIGURE 2

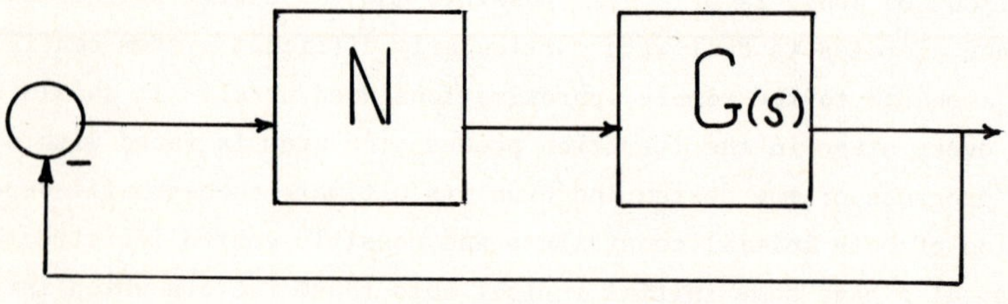

FIGURE 3

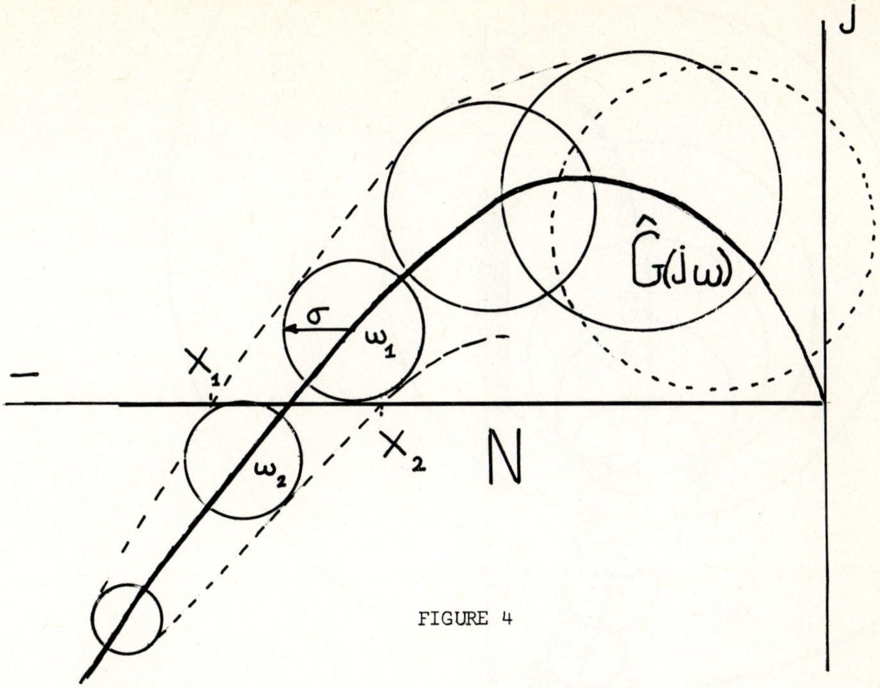

FIGURE 4

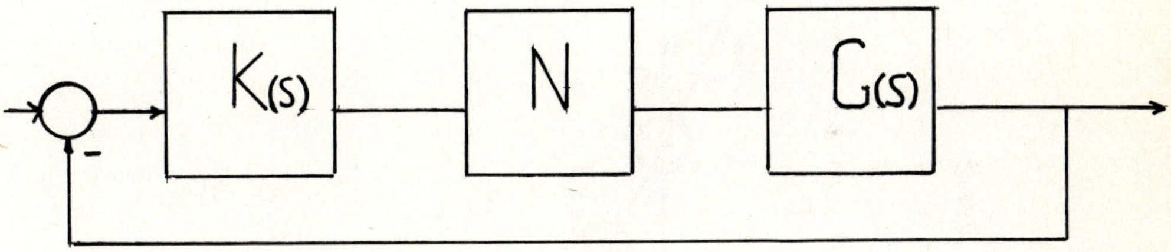

FIGURE 5

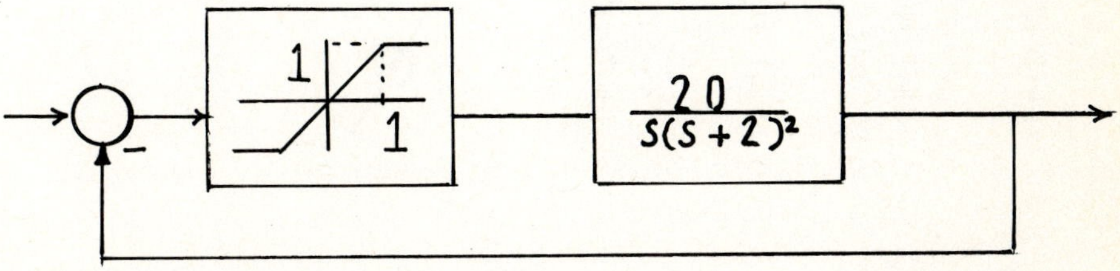

FIGURE 6

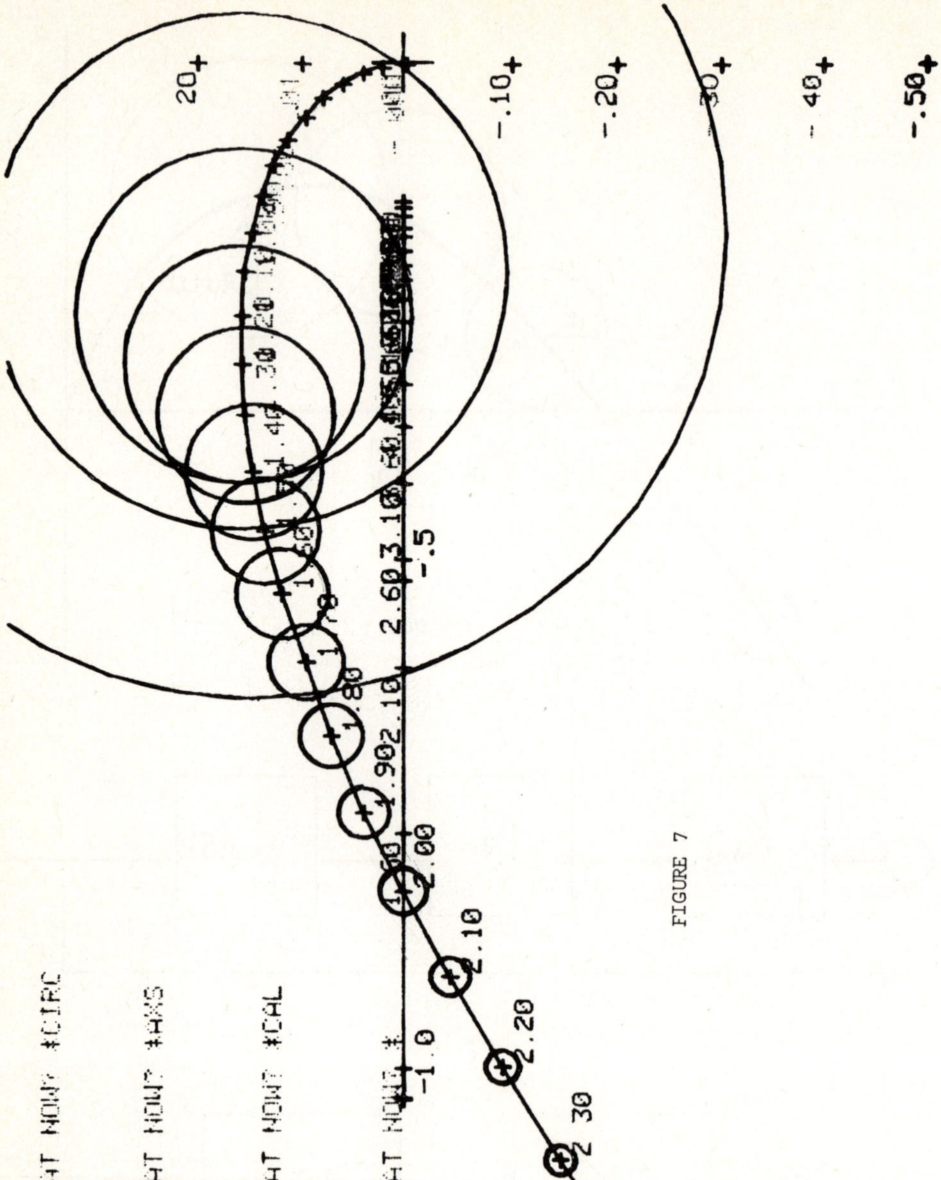
WHAT NOW? #CIRC
WHAT NOW? #AXS
WHAT NOW? #CAL
WHAT NOW? #
-1.0
-.5
-.10
-.20
-.40
-.50
2.00
2.10
2.20
2 30

FIGURE 7

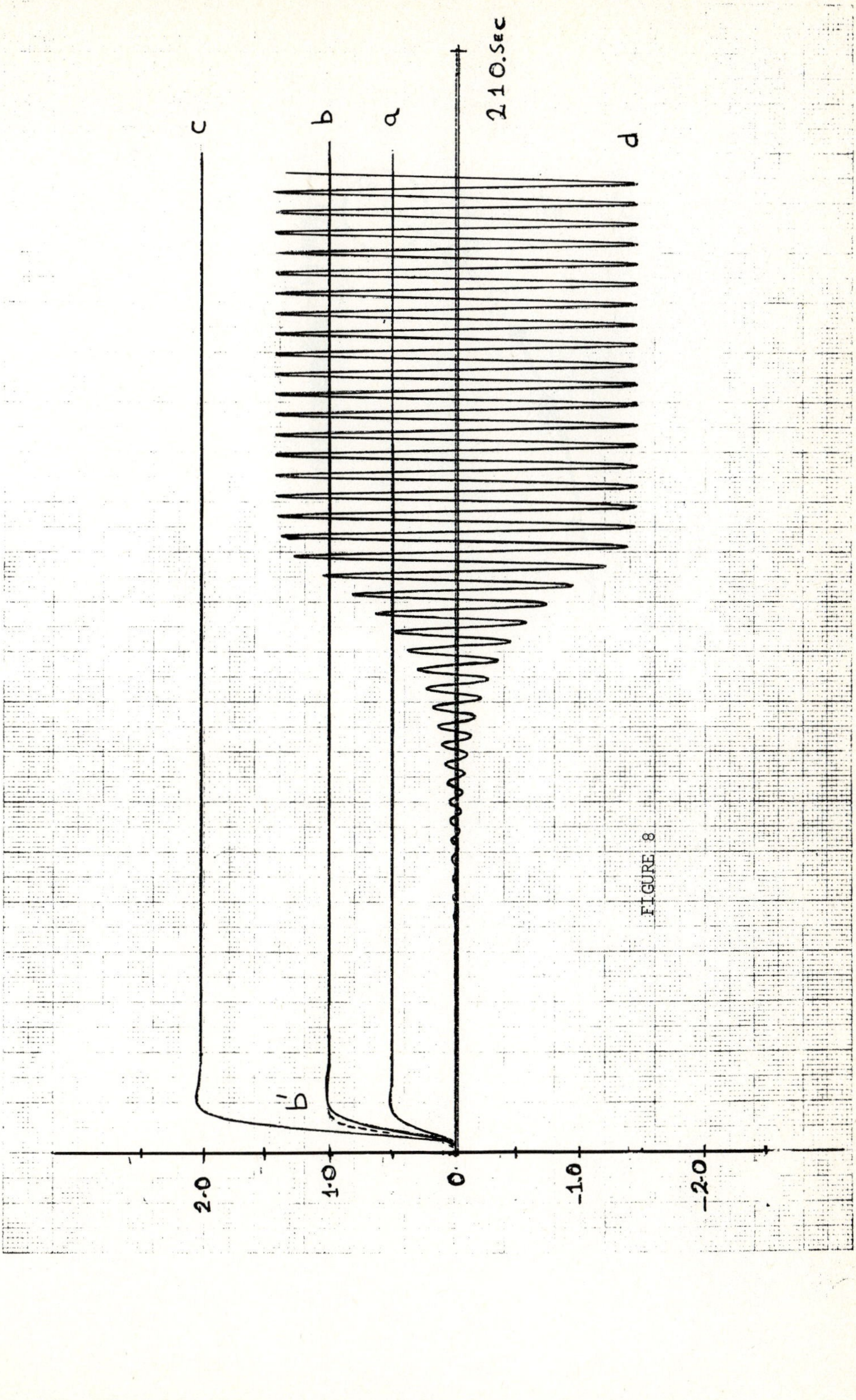

FIGURE 8

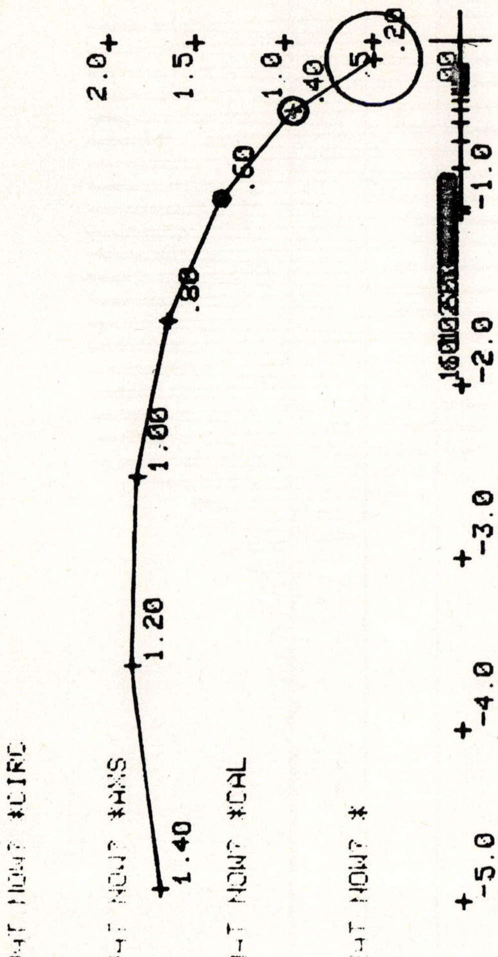
WHAT NOW? #CIRC
NOW? #AXS
NOW? #CAL
NOW? #
1.40
1.20
1.00
.80
.60
.40
.20
2.0
1.5
1.0
-5.0
-4.0
-3.0
-2.0
-1.0

FIGURE 9

References

1. Zakian, V., Al Naib, U., "Design of dynamical and control systems by the method of inequalities". Proc. I.E.E. Vol. 120, No. 11. Nov. 1973. pp 1421-1427.

2. Mees, A., Bergen, A. R.; "Describing function revisited"; Proc. I.E.E.E. Vol. AC 20, 1975, pp 473 - 478

3. Gelb, A., Vander Velde, W.; "Multiple input describing functions and non linear systems design". McGraw Hill 1958.

4. Al-Janabi, T. H.; "The numerical design of multivariable non linear control systems". Ph.D. Thesis, U.M.I.S.T., 1975.

SUR L'APPROXIMATION DU CONTRÔLE OPTIMAL DES SYSTÈMES GOUVERNES PAR DES EQUATIONS DIFFERENTIELLES AVEC RETARD PAR LA METHODE DE DIFFERENCES FINIES

Irena Lasiecka, Andrzej Hatko

Institut de l'Organisation, Gestion

et du Contrôle

Varsovie, KRN 55

1. Introduction

Le problème qui va nous préocuper est l'approximation numérique du problème de la minimization d'une fonctionelle intégrale pour les systèmes gouvernés par des équations différentielles avec retard.

On considère la méthode de différences finies qui est relativement simple au point de vue de l'algorithme numérique [1], [2].

Le but principal de ce travail est d'estimer à priori la différence au sens de la norme L^2 entre la solution optimale d'un problème exact et d'un probleme approximé.

2. Position du problème.

Soit $x \in H^1 [OT; R^n]$; $\varphi \in H^1 [-h,0; R^n]$; $u \in H^1 [OT; R^m]$

$A: L^2[OT; R^n] \times L^2[OT; R^n] \times L^2[OT; R^m] \to L^2[OT; R^n]$

$\Phi: L^2[OT; R^n] \times L^2[OT; R^m] \to L^2[OT; R^n]$

$\Omega \subset H^1[OT; R^m]$ - un convexe fermé à l'intérieur non-vide.

On admet que les hypothèses suivantes sont verifiées

H1 - A est un opérateur borné dans $L^2 [OT; R^n]$

H2-A est un opérateur strictement monotonique c'est à dire-il existe $\alpha > 0$ tel que pour chaque $x_1, x_2, y \in L^2 [OT; R^n]$; $u \in L^2[OT; R^m]$

$$\int_{kh}^{(k+1)h} (A(x_1(t), y(t), u(t)) - A(x_2(t), y(t), u(t)), x_1(t) - x_2(t))dt \geq$$

$$\geq \alpha \int_{kh}^{(k+1)h} | x_1(t) - x_2(t)|^2 dt$$

pour $k = 0,1 \ldots m-1$ où on admet que $m = \frac{T}{h}$ est un entier.

H3 - La fonctionnelle $J(x,u) = \int_0^T \Phi(x(t), u(t))\, dt$ est radialement non bornée- on dit que $J(x,u)$ est radialement non borné si $\Phi(x,u)$ tend vers l'infini quand $\|u\|$ tend vers l'infini (par $\|\cdot\|$ on note la norme au sens de l'espace L^2).

On considère alors le problème suivant:

PROBLÈME Q_0

Trouver $(x^0, u^0) \in H^1 [OT; R^n] \times H^1 [OT; R^m]$ minimizant $J(x,u)$ sous les contraintes.

(1) $\frac{dx(t)}{dt} + A(x(t), x(t-h), u(t)) = 0$[1]; $t \in [0,T]$ où $h > 0$

---

(1) Les derivations sont prises au sens des distributions.

(2) $x(Q) = \varphi(Q) \quad Q \in [-h,0]$

(3) $u \in \Omega$

On suppose tout au long du travail que le problème Q_0 admet une solution. On associe au problème Q_0 une famille de problèmes "approchés" et au lieu du problème Q_0 on resout le problème "approché".

Pour cela on introduit [3] une famille d'espaces

$E_\tau[-h;\ T+\tau\ ;\ R^n] \subset L^2[-h;\ T+\tau\ ;\ R^n]$ où τ est un pas en temps destiné à tendre vers zéro.

En plus on admet qu'il est possible de choisir τ de façon que $l \overset{dt}{=} \frac{h}{\tau}$ et $k \overset{dt}{=} \frac{T}{\tau}$ sont des entiers (cette condition n'est pas indispensable au point de vue du résultat final).

On définit par $E_\tau[-h;\ T+\tau\ ;\ R^n]$ l'espace des fonctions de la forme:

$$x_\tau(t) \overset{dt}{=} \sum_{r=-l}^{k} x_\tau(r\tau)\, W_r(t)$$

où $W_r(t)$ est la fonction caractéristique de l'intervalle $[r\tau\ ;\ (r+1)]$ et $x_\tau(r\tau) \in R^n$ (renvoyez à [3]).

Donc les fonctions de $L^2[-h, T;\ R^n]$ sont approximées par des fonctions en escalier.

Soit $P_\tau : H^1[-h; T;\ R^n] \to E_\tau[-h;\ T+\tau\ ;\ R^n]$ un opérateur de la projection tel que:

$\exists C > 0 \quad \| P_\tau x - x \| \leq C\tau$ pour chaque $x \in H^1[-h;\ T;\ R^n]$

Un ensemble fermé et convex $\mathcal{P}_\tau \Omega \subset E_\tau[0;\ T\ ;\ R^m]$ (2) est dit

---

(2) $E_\tau[0;\ T+\tau\ ;\ R^m]$ designe le decoupage de $E_\tau[-h\ ;\ T+\tau\ ;\ R^m]$

l'approximation de Ω s'il verifie l'axiomes suivants:

(4) $\exists\, c_1 > 0\ \forall u \in \Omega\ \exists u_\tau \in P_\tau \Omega$ tel que $\|u-u_\tau\| \leq c_1 \tau$

(5) $\exists\, c_2 > 0\ \forall\, u_\tau \in P_\tau \Omega\ \exists\, u \in \Omega$ tel que $\|u - u_\tau\| \leq c_2 \tau$ (cf. [9])

Soit $x_\tau \in E_\tau[0; T+\tau; R^n]$; ; $u_\tau \in E_\tau[0; T+\tau ; R^m]$

On définit un opérateur de differenees finiès ∇_τ par:

$$\nabla_\tau x_\tau(t) \overset{dt}{=} \frac{x_\tau(t+\tau)-x_\tau(t)}{\tau}$$

On est maintenant en mesure d'énoncer pour chaque τ un problème "approché" associé au problème Q_o.

PROBLÈME Q_τ

Trouver $(x_\tau^o, u_\tau^o) \in E_\tau[0; T+\tau\ R^n] \times E_\tau[0; T : R^m]$

minimizant $J(x_\tau, u_\tau) = \int_0^T \Phi(x_\tau(t), u_\tau(t))dt$ sous les contraintes

(6) $\nabla x_\tau(t) + A(x_\tau(t),\ x_\tau(t-h), u_\tau(t)) = 0$

(7) $x_\tau(Q) = P_\tau \varphi(Q)$ $Q \in [-h, 0)$; $x_\tau(0) = \varphi(0)$

(8) $u_\tau \in P_\tau \Omega$

En utilisant les hypothèses H1, H3 d'un raisonnement analogue à celui de [4] on prouve que le problème Q_τ admet la solution optimale. Plus que ça, par les hypothèses H1, H2, H3 on montre que (x^o, u^o) et (x_τ^o, u_τ^o) appartiennent à certain ensemble $G \subset L^2[OT; R^m] \times L^2[OT; R^m]$ qui peut être estimé à priori. Notons, que les problèmes Q_o et Q_τ on peut considerer comme des problèmes variationnelles de la minimization d'une fonctionnelle dans une espèce d'Hilbert.

Cependant on associe aux problèmes Q_o et Q_τ les fonctionnelles de Lagrange L et L_τ respectivement

Soit $\lambda \in L^2[OT; R^m]$

on pose $y(t) \overset{dt}{=} x(t-h)$

$L : H^1[-h,T\,;\,R^n] \times H^1[0,T;R^m] \times L^2[0,T;R^n] \rightarrow R^1$ est défini la formule :

(9) $L(x,u,\lambda) \overset{dt}{=} J(x,u) + \langle \lambda, \frac{dx}{dt} + A(x,y,u) \rangle$ où $\langle\ ,\ \rangle$ désigne la dualité dans $L^2[0,T, R^n]$.

Si on admet certaines conditions de la regularité posées sur A et Φ on a [5] qu'il existe $\lambda^0 \in H^1[0,T\,;\,R^n]$ tel que :

(10) $\langle \delta_x L(x^0,u^0,\lambda^0), \delta x \rangle + \langle \delta_y L(x^0,u^0,\lambda^0), \delta y \rangle = 0$

pour chaque $\delta x \in H^1[-h,T\,;\,R^n]$ tel que

$\delta x(Q) = 0 \quad Q \in [-h,0]$. (on se souvient que $\delta y(t) \overset{dt}{=} \delta x(t-h)$)

(11) $\delta_\lambda L(x^0,u^0,\lambda^0) = 0$.

(12) $\langle \delta_u L(x^0,u^0,\lambda^0), u-u^0 \rangle \geq 0 \quad \forall u \in \Omega$

On fait en plus sur L l'hypothèse

H4 $$\left\langle \begin{bmatrix} L_{xx}(\tilde{x},\tilde{u},\tilde{\lambda}) & ,L_{ux}(\tilde{x},\tilde{u},\tilde{\lambda}) \\ L_{xu}(\tilde{x},\tilde{u},\tilde{\lambda}) & ,L_{uu}(\tilde{x},\tilde{u},\tilde{\lambda}) \end{bmatrix} \begin{bmatrix} x \\ u \end{bmatrix}, \begin{bmatrix} x \\ u \end{bmatrix} \right\rangle \geq \gamma \|u\|^2$$

ou $\gamma > 0$; $(\tilde{x},\tilde{u},\tilde{\lambda})$ appartiennent à un certain voisinage de (x^0, u^0, λ^0) (cf [6,7]),

Par une mèthode analogue on introduit :

$L_\tau : E_\tau[-h;\ T+\tau\ ,R^n] \times E_\tau[0;T+\tau\ ;R^m] \times E_\tau[0;T+\tau\ ;R^n] \rightarrow R^1$
defini par

(13) $L_\tau(x_\tau, u_\tau, \lambda_\tau) \overset{dt}{=} J(x_\tau, u_\tau) + \langle \lambda_\tau, \nabla x_\tau + A(x_\tau, y_\tau, u_\tau) \rangle$
ou $y_\tau(t) \overset{dt}{=} x_\tau(t-h)$.

De même on a bien qu'il existe $\lambda_\tau^0 \in E_\tau[Q;T+\tau\ ;R^n]$ tel que :

(14) $\langle \delta_x L_\tau(x_\tau^0,u_\tau^0,\lambda_\tau^0), \delta x_\tau \rangle + \langle \delta_y L_\tau(x_\tau^0,u_\tau^0,\lambda_\tau^0), \delta y_\tau \rangle = 0$
pour tout $\delta x_\tau \in E_\tau[0;T+\tau\ ;R^n]$ tel que $\delta x_\tau(Q) = 0$
$Q \in [-h,0]$.

(15) $\delta_{\lambda_\tau} L_\tau(x_\tau^0, u_\tau^0, \lambda_\tau^0) = 0$

(16) $\langle \delta_{u_\tau} L_\tau(x_\tau^0, u_\tau^0, \lambda_\tau^0), u_\tau - u_\tau^0 \rangle \geq 0 \quad \forall u_\tau \in \mathcal{R}_\tau \Omega$

Respectivement sur L_τ on fait l'hypothèse :

H4' $$\left\langle \begin{bmatrix} L_{\tau\, xx}(\tilde{x}_\tau, \tilde{u}_\tau, \tilde{\lambda}_\tau), & L_{\tau\, ux}(\tilde{x}_\tau, \tilde{u}_\tau, \tilde{\lambda}_\tau) \\ L_{\tau\, xu}(\tilde{x}_\tau, \tilde{u}_\tau, \tilde{\lambda}_\tau), & L_{\tau\, uu}(\tilde{x}_\tau, \tilde{u}_\tau, \tilde{\lambda}_\tau) \end{bmatrix} \begin{bmatrix} x_\tau \\ u_\tau \end{bmatrix}, \begin{bmatrix} x_\tau \\ u_\tau \end{bmatrix} \right\rangle \geq \gamma \| u_\tau \|^2$$

où $(\tilde{x}_\tau, \tilde{u}_\tau, \tilde{\lambda}_\tau)$ appartiennent à un certain voisinage de $(x_\tau^0, u_\tau^0, \lambda_\tau^0)$.

Notons que les hypothèses H4 et H4' permettent d'obtenir l'unicité des solutions (x^0, u^0) et (x_τ^0, u_τ^0).

D'après les hypothèses H4 et H4' on constate que (x^0, u^0, λ^0) et $(x_\tau^0, u_\tau^0, \lambda_\tau^0)$ sont les points de la selle de L et L_τ respectivement. (la démonstration de ce point est classique).

Il est facile de remarquer que le problème Q_τ admet une solution u_τ^0 qui n'appartient pas à Ω - un ensemble des contrôles admissibles .

Donc à l'aide de u_τ^0 on construit effectivement un autre contrôle - disons u_τ^* proche à u^0 et en même temps appartenant à Ω.

L'existence de tel u_τ^* est assurée par la condition (5).

3. L'estimation de la difference entre la solution exacte et "approchée".

Le bût général de ce travail est de faire les estimations de $\|u_\tau^* - u^o\|$; $\|x_\tau^* - x^o\|$ et $J(x_\tau^*, u_\tau^*) - J(x^o, u^o)$ où x_τ^* est une solution de l'équation (1) correspondante au u_τ^*. On va distinguer deux cas selon des propriétés de Ω.

1.er cas - où $\Omega = H^1[OT; R^m]$ (on dit le problème sans contraintes).

2ième cas - où Ω est un sousensemble convex et fermé dans $H^1[OT; R^m]$ (on dit le problème aux contraintes).

On énonce donc le résultat principal du travail sous la forme des Théorèmes (1) et (2). D'abord on formule le Théorème (1) qui concerne un problème sans contraintes.

THÉORÈME (1)

Soit

(a) (x^o, u^o) et (x_τ^o, u_τ^o) des solutions de problème Q_o et Q_τ respectivement; et u_τ^* est définit par (5).

(b) Les hypothèses H1, H2, H3, H4' sont verifiées.

(c) Les opérateurs $A, A_x, A_y, A_u, \Phi_x, \Phi_u$ satisfont la condition de Liphschitz sur G avec des constantes $L_o, L_1, L_2, L_3, L_4, L_5$ respectivement.

(d) Les opérateurs $A_x, A_y, A_u, \Phi_x, \Phi_u, A_{xx}, A_{xy}, A_{xu}, \Phi_{xx}, \Phi_{xu}$ sont bornés au sense de la norme $L_\infty(G)$ par $M > 0$.

(e) $\Omega = H^1[OT; R^m]$

(f) $\tau < \dfrac{\alpha}{2L_0^2}$

donc,

(g) $\| u^0 - u_\tau^* \| \leq d_1 \tau$

(h) $\| x^0 - x_\tau^* \| \leq d_2 \tau$

(i) $| J(x_\tau^*, u_\tau^*) - J(x^0, u^0) | \leq d_3 \tau$

où d_1, d_2, d_3 dependent de: $\left\|\frac{du^0}{dt}\right\|$, $\left\|\frac{d\varphi}{dt}\right\|$, $\|\varphi\|$, α, γ, L_0, L_1, L_2 L_3, L_4, L_5, M.

Le schéma géneral de la preuve est suivant:

I. Par l'hypothèse H4' et les conditions (10,11,12) on trouve d'abord l'estimation de la norme de difference $\|u^0 - u_\tau^0\|$ en termes de $\|x^0 - x_\tau^0\|$ et $\|\lambda^0 - \lambda_\tau^0\|$. Alors on déduit immédiatement du Théorème du Point de la Selle que:

(17) $J(x_\tau^0, u_\tau^0) = L_\tau(x_\tau^0, u_\tau^0, \lambda_\tau^0) \leq L_\tau(\tilde{x}_\tau^0, \tilde{u}_\tau^0, \lambda_\tau^0)$ où

$\tilde{x}_\tau^0 = P_\tau x^0$ et $\tilde{u}_\tau^0$

est une approximation de u^0 appartenante à $P_\tau \Omega$ (celui là existe d'après la condition (4)).

D'autre part en employant la formule de Taylor et l'Hypothèse H4[1] et en désignant $\tilde{\lambda}_\tau^0 \overset{dt}{=} P_\tau \lambda^0$ on a bien que:

(18) $J(x_\tau^0, u_\tau^0) = L_\tau(x_\tau^0, u_\tau^0, \tilde{\lambda}_\tau^0) \geq L_\tau(\tilde{x}_\tau^0, \tilde{u}_\tau^0, \tilde{\lambda}_\tau^0) +$

$+ \langle \delta_{x_\tau} L_\tau(\tilde{x}_\tau^0, \tilde{u}_\tau^0, \tilde{\lambda}_\tau^0), x_\tau^0 - \tilde{x}_\tau^0 \rangle + \langle \delta_{y_\tau} L_\tau(\tilde{x}_\tau^0, \tilde{u}_\tau^0, \tilde{\lambda}_\tau^0), y_\tau^0 - \tilde{y}_\tau^0 \rangle +$

$+ \langle \delta_{u_\tau} L_\tau(\tilde{x}_\tau^0, \tilde{u}_\tau^0, \tilde{\lambda}_\tau^0), u_\tau^0 - \tilde{u}_\tau^0 \rangle + \gamma \|u_\tau^0 - \tilde{u}_\tau^0\|^2$

D'où il vient:

(19) $\gamma \| u_\tau^0 - \tilde{u}_\tau^0 \|^2 \leq L_\tau(\tilde{x}_\tau^0, \tilde{u}_\tau^0, \lambda_\tau^0) - L_\tau(\tilde{x}_\tau^0, \tilde{u}_\tau^0, \tilde{\lambda}_\tau^0) +$

$+ \langle \delta_{x_\tau} L_\tau(\tilde{x}_\tau^0, \tilde{u}_\tau^0, \tilde{\lambda}_\tau^0), x_\tau^0 - \tilde{x}_\tau^0 \rangle + \langle \delta_{y_\tau} L_\tau(\tilde{x}_\tau^0, \tilde{u}_\tau^0, \tilde{u}_\tau^0), y_\tau^0 - \tilde{y}_\tau^0 \rangle +$

$+ | \delta_{u_\tau} L_\tau(\tilde{x}_\tau^0, \tilde{u}_\tau^0, \tilde{\lambda}_\tau^0) | \; \| u_\tau^0 - \tilde{u}_\tau^0 \|$

Pour obtenir le résultat voulu il est necessaire d'éstimer des expressions

(20) $L_\tau(\tilde{x}_\tau{}^o,\tilde{u}_\tau{}^o,\lambda_\tau{}^o) - L_\tau(\tilde{x}_\tau{}^o,\ \tilde{u}_\tau{}^o,\tilde{\lambda}_\tau{}^o)$

(21) $\langle \delta_{x_\tau} L_\tau(\tilde{x}_\tau{}^o,\tilde{u}_\tau{}^o,\tilde{\lambda}_\tau{}^o), x_\tau{}^o-\tilde{x}_\tau{}^o\rangle + \langle \delta_{y_\tau} L_\tau(\tilde{x}_\tau{}^o,\tilde{u}_\tau{}^o,\tilde{\lambda}_\tau{}^o),\ y_\tau{}^o-\tilde{y}_\tau{}^o\rangle$

(22) $\delta_{u_\tau} L_\tau(\tilde{x}_\tau{}^o,\tilde{u}_\tau{}^o,\tilde{\lambda}_\tau{}^o)$.

Par la définition de L_τ on estime (20) d'où il révient que:

$$L_\tau(\tilde{x}_\tau{}^o,\tilde{u}_\tau{}^o,\lambda_\tau{}^o)-L_\tau(\tilde{x}_\tau{}^o,\tilde{u}_\tau{}^o,\tilde{\lambda}_\tau^o) \leq \tau C^{(1)} \cdot \|\lambda^o - \tilde{\lambda}_\tau^o\|$$

Ensuite par la condition (10) et (14) on a que :

$$\langle \delta_{x_\tau} L_\tau(\tilde{x}_\tau{}^o,\tilde{u}_\tau{}^o,\tilde{\lambda}_\tau{}^o), x_\tau{}^o-\tilde{x}_\tau{}^o\rangle + \langle \delta_{y_\tau} L_\tau(\tilde{x}_\tau{}^o,\tilde{u}_\tau{}^o,\tilde{\lambda}_\tau{}^o), y_\tau{}^o-\tilde{y}_\tau{}^o\rangle \leq$$
$$\leq \tau C \|x^o - x_\tau^o\|.$$

En employant la condition (16) qui d'après (e) est de la forme

$\delta_{u_\tau} L_\tau(\tilde{x}_\tau{}^o,\tilde{u}_\tau{}^o,\tilde{\lambda}_\tau{}^o) = 0$ on obtient que:

$$|\delta_u L_\tau(\tilde{x}_\tau{}^o,\tilde{u}_\tau{}^o,\tilde{\lambda}_\tau{}^o)| \leq \tau C$$

En effet il en résulte que:

$$\gamma\|u^o - u_\tau^o\|^2 \leq \tau C\ (\|x^o-x_\tau^o\| + \|\lambda_\tau^o - \lambda^o\| + \|u_\tau{}^o - u^o\|)$$

II. $\|\lambda^o - \lambda_\tau{}^o\|$ on estime en termes de $\|x^o - x_\tau{}^o\|$ et $\|u^o-u_\tau{}^o\|$ en utilisant l'équation de l'état adjoint ce qui découle d'une façon immédiate des conditions (10) et (14). D'où on a alors

$$\|\lambda^o - \lambda_\tau^o\| \leq C(\|x^o-x_\tau{}^o\| + \|u^o - u_\tau{}^o\|)$$

III D'après l'hypothèse H2 et en utilisant l'équation de l'état on estime que: $\|x^o - x_\tau{}^o\| \leq C\|u^o - u_\tau{}^o\|$

IV En appliquant les résultats données par I, II, III on déduit:

(23) $\|u^o - u_\tau{}^o\| \leq C\tau$.

V Parce que $\|u^o - u_\tau^*\| \leq \|u^o-u_\tau{}^o\| + \|u_\tau{}^o-u_\tau^*\|$ donc en employant (23) et la condition (4) on a:

$\|u^o - u_\tau^*\| \leq C\tau$ ce qui est le résultat voulu (g).

---

(1) C designe les constantes diverses.

VI. Ensuite de facon général on estime $\|x^o - x_\tau^*\|$ (où x_τ^* est une solution de l'équation (1) correspondante au u_τ^*).

D'où on obtient (h).

VII À la fin on estime $|J(x_\tau^*, u_\tau^*) - J(x^o, u^o)|$ en utilisant les certaines conditions de la régularité faites sur J(x,u) $(\varphi_x, \varphi_u \in L_\infty(G))$.

Le Théorème (2) formulé dessous traite d'un problème aux contraintes (le deuxième cas). Il faut rémarquer que dans ce cas les conditions (12), (16) ne peuvent pas être remplacées par : $\delta_u L(x^o, u^o, \lambda^o) = 0$ et $\delta_{u_\tau} L_\tau(x_\tau^o, u_\tau^o, \lambda_\tau^o) = 0$. Pour cela on obtient un résultat plus faible par rapport au cas précédent- le rang de la convergence est égal $O(\tau^{\frac{1}{2}})$ (au contraire d'un problème sans contraintes, où le même rang est égal $O(\tau)$).

THÉORÈME 2

Que : (x^o, u^o), (x^o, u^o), u_τ^*, A, φ_x, φ_y, A_x, A_y, A_u satisfont les conditions du Théorème (1). En plus on admet que $\Omega \neq H^1[OT; R^m]$

Donc:

$\|u^o - u_\tau^*\| \leq C_4 \tau^{\frac{1}{2}}$

$\|x^o - x_\tau^*\| \leq C_5 \tau^{\frac{1}{2}}$

$J(x^*, u^*) - J(x^o, u^o) \leq C_6 \tau^{\frac{1}{2}}$

où C_4, C_5, C_6 dependent de : L_o, L_4, L_5, $\|\varphi\|$, $\|\frac{d\varphi}{dt}\|$, $\|\frac{du^o}{dt}\|$, M, α, γ.

Le Théorème (2) se démontre par une méthode analogue au Théorème (1). (Les détails qui concernent la démonstration du Théorème (2) sont présentés à [8]).

Remarque :

Les Théorèmes (1) et (2) restent valables dans, le cas ou l'opèrateur A et la fonctionelle J dèpendent d'une facon explicité du temps. La démonstration du Théorème dans ce cas se fait sans difficultés.

4. Exemple

Un exemple numérique qui illustre la méthode de l'approximation presenté avant, se trouve dans ce chapitre.

Pour pouvoir comparer la solution optimale determinée d'un facon analytique avec celle obtenue par l'approximation, on se limite dans ce cas-ci au probléme d'optimisation dans le cas d'une fonctionelle quadratique et d'équations d'etat lineaires.

Étant donnés :

$$z \in H^1\left[0,2;R^1\right] \quad ; \quad w \in H^1\left[0,2;R^1\right] \quad -1,0;R^1$$

definiès respectivement par :

$$z(t) = -\frac{1}{4}\, t^2 + \frac{7}{8}\, t \quad \text{pour} \quad t \in [0,2] \tag{24}$$

$$w(t) = \begin{cases} t & t \in [0,1) \\ -t+2 & t \in [1,2] \end{cases} \tag{25}$$

$$\varphi(Q) = -\frac{1}{4}\, Q^2 - \frac{9}{8}\, Q \qquad Q \in [-1,0] \tag{26}$$

On considère alors le problème suivant :

minimizer $\int_0^2 \left[|x(t) - z(t)|^2 + |u(t) - w(t)|^2 \right] dt$ sous les contraintes

$$\frac{dx(t)}{dt} + x(t) - x(t-1) - u(t) = 0 \qquad t \in [0,2]$$

$$x(Q) = \varphi(Q) \qquad Q \in [-1,0]$$

Il est facile à verifièr que la résolution analytique x^0, u^0 de ce probleme se presente sous la forme :

$x^o = w$

$u^o = z$

et la valeur optimale de la fonctionnelle est égale zero.

Étant donné N.

Soit $\tau \overset{df}{=} \frac{2}{N-1}$

Au problème exacte on associe un problème"approché" Q_τ qui prend dans ce cas la forme:

minimizer $\int_0^2 \left[|x(t) - P_\tau z(t)|^2 + |u(t) - P_\tau w(t)|^2 \right] dt$

sous les contraintes

(27) $\nabla x_\tau(t) + x_\tau(t) - x_\tau(t-1) - u_\tau(t) = 0 \qquad t \in [0,2]$

(28) $x_\tau(Q) = P_\tau \varphi(Q) \qquad Q \in [-1,0)$

(29) $x_\tau(0) = \varphi(0)$

où $P_\tau z(t) \overset{dt}{=} \sum_{r=0}^{N-1} \frac{1}{\tau} \int_{r\tau}^{(r+1)\tau} z(t)dt \cdot W_r(t)$

Le problème de la minimization d'une fonctionnelle on resout pour chaque N par le méthode iterative du type gradient conjugué.
La résolution de l'équation de l'état correspondante au contrôle obtenu dans les iterations successives on calcule par le suivant algorithm:

$$\frac{x_{r+1} - x_r}{\tau} + x_r - x_{r-\frac{N}{2}} - u_r = 0 \qquad r = 0,1 \quad N-1$$

$$x_r = \varphi_r \qquad r = -\frac{N}{2}, \ldots -1 .$$

Les calculs sont effectués sur l'ordinateur ODRA-1325

Les résultats obtenus sont donnés sur le tableau suivant:

Tabl. 1

N	Numeros des itérations	g	$J\ x_N,\ u_N$	$u_N - u^o\ ^2$
10	1	$0.32502.10^{3}$	$0.32692.10^{1}$	$0.25851.10^{2}$
	2	$0.30649.10^{1}$	$0.91566.10^{2}$	$0.22733.10^{2}$
	3	$0.56083.10^{-2}$	$0.33678.10^{-2}$	022728.10^{2}
	5	$0.36797.10^{-8}$	$0.33629.10^{-2}$	$0.22728.10^{2}$
	10	$0.32581.10^{-14}$	033629.10^{-2}	$0.22728.10^{2}$
50	1	$0.63649.10^{2}$	$0.27087.10^{1}$	$0.66636.10^{1}$
	2	0.47066	$0.39977.10^{-2}$	$0.4102.10^{1}$
	3	$0.67419.10^{-3}$	$0.98461.10^{-4}$	$0.40883.10^{1}$
	5	$0.15849.10^{-9}$	$0.96367.10^{-4}$	$0.40883.10^{1}$
	10	$0.10060.10^{-15}$	$0.96367.10^{-4}$	$0.40983.10^{1}$
100	1	$0.35294.10^{2}$	$0.26720.10^{1}$	$0.47802.10^{1}$
	2	0.25591	$0.37332.10^{-2}$	$0.22556.10^{1}$
	3	$0.35173.10^{-3}$	$0.30724.10^{-4}$	$0.22521.10^{1}$
	5	$0.73443.10^{-1}$	$0.28839.10^{-4}$	$0.22521.10^{1}$
	10	$0.7754.10^{-1}$	$0.28839.10^{-4}$	$0.22521.10^{1}$
130	1	$0.24417.10^{2}$	$0.26588.10^{1}$	$0.40673.10^{1}$
	2	0.17576	$0.36437.10^{-2}$	$0.15560.10^{1}$
	3	$0.23753.10^{-3}$	$0.15473.10^{-4}$	$0.15526.10^{1}$
	5	$0.47404.10^{-1o}$	$0.13663.10^{-4}$	$0.15526.10^{1}$
	10	$0.21749.10^{-15}$	$0.13663.10^{-4}$	$0.15526.10^{1}$
170	1	$0.18665.10^{2}$	$0,26519.10^{1}$	$0.36924.10^{1}$
	2	0.13385	$0.35991.10^{-2}$	$0.1188.10^{1}$
	3	$0.17937.10^{-3}$	$0.97128.10^{-5}$	$0.11846.10^{1}$
	5	$0.34927,10^{-1o}$	$0.79413.10^{-5}$	$0.11846.10^{1}$
	10	$0.20042.10^{-16}$	$0.79413.10^{-5}$	$0.11846.10^{1}$

Dans les trois colonnes qui suivent, la première et la deuxième donnent la valeur obtenue pour la norme du gradient de la fonctionnelle g et pour $J(x_N, u_N)$ respectivement dans les iterations successives toutes les 40 valeurs de N en partant de la valeur N = 10. Dans la dernière colonne on donne la difference au sens de la norme L^2 entre la solution optimale u^o et approximée u_N.
On voit de ce tableau que les valeurs obtenues pour $J(x_N, u_N)$ et $\| u_N - u^o \|^2$ sont a peu près les mêmes quand on ne fait que 3 itèrations. Les valeurs de $J(x_N, u_N)$ et $\| u_N - u^o \|^2$ calculées après les cinq iterations diminuent de 99,6 % et 94,7 % respectivement losrsque N passe de 10 a 170.

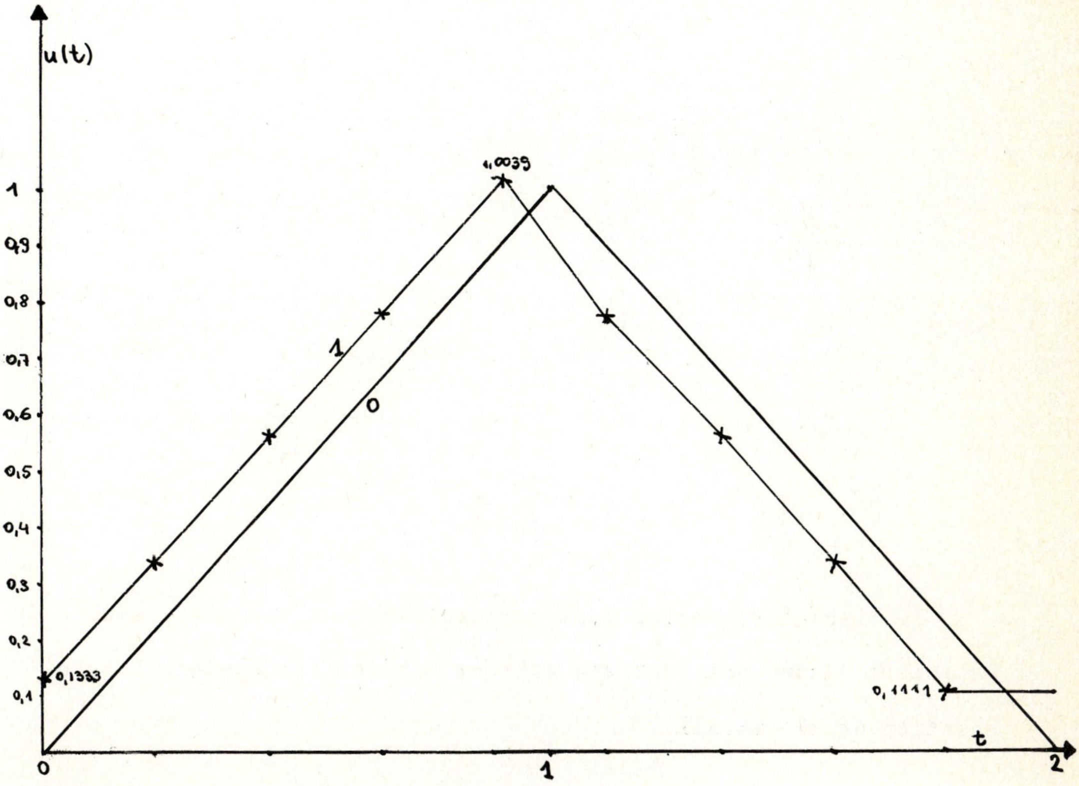

Le graphique présenté au dessous indique la disposition des points u_N obtenue aprés 10 iterations en comparaison avec les valeurs exactes u^o .

La liqne notée par 0 est la solution optimale u^o . Les points dressés sur la liqne 1 representent les valeurs u_N qu'on obtient pour $N = 10$.

Soit u_N la fonction construite a l'aide de u_N de façon suivant.

$$u_N^*(t) = \frac{u_{r+1}-u_r}{\tau}\left(t - r\tau\right) + u_r \qquad \text{pour} \quad t \in \left[r\tau, (r+1)\,\tau\right]$$

alors, la liqne 1 donne le graphique de u_N correspondante à valeur $N = 10$.

Je tiens à remercier tout particulièrement le Proffesseur Kazimierz Malanowski pour son aide et supervision pendant la préparation de ce travail.

Bibliographie

[1] W.E. Bosarge,„ Iterative continuation and the solution of nonlinear two-point boundary value problems". Numer.Math.17 1971

[2] J.M.Ortega, W.C.Rheinboldt,„ Iterative solution of nonlinear equations in several variables." Academic Press,New York 1970.

[3] P.A.Raviart,„ Sur l'approximation de certaines équations d'évolution linéaires et non linéaires." J.de Math.pures et appl. 46 1967.

[4] J.W.Daniel,„ The approximate minimization of functionals," Prentice-Hall Englewood Cliffs N.J. 1971

[5] D.G.Luenberger,„ Optimization by vector space methods," John Wiley, New York 1969.

[6] R.S. McKnight,„ W.E.Bosarge, „The Ritz-Galerkin procedure for parabolic control problems." SIAM J.Control 11,3 1973

[7] W.E.Bosarge, O.G.Johnson, R.S.McKnight, W.P.Timlake,„ The Ritz-Galerkin procedure for nonlinear control problems." SIAM J.Numer.Anal.10 1973.

[8] I. Lasiecka,„Finite difference approximation of optimal control for systems described by nonlinear differential equation with delay." Control and Cybernetics (to be published).

[9] K. Malanowski " On discrete approximation for a class of sets of admissible control" , Control and Cybernetics , (to be published)

CANONICAL REALIZATIONS OF TRANSFER OPERATORS *

N. Levan
Department of System Science
4532 Beolter Hall
University of California
Los Angeles, California 90024 USA

Abstract. We wish to present in this paper the realization of a class of transfer operators of infinite dimensional state space discrete-time systems. The realization will be carried out on a functional state space constructed from a given transfer operator. Our method here is based on the canonical model theory of contraction Hilbert Space operators of Nagy and Foias.

It will be shown that the state space in this case has beside the output component, a second component which characterizes the energy dissipated in the system. Furthermore, the realization will be automatically canonical when one uses cyclic subspaces of the restricted shift operator on a Nagy-Foias Space. Relationships between the realization here and the scattering synthesis of passive networks will also be discussed.

I. Introduction. We study in this paper the realization of linear discrete-time systems whose transfer functions are contractive analytic functions from the unit disc to the operators from one Hilbert Space to another Hilbert Space. Our method here is based on the operator model theory of Nagy and Foias.

A model of an operator is, plainly speaking, another operator (or operators) which is simpler in some suitable sense, and at the same time, has richer structure. In their theory, Nagy and Foias have shown that every Hilbert Space contraction is unitarily equivalent to a shift operator compressed to a functional space, called a Nagy-Foias Space.

In this paper we shall show that given a contractive analytic transfer function, the Nagy-Foias space constructed from this function is a state-energy type space, and the realization on this space, using cyclic subspaces of the compressed shift operator will naturally be canonical.

In Section 2 we present the basic mathematical preliminaries and background motivation. Structures of a Nagy-Foias space will be discussed in detail in Section 3. Section 4 is devoted to the realization problem. Relationships between the Nagy-Foias theory and scattering realization of networks will also be discussed here.

---

*This work was supported by National Science Foundation under Grant # ENG 75-11876.

II. Mathematical Preliminaries. In this paper we shall be, unless otherwise stated, dealing with linear bounded operators on separable Hilbert spaces. Norm and inner product are denoted by $|| \ ||$ and by $[\ ,\ ]$ respectively.

We briefly present in this section some basic notations and definitions which will be needed throughout the paper. The Nagy-Foias Space will then be constructed while its structures and applications to system realizations will be dealt with in Section 3.

Given a Hilbert Space H_1 the space of power series: $f(z) = \sum_{n=0}^{\infty} f_n z^n$, where the f_n's are in H, $\sum_{n=0}^{\infty} ||f_n||^2 < \infty$, and $|z| < 1$, is denoted by $H^2(H)$. The norm and inner product in $H^2(H)$ are defined by $||f||^2 = \sum_{n=0}^{\infty} ||f_n||_H^2$, and $[f,g] = \sum_{n=0}^{\infty} [f_n,g_n]_H$, where $g(z) = \sum_{n=0}^{\infty} g_n z^n$. Clearly, $H^2(H)$ can be identified with the space of square summable H – sequences $\{f_0,f_1,f_2,\ldots;\ f_i \in H\}$.

We can associate with each $f(z)$ in $H^2(H)$ its "boundary function" $f(e^{it}) = \sum_{n=0}^{\infty} f_n e^{int}$, and consequently, $H^2(H)$ can be identified with the space $L_+^2(H)$ of Fourier series with non-negative powers of e^{it}. This space, and therefore $H^2(H)$ also, are in turn, a subspace of the space $L^2(H)$ of Fourier Series with all powers of e^{it}. We have the orthogonal decomposition $L^2(H) = L_-^2(H) \oplus L_+^2(H)$, where $L_-^2(H)$ is the set of Fourier Series with negative powers of e^{it}.

Given two Hilbert Spaces H_1 and H_2, a function $\theta(z)$ from the unit disc to the operators from H_1 to H_2 is denoted by $\{\theta(z), H_1,H_2\}$. Such a function is bounded analytic when

$$\theta(z) = \sum_{n=0}^{\infty} \theta_n z^n,\ \theta_n:\ H_1 \to H_2,\ |z| < 1$$

and

$$||\theta(z)h_1|| \leq M||h_1||,\ h_1 \text{ in } H_1.$$

θ is said to be contractive when $M = 1$, and purely contractive, if in addition, $||\theta(0)h_1|| < ||h_1||$ for any h_1 in H_1.

Given a bounded analytic function $\{\theta(z), H_1, H_2\}$ we can associate with it the following spaces and operators:

a) The space $H^2(H_1) \subset L^2(H_1)$ and $H^2(H_2) \subset L^2(H_2)$.

b) As in the above, we can associate with $\theta(z)$ its "boundary function" $\theta(e^{it})$ defined almost everywhere. Hence, we have the following operators

$$\theta_z:\ H^2(H_1) \to H^2(H_2)$$

$$(\theta_z f)(z) = \theta(z)f(z)$$

$$\theta_t:\ L^2(H_1) \to L^2(H_2)$$

$$(\theta_t f)(t) = \theta(e^{it})f(t)$$

$$\theta_t^*:\ L^2(H_2) \to L^2(H_1)$$

$$(\theta_t^* f)(t) = \theta(e^{it})^* f(t),\quad \theta(e^{it})^* = \sum_{n=0}^{\infty} \theta_n^* e^{-int}$$

If $\theta(z)$ is contractive, then we can, in addition, define the operator

$$\Delta_t:\ L^2(H_1) \to L^2(H_1)$$

$$(\Delta_t f)(t) = \Delta(t)f(t)$$

where $\Delta(t)$ is the unique positive square root $[I - \theta(e^{it})^*\theta(e^{it})]^{\frac{1}{2}}$, and it is bounded between 0 and 1.

To each contractive analytic function $\{\theta(z), H_1, H_2\}$ there corresponds a Nagy-Foias Space which can be constructed as follows [1]:

First let H be the Hilbert space of pairs of functions $(v(z),\Delta(t)u(t))$, for v in $H^2(H_2)$ and u in $L^2(H_1)$

$$H = H^2(H_2) \oplus \overline{\Delta(t)L^2(H_1)} \tag{2-1}$$

where $\overline{}$ indicates the closure. The inner product and norm in H are defined in the usual way

$$[(v_1,\Delta u_1),(v_2,\Delta u_2)]_H = [v_1,v_2]_{H^2(H_2)} + [\Delta u_1,\Delta u_2]_{L^2(H_1)} \tag{2-2}$$

and

$$||(v,\Delta u)||_H^2 = ||v||^2_{H^2(H_1)} + ||\Delta u||^2_{L^2(H_1)} \tag{2-3}$$

The space H is clearly a subspace of the space

$$K = L^2(H_2) \oplus \overline{\Delta(t)L^2(H_1)} \tag{2-4}$$

$$= \{(f_-,0),\ f_- \in L^2_-(H_2)\} \oplus H \tag{2-5}$$

Next, in H, consider the set

$$M = \{(\theta(z)w(z),\ \Delta(t)w(e^{it})),\ w \in H^2(H_1)\} \tag{2-6}$$

It is clear that

$$||(\theta w,\Delta w)||^2 = ||\theta w||^2 + ||w||^2 - ||\theta w||^2 = ||w||^2$$

Hence the map $w \to (\theta w, \Delta w)$ is an isometry, and therefore M is closed in H.

Definition

The orthogonal complement $M^{\perp}$ of M in H:

$$M^{\perp} = H^2(H_2) \oplus \overline{\Delta(t)L^2(H_1)} \ominus \{(\theta w, \Delta w),\ w \in H^2(H_1)\} \qquad (2\text{-}7)$$

is called the Nagy-Foias space of $\{\theta(z), H_1, H_2\}$.

It is not at all clear from this definition what are the meanings and structures of $M^{\perp}$. In the next section we shall investigate these from a system-network viewpoint.

In what follows we will be concerned with linear, fixed, discrete-time system which has the state space description:

$$\begin{cases} x_n = Ax_n + Bu_n & (2\text{-}8) \\ v_n = Cx_n + Du_n & (2\text{-}9) \end{cases}$$

where $n = 0, 1, 2, \ldots$, A, B, C, and D are appropriate operators; $\{x_n\}$, $\{u_n\}$ and $\{v_n\}$ are the states, inputs and outputs respectively. The $\{x_n\}$ and $\{v_n\}$ are chosen to be square-summable H_2-sequences, and $\{u_n\}$ is taken to be square-summable H_1-sequences.

Given a square-summable sequence $\{f_0, f_1, \ldots\}$, its discrete Fourier transform is the map

$$\{f_n\} \to \sum_{n=0}^{\infty} f_n z^n = f(z) \text{ say} \qquad (2\text{-}10)$$

Hence taking the transforms of equations (2-8) and (2-9) with $x_0 = 0$, we find

$$v(z) = \left[D + zC[I - zA]^{-1}B\right]u(z) \qquad (2\text{-}11a)$$

$$= \theta(z)\, u(z) \text{ say} \qquad (2\text{-}11b)$$

The function $\theta(z)$ is called the transfer operator of the system. Again, we assume that $\theta(z)$ is contractive. Given a discrete-time signal $\{\ldots f_{-2}, f_{-1}, f_0, f_1, f_2, \ldots\}$, if 0 is taken to be the present instant of time, then the sequence $\{f_0, f_1, \ldots\}$ is the present-future segment of the signal, while the sequence $\{\ldots f_{-3}, f_{-2}, f_{-1}\}$ is the past segment of the signal. It then follows that, the space $H^2(H_1)$ $\left(\text{or } L_+^2(H_1)\right)$ is just the space of (transform of) present-future inputs, and $L_-^2(H_1)$ is the space of (transform of) past inputs, while $L^2(H_1)$ is the space of allowable inputs over all (discrete) time. Similarly for the output spaces $H^2(H_2)$, $L_-^2(H_2)$ and $L^2(H_2)$.

The boundary function $\theta(e^{it})$ is called the system frequency operator. For our purpose, $\theta(e^{it})$ is represented by the matrix operator

$$\theta(e^{it}) = \begin{bmatrix} \theta_{11} & \theta_{12} \\ \theta_{21} & \theta_{22} \end{bmatrix} \tag{2-12}$$

with respect to the decompositions $L^2(H_i) = L_-^2(H_i) \oplus L_+^2(H_i)$, i = 1, 2. Clearly $\theta_{11} = P_2^-\theta(e^{it})P_1^-$, $\theta_{12} = P_2^-\theta(e^{it})P_1^+$, $\theta_{21} = P_2^+\theta(e^{it})P_1^-$ and $\theta_{22} = P_2^+\theta(e^{it})P_1^+$, $P_i^{\pm}$, i = 1, 2 are projection operators from $L^2(H_i)$ onto $L_{\pm}^2(H_i)$. We note that θ_{22} can actually be identified with $\theta(z)$.

The system is said to be causal when $\theta(e^{it})$ is lower triangular [2], that is when $\theta_{12} = 0$, and this certainly is the case when θ is analytic.

The space H (equation 2-1) whose elements are pairs of functions $(v(z), \Delta(t)u(t))$ can be considered as the space of '<u>present-future output - "input" pairs</u>' of a system whose transfer operator is $\theta(z)$. The function Δu is not really an input although it is in the input space $L^2(H_1)$, while u itself is an input. However, since

$$||\Delta u||^2_{L^2(H_1)} = ||u||^2_{L^2(H_1)} - ||\theta u||^2_{L^2(H_2)} \tag{2-13}$$

which can be regarded as the amount of energy which the system absorbed from the outside world. Thus, in some sense, Δu characterizes the passage of energy through the system.

We conclude the section by noting that, in a pair $(v, \Delta u)$, v and u are quite arbitrary, in the sense that the output v may or may not result from the input u. When $(v = \theta w, \Delta u = \Delta w)$ which is an element in M(equation 2-6), then the present-future output θw in this case is the response of the system to the present-future input w.

III. <u>Structures of A Nagy-Foias Space</u>. The Nagy-Foias Space $M^{\perp}$ of a contractive analytic function $\{\theta(z), H_1, H_2\}$ was introduced by Nagy and Foias [1] in the study of models of Hilbert space contractions (i.e. operators with norms less than or equal to 1). They showed that such an operator is unitarily equivalent to the compressed shift operator on $M^{\perp}$.

Here, in this section we shall investigate the structures of $M^{\perp}$, and in particular, its roles in the realization of $\theta(z)$.

Let $(v,\Delta u)$ be an element in H, its component $P_{M^{\perp}}(v,\Delta u)$, where $P_{M^{\perp}}$ is the orthogonal projection onto $M^{\perp}$, is clearly

$$P_{M^\perp}(v,\Delta u) = (v,\Delta u) - (\theta w,\Delta w) \tag{3-1}$$

where w is in $H^2(H_1)$ and is such that $P_{M^\perp}(v,\Delta u) \perp M$, that is

$$[(v-\theta w, \Delta u-\Delta w), (\theta\tilde{w},\Delta\tilde{w})]_H = 0, \ \tilde{w} \text{ in } H^2(H_1)$$

or

$$[\theta^* v + \Delta^2 u - w, \tilde{w}]_{L^2(H_1)} = 0$$

Hence, as a function in $L^2(H_1)$

$$(\theta^* v + \Delta^2 u - w) \perp H^2(H_1)$$

For this to be ture, we must have

$$P_1^+[\theta^* v + \Delta^2 u - w] = 0$$

Therefore

$$w = P_1^+[\theta^* v + \Delta^2 u] \tag{3-2}$$

From which it is evident that $(v,\Delta u)$ in H is in $M^\perp$ if and only if

$$P_1^+[\theta^* v + \Delta^2 u] = 0 \tag{3-3}$$

This condition can also be expressed differently as follows. To each pair $(v,\Delta u)$ in H there corresponds the pair $(\theta_{21}u_-,\Delta u_-)$ in $M^\perp$ and $(\theta_{22}u_+,\Delta u_+)$ in M, where u_- is in $L_-^2(H_1)$, u_+ in $L_+^2(H_1)$ and $u_- + u_+ = u$. From these pairs, we form the pair

$$(\hat{v},\Delta u) = (\theta_{21}u_- + \theta_{22}u_+,\Delta u) \tag{3-4}$$

which is an output-input pair in which the present-future output $\hat{v}$ resulted from the input (over all time) u. We can therefore write

$$(v,\Delta u) = (\theta_{21}u_-,\Delta u_-) + (\theta_{22}u_+,\Delta u_+) + (v-\hat{v}, 0)$$

hence

<u>Lemma 1</u>

$(v,\Delta u)$ in H is in $M^\perp$ if and only if

$$P_1^+(\theta^* v + \Delta^2 u) = 0 \tag{3-3}$$

or alternately

$$(v,\Delta u) = (\theta_{21}u_-,\Delta u_-) + P_{M^\perp}(v-\hat{v}, 0) \tag{3-5}$$

We note that $(v-\hat{v})$ can be regarded as the error between the two outputs v and $\hat{v}$.

In what follows we shall concentrate on two special subspaces of $M^{\perp}$:

$$M_1^{\perp} = \text{closure } \{(\theta_{21}u_-, \Delta u_-),\ u_- \text{ in } L_-^2(H_1)\} \tag{3-6}$$

and

$$M_2^{\perp} = \text{closure } \{P_{M^{\perp}}(y,0),\ y \text{ in } H^2(H_2)\} \tag{3-7}$$

Plainly speaking, in $M_1^{\perp}$, the set of present-future outputs comes entirely from past inputs, while in $M_2^{\perp}$, the inputs are not specified.

To proceed further we define

Definition 1

The shift operator S on H is defined by

$$S(v,\Delta u) = (zv,\ e^{it}\Delta u),\ (v,\Delta u) \text{ in } H \tag{3-8}$$

and its adjoint S^* is given by

$$S^*(v,\Delta u) = \left(\frac{v(z)-v(0)}{z},\ e^{-it}\Delta u\right),\ (v,\Delta u) \text{ in } H \tag{3-9}$$

We note that S is an isometry on H, and M is invariant [1] under S, as a consequence, $M^{\perp}$ is invariant under S^*.

Since $M^{\perp}$ is not invariant under S, the restriction of S onto $M^{\perp}$ is called the compressed shift and is defined by

Definition 2

The operator $T\colon\ M^{\perp} \to M^{\perp}$ given by

$$T(v,\Delta u) = P_{M^{\perp}}(zv,\ e^{it}\Delta u),\ (v,\Delta u) \text{ in } M^{\perp} \tag{3-10}$$

is called the compressed shift operator on the Nagy-Foias space $M^{\perp}$.

The adjoint operator T^* is just S^* restricted to $M^{\perp}$:

$$T^*(v,\Delta u) = \left(\frac{v(z)-v(0)}{z},\ e^{-it}\Delta u\right),\ (v,\Delta u) \text{ in } M^{\perp}, \tag{3-11}$$

Now we assume that θ is causal, that is $\theta_{12} = 0$, in this case $M_1^{\perp}$ and $M_2^{\perp}$ have very nice structures, and as we shall see, they play a cruicial role in the realization of θ as well as in the Nagy-Foias model theory.

Structures of $M_1^{\perp}$

As we have defined above

$$M_1^{\perp} = \text{closure } \{(\theta_{21}u_-, \Delta u_-),\ u_- \text{ in } L_-^2(H_1)\}$$

setting

$$u_- = e^{-it}\alpha_1 + e^{-2it}\alpha_2 + \ldots + e^{-int}\alpha_n + \ldots,\ \alpha_i \text{ in } H_1, \tag{3-12a}$$

and

$$\theta(e^{it}) = \theta_0 + \theta_1 e^{it} + \theta_2 e^{2it} + \ldots \theta_n e^{int} + \ldots\ ,\quad \theta_i: H_1 \to H_2 \tag{3-12b}$$

Then $\theta_{21}u_-$ can be calculated as follows

$$\begin{aligned}\theta_{21}u_- &= P_2^+\{\theta(e^{it})u_-\}\\ &= (\theta_1\alpha_1 + \theta_2\alpha_2 + \theta_3\alpha_3 + \ldots)\,1\\ &+ (\theta_2\alpha_1 + \theta_3\alpha_2 + \theta_4\alpha_2 + \ldots)\,e^{it}\\ &+ \ldots\end{aligned} \tag{3-13}$$

Hence the matrix of θ_{21} with respect to the orthonormal basis $\{1, e^{it}, e^{2it}, \ldots\}$ is

$$[\theta_{21}] = \begin{bmatrix} \theta_1 & \theta_2 & \theta_3 & \cdots \\ \theta_2 & \theta_3 & \theta_4 & \cdots \\ \theta_3 & \theta_4 & \theta_5 & \cdots \\ \vdots & \vdots & \vdots & \end{bmatrix} \tag{3-14}$$

This infinite matrix is called the Hankel matrix of the Hankel operator generated by θ, denoted by $\mathcal{H}_\theta$. Hence $\theta_{21} \equiv \mathcal{H}_\theta$.

Let $\big(\theta(z)w(z), \Delta(t)w(e^{it})\big)$ be an element of M, and since M is not invariant under S^*, $S^*(\theta w, \Delta w)$ will just be in H. What are the elements of M which are mapped entirely into $M^\perp$ under S^*? We have

$$S^*(\theta w, \Delta w) = \Big(z^{-1}\big(\theta(z)w(z) - \theta(0)w(0)\big),\ e^{-it}\Delta w(e^{it})\Big),\quad w \text{ in } H^2(H_1) \tag{3-15}$$

Using (3-3) of Lemma 1, we find

$$P_{M^\perp} S^*(\theta w, \Delta w) = S^*(\theta w, \Delta w) - \Big(\theta e^{-it}\big(w(z) - w(0)\big),\ \Delta e^{-it}\big(w(z) - w(0)\big)\Big) \tag{3-16}$$

Thus for $S^*(\theta w, \Delta w)$ to be in $M^\perp$, we must have

$$w(z) = w(0) = \alpha \quad \text{say}$$

that is $w(z)$ must be a constant function in $L^2(H_1)$, in other words w is in H_1. Conversely, for α in H_1,

$$S^*(\theta\alpha, \Delta\alpha) = \Big(\frac{\theta(z) - \theta(0)}{z}\,\alpha,\ e^{-it}\Delta\alpha\Big) \tag{3-17}$$

Using (3-12a) and (3-13) we find

$$\frac{\theta(z)-\theta(0)}{z}\alpha = (\theta_1 + \theta_2 z + \theta_3 z^2 + \dots)\alpha$$

$$= \theta_{21}\, e^{-it}\, \alpha \tag{3-18}$$

Therefore

$$S^*(\theta\alpha, \Delta\alpha) = (\theta_{21}\, e^{-it}\alpha,\ \Delta e^{-it}\alpha) \tag{3-19}$$

which shows that $S^*(\theta\alpha,\Delta\alpha)$ is in $M_1^\perp$ and therefore it is in $M^\perp$.

Lemma 2

S^* sends $(\theta w, \Delta w)$ into $M_1^\perp$ if and only if $w(z) = \alpha$ in H_1, in which case

$$S^*(\theta\alpha,\Delta\alpha) = \left(\frac{\theta(z)-\theta(0)}{z},\ \Delta e^{-it}\right)\alpha$$

$$= (\theta_{21}\, e^{-it}\alpha,\ \Delta e^{-it}\alpha)$$

This lemma suggests that in $M_1^\perp$ we can have a sequence of subspaces which are constructed as follows.

First set

$$\phi_0(z) = \theta(z) \tag{3-20}$$

Then for $n \geq 0$, define

$$\phi_{n+1}(z) = \frac{\phi_n(z)-\phi_n(0)}{z} \tag{3-21}$$

It is clear that $\phi_n(0) = \theta_n$, the nth coefficient in the power series of $\theta(z)$, and for α in H_1, $\phi_n(z)\alpha = \theta_{21}e^{-int}\alpha$. Next, we define

$$K_n = [\phi_n(z),\ \Delta e^{-int}],\quad n = 0, 1, 2, \dots \tag{3-22a}$$

Then for α in H_1, $K_0\alpha = [\theta\alpha,\Delta\alpha]$ is an element of M, while for $n \geq 1$

$$K_n\alpha = [\phi_n,\ \Delta e^{-int}]\alpha \tag{3-22b}$$

$$= (\theta_{21}e^{-int}\alpha,\ \Delta e^{-int}\alpha) \tag{3-22c}$$

which are in $M_1^\perp$.

It follows at once from the above that

$$(\theta_{21}u_-, \Delta u_-) = K_1\alpha_1 + K_2\alpha_2 + \dots + K_n\alpha_n + \dots \tag{3-23}$$

and therefore

$$M_1^\perp = \overline{\text{span}}\{K_1\alpha, K_2\alpha, \dots, K_n\alpha, \dots\} \tag{3-24}$$

Furthermore, for $n \geq 1$

$$T^*K_n\alpha = \left(\frac{\phi_n(z)-\phi_n(0)}{z}, \Delta e^{-i(n+1)t}\right)\alpha$$

$$= K_{n+1}\,\alpha \qquad (3\text{-}25)$$

Hence, for $n \geq 1$

$$K_{n+1}\alpha = T^{*n}K_1\alpha \qquad (3\text{-}26)$$

Thus we have shown that

<u>Theorem 1</u>

$\overline{K_1\alpha}$ is a cyclic subspace [1] of $T^*\big|_{M_1^\perp}$, the restriction of T^* to $M_1^\perp$.

<u>Structures of $M_2^\perp$</u>

The subspace $M_2^\perp$ was defined above:

$$M_2^\perp = \text{closure } \{P_{M^\perp}(y,0),\ y \text{ in } H^2(H_2)\}$$

straightforward calculations give

$$P_{M^\perp}(y,0) = (I - \theta(z)\theta_{22}^*, -\Delta\theta_{22}^*)y$$

$$= k\,y \quad \text{say, for } y \text{ in } H^2(H_2) \qquad (3\text{-}27)$$

We note that for any β in H_2,

$$\theta_{22}^*\beta = P_1^+\{\theta(e^{it})^*\beta\} = \theta(0)^*\beta \qquad (3\text{-}28)$$

Therefore

$$P_{M^\perp}(\beta,0) = k\beta = \left(I-\theta(z)\theta(0)^*, -\Delta\theta(0)^*\right)\beta = k_0\beta \quad \text{say, for } \beta \text{ in } H_2 \qquad (3\text{-}29)$$

Next, let us calculate $P_{M^\perp}(\beta e^{it},0)$, we have

$$P_{M^\perp}(\beta e^{it},0) = k\beta e^{it} = (I-\theta\theta_{22}^*, -\Delta\theta_{22}^*)\beta e^{it} \qquad (3\text{-}30)$$

where

$$\theta_{22}^*\beta e^{it} = P_1^+\{\theta(e^{it})^*\beta e^{it}\}$$

$$= P_1^+\{\theta(0)^*\beta e^{it} + \theta_1^*\beta + \theta_2^*\beta\, e^{-it} + \ldots\}$$

$$= \theta(0)^*\beta e^{it} + \theta_1^*\beta$$

Therefore

$$P_{M^\perp}(\beta e^{it},0) = k\beta e^{it}$$

$$= \left(I-\theta\theta(0)^*,\ -\Delta\theta(0)^*\right)\beta e^{it} - (\theta\theta_1^*\beta,\Delta\theta_1^*\beta)$$

$$= k_0\beta e^{it} - (\theta\theta_1^*\beta,\Delta\theta_1^*\beta)$$

we have

$$P_{M^\perp}(\beta e^{it},0) = k\beta e^{it} = P_{M^\perp}(k_0\beta e^{it}) = k_1\beta \quad \text{say} \tag{3-31}$$

But, $k_0\beta e^{it} = e^{it}\,k_0\,\beta$, hence

$$k_1\beta = P_{M^\perp}(e^{it}k_0\beta) = T\,k_0\,\beta \tag{3-32}$$

Similarly, setting

$$k_2\beta = k\beta e^{2it} = P_{M^\perp}(\beta e^{2it},0) \tag{3-33}$$

we find

$$k_2\beta = P_{M^\perp}(e^{it}k\beta e^{it}) = T(k_1\beta) = T^2(k_0\beta)$$

Thus, in general for $n \geq 0$, and β in H_2

$$k_n\beta = k\beta e^{int} = P_{M^\perp}(\beta e^{int},0) = T^n(k_0\beta) \tag{3-34}$$

Consequently

$$M_2^\perp = \overline{\text{span}}\{k_0\beta,\ k_1\beta,\ k_2\beta,\ \ldots,\ k_n\beta,\ \ldots\} \tag{3-35}$$

We have therefore shown that

<u>Theorem 2</u>

$\overline{k_0\beta}$ is a cyclic subspace [1] of $T\big|_{M_2^\perp}$, the restriction of T to $M_2^\perp$.

Our derivations of $K_n\alpha$ and $k_n\beta$ above were motivated by the work of D. N. Clark [3] on one dimensional perturbations of the restricted shift on a Nagy-Foias space associated with a scalar inner function θ. In his work Clark was using only K_1 (which he denoted by K_0) and k_0. It is a pleasure to thank Doug. Clark for introducing this work to me. Clark's work was subsequently generalized by Fuhrmann [4], Ball and Lubin [5].

To proceed further, we now consider the necessary and sufficient condition for an element $(v,\Delta u)$ in $M^\perp$ to be orthogonal to $K_n\alpha$.

The case $n = 0$ is trivial since $K_0\alpha$ is in M. For $n \geq 1$, we have:

$$[(v,\Delta u), K_n\alpha] = [(v,\Delta u), (P_2^+\theta(e^{it})e^{int}\alpha, \Delta e^{-int}\alpha)]$$

$$= [(Z^n v, \Delta e^{int}u), (\theta\alpha, \Delta\alpha)]$$

$$= [e^{int}(\theta^* v + \Delta^2 u, \alpha]_{L^2(H_1)}, \quad \alpha \text{ in } H_1$$

Now since $(v,\Delta u)$ is in $M^\perp$, $\theta^* v + \Delta^2 u$ has an expansion in negative powers of e^{it}

$$\theta^* v + \Delta^2 u = \gamma_1 e^{-it} + \gamma_2 e^{-2it} + \ldots + \gamma_n e^{-int} + \ldots, \quad \gamma_i \text{ in } H_1$$

Consequently,

$$[(v,\Delta u), K_n\alpha] = [\gamma_n,\alpha]_{H_1}, \quad \gamma_n,\alpha \quad \text{in } H_1, \quad n \geq 1 \tag{3-36}$$

Hence $(v,\Delta u)$ in $M^\perp$ is orthogonal to $K_n\alpha$ if and only if $\gamma_n = 0$.

Similarly, for $(v,\Delta u)$ in $M^\perp$ and for $n \geq 0$

$$[(v,\Delta u), k_n\beta] = [(v,\Delta u), P_{M^\perp}(z^n\beta, 0)]$$

$$= [v, z^n\beta]$$

$$= [v_n,\beta] \tag{3-37}$$

where v_n is the coefficient of z^n in the power series expansion of v_n. Hence $(v,\Delta u)$ in $M^\perp$ is orthogonal to $k_n\beta$ if and only if $v_n = 0$. It follows from the above that

<u>Lemma 3</u>

For $(v,\Delta u)$ in $M^\perp$,

$(v,\Delta u) \perp M_1^\perp$ if and only if $v = 0$

$(v,\Delta u) \perp M_2^\perp$ if and only if $\theta^* v + \Delta^2 u = 0$

The orthogonal complement (in $M^\perp$) of $M_1^\perp$ will be denoted by M_1, while that of $M_2^\perp$ will be denoted by M_2.

For any $(v,\Delta u)$ in $M^\perp$, straightforward calculations give

$$T^{*n}(v, \Delta u) = (\frac{v}{z^n}, e^{-int}\Delta u) - (\frac{v_0}{z^n} + \frac{v_1}{z^{n-1}}) + \ldots + V_{n-1} \quad n \geq 1, \tag{3-38}$$

and

$$T^{n}T^{*n}(v,\Delta u) = (v,\Delta u) - \sum_{\ell=0}^{n-1} k_{\ell}v_{\ell}, \quad n \geq 1 \tag{3-39}$$

Hence

$$(I-T^{n}T^{*n})\ (v,\Delta u) = \sum_{\ell=0}^{n-1} k_{\ell}v_{\ell},\ n \geq 1 \tag{3-40}$$

Similarly

$$(I-T^{*n+1}T^{n+1})(v,\Delta u) = \sum_{\ell=0}^{n} K_{\ell+1}\ \gamma_{\ell+1},\ n \geq 0, \tag{3-41}$$

It then follows that

Lemma 4

For $(v,\Delta u)$ in $M^{\perp}$:

(i) $T^{*n}(v,\Delta u) = (\frac{v}{z^{n}},\ e^{-int}\Delta u),\quad n \geq 1$

or equivalently

$(I-T^{n}T^{*n})\ (v,\Delta u) = 0\ ,\quad n \geq 1$

if and only if $(v,\Delta u) \perp k_{\ell}\ \beta$ for all $\ell = 0, 1, 2, \ldots, n-1$.

(ii) $T^{n+1}(v,\Delta u) = \left(z^{n+1}v,\ e^{i(n+1)t}\Delta u\right),\quad n \geq 0$

or equivalently

$(I-T^{*n+1}T^{n+1})\ (v,\Delta u) = 0,\quad n \geq 0$

if and only if $(v,\Delta u) \perp K_{\ell}\ \alpha$ for all $\ell = 0, 1, \ldots, n+1$.

This lemma is a generalization of Clark's results for the scalar case [3].

The following results can be easily verified:

Lemma 5

$T^{n+1}K_{1}\alpha \quad = -\ k_{n}\theta(0)\alpha\ ,\qquad n \geq 0$

and

$T^{*n+1}k_{0}\beta = -\ K_{n+1}\theta(0)^{*}\ \beta,\quad n \geq 0$

In the next Section, we shall use the above results for system operators realizations.

IV. Realizations of Transfer Operators. In this Section we shall discuss the realization of a given contractive analytic transfer operator $\{\theta(z), H_1, H_2\}$. Thus, our problem is to find operators A, B, C and D such that

$$\theta(z) = D + z\, C[I-zA]^{-1}B, \quad |z| < 1, \tag{4-1}$$

We shall use results of previous Sections. First, let us see what are the meanings of a Nagy-Foias space associated with a given transfer operator $\{\theta(z), H_1, H_2\}$.

As we have seen above, the Hilbert Space H (equation 2-1) is the space of present-future outputs (together with elements of the form Δu, which characterize the net energy absorbed by the system) while its subspace M (equation 2-6) is the set of all present - future outputs, resulted entirely from present - future inputs. Thus, the orthogonal complement $M^\perp$ of M (in H), can be regarded [6] as a state-energy type space.

The subspace $M_1^\perp$ (equation 3-6) consists of all present - future outputs $\theta_{21}u_-$ resulted entirely from past inputs u_-, further more we have

$$||(\theta_{21}u_-,\Delta u_-)||^2 = ||\theta_{21}u_-||^2 + ||\Delta u_-||^2 = ||u_-||^2 - ||\theta_{11}u_-||^2$$

which can be regarded as energy stored in the system - due to inputs in the past.

To proceed with the realization problem, we first observe, from Lemma 4, with $n = 1$, that

$$(I-T^*T)\,(v,\Delta u) = 0 \iff (v,\Delta u) \perp K_1\alpha, \tag{4-2}$$

and

$$(I-TT^*)\,(v,\Delta u) = 0 \iff (v,\Delta u) \perp k_0\beta, \tag{4-3}$$

Also using Lemma 5 with $n = 0$ we get,

$$(I-T^*T)\,K_1\alpha = K_1[I-\theta(0)^*\,\theta(0)]\alpha \quad , \tag{4-4}$$

and

$$(I-TT^*)\,k_0\beta = k_0[I-\theta(0)\,\theta(0)^*] \quad , \tag{4-4}$$

Now, since $\{\theta(z), H_1, H_2\}$ is purely contractive, that is $||\theta(0)|| < 1$, it can be shown that [1] the ranges of $[I-\theta(0)^*\,\theta(0)]$ and of $[I-\theta(0)\,\theta(0)^*]$ are dense in H_1 and in H_2 respectively. Hence

$$[I-T^*T]\,\overline{K_1 H_1} = \overline{K_1 H_1} \tag{4-6}$$

$$[I-TT^*]\ \overline{k_0 H_2} = \overline{k_0 H_2} \tag{4-7}$$

From which it follows that

$$[I-T^*T]^p \overline{K_1 H_1} = \overline{K_1 H_1} \tag{4-8}$$

$$[I-TT^*]^p \overline{k_0 H_2} = \overline{k_0 H_2} \tag{4-9}$$

for $p = 1, 2, \ldots$

The operators $[I-T^*T]$ and $[I-TT^*]$ are both positive and bounded between 0 and 1, consequently we can define its positive square roots $[I-T^*T]^{\frac{1}{2}}$ and $[I-TT^*]^{\frac{1}{2}}$. Furthermore, it follows from (4-8) and (4-9) that the range of $[I-T^*T]^{\frac{1}{2}}$ is dense in $\overline{K_1 H_1}$ while that of $[I-TT^*]^{\frac{1}{2}}$ is dense in $\overline{k_0 H_2}$.

The following model theorem of Nagy-Foias gives a solution to the realization problem.

Nagy-Foias Model Theorem [1]

Let $\{\theta(z), H_1, H_2\}$ be a purely contractive analytic function, and let T be be the restricted shift operator on the Nagy-Foias space $M^{\perp}$ generated by $\theta(z)$, then

$$\theta(z) = U\ \theta_T(z)\ V \tag{4-10}$$

where $\theta_T(z)$ is called the characteristic operator function of T and is defined by

$$\theta_T(z) = -T + z(I-TT^*)^{\frac{1}{2}}\,[I-zT^*]^{-1}\,(I-T^*T)^{\frac{1}{2}} \tag{4-11}$$

and maps $\overline{K_1 H_1}$ into $\overline{k_0 H_2}$.

U and V are unitary maps

$$U:\ \overline{k_0 H_2} \to H_2 \quad \text{and} \quad V:\ H_1 \to \overline{K_1 H_1}$$

It follows at once from this theorem that the operators A, B, C, and D which realize $\theta(z)$ are

$$A = T^*$$
$$B = (I-T^*T)^{\frac{1}{2}}\,V$$
$$C = U(I-TT^*)^{\frac{1}{2}}\Big|_{\overline{k_0 H_2}}$$

$$D = U\,\theta(0)V = -\,UT\Big|_{\overline{K_1 H_1}}\,V$$

Moreover, since $\overline{K_1 H_1}$ and $\overline{k_0 H_2}$ are cyclic for $T^*\Big|_{M_1^\perp}$ and $T\Big|_{M_2^\perp}$ (Theorems 1 and 3 of Section 3), we conclude that the Nagy-Foias realization is both controllable and observable.

We note that the Nagy-Foias model theory was developed via the unitary dilations of contraction operators [1], this is why they used the two operators $(I-T^*T)$ and $(I-TT^*)^{\frac{1}{2}}$, since the operator

$$\Sigma = \begin{bmatrix} -T & (I-TT^*)^{\frac{1}{2}} \\ (I-T^*T)^{\frac{1}{2}} & T^* \end{bmatrix} \tag{4-12}$$

is unitary and is a unitary dilation of $-T$.

One can of course obtain other realization schemes, using T, T^*, $(I-T^*T)^p$ and $(I-TT^*)^p$. For instance, if we form

$$\Phi(z) = -\,T + z(I-TT^*)\,[I-zT^*]^{-1}\,(I-T^*T)$$

Then we have

$$\tau_2\,\theta(z) = \Phi(z)\tau_1$$

where τ_1 and τ_2 are bounded invertible operators: τ_1: $H_1 \to \overline{K_1 H_1}$ and τ_2: $H_2 \to \overline{k_\theta H_2}$. In this case θ and Φ are said to be quasi - similar [1].

Finally, we note that if we set $z = \frac{p-1}{p+1}$, then $\theta(z = \frac{p-1}{p+1})$ can be taken to be the scattering operator of a linear passive multiport network [7], and the Nagy-Foias characteristic operator function $\theta_T(z)$ (equation 4-11) can be gotten by cascade loading the lossless network whose scattering operator is Σ (equation 4-12) in unit inductors.

Thus, for multiport passive networks, the Nagy-Foias model theorem results in the cascade load synthesis procedure. For a complete discussion of this, we refer to [8].

REFERENCES

[1] B. Sz-Nagy and C. Foias, "Harmonic Analysis of Operators on Hilbert Space", North Holland-American Elsevier, Amsterdam, New York, 1970.

[2] R. Saeks, "Causality in Hilbert Space", Siam Review, Vol. 12, pp. 357-383, 1970.

[3] D. N. Clark, "One Dimensional Perturbations of Restrictive Shifts", J d' Analyse Mathematique, Vol. 25, pp. 169-191, 1972.

[4] P. A. Fuhrmann", On a Class of Finite Dimensional Contractive Perturbations of Restrictive Shifts of Finite Multiplicity", Israel J. of Maths., Vol. 16, pp. 162-175, 1973.

[5] J. A. Ball and A. Lubin, "On a Class of Contractive Perturbations of Restricted Shifts", To be published.

[6] A. V. Balakrishnan, "State Space Theory of Linear Time-Varying Systems" pp. 95-125 of "System Theory", L. A. Zadeh and E. Pollack Editors, McGraw-Hill, New York, 1969.

[7] R. W. Newcomb, "Linear Multiport Synthesis", McGraw-Hill, New York, 1966.

[8] N. Levan, "The Nagy-Foias Operator Models, Networks and Systems", IEEE Circuits and Systems Transactions, to appear.

ON OPTIMAL CONTROL PROBLEMS WITH BOUNDED STATE VARIABLES AND CONTROL APPEARING LINEARLY

H. Maurer

Mathematisches Institut der Universität Würzburg
87 Würzburg, Am Hubland, W.-Germany

Summary: Necessary conditions for the junction of interior arcs and boundary arcs of an optimal control are given. These conditions are based on necessary conditions for the switching function holding at junction points or contact points with the state boundary. The junction theorems obtained are similar to junction theorems in singular control problems.

1. Formulation of the problem

We consider the following control problem with control appearing linearly: determine the scalar, piecewise continuous control $u(t)$, $t \in [0,T]$, which minimizes the functional

(1) $$J(u) = G(x(T))$$

subject to

(2) $$\dot{x} = f_1(x) + f_2(x)u$$

(3) $$x(0) = x_o \ , \ \psi(x(T)) = 0$$

(4) $$|u(t)| \leq K(t) \ , \ K(t) > 0 \ , \ 0 \leq t \leq T$$

and the scalar state inequality constraint of order p

(5) $$S(x) \leq 0 \ .$$

The state x is an n-vector. The functions $G:\mathbb{R}^n \to \mathbb{R}$, $\psi: \mathbb{R}^n \to \mathbb{R}^k$, $k < n$, are differentiable and for simplicity the functions $f_1, f_2 : \mathbb{R}^n \to \mathbb{R}^n$, $S : \mathbb{R}^n \to \mathbb{R}$ are assumed to be analytic in a suitable domain; $K(t)$ is analytic in $[0,T]$.

Along a trajectory $x(t)$ of (2) the i-th time derivative of $S(x(t))$ is denoted by S^i, $i \geq 0$. Then by definition of the order p of the state constraint (5) S^p is the first derivative containing the control u explicitly and we have

(6) $$S^i = S^i(x) \ , \ i=0,..,p-1 \ , \quad S^p = S^p(x,u) = a(x) + b(x)u \ .$$

A subarc of $x(t)$ with $S(x(t)) < 0$ is called an interior arc and a subarc of $x(t)$ with $S(x(t)) = 0$ for $t_1 \leq t \leq t_2$, $t_1 < t_2$, is called a boundary arc. Here t_1 and t_2 are called entry- and exit-point or simply junction points of the boundary arc. If $S(x(t_1)) = 0$ and $S(x(t)) < 0$ for $t \neq t_1$ in a neighborhood of t_1 then t_1 is called a contact point of $x(t)$ with the boundary.

The boundary control is determined by $S^p(x,u) = 0$ which gives

$$u = u(x) = -a(x)/b(x) . \tag{7}$$

Let $u(t) = u(x(t))$ and $b(t) = b(x(t))$. It is assumed that along a boundary arc in $[t_1, t_2]$ the following condition holds:

$$b(t) \neq 0 \quad \text{for} \quad t_1 \leq t \leq t_2 , \quad |u(t)| < K(t) \quad \text{for} \quad t_1 < t < t_2 . \tag{8}$$

2. Necessary conditions of the Minimum-Principle

The necessary conditions for an extremal arc of (1)-(5) are developed in [1],[2]. It can be shown that the function η^* of bounded variation in [2,Th.2.3] has a continuous derivative η on the interior of a boundary arc for p-th order state constraints. Define the Hamiltonian

$$H(x,u,\lambda,\eta) = \lambda^T f_1(x) + \lambda^T f_2(x)u + \eta S(x) \tag{9}$$

where $\lambda \in \mathbb{R}^n$, $\eta \in \mathbb{R}$ and where the superscript T denotes the transpose.

1. There exists a scalar function $\eta(t) \geq 0$ which satisfies $\eta(t)S(x(t)) = 0$, $t \in [0,T]$, and which is continuous on the interior of a boundary arc. The adjoint variable $\lambda(t)$ satisfies

$$\dot{\lambda}^T = - \lambda^T(f_1+f_2u)_x - \eta S_x , \quad \lambda^T(T) = G_x(x(T))+\sigma^T\psi_x(x(T)) , \quad \sigma \in \mathbb{R}^k . \tag{10}$$

2. The jump condition at a contact point or junction point t_1 is

$$\lambda^T(t_1^+) = \lambda^T(t_1^-) - \nu_1 S_x(x(t_1)) , \quad \nu_1 \geq 0 . \tag{11}$$

3. The optimal control $u(t)$ minimizes $H(x(t),u,\lambda(t),\eta(t))$ over u with $|u| \leq K(t)$.

The coefficient of u in (9) is called the switching function

$$\Phi(t) = \lambda^T(t)f_2(x(t)) . \tag{12}$$

Then the optimal control $u(t)$ is given on an __interior__ __arc__ by

(13) $$u(t) = - K(t)\,\mathrm{sgn}\,\Phi(t)$$

where for simplicity $\Phi(t)$ is assumed to have only isolated zeros, i.e. $u(t)$ is a nonsingular control. On a __boundary__ __arc__ in $[t_1,t_2]$ the optimal control is the boundary control (7). The assumption (8) and the Minimum-Principle then imply

(14) $$H_u(t) = \Phi(t) = 0 \quad \text{for} \quad t_1^+ \leq t \leq t_2^- \ .$$

Thus the boundary control behaves like a __singular__ control in singular control problems. We can expect therefore necessary conditions for junctions between interior and boundary arcs which are similar to those in McDanell, Powers [3] .

3. Relations for the switching function at contact or junction points

Let t_1 be a contact point or a junction point and let $u^{(r)}(t)$, $r \geq 0$, be the lowest order derivative of the control $u(t)$ which is __discontinuous__ at t_1 . Furthermore let the integer q be the order of a singular arc, i.e. $\Phi^{(2q)}$ is the lowest order time derivative of Φ which contains the control u explicitly. Under the assumption $p \leq 2q+r$ one can show the following relations for the switching function by using the jump condition (11):

(15) $$\Phi^{(i)}(t_1^+) = \Phi^{(i)}(t_1^-) \quad , \quad i = 0,..,p-2 \quad ,$$

(16) $$\Phi^{(p-1)}(t_1^+) = \Phi^{(p-1)}(t_1^-) - \nu_1(-1)^{p-1}b(t_1) \ .$$

The assumption $p \leq 2q+r$ always holds for $p \leq 2$. The relation (16) implies that $\nu_1 > 0$ is equivalent to the discontinuity of $\Phi^{(p-1)}(t)$ at t_1 . Now let t_1 be an __entry-point__ of a boundary arc. Then we get $\Phi^{(i)}(t_1^+) = 0$ for $i \geq 0$ by virtue of (14) and hence (15),(16) yield

(17) $$\Phi^{(i)}(t_1^-) = 0 \quad , \ i = 0,..,p-2 \quad ,$$

(18) $$\nu_1 = (-1)^{p-1}\,\Phi^{(p-1)}(t_1^-)/b(t_1) \geq 0 \ .$$

The relations (17),(18) remain valid at an exit-point t_2 with t_1^- resp. ν_1 replaced by t_2^+ resp. $-\nu_2$.

4. Junction Theorems

Based on (17),(18) the following theorem can be proved using ideas similar to those in [3,Th.1]

Theorem 1: Let t_1 be a point where an interior nonsingular arc and a boundary arc of an optimal control u are joined and assume that u is piecewise analytic in a neighborhood of t_1. Let $u^{(r)}$, $r \geq 0$, be the lowest order derivative of u which is discontinuous at t_1 and let $p \leq 2q+r$. If $\nu_1 > 0$ then $p+r$ is an even integer.

Corollary 1: Under the assumptions of Theorem 1 the following statements are valid.

(i) If $p+r$ is odd then $\nu_1 = 0$ holds at a junction point t_1.

(ii) If $p+r$ is odd and if $\nu_1 > 0$ then t_1 cannot be a junction point but can only be a contact point with the boundary.

The next theorem treats the case $\nu_1 = 0$, i.e. $\Phi^{(p-1)}(t_1^-) = 0$, and is dual to a result for singular control problems [3,Th.2].

Theorem 2: Let t_1 be a point where an interior nonsingular arc and a boundary arc of an optimal control u are joined and assume that u is piecewise analytic in a neighborhood of t_1. Let $\Phi^{(p+m)}(t_1^-)$, $m \geq 0$, be the lowest order nonvanishing derivative of Φ and let $u^{(r)}$, $r \geq 0$, be the lowest order derivative of u which is discontinuous at t_1. If $p+m < 2q+r$ then $p+r+m$ is an odd integer.

In the 'normal case' $r = 0$ the preceding junction theorems allow a rough classification of the behaviour of the extremals with respect to the order p. For $p = 1$ numerical examples show that the extremals contain in general only boundary arcs. Hence $\nu_1 = 0$ at a junction point t_1 by Corollary 1 and the integer m in Theorem 2 is even. Usually we have $m = 0$ and thus $\Phi(t_1) = \Phi(t_2) = 0$, $\dot{\Phi}(t_1^-) \neq 0$, $\dot{\Phi}(t_2^+) \neq 0$ at an entry-point t_1 or exit-point t_2. If p is even then contact points and boundary arcs are possible for $\nu_1 > 0$. If p is odd and $p \geq 3$ then only contact points with the boundary are possible for $\nu_1 > 0$. A similar result holds for a regular Hamiltonian, cf. [1].

Proofs, further junction theorems and numerical examples will appear elsewhere [4]. The duality of control problems with bounded state variables and singular control problems is also displayed by similar numerical algorithms for both problems [5],[6].

References

[1] Jacobson,D.H., Lele,M.M., Speyer,J.L.: New Necessary Conditions of Optimality for Control Problems with State-Variable Inequality Constraints. J. of Math. Analysis and Appl. 35 (1971), 255-284.

[2] Norris,D.O.: Nonlinear Programming Applied to State-Constrained Optimization Problems. J. of Math. Analysis and Appl. 43 (1973), 261-272.

[3] McDanell,J.P., Powers,W.F.: Necessary Conditions for Joining Singular and Nonsingular Subarcs. SIAM J. on Control 9 (1971), 161-173

[4] Maurer,H.: On Optimal Control Problems with Bounded State Variables and Control Appearing Linearly. Submitted to SIAM J. on Control.

[5] Maurer,H., Gillessen,W.: Application of Multiple Shooting to the Numerical Solution of Optimal Control Problems with Bounded State Variables. To appear in COMPUTING.

[6] Maurer,H.: Numerical Solution of Singular Control Problems Using Multiple Shooting Techniques. To appear in JOTA 18 , No.2 (1976).

ON THE OPTIMAL CONTROL OF VARIATIONAL INEQUALITIES

F. Patrone
Istituto Matematico, Università di Genova
via L. B. Alberti 4, 16132 GENOVA (ITALY)

I. An existence theorem.

We are given a real and reflexive Banach space V, with dual V', an operator $A:V \longrightarrow V'$ and a function $j:V \longrightarrow (-\infty,+\infty]$.

We consider variational inequalities of the following type: find $y \in V$ such that:

$$(I.1) \qquad \langle Ay, z-y \rangle + j(z) - j(y) \geq \langle g, z-y \rangle \qquad \forall z \in V,$$

where $g \in V'$ and $\langle \cdot, \cdot \rangle$ is the pairing between V' and V.

We assume A a pseudo-monotone operator, i.e:

$$(I.2) \begin{cases} \text{a) } A \text{ is bounded (on bounded subsets)} \\ \text{b) } y_n \rightharpoonup y \text{ and } \overline{\lim_{n\to\infty}} \langle Ay_n, y_n - y \rangle \leq 0 \text{ imply} \\ \qquad \underline{\lim}_{n\to\infty} \langle Ay_n, y_n - z \rangle \geq \langle Ay, y-z \rangle \quad \forall z \in V. \end{cases}$$

j is a convex, proper (i.e. $j \not\equiv +\infty$) and lower semicontinuous function.

It is well known that inequalities of the type (I.1), for any given $g \in V'$, have a solution (see J.L.Lions [4]), if the following coercivity hypothesis is satisfied:

$$(I.3) \begin{cases} \exists\, z_0 \in V \text{ such that } j(z_0) < +\infty \text{ and} \\ \dfrac{\langle Az, z-z_0 \rangle + j(z)}{\|z\|} \longrightarrow +\infty \quad \text{when } \|z\| \longrightarrow \infty . \end{cases}$$

We shall deal with the following control problem. The space of controls is U, a real and reflexive Banach space; the set of admissible controls is U_{ad}, a closed convex (non empty) subset of U; $B:U \longrightarrow V'$ is a map such that $u_n \xrightarrow[U]{} u$ implies $Bu_n \xrightarrow[V']{} Bu$.

For any given $u \in U_{ad}$, the state $y=y(u)$ is given by the solution (not necessarily unique) of the following inequality($f \in V'$ is fixed):

(I.4) $\quad \langle Ay, z-y\rangle + j(z) - j(y) \geq \langle f+Bu, z-y\rangle \qquad \forall z \in V.$

The cost is assumed to be quadratic:

$J(u,y(u)) = \|y(u)-z_d\|_V^2 + \nu\|u\|_U^2$, $\nu \geq 0$ (a linear and continuous observation operator might as well be considered).

The control problem is: minimize J on U_{ad}.

We have the following theorem:

Theorem I.I Under the hypotheses above, if we have:

(I.5) $\quad$ either $\nu > 0$ or U_{ad} is bounded,

there exists an optimal pair $(u,y(u))$.

We give here a brief sketch of the proof. We take a minimizing sequence $\{u_n, y_n\}_{n \in N}$, where y_n is selected among the solutions of (I.4) corresponding to u_n. It follows from hypothesis (I.5) and the coercivity hypothesis (I.3) that we can extract a subsequence $\{u_{n_i}, y_{n_i}\}_{i \in N}$ such that $u_{n_i} \rightharpoonup u$ and $y_{n_i} \rightharpoonup y$ in V. The compactness of B, the lower semicontinuity of j and the pseudo-monotonicity of A allow us to pass to the limit in the inequality and to prove that y is a solution corresponding to u of (I.4). To conclude, the weak lower semicontinuity of J on U×V assure us that $\{u,y\}$ is an optimal control-state pair. ■

Remark I A particular case of inequality (I.I) is:

(I.6) $\quad \langle Ay, z-y\rangle \geq \langle g, z-y\rangle \qquad \forall z \in K,$

where K is a closed convex subset of V. It suffices to define $j(z) = \delta_K(z)$ where δ_K is the indicator function of K, i.e.:

$$\delta_K(z) = \begin{cases} 0 & \text{if } z \in K \\ +\infty & \text{if } z \notin K \end{cases}$$

Remark 2 Generally we have not uniqueness of the optimal control. A very simple counterexample is the following (V=U=R). The inequality is given by ($Bu = u + \sqrt{2}$):

$$(I.7) \qquad y\cdot(z-y) \geq (u+\sqrt{2})\cdot(z-y) \qquad \forall z\in[0,I]$$

The (unique) solution of (I.7) is given by $y(u)=Pr_K(u+\sqrt{2})$, the projection of $u+\sqrt{2}$ on $K=[0,I]$.

If we set $U_{ad}=U$ and $J(u)=u^2+(y(u))^2$, both $u=0$ and $u=-\frac{\sqrt{2}}{2}$ are optimal controls.

Former results on the optimal control of variational inequalities may be found in J.P.Yvon [6], or in J.L.Lions [3] (here a special case is considered); see also R.Kluge [2], and the bibliography listed there for further references.

2. Further results on the control of variational inequalities.

The existence theorem previously given can be extended to more general cost functionals, provided that the operator A is actually a monotone hemicontinuous and bounded operator. To do this, we need the following theorem of F.E.Browder [I] on the sequential lower semicontinuity of certain types of functionals.

Theorem 2.I We are given three real Banach spaces X, X_1, X_2 ; a map $h:X_1\ X_2 \longrightarrow R$ such that:

a) $h(x_1,x_2)$ is convex and strongly continuous in x_1 when x_2 is fixed

b) $h(x_1,x_2)$ is strongly continuous in x_2, when x_1 is fixed, and uniformly continuous in x_2 when x_1 varies in bounded subsets.

We are given also $L:X\longrightarrow X_1$, a linear and continuous map, and $M:X\longrightarrow X_2$ a map sequentially continuous from the weak to the strong topology. Then, setting $J(u)=h(Lu,Mu):X\longrightarrow R$, J is weakly sequentially lower semicontinuous.

We note that in Browder's theorem it is required that M is a linear map, but this hypothesis can be suppressed, as can be easily verified.

If we require that A is a strongly monotone operator, we can apply Browder's theorem to our control problem: to do this we note that under this assumption the map $G:U\longrightarrow V$, which associates to the control u the corresponding (unique) state y(u), is sequentially continuous from weak to strong. If we define X=U, $X_2=V$ and M=G, while h, X_1 and L are as in Browder's theorem, we obtain that the map J(u)=h(Lu,Mu) is sequentially weakly lower semicontinuous on U. This fact gives us an existence theorem if we assume that the set of admissible controls U_{ad} is bounded.

To conclude, we give here briefly a result on the sensitivity of the control problem. We consider variational inequalities of type (I.6), where A is a strictly monotone hemicontinuous and bounded operator. We perturb A,K,z_d and f. More precisely, we consider a sequence of problems like this:

(P_n) : minimize $J_n(v)=\|y_n(v)- z_{d_n}\|_V^2 +\nu\|v\|_U^2$

for $v\in U_{ad}$ and $y_n(v)$ the solution of:

$$\langle A_n y_n(v),z-y_n(v)\rangle \geq \langle f_n+Bv,z-y_n(v)\rangle \quad \forall\ z\in K_n$$

The initial problem is:

(P_o) : minimize $J_o(v)=\|y_o(v)-z_{d_o}\|_V^2 +\nu\|v\|_U^2$

for $v\in U_{ad}$ and $y_o(v)$ the solution of:

$$\langle A_o y_o(v),z-y_o(v)\rangle \geq \langle f_o+Bv,z-y_o(v)\rangle \quad \forall\ z\in K_n$$

If we assume that $f_n\longrightarrow f_o$, $z_{d_n}\longrightarrow z_{d_o}$, and that A_n,K_n and A_o,K_o verify the hypotheses introduced by U.Mosco in [5] to prove the strong convergence of solutions of perturbed variational inequalities, then we obtain the following result:

Theorem 2.2 Given a sequence of optimal controls u_n for P_n, we can extract a subsequence u_{n_i} weakly converging to an optimal control $\bar{u}$ for P_o; the corresponding states y_{n_i} converge strongly to $\bar{y}=y_o(\bar{u})$.

Let us give here a brief sketch of the proof. Our aim is to prove the boundedness of u_n, and this is obvious if we prove that $J_n(u_n)$ is such. To prove this fact, we note that $J_n(u_n) \leq J_n(u_o)$, where u_o is an optimal control for P_o. But $J_n(u_o)$ is bounded because the states $y_n' = y_o(u_n)$ converge strongly to y_o; more precisely we obtain that:

$$\forall \varepsilon > 0, \ \exists n(\varepsilon) \ \text{s.t.} \quad |J_n(u_o) - J_o(u_o)| < \varepsilon \ . \tag{2.I}$$

This fact implies the boundedness of $J_n(u_n)$, so of u_n. If we consider a subsequence $u_{n_i} \longrightarrow \bar{u}$, we have $Bu_{n_i} \longrightarrow B\bar{u}$: from this follows that $y_{n_i}(u_{n_i}) \longrightarrow y_o(\bar{u})$. To conclude, the weak lower semicontinuity of the costs J imply that $\underline{\lim}_{n\to\infty} J_n(u_n) \geq J_o(\bar{u})$. From this, and from (2.I), we get that $J_o(\bar{u}) \leq J_o(u_o) + \varepsilon$ for every $\varepsilon > 0$, so $J_o(\bar{u}) \leq J_o(u_o)$. But $J_o(u_o) \leq J_o(\bar{u})$ for definition of u_o: this means that $\bar{u}$ is actually an optimal control for P_o. ∎

References

1. BROWDER,F.E. "Remarks on the Direct Method of the Calculus of Variations", Arch. Rat. Mech. Anal., vol. 20, pp. 251-258, 1965.

2. KLUGE,R. "Optimal Control with Minimum Problems and Variational Inequalities", Lecture Notes in Computer Science, Optim. Techn. IFIP Technical Conference, Springer, Berlino-Heidelberg-New York, 1975.

3. LIONS,J.L. "Optimal Control of Systems Governed by Partial Differential Equations", Springer, Berlino-Heidelberg, 1971.

4. LIONS,J.L. "Quelques méthodes de résolution des problèmes aux limites non linéaires", Dunod, Gauthier-Villars, 1969.

5. MOSCO,U. "Convergence of Convex Sets and of Solutions of Variational Inequalities", Adv. in Math., vol. 3, pp. 510-585, 1969.

6. YVON,J.P. "Optimal Control of Systems Governed by Variational Inequalities", Lecture Notes in Computer Science, 5 Conference on Optim. Techn., PartI, Springer, Berlin-Heidelberg-New York, 1973.

MODELLING AND CONTROL FOR DISTRIBUTED PARAMETER SYSTEMS

A. J. Pritchard

Control Theory Centre

University of Warwick

England

P. E. Crouch

Department of Engineering and Applied Physics

Harvard University

U.S.A.

1. INTRODUCTION

For most systems it is usual to first perform the identification in order to obtain a mathematical model, and then design the control action on the basis of this model. However, it may be that the performance index which is used to design the control is not particularly sensitive to variations which result from updating the model by identification. For distributed parameter systems the identification and control problems are particularly difficult because of the limitation on the number of sensors, and the places at which control can be applied. In this paper a theory is developed from which it is possible to determine whether the identification should continue, the class of controls expanded, or whether the existing model and controls are satisfactory.

Another application of the theory is to control problems which are subject to perturbations. The optimal control problem for linear systems with quadratic performance indices is well understood (Curtain and Pritchard (1), Lions (2)). If the control can be implemented it will be of the feedback variety and will be independent of the initial state. This is regarded as particularly desirable since the control although designed from the mathematical model depends in an important way on the actual state of the system, and will, therefore take some account of a lack of perfect identification through perturbations to the system. For distributed parameter systems the class of controls which can be practically implemented is very small, and the best control within this class will in general depend on the initial state of the system. In our theory we will reformulate the performance index so that the optimal control from the limited class for the new index will not depend on the initial state, and will take some account of the perturbation to the system dynamics.

In order to develop our ideas we will assume that the basic model for the system is determined by a semigroup (Hille and Phillips (3)). Thus our results will also be valid for systems governed by ordinary differential equations, and certain differential-delay equations, although our main interest is in systems governed by partial differential equations. We could easily extend our results to systems determined by mild evolution operators (1), but for ease of presentation we will only consider the autonomous case.

2. PRELIMINARIES

The mathematical model of the control process is assumed to take the form

$$(2.1) \qquad \dot{z} = A_o z + Bu \quad , \quad z(o) = z_o$$

where z is an element of a real Hilbert space H, and u is a control with $u(.) \in L_2(o, T, \mathcal{U})$, where $\mathcal{U}$ is also a real Hilbert space. We assume that the operator A_o is a closed linear operator, defined on a dense domain $D(A_o) \subset H$, and A_o is the infinitesimal generator of a strongly continuous semigroup T_t. We take $B \in \mathcal{L}(\mathcal{U}, H)$, although this is a very restrictive assumption and does not allow us to consider control action on the boundary. Our results can be extended to the case where B is unbounded following the results in Curtain and Pritchard (4).

The first problem is to define a solution of (2.1). We use the concept of a mild solution

$$z(t) = T_t z_o + \int_o^t T_{t-s} B\, u(s) ds \tag{2.2}$$

It can be shown for $z_o \in H$, $u \in L_2(o, T; \mathcal{U})$ that $z \in C(o, T; H)$, however we are not able to differentiate (2.2) to obtain a strict solution of (2.1) unless for example $z_o \in D(A_o)$, and $u \in C'(o, T; \mathcal{U})$.

The controls will be restricted by practical considerations and from this limited class a feedback control u_o is taken of the form

$$u_o = - Fz \tag{2.3}$$

where $F \in \mathcal{L}(H, \mathcal{U})$. Again we can extend the results to F depending on time, or F unbounded (4), and this would allow us to consider point sensors.

With u_o given by (2.3) we may write (2.1) in the form

$$\dot{z} = Az + B\bar{u} \quad , \quad z(o) = z_o \tag{2.4}$$

where
$$A = A_o - BF \quad , \quad D(A) = D(A_o)$$
$$\bar{u} = u - u_o$$

Of course we must interpret (2.4) in the mild sense, that is

$$(2.5) \qquad z(t) = S_t z_o + \int_o^t S_{t-s} B \bar{u}(s)$$

where

$$(2.6) \qquad S_t z_o = T_t z_o - \int_o^t T_{t-s} BF S_s z_o \, ds$$

It is easily shown that S_t is well defined by (2.6) and S_t is a strongly continuous semigroup (1, 3).

Now we wish to consider unbounded perturbations of the operator A_o, so that the perturbed system can be written formally as

$$(2.7) \qquad \dot{z} = A_o z + A_1 z + Bu \quad , \quad z(o) = z_o$$

or equivalently

$$(2.8) \qquad \dot{z} = Az + A_1 z + B\bar{u} \quad , \quad z(o) = z_o \quad .$$

In order to explain what is meant by a solution of (2.7), (2.8), we need to impose conditions on the unbounded operator A_1. We will assume

$$(2.9) \qquad D(A_o) \subset D(A_1)$$

$$(2.10) \qquad ||T_t A_1 z_o|| \leq \frac{M}{t^\alpha} ||z_o|| \quad \text{for all } t \in (o, T], \; z_o \in D(A_o)$$

where M and α are constants with $0 \leq \alpha < 1$. Then for each $t \in (o, T]$, $T_t A_1 \in \mathcal{L}(D(A_o), H)$ and since $D(A_o)$ is dense in H we may extend the operator $T_t A_1$ to $\overline{T_t A_1}$ where $\overline{T_t A_1} \in \mathcal{L}(H)$. We show in the appendix that we are then able to define a strongly continuous semigroup U_t, by

$$(2.11) \qquad U_t z_o = T_t z_o - \int_o^t T_{t-s} BF U_s z_o \, ds + \int_o^t \overline{T_{t-s} A_1} \, U_s z_o \, ds$$

or

$$(2.12) \qquad U_t z_o = S_t z_o + \int_o^t \overline{S_{t-s} A_1} \, U_s \, z_o \, ds$$

The mild solution of either (2.7) or (2.8) is then given by

$$(2.13) \qquad z(t) = U_t \, z_o + \int_o^t U_{t-s} \, B\bar{u}(s) \, ds$$

or

$$(2.14) \qquad z(t) = S_t \, z_o + \int_o^t \overline{S_{t-s} A_1} \, z(s) \, ds + \int_o^t S_{t-s} \, B\bar{u}(s) \, ds \, .$$

Associated with the control problem will be a cost functional

$$(2.15) \qquad J(u) = \langle z(T), G \, z(T) \rangle_H + \int_o^T (\langle z(s), W \, z(s) \rangle_H + \langle u(s), R \, u(s) \rangle_{\mathcal{U}}) \, ds$$

where $R \in \mathcal{L}(\mathcal{U})$, $G, W \in \mathcal{L}(H)$. R, G, W are self adjoint and $G, W \geq 0$, $R > 0$.

3. PERTURBATION RESULTS

In this section we derive a number of perturbation results for the cost functional (2.15). We will distinguish between the cost of a control u for the mathematical model (2.2) and the cost for the perturbed system (2.13) or (2.14) by writing $J_M(u)$ and $J_p(u)$ respectively.

A fundamental role will be played by a strongly continuous linear operator Q(t), defined by

$$(3.1) \qquad Q(t) \, z_o = S^*_{T-t} \, G \, S_{T-t} \, z_o + \int_t^T S^*_{s-t} (W+F^*RF) \, S_{s-t} \, z_o \, ds$$

where S^*_t is the dual semigroup of S_t and is known to be strongly continuous since H is a Hilbert space. It is shown in (1) that Q satisfies the differential equation

$$(3.2) \qquad \frac{d}{dt} < Q(t)z_1, z_2>_H + < Az_1, z_2>_H + <z_1, Az_2>_H + <z_1, (W+F^*RF)z_2>_H = 0$$

with $Q(T) = G$, and $z_1, z_2 \in D(A_o)$.

Theorem 3.1

For the perturbed system defined by (2.14), the model defined by (2.2) and Q defined by (3.1)

a) the cost of the control u_o for the model is

$$(3.3) \qquad J_M(u_o) = < z_o, Q(o) z_o >_H ,$$

b) the difference in the cost of the perturbed system and the model will control u_o, is

$$(3.4) \qquad J_p(u_o) - J_M(u_o) = 2 \int_o^T (< z(t), \overline{Q(t)A_1} z(t)>_H) dt$$

c) the difference in the cost of the perturbed system with control $\bar{u} + u_o$, and the model with control u_o is

$$(3.5) \qquad J_p(\bar{u}+u_o) - J_M(u_o) = 2 \int_o^T (< z(t), \overline{Q(t)A_1} z(t)>_H + <\bar{u}(t), B^*Q(t)z(t)>_{\mathcal{U}} - <\bar{u}(t), RFz(t)>_{\mathcal{U}}) dt$$

Proof

a) The solution of (2.2) with $u = u_o$ is given by (2.5) with $\overline{u} = 0$, that is

$$z(t) = S_t z_o$$

From (3.1)

$$< z_o, Q(o) z_o > = <z(T), G z(T)>_H + \int_o^T [<z(s), (W+F^*RF) z(s)>_H] ds$$

$$= <z(T), G z(T)>_H + \int_o^T [<z(s), Wz(s)>_H + <u_o(s), Ru_o(s)>_{\mathcal{U}}]ds$$

$$= J_M(u_o)$$

b) This follows from part (c) with $\overline{u} = 0$.

c) It is easy to show that

$$|| Q(t)A_1 z_o || \leq \frac{\overline{M}}{(T-t)^\alpha} || z_o || \quad \text{for all } t \in [o,T) \; z_o \in D(A_o)$$

Thus $\overline{Q(t)A_1}$ is the bounded extension of $Q(t)A_1$ for all $t \in [o, T)$. With this interpretation the proof is by direct substitution and is a special case of a similar result in Pritchard and Crouch (5) and so will not be repeated here.

We now use this theorem to obtain bounds for $J_p(u_o) - J_M(u_o)$, and $J_p(\overline{u}+u_o) - J_n(u_o)$. In particular we need to establish the conditions under which there exists positive constants β, γ, such that

$$(3.6) \qquad | J_p(u_o) - J_M(u_o) | \leq \beta J_p(u_o)$$

$$(3.7) \qquad J_p(\overline{u}+u_o) - J_M(u_o) \geq - \gamma J_p(\overline{u}+u_o) .$$

Lemma 3.1

If either

a) $G \equiv 0$, $\alpha < 1$, $W + F^*RF > wI$ for $w > 0$.

or b) $G > gI$, $g > 0$; $W + F^*RF > wI$, $w > 0$, $\alpha < \frac{1}{2}$.

then there exists β such that (3.6) holds.

Proof

Assume (a) holds and let $W + F^*RF < kI$. Substituting for the expression (3.1) for $Q(t)$ with $G = 0$ in (3.4), yields

$$J_p(u_o) - J_M(u_o) = 2 \int_o^T \int_t^T < S_{\sigma-t} z(t), (W+F^*RF)\, \overline{S_{\sigma-t} A_1}\, z(t) >_H d\sigma\, dt .$$

Now S_t is a strongly continuous semigroup, hence there exists a constant M_s such that

$$|| S_t || \leqslant M_s \text{ for } t \in [0, T] .$$

In the appendix we have shown that

$$|| \overline{S_{\sigma-t} A_1} || \leqslant \frac{M_1}{(\sigma-t)^\alpha} .$$

Thus

$$|J_p(u_o) - J_M(u_o)| \leqslant 2 \int_o^T \int_t^T M_s M_1 k \frac{||z(t)||^2}{(\sigma-t)^\alpha}\, d\sigma\, dt$$

$$\leqslant \frac{2T^{1-\alpha} M_s M_1 k}{1 - \alpha} \int_o^T ||z(t)||^2\, dt$$

$$\leqslant \beta\, J_p(u_o)$$

where $\beta = \dfrac{2T^{1-\alpha} M_s M_1 k}{w(1-\alpha)}$

Now assume (b) holds, then from (2.14) we have

$$z(T) = S_{T-t} z(t) + \int_t^T \overline{S_{T-s} A_1}\, z(s)\, ds \tag{3.8}$$

Substituting from (3.8) and (3.1) in (3.4) yields

$$J_p(u_o) - J_M(u_o) = 2 < z(T),\ G \int_o^T \overline{S_{T-\sigma} A_1}\, z(\sigma)\, d\sigma >_H$$

$$- 2 \int_o^T < \int_t^T \overline{S_{T-\sigma} A_1}\, z(\sigma)\, d\sigma,\ G\, \overline{S_{T-t} A_1}\, z(t) >_H dt$$

$$+ 2 \int_o^T \int_t^T < S_{\sigma-t} z(t),\ (W+F^*RF)\, \overline{S_{\sigma-t} A_1}\, z(t) >_H d\sigma\, dt$$

Hence

$$|J_p(u_o) - J_M(u_o)| \leq 2\, ||z(T)||\, ||G|| \int_o^T \frac{M_1}{(T-\sigma)^\alpha} ||z(\sigma)|| d\sigma$$

$$+ 2 \int_o^T \int_t^T \frac{M_1}{(T-\sigma)^\alpha} ||z(\sigma)|| d\sigma ||G|| \frac{M_1}{(T-t)^\alpha} ||z(t)|| dt$$

$$+ 2 \int_o^T \int_t^T \frac{M_s k M_1 ||z(t)||^2}{(\sigma - t)^\alpha}\, d\sigma\, dt$$

$$\leq N_1 ||z(T)||\, ||z(.)||_T + N_2 ||z(.)||^2_T$$

where $N_1 = \dfrac{2||G||M_1\, T^{\frac{1}{2}-\alpha}}{\sqrt{1 - 2\alpha}}$, $N_2 = w\beta + \dfrac{2M_1{}^2||G||T^{1-2\alpha}}{1 - 2\alpha}$

and $||z(.)||_T = ||z(.)||_{L_2(o,\, T;\, H)}$

So that

$$|J_p(u_o) - J_M(u_o)| \leq \beta_1 J_p(u_o)$$

if $\quad N_1||z(T)||\,||z(.)||_T + N_2||z(.)||^2_T \leq \beta_1(g||z(T)||^2 + w||z(.)||^2_T)$

That is

$$4(\beta_1 w - N_2)\,\beta_1\, g > N_1^2$$

Clearly such a β_1 can be found if $w > o$.

Lemma 3.2

If either

a) $\quad G \equiv 0, \;\; \alpha < 1, \; W > wI$

or b) $\quad G > gI, \; \alpha < \frac{1}{2}, \; W > wI$

there then exists a γ such that (3.7) holds.

Proof

The condition that

$$J_p(\bar{u} + u_o) - J_M(u_o) \geq -\gamma\, J_p(\bar{u} + u_o)$$

is by (3.5) equivalent to

$$(3.9) \qquad \gamma < z(T), Gz(T)>_H + \int_o^T [2 < z(s), \overline{Q(s)A_1}\, z(s)>_H + (1+\gamma) < \bar{u}(s), R\bar{u}(s)>_{\mathcal{U}}$$

$$+ 2 < \bar{u}(s), (B^*Q(s) - RF - \gamma RF)z(s)>_{\mathcal{U}} + \gamma < z(s), (W + F^*RF)z(s)>_H]\, ds \geq 0$$

From (2.14) we have

$$(3.10) \qquad z(T) = S_{T-t}z(t) + \int_t^T [\overline{S_{T-t}A_1}\, z(s) + S_{T-s}\, B\bar{u}(s)]\, ds$$

a) Using (3.1) with $G = 0$, as in Lemma 3.1, we can show

$$\left| \int_o^T < z(s), \overline{Q(s)A_1}\, z(s) > ds \right| \leq \beta w ||z(.)||^2_T$$

Now (3.9) with G = 0 may be written in the form

$$\int_o^T \{ (1+\gamma) < [\overline{u}(s) + \frac{1}{1+\gamma} R^{-1}(B^*Q(s)-RF-\gamma RF)z(s)],\ R[\overline{u}(s) + \frac{1}{1+\gamma} R^{-1}(B^*Q(s)-RF-\gamma RF)z(s)>_{\mathcal{U}}]$$

$$+ \gamma < z(s),\ (W+F^*RF)z(s)>_H - \frac{1}{1+\gamma} < z(s),\ (B^*Q(s)-(\gamma+1)RF)^*R^{-1}(B^*Q(s)-(\gamma+1)RF)z(s)>_H$$

$$+ \quad 2 < z(s),\ \overline{Q(s)A_1}\, z(s) >_H \} \ ds \quad \geq 0.$$

So that this will certainly hold if

(3.11)

$$\gamma \int_o^T < z(s),\ (W+F^*RF)z(s)>_H ds - \frac{1}{1+\gamma} \int_o^T < z(s),\ (B^*Q(s)-(\gamma+1)RF)^*R^{-1}(B^*Q(s)-(\gamma+1RF)z(s)>_H ds$$

$$- \beta w ||z(.)||^2_T \quad \geq 0 .$$

Clearly we can find such a γ if w is sufficiently large.

b) Using (3.1) and (3.10) we obtain

$$\int_o^T < z(s),\ \overline{Q(s)A_1}\, z(s)>_H ds = \int_o^T < z(T),\ G\ \overline{S_{T-t}A_1}\, z(t)>_H\, dt$$

$$+ \int_o^T < z(t),\ \int_t^T S^*_{\sigma-t}(W+F^*RF)\ \overline{S_{\sigma-t}A_1}\, z(t)>_H\, ds\, dt$$

$$- \int_o^T < \int_t^T \overline{S_{T-\sigma}A_1}\, z(\sigma)d\sigma,\ G\ \overline{S_{T-t}A_1}\, z(t)>_H\, dt$$

$$- \int_o^T < \int_t^T < S_{T-\sigma}B\overline{u}(\sigma)d\sigma,\ G\ \overline{S_{T-t}A_1}\, z(t)>_H\, dt$$

The terms in the above expression may be estimated by

$\frac{N_1}{2}||z(T)||\,||z(.)||_T$, $\frac{N_2}{2}||z(.)||^2_T$ as in Lemma 3.1, and the last term may be written $\int_0^T < \bar{u}(t),\ B^* r(t)>_{\mathcal{U}}\ dt$

where

$$r(t) = \int_0^t S^*_{T-t}\ G\ \overline{S_{T-\sigma}A_1}\ z(\sigma)\ d\sigma$$

Completing the square with respect to $\bar{u}$, we find that (3.9) will certainly hold if

$$||z(T)||^2 - N_1||z(T)||\,||z(.)||_T - \frac{N_2}{2}||z(.)||^2_T + \gamma w||z(.)||^2_T$$

$$- \frac{1}{1+\gamma}\int_0^T <\{(B^*Q(t)-(1+\gamma)RF)z(t)-B^*r(t)\},R^{-1}\{(B^*Q(t)-(1-\gamma)RF)z(t)-B^*r(t)> dt \geq 0$$

So we can find such a γ if w or g is sufficiently large.

The above lemmas are essentially technical. Their main interest lies in the conditions which must be imposed on α and this indicates the class of unbounded perturbation operators which it is possible to consider. In applications, where it is necessary to obtain sharp estimates for β, γ, the explicit forms of Q, A_1 etc should be used rather than the crude estimates carried out in the lemmas.

We now show how (3.6) and (3.7) can be used to answer some of the problems posed in the introduction. From (3.6) we have

$$(3.12) \qquad \frac{1}{1+\beta} \leq \frac{J_p(u_o)}{J_M(u_o)} \leq \frac{1}{1-\beta}$$

and from (3.7)

$$(3.13) \qquad \frac{J_p(\bar{u} + u_o)}{J_M(u_o)} \geq \frac{1}{1+\gamma}$$

Combining (3.11) and (3.12) we find

$$\frac{J_p(\bar{u} + u_o)}{J_p(u_o)} \geq \frac{1-\beta}{1+\gamma}$$

Since $\bar{u}$ is any control we may take $\bar{u} + u_o$ to be the optimal control u^* for the perturbed system, then from the above we have

$$(3.14) \qquad 1 \geq \frac{J_p(u^*)}{J_p(u_o)} \geq \frac{1-\beta}{1+\gamma}$$

Now β, γ depend only on the model, the parameters M, α (which determine a class of perturbation operators), and the control u_o. They do not depend on any knowledge of u^*, or the initial state z_o. Moreover, if the perturbation is zero, then $\beta = 0$, and if then the limited class of controls is wide enough to include the optimal control u^*, then $\gamma = 0$ where $u_o = u^*$. We propose the lower bound $\frac{1-\beta}{1+\gamma}$ as a criteria for answering the following questions.

a) For a system which is imperfectly identified and for which there is limited control action, should the identification continue or the class of controls expanded? - If we think of the operator A_1 as being the unidentified part of the system, then the largest value of $\frac{1-\beta}{1+\gamma}$ as the controls vary over the limited class defines a number ρ which reflects the lack of perfect identification and limited control action. Obviously $\rho \leq 1$, and

$$J_p(u_o) \leq \frac{1}{\rho} J_p(u^*)$$

so that the cost is at most a factor $1/\rho$ of the best that can possibly be achieved assuming perfect identification and no constraints on the control action. If this is not regarded as satisfactory the exact forms of β, and γ will indicate whether or not the identification should continue or the class of controls expanded.

b) For a system working in the presence of perturbations, and with limited control action, what bounds should be placed on the perturbation so that the system operates in a satisfactory manner? - If the system is regarded as working satisfactorily for $\rho \geq \bar{\rho}$, then the value of $\rho = \bar{\rho}$ will give an upper bound on the class of perturbations.

4. EXAMPLE - TRAFFIC FLOW PROBLEM

We will apply the theory of § 3 to a traffic flow problem for which the basic dynamics are governed by the Lighthill-Whitham model (6). If the flow is heavy we may treat a stream of traffic as a continuum so that the number of cars passing a position x per unit time is $q(x, t)$, and the number of cars per unit length is $\rho(x, t)$. Then the flow velocity $v = q/\rho$, and conservation requires that at points where there is no entry or exit

$$\rho_t + (\rho v)_x = 0$$

Lighthill and Whitham assume there is a relationship between q and ρ of the form

$$q = Q(\rho) - \nu\rho_x$$

where Q is convex, zero at $\rho = 0$, and $\rho = \rho_j$ with a maximum at ρ_m, although the exact form of Q is in general not known. Actual observations indicate that the typical values of ρ_1, ρ_m, $Q(\rho_m)$ are 225 vehicles per mile, 80 vehicles per mile, 1500 vehicles per hour. ν is a diffusion parameter and the term $-\nu\rho_x$ is a way of introducing a driver's awareness of conditions ahead.

There is a steady state solution of the form

$$v = v_o = \frac{Q(\rho_o)}{\rho_o} , \qquad \rho = \rho_o$$

and if we wish to maximise the flow rate q we would set $\rho_o = \rho_m$, $v_o = \dfrac{Q(\rho_m)}{\rho_m}$. Now suppose that at a given time there is a deviation from the steady state, then the linearized equations governing the perturbation in velocity, are

$$(4.1) \qquad v_t = - c_o v_x + \nu v_{xx} , \qquad v(x,o) = \overline{v}(x)$$

where $c_o = Q'(\rho_o)$. We have attempted to choose $\rho_o = \rho_m$ which maximizes $Q(\rho)$, and so we would expect $c_o = 0$. However, since the exact form of $Q(\rho)$ is not known this is unlikely to be the case. We may therefore regard c_o as an unidentified parameter with nominal value zero, and the term $c_o v_x$ as a perturbation of the system.

The boundary conditions for (4.1) are difficult to formulate, but since our main aim is to illustrate the theory of § 3 we will assume the boundary conditions

$$v = 0 \quad \text{at } x = 0, \ x = L$$

Thus the speed at entry and exit is fixed at v_o. If we assume that the system can be controlled, we have

(4.2) $$v_t = -c_o v_x + \nu v_{xx} + g$$

$$v = 0 \text{ at } x = 0, L, \quad v(x,o) = \overline{v}(x)$$

The control action is to be designed so that the system operates close to the steady state with a minimum use of control energy. One was of expressing this requirement is to choose the control to minimize

$$J(g) = \int_o^{T_1}\int_o^{L} (v^2 + \lambda g^2)\, dx\, dt$$

The problem can be transformed into an equivalent problem

(4.3) $$v_t = a v_x + v_{xx} + u$$

$$v = 0 \text{ at } x = 0, 1, \quad v(x,o) = \overline{v}(x)$$

$$J(u) = \int_o^{T}\int_o^{1} (v^2 + b\, u^2)\, dx\, dt$$

where $a = -\dfrac{c_o L}{\nu}$, $b = \dfrac{\nu^2 \lambda}{L^4}$, $u = \dfrac{L^2 g}{\nu}$

If we take $H = L_2(0,1) = \mathcal{U}$, then (4.3) is equivalent to the abstract evolution problem

(4.4) $$\dot{z} = A_o z + A_1 z + Bu, \quad z(o) = z_o$$

where $A_o z = z_{xx}, \quad z \in D(A_o)$

$$D(A_o) = \{z \in L_2(0,1),\ Az \in L_2(0,1),\ z = 0 \text{ at } x = 0, 1\}$$

$$A_1 z = a\, z_x \qquad z \in D(A_o)$$

$$B = I \quad (\text{identity in } \mathcal{L}(H))$$

$$J(u) = \int_o^T (||z||^2 + b||u||^2)\, dt$$

It is easy to show that A_o generates a strongly continuous semigroup T_t, where

$$T_t\, z_o = \sum_1^\infty 2\, e^{-n^2\pi^2 t} \sin n\pi x \int_o^1 \sin n\pi y\; z_o(y)\, dy$$

Hence

$$||T_t|| = ||T^*_t|| \leq e^{-\pi^2 t}$$

Moreover

$$||T_t A_1 z_o||^2 = a^2 \sum_1^\infty 2\, e^{-2n^2\pi^2 t}\, n^2\pi^2 \left(\int_o^1 \cos \pi y\; z_o(y) dy\right)^2, \quad z_o \varepsilon\, D(A_o)$$

and so

$$||T_t A_1 z_o|| \leq \frac{|a|}{\sqrt{2e}\; t^{\frac{1}{2}}}$$

Since (4.4) is asymptotically stable in the Liapunov sense, and control action is difficult to implement we will make comparisons with the case $u_o = 0$, so that $F = 0$. Then from (3.1)

$$Q(t)\, z_o = \int_t^T T^*_{s-t}\; T_{s-t}\; z_o\, ds$$

It is easy to show that

$$\left| 2 \int_o^T \langle z(t), \overline{Q(t)A_1}\; z(t)\rangle_H\, dt \right| \leq \frac{4|a|}{\sqrt{2e}} \left(\int_o^T e^{-\pi^2 y^2}\, dy\right) ||z(.)||^2_T$$

Hence from (3.4)

$$\frac{1}{1-\beta} \geq \frac{J_p(o)}{J_M(o)} \geq \frac{1}{1+\beta}$$

where $\quad = \frac{4|a|}{\sqrt{2e}} \int_o^T e^{-\pi^2 y^2}\, dy$.

γ is obtained from (3.11) with F = 0, so that we require

$$\gamma||z(.)||^2_T - \beta||z(.)||^2_T - \frac{1}{(1+\gamma)b}||(Qz)(.)||^2_T \geq 0 .$$

Now

$$||(Qz)(t)|| = ||\int_t^T T^*_{s-t}\, T_{s-t}\, z(t)\, ds\, ||$$

$$= ||\int_t^T T_{2(s-t)}\, z(t)\, ds\, ||$$

$$\leq \frac{1 - e^{-2\pi^2 T}}{2\pi^2}||z(t)|| = \alpha||z(t)||$$

Hence

$$\gamma = \frac{\beta+1 + \sqrt{(\beta+1)^2 + 4\alpha^2/b}}{2}$$

and

$$1 \geq \frac{J_p(u^*)}{J_p(o)} \geq \frac{1-\beta}{1+\gamma}$$

For some numerical results we take b = 1, and estimate α and β by their values for $T = \infty$, then

$$1 > \frac{J_p(u^*)}{J_p(o)} > \frac{1 - .484|a|}{1 + .484|a| + .00257}$$

We see therefore that if a = 0 so that the system is perfectly identified, then

$$1 \geq \frac{J_p(u^*)}{J_p(o)} \geq .9974$$

and it would seem that it is not worth introducing sensors and applying control action. On the other hand if the process is regarded as working satisfactorily for

$$J_p(o) \leq 2\, J_p(u^*)$$

then this will allow $|a| < .687$. So that identification need only continue until this bound is met.

5. REFERENCES

(1) R. F. CURTAIN, A.J. PRITCHARD "The infinite dimensional Riccati equation for systems described by evolution operators", to appear in SIAM J. Control, 1975.

(2) J. L. LIONS "Optimal control of systems governed by partial differential equations", Springer-Verlag, 1971.

(3) E. HILLE, R.S. PHILLIPS "Functional analysis and semigroups", AMS, Vol. 31, 1957.

(4) R. F. CURTAIN, A.J. PRITCHARD "Boundary value control and filtering with point observations", Control Theory Centre Report No. 42, 1975.

(5) A. J. PRITCHARD, P.E. CROUCH "Sensitivity analysis for linear quadratic control problem in Hilbert space", Control Theory Centre Report No. 27, 1974.

(6) M. J. LIGHTHILL, G.B. WHITHAM "Theory of traffic flow on long crowded raods", Proc. Roy. Soc. A.229, pp. 281-345, 1955.

(7) G. B. WHITHAM "Linear and nonlinear waves", Wiley, 1974.

6. APPENDICES

Appendix I

If A_1 satisfies (2.5), (2.10) and S_t is defined by (2.6), that is

$$S_t z_o = T_t z_o + \int_o^t T_{t-s} BF S_s z_o \, ds \tag{6.1}$$

where BF ε $\mathcal{L}(H)$, then

$$||S_t A_1 z_o|| \leq \frac{M_1}{t^\alpha} ||z_o|| \quad , \qquad z_o \varepsilon D(A_o) . \tag{6.2}$$

Proof

Using the usual iterative scheme

$$S^o{}_t = T_t \quad ,$$

$$S^n{}_t z_o = \int_o^t T_{t-s} BF S_s^{n-1} z_o \, ds \ ,$$

$$S_t = \sum_{n=0}^{\infty} S^n{}_t \quad ,$$

it is easy to show that (6.1) defines, uniquely, a strongly continuous semigroup with $||S_t|| \leqslant M_s$ $t \varepsilon [o, T]$. Moreover it can also be shown that S_t satisfies

$$S_t z_o = T_t z_o + \int_o^t S_{t-s} BF T_s z_o \quad .$$

Hence

$$S_t A_1 z_o = T_t A_1 z_o + \int_o^t S_{t-s} BF T_s A_1 z_o \qquad z_o \varepsilon D(A) \quad t > 0$$

Then

$$||S_t A_1 z_o|| \leq \frac{M}{t^\alpha} ||z_o|| + \frac{M_s ||BF|| MT}{T^\alpha (1-\alpha)}$$

$$\leq \frac{M_1}{t^\alpha} \quad \text{where} \quad M_1 = M + \frac{M_s ||BF|| MT}{1 - \alpha} ,$$

Appendix II

U_t defined by (2.12) uniquely determines a strongly continuous semigroup.

Proof

The proof is straightforward, using (6.2) and the iterative scheme

$$U^o{}_t = S_t$$

$$U^n{}_t z_o = \int_o^t \overline{S_{t-s} A_1} \; U_s^{n-1} z_o \, ds \; ,$$

ON BANG – BANG CONTROL POLICIES

Roberto GONZALEZ (*) and Edmundo ROFMAN (**)
Instituto de Matemática "Beppo Levi"
Universidad Nacional de Rosario
ARGENTINA

(Work included in the cooperation program with I.R.I.A. – Rocquencourt – FRANCE)

ABSTRACT: In this paper it is proposed a method for the determination of the optimal distribution of N switching points for a bang–bang control applied to a differential system.

After pointing the necessary conditions to be verified by such switching points it is showed the existence of optimal policy for a fixed number N of them.

Once characterized these points thru the application of the Pontryagin principle the problem, considered till now in the space of step functions, is put into the L_1 space, in order to show the existence of a minimizing succession of the amplified problem and analized its correspondence to an optimal policy.

After reducing the problem into one of optimization on a convex K of $\mathbb{R}^n$ there are added considerations which let us, with the proposed method, obtain the optimal also with a number of switching points n less than the predetermined N .

Now it is proved that the function to optimize is of C^2 class in K and the applied methods are these of the projected gradient and the conjugated gradient conveniently penalized.

Finally, the obtained algorithms are applied in one example: the shut down policy of a nuclear reactor where the optimun is obtained with a finite number of switchings; this number remains constant although increasing values of N are proposed.

§1. STATEMENT OF THE PROBLEM. NECESSARY CONDITIONS OF OPTIMALITY. EXISTENCE OF MINIMUN IN THE CASE OF FIXED NUMBER OF SWITCHING POINTS.

Given the dinamical system governed by the differential equation:

$$\dot{x} = F(t)x + G(t)u \qquad x \in R^{\nu} \ , \ u \in R^{1} \tag{1}$$

with initial condition $x(0) = x_0$

and the cost functional

$$J(u(.)) = \int_0^T l(x(s), u(s), s)ds + g(x(T)) \tag{2}$$

we try to find the control function u(.) that minimices J .

(*) Reasercher of the "Consejo de Investigaciones de la Universidad Nacional de Rosario" for the project: "Optimization and Control. Theory and applications".

(**) Director of the above referred project.

The control $u(.)$ belongs to the family $\mathcal{U}_{ad}$ that satisfy the following restrictions.

a) $u(t) = v_1$ or $u(t) = v_2$ $\forall\, t \in [0,T]$

b) $u(.)$ is a step-function with n switchings

c) $u(0) = v_1$.

We denote with $\theta_1, \theta_2, \ldots, \theta_n$ the switching points and this set with the vector $\theta = (\theta_1, \ldots, \theta_n)'$

(3) θ satisfies the restrictions: $0 < \theta_1 < \theta_2 < \ldots < \theta_n < T$.

Then, if we fix θ , we know the value of $u(t)$ $\forall\, t \in [0,T]$ and we can think of $J(u(.))$ as a function $J(\theta)$ of $\theta \in \mathring{\Omega}$, where $\mathring{\Omega} = \{\theta \in R^n \,/\, 0 < \theta_1 < \ldots < \theta_n < T\}$. If the minimun of the problem exists, the necessary conditions will be (because θ belongs to an open set):

$$\frac{\partial J}{\partial \theta_1} = 0 \quad ; \ldots ; \quad \frac{\partial J}{\partial \theta_n} = 0 \quad .$$

We can modify the restrictions (3) in the following form:

(3') $$0 \leqq \theta_1 \leqq \theta_2 \leqq \ldots \leqq \theta_n \leqq T$$

and analyze the meaning of a point in the boundary of $\mathring{\Omega}$.

a) $\theta_1 = 0$ means that the first step has the value $u = v_2$.

b) $\theta_i = \theta_{i+1}$

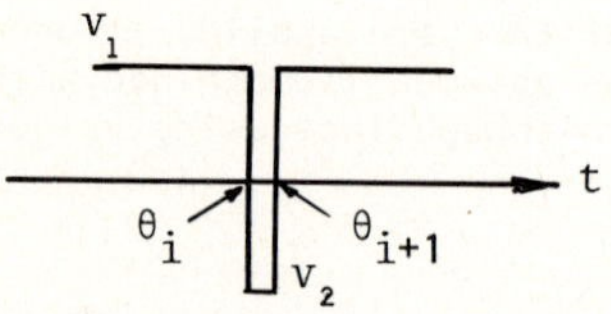

there are two simultaneous commutations that could be eliminated and it remains a new control function with n-2 commutations.

c) $\theta_i = \theta_{i+1} = \theta_{i+2}$

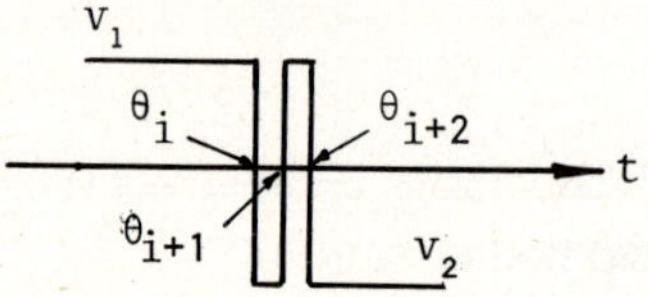

there are three simultaneous switchings, we could eliminate two and obtain a new control with n-2 commutations.

d) $\theta_n = T$ a commutation at the end that could be eliminated and the new control has $n-1$ switchings.

With this meaning we can define the function $J(\theta)$ in the set $\Omega = \{\theta \in R^n \,/\, 0 \leqq \theta_1 \leqq \theta_2 \leqq \ldots \leqq \theta_n \leqq T\}$. In this compact set, under suitable conditions on l, g, F, G, (it will be enough the continuity of l and g, and that F, G be integrable), is $J(\theta)$ continuos; then there is an optimal control in Ω that provides the minimun value of J and has $n' \leqq n$ switching points.

§2. THE RELATION BETWEEN THE NECESSARY CONDITION $\frac{\partial J}{\partial \theta_i} = 0$ AND PONTRYAGIN'S MAXIMUN PRINCIPLE.

We shall see in the following number that $\frac{\partial J}{\partial \theta_i}$ has the form:

$$\frac{\partial J}{\partial \theta_i} = l(x(\theta_i), u(\theta_i -), \theta_i) - l(x(\theta_i), u(\theta_i +), \theta_i) + p(\theta_i) G(\theta_i) \left[u(\theta_i +) - u(\theta_i -) \right]$$

where $p(t)$ satisfies:

$$\begin{cases} \dfrac{dp}{dt} = - p\,F(t) + \dfrac{\partial l}{\partial x}(x(t), u(t), t) \\ p(T) = - \dfrac{\partial g}{\partial x}\Big|_{x(T)} \end{cases}$$

and $u(\theta_i -) = \lim\limits_{\substack{\varepsilon \to 0 \\ \varepsilon > 0}} u(\theta_i - \varepsilon)$; $u(\theta_i +) = \lim\limits_{\substack{\varepsilon \to 0 \\ \varepsilon > 0}} u(\theta_i + \varepsilon)$.

If we define:

$$H(x,p,u,t) = p(t)[F(t)x(t) + G(t)u(t)] - l(x(t),u(t),t)$$

we can write:

$$\frac{\partial J}{\partial \theta_i} = H(x(\theta_i),p(\theta_i),u(\theta_i+),\theta_i) - H(x(\theta_i),p(\theta_i),u(\theta_i-),\theta_i) \quad .$$

We also know that in the problem $\min\limits_{u \in \mathcal{U}_{ad}} J(u(.))$, where $\mathcal{U}_{ad}$ is the set of step functions with values v_1, v_2, if $\bar{u}(.)$ is the optimal control in this set, it must satisfie the maximun principle:

$$H(x(t),p(t),\bar{u}(t),t) = M(x(t),p(t),t) \qquad \text{a.e.}$$

where, by definition

$$M(x,p,t) = \max_{u=\begin{cases}u_1\\u_2\end{cases}} H(x,p,u,t)$$

$M(x(t), p(t), t)$ is continuous in t , then:

$$\lim_{\varepsilon \to 0+} M(x(t+\varepsilon), p(t+\varepsilon), t) = \lim_{\varepsilon \to 0-} M(x(t+\varepsilon), p(t+\varepsilon), t+\varepsilon)$$

but

$$\lim_{\varepsilon \to 0+} M(x(t+\varepsilon),p(t+\varepsilon),t+\varepsilon) = \lim_{\varepsilon \to 0+} H(x(t+\varepsilon),p(t+\varepsilon),u(t+\varepsilon),t+\varepsilon) = H(x(t),p(t),u(t+),t)$$

and also

$$\lim_{\varepsilon \to 0-} M(x(t+\varepsilon),p(t+\varepsilon),t+\varepsilon) = \lim_{\varepsilon \to 0-} H(x(t+\varepsilon),p(t+\varepsilon),u(t+\varepsilon),t+\varepsilon) = H(x(t), p(t), u(t-), t)$$

from where it follows:

$$\frac{\partial J}{\partial \theta_i} = H(x(\theta_i),p(\theta_i),u(\theta_i+),\theta_i) - H(x(\theta_i),p(\theta_i),u(\theta_i-),\theta_i) = 0$$

then, the maximun principle implies the necesary conditions of optimality: $\dfrac{\partial J}{\partial \theta_i} = 0$.

§3. COMPUTATION $\frac{\partial J}{\partial \theta_i}$

The equation of the system's evolution is

$$x(t) = \Phi(t,0)x_o + \sum_{k=1}^{j-1} \int_{\theta_k}^{\theta_{k+1}} \Phi(t,s)G(s)u(s)ds + \int_{\theta_j}^{t} \Phi(t,s)G(s)u(s)ds$$

where $j \,/\, \theta_j \leqq t \leqq \theta_{j+1}$

and we can calculate $\dfrac{\partial}{\partial \theta_i} x(t)$.

We suppose $G(t)$ is continuous in the interval $[0,T]$

a) $t > \theta_i$

$$\frac{\partial}{\partial\theta_i}x(t) = \Phi(t,\theta_i)G(\theta_i)(v_2 - v_1)(-1)^i \quad .$$

It must be remembered that $\Phi(t,s)$ is the solution of the matrix differential equation

$$\frac{d}{dt}\Phi(t,s) = F(t)\Phi(t,s) \qquad \Phi \cdot \nu\times\nu \quad \text{matrix}$$

with initial condition: $\Phi(s,s) = I$.

b) $t < \theta_i$

$x(t)$ does not depend on Θ_i , then $\frac{\partial}{\partial\theta_i}x(t) = 0$.

In this form, we can say that

$$\begin{cases} \frac{d}{dt}\left(\frac{\partial}{\partial\theta_i}x(t)\right) = F(t)\left(\frac{\partial}{\partial\theta_i}x(t)\right) & t > \theta_i \\ \text{with initial conditions:} \\ \left.\frac{\partial}{\partial\theta_i}x(t)\right|_{t=\theta_i} = G(\theta_i)(v_2 - v_1)(-1)^i \end{cases}$$

$$J(\theta) = g(x(T)) + \int_0^T 1(x(s),\, u(s),\, s)d\tau$$

if 1 and g are continuously differentiable, we have:

$$\frac{\partial J}{\partial\theta_i} = 1(x(\theta_i), u(\theta_i -), \theta_i) - 1(x(\theta_i), u(\theta_i +), \theta_i) + \\ + g'(x(T)).\frac{\partial}{\partial\theta_i}x(T) + \int_{\theta_i}^T \frac{\partial 1}{\partial x}(x(s),u(s),s).\frac{\partial}{\partial\theta_i}x(s)ds$$

if we introduce the adjoint vector $p(t)$ that satisfies

$$\begin{cases} -\frac{dp}{dt}(t) = p(t)F(t) - \frac{\partial 1}{\partial x}(x(t),u(t),t) \\ p(T) = -g'(x(T)) \end{cases}$$

we can write:

$$\frac{\partial J}{\partial\theta_i} = 1(x(\theta_i), u(\theta_i -), \theta_i) - 1(x(\theta_i), u(\theta_i +), \theta_i) + \\ + g'(x(T))\frac{\partial}{\partial\theta_i}x(T) + \int_{\theta_i}^T \left(p(t)F(t) + \frac{dp}{dt}(t)\right)\frac{\partial}{\partial\theta_i}x(t)dt$$

and integrating by parts we obtain:

$$\frac{\partial J}{\partial\theta_i} = 1(x(\theta_i),u(\theta_i -),\theta_i) - 1(x(\theta_i),u(\theta_i +) + p(\theta_i)G(\theta_i)(v_1 - v_2)(-1)^i$$

§4. CONTINUITY OF $\frac{\partial J}{\partial\theta_i}$

(1) $$\frac{\partial J}{\partial\theta_i} = 1(x(\theta_i),u(\theta_i -),\theta_i) - 1(x(\theta_i),u(\theta_i +),\theta_i) + p(\theta_i)G(\theta_i)(v_1 - v_2)(-1)^i \quad .$$

We suppose that $F(.),G(.)$ are continuous and $1,g$ are continuously differentiable. $\frac{\partial J}{\partial\theta_i}$ is a continuous function of $x(\theta_i),p(\theta_i)$ and $\theta_i)$. The values $u(\theta_i +)$ and $u(\theta_i -)$ are constant (v_1 and v_2) , then we must prove only the continuity of $x(\theta_i)$, $p(\theta_i)$ to obtain the continuity of $\frac{\partial J}{\partial\theta_i}$.

We can easily see that the transformation $\theta : \longrightarrow u(.)$ defined by

$$(2)\quad \begin{cases} u(t) = v_1 & 0 \leqq t < \theta_1 \\ u(t) = v_1 + (v_2 - v_1)\left[\frac{1-(-1)^i}{2}\right] & \theta_i \leqq t < \theta_{i+1} \\ u(t) = v_1 + (v_2 - v_1)\left[\frac{1-(-1)^n}{2}\right] & \theta_n \leqq t \leqq T \end{cases}$$

is continuous from $\Omega \longrightarrow L_1(0,T)$.

$$(3)\qquad x(t) = \Phi(t,0)x_0 + \int_0^t \Phi(t,s)G(s)u(s)\,ds$$

and this formula defines a continuous transformation from $L_1(0,T) \longrightarrow C(0,T\,;R^\nu)$ because, if $u_1(.)$, $u_2(.)$ are two controls in $L_1(0,T)$ and $x_1(.)$, $x_2(.)$ the system's evolution, it is:

$$(4)\qquad \|x_1(t) - x_2(t)\| \leqq M\int_0^T |u_1(s) - u_2(s)|\,ds = M\|u_1 - u_2\|_{L_1(0,T)}$$

where

$$(5)\qquad M = \sup_{\substack{t\in[0,T]\\ s\in[0,t]}} \|\Phi(t,s)\| \,.\, \sup_{s\in[0,T]} \|G(s)\|$$

and M is finite due to the continuity of G and Φ .

Obviously

$$(6)\qquad x(.) \longrightarrow x(\theta_i) \text{ is continuous from } C(0,T\,;R^\nu) \longrightarrow R^\nu \quad .$$

Then,

$$(7)\qquad \theta \longrightarrow x(\theta_i) \quad \text{is continuous from } R^\nu \longrightarrow R^\nu \quad .$$

From the differential equation of $p(t)$ (p is a row vector)

$$(8)\qquad -\frac{d}{dt}\,p(t) = p(t)F(t) - \frac{\partial l}{\partial x}$$

we obtain the complete solution using the homogeneous solution and "variation of constants" method

$$(9)\qquad p(t) = p(T)\Phi(T,t) + \int_T^t \frac{\partial l}{\partial x}\,(x(s),u(s),s)\,.\,\Phi(s,t)\,ds$$

$$(10)\qquad p(T) = -\,g'(x(T))$$

thus, $-\,g'(x(T))\Phi(T,\theta_i)$ defines a continuous function from $\theta \in R^n \longrightarrow R^\nu$, because g' is continuous, $x(T)$ is a continuous function of θ and Φ is absolutely continuous in its both arguments. The transformation $R^n \longrightarrow L_1(0,T\,;R^\nu)$ given by:

$$\theta \longrightarrow u(.) \longrightarrow \frac{\partial l}{\partial x}\,(x(t),u(t),t) \in L_1(0,T\,;R^\nu)$$
$$\searrow x(.) \nearrow$$

is continuous from $R^n \longrightarrow L_1(0,T\,;R^\nu)$.

To prove it, let $\theta, \theta_\varepsilon$ be two set of switching points such that $\|\theta-\theta_\varepsilon\| \longrightarrow 0 \quad \varepsilon \to 0$.

Then, $u(t) \longrightarrow u_\varepsilon(t)$ a.e.

and
$$\max_{t\in[0,T]} \|x(t) - x_\varepsilon(t)\| \longrightarrow 0$$

now, $\frac{\partial l}{\partial x}\,(x_\varepsilon(s),u_\varepsilon(s),s) \longrightarrow \frac{\partial l}{\partial x}\,(x(s),u(s),s)$ a.e. , and

$\int_0^T \|\frac{\partial l}{\partial x}\,(x_\varepsilon(s),u_\varepsilon(s),s) - \frac{\partial l}{\partial x}\,(x(s),u(s),s)\|\,ds \longrightarrow 0$ as we can see applying Lebesgue's theorem.

The formula $\int_T^t \frac{\partial l}{\partial x}(x(s),u(s),s)\Phi(s,t)ds$ defines a continuous transformation from $L_1(0,T;R^\nu) \longrightarrow C(0,T;R^\nu)$. Then, taking into account all these results we proof that $p(\theta_i)$ and also $\frac{\partial J}{\partial \theta_i}$ are continuous functions of θ .

§5. NUMERICAL SOLUTION OF THE PROBLEM BY THE APPLICATION OF THE PROJECTED GRADIENT METHOD.

We have seen that the problem of finding an optimal bang—bang policy with $n' \leqq n$ switchings was reduced to the finite dimensional problem.

(1) $$\min_\Omega J(\theta) \qquad \Omega = \left\{\theta \in R^n \;/\; 0 \leqq \theta_1 \leqq \theta_2 \leqq \cdots \leqq \theta_n \leqq T\right\}$$

Ω is a convex and compact subset where J is continuous, and this implicates the existence of a minimun.

We write the $n+1$ restrictions defining Ω in vector form.

$f(\theta) \leqq 0 \qquad f \in R^{n+1}$ and $f_1 = -\theta_1$, $f_2 = \theta_1 - \theta_2, \ldots, f_n = \theta_{n-1} - \theta_n$, $f_{n+1} = \theta_n - T$.

DEFINITION.

$$I_\varepsilon(\theta) = \{i \;/\; f_i(\theta) + \varepsilon \geqq 0\} \quad .$$

DEFINITION.

Given a set of integers, $I \subset \{1,\ldots, n+1\}$, the projection of $y \in R^n$ on the subspace generated by the vectors ∇f_i, $i \in I$ is the vector $F_I \bar{\mu}$ that minimizes $\|y - F_I \mu\|^2$, where

$$F_I = \left[\nabla f_{i_1}, \ldots, \nabla f_{i_m}\right] \;,\quad I = \left\{i_1, \ldots, i_m\right\} \quad \text{and} \quad \mu \in R^m \quad .$$

It is easily shown that

$$\bar{\mu} = \left(F_I' F_I\right)^{-1} F_I' y \qquad (F' \text{ is transpose of } F) \quad .$$

Then, the projection of y is:

$$P_I y = F_I\left(F_I' F_I\right)^{-1} F_I' y$$

and we can define the projection matrix

$$P_I = F_I\left(F_I' F\right)^{-1} F_I' \quad .$$

In the same form we define the projection on the subspace orthogonal to ∇f_i , $i \in I$ and the corresponding matrix is:

$$P_I^\perp = I - P_I \quad .$$

In the definition of P_I , we have supposed that the vectors ∇f_i, $i \in I$ are linearly independent, and then the matrix $F_I' F_I$ is invertible.

It is known (Kuhn—Tucker's theorem) that if $\bar{\theta}$ is a solution of the problem (1), then

$$\nabla J(\bar{\theta}) = F_{I_0(\bar{\theta})}\bar{\mu} \qquad \bar{\mu}_1 \leqq 0, \ldots, \bar{\mu}_m \leqq 0$$

and

$$\bar{\mu} = \left(F'_{I_0(\bar{\theta})} F_{I_0(\bar{\theta})}\right)^{-1} F'_{I_0(\bar{\theta})} \nabla J(\bar{\theta})$$

We define a point as desirable if:

a) $\theta \in \Omega$ b) $\nabla J(\theta) = F_{I_0(\theta)}\,\mu(\theta) \qquad \mu(\theta) \leqq 0$.

It is posible to apply the following algorithm: "Gradient Projected".

ALGORITHM.

Step 0 : Select $\theta_o \in \Omega$;
$\varepsilon' > 0 \ / \ \forall\, \varepsilon > 0 \ , \ \varepsilon \leqq \varepsilon'$ $\left\{\nabla f_i(\theta) \ / \ i \in I_\varepsilon(\theta), \ \theta \in \Omega\right\}$ is a set of linearly independent vectors.
Choose $\beta \in (0,1)$, $\overline{\varepsilon} \in (0,\varepsilon')$, $\varepsilon'' \in (0,\overline{\varepsilon})$.
Set $i = 0$.

Step 1 : Set $\theta = \theta_i$.

Step 2 : Set $\varepsilon_o = \overline{\varepsilon}$ and $j = 0$.

Step 3 : Compute $h_{\varepsilon_j} = P^{\perp}_{I_{\varepsilon_j}(\theta)} \nabla J(\theta)$.

Step 4 : If $\|h_{\varepsilon_j}\| > \varepsilon_j$ $h(\theta) = - h_{\varepsilon_j}$ and go to step 12; else, go to 5.

Step 5 : If $\varepsilon_j \leqq \varepsilon''$, compute $h_o(\theta) = P^{\perp}_{I_o(\theta)} \nabla J(\theta)$ and
$\mu_o(\theta) = \left(F'_{I_o(\theta)} F_{I_o(\theta)}\right)^{-1} F'_{I_o(\theta)} \nabla J(\theta)$ and go to 6; else, go to 7 .

Step 6 : If $\mu_o(\theta) \leqq 0$ and $\|h_o(\theta)\| = 0$ set $\theta_{i+1} = \theta$ and stop, else, go to 7.

Step 7 : Compute $\mu_{\varepsilon_j}(\theta) = \left(F'_{I_{\varepsilon_j}}(\theta) \ F_{I_{\varepsilon_j}}(\theta)\right)^{-1} F'_{I_{\varepsilon_j}}(\theta) \ \nabla J(\theta)$.

Step 8 : If $\mu_{\varepsilon_j}(\theta) \leqq 0$, set $\varepsilon_{j+1} = \beta\, \varepsilon_j$, set $j = j+1$ and go to step 3 ;
else, go to step 9 .

Step 9 : Assuming that $I_{\varepsilon_j}(\theta) = \left\{k_1, \ldots, k_{m'}\right\}$ and that $k_1 < k_2 < \ldots < k_{m'}$,
set $y^{k_\alpha}_{\varepsilon_j}(\theta) = \mu^{\alpha}_{\varepsilon_j}(\theta)$ for $\alpha = 1,2,\ldots, m'$ (where $\mu^{\alpha}_{\varepsilon_j}(\theta)$ is the αth component of the vector $\mu_{\varepsilon_j}(\theta)$) .

Step 10 : Find the smallest $k \in I_{\varepsilon_j}(\theta)$ such that the vector $\overline{h}_{\varepsilon_j}(\theta) =$
$= P^{\perp}_{I_{\varepsilon_j}(\theta) - k} \nabla J(\theta)$ satisfies the relation

$$\|\overline{h}_{\varepsilon_j}(\theta)\| = \max\left\{\|P^{\perp}_{I_{\varepsilon_j}(\theta) - 1} \nabla J(\theta)\| \ / \ 1 \in I_{\varepsilon_j}(\theta) \ , \ y^1_{\varepsilon_j}(\theta) > 0\right\}$$

and set $h(\theta) = - \overline{h}_{\varepsilon_j}(\theta)$.

Step 11 : If $\|h(\theta)\| \leqq \varepsilon_j$ set $\varepsilon_{j+1} = \beta\, \varepsilon_j$, set $j = j+1$, and go to step 3;
else, go to step 12.

Step 12 : Compute $\lambda(\theta) > 0$ to be the smallest scalar satisfying $J(\theta+\lambda(\theta)h(\theta)) =$
$= \min\{J(\theta + \lambda h(\theta)) \ / \ \lambda \geqq 0 \ , \ (\theta + \lambda\, h(\theta)) \in \Omega\}$.

Step 13 : Set $\theta_{i+1} = \theta_i + \lambda(\theta)h(\theta)$, set $i = i+1$, and go to step 1 .

We have shown that J is continuously differentiable, then is valid the following theorem:

THEOREM. *The sequence* θ_i *given by the algorithm is finite and its last element is desirable or is infinite and each accumulation point of the sequence is desirable.*
(The proof of this Th. is, essentially, the same that we find in [1] , pag. 195) .

§6. NUMERICAL SOLUTION OF AN EXAMPLE : THE SHUT DOWN OF A NUCLEAR REACTOR.

The problem is the reduction of the power of a nuclear reactor in a fixed time. The functional to minimize is the xenon poisoning.

The model is ruled by the differential equations:

$$(1)\qquad \begin{cases} \dot{I} = -aI + b\phi \\ \dot{x} = aI + c\phi - (d + e\phi)x \\ \dot{\phi} = U\phi . \end{cases}$$

I is the iodine concentration and x the xenon concentration.

ϕ is the flux of neutrons.

U is the control and it can only assume two values.

The control is applied in the interval $[0,T]$ in such a form that $\phi(T) = \phi_f$ (a fixed value). After that $(t > T)$, the flux is held constant.

If we define $x_M = \max_{t>T} x(t)$, is possible to state the problem in the following form:

Find $U(t)$, $0 \leqq t \leqq T$, where $U(t)$ is a step function with n' switching ($n' \leqq n$, n fixed), that takes only the values $V1$, $V2$ and such that the corresponding response of the system (1) satisfies $\phi(T) = \phi_f$ and gives the minimun value of $x_M(u(0,T))$.

This problem differs from the models studied in the nonlinearities of the equations (1) and in the fixed final condition.

In this case, it can be shown that $J(\theta) = x_M$ (where $\theta = (\theta_1,\dots,\theta_n)$ is the set of switching points) is a continuous function both with the derivatives of J, and then the theorem remains valid.

The final condition could be introduced in the functional through a penalization function. Another method is the following, we use the propertie that $\phi(T) = \phi_f$ implies that $\forall\, u(0,T) \,/\, \phi(T) = \phi_f$

$$f^\circ(\theta) = \sum_{i=1}^{[(N+1)/2]} \left(\theta_{2i} - \theta_{2i-1}\right) = \text{constant}$$

and consider this relation as an additional restriction. In the projected gradient algorithm,, the matrix $F_{I_\varepsilon(\theta)}$ is enlarged in the following form:

$$F_{I_\varepsilon(\theta)} \longrightarrow \tilde{F}_{I_\varepsilon(\theta)} = \left[\nabla f^\circ ,\ \nabla f_{i_1},\dots,\ \nabla f_{i_m}\right] \quad \{i_1,\dots,i_m\} = I_\varepsilon(\theta)$$

and the points constructed by the algorithm satisfie $\phi(T) = \phi_f$,provided the initial point (θ_0) satisfies that condition.

FORMULAS OF J AND ∇J :

We have defined $J = x_M$; to compute it, we solve the equations:

$$\begin{cases} \dot{I} = -I + b\phi_f \\ \dot{x} = I + c\phi_f - (d + e\phi_f)x \end{cases} \qquad \text{for } t > T$$

with initial conditions $x(T)$, $I(T)$ and find the value $x_M = \max_{t \geq T} x(t)$. (We set a=1 making a change of variables).

If $\phi_f = 0$, it is:

$$\begin{cases} x_M = \exp(-t^*)\,.\,\dfrac{I(T)}{d-1} + \exp(-dt^*)\left[x(T) - \dfrac{I(T)}{d-1}\right] & \text{if} \quad \dfrac{x(T)}{I(T)}\,d < 1 \quad \text{and} \\ x_M = x(T) \quad \text{if} \quad \dfrac{x(T)}{I(T)}\,d \geqq 1 & \text{where} \\ t^* = \dfrac{1}{d-1}\ \ln\left[d + \dfrac{x(T)}{I(T)}\,d(1-d)\right] & . \end{cases}$$

To compute $\frac{\partial J}{\partial \theta_i} = p_3(\theta_i)x_3(\theta_i)(v_2 - v_1)(-1)^i$, we integrate backwardly the adjoint equations:

$$\begin{cases} -\dfrac{dp_1}{dt} = -\ p_1 + p_2 \\ -\dfrac{dp_2}{dt} = -\ p_2(d + e\phi) \\ -\dfrac{dp_3}{dt} = \ p_1 + cp_2 - p_2.e.x + p_3U \end{cases}$$

with final conditions:

$$\begin{cases} p_1(T) = -\dfrac{\partial x_M}{\partial x(T)} \\ p_2(T) = -\dfrac{\partial x_M}{\partial I(T)} \\ p_3(T) = -\dfrac{\partial x_M}{\partial \phi(T)} \end{cases} .$$

The projected gradient algorithm (with the shown modification) was used to solve numerically the problem. The values of J and ∇J were computed integrating the differential equations of x, I, ϕ, p with a 4th order Runge — Kutta method.

NUMERICAL VALUES.

$a = 0.1$	$\phi_f = 0.674 \times 10^{-2}$
$b = 1.0$	$x_o = 2.0$
$c = 1.0$	$I_o = 10.0$
$d = 0.05$	$\phi_o = 1.0$
$e = 0.95$	$T = 10.0$
$v_1 = -2.0$	
$v_2 = 0.0$	

NUMERICAL RESULTS.

The optimal values obtained for $n = 2$ are:

$$\bar{\theta}_1 = 0.6561$$
$$\bar{\theta}_2 = 8.156$$
$$\bar{x}_M = 4.6351 \quad , \quad (\text{for } t_{\bar{x}_M} = 19.86)$$

The following is a sample of the sequence produced by the algorithm, that shows rate of convergence.

θ_1	θ_2	x_M
1.000	8.500	4.82751
0.625	8.125	4.63729
0.659	8.154	4.63514
0.656	8.156	4.63513

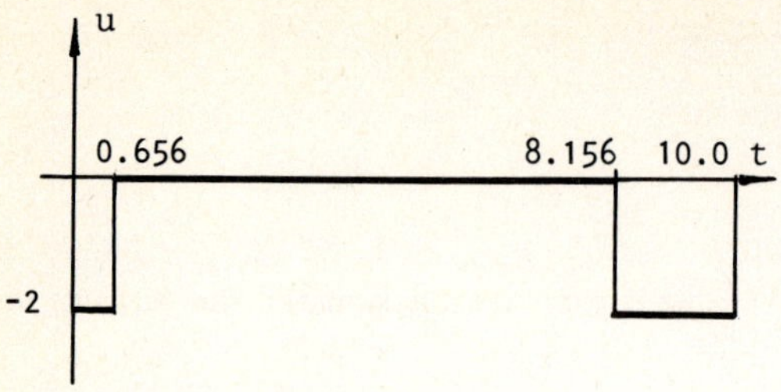

Control u

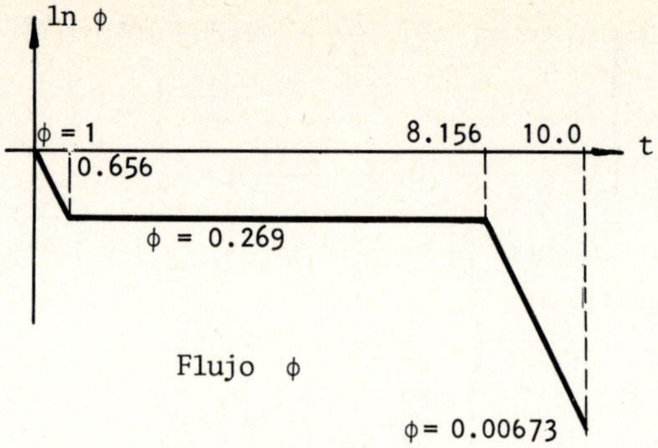

Flujo φ

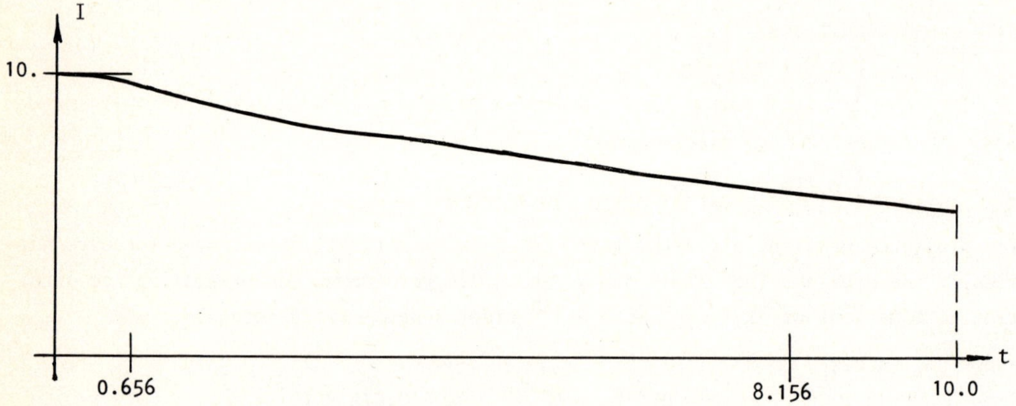

Iodo I

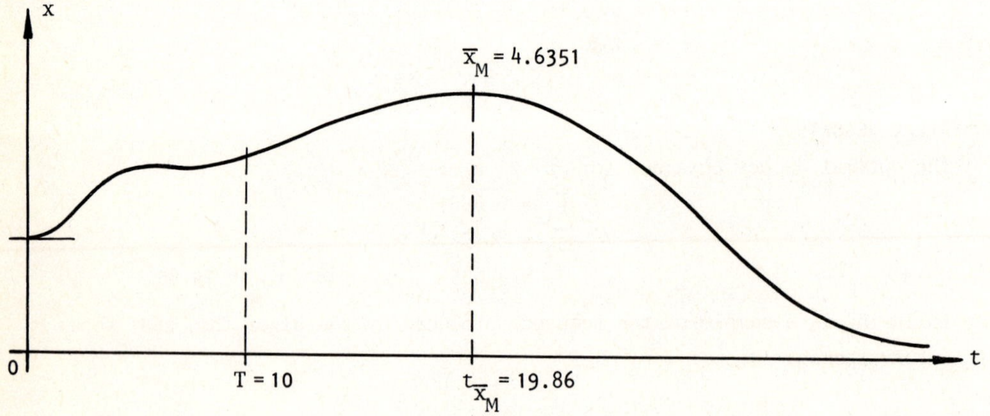

Xenon: x

For $n \geq 2$ the sequences obtained were convergent to the optimal point for $n=2$, as is shown in the case $n=6$. Then, in the set of control with 6 or less switchings, the optimal one is a policy with two switchings.

θ_1	θ_2	θ_3	θ_4	θ_5	θ_6	X_M
0.250	1.000	1.250	2.000	2.250	8.250	4.7485
0.401	0.999	1.213	2.115	2.115	8.115	4.6634
0.488	0.993	1.195	2.115	2.115	8.191	4.6536
0.512	1.028	1.142	2.115	2.115	8.126	4.6470
0.568	1.026	1.132	2.115	2.115	8.172	4.6427
0.585	1.050	1.102	2.115	2.115	8.137	4.6396
0.645	1.059	1.085	2.115	2.115	8.171	4.6370
0.649	1.072	1.072	2.115	2.115	8.149	4.6352
0.656	1.072	1.072	2.115	2.115	8.156	4.6351

The switchings 2-3 and 4-5 are simultaneous and could be eliminated and the new policy is the optimal one for the problem with $n=2$.

§7. FORM AND PROPERTIES OF THE SECOND DERIVATIVES OF J.

$$\frac{\partial J}{\partial \theta_i} = l(x(\theta_i),u(\theta_i-),\theta_i) - l(x(\theta_i),u(\theta_i+),\theta_i) + p(\theta_i)G(\theta_i)(v_1 - v_2))(-1)^i$$

a) $\dfrac{\partial^2 J}{\partial \theta_i^2}$

First, we find the formulas of $\dfrac{\partial x(\theta_i)}{\partial \theta_i}$, $\dfrac{\partial p(\theta_i)}{\partial \theta_i}$

$$x(\theta_i) = \Phi(\theta_i,0)x_o + \int_o^{\theta_i} \Phi(\theta_i,s)G(s)u(s)ds$$

then,

$$\frac{\partial x(\theta_i)}{\partial \theta_i} = F(\theta_i)x(\theta_i) + G(\theta_i)u(\theta_i-) \quad .$$

Also,

$$p(\theta_i) = p(T)\Phi(T,\theta_i) - \int_{\theta_i}^{T} \frac{\partial l}{\partial x}(x(s),u(s),s)\Phi(s,\theta_i)ds$$

with

$$p(T) = -\frac{\partial g}{\partial x}(x(T))$$

$$\frac{\partial p(\theta_i)}{\partial \theta_i} = -p(T)\Phi(T,\theta_i)F(\theta_i) + \int_{\theta_i}^{T} \frac{\partial l}{\partial x}(x(s),u(s),s)\Phi(s,\theta_i)F(\theta_i)ds +$$
$$+ \frac{\partial l}{\partial x}(x(\theta_i),u(\theta_i+),\theta_i) - \left(\frac{\partial}{\partial \theta_i}x(T)\right)' \frac{\partial^2 g}{\partial x^2}(x(T))\Phi(T,\theta_i) -$$
$$- \int_{\theta_i}^{T} \left(\frac{\partial}{\partial \theta_i}x(s)\right)' \frac{\partial^2 l}{\partial x^2}(x(s),u(s),s)\Phi(s,\theta_i)ds$$

but, we know that, for $t > \theta_i$

$$\frac{\partial}{\partial \theta_i}x(t) = \Phi(t,\theta_i)G(\theta_i)(v_2 - v_1)(-1)^i$$

then,

$$\frac{\partial p(\theta_i)}{\partial \theta_i} = -p(\theta_i)F(\theta_i) + \frac{\partial l}{\partial x}(x(\theta_i),u(\theta_i+),\theta_i) -$$
$$- (-1)^i(v_2 - v_1)G'(\theta_i)\Phi'(T,\theta_i)\frac{\partial^2 g}{\partial x^2}(x(T))\Phi(T,\theta_i) -$$

$$- \int_{\theta_i}^{T} (-1)^i (v_2 - v_1) G'(\theta_i) \Phi'(s,\theta_i) \frac{\partial^2 l}{\partial x^2}(x(s), u(s), s).\Phi(s,\theta_i) ds \quad .$$

Now, it is possible to compute $\frac{\partial^2 J}{\partial \theta_i^2}$

$$\frac{\partial^2 J}{\partial \theta_i^2} = \frac{\partial}{\partial \theta_i} l(x(\theta_i), u(\theta_i -), \theta_i) - \frac{\partial l}{\partial \theta_i}(x(\theta_i), u(\theta_i +), \theta_i) +$$

$$+ \left[\frac{\partial}{\partial x} l(x(\theta_i), u(\theta_i -), \theta_i) - \frac{\partial l}{\partial x}(x(\theta_i), u(\theta_i +), \theta_i)\right]\left[F(\theta_i) x(\theta_i) + G(\theta_i) u(\theta_i -)\right] +$$

$$+ p(\theta_i)\left(\frac{dG(\theta_i)}{d\theta_i}\right)(v_1 - v_2)(-1)^i + \left[- p(\theta_i) F(\theta_i) + \frac{\partial l}{\partial x}(x(\theta_i), u(\theta_i +), \theta_i)\right] \quad .$$

$$.G(\theta_i)(v_1 - v_2)(-1)^i + (-1)^i (v_1 - v_2) G'(\theta_i) \Phi'(T,\theta_i) \frac{\partial^2 g}{\partial x^2}(x(T)) \Phi(T,\theta_i) G(\theta_i)(v_1 - v_2)(-1)^i +$$

$$+ (v_1 - v_2)(-1)^i G'(\theta_i) \int_{\theta_i}^{T} \Phi'(s,\theta_i) \frac{\partial^2 l}{\partial x^2}(x(s), u(s), s) \Phi(s,\theta_i) ds . G(\theta_i)(v_1 - v_2)(-1)^i \quad .$$

b) If $j > i$

$$\frac{\partial}{\partial \theta_j} x(\theta_i) = 0$$

then:

$$\frac{\partial}{\partial \theta_j}\left(\frac{\partial J}{\partial \theta_i}\right) = \frac{\partial}{\partial \theta_j} p(\theta_i) \; . \; G(\theta_i)(v_1 - v_2)(-1)^i \quad ,$$

and we must only know $\frac{\partial}{\partial \theta_j} p(\theta_i)$ to find the form of $\frac{\partial^2 J}{\partial \theta_j \, \partial \theta_i}$.

From the integral formula of p we obtain:

$$\frac{\partial}{\partial \theta_j} p(\theta_i) = \left(\frac{\partial}{\partial \theta_j} x(T)\right)' . \frac{\partial^2 g}{\partial x^2}(x(T)) . \Phi(T,\theta_i) +$$

$$+ \left[\frac{\partial l}{\partial x}(x(\theta_j), u(\theta_j +), \theta_j) - \frac{\partial l}{\partial x}(x(\theta_j), u(\theta_j -), \theta_j)\right] \Phi(\theta_j, \theta_i) -$$

$$- \int_{\theta_j}^{T} (-1)^j (v_2 - v_1) G'(\theta_j) \Phi'(s,\theta_j) \frac{\partial^2 l}{\partial x^2}(x(s), u(s), s) \Phi(s,\theta_i) d\tau$$

and then,

$$\frac{\partial}{\partial \theta_j}\left(\frac{\partial J}{\partial \theta_i}\right) = (-1)^j (v_2 - v_1) G'(\theta_j) \Phi(T,\theta_j) \frac{\partial^2 g}{\partial x^2}(x(T)) . \Phi(T,\theta_i) G(\theta_i)(v_1 - v_2)(-1)^i +$$

$$+ \left[\frac{\partial l}{\partial x}(x(\theta_j), u(\theta_j +), \theta_j) - \frac{\partial l}{\partial x}(x(\theta_j), u(\theta_j -), \theta_j)\right] \Phi(\theta_j, \theta_i) G(\theta_i)(v_1 - v_2)(-1)^i -$$

$$- (-1)^j (v_2 - v_1) G'(\theta_j) \int_{\theta_j}^{T} \Phi'(s,\theta_j) \frac{\partial^2 l}{\partial x^2}(x(s), s) \Phi(s,\theta_i) ds \; . \; G(\theta_i)(v_1 - v_2)(-1)^i$$

c) If $i > j$

$$\frac{\partial}{\partial \theta_j} x(\phi_i) = \Phi(\theta_i, \theta_j) G(\theta_j)(v_2 - v_1)(-1)^j$$

and

$$\frac{\partial}{\partial \theta_j} p(\theta_i) = - \left(\frac{\partial}{\partial \theta_j} x(T)\right)' \frac{\partial^2 g}{\partial x^2}(x(T)) \Phi(T,\theta_i) - \int_{\theta_i}^{T} \left(\frac{\partial}{\partial \theta_j} x(s)\right)' \frac{\partial^2 l}{\partial x^2}(x(s), u(s), s) \Phi(s,\theta_i) ds$$

then,

$$\frac{\partial}{\partial \theta_j}\left(\frac{\partial J}{\partial \theta_i}\right) = \left[\frac{\partial l}{\partial x}(x(\theta_i), u(\theta_i -), \theta_i) - \frac{\partial l}{\partial x}(x(\theta_i), u(\theta_i +), \theta_i)\right] \Phi(\theta_i, \theta_j) \quad .$$

$$.G(\theta_j)(v_2 - v_1)(-1)^j + (-1)^j (v_2 - v_1) G'(\theta_j) \Phi'(T,\theta_j) \frac{\partial^2 g}{\partial x^2}(x(T)) \Phi(T,\theta_i) G(\theta_i)(v_1 - v_2)(-1)^i -$$

$$- (-1)^j (v_2 - v_1) G'(\theta_j) \int_{\theta_i}^{T} \Phi'(s,\theta_j) \frac{\partial^2 l}{\partial x^2}(x(s),u(s),s)\Phi(s,\theta_i)ds.G(\theta_i)(v_1 - v_2)(-1)^i \quad .$$

It can be proved, in the same form we have done for the first derivatives, that the second derivatives are continuous provided $g \in C^2$, $l \in C^2$, $G \in C^1$ and $F \in C$. This continuity is important to obtain superlinear convergence when it is applied the conjugate gradient method.

§8. SOLUTION OF THE PROBLEM USING A MIXED METHOD OF PENALIZATION AND CONJUGATE GRADIENTS.

The problem of minimum with restrictions:

(1) $$\min_{\Omega} J(\theta) \quad ; \quad \Omega = \{\theta \,/\, 0 \leqq \theta_1 \leqq \dots \leqq \theta_n \leqq T \quad , \quad \theta \in R^n\}$$

is transformed into another that could be solved using the methods of optimization without restrictions. This is done applying penality functions.

The new problem is:

$$\min_{\mathring{\Omega}} J_\beta(\theta) \quad ; \quad \mathring{\Omega} = \left\{\theta \in R^n \,/\, 0 < \theta_1 < \theta_2 < \dots < \theta_n < T\right\}$$

and

$$J_\beta(\theta) = J(\theta) + \beta \sum_{i=1}^{n+1} - \varphi_i^{-1}(\theta)$$

$$\varphi_1 = -\theta_1 \quad ; \quad \varphi_i = -\theta_i + \theta_{i-1} \quad i = 2,\dots,n \quad ; \quad \varphi_{n+1} = \theta_n - T \quad .$$

ALGORITHM.

Step 0: Choose $\theta_0 \in \mathring{\Omega}$; $\beta_0 > 0$; $\varepsilon_0 > 0$ and set $i = 0$.

Step 1: Apply the conjugate gradient method to the minimization of J_β until it is obtained a point θ_{i+1} such that $\|\nabla J_\beta(\theta_{i+1})\| < \varepsilon_i$

Step 2: Let $\beta_{i+1} = \frac{\beta_i}{2}$, $\varepsilon_{i+1} = \frac{\varepsilon_i}{2}$, $i = i+1$ and go to 1 .

REMARK.

The conjugate gradient method could be applied in $\mathring{\Omega}$, modifying the one dimensional search (along the conjugate directions) in such a form that the point are always chosen in $\mathring{\Omega}$.

It is known (Kuhn–Tucker's theorem) that a necessary condition for optimality of $\overline{\theta}$ in problem 1 is:

$$\nabla J(\overline{\theta}) + \sum_{i \in I(\overline{\theta})} \mu_i \nabla\varphi_i(\overline{\theta}) = 0 \qquad \mu_i \geqq 0$$

$$I(\theta) = \left\{i \,/\, \varphi_i(\theta) = 0\right\} \quad .$$

DEFINITION.

A point θ is desirable if:

i) $\theta \in \Omega$

ii) $\nabla J(\theta) + \sum_{i \in I(\theta)} \mu_i \nabla\varphi_i(\theta) = 0 \quad ; \quad \mu_i \geqq 0 \quad i \in I(\theta)$

The algorithm has the property:

LEMA: If $J(\theta)$ is continuously differentiable, then the algorithm produces a sequence of different points and all the accumulation points are desirable, or the sequence has a finite number of different points and the last (infinitely repeated) is desirable.

PROOF.

i) $\mathring{\Omega}$ is relatively compact, then $\{\theta_i\}$ has accumulation points.

Let be $\theta_{i_k} \longrightarrow \tilde{\theta}$ $k = 1,2,\ldots$

$$\overline{\mathring{\Omega}} = \Omega \quad , \quad \text{then } \tilde{\theta} \in \Omega \quad .$$

It will be proved that $\tilde{\theta}$ satisfies the Kuhn–Tucker conditions.

From the step 1 of the algorithm it follows:

$$\text{(1)} \qquad \nabla J(\theta_{i_k}) + \sum_{j=1}^{n+1} \left[\beta_{i_k} / \varphi_j^2(\theta_{i_k})\right] \nabla\varphi_j(\theta_{i_k}) \longrightarrow 0 \qquad .$$

It is easily seen that, $\forall\, \theta \in \Omega, I(\theta)$ has at most n elements and $\{\nabla\varphi_i(\theta),\ i \in I(\theta)\}$ is a set of linearly independent vectors.

Then,

$$\nabla J(\theta_{i_k}) + \sum_{j \in I(\tilde{\theta})} \left[\beta_{i_k} / \varphi_j^2(\theta_{i_k})\right] \nabla\varphi_j(\theta_{i_k}) \longrightarrow 0 \qquad .$$

We define: ψ_{i_k} (matrix $n \times m$)

$$\psi_{i_k} = \left[\nabla\varphi_{j_1}, \nabla\varphi_{j_2}, \ldots, \nabla\varphi_{j_m}\right] \qquad \{j_1, \ldots, j_m\} = I(\tilde{\theta})$$

and then:

$$\begin{pmatrix} \beta_{i_k} / \varphi_{j_1}^2(\theta_{i_k}) \\ \vdots \\ \beta_{i_k} / \varphi_{j_m}^2(\theta_{i_k}) \end{pmatrix} - (\psi'_{i_k}\,\psi_{i_k})^{-1}\,\psi'_{i_k}\,\nabla J(\theta_{i_k}) \longrightarrow 0$$

because, $\psi_{i_k} \longrightarrow \psi = \left[\nabla\varphi_{j_1}(\tilde{\theta}), \ldots, \nabla\varphi_{j_m}(\tilde{\theta})\right]$, this matrix is of maximun rank and then $\psi'_{i_k}\,\psi_{i_k} \longrightarrow \psi'\psi$ invertible matrix.

Then: $\beta_{i_k} / \varphi_{j_s}^2(\theta_{i_k}) \longrightarrow \mu_{j_s} \geq 0 \qquad j_s \in I(\theta)$

and also, taking limits in (1):

$$\nabla J(\tilde{\theta}) + \sum_{j \in I(\tilde{\theta})} \mu_j \nabla\varphi_j(\tilde{\theta}) = 0 \qquad .$$

Thus, $\tilde{\theta}$ is desirable.

ii) If $\exists N \,/\, \forall\, i \geq N, \theta_i = \theta_N$, from Step 1 it follows:

$$\left\| \nabla J(\theta_N) + \beta_i \nabla\left(\sum_{j=1}^{n+1} - \varphi_j^{-1}(\theta_i)\right)\right\| \longrightarrow 0 \qquad i \to \infty$$

but $\beta_i \longrightarrow 0$

then: $\nabla J(\theta_N) = 0$, $\theta_N \in \mathring{\Omega}$ and θ_N is desirable.

NUMERICAL RESULTS.

We have applied this method to the problem of the shutdown of the nuclear reactor. The results are the same obtained with the projected gradient method: the optimal policy in the set of step functions is a policy with two switchings.

The following table shows the convergence for $n = 2$

θ_1	θ_2	J_β	β
1.000	8.500	6.62755	1.0000
0.913	8.413	5.49471	0.4000
0.792	8.292	4.98835	0.1600
0.720	8.220	4.77629	0.0640
0.684	8.184	4.69171	0.0256
0.667	8.167	4.65213	0.0076
0.659	8.159	4.64021	0.0023
0.658	8.158	4.63601	0.0004
0.6570	8.157	4.63530	0.00008
0.6564	8.1564	4.63516	0.000016

§9. PROOF THAT THE SOLUTIONS OF THE PROBLEM WITH FIXED NUMBER OF SWITCHING ARE MINIMIZATING SEQUENCE FOR THE PROBLEM WITH MEASURABLE CONTROLS.

The set of problems with step function controls (as were stated in §1) could be considered as a set of approximations to the problem:

$$\min_{\mathcal{U}_{ad}} J(u(.))$$

with

$$\mathcal{U}_{ad} = \left\{ u(.) \text{ measurable in } [0,T] \;/\; u(t) = v_1 \text{ or } u(t) = v_2 \quad \text{a.e.} \right\}$$

$$J(u(.)) = \int_0^T l(x(s), u(s), s)ds + g(x(T))$$

and $x(s)$ satisfies:

$$\begin{cases} \dfrac{dx}{dt}(t) = F(t)x(t) + G(t)u(t) \\ x(0) = x_0 \end{cases} .$$

Under the assumptions that l,g are continuous and F,G are integrable, it can be proved that J is a continuous functional for $u \in \mathcal{U}_{ad}$ (with the $L_1(0,T)$ topology) If we denote with $\bar{u}_n$ the optimal solution with at most n switchings, we shall prove that $\bar{u}_n$ is a minimizating sequence for the new problem.

Let w_n be a minimizating sequence:

$$\lim_{n\to\infty} J(w_n) = \inf_{\mathcal{U}_{ad}} J(u) \qquad w_n \in \mathcal{U}_{ad} .$$

But, for the continuity of J, it is possible to find $\tilde{w}_n$ ($\tilde{w}_n$ a step function / $\tilde{w}_n \in \mathcal{U}_{ad}$) such that $|J(\tilde{w}_n) - J(w_n)| < \frac{1}{2^n}$

Let $k(n)$ be the number of switchings of $\tilde{w}_n$. By definition, $\forall\, k(n)$

$$J\left(\bar{u}_{k(n)}\right) \leq J(\tilde{w}_n) .$$

Then, using the property that $J(\bar{u}_n)$ is non-incresing.

$$\lim_{m\to\infty} J(\bar{u}_m) \leqq \lim_{n\to\infty} J(\tilde{w}_n) = \lim_{n\to\infty} J(w_n) = \inf_{\mathcal{U}_{ad}} J$$

but:
$$J(\bar{u}_m) \geqq \inf_{\mathcal{U}_{ad}} J$$

then
$$\lim_{m\to\infty} J(\bar{u}_m) = \inf_{u\in\mathcal{U}_{ad}} J(u) \quad .$$

§10. OTHERS RESULTS.

The bang–bang problem with restricted number of switchings could be analysed in global form (i.e., the initial state $x_o \in R^n$ or $x_o \in \Omega$) and is reduced to a sequence of stopping–times problems. Also in the problem with measurable control ($u=0$ or $u=1$ a.e.) it is possible to prove existence theorems and to analyse the optimal cost function with the hamiltonian technique. These are the objects of forthcoming papers.

REFERENCES

1. E. POLAK. "*Computational methods in optimization*". Academic Press 1971.

2. E.B. LEE, L. MARKUS. "*Foundations of Optimal Control Theory*". Wiley. 1968.

3. L.S. PONTRYAGIN, V.G. BOLTYANSKII, R.V. GAMKERELIDZE, E.F. MISCHENKO. "*Mathematical Theory of Optimal Processus*". New York, Wiley 1962.-

Los originales de este trabajo fueron preparados en el *Instituto de Matemática "Beppo Levi"* por la Sra. H. I. Warecki de MUTY.

OPTIMAL CONTROL PROBLEMS IN SOBOLEV SPACES WITH WEIGHTS . NUMERICAL APPROACHES .APPLICATIONS TO PLASMA OPTIMAL CONTROL AND TIME.DELAY PROBLEMS

Claudia Simionescu
University of Braşov
Department of Mathematics
2200 Braşov- R.S.Romania

Abstract

We consider systems for which the state , $y(t,u)$, is given by the solution of the equation

$$(P(t,D)y,v)_H + a(t;y,v) = (g+Bu,v)_H \qquad v \in V$$

in the sense of scalar distributions , where

$$P(t,D)= \sum_{r=-1}^{s} B_r(t)D^r \qquad s=1,2$$

A generalized class of optimal control problems with a quadratic performance criterion is considered and existence and unicity results are obtained.

The optimality system is studied, approximations for the solutions are given and applications to time-delay problems and plasma optimal control problems are shown .

Similar control problems for particular cases have been considered by the author in [11], [12], [13]. The definitions and nonations are the same as in , .

Since the technique of proving the existence and uniqueness of the solution in all cases is the same as the one we have been used in [11], [12], and [13], we shell resume ourselves to only give the results and comment them.

1. STATEMENT OF THE PROBLEM.

Let V and H be two Hilbert spaces, $V \subset H$, V dense in H and let the injection $V \longrightarrow H$ be continuous. We identify H to its dual so that if V' denotes the dual of V, we have

$$V \subset H \equiv H' \subset V'$$

If in general we denote by $\mathcal{E}_t(\mathcal{L}(E,F))$ the space of all linea bounded operators from $E \longrightarrow F$ which have continuous derivatives of every order with respect to t ; then, for $B_r(t) \in \mathcal{E}_t(\mathcal{L}(H,H))$ we shell be considering integro-differential operators of the form:

$$P(t,D) = \sum_{r=-n}^{\ell} B_r(t)D^r \qquad n \in \mathcal{N} \text{ (parabolic) } (1$$

and

$$P(t,D) = \sum_{r=-n}^{\ell} B_r(t)D^r \qquad n \in \mathcal{N} \text{ (hyperbolic) } \quad (2)$$

Let a(t;u,v) be a sesqui-linear Hermitian continuous form on VxV such that for every $u,v \in V$:

$$|a(t;u,v)| \leq k \|u\| . \|v\| \quad , \quad k > 0 \qquad (3)$$

$$a(t;u,u) \geqslant \alpha(t) \|u\|_V^2 \qquad \alpha(t) > 0 , \alpha(t) \in \mathcal{E}_t , t \in R \quad (4)$$

We suppose that the mapping $t \longrightarrow a(t;u,v)$ is measurable.

Let us consider now $\mathcal{U}$ - a hilbert space (the space of controls) , $\mathcal{U}_{ad}$ - a closed convex subset of $\mathcal{U}$, (admissible controls) $\mathcal{H}$ - another hilbert space and (S) a system governed by one of the integro- differential operators (1) or (2) , in the way that the state of the system , for each $u \in \mathcal{U}$ and $t \in R$, is given by the solution $y=y(t,u)$ of the equation

$$(P(t,D)y,v)_H + a(t;y,v) = (g+Bu,v)_H \quad , \quad v \in V \tag{5}$$

in the sense of scalar distributions.

<u>OPTIMAL CONTROL PROBLEM.</u> If $\mathcal{U}$ is the space of controls, $y(t,u)$ is the solution for (5) , C - the observation operator and $N \in \mathcal{L}(\mathcal{U},\mathcal{U})$, find $u \in \mathcal{U}_{ad}$, such that

$$J(u) = \inf_{v \in \mathcal{U}_{ad}} J(v)$$

where J(u) is a quadratic performance criterion:

$$J(u) = \| Cy - z_d \|^2_{\mathcal{H}} + (Nu,u)_{\mathcal{U}} \tag{6}$$

2.RESULTS IN SOBOLEV SPACES WITH WEIGHTS.

We are going to establish a first result regarding this optimal control problem by considering Sobolev spaces with weights. This kind of spaces have been extensively used in obtaining existence and uniqueness results for boundary and mixt boundary value problems of parabolic or hyperbolic type in [3] , [5], [15], [16] .

We consider E a Hilbert space and q(t) a real function with continuous derivatives , that satisfies the condition:

(A) there exists a constant $p_o > 0$, such that $q'(t) \geq p_o$, $t \in R$

For $k \in Z$ (an integer) , $\mathcal{D}^k(q;E)$ denotes the Hilbert space obtained by completing $\mathcal{D}(E)$ - (the space of functions defined on the real line and taking values in E , with derivatives of all orders and compact support) , with respect to the structure defined by the Hermitian product

$$(\varphi,\psi)_{E;q,k} = (e^{-q(t)}D^k\varphi , e^{-q(t)}D^k\psi)_{L^2(E)} = \int_{-\infty}^{\infty}(e^{-q(t)}D^k\varphi, e^{-q(t)}$$

The norm in $\mathcal{D}^k(q;E)$ is then:

$$\|\varphi\|_{E;q,k} = \| e^{-q(t)}D^k\varphi\|_{L^2(E)} = \left[\int_{-\infty}^{\infty}\|e^{-q(t)}D^k\varphi\|_E^2 \, dt\right]^{\frac{1}{2}}$$

For each integer $k \in Z$ and each real function q(t) verifyi (A) , $\mathcal{D}^k(q;E)$ is a space of distributions with values in E ,

$$\mathcal{D}^k(q;E) \subset \mathcal{D}'(E)$$

If $k \geq 0$, $\mathcal{D}^k(q;E)$ is the space of all (classes of) measurable functions from R into E such that $e^{-q(t)}D^h\varphi(t) \in L^2(E)$ for al $0 \leq h \leq k$.

If $k > 0$, $\mathcal{D}^{-k}(q;E)$ is a space of distributions having the property : for $T \in \mathcal{D}^{-k}(q;E)$, there exists k+1 functions $g_h \in L^2(E)$ such that

$$T = e^q g_o + D(e^q g_1) + \ldots + D^k(e^q g_k)$$

<u>THEOREM 2.1.</u> There exists a positive function $G(t)$, $G(t) \in C^1$, $(\forall)\, t \in R$, and for each integer k , the positive functions g_k , $g_k(t) \in C^1$ $t \in R$, such that if:

a) $p(t) \in C^1$, with $p'(t) \geqslant g_k(t)$ for all $t \in R$

and $p'(t) + G'(t) \geqslant p_o$ for all $t \in R$,

b) there exists a positive function $b(t)$, $b(t) \in \mathcal{E}_t$ such that

$$(B_1(t)u,u)_H \geqslant b(t) \, \| u \|^2_H \quad \text{for every } u \in H \text{ , } t \in R$$

c) $B \in \mathcal{L}(\mathcal{U} , \mathcal{D}^k(p,H))$

d) $g \in \mathcal{D}^k(p;H)$

e) N is Hermitian and for every $u \in \mathcal{U}$, $(Nu,u)_{\mathcal{U}} \geqslant \alpha \| u \|^2_{\mathcal{U}}$

then , the stated optimal control problem for (1) has a unique solution $u^* \in \mathcal{U}_{ad} \subset \mathcal{U}$.

In this case , the state space of the problem is

$$\mathcal{D}^k(p+G;V) \cap \mathcal{D}^k(p;H)$$

with the norme $\quad \| \cdot \|^2_{V;p+G,k} + \| \cdot \|^2_{H;p,k}$

Also, for each integer $k \in Z$ there exists a positive function $G_k(t) \in C^1$, such that if $p+G_k$ and $p+G+G_k$ are verifying (A) , then , for $u \in \mathcal{U}_{ad}$ we have:

$$\| y \|^2_{V;p+G+G_k,k} + \| y \|^2_{H;p+G_k,k} \leqslant (g+Bu \text{ , } y)_{H;p+G_k,k} \tag{7}$$

Since

$$C : \mathcal{D}^k(p+G,V) \cap \mathcal{D}^k(p,H) \longrightarrow \mathcal{H}$$

the adjoint :

$$C^* \in \mathcal{L} (\mathcal{H}' ; [\mathcal{D}^k(p+G;V) \cap \mathcal{D}^k(p,H)]')$$

or

$$C^* \in \mathcal{L} (\mathcal{H}' ; \mathcal{D}^{-k}(-(p+G);V) \cup \mathcal{D}^{-k}(-p,H))$$

We denote by $\widetilde{P}(t,D)$ the formal adjoint of (1), with Λ the canonical isomorphism $\mathcal{H} \to \mathcal{H}'$ and with $a^*(t;u,v)$ the adjoint defined by $a^*(t;u,v) = \overline{a(t;v,u)}$.

Thanks to the assumptions we made , the equation

$$(\widetilde{P}(t,D)p(t,u^*),v)_H + a^*(t;p(t,u^*),v) = (C^*\Lambda(Cy(t,u^*)-z_d),v)_H ;\ v\in V \quad (8)$$

has a unique solution

$$p(t,u^*)\in\left[\mathcal{D}^{-k}(-(p+G)+F;V)\cap\mathcal{D}^{-k}(-(p+G);H)\right]\cup\left[\mathcal{D}^{-k}(-p+E;V)\cap\mathcal{D}^{-k}(-p;H)\right]$$

where F and E are positive functions of class C^1 , such that $F-(p+G)$ and $-p+E$ are satisfying the condition a) . Since $a(t;u,v)$ has been considered real and Hermitian , we can write then the optimality system in the following manner:

$$(9)\quad\begin{cases} a(t;y(t,u^*),v)+\left(\sum_{r=-n}^{1} B_r(t)D^r y(t,u^*),v\right)_H = (g+Bu^*,v)_H \quad v\in V \\ y(t;u^*)\in\mathcal{D}^{k}(p+G;V)\cap\mathcal{D}^{k}(p;H) \\ a(t;v,p(t,u^*)) + \left(\widetilde{\sum_{r=-n}^{1} B_r(t)D^r} p(t,u^*),v\right)_H = (C^*\Lambda(Cy(t,u^*)-z_d),v)_V ;\ v\in V \\ p(t,u^*)\in\left[\mathcal{D}^{-k}(-(p+G)+F;V)\cap\mathcal{D}^{-k}(-(p+G),H)\right]\cup\left[\mathcal{D}^{-k}(-p+E;V)\cap\mathcal{D}^{-k}(-p;H)\right] \\ \left(\Lambda_{\mathcal{U}}^{-1}B^*p(t,u^*)+Nu^*,\ u-u^*\right)_{\mathcal{U}} \geq 0 \qquad u\in\mathcal{U}_{ad} \end{cases}$$

<u>THEOREM 2.2.</u> For each integer k , there exists a positive function $g_k(t)$, $g_k(t)\in C^1$ for $t\in R$, such that if:

a') $p(t)\in C^1$ and for every $t\in R$, $p'(t)\geq g_k(t)$

b') $(\exists)$ $b(t)>0$, $b(t)\in\mathcal{E}_t$ such that

$$(B_2(t)u,u)_H \geq b(t)\ \|u\|^2_H \qquad (\forall)\ u\in H ,\ t\in R$$

c') $B\in\mathcal{L}(\mathcal{U},\ \mathcal{D}^{k}(p;H))$

d') $g\in\mathcal{D}^{k}(p;H)$

e') N is hermitian and $(Nu,u)_{\mathcal{U}} \geq \alpha \|u\|^2_{\mathcal{U}}$ $u \in \mathcal{U}$,

then we have a unique solution $u^* \in \mathcal{U}_{ad}$ for the optimal control problem where $P(t,D)$ has the hyperbolic form.

The state space is given by $\mathcal{D}^k(p;V) \cap \mathcal{D}^{k+1}(p;H)$,
$C \in \mathcal{L}(\mathcal{D}^k(p;V) \cap \mathcal{D}^{k+1}(p;H), \mathcal{H})$, $C^* \in \mathcal{L}(\mathcal{H}', \mathcal{D}^{-k}(-p;V') \cup \mathcal{D}^{-k-1}(-p;H))$

Hence, the optimality system is going to take the form:

$$
(10)\quad \begin{cases}
a(t;y(t,u^*),v) + (\sum_{r=-n}^{l} B_r(t)D^r y(t,u^*),v)_H = (g+Bu^*, v)_H \quad v \in V \\
y(t,u^*) \in \mathcal{D}^k(p;V) \cap \mathcal{D}^{k+1}(p;H) \\
a(t;v,p(t,u^*)) + (\widetilde{\sum_{r=-n}^{l} B_r(t)D^r} p(t,u^*),v)_H = (C^*\Lambda(Cy(t,u^*)-z_d),v)_V ; v \in V \\
p(t,u^*) \in [\mathcal{D}^{-k}(-p;V) \cap \mathcal{D}^{-k+1}(-p;H)] \cup [\mathcal{D}^{-k}(-p;V) \cap \mathcal{D}^{-k+1}(-p;H)] \\
(\Lambda_{\mathcal{U}}^{-1} B^* p(t,u^*) + Nu^*, u-u^*)_{\mathcal{U}} \geq 0 \qquad u \in \mathcal{U}_{ad}
\end{cases}
$$

3. OPTIMAL CONTROL IN DISTRIBUTIONS.

We denote with $\mathcal{D}'(A)$ the space of distributions on the real li line haming the values in H, $\mathcal{D}_-(A)$ - the space of indefinite differentiable functions on the real line with the support limited at right, having the values in H and with the usual Schwartz topology, $\mathcal{D}'_+(A)$ - the dual of $\mathcal{D}_-(A)$, $\mathcal{D}'_+(A) = \mathcal{L}(\mathcal{D}_-(A), H)$.

Because $a(t;u,v)$ satisfies to the conditions (3), (4), we can extend it to a sesqui-linear continuous form on $\mathcal{D}'(V) \times V$, $a(t;U,v)$. We can also extend canonicaly the Hermitian product $(\ ,\ )_H$ to $\mathcal{D}'(H) \times H$.

Let be then a system for which the state is a distribution, solution for the equation

$$a(t;U,v) + (P(t,D)U,v)_H = (T+Bu,v)_H \qquad v \in V$$

in the sense of scalar distributions.

THEOREM 3.1. If:

a'') $B \in \mathcal{L}(\mathcal{U}, \mathcal{D}'_+(H))$

b'') $C \in \mathcal{L}(\mathcal{D}'_+(V), \mathcal{H})$

c'') $N \in \mathcal{L}(\mathcal{U}, \mathcal{U})$ Hermitian and $(Nu,u)_{\mathcal{U}} \geq k \|u\|^2_{\mathcal{U}}$; $u \in \mathcal{U}$

d'') $(B_1(t)u,u) \geq b(t) \|u\|^2$; $u \in H$, $t \in R$ for parab.

or $(B_2(t)u,u) \geq b(t) \|u\|^2$; $u \in H$, $t \in R$ for hyperb.

e'') $T \in \mathcal{D}'_+(H)$

then the optimal control problem has a unique solution $u^* \in \mathcal{U}$.

The state space is $\mathcal{D}'_+(V)$ and if $\mathrm{supp}(T+Bu) \subset \{t \; ; \; t \geq a\}$ then the same for U , $\mathrm{supp}\,U(t,u) \subset \{t \; ; \; t \geq a\}$.

Denoting by $X(t,u)$ the adjoint state , the optimality system gets the form:

$$
(11) \quad \begin{cases}
a(t;U(t,u^*),v) + (P(t,D)U(t,u^*),v)_H = (T+Bu^*,v)_H \; ; \quad v \in V \\
U(t,u^*) \in \mathcal{D}'_+(V) \\
a^*(t;X(t,u^*),v) + (\tilde{P}(t,D)X(t,u^*),v)_H = (C^* \Lambda (CU(t,u^*)-z_d), v)_V ; \; v \in V \\
X(t,u^*) \in \mathcal{D}_-(V) \\
(\Lambda_{\mathcal{U}}^{-1} B^* X(t,u^*) + Nu^* , u-u^*)_{\mathcal{U}} \geq 0 \qquad u \in \mathcal{U}_{ad} \subset \mathcal{U}
\end{cases}
$$

Another exemple of control problem in non normed spaces is given in ([6]) . See also [1].

If $\mathcal{D}'_{a+}(A)$ has been a subspace of $\mathcal{D}'_+(A)$ and namely the space of distributions having the support limited at left by $a+ > 0$ then, we denote by $\mathcal{D}'f_{a+}(A)$ the space of distributions of finite order , on the real line, with the values in A and having the support limited at left by $a+ > 0$.

This means that for $0 \leq h \leq m$, we can find a class of continuous functions $\{g_h(t)\}$ with the values in A , such that we can write:

$$T = g_o + Dg_1 + \dots + D^m g_m \qquad (g_i = 0 \text{ for } t < a)$$

Obviously

$$\mathcal{D}'f_+(A) = \bigcup_{a \in R} \mathcal{D}'f_{a+}(A)$$

THEOREM 3.2. If:

aIV) $B \in \mathcal{L}(\mathcal{U}, \mathcal{D}'f_+(H))$

bIV) $C \in \mathcal{L}(\mathcal{D}'f_+(V), \mathcal{H})$

cIV) $N \in \mathcal{L}(\mathcal{U}, \mathcal{U})$, Hermitian and

$$(Nu,u)_{\mathcal{U}} \geq k \|u\|^2_{\mathcal{U}} \quad ; \; u \in \mathcal{U} \, , \; k > 0$$

dIV) $(B_i(t)u,u) \geq b(t) \|u\|^2 \quad u \in H \, , \; t \in R \, , \; i = 1 \text{ or } 2$

eIV) $T \in \mathcal{D}'_+f(H)$

then the optimal control problem has a unique solution $u^* \in \mathcal{U}$.

The state space is a space of distributions of finite order, $\mathcal{D}'^f{}_+(V)$ and the support of $U(t,u)$ is on the half line $t \geq a$ a-real, if the support of T+Bu is the same .

We obtaine the following optimality system:

$$(12) \begin{cases} a(t;U(t,u^*),v) + (P(t,D)U(t,u^*),v)_H = (T + Bu^* \, , \, v)_H \; ; \quad v \in V \\ U(t,u^*) \in \mathcal{D}'^f{}_+(V) \\ a^*(t;X(t,u^*),v) + (\widetilde{P}(t,D)X(t,u^*),v)_H - (C^*\Lambda(CU(t,u^*)-z_d),v)_V ; \quad v \in V \\ X(t,u^*) \in \mathcal{D}^f{}_-(V) \\ (\Lambda_{\mathcal{U}}^{-1} B^*X(t,u^*) + Nu^* \, , \, u - u^*)_{\mathcal{U}} \geq 0 \qquad u \in \mathcal{U}_{ad} \subset \mathcal{U} \end{cases}$$

Observation.

In general, if Ω is a domain in R^n, $\Omega \subset R^n$, the space of distributions of order zero on Ω is the space of Radon measures on Ω and we put $\mathcal{M}(\Omega) = \mathcal{D}'^0(\Omega) = \mathcal{K}'(\Omega)$, $\mathcal{K}(\Omega)$ being the vector space of all continuous functions on Ω whose support is contained in some compact subset of Ω, equiped with the finest locally convex topology for which the canonical injection $\mathcal{K}(\Omega) \to \mathcal{K}(\Omega)$ is continuous ($K \subset \Omega$, K- compact). If $\varphi \in \mathcal{K}(\Omega)$ and $\mu \in \mathcal{M}(\Omega)$, then the value

$$<\mu, \varphi> = \int_\Omega \varphi(x)\, d\mu(x)$$

is called the integral of φ with respect to the measure μ.

Hence if the state of a system is represented as a Radon measure on an interval, solution for the equation

$$a(t;\mu(t,u),v) + (P(t,D)\mu(t,u),v)_H = (T + Bu^*,v)_H \qquad v \in V$$

in the distributional sense, then, by the THEOREM 3.2. we get an existence and uniqueness result for the optimal control problem and (11) becomes:

$$(13)\quad \begin{cases} a(t;\mu(t,u^*),v) + (P(t,D)\mu(t,u^*),v)_H = (T+Bu^*,v)_H & v \in V \\ a^*(t;\varphi(t,u^*),v) + (\tilde{P}(t,D)\varphi(t,u^*),v)_H = (C^*\Lambda(C\mu(t,u^*)-z_d),v)_V & v \in V \\ \mu(t,u^*) \in \mathcal{M}_+(V) & \\ \varphi(t,u^*) \in \mathcal{K}_-(V) & \\ (\Lambda_{\mathcal{U}}^{-1} B^*\varphi(t,u^*) + Nu^*, u-u^*)_{\mathcal{U}} \geq 0 & u \in \mathcal{U}_{ad} \subset \mathcal{U} \end{cases}$$

4. INTERPRETATION OF RESULTS.

There are manyways of giving an interpretation to the system (9) . All depends on the different assumptions regarding V,H,C,B. For instance for $V=H^1_0(\Omega)$ and $H=L^2(\Omega)$, or $V=H^1(\Omega)$ and $H=L^2(\Omega)$ we obtaine well known optimal control problems .

We are going to consider here, that interpretation, which is more connected to our further applications.

Since $a(t;u,v)$ is continuous on VxV and (3) is satisfied , we can find an operator $\mathcal{A}(t)\in \mathcal{E}_t(\mathcal{L}(V,V))$, such that

$$a(t;u,v) = (\mathcal{A}(t)u,v)_V \qquad (14)$$

From (4) there is obvious that we have

$$(\mathcal{A}(t)u,u)_V \geq \alpha(t)\ \|u\|^2_V \ , \quad \alpha(t)>0\ , \quad \alpha(t)\in\mathcal{E}_t, t\in R \qquad (15)$$

We supposed $a(t;u,v)$ real and Hermitian ; then $a(t;u,v) = a(t;v,u) = a^*(t;u,v)$ and hence $\mathcal{A}(t)$ is autoadjoit $\mathcal{A}(t) = \mathcal{A}^*(t)$, $t\in R$

We introduce a continuous linear operator $\mathcal{J}\in\mathcal{L}(H,V)$ by the equality:

$$(f,v)_H = (\mathcal{J}f,v)_V \qquad f\in H\ ,\ v\in V$$

Then, the equations in (9) can be written as:

$$(\mathcal{A}(t)y,v)_V + (\sum_{r=-n}^{1} B_r(t)D^r y,v)_H = (\mathcal{J}g,v)_V + (Bu^*,v)_H \qquad v\in V$$

$$(\mathcal{A}(t)p,v)_V + (\widetilde{\sum_{r=-n}^{1} \mathcal{J} B_r(t)D^r} p,v)_V = (C^*\Lambda(Cy(t,u^*)-z_d),v)_V \qquad v\in V$$

and so , the system (9) becomes:

$$
(16)\quad \begin{cases}
\sum_{r=-n}^{1} B_r(t) D^r y(t,u^*) = Bu^* & \text{in } H \\
\mathcal{A}(t) y(t,u^*) = \gamma g(t) & \text{in } V \\
\mathcal{A}(t) p(t,u^*) + \sum_{r=-n}^{1} \widetilde{\gamma B_r(t)} D^r p(t,u^*) = C^* \Lambda (Cy(t,u^*) - z_d) & \text{in } V \\
e^{-(p+G)} D^k y \in L^2(V) \\
e^{-p} D^k y \in L^2(H) \\
p(t,u^*) \in \left[\mathcal{D}^{-k}(F-p-G;V) \cap \mathcal{D}^{-k}(-p-G;H)\right] \cup \left[\mathcal{D}^{-k}(E-p;V) \cap \mathcal{D}^{-k}(-p;H)\right] \\
(\Lambda_{\mathcal{U}}^{-1} B^* p(t,u^*) + Nu^* , u - u^*)_{\mathcal{U}} \geq 0 & u \in \mathcal{U}_{ad} \subset \mathcal{U}
\end{cases}
$$

and the same for (10) :

$$
(17)\quad \begin{cases}
\sum_{r=-n}^{2} B_r(t) D^r y = Bu^* & \text{in } H \\
\mathcal{A}(t) y(t,u^*) = \gamma g(t) & \text{in } V \\
\mathcal{A}(t) p(t,u^*) + \sum_{r=-n}^{2} \widetilde{\gamma B_r(t)} D^r p(t,u^*) = C^* \Lambda (Cy(t,u^*) - z_d) & \text{in } V \\
e^{-p} D^k y \in L^2(V) \\
e^{-p} D^{k+1} y \in L^2(H) \\
p(t,u^*) \in \left[\mathcal{D}^{-k}(-p;V) \cap \mathcal{D}^{-k+1}(-p;H)\right] \cup \left[\mathcal{D}^{-k}(-p;V) \cap \mathcal{D}^{-k+1}(-p;H)\right] \\
(\Lambda_{\mathcal{U}}^{-1} B^* p(t,u^*) + Nu^* , u - u^*)_{\mathcal{U}} \geq 0 & u \in \mathcal{U}_{ad} \subset \mathcal{U}.
\end{cases}
$$

We must mentione that , as usual([9], [10]), by $D^r\varphi$, $r > 0$ we understand the derivative in the distributional sense , and for $\varphi \in \mathcal{D}_-(E)$,

$$D^{-r}\varphi(t) = (Y^{*(r)} * \varphi)(t) = (Y_r * \varphi)(t) = \frac{1}{\Gamma(r)}\left[Pf(y^{r-1})_{y>0} * \varphi\right](t) =$$

$$= \frac{1}{\Gamma(r)}\int_0^{\infty} y^{r-1}\varphi(t-y)dy \qquad \text{in} \quad E$$

where Y is the Heaviside function , and

$$Y_m = \frac{1}{\Gamma(m)} Pf(x^{m-1})_{x>0} \qquad m>0$$

$$Y_{-1} = \delta^{(1)} = (-1)\varphi^{(1)}(0) \qquad m=-1 \leq 0$$

$$Y^{*m} = Y_m$$

$$\left.\begin{array}{l} I^m T = Y_m * T \\ D^m T = Y_{-m} * T \end{array}\right\} \quad \text{for } T \in \mathcal{D}'_+$$

This way , if $B_r(t) \in \mathcal{C}_t(\mathcal{L}(H,H))$, for every $\varphi \in \mathcal{D}_-(H)$ we have:

$$P(t,D)\varphi(t) = B_{-n}(t)\frac{1}{\Gamma(n)}\int_0^{\infty} y^{n-1}\varphi(t-y)dy + \ldots + \frac{B_{-2}(t)}{2}\int_0^{\infty} y\,\varphi(t-y)dy +$$

$$+ B_{-1}(t)\int_0^{\infty}\varphi(t-y)dy + B_0(t)\varphi(t) + B_1(t)\frac{d\varphi}{dt}$$

with the values in H.

In the same time , for $\varphi \in \mathcal{D}_-(V)$

$$\widetilde{P}(t,D)\varphi(t) = \sum_{r=-n}^{1}(-1)^r D^r\left[B_r(t)\varphi(t)\right]$$

If we consider for instance n = 2 , we obtaine the following form for the optimality system:

$$
(18)\quad \begin{cases}
B_1(t)\dfrac{dy}{dt} + B_o(t)y + B_{-1}(t)\displaystyle\int_0^{\infty} y(t-x)dx + \dfrac{B_{-2}(t)}{2}\displaystyle\int_0^{\infty} xy(t-x)dx = Bu^{x} \quad \text{in } H \\
\mathcal{A}(t)y = \gamma g(t) \quad \text{in } V \\
-A_1(t)\dfrac{dp}{dt} + (A_o(t) - A_1^{(1)}(t) + \mathcal{A}(t))p - A_{-1}(t)\displaystyle\int_0^{\infty} p(t-x)dx + \\
\quad + d_1^{-1}\displaystyle\int_0^{\infty} dx \int_0^{\infty} A_{-1}^{(1)}(t-x)p(t-x-z)dz + \dfrac{A_{-2}(t)}{2}\displaystyle\int_0^{\infty} xp(t-x)dx + \\
\quad + d_1^{-1}\displaystyle\int_0 dx \int_0 A_{-2}^{(1)}(t-x)p(t-x-z)dz + \dfrac{d_2^{-2}}{2}\displaystyle\int_0^{\infty} x\, B_{-2}^{(2)}(t-x)dx \int_0^{\infty} zp(t-x-z)dz \\
\qquad = C^{x}\Lambda(C(y(t,u^{x})-z_d)) \quad \text{in } V \\
e^{-(p+G)}D^k y \in L^2(V) \\
e^{-p}D^k y \in L^2(H) \\
p(t,u^{x}) \in [\mathcal{D}^{-k}(F-p-G;V) \cap \mathcal{D}^{-k}(-p-G;H)] \cup [\mathcal{D}^{-k}(E-p;V) \cap \mathcal{D}^{-k}(-p;H)] \\
(\Lambda_{\mathcal{U}}^{-1} B^{x}p(t,u^{x}) + Nu^{x},\ u-u^{x})_{\mathcal{U}} \geq 0 \qquad u \in \mathcal{U}_{ad} \subset \mathcal{U}.
\end{cases}
$$

where $d_i^{\,j}$ are integers , $\gamma B_r(t) = A_r(t) \quad t \in R$ and $A_r^{(i)}(t)$ means the i-th derivative of $A_r(t)$.

4. NUMERICAL APPROACH.

From all the numerical methods used in control theory , we choosed the one which seems to us that works the best ; this is the Galerkin method , and we shell apply it to the systems (9) and (10).

Let V be separable and we consider an internal Hilbert approximation for V , $\{V_h, P_h, R_h\}_{h \in \mathbb{N}}$ which is stable and convergent ([5], [14]) and such that

$$\overline{\bigcup_{h=1}^{\infty} V_h} = V$$

For instance , we take w_1 , w_2 , ..., a free and total sequence in V and we denote by V_h the subspace of V generated by w_1 , w_2 ,...,w_h. P_h is the canonical injection $V_h \longrightarrow V$ and R_h is the ortogonal projection of V on V_h .

This way , to each V_h , $k \in Z$, it corresponds the space $\mathcal{D}^k(p;V_h)$ where the weight p satisfies (A). Obvious that:

$$\mathcal{D}^k(p;V_h) \subset \mathcal{D}^k(p;V) \qquad \text{and} \qquad \bigcup_{h=1} \mathcal{D}^k(p;V_h) = \mathcal{D}^k(p;V)$$

For each h , we define y_h as the solution of the equation:

$$(P(t,D)y_h , v_h)_H + a(t;y_h , v_h) = (f + Bu , v_h)_H ; \; v_h \in V_h \qquad (19)$$

If $f+Bu \in \mathcal{D}^k(p;H)$ then , we find a unique solution $y \in \mathcal{D}^k(p+G;V_h) \cap \mathcal{D}^k(p;H)$ such that:

$$\|y_h\|^2_{V_h,p+G+G_k,k} + \|y_h\|^2_{H,p+G_k,k} \leq (f+Bu,y_h)_{H,p+G_k,k}$$

where G can be choosen the same for all h (by taking the supremum) in the way that $p+G_k$, $p+G+G_k$ are satisfying (A).

Let us consider then , the sequence pf approximative solutions $\{y_h\}_{h \in \mathbb{N}}$. If y is the solution of the first equation in (9) then, in particular , we have:

$$(P(t,D)y,v_h)_H + a(t;y,v_h) = (f+Bu,v_h)_H \qquad v_h \in V_h$$

and for each $h \in \mathcal{N}$, $y_h - y$ is satisfying the homogenuous eqation

$$(P(t,D)\,[y_h-y]\,,v_h)_H + a(t;y_h-y,v_h) = 0 \qquad v_h \in V_h$$

and

$$\| y_h-y \|^2_{V_h,p+G+G_k,k} + \| y_h-y \|^2_{H,p+G_k,k} \leq 0$$

Since P_h is a stable continuous injection from $V_h \longrightarrow V$ the application $\varphi_h \longrightarrow P_h \varphi_h$ from $\mathcal{D}(V_h) \longrightarrow \mathcal{D}(V)$ can be extended to a continuous application of

$$\mathcal{D}^k(p;V_h) \longrightarrow \mathcal{D}^k(p;V)$$

Also , the approximation $\{ V_h,P_h,R_h \}_{h \in \mathcal{N}}$ was supposed convergent, so, we have:

$$\| y_h-y \|^2_{V,p+G+G_k,k} + \| y_h-y \|^2_{H,p+G_k,k} \leq 0$$

and from here $y_h \longrightarrow y$ when $h \to \infty$ in the norm of $\mathcal{D}^k(p+G+G_k,V) \cap \mathcal{D}^k(p+G_k,H)$ for the P(t,D) parabolic and in the norm of $\mathcal{D}^k(p+G_k,V) \cap \mathcal{D}^{k+1}(p+G_k,H)$ for thehyperbolic form of P(t,D).

If $G_k(t)$ is such that $\exp(G_k(t))$ is bounded over the whole real line , then there is a continuous embeading of $\mathcal{D}^k(p+G+G_k;V)$ in $\mathcal{D}^k(p+G;V)$, of $\mathcal{D}^k(p+G_k;H)$ in $\mathcal{D}^k(p;H)$, of $\mathcal{D}^k(p+G_k;V)$ in $\mathcal{D}^k(p,V)$ and $\mathcal{D}^{k+1}(p+G_k;H)$ in $\mathcal{D}^{k+1}(p;H)$ for all $k \in Z$.
Hence $y_h \longrightarrow y$ in the norm of the state spaces.

The approximative adjoint state $p_h(t,u)$ is defined as the solution of

$$(\widetilde{P}(t,D)p_h, v_h)_H + a(t;p_h,v_h) = (C^{x}\Lambda (Cy_h - z_d), v_h)_{V_h} \qquad v_h \in V_h \qquad (20)$$

and the convergence of $\{p_h\}_{h\in\mathcal{H}}$ is obtained in the same way.

Now, if

$$J_h(u) = \| Cy_h - z_d \|^2_{\mathcal{H}} + (Nu,u)_{\mathcal{U}} \qquad u \in \mathcal{U}_{ad} \subset \mathcal{U}$$

we denote by $u_h \in \mathcal{U}_{ad} \subset \mathcal{U}$ the corresponding optimal control

$$J_h(u_h) = \min_{u\in \mathcal{U}_{ad}} J_h(u)$$

By an already classical result ([3], [7], [14]) we have , under our hypothesis that $u_h \longrightarrow u$ strongly in $\mathcal{U}$.

Remarque. Discretization for $\mathcal{U}$ in $\{\mathcal{U}_\xi\}$ can also be considered.

6. APPLICATIONS.

There are many interesting control problems that are arising in different fields of research and to which the above results can be applied.

Let us consider first the general form of a linear bounded delay-differential equation of closed-cycle type:

$$\dot{x}(t) + \int_{-\infty}^{t} d_\tau S(t-\tau)x(\tau) + \int_{-\infty}^{t} d_\tau K(t-\tau)x(\tau) = f(t) \qquad (21)$$

where $S(t-\tau)$ and $K(t-\tau)$ are the saltus matrix and the continuous kernel corresponding to the kernel $G(t-\tau)$ If $S \equiv 0$ and K is differentiable and Riemann integrable , then the equation (21) becomes an integro-differential equation of the form:

$$\dot{x}(t) + \int_{-\infty}^{t} K_{\tau}(t-\tau).x(\tau)\,d\tau = f(t)$$

This is leading us to consider the equation

$$\dot{x}(t) + \frac{B_2(t)}{2}\int_{0}^{\infty} \tau\; x(t-\tau)\,d\tau = B(t)u(t) + f(t) \qquad (22)$$

in H , with the initial condition $x(t) = \mathcal{A}^{-1}(t)\,\gamma g(t)$ in V , and the corresponding hereditary process of closed-cycle type described by it. Each pair $(\mathcal{A}^{-1}(t)\ g(t)\ ,\ u(t))$ where $\mathcal{A}(t)$, γ and u are defined as in 2 . is called optimal. The optimal control problem is then: find an admissible pair $(\mathcal{A}^{-1}(t)\,\gamma g(t)\ ,\ u(t))$, with the correspon-ding trajectory x(t,u) such that the quadratic performance criterion (6) gets minimized.

Under the hypothesis a) , c),d), e) the THEOREM 2.1. is giving us an existence and uniqueness result . There is evident that we can take $H=R^n$ and V a Hilbert subspace of R^n , dense in R^n and this way , already classical results are included in our statement. The optimality system is easy to be written.

6.2. Areas of most importance as plasma phisics , radiative transfer , rarefied gas dynamics , neutron transport thaory are employing a form of transport equation.

In the kinetic theory a basic problem concerns the rate and manner in which gas molecules approach their equilibriums distribution .

A general problem can be envisaged as a coupled time and space relaxation due to an initial perturbation.

If the problem is considered linear , then, discussing separately the time and the space relaxation due to an initial perturbation , we can consider the two equations:

$$\frac{\partial h(c,t)}{\partial t} = - \hat{I}_0 \left[h(c,t) \right] \tag{19}$$

$$c \nabla h(r,c) = - \hat{I}_0 \left[h(r,c) \right] \tag{20}$$

The equation (19) describes the time-dependent perturbation of a single gas due to spatially uniform anisotropic disturbances and equation (20) describes space-dependent perturbation due to non-uniform anisotropic stationary disturbances.

The time dependent problem is closely related to the pulsed neutron problem in an infinite medium . This arises when a small admixture of an alien gas is introduced into a host gas , the latter being in an equilibrium condition. However we should point out that deviation from the equilibrium neutron distribution are introduced by the injection of particles , whereas in the gas problem the deviation is brought about by the introduction of energy.

If in (19) we set:

$$\hat{I}_0(h) = V(c)h(x,c) - \int_0^\infty dc' e^{-c'^2} K(c,c')h(x,c') \tag{21}$$

$c=v(m/2kT)^{\frac{1}{2}}$ - the reduced velocity

and if we shell require now that the kernel $K(c,c')$ takes the form:

$$K(c,c') = \frac{c' e^{c'^2} A(c,t)h(t-c',t)}{h(c',t)} \tag{22}$$

such that $A(c,t) \in \mathcal{C}_t(R)$ for each c and every $t \in R$,we can consider the equation

$$\frac{\partial h(c,t)}{\partial t} + V(c)h(c,t) - \int_0^\infty e^{-c'^2} K(c,c')h(c',t)dc' = u \text{ in } R \tag{22'}$$

If we want now to control even the injection of particles even the inintroduction of the energy , we can formulate the problem: for a kernel given by (22) find a control $u \in \mathcal{U}_{ad} \subset \mathcal{U}$ such that if $h=h(c,t,u)$ is the solution in (22') , the quadratic performance cr terion $J(u)$ is minimized.

The THEOREM 2.1. is giving us and in this case an existence and uniqueness result, for $H = R$.

REFERENCES

[1]. Balakrishnan,A.V.:"Foundations of the state space theory of continuous systems " I Journal of Computer and System Sciences, I,1967

[2]. Kailath,T:"A view of Three Decades of Linear Filtring Theory" - I.E.E.E. Transaction on Information Theory, vol.IT-20, 2,1974

[3]. Lions,J.L.:"Thèse", Acta Mathematica,95 (1955),13-153

[4]. -"- :"Boundary value problems", Univ. of Kansas, 1957

[5]. -"- :"Equations differentielles operationnelles et problèmes aux limites",Springer Verlag, 1961

[6]. -"- :"Contrôle optimal de systèmes gouvernés par des équations aux dérivées partielles", Dunod,Gauthier-Villars,1968

[7]. -"- :"Some aspects of the optimal control of distributed parameter systems"-SIAM-series, Philadelphia,Pe.19103,1972

[8]. Oguztöreli,M.N.:"Time-Lag Control Systems"-Acad.Press 1966

[9]. Schwartz,L:"Théorie des distributions" Paris,Hermann,1950

[10]. -"- :"Distributions à valeurs vectorielles"-Annales de l'Institut Fourier 1958

[11]. Simionescu,Cl.:"O problemă de control optimal" Buletinul Universității din Brașov,Seria C,vol.XV,1973

[12]. -"- :"Optimal control problems in distributions" Proceedings of the UNESCO-Seminar in modeling and optimization techniques București-Brașov, 3-5 June 1975 (under print)

[13]. -"- :" Optimal control problems in Sobolev spaces with weights" SIAM Journal on Control,14-1,Ja-1976

[14]. Temam,R:"Résolution approchée d'équations aux dérivées partielles" Press Universitaires de France,1970

[15]. Trèves,F.:"Domination et problèmes aux limites de type mixte", C.R.Acad.Sci.Paris. (245) 1957,pg. 2454, 1957

[16]. -"- :"Thèse"- Acta Mathematica 101 (1959) pg.1-139

[17]. Williams,M.M.R.:"Mathematical Mathods in Particle Transport Theory"-London-Butterwarth.

ON OPTIMAL PARAMETRIC CONTROL OF PARABOLIC SYSTEM

Jan Sokołowski

Institute for Organization, Management
and Control Sciences
00-818 Warszawa, ul. KRN, Poland

Introduction

In the paper we consider a parametric optimization problem for an abstract parabolic equation. Problems of such a type are investigated in the case of elliptic equations in [2], [10] and parabolic equations in [2], [3], [5], [11]. In this paper two types of observations : in spaces $C(0,T;H)$ and $C(0,T;V)$ are considered.

For both cases sufficient conditions of existence of an optimal control are given. Using the so called generalized adjoint state equation necessary conditions of optimality are formulated. Proofs of presented results are given in [7], [8], [9].

Let there be given Hilbert spaces V, H with

/1.1/ $\quad V \subset H \quad , \quad V$ dense in H

By U_{ad} we denote the set of admissible controls which is assumed to be a convex subset of a Hilbert space U.

Let there be given a family of bilinear forms on V :

/1.2/ $\quad a_u(t;y,z)$

where $u \in U_{ad}$

$t \in [0,T]$

$y,z \in V$

We assume :

(i) family /1.2/ is continuous on V, that is

/1.3/ $$|a_u(t;y,z)| \leq M \|y\|_V \|z\|_V , \forall t \in [0,T] ,$$

$$\forall u \in U_{ad} , \qquad \forall y,z \in V$$

(ii) mapping

$$t \longmapsto a_u(\cdot\,;y,z)$$

is measurable for all $y,z \in V$ and all $u \in U_{ad}$ with respect to Lebesque measure on the interval $[0,T]$

(iii)

/1.4/ $$a_u(t;y,y) \geqslant \alpha \|y\|_V^2 , \qquad \alpha > 0 ,$$

$$\forall u \in U_{ad} , \quad \forall t \in [0,T] , \qquad \forall y \in V$$

Following [3] we denote by $W(0,T) \subset L^2(0,T;V)$ a Hilbert space with the scalar product :

/1.5/ $$(y,z)_{W(0,T)} = \int_0^T \left\{ \left(\frac{dy}{dt} , \frac{dz}{dt} \right)_{V'} + (y,z)_V \right\} dt$$

where $(\cdot\,,\cdot)_V$ denotes scalar product in V .

Let there be given elements $y_0 \in H$ and $f \in L^2(0,T;V')$.

For given control $u \in U_{ad}$ we define state trajectory $y_u \in W(0,T)$ as the solution of an abstract parabolic equation of the form :

/1.6/ $$\left(\frac{dy_u}{dt}(t), z \right)_{V'V} + a_u(t;y_u(t),z)$$

$$= (f(t),z)_{V'V} , \qquad \forall z \in V, \quad \text{a.e. in }]0,T[$$

/1.7/ $\quad y_u(0) = y_0$

where $(\cdot\,,\,\cdot)_{V'V}$ denotes scalar product between V' and V. Under the above assumptions problem /1.6/, /1.7/ has the unique solution [4] which continuously depends on the data $y_0 \in H$ and $f \in L^2(0,T;V)$.

2. Observation in $C(0,T;H)$

Let us define on the set $U_{ad} \subset U$ the cost functional

/2.1/ $$J(u) = \frac{1}{2}\, \| y_u(T) - z_d \|^2_H + \frac{\varepsilon}{2}\, \| u \|^2_U$$

where

$$y_u(T) = y_u \Big|_{t=T}$$

and $z_d \in H$ is a given element.

We shall consider the following minimization problem :

/2.2/ $$\text{find} \quad \inf_{u \in U_{ad}} J(u)$$

Lemma :

If we assume that :

(i) bilinear form /1.2/ is Lipschitzian with respect to u, that is

/2.3/ $$| a_{u1}(t;y,z) - a_{u2}(t;y,z) | \leqslant C\, \| u_1 - u_2 \|_U \|y\|_V \, \|z\|_V$$

$$\forall t \in [0,T], \quad \forall u_1, u_2 \in U_{ad}, \quad \forall y,z \in V$$

(ii) set $U_{ad} \subset U$ is compact

then there exists a solution $u \in U_{ad}$ to the problem /2.2/.

Proof is given in [8] .

Let us assume that at the point $u \in U_{ad}$ there exists Frechet derivative $\tilde{A}_u(\hat{u})$ of the operator $\tilde{A}(u) \in \mathcal{L}(L^2(0,T;V) ; L^2(0,T;V'))$ which is defined by the equality

$$/2.4/ \qquad ((\tilde{A}(u))(t)\, y,z)_{V'V} = a_u(t;y,z)$$
$$\forall u \in U_{ad} \quad , \quad \forall t \in [0,T] \quad , \quad \forall y,z \in V$$

To obtain a simple form of necessary conditions of optimality we introduce adjoint state $p_u \in W(0,T)$ which is defined as the solution of adjoint state equation :

$$/2.5/ \qquad -\left(\frac{dp_u}{dt}(t),z\right)_{V'V} + a_u\left(t,z,p_u(t)\right) = 0$$
$$\forall z \in V \quad , \quad \text{a.e. in }]0,T[$$

$$/2.6/ \qquad p_u(T) = -\, y_u(T) + z_d$$

Optimal control $\hat{u} \in U_{ad}$ is characterized [2] , [8] by inequality :

$$/2.7/ \qquad \int_0^T \left(\langle \tilde{A}_u(\hat{u}), \; u-u \rangle \; y_{\hat{u}}(t), \; p_{\hat{u}}(t) \right)_{V'V} dt \geqslant 0 \quad ,$$
$$\forall u \subset U_{ad}$$

3. <u>Observation in C (0,T;V)</u>

In order to consider the case where cost dunctional is defined on $C(0,T;V)$ we have to use some representation of bilinear form /1.2/ . To do that we need some additional definictions.

Let there be given a Hilbert space S and linear, bounded operator $\gamma \in \mathcal{L}(V,S)$. We assume

/3.1/ operator γ maps V onto S

/3.2/ kernel $\ker \gamma = V_0$ is dense in H

It is easy to show that for given element $\varphi \in L^2(0,T;S')$ linear functional

$$/3.3/ \qquad V \ni z \longmapsto (\varphi(t), \ \gamma z)_{S'S} \in R^1$$

is continuous a.e. in $]0,T[$.

With the form /1.2/ we associate the so called formal operator :

$$/3.4/ \qquad A(u) \in \mathcal{L}(L^2(0,T;V) \ ; \ L^2(0,T;V_0'))$$

whichis defined by the formula

$$/3.5/ \qquad \int_0^T a_u(t,y(t),z(t))\,dt = \int_0^T ((A(u))(t)\,y(t), \ z(t))_{V_0' \, V_0} \, dt$$

$$\forall z \in L^2(0,T;V_0), \qquad \forall y \in L^2(0,T;V)$$

Furthermore we assume that the domain x/ of operator $A(u)$ considered as an unbounded operator in $L^2(0,T;H)$ is a space $L^2(0,T;D)$ where $D \subset V$ is a given Hilbert space.

It can be shown [1] that there exists the unique operator, called Neumann operator :

$$/3.6/ \qquad \sigma(u) \in \mathcal{L}(L^2(0,T\ ;D) \ ; \ L^2(0,T;S'))$$

such that the following representation takes place :

$$/3.7/ \qquad \int_0^T a_u(t \ ; \ y(t), \ z(t))\,dt$$

$$= (A(u)\,y,z)_{\mathcal{H}} + ((\sigma(u)\,y \ , \ \tilde{\gamma} z))$$

$$\forall y \in L^2(0,T\ ;D) \ , \qquad \forall z \in L^2(0,T;V)$$

where

$$\mathcal{H} = L^2(0,T;H)$$

$$(\tilde{\gamma} z)(t) = \gamma z(t) \qquad \text{a.e. in }]0,T[$$

and $\tilde{\gamma} \in \mathcal{L}(L^2(0,T;V); \ L^2(0,T;S'))$

$((\cdot \ , \ \cdot))$ denotes scalar product between $L^2(0,T;S)$

---

x/ It is the set of elements $y \in L^2(0,T;V)$ such that $A(u)y \in L^2(0,T;H)$.

and $L^2(0,T;S)$.

Let us assume that there exist Hilbert spaces:

/3.8/ (i) $W^1(0,T) \subset C(0,T;V)$,

injection $W^1(0,T) \to C(0,T;V)$ is continuous

/3.9/ (ii) $Y \subset L^2(0,T;S')$

injection $Y \longrightarrow L^2(0,T;S')$ is continuous

such that for any given

/3.10/ $y_0 \in V$

/3.11/ $f \in L^2(0,T;H)$

/3.12/ $\varphi \in Y$

the following state equation has the unique solution $y_u \in W^1(0,T)$:

/3.13/
$$\left(\frac{dy_u}{dt}(t), z \right)_{V'V} + a_u(t; y_u(t), z)$$
$$= (f(t), z)_H + (\varphi(t), \gamma z)_{S'S} , \quad \forall z \in V$$
a.e. in $]0,T[$

/3.14/ $y_u(0) = y_0$

<u>Remark</u>

Problem of existence of the spaces /3.8/, /3.9/ is discussed in [9] . Using /3.7/ we obtain another representation [9] of the system /3.13/, /3.14/ namely

/3.15/
$$\frac{dy_u}{dt} + A(u)y = f$$
$$\sigma(u)\, y_u = \varphi$$
$$y_u(0) = y_0$$

We introduce an optimization problem similar to that in previous section. We define cost functional

/3.16/ $$J(u)= \frac{1}{2} \|y_u(T)- z_d\|_V^2 + \frac{\varepsilon}{2} \|u\|_U^2 , \quad \varepsilon \geqslant 0$$

where $z_d \in V$ is a given element.

We consider the problem of minimization of the cost functional on a given convex set $U_{ad} \subset U$.

If mappings

/3.17/ $$U \supset U_{ad} \ni u \longmapsto A(u) \in \mathcal{L}(W^1(0,T);L^2(0,T;H))$$

/3.18/ $$U \supset U_{ad} \ni u \longmapsto \sigma(u) \in \mathcal{L}(W^1(0,T);Y)$$

are locally Lipschitzian and the set $U_{ad} \subset U$ is compact then there exists an optimal control $\hat{u} \in U_{ad}$ such that

/3.19/ $$J(\hat{u}) \leqslant J(u), \qquad \forall u \in U_{ad}$$

Proof is given in [9] .

To obtain a simple form of necessary conditions of optimality we introduce [8] the so called generalized adjoint state $(p,r) \in L^2(0,T;H) \times Y$ defined at the given point $u \in U_{ad}$ as a solution of generalized adjoint state equation of the form :

/3.20/ $$\left(p, \frac{dw}{dt} + A(u)w\right)_{\mathcal{H}} + \left(r, \sigma(u)w\right)_Y = -\left(y_u(T)- z_d, w(T)\right)_V , \qquad \forall w \in W_0^1(0,T)$$

where

$$W_0^1(0,T) = \{ w \in W^1(0,T) \mid w|_{t=0} = 0 \}$$

It can be shown [8] , that there exists the unique solution of /3.20/ for each $u \in U_{ad}$.

If $\hat{u} \in U_{ad}$ is an optimal control, mappings /3.17/, /3.18/ are

Frechet differentiable and $(\hat{p},\hat{r})$ is a generalized adjoint state at $\hat{u} \in U_{ad}$ then necessary conditions of optimality takes on the form :

/3.21/
$$\left(\langle A_u(\hat{u});\ u-u\rangle\ y_{\hat{u}}, \hat{p}\right)_{\mathcal{H}}$$
$$+\left(\langle \sigma_u(\hat{u});\ u-\hat{u}\rangle\ y_{\hat{u}}, \hat{r}\right)_Y + \varepsilon(\hat{u}, u-\hat{u})_U \geqslant 0 \quad , \quad \forall u \in U_{ad}$$

where $A_u(\hat{u})$, $\sigma_u(\hat{u})$ denotes Frechet derivatives of mappings /3.17/ /3.18/ taken at optimal point $\hat{u} \in U_{ad}$.

Example :

Let Ω be an open region in R^n with smooth boundary $\Gamma = \partial\Omega$

We introduce the following functional spaces :

(i) $V = H^1(\Omega)$, $H = L^2(\Omega)$

(ii) $W^1(0,T) = H^{2,1}(Q)$

where $Q = \Omega \times]0,T[$, $T > 0$

(iii) $Y = H^{1/2,1/4}(\Sigma)$

where $\Sigma = \Gamma \times]0,T[$

(iv) $D = H^1(\Omega;\Delta)$

where
$$H^1(\Omega;\Delta) = \{ y \in H^1(\Omega) \mid y \in L^2(\Omega) \}$$
$$S = H^{1/2}(\Gamma)$$

We define the set of admissible controls $U_{ad} \subset U = H^2(0,T)$ as the set of solutions of ordinary differential equation

$$\begin{cases} \dfrac{du}{dt} = -a_1 u + v_1 \\ \dfrac{dv_1}{dt} = -a_2 v_1 + v \end{cases}$$
$$u(0) = v_1(0) = 0$$

for all $v \in L^2(0,T)$ such that

$$0 \leqslant v(t) \leqslant 1 \qquad \text{a.e. in }]0,T[$$

where $a_1, a_2 > 0$ are given constants.

Let there be given real functions $F(.)$, $g(.)$ such that

(i) $F(.)\ ,\ g(.) \in C^2[0,1]$

(ii) $F(r) \geqslant \alpha > 0\ ,\quad \forall r \in [0,1]$

$g(r) \geqslant 0\ ,\quad \forall r \in [0,1]$

We introduce the state equation of the form :

$$\int_\Omega \frac{\partial y_u}{\partial t} z\, d\Omega + F(u(t)) \sum_{i=1}^{n} \int_\Omega \frac{\partial y_u}{\partial x_i} \frac{\partial z}{\partial x_i} d\Omega$$

$$+ g(u(t)) \int_\Gamma y_u z\, d\Gamma = \int_\Omega f z\, d\Omega + \int_\Gamma \varphi z\, d\Gamma$$

$$\forall z \in H^1(\Omega), \qquad \text{a.e in }]0,T[$$

$$y_u(x,0) = y_0(x)\ , \qquad x \in \Omega$$

It can be shown that for given

$\varphi \in H^{1/2,1/4}(\Sigma)$

$f \in L^2(Q)$

$y_0 \in H^1(\Omega)$

$u \in U_{ad}$

There exists the unique solution $y_u \in H^{2,1}(Q)$ of the above problem.

In this case operators /3.17/, /3.18/ have the form

$$A(u)y = F(u)\,\Delta y\ , \qquad y \in D = H^1(\Omega;\Delta)$$

$$\sigma(u)y = \frac{\partial y}{\partial n} + g(u)\, y$$

where $\frac{\partial}{\partial n}$ denotes normal derivative to the boundary $\Gamma = \partial\Omega$

We introduce the cos t functional of the form

$$J(u) = \frac{1}{2} \| y_u(T) - z_d \|^2_{H^1(\Omega)} + \frac{\varepsilon}{2} \| u \|^2_{H^2(0,T)}$$

where $\varepsilon \geqslant 0$ and $z_d \in H^1(\Omega)$ is a given element.

For any given $\varepsilon \geqslant 0$ there exists [9] an optimal control $\hat{u} \in U_{ad}$ which is characterized by the following inequality :

$$\int_Q \left[\frac{dF}{du}(\hat{u})(u-u)\right] \Delta y_{\hat{u}} \, \hat{p} \, dQ$$

$$+ \int_\Sigma \left[\frac{dF}{du}(\hat{u}) \frac{\partial y_{\hat{u}}}{\partial n} + \frac{dg}{du}(\hat{u}) \, y_{\hat{u}} \right] (u-\hat{u}) \hat{r} \, d\Sigma$$

$$+ \varepsilon(\hat{u}, u-\hat{u})_{H^2(0,T)} \geqslant 0 , \qquad \forall u \in U_{ad}$$

where $(\hat{p},\hat{r}) \in L^2(Q) \times (H^{1/2,1/4}(\Sigma))'$ is the unique solution [9] of the problem :

$$\int_Q \left(\frac{dw}{dt} + F(\hat{u}) \Delta w \right) \hat{p} \, dQ$$

$$+ \int_\Sigma \left(F(\hat{u}) \frac{\partial w}{\partial n} + g(\hat{u}) w \right) \hat{r} \, d\Sigma$$

$$= - \left(y_u(.\,,T) - z_d \, , \, w(.\,,T) \right)_{H^1(\Omega)} \qquad \forall w \in H^{2,1}(Q)$$

such that $w(.,\, 0) = 0$

where we use the same notation for the scalar product in $L^2(\Sigma)$ and the scalar product between $H^{1/2,1/4}(\Sigma)$ and $(H^{1/2,1/4}(\Sigma))'$.

References

[1] Aubin J.P. : Approximation of Elliptic Boundary Value Problems, Wiley, Interscience 1972

[2] Chavent G. : These, Paris 1971.

[3] Lions J.P. : Contrôle optimale de systèmes gouvernés par des équations aux derivées partielles, Dunod, Paris 1968.

[4] Lions J.P., Magenes E. : Problèmes aux limites non homogenes et applications, Vol. 1,2, Dunod, Paris 1968.

[5] Fleming W.H. : Optimal control of partially observable diffussion, SIAM J. Control 6, 1968.

[6] Ladyzhenskaja O.A. : Krajewyje zadachi matematicheskoj fizyki, Nauka, 1973 /in Russian/.

[7] Sokołowski J. : On parametric optimal control for a class of linear and quasilinear equations of parabolic type, Control and Cyrernetics 1, 1975.

[8] Sokołowski J. : On parametric otpimal control for weak solutions of abstract parabolic equations, Control and Cybernetics /to appear/.

[9] Sokołowski J. : Thesis, Warsaw 1975.

[10] Tartar L. : Problèmes de contrôle des coefficients dans des équations aux derivées partielles, in Lectures Notes /Control Theory/ Vol. 107, Springer Verlag 1975.

[11] Zollezi T. : Necessary conditions for optimal controls of elliptic or parabolic problems, SIAM J. Control, Vol. 10, No 4, 1972.

ON THE CONVERGENCE OF BALAKRISHNAN'S METHOD

T. Zolezzi
Centro di Studio
per la Matematica e la Fisica Teorica del C.N.R.
GENOVA

We show that the epsilon method of Balakrishnan gives, in a sense, a constructive proof of an existence theorem of Cesari for the following general problem of optimal control: minimize

$$\int_a^b f(t,x,u)dt$$

over the set of all pairs (u,x), u measurable, x absolutely continuous on $[a,b]$, such that

$$\dot{x} = g(t,x,u) \text{ a.e.},$$

(1) $(a,x(a),b,x(b)) \in T$,

(2) $(t,x(t)) \in G$

(3) $u(t) \in V(t,x(t))$ q.e.

Let

$$I_n(u,x) = \int_a^b [f(t,x,u) + n|\dot{x}-g(t,x,u)|]dt \text{ , } n = 1,2, \ldots$$

We consider the following problem P_n: minimize

$$I_n(u,x)$$

over the set of all pairs (u,x) as above such that (1),(2),(3) hold.

Under linearity and convexity assumptions (on g,f respectively) it is known that optimal solutions of P_n approximate optimal solutions of P_o. See Balakrishnan, SIAM J. Control 6 (1968), also in "Control Theory and the calculus of variations", UCLA 1968. In this work these results are extended to the above optimal control problem. In the simplest case, the (Q)-property of Cesari (as weakened by Berkowitz) has an important role in the proof. The type of convergence is the following: given any sequences $\varepsilon_n \longrightarrow 0$ and $\{u_n,x_n\}$ such that

$$I_n(u_n,x_n) \leq \inf I_n + \varepsilon_n$$

for some subsequence we have

$$x_n \longrightarrow x_o \text{ uniformly ,}$$
$$\dot{x}_n \longrightarrow \dot{x}_o \text{ in } L^1 \text{ ,}$$

$$\int_{a_n}^{b_n} f(f,x_n,u_n)dt \longrightarrow \int_{a_o}^{b_o} f(t,x_o,u_o)dt \text{ ,}$$

(u_o,x_o) an optimal pair for the original problem,

$$\inf I_n \longrightarrow \min I_o .$$

This approach avoids any use of generalized controls. Only an approximate minimization is required for every P_n.
When the state equations are linear at least in the control variables, and the cost meets suitable convexity conditions, then we get strong convergence

$$u_n \longrightarrow u_o \text{ in } L^p .$$

If the original problem has a piecewise smooth optimal solution, then by pointwise minimization of
$u \longrightarrow f(t,x(t),u) + n|x(t) - g(t,x(t),u)|$ over the control region $V(t,x(t))$, and then an approximate minimization of I_n with respect of x, we get a sequence (u_n, x_n) such that

$$u_n \longrightarrow u_o \text{ piecewise uniformly.}$$

Complete statements and the details will appear elsewhere.

- Minimum Variance Control of Discrete - Time Linear Stochastic System, using instantaneous Output Feedback.

P. BLANVILLAIN [+] G. FAVIER [++]

Abstract

This paper considers the problem of determining the linear output feedback control which minimizes a quadratic performance index, for a linear, discrete - time stochastic system. Both the finite and infinite - time versions of this problem are solved. For the finite terminal time case, the two-point boundary value problem that specifies the optimal feedback gain matrices is derived, and an algorithm is proposed for solving it. For the infinite-time case, two coupled non linear matrix equations must be solved to realize the optimal control ; an algorithm is also proposed for solving those equations. A numerical example is treated comparing this control policy to the optimal Kalman - type control policy.

+ , ++ : The authors are with the "Laboratoire d'Automatique et d'Analyse des Systèmes", Centre National de la Recherche Scientifique, 7 Avenue du Colonel Roche, 31400 TOULOUSE, France.

I - Introduction -

The object of this paper is to solve a linear regulator problem for which the control is constrained to be, at any instant, a linear transformation of the measured system output.

As it is well known, [1] , [2] , the optimal control policy, when the admissible control is not constrained, is provided by the celebrated separation theorem [3] . However, the optimal controller, composed in this case of a Kalman filter together with the optimal control gains for the associated deterministic problem, might be much too complex to be implemented in practice.

It is thus interesting to look for a simple control policy, such as the one proposed, if we agree with the following statements :

- We don't require state estimation, but we rather want a good regulation
- We are ready to trade optimality for engineering simplicity.

Several authors have taken this approach, in a deterministic as well as in a stochastic framework. Athans and Levine [4] and Kosut [5] are to be mentionned for the deterministic case, Mac Lane [6] , Axaster [7] , and more recently Ben Zurion Kurtaran [8] , are to be mentionned for the stochastic problem. However, whereas noise - free measurements were considered in [6] , [7] , [8] , we will consider in this paper, that the measures are corrupted by additive white noise. Also we will not directly penalize the control, but rather use a minimum variance criterion on the state of the plant.

The organization of the paper is as follows : in section II the finite terminal - time problem is solved and an algorithm is proposed for solving the associated two point boundary value problem ; in section III the infinite terminal-time problem is solved and an algorithm is proposed for solving the associated non linear matrix equations. In section IV a numerical example is presented.

II - Finite terminal - time problem -

Consider the following linear, discrete - time stochastic system :

$$x_{k+1} = A_k x_k + B_k u_k + W_k \qquad k = 0, 1, \ldots, N-1 \qquad (1)$$
$$y_k = C_k x_k + V_k$$

where :

x the state , is an n - dimensional vector

u the input , is an r - dimensional vector

y the output , is an m - dimensional vector

and where A_k, B_k, C_k are time - varying matrices of appropriate dimensions. The following assumptions are also made on the initial state of the plant and the white - noise sequences w_k and v_k :

$$x_0 \text{ is } N(0, X_0) \quad ; \quad E\{v_k\} = 0 \quad ; \quad E\{w_k\} = 0$$
$$E\{w_k w'_1\} = \Sigma_k \delta(k-1) \quad ; \quad E\{v_k v'_1\} = \Theta_k \delta(k-1) \qquad (2)$$
$$E\{w_k v'_1\} = 0 \quad ; \quad E\{w_k x'_0\} = 0 \quad ; \quad E\{v_k x'_0\} = 0$$

The control law to be implemented is the following linear transformation on the output y_k :

$$u_k = L_k y_k \qquad (3)$$

where L_k is an r x m time - varying gain matrix to be optimized.

The closed - loop system becomes accordingly :

$$x_{k+1} = \Gamma_k x_k + B_k L_k v_k + w_k$$
$$\Gamma_k = A_k + B_k L_k C_k \qquad k = 0, 1, \ldots, N-1 \tag{4}$$

The problem is then to find the sequence $\left\{L_k\right\}_{k=0}^{k=N-1}$

that minimizes the following cost criterion :

$$J = E\left\{x'_N S_{xN}\right\} + \sum_{k=1}^{N-1} E\left\{x'_k Q_k x_k\right\} \tag{5}$$

where $S > 0$ and $Q_k > 0 \qquad k = 1, \ldots, N-1$

Defining now $\Lambda_k = E\left\{x_k x'_k\right\}$, the second moment of x_k, the cost can then be rewritten :

$$J = \mathrm{tr}\ (S\Lambda_N) + \sum_{k=1}^{N-1} \mathrm{tr}\ (Q_k \Lambda_k) \tag{6}$$

Now from (4), Λ_k is known to satisfy :

$$\Lambda_{k+1} = \Gamma_k \Lambda_k \Gamma'_k + B_k L_k \Theta_k L'_k B'_k + \Sigma_k, \Lambda_0 = X_0 \tag{7}$$

so that we can formulate our problem as follows :

<u>Problem statement</u> : Given the plant parameters A_k, B_k, C_k, Σ_k, Θ_k, X_0, and the weighting matrices S and Q_k, find the time-varying gain matrix L^*_k such that :

$$J\left[L^*_k\right] \leqslant J\left[L_k\right] \tag{8}$$

for all real r x m time - varying matrices L_k, where J is given by (6) subject to the dynamic constraint (7) on Λ_k.

This constrained dynamic optimization problem may be solved via the matrix minimum principle [9]. The results are given in the following theorem :

Theorem 1 - (Necessary conditions) - Any matrix of time-varying gains L^*_k which satisfies (8) for the cost functional (6) subject to the dynamic constraint equation (7), also satisfies the following equations :

$$L^*_k = -(B'_k \Pi^*_{k+1} B_k)^{-1} B'_k \Pi^*_{k+1} A_k \Lambda^*_k C'_k (C_k \Lambda^*_k C'_k + \Theta_k)^{-1} \tag{9}$$

$$\Lambda^*_{k+1} = \Gamma^*_k \Lambda^*_k \Gamma'^*_k + B_k L^*_k \Theta_k L'^*_k B'_k + \Sigma_k \quad , \Lambda^*_0 = X_0 \tag{10}$$

$$\Pi^*_k = \Gamma'^*_k \Pi^*_{k+1} \Gamma^*_k + Q_k \quad , \Pi^*_N = S \tag{11}$$

$$\Gamma^*_k = A_k + B_k L^*_k C_k \qquad k = 0, 1, \ldots, N-1. \tag{12}$$

Proof : The derivation of this non - linear two - point boundary value problem is given in appendix A.

It is interesting to note the particular structure of the optimal gain L^*_k : namely the first part $(B'_k \Pi^*_{k+1} B_k)^{-1} B'_k \Pi^*_{k+1}$ has a typical control - type form, whereas the second part $A_k \Lambda^*_k C'_k (C_k \Lambda^*_k C'_k + \Theta_k)^{-1}$ has a typical filtering-type form.

Also, since the conditions are only necessary, it may exist a sequence $\left\{ \hat{L}_k \right\}_{k=0}^{N-1}$ satisfying (9) - (12) which is not globally optimal.

Finally, existence and uniqueness of solutions of (9) - (10) are still largely open questions.

A feasible algorithm is now proposed, which has proven to have reasonable convergence properties.

<u>Notation</u> : L^n_k

n = number of iteration

k = index of time

1 - Make an initial guess on $\left\{L^{\circ}_k\right\}_{k=0}^{N-1}$ which determines also $\left\{\Gamma^{\circ}_k\right\}_{k=0}^{N-1}$.

2 - Compute forward in time, for $n = 0$ and $k = 0, 1, \ldots, N-2$

$$\Lambda^{n+1}_{k+1} = \Gamma^n_k \Lambda^{n+1}_k \Gamma'^n_k + B_k L^n_k \Theta_k L'^n_k B'_k + \Sigma_k \quad , \quad \Lambda^{n+1}_0 = X_0 \tag{13}$$

giving the sequence $\left\{\Lambda^1_k\right\}_{k=0}^{N-1}$

3 - Compute backwards in time, for $n = 0$ and $k = N-1, \ldots, 1$

$$\Pi^{n+1}_k = \Gamma'^n_k \Pi^{n+1}_{k+1} \Gamma^n_k + Q_k \quad , \quad \Pi^{n+1}_N = S \tag{14}$$

giving the sequence $\left\{\Pi^1_k\right\}_{k=1}^{N}$

4 - Compute for $n = 0$, $k = 0, \ldots, N-1$

$$L^{n+1}_k = -(B'_k \Pi^{n+1}_{k+1} B_k)^{-1} B'_k \Pi^{n+1}_{k+1} A_k \Lambda^{n+1}_k C'_k (C_k \Lambda^{n+1}_k C'_k + \Theta_k)^{-1} \tag{15}$$

$$\Gamma^{n+1}_k = A_k + B_k L^{n+1}_k C_k$$

giving the sequences $\left\{L^1_k\right\}_{k=0}^{N-1}$ and $\left\{\Gamma^1_k\right\}_{k=0}^{N-1}$

5 - Iterate steps 2, 3, 4 for $n = 1, \ldots$ giving

$\left\{\Lambda^n_k\right\}_{k=0}^{N-1}$, $\left\{\Pi^n_k\right\}_{k=1}^{N}$, $\left\{L^n_k\right\}_{k=0}^{N-1}$, $\left\{\Gamma^n_k\right\}_{k=0}^{N-1}$ until

the desired degree of convergence is reached.

III - Infinite terminal - time problem -

Consider now the following linear, discrete - time, time - invariant, stochastic system :

$$x_{k+1} = A x_k + B u_k + w_k \qquad (16)$$
$$y_k = C x_k + v_k$$

where x, y, u are as in section II, and A, B, C are constant matrices of appropriate dimension. We will also make the following assumptions on the initial state of the plant and the stationary white - noise sequences w_k and v_k :

$$x_0 \text{ is } N(0, X_0) \; ; \; E\{v_k\} = 0 \; ; \; E\{w_k\} = 0$$
$$E\{w_k w'_\ell\} = \Sigma \delta(k-\ell) \; ; \; E\{v_k v'_\ell\} = \Theta \delta(k-\ell) \qquad (17)$$
$$E\{w_k v'_\ell\} = 0 \; ; \; E\{w_k x'_0\} = 0 \; ; \; E\{v_k x'_0\} = 0$$

The control law to be implemented is the following linear transformation on the output y_k :

$$u_k = L y_k \qquad (18)$$

where L is an r x m constant gain matrix to be optimized.

The closed - loop system becomes accordingly :

$$x_{k+1} = \Gamma x_k + B L v_k + w_k \qquad (19)$$
$$\Gamma = A + B L C$$

The following cost is to be minimized :

$$J = \lim_{N \to \infty} \frac{1}{N} \sum_{k=0}^{N} E\left\{ x'_k \, Q \, x_k \right\} \tag{20}$$

or equivalently, assuming the system has reached steady - state :

$$J = E\left\{ x'_k \, Q \, x_k \right\} \tag{21}$$

Important remark - Note that we are tacitly assuming that the initial system (16) is output stabilizable (i. e. : there exists a constant r x m matrix L_0 such that $A + B L_0 C$ is stable). If it was not so, our problem would be meaningless since J would become infinite.

Defining now $\Lambda = E\left\{ x'_k \, x_k \right\}$ the steady - state covariance of the system state, J can simply be rewritten as :

$$J = \mathrm{tr}\ (\Lambda\ Q) \tag{22}$$

Now from (19), Λ is known to satisfy :

$$\Lambda = \Gamma \Lambda \Gamma' + B L \Theta L' B' + \Sigma \tag{23}$$

so that we can formulate our problem as follows :

Problem statement : Given the output stabilizable plant (A, B, C), the noise statistics Σ and Θ, and the weighting matrix Q, find the constant gain matrix L^* such that :

$$J\ (L^*) \leq J\ (L) \tag{24}$$

for all real r x m constant matrices L , where J is given (22).

This parameter optimization problem may be solved using Lagrange multipliers technique. The results are given in the following theorem :

Theorem 2 - (Necessary conditions) - Any constant gain matrix L which satisfies (24) for the cost criterion (22), also satisfies the following equations :

$$L^* = -(B'\Pi^* B)^{-1} B'\Pi^* A \Lambda^* C' (C\Lambda^* C' + \Theta)^{-1} \qquad (25)$$

$$\Lambda^* = \Gamma^* \Lambda^* \Gamma'^* + B L^* \Theta L'^* B' + \Sigma \qquad (26)$$

$$\Pi^* = \Gamma'^* \Pi^* \Gamma^* + Q \qquad (27)$$

$$\Gamma^* = A + B L^* C \qquad (28)$$

Proof : The simple proof of this theorem is omitted. Π is an n x n Lagrange multiplier matrix.

Again note the particular structure of L^*. Also, since the conditions are only necessary, il may exist a matrix $\hat{L}$ satisfying (25) - (28) which is not globally optimal. Finally, note that plugging (25) and (28) in (26) and (27) reduces the problem to solving two coupled non linear equations in Λ^* and Π^*.

We will now propose a feasible algorithm, for solving (25) - (28) which has proven to have reasonable convergence properties.

1 - Choose a stabilizing $L°$, for $\Gamma° = A + BL°C$

2 - Solve the linear algebraic matrix equation :

$$\Lambda^{n+1} = \Gamma^n \Lambda^{n+1} \Gamma'^n + B L^n \Theta L'^n B' + \Sigma$$

for n = 1 giving Λ^1.

3 - Solve the linear algebraic matrix equation :

$$\Pi^{n+1} = \Gamma'^{\,n}\, \Pi^{n+1}\, \Gamma^{n} + Q$$

for n = 1 giving Π^{1}.

4 - Compute :

$$L^{n+1} = -\left(B'\Pi^{n+1} B\right)^{-1} B'\Pi^{n+1} A\Lambda^{n+1} C' \left(C\Lambda^{n+1} C' + \Theta\right)^{-1}$$

$$\Gamma^{n+1} = A + B L^{n+1} C$$

5 - Iterate steps 2, 3, 4 for n = 2, ... giving sequences $\{\Lambda^n\}$, $\{\Pi^n\}$, $\{L^n\}$, $\{\Gamma^n\}$ until the desired degree of convergence is reached.

Remark - Consider a time - invariant system and weighting matrix over a finite - horizon $[0, N]$, but allow as in section II time varying gains. It does not follow that, as $N \to \infty$, the time - varying gains of (9) - (12), go to the steady - state solution (25) - (28). From what we have previously seen, it is clear that it will be so if and only if the system is output stabilizable. Now, if this condition is not satisfied, section II gives a method of finding time-varying gains to stabilize the given system.

IV - A Numerical Example -

The results of sections II and III are applied to a fifth order academic example, simulated on an IBM 370/168 .

The actual values of A, B, C, Σ , Θ , Q, S and X_0 are :

$$A = \begin{bmatrix} 0.9 & 0. & -0.66 & 0. & 0.071 \\ 0. & 0.82 & -0.078 & 0. & 0.016 \\ 0. & 0. & 0.74 & 0. & 0.027 \\ 0. & 0. & 0. & 0.67 & 0. \\ 0. & 0. & 0. & 0. & 0.6 \end{bmatrix}$$

$$B = \begin{bmatrix} 0.62 & 0. \\ 0.086 & 0.27 \\ 0.17 & 0. \\ 0. & 0.5 \\ 0.4 & 0. \end{bmatrix}$$

$$C = \begin{bmatrix} 1.5 & 0. & -3. & 0. & 2.1 \\ 0. & 1. & -1. & 2. & 0.2 \end{bmatrix}$$

$$\Sigma = \begin{bmatrix} 0.11 & 0.0027 & -0.027 & 0. & 0.0024 \\ 0.0027 & 0.083 & -0.0032 & 0. & 0.0005 \\ -0.027 & -0.0032 & 0.076 & 0. & -0.0008 \\ 0. & 0. & 0. & 0.071 & 0. \\ 0.0024 & 0.0005 & -0.0008 & 0. & 0.067 \end{bmatrix}$$

$$\Theta = \begin{bmatrix} 20. & 0. \\ 0. & 10. \end{bmatrix}$$

$$Q = 0.1 \quad I\,5 \quad , \qquad S = I\,5 \quad , \qquad X_0 = I\,5$$

1 - <u>Finite - time case (T = 10 s) : Time Varying Gains</u> :

We chose $\left\{ L_k^{\circ} \right\}_{k=0}^{N-1} = 0$

Convergence of the algorithm is obtained when the whole sequence $\left\{ L_k^{n} \right\}_{k=0}^{N-1}$ reaches some stationary state.

As it was noted in section II, the optimal gain L_k has a typical behaviour as described by Fig. 1.

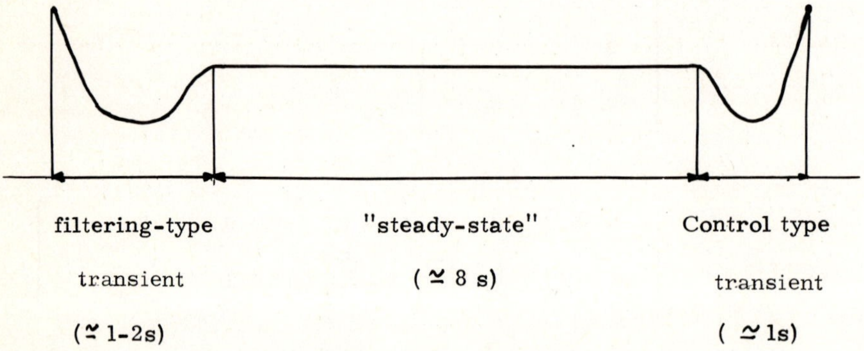

Figure 1

Fig. 2 (a. b. c. d) shows the different components $L^{*}ij$ (i = 1, 2 ; j = 1, 2) of $\left\{ L^{*}_{k} \right\}$.

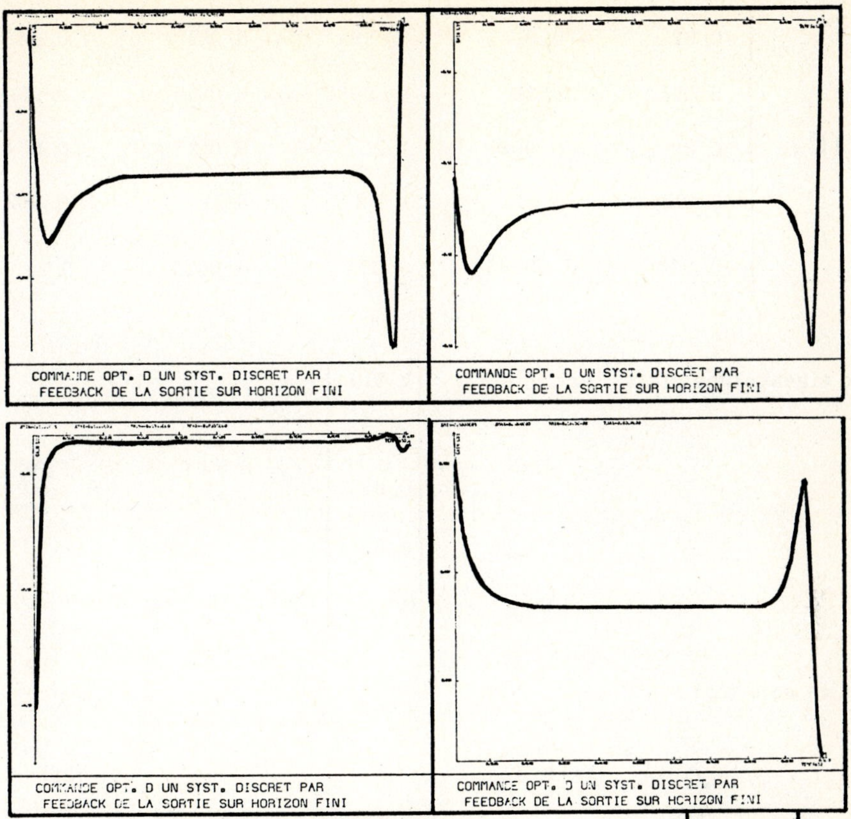

Figure 2

2 - Infinite - time case : Constant gains :

We chose for the initialisation : $L^\circ = 0$ that stabilizes

$\Gamma^\circ = A + B L^\circ C.$

The results are :

$$L^* = \begin{bmatrix} -0.149 & -0.048 \\ 0.017 & -0.037 \end{bmatrix}$$

$$\Lambda^* = \begin{bmatrix} 1.36 & 0.068 & -0.35 & 0.0066 & -0.18 \\ 0.068 & 0.25 & -0.025 & -0.0075 & -0.0057 \\ -0.035 & -0.025 & 0.33 & -0.015 & 0.19 \\ 0.0066 & -0.0075 & -0.015 & 0.13 & -0.019 \\ -0.18 & -0.0057 & 0.19 & -0.019 & 0.30 \end{bmatrix}$$

$$\Pi^* = \begin{bmatrix} 0.32 & -0.017 & -0.39 & -0.012 & -0.048 \\ -0.017 & 0.29 & -0.0086 & -0.015 & 0.00086 \\ -0.39 & -0.0086 & 1.25 & 0.012 & 0.074 \\ -0.012 & -0.015 & 0.012 & 0.17 & 0.0027 \\ -0.048 & 0.00086 & 0.074 & 0.0027 & 0.14 \end{bmatrix}$$

The eigenvalues of Γ^* are :

$$\begin{bmatrix} 0.916 \\ 0.813 \\ 0.687 \\ 0.630 \\ 0.451 \end{bmatrix}$$

and we note that :

$$\sum_{i=1}^{5} \lambda^*_{ii} = 2.37 \qquad (29)$$

Let's compare now these results with the L Q G design :

($Q_{LQG} = 0.5\ I5$, $R_{LQG} = 0.\ I2$)

The separation theorem gives us :

- Filtering Part : Optimal gain G^*

$$\begin{cases} G = -A\ \tilde{\Lambda}^*_{LQG}\ C'\ (C\ \tilde{\Lambda}^*_{LQG}\ C' + \Theta)^{-1} \\ \tilde{\Lambda}^*_{LQG} = (A + G^* C)\ \tilde{\Lambda}^*_{LQG}\ A' + \Sigma \end{cases}$$

- Control Part : Optimal gain K^*

$$\begin{cases} K^* = -(B'\Pi^* B + R_{LQG})^{-1} B'\Pi^* A \\ \Pi^* = A'\Pi^*(A+BK^*) + Q_{LQG} \end{cases}$$

where $\tilde{\Lambda}^*_{LQG}$ is the steady - state error covariance matrix.

The obtained results were :

$$\tilde{\Lambda}^*_{LQG} = \begin{bmatrix} 0.94 & 0.05 & -0.16 & -0.008 & -0.01 \\ 0.05 & 0.25 & -0.016 & -0.007 & 0.0008 \\ -0.16 & -0.016 & 0.15 & 0.0024 & 0.0056 \\ -0.008 & -0.007 & 0.0024 & 0.12 & -0.00005 \\ -0.01 & 0.0008 & 0.0056 & -0.00005 & 0.10 \end{bmatrix}$$

Moreover the steady-state value of $\hat{\Lambda}^*_{LQG} = E\left[\hat{x}\,\hat{x}^t\right]$ is given by :

$$\hat{\Lambda}^*_{LQG} = (A+BK^*)\,\hat{\Lambda}^*_{LQG}\,(A+BK^*)' + G^*(C\tilde{\Lambda}^*_{LQG}\,C'+\Theta)\,G'^*$$

It was obtained :

$$\hat{\Lambda}^*_{LQG} = \begin{bmatrix} 0.37 & 0.013 & -0.21 & 0.008 & -0.18 \\ 0.013 & 0.007 & -0.0003 & 0.002 & 0.007 \\ -0.21 & -0.0003 & 0.17 & -0.007 & 0.18 \\ 0.008 & 0.002 & -0.007 & 0.003 & -0.006 \\ -0.18 & 0.007 & 0.18 & -0.006 & 0.2 \end{bmatrix}$$

From where, for the LQG design, we have :

$$\Lambda^*_{LQG} = \hat{\Lambda}^*_{LQG} + \tilde{\Lambda}^*_{LQG}$$

$$\Lambda^*_{LQG} = \begin{bmatrix} 1.31 & 0.06 & -0.38 & -0.0001 & -0.19 \\ 0.062 & 0.25 & -0.017 & -0.005 & 0.008 \\ -0.38 & -0.017 & 0.32 & -0.004 & 0.18 \\ -0.0001 & -0.005 & -0.004 & 0.13 & -0.006 \\ -0.19 & 0.008 & 0.18 & -0.006 & 0.31 \end{bmatrix}$$

and $$\sum_{i=1}^{5} (\lambda^*_{ii})_{LQG} = 2.32 \qquad (30)$$

Moreover, the eigenvalues of $(A+G^*C)$ are :

$$\begin{bmatrix} 0.808 \\ 0.725 + 0.09\,j \\ 0.725 - 0.09\,j \\ 0.635 \\ 0.587 \end{bmatrix}$$

and the eigenvalues of $(A + BK^*)$ are :

$$\begin{bmatrix} 0.923 \\ 0.781 \\ 0.589 \\ 0.238 \\ 0.169 \end{bmatrix}$$

By comparaison of the traces (29) and (30) we show that the performances of the linear output feedback control may be quite close to the optimum.

It is interesting to note that, as expected, the values for the infinite-time constant gain problem correspond to the "steady-state" values of the gains in the finite-time case.

V - Conclusion -

The control laws derived in this paper are to be thought as sub-optimal control laws for the linear regulator problem. Their interest relies on their easiness of implementation. An academic example has shown that their performances may be quite close to the optimum. Of course, this is not always the case and dynamic compensation may have to be used.

In case of a time - invariant, non output stabilizable system, section II provides a way of finding the time-varying gains that stabilizes the system.

Appendix A - Demonstration of theorem 1 -

From (6) J is given by :

$$J = \operatorname{tr}\,(S \Lambda_N) + \sum_{k=1}^{N-1} \operatorname{tr}\,(Q_k \Lambda_k) \qquad (6)\ (6)$$

Also from (7) Λ_{k+1} is given by :

$$\Lambda_{k+1} - \Lambda_k = \Gamma_k \Lambda_k \Gamma'_k - \Lambda_k + B_k L_k \Theta_k L'_k B'_k + \Sigma_k \qquad (7\text{bis})$$

Consider now the scalar hamiltonian :

$$H(\Lambda_k, \Pi_{k+1}, L_k) = \operatorname{tr}\,(\Lambda_k Q_k) + \operatorname{tr}\left\{(\Gamma_k \Lambda_k \Gamma'_k - \Lambda_k + B_k L_k \Theta_k L'_k B'_k + \Sigma_k)\Pi_{k+1}\right\}$$

From [9] we then obtain :

(i) Canonical equations :

$$\Lambda^*_{k+1} - \Lambda^*_k = \left.\frac{\partial H}{\partial \Pi_{k+1}}\right|_* \quad \text{which gives eq.} \quad (10)$$

$$\Pi^*_{k+1} - \Pi^*_k = -\left.\frac{\partial H}{\partial \Lambda_k}\right|_* \quad \text{which gives eq.} \quad (11)$$

(i i) Boundary conditions :

At $k = 0$ $\quad \Lambda^*_0 = X_0$

At $k = N$ $\quad \Pi^*_N = S$

(i i i) Minimization of the halmitonian :

Since the L'_k s are not constrained :

$$\left.\frac{\partial H}{\partial L_k}\right|_* = 0 \text{ which yields eq. (9)}$$

References

[1] P. DORATO, A. H. LEVIS "Optimal Linear Regulators : the discrete-time case" - IEEE Trans. Aut. Contr. 1971 pp. 613 - 620.

[2] E. TSE "The optimal Control of Stochastic Linear Systems" - IEEE Trans. Aut. Contr. 1971 pp. 776 - 785.

[3] W. M. WONHAM "On the separation theorem of Stochastic Control" SIAM J. Control, 1968 pp. 312 - 326

[4] W. S. LEVINE, T. L. JOHNSON, M. ATHANS "Optimal Limited State Variable Feedback Controllers for Linear Systems" - IEEE Trans. Aut. Contr. 1971 pp. 785 - 793.

[5] R. L. KOSUT "Suboptimal Control of Linear Time - invariant Systems subject to control Structure constraints" - IEEE Trans. Aut. Contr. 1970 pp. 557 - 563.

[6] P. J. Mc LANE "Linear Optimal Stochastic Control using instantaneous output feedback" Int. J. Control 1971 pp. 383 - 396.

[7] S. AXASTER "Suboptimal time-variable feedback control of linear dynamic systems with random inputs" - Int. J. Control, 1966 - pp. 549 - 566.

[8] B. Z. KURTARAN and M. SIDAR " Optimal Instantaneous output feedback controllers for linear stochastic Systems" Int. J. Contr., 1974, pp. 797-816.

[9] M. ATHANS " The Matrix Minimum Principle" - Information and Control, 1967, pp. 592 - 606.

FINDING A FEASIBLE CONTROL FOR REAL PROCESS UNDER UNCERTAINTY

M. Brdyś

Technical University of Warsaw
Institute of Automatic Control
ul. Nowowiejska 15/19
00-665 Warsaw, Poland

Abstract

In the paper the problem of finding a feasible control for real process is discussed. It is assumed that the available mathematical model of the process differs from its real mathematical description but there is some consistence between them. Under such assumption the feasible control for real process is found by using the process mathematical model only. To show that this approach makes sense the existence theorems are given.

From the point of view of the control the problem is considered in two aspects. First, we want only to find the feasible control for the process; a numerical algorithm with convergence analysis is given. Second, we want to generate on-line a feasible control for the real process when its constraints are violated. In this case numerical algorithms with convergence analysis are also given.
Finally, numerical example is presented.

1. Introduction

One of the important aspects of the practice of control of real processes is connected with the problem of finding control satisfying the real process constraints /feasible control/.

In the practical situation the mathematical model of the process differs from its exact but unknown mathematical description for several reasons, e.g. simplifying assumptions made when the model is being built, disturbances which cannot be exactly foreseen, etc. It means that the control for which the model constraints are fulfilled/feasible control for the model/ is not feasible for the real process in general. In many practical cases we are not allowed to apply non-feasible control to the real process. The possibilities of adaptation of the model cannot therefore be considered at this stage.

It means that for generating the feasible control for the real process its mathematical model can only be used.

In the paper this problem of finding feasible control for real process is discussed and numerical algorithms are proposed.

2. Problem formulation and concept of the solution

Let us suppose that the mathematical model of the process has the following form:

process equation -

$$F : C \times \mathcal{B} \longrightarrow Y, \qquad (1)$$

$$F(c,\beta) = y,$$

constraints -

$$P : C \times Y \times \mathcal{A} \quad Z, \qquad (2)$$

$$P(c,y,\alpha) \in S \subset Z,$$

$$c \in G \subset C, \quad y = F(c,\beta).$$

Variables α, β which are arguments of mappings F,P represent the unknown parameters or disturbances in the process, the sets of their values being estimated by $\mathcal{A}, \mathcal{B}$.

Variables c,y are respectively control and output of the process.

Definition 1

We say that control $\bar{c}$ is $(\bar{\alpha},\bar{\beta})$ - feasible control for the model if and only if the following relations are fulfilled

$$P(\bar{c}, \bar{y}, \bar{\alpha}) \in S, \qquad (3)$$

$$\bar{c} \in G, \quad \bar{y} = F(\bar{c}, \bar{\beta}).$$

and $\bar{\alpha}, \bar{\beta}$ are fixed in sets $\mathcal{A}$ and $\mathcal{B}$.

This means that the control $\bar{c}$ satisfies constraints in the model when variables α, β take the values $\bar{\alpha}, \bar{\beta}$ in sets $\mathcal{A}, \mathcal{B}$.

As it was mentioned above the real mathematical description of the process is different from the process model and it is assumed that this real process description has the following yet unknown form:

process equation -

$$F_* : C \times \mathcal{B}_* \longrightarrow Y, \qquad (4)$$

$$F(c,\beta_*) = y,$$

constraints -

$$P_* : C \times Y \times \mathcal{A}_* \longrightarrow Z ,$$

$$P_*(c, y, \alpha_*) \in S \subset Z , \tag{5}$$

$$c \in G \subset C , \quad y = F_*(c, \beta_*)$$

In the deterministic case the sets $\mathcal{A}_*, \mathcal{B}_*$ are single-point ones and mappings P_* , F_* do not depend on variables α_*, β_* .
Such situation takes place for example when the values of some parameters in the process are not given but it is known that they belong to sets $\mathcal{A}$ and $\mathcal{B}$.

In the presence of disturbances, α_* and β_* are stochastic processes and sets $\mathcal{A}_*$, $\mathcal{B}_*$ are the realization sets of them. In this case sets $\mathcal{A}$, $\mathcal{B}$ in the model approximate $\mathcal{A}_*, \mathcal{B}_*$.

Definition 2

We say that control $\bar{c}$ is a feasible control for the process if and only if the following relations are fulfilled:

$$P_*(\bar{c} , \bar{y}, \bar{\alpha}_*) \in S , \tag{6}$$
$$\bar{c} \in G , \bar{y} = F_*(\bar{c} , \bar{\beta}_*) ,$$

where $\bar{\alpha}_*$, $\bar{\beta}_*$ are the actual values of variables α_*, β_* in the process. This means that control $\bar{c}$ satisfies constraints in the process when variables α_* , β_* take values $\bar{\alpha}_*, \bar{\beta}_*$.
Because we do not know the actual values of variables α_*, β_* in the process and only the model is given, some connection between the mathematical model and real mathematical description of the process have to be assumed.

Definition 3

We say that the real mathematical description of the process is consistent with the process mathematical model if and only if the following condition is satisfied:

$$\forall(\beta_* \in \mathcal{B}_*) \, \exists(\beta \in \mathcal{B}) \, \forall(c \in G) \quad F_*(c, \beta_*) = F(c, \beta)$$

and

$$\forall(\alpha_* \in \mathcal{A}_*) \, \exists(\alpha \in \mathcal{A}) \, \forall(c \in G) \, \forall(y \in F_*(G \times \mathcal{B}_*)) \; P_*(c, y, \alpha_*) = P(c, y, \alpha) \tag{C}$$

It is very easy to see that the control which is (α,β)-feasible for the model for every $(\alpha,\beta)\in\mathcal{A}\times\mathcal{B}$ is also feasible for the process for every $(\alpha_*,\beta_*)\in\mathcal{A}_*\times\mathcal{B}_*$ if the consistency condition (C) is satisfied. Thus feasible control for the process could be found as a control which is (α,β) -feasible for the model, independently on what are the values of variables α, β (we will denote such control as $(\mathcal{A},\mathcal{B})$- feasible control). To show that this approach makes sense the existence theorem for such control under reasonable assumptions should be given. That is the subject of our considerations in the next section.

3. An existence theory

We start with the definition of convexlike and concavelike mappings.

Definition 4

We say that the mapping $f: X \times Y \to Z$ is convexlike (concavelike) on set X (Y) with respect to set S if and only if the following relation is fulfilled:

$$\forall(x_1, x_2 \in X)\forall(t \in [0,1])\exists(x \in X)\forall(y \in Y)\ tf(x_1,y)+(1-t)f(x_2,y)-f(x,y) \in S$$

$$(\forall(y_1,y_2 \in Y)\forall(t \in [0,1]\ \exists(y \in Y)\forall(x \in X)\ f(x,y)-tf(x,y_1)-(1-t)f(x,y_2) \in S)$$

The definition of a convexlike (concavelike) function (that is the case when $Z = R$ and $S = \{z \in Z : z \geqslant 0\}$ is given in [6].

Proposition 1

If the mapping f is convexlike on X and concavelike on Y with respect to S, Z is the linear space, then the function μ defined as follows:

$$\mu : X \times Y \to R$$

$$\mu(x, y) \triangleq z'(f(x,y)),$$

where z' is linear functional on Z such that

$$\forall(z \in S)\ z'(z) \geqslant 0;$$

is convexlike on X and concavelike on Y.

The proof of this proposition instantly results from the Definition 4 and the property of z'.

Let us define mapping $\tilde{P}$ as follows:

$$\tilde{P} : C \times \Gamma \to Z,$$

$$\tilde{P}(c,\gamma) \triangleq P(c,F(c,\beta),\alpha), \tag{7}$$

where

$$\Gamma = \mathcal{A} \times \mathcal{B}, \quad \gamma = (\alpha,\beta).$$

Now we can formulate the basic existence theorem.

Theorem 1

Let us assume that G, Γ are compact topological Hausdorff spaces,

Z is Banach space, S is a closed convex cone and

1) mapping $\tilde{P}$ is convexlike on and concavelike on G with respect to S,

2) $\forall (z^* \in \tilde{S}) \; \forall (\gamma \in \Gamma)$ function $G \ni c \longrightarrow z^*(\tilde{P}(c,\gamma)) \in R$ is upper semicontinuous,
where $\tilde{S} = S^* \cap K(\Theta ; 1)$,
$S^* \subset Z^*$ is a cone conjugate to S, where Z^* is a space conjugate to Z,
by $K(\Theta ; 1)$ we denote the closed unit ball in Z^*.

3) $\forall (z_1^*, z_2^* \in \tilde{S}) \forall (c \in G) \forall (t \in [0,1]) \inf_{\gamma \in \Gamma} [t \; z_1^*(\tilde{P}(c,\gamma)) + (1-t) z_2^*(\tilde{P}(c,\gamma))] =$
$= t \inf_{\gamma \in \Gamma} z_1^*(\tilde{P}(c,\gamma)) + (1-t) \inf_{\gamma \in \Gamma} z_2^*(\tilde{P}(c,\gamma))$,

4) $\forall (\gamma \in \Gamma) \; \exists (c \in G) \; \tilde{P}(c,\gamma) \in S$,

5) $\forall (z^* \in \tilde{S}) \forall (c \in G)$ function $\Gamma \ni \gamma \longrightarrow z^*(\tilde{P}(c,\gamma)) \in R$ is lower semicontinuous.

Then there exists the $(\mathcal{A}, \mathcal{B})$ - feasible control.

Proof

To simplify the notation we denote:

$$\inf_{z^* \in \tilde{S}} = \inf_{z^*} , \quad \inf_{\gamma \in \Gamma} = \inf_{\gamma} , \quad \sup_{c \in G} = \sup_{c}$$

On the base of the assumption 4) and the definition of S^* we have

$$\forall (\gamma \in \Gamma) \; \exists (c \in G) \; \forall (z^* \in S^*) \; z^*(\tilde{P}(c,\gamma)) \geqslant 0 ,$$

or (what is equivalent)

$$\inf_{\gamma} \; \sup_{c} \; \inf_{z^*} \; z^*(\tilde{P}(c,\gamma)) \geqslant 0 . \tag{8}$$

Because for any function $f: X \times Y \longrightarrow R$ there is ([2])

$$\sup_{x \in X} \; \inf_{y \in Y} \; f(x,y) \leqslant \inf_{y \in Y} \; \sup_{x \in X} \; f(x,y) ,$$

hence from (8) it follows that

$$\inf_{z^*} \; \inf_{\gamma} \; \sup_{c} \; z^*(\tilde{P}(c,\gamma)) \geqslant 0 . \tag{9}$$

By Proposition 1 and assumption 1) the function

$\mu_1 : G \times \Gamma \to R$, $\mu_1(c,\gamma) \triangleq z^*(\tilde{P}(c,\gamma))$, where z^* is any element from $\tilde{S}$, is convexlike on Γ and concavelike on G.

Besides by assumption 5) for every $c \in G$ the function $\mu_1(c, \cdot)$ is lower semicontinuous on Γ.

Thus we can apply Kneser-Fan Theorem (see [6] , [12]), by virtue of which the following equality is fulfilled:

$$\sup_c \inf_\gamma z^*(\tilde{P}(c,\gamma)) = \inf_\gamma \sup_c z^*(\tilde{P}(c,\gamma)). \tag{10}$$

Combining relations (9) and (10) we have

$$\inf_{z^*} \sup_c \inf_\gamma z^*(\tilde{P}(c,\gamma)) \geqslant 0 \tag{11}$$

Let us consider the function μ_2, where

$$\mu_2 : G \times \tilde{S} \to R , \tag{12}$$

$$\mu_2(c, z^*) \triangleq \inf_\gamma z^*(\tilde{P}(c,\gamma)) .$$

By assumption 2) the function $\mu_2(\cdot , z^*)$ is upper semicontinuous on G , for every $z^* \in \tilde{S}$.

Taking into account assumption 1) and the property of $z^* \in \tilde{S}$ we conclude what follows:

$$\forall(c_1,c_2 \in G)\forall(t \in [0,1])\exists(c \in G)\forall(z^* \in \tilde{S}) \inf_\gamma z^*(\tilde{P}(c,\gamma)) \geqslant t \inf_\gamma z^*(\tilde{P}(c_1,\gamma)) + (1-t)\inf_\gamma z^*(\tilde{P}(c_2,\gamma)). \tag{13}$$

It means that the function μ_2 is concavelike on G.

Further from assumption 3) it directly follows that the function μ_2 is convexlike on $\tilde{S}$.

Now we can again apply Kneser-Fan Theorem with respect to the function μ_2 and conclude what follows:

$$\inf_{z^*} \sup_c \mu_2(c,z^*) = \sup_c \inf_{z^*} \mu_2(c,z^*). \tag{14}$$

Combining relations (11) and (14) we have

$$\sup_c \inf_\gamma \inf_{z^*} z^*(\tilde{P}(c,\gamma)) \geqslant 0 . \tag{15}$$

Since the function

$$G \ni c \quad \inf_\gamma \inf_{z^*} z^*(\tilde{P}(c,\gamma)) \in R ,$$

is upper semicontinuous (see assumption 2)) and G is a compact set, the supremum on the left side of inequality (15) is attained. It means that

$$\exists (c \in G) \forall (\gamma \in \Gamma) \forall (z^* \in \tilde{S}) \; z^*(\tilde{P}(c,\gamma)) \geqslant 0 ,$$

and this is, by the property of $\tilde{S}$, equivalent to

$$\exists (c \in G) \forall (\gamma \in \Gamma) \; \tilde{P}(c,\gamma) \in S , \tag{16}$$

and the proof is completed.

Remark 1

Assumption 4) seems to be a natural one. It is strictly connected with the quality of the mathematical model. A mathematical model is better than another one when its sets $\mathcal{A}, \mathcal{B}$ approximate more exactly the unknown constant values of the parameters in the process (when the deterministic case takes place) or the sets $\mathcal{A}_*, \mathcal{B}_*$ (in the stochastic case).

Remark 2

Fulfilment of assumption 3) mainly depends on the character of the mapping $\tilde{P}(c, \cdot)$, where $c \in G$. We will give below the example of such situation where this assumption is fulfilled.

Example 1

Assume that $Z = R^n$, $S = \{z \in R^n : z \leqslant 0\}$,

$\forall (z^* \in \tilde{S}) \forall (c \in G)$ number $\inf_{\gamma} z^*(\tilde{P}(c,\gamma))$ is finite, mapping $\tilde{P}$ has the following structure:

$$\tilde{P}(c,\gamma) = (\tilde{P}_1(c,\gamma_1), \ldots, \tilde{P}_n(c,\gamma_n)) \tag{17}$$

where $\Gamma = \Gamma_1 \times \ldots \times \Gamma_n$, $\gamma = (\gamma_1, \ldots, \gamma_n)$.

By direct computation we can test that the assumption 3) of Theorem 1 is satisfied here.

Remark 3

If $Z = R^n$, $S = \{z \in R^n : z \leqslant 0\}$, sets $G, \mathcal{A}, \mathcal{B}$ are convex , mapping $\tilde{P}$ has the following structure

$$\tilde{P}(c,\gamma) = (\tilde{P}_1(c,\gamma), \ldots, \tilde{P}_n(c,\gamma)) \tag{18}$$

and for every i, $1 \leq i \leq n$ functions $\tilde{P}(\circ,\gamma)$ and $\tilde{P}_i(c,\circ)$ are convex on G and concave on Γ respectively, then the assumption 1) of Theorem 1 is satisfied. Truthfulness of the above fact can be easily verfied by direct computation.

To apply Theorem 1 we must test whether $G, \mathcal{A}, \mathcal{B}$ are compact sets. However, especially the set G is not always compact in practice. So our important task is to omit this requirement. To do it we will only consider the mappings $\tilde{P}$ which have the following form:

$$\tilde{P}(c,\gamma) = \phi(\tilde{P}_1(c), \tilde{P}_2(\gamma)) \tag{19}$$

where

$$\tilde{P}_1 : C \longrightarrow Z, \quad \tilde{P}_2 : \Gamma \longrightarrow Z, \quad \phi : Z \times Z \longrightarrow Z.$$

For mappings $\tilde{P}$ which belong to the class defined by (19) we can formulate.

<u>Theorem 2</u>

Assume that Z is Banach space, S is a closed convex cone in Z and

1) mapping $\tilde{P}$ has the form given by (19), where mappings $\phi(\circ, z)$ and $\phi(z, \cdot)$ are weakly contunuous on Z for arbitrary $z \in Z$,

2) mapping $\tilde{P}$ is convexlike on Γ and concavelike on G with respect to S,

3) sets $\tilde{P}_1(G)$ and $\tilde{P}_2(\Gamma)$ are weakly compact,

4) $\forall(z_1^*, z_2^* \in \tilde{S})\forall(c \in G)\forall(t \in [0,1]) \inf_{\gamma\in\Gamma}\left[t\, z_1^*(\tilde{P}(c,\gamma)) + (1-t)\, z_2^*(\tilde{P}(c,\gamma))\right] =$

$$= t \inf_{\gamma\in\Gamma} z_1^*(\tilde{P}(c,\gamma)) + (1-t) \inf_{\gamma\in\Gamma} z_2^*(\tilde{P}(c,\gamma)),$$

5) $\forall(\gamma\in\Gamma)\ \exists(c\in G)\ \tilde{P}(c,\gamma)\in S$,

then there exist the control which is $(\mathcal{A},\mathcal{B})$- feasible.

The proof of this theorem is given in [3].

<u>Remark 4</u>

Compactness requirements for the sets $G, \mathcal{A}, \mathcal{B}$ were in fact omitted. They were substituted by assumption 3). Interpretation of the other assumptions of Theorem 2 are the same as in Theorem 1 (see Remarks 1, 2, 3).

From the above existence theorems it follows that the class of problems in which the control $(\mathcal{A}, \mathcal{B})$ - feasible exists is quite wide.

It means that our approach to the problem of finding feasible control for real process using only its mathematical model makes sense.

5. Numerical methods for feasible control generation

In this section the algorithms for finding feasible control are presented. The feasible control is found as a control which is $(\mathcal{A},\mathcal{B})$-feasible. It is assumed in this section that the model is consistent with the real mathematical description of the process, mapping $\tilde{P}$ has the representation given by (19), $S = \{z \in R^N : z \leq 0\}$, C is Hilbert space, $\mathcal{A}$ and $\mathcal{B}$ are compact topological Hausdorff spaces, G is defined by a finite set of functional inequalities which are considered to be included in the set of mappings $\tilde{P}_i$ and for all $1 \leq i \leq N$, $\tilde{P}_i$ is continuous on $C \times \Gamma$, $\tilde{P}_i(\cdot, \gamma)$ is convex on C for each fixed $\gamma \in \Gamma$, $C \times \Gamma \ni (e,\gamma) \rightarrow \tilde{P}'_i(c,\gamma;e) \in R$ is continuous function for each fixed $c \in G$,

where

$$\tilde{P}'_i(c,\gamma;e) = \lim_{t \to 0^+} \frac{\tilde{P}_i(c + te,\gamma) - \tilde{P}_i(c,\gamma)}{t} .$$

As it was mentioned above our task of finding a feasible control is equivalent under this representation of $\tilde{P}$ to finding $\bar{c}$ such that

$$P_{max}(\bar{c}) \leq 0 \tag{24}$$

where $P_{max} : C \rightarrow R$, $P_{max}(c) \triangleq \max_{1 \leq i \leq N} \bar{P}_i(c)$

where for all i, $1 \leq i \leq N$, $\bar{P}_i(c) = \max_{\gamma \in \Gamma} \tilde{P}_i(c,\gamma)$.

Under above assumptions the following hold (see [7], [9], [13]):

(i) P_{max} is convex and continuous on C,

(ii) for each point $c \in G$ and direction $e \in C$ the directional derivative $P'_{max}(c;e)$ of P_{max} exists and is given by the formula

$$P'_{max}(c;e) = \max_{i \in I(c)} \max_{\gamma \in \Gamma_i(c)} \tilde{P}'_i(c,\gamma;e) \tag{25}$$

where

$$I(c) = \{i : P_{max}(c) = \bar{P}_i(c)\}, \quad \Gamma_i(c) = \{\gamma \in \Gamma : \bar{P}_i(c) = \tilde{P}_i(c,\gamma)\}.$$

In order to solve problem (24) a descent technique is applied. More precisely, let us consider the following algorithm.

Algorithm 1

Let be given an accuracy parameter of the solution ε, $\varepsilon > 0$, a starting point c_o, $c_o \in C$ and a sequence $\{\rho_n\}_{n\in\mathcal{N}}$ such that $\forall (n\in\mathcal{N})$ $\rho_n > 0$, $\lim_{n\to\infty} \rho_n = 0$, $\sum_{n=1}^{\infty} \rho_n = +\infty$. Let be given also c_n such that $P_{max}(c_n) > \varepsilon$.

Step 1. Find a direction e_n as a solution of the problem

$$\min_{e\in K(\Theta;1)} P'_{max}(c,e), \tag{26}$$

where

$K(\Theta; 1) = \{e \in C : \|e\| \leq 1\}$, and $P'_{max}(c; e)$ is given by (25).

Step 2. Set $c_{n+1} = c_n + \rho_n e_n$.

Step 3. If $P_{max}(c_{n+1}) \leq \varepsilon$ then stop. In the opposite case return to Step 1.

We will start to analyse the properties of Algorithm 1 with the following propositions.

Proposition 2

Let be given a positevely homogeneous functional f on a linear normed space C. If $\bar{c}$ is a solution of the problem

$$\min_{c\in X} f(c),$$

where

$X = \{c \in C : \|c\| \leq 1, \varphi_k(c) \leq 0 \text{ for } k, 1 \leq k \leq r\}$,

where

$\forall (1 \leq k \leq r)$, φ_k is positively homogeneous

and

$f(\bar{c}) < 0$, then $\|\bar{c}\| = 1$

Proof

Assume the contrary, i.e., $\|\bar{c}\| < 1$ and set $\tilde{c} = \frac{\bar{c}}{\|\bar{c}\|}$. Since

$\forall (1 \leq k \leq r) \quad \varphi_k(\tilde{c}) = \frac{1}{\|\bar{c}\|} \varphi_k(\bar{c}) \leq 0$ and $\|\tilde{c}\| = 1$, hence $\tilde{c} \in X$.

On the other hand $f(\bar{c}) \leq f(\tilde{c})$ which implies that $f(\bar{c}) \geq 0$ what contradicts the assumption. Thereby $\|\bar{c}\| = 1$ and the proof is completed.

Proposition 3

Let be given a positively homogeneous, convex functional f on Hilbert space C. If $c_1, c_2 \in C$ are such that

$$f(c_1) = f(c_2) = \min_{c \in \{c \in C : \|c\| = 1\}} f(c) = d,$$

and $\|c_1\| = \|c_2\| = 1$, $d < 0$

then $c_1 = c_2$.

The proof of this proposition is given in [3].

We can formulate now the following lemma.

Lemma 1

If there exists $\bar{c}$ such that $P_{max}(\bar{c}) \leq 0$ (i.e., a problem (1) has a solution) then the problem (3) has a unique solution e_n and $\|e_n\| = 1$

Proof

By (i), $P'_{max}(c_n, \cdot)$ is weakly lower semicontinuous on C (see [7], [5]). $K(\Theta; 1)$ is a weakly compact set ([5]).

Therefore, the solution e_n of the problem (3) exists. Further, from the convexity of P_{max} it follows

$$P_{max}(\bar{c}) \geq P_{max}(\tilde{c}) + P'_{max}(\tilde{c}; \bar{c} - \tilde{c}),$$

which implies that

$$P'_{max}(\tilde{c}; e) < 0, \text{ where } e = \bar{c} - \tilde{c}.$$

Hence, the fact that e_n is unique and $\|e_n\| = 1$ follows from the above Propositions.

The task solved in Step 1 of Algorithm 1 can sometimes be formulated in more suitable way. If for example $C = R^k$, $\Gamma_i(c_n) = \{\gamma_i\}$ for all i, $1 \leq i \leq N$ and we get $\|e\| = \max_i |e^i|$, where $e = (e^1, \dots, e^k) \in R^k$ then the problem (26) is equivalent to the following one

$$\min_{(e,\,x)} x \tag{27}$$

where

$$\tilde{P}_i'(c_n, \gamma_i ; e) \leq x ,$$

$$-1 \leq e^i \leq 1 , \quad i = 1,..,k .$$

When in addition $\tilde{P}_i'(c_n, \gamma_i ; \cdot)$ are linear functions for all i, $1 \leq i \leq N$, then the problem (27) is a linear programming one.
The coefficients ρ_n were chosen in Step 2 in such a way as to guarantee that the sequence $\{c_n\}$ generated by Algorithm 1 converge to the solution of the problem (24) .
The choice ρ_n by directional minimization seems to be more attractive since it guarantees a bigger progress in every iteration. However, on the other hand it is well known that the steepest descent procedure can generate a sequence which does not converge to the solution of the minimization problem with a cost function which is not differentiable in the Fréchet sense, when a directional minimization is used.
Because of that and in view of the fact that P_{max} is in general not Fréchet differentiable functional the directional minimization in Step 2 of the Algorithm should be avoided. Now we will investigate the properties of the direction e_n generated in Step 1. A satisfactory information about it is given by the following lemma.

<u>Lemma 2</u>

Let be given a convex, continuous functional f on Hilbert space C and an element $c \in C$ such that $0 \notin \partial f(c)$.[1/]
Then the solution $\bar{e}$ of the problem

$$\min_{e \in K(\Theta;1)} f'(c; e) \tag{28a}$$

exists and $\bar{e} = -\dfrac{c^*}{\|c^*\|}$, where $c^* \in \partial f(c)$.

1/ $\partial f(c)$ is a subdifferential of f at c.

Proof

The first part of the Lemma results from the facts that $f'(c;\cdot)$ is a weakly lower semicontinuous function and $K(0;1)$ is a weakly compact set. Consider now the following problem

$$\max_{c^*\in\partial f(c)} \min_{e\in K(\Theta;1)} \langle e, c^*\rangle^{2/} \tag{28}$$

Since $\forall(e\in K(\Theta;1))\forall(c^*\in\partial f(c))\ \langle e,c^*\rangle \geqslant -\|c^*\|$,

$$\langle e, c^*\rangle\Big|_{e=\frac{-c^*}{\|c^*\|},\ c^*\neq 0} = -\|c^*\| \text{ and by the assumption}$$

that $0\notin\partial f(c)$, it follows that any solution of the problem (28) has the form $\left(c^*, \frac{-c^*}{\|c^*\|}\right)$.

A functional f is convex and continuous from the assumption, what implies that $\partial f(c)$ is weakly compact ([7], [8]). Hence by Ky Fan-Sion Theorem (see [4]) a saddle point $(\hat{e}, \hat{c}^*)$ of a function $C\times C\ni(e, c^*)\to\langle e, c^*\rangle\in R$ on $K(\Theta;1)\times\partial f(c)$ exists. Since this saddle point solves a problem (28), then $\hat{e} = -\frac{\hat{c}^*}{\|\hat{c}^*\|}$. From the saddle point existency and by the property of a saddle point we have that $(\bar{e}, \bar{c}^*)$, where $\bar{c}^*$ is a part of the solution of the following problem

$$\min_{e\in K(\Theta;1)} \max_{c^*\in\partial f(c)} \langle e, c^*\rangle, \tag{29}$$

is also a saddle point of $\langle\cdot,\cdot\rangle$ on $K(\Theta;1)\times\partial f(c)$.
Hence and by the expression ([8], [9])

$$\min_{e\in K(\Theta;1)} f'(c;e) = \min_{e\in K(\Theta;1)} \max_{c^*\in\partial f(c)} \langle e, c^*\rangle, \tag{30}$$

we have $\bar{e} = -\frac{\bar{c}^*}{\|\bar{c}^*\|}$ what completed the proof.

The convergence property of Algorithm 1 is formulated in the following theorem.

Theorem 3

Assume that the solution set of the problem (1) is not empty. Then

---

2/ $\langle\cdot,\cdot\rangle$ is a scalar product.

(a) for every sequence $\{c_n\}_{n\in\mathcal{N}}$ generated by Algorithm 1 there exists the natural number M such that $P_{max}(c_M)\leqslant\varepsilon$.

(b) if, in addition, there exists $\bar{c}$ such that $P_{max}(\bar{c})<0$ then there exists the natural number J for which $P_{max}(c_J)<0$, i.e., $(\mathcal{A},\mathcal{B})$ - feasible control after a finite number of steps is attained.

Proof

Note that P_{max} satisfies the assumptions of Lemma 2.
Hence by the property of e_n we can look at Algorithm 1 as at the well known subdifferential minimization technique which is applied to P_{max}. Thereby the proof is completed by [1], [10].

Remark 5

For given c we can treat $P_{max}(c)$ as a certain measure of feasibility of c. The open problem is to find another measures.
The very important property of Algorithm 1 is that the constructive method of finding a subgradient is used in it. Bertsekas and Mitter presented in [2] so called ε- Subgradient Method for optimization problems with nondifferentiable cost functionals. Since the directional minimization in ε- subgradient direction can be used in this method then it could be more efficient than Algorithm 1 when applied to problem (1). But on the other hand constructive possibilities of the ε- subgradient generation seem to be more limited than the possibility of subgradient generation by formula (28a).

Suppose now that the constraints of the process are violated because of some reasons. The following task then arises: starting from the actual nonfeasible control, generate a sequence $\{c_n\}_{n\in\mathcal{N}}$ convergent to a feasible control such that every element in this sequence is better for the real constraints from the previous one. The important property of such sequence is that all its elements can be applied in the process successively during their generation. This task is the subject of our further considerations. First we have to define more precisely what it means that a certain control is better from another one. It is of course not possible to give a universal definition. However it seems that the following one will be useful in many practical situations.

Definition 5 (Control Quality Definition)

We say that a control c_1 nonfeasible for the process is better from c_2 which is also nonfeasible if and only if the following relations are satisfied:

$$\forall(i\in\tilde{I}_1(c_2))\ \tilde{P}_i(c_1,\gamma_a)\leqslant\tilde{P}_i(c_2,\gamma_a)\ , \tag{31}$$

and

$$\forall(i\in\tilde{I}_2(c_2))\ \tilde{P}_i(c_1,\gamma_a)\leqslant 0\ ,$$

where

$$\forall(c\in C)\ \tilde{I}_1(c)=\{i: 1\leqslant i\leqslant N, \tilde{P}_i(c_2,\gamma_a)>0\}\text{ and }\tilde{I}_2(c)=\{i: 1\leqslant i\leqslant N, \tilde{P}_i(c_2,\gamma_a)\leqslant 0\}\ ,$$

and γ_a is the actual value of the variable γ in the process.

By the essential assumption of this paper the adaptation procedure for recognizing γ_a is excluded. Hence we are not able to generate such sequence as it was mentioned above if the Definition 5 is applied. However the possibility of doing it by using the model only, would if the Control Quality Definition (31) is modified as follows:

$$\forall(i\in\bar{I}_1(c_2))\ \bar{P}_i(c_1)\leqslant\bar{P}_i(c_2)\text{ and }\forall(i\in\bar{I}_2(c_2))\ \bar{P}_i(c_1)\leqslant 0\ , \tag{32}$$

where

$$\forall(c\in C)\ \bar{I}_1(c)=\{i: 1\leqslant i\leqslant N, \bar{P}_i(c_2)>0\}\text{ and }\bar{I}_2(c)=\{i: 1\leqslant i\leqslant N, \bar{P}_i(c_2)\leqslant 0\}\ .$$

This modified Definition will be valid in the next part.

Further, the set of the assumption from page 14 will be broadened by supplementing the following one:

$$\exists(\bar{c}\in C)\ P_{max}(\bar{c})<0. \tag{33}$$

We are now starting construction of an algorithm for the generation of the above sequence with some set definitions:

$$\forall(c\in C)\ \bar{I}_3(c)\triangleq\{i: 1\leqslant i\leqslant N, \bar{P}_i(c)\geqslant 0\}, \tag{34}$$

and

$$\bar{K}(c)=\{e\in C:\|e\|\leq 1,\exists(t(c)\in R)\forall(t\in(0,t(c)))\forall(i\in\bar{I}_3(c))\quad \bar{P}_i(c+te)<\bar{P}_i(c)\} \quad (35)$$

Roughly speaking, for a given control c, $\bar{K}(c)$ consist of all improvement directions for these constraints $\bar{P}_i$ which are violated by c or can be violated by $c+\Delta c$, where Δc is arbitrarily small. Consider the possibility of choosing a direction from $\bar{K}(c)$ by solving the following problem:

$$\min_{e\in K(c)} \max_{i\in\bar{I}_3(c)} \bar{P}_i'(c;e). \quad (36)$$

An exhaustive information about it is given by the following lemma.

<u>Lemma 3</u>

Suppose that the set $\bar{I}_3(c)$ is not empty. Then the problem (36) is equivalent to the following one:

$$\min_{e\in K(\Theta;1)} \max_{i\in\bar{I}_3(c)} \bar{P}_i'(c;e) \quad (37)$$

The problem (37) has a unique solution with a unit norm.

<u>Proof</u>

If $e\in K(\Theta;1)-\bar{K}(c)$ then $\exists(i\in\bar{I}_3(c))\ \bar{P}_i'(c;e)\geq 0$.
If $e\in\bar{K}(c)$ then $\forall(i\in\bar{I}_3(c))\ \bar{P}_i'(c;e)\leq 0$. Thus we conclude that every solution of the problem (36) solves the problem (37). Suppose now that $\bar{e}$ solves (37). Since $\bar{K}(c)\subset K(\Theta;1)$ then $\forall(e\in\bar{K}(c))\max_{i\in\bar{I}_3(c)}\bar{P}_i'(c;\bar{e})\leq\max_{i\in\bar{I}_3(c)}\bar{P}_i'(c;e)$ and in order to show that $\bar{e}$ solves (36) it is enough to show that $\bar{e}\in\bar{K}(c)$. Assume the contrary, i.e., $\bar{e}\notin\bar{K}(c)$. Then $\max_{i\in\bar{I}_3(c)}\bar{P}_i'(c;e)\geq 0$.
On the other hand, from convexity of the functions $\bar{P}_i$, assumption (33), definition of the set $\bar{I}_3(c)$ and assumption of Lemma it follows that $\forall(i\in\bar{I}_3(c))\ \bar{P}_i'(c;\frac{\bar{c}-c}{\|\bar{c}-c\|})<0$, which is not possible. Hence the first part of Lemma is proved. Further, functional $C\ni e\to\max_{i\in\bar{I}_3(c)}\bar{P}_i'(c;e)\in R$ is a positively homogeneous, convex and continuous one and from the above consideration it follows that

$$\min_{e\in K(\Theta;1)}\max_{i\in\bar{I}_3(c)}\bar{P}_i'(c;e)<0.$$

Thus the second part of Lemma results immediately from Proposition 2 and Proposition 3. This completes the proof.

Lemma 3 provides that the problem (36) makes sense and enables us to substitute it by a simpler one.

Now we can formulate an algorithm.

Algorithm 2

Let be given an accuracy of the solution ε, $\varepsilon > 0$ and a starting nonfeasible point c_o, $c_o \in C$. Let be given also c_n such that $P_{max}(c_n) > \varepsilon$.

Assumption: $\{c \in C: \bar{P}_i(c) \leq \bar{P}_i(c_o)\}$ are bounded sets for all i, $1 \leq i \leq N$.

Step 1. Find a direction e_n as the solution of the problem

$$\min_{e \in K(\theta;1)} \max_{i \in \bar{I}_3(c_n)} \bar{P}_i'(c; e).$$

Step 2. For all $i \in \bar{I}_3(c_n)$ determine numbers $\bar{\rho}_i$ as the solutions of the problems

$$\min_{\rho \in [0,\infty)} \bar{P}_i(c_n + \rho e_n).$$

Step 3. Set $\tilde{c}_{n+1} = c_n + \tilde{\rho}_n e_n$, where $\tilde{\rho}_n = \inf_{i \in \bar{I}_3(c_n)} \{\bar{\rho}_i\}$.

Step 4. If $\forall (i \in (1:N) - \bar{I}_3(c_n))\, \bar{P}_i(\tilde{c}_{n+1}) \leq 0$ then 1/
set $\rho_n = \tilde{\rho}_n$ and go to Step 6. In opposite case go to Step 5.

Step 5. Select the greatest number ρ_n such that

$$\forall (i \in (1:N) - \bar{I}_3(c_n))\; \bar{P}_i(c_n + \rho_n e_n) \leq 0.$$

Step 6. Set $c_{n+1} = c_n + \rho_n e_n$

Step 7. If $P_{max}(c_{n+1}) \leq \varepsilon$ then stop. In the opposite case return to Step 1.

1/ $(1 : N) = \{i: 1 \leq i \leq N\}$

Theorem 4

Algorithm 2 is well defined and generates the sequence which satisfies the quality requirement expressed by inequalities (32)

Proof

The proof directly follows from the Algorithm construction and Lemma 3. Unfortunately Algorithm 2 can sometimes generate a sequence which is not convergent to the point satisfing relation (24). This phenomenon is strictly connected with the fact mentioned already in this section, i.e., that $\max_{i\in\bar{I}_3(c)} \bar{P}_i(c)$ is not Fréchet differentiable in general. In spite of such disadvantage this heuristic algorithm can be useful in practise. The algorithm which has a convergence property is formulated below.

Algorithm 3

Let be given an accuracy of solution ε, $\varepsilon > 0$, a starting nonfeasible point c_o, $c_o \in C$ and a sequence $\{\varrho_n\}_{n\in\mathcal{N}}$ such that $\forall(n\in\mathcal{N})\ \varrho_n > 0$, $\lim_{n\to\infty} \varrho_n = 0$, $\sum_{n=1}^{\infty} \varrho_n = +\infty$. Let be given also c_n such that $P_{max}(c_n) > \varepsilon$.

Step 1. Find a direction e_n as solution of the problem

$$\min_{e\in K_1(c_n)} \sum_{i\in\bar{I}_1(c_n)} \bar{P}_i'(c_n; e), \tag{38}$$

where

$$\forall(c\in C)\ K_1(c) = \{e\in C: \|e\| \leqslant 1, \forall(i\in\bar{I}_3(c))\ \bar{P}_i'(c_n; e) \leqslant 0\}. \tag{39}$$

Step 2. Set $c_{n+1} = c_n + \varrho_n e_n$.

Step 3. If $P_{max}(c_{n+1}) \leqslant \varepsilon$ then stop. In the opposite case return to Step 1.

Theorem 5

Algorithm 3 is well defined.

Proof

Only Step 1 requires the proof. Since $\forall(n\in\mathbb{N})\{e\in C:\bar{P}_i'(c_n;\ e)\leqslant 0\}$ are a convex, closed sets, $\{e\in C:\ \|e\|\leqslant 1\}$ is a weakly compact set then $K_1(c_n)$ is a weakly compact one.

Besides $\forall(c\in C)\,C\ni e\longrightarrow\sum_{i\in\bar{I}_1(c)}\bar{P}_i'(c;\ e)\in R$ is a weakly lower semicontinuous functional. Then the proof is completed by application of Weierstrass Theorem.

The convergence property of that Algorithm is expressed by the following theorem.

Theorem 6

Assume that

$$\forall(c\in C)\forall(i\in\bar{I}_3(c))\forall(c^*\in K_2(c))\ \bar{P}_i'(c;\ -c^*)\leqslant 0. \tag{40}$$

where

$$(c^*\in K_2(c))\leftrightarrow\exists(e\in K_1(c))\qquad (e,c^*)\text{ is a saddle point of a functional}$$

$$\langle\cdot,\ \cdot\rangle \text{ on } K_1(c)\times K_3(c)\ ,$$

where

$$K_3(c)=\{c^*\in C:\ c^*=\sum_{i\in\bar{I}_1(c)}c^{*i}\ ,\ \forall(i\in\bar{I}_1(c))\ c^{*i}\in\partial\bar{P}_i(c)\} \tag{41}$$

Then for every sequence $\{c_n\}_{n\in\mathbb{N}}$ generated by Algorithm 3 there exists the natural number M such that $P_{max}(c_M)\leqslant\varepsilon$.

Proof

Suppose that $P_{max}(c_n)>\varepsilon$. Note that $\sum_{i\in\bar{I}_1(c_n)}\bar{P}_i'(c_n;\ \cdot)$ is a positively homogeneous convex functional and the inequality (33) holds. Hence from the above and by definitions of the sets

$$K_1(c_n)\ ,\ \bar{I}_1(c_n),\ \bar{I}_3(c_n)$$

we have

$$\exists(\bar{c}\in C)\qquad\sum_{i\in\bar{I}_1(c_n)}\bar{P}_i'\left(c_n;\frac{\bar{c}-c}{\|\bar{c}-c\|}\right)<0. \tag{42}$$

Making use of the property of the subgradient we can write the following

$$\min_{e\in K_1(c_n)} \sum_{i\in \bar{I}_1(c_n)} \bar{P}_i'(c_n; e) = \min_{e\in K_1(c_n)} \sum_{i\in \bar{I}_1(c_n)} \max_{c_n^{*i}\in\partial \bar{P}_i(c_n)} \langle e, c_n^{*i}\rangle =$$

$$= \min_{e\in K_1(c_n)} \max_{c_n^{*}\in K_3(c_n)} \langle e, c_n^{*}\rangle . \tag{43}$$

Since $\forall(i\in(1:N))\ \partial\bar{P}_i(c_n)$ are weakly compact, convex sets ([7] , [8]), then $K_3(c_n)$ is also weakly compact and convex([5]) . As it was noted in the proof of Theorem 5, $K_1(c_n)$ is a weakly compact convex set too. Hence by application of Ky Fan-Sion Theorem we conclude that a saddle point of $\langle\cdot,\cdot\rangle$ on $K_1(c_n)\times K_3(c_n)$ exists. Let $(\hat{e}_n, \hat{c}_n^*)$ be this point. It also solves the following problem:

$$\max_{c_n^*\in K_3(c_n)} \min_{e\in K_1(c_n)} \langle e, c_n^*\rangle . \tag{44}$$

Note that $c_n^*\neq 0$, since (42) holds. Now, by application of Schwartz inequality and from the inequality (40) the saddle point representation is received in the form:

$$(\hat{e}_n, \hat{c}_n^*) = \left(-\frac{\hat{c}_n^*}{\|\hat{c}_n^*\|}, \hat{c}_n^*\right), \tag{45}$$

where

$$\hat{c}_n^*\in K_3(c_n) .$$

Since $(e_n, \bar{c}_n^*)$ where $\bar{c}_n^*$ is a part of any solution of the problem (44) is the saddle point then

$$e_n = \frac{-c_n^*}{\|c_n^*\|}, \tag{46}$$

where c_n^* is a certain element from the set $K_3(c_n)$.

Consider now a functional $\bar{\Psi}$ where

$$\bar{\Psi}(c) \triangleq \sum_{i=1}^{N} \max\{0, \bar{P}_i(c)\} . \tag{47}$$

By the well known rule of computing of a subgradient ([9]) we conclude that an element $\sum_{i\in\bar{I}_1(c_n)} c_n^{*i}$ where $c_n^{*i}\in\partial\bar{P}_i(c_n)$, belong to $\partial\bar{\Psi}(c_n)$.

Hence, taking into account the definition of $K_3(c)$ and (46) we have:

$$e_n = - \frac{c_n^*}{\|c_n^*\|} , \qquad (48)$$

where $c_n^* \in \partial \bar{\Psi}(c_n)$.

The important conclusion from the above is that Algorithm 3 generates a minimizing sequence for $\bar{\Psi}([1], [10])$.
This together with the definition of $\bar{\Psi}$, completed the proof.

Remark 6

Not all elements of the sequence generated by Algorithm 3 satisfy the quality requirement. There are two reasons of such situation. First, the direction e_n is not the direction of improvement for every $\bar{P}_i$ in c_n. We have only that $\bar{P}_i'(c_n; e_n) \leq 0$, for all $i, i \in \bar{I}_3(c_n)$. Second, even if e_n is a direction of improvement for all $\bar{P}_i$ in c_n then the choice of a coefficient ρ_n in Algorithm 3 would not quarantee satisfaction of the relations (32) . On the other hand, as it was shown in the proof of Theorem 6, every sequence which is generated by Algorithm 3, minimizes the functional $\bar{\Psi}$. From the form of $\bar{\Psi}$ then follows that these drawbacks of Algorithm 3 do not considerably limit the possibility of its application. The positive property of Algorithm 3 is that we can favour some constraints, depending on their practical validity. It should be done by introducing weight coefficients w_i and substituting the problem (38) in Step 1 by the following one:

$$\min_{e \in K_1(c_n)} \sum_{i \in \bar{I}_1(c_n)} w_i \bar{P}_i'(c_n; e) .$$

It seems that a good idea might be to construct on the base of Algorithm 2 and Algorithm 3 a new one. Numerical experiments relating to this concept were done in [11] .

Acknowledgment

The author wishes to thank Prof. W.Findeisen for many stimulating discussions about the problem considered in this paper.
He is also very grateful for the discussions provided by Dr.J.Szymanowski, Dr.K.Malinowski, Dr.A.Woźniak and other members of the academic staff of the Institute of Automatic Control, Technical University of Warsaw.

An numerical example

The following example of the process has been considered:

$$\tilde{P}_1(c,\gamma) = c_1^2 + c_2^2 - \gamma_1 ,$$

$$\tilde{P}_2(c,\gamma) = c_1 - \gamma_1 c_2^2 + \gamma_2 ,$$

$$\tilde{P}_3(c,\gamma) = - c_2 + \gamma_2 c_1 ,$$

where

$$\gamma = (\gamma_1, \gamma_2) ; \Gamma = [2,3] \times [-1,1] , \gamma_1 \in [2,3], \gamma_2 \in [-1,1] , c = (c_1, c_2) .$$

$(\mathcal{A}, \mathcal{B})$- feasible controls set for this example is shown on Fig.1. Three possibilities of a sequence $\{\rho_n\}_{n \in \mathcal{N}}$ choosing in Algorithm 2 and Algorithm 3 have been tested. It has been assumed in general that $\forall (n \in \mathcal{N})\ \rho_n = \frac{a}{n}$, where a has depended on a variable KROK as follows:

$$a = \begin{cases} \|c_o\| , & \text{where KROK} = 1 , \\ \hat{\rho}, & \text{where } \hat{\rho} \text{ solves a problem: } \min_{\rho \in [0,\infty)} P_{max}(c_o + \rho e_1) ; \\ & \text{where KROK} = 0, \\ P_{max}(c_o) , & \text{where KROK} = -1. \end{cases}$$

The best numerical results have been attained for $a = \hat{\rho}$.

Next, two possibilities of choosing of a starting value of a step coefficient ρ_o in a directional minimization procedure used in Step 1 of Algorithm 2, have been tested.

The value of a coefficient ρ_o has depended on a variable KROK as vallows:

$$\rho_o = \begin{cases} \|c_n\| , & \text{where KROK} = 1 \text{ or } 0 , \\ P_{max}(c_n) , & \text{where KROK} = -1. \end{cases}$$

On the base of numerical results we can say that one and the other possibility is useful. Finally, two norms have been considered in Step 1 of all Algorithms: $\|e\| = (\sum_{i=1}^{k} e_i^2)^{1/2}$ (NL) and $\|e\| = \max_{1 \le i \le k} |e_i|$ (LN), where $e = (e_1, \dots, e_k)$.

It seems that the proper choice of a norm is strictly depended on the problem.

The computational results are given in Table 1.

The sequence generated by Algorithm 1 when $c_o = (5,-7)$ is shown on Fig. 1.

Table 1

Starting point	Algorithm	K R O K											
		1				0				-1			
		LN		NL		LN		NL		LN		NL	
-0.5,3.0	1	3	-	2	145	3	-	2	145	3	72	1	30
	2	2	-	2	-	2	-	2	-	3	70	2	89
	3	6	-	4	127	3	-	2	-	10	68	7	77
5.0,-7.0	1	-	-	5	-	-	-	5	186	-	-	2	61
	2	-	-	-	-	-	-	5	138	-	-	-	-
	3	9	-	7	-	3	44	3	61	-	-	11	118
10.0,0.0	1	3	52	-	-	3	-	-	-	5	136	3	74
	2	2	19	12	652	2	-	7	-	5	57	4	112
	3	12	62	9	89	5	-	11	-	31	176	11	114
-9.5,5.0	1	4	75	-	-	4	87	5	206	-	-	3	102
	2	3	36	3	94	3	41	4	165	4	74	2	77
	3	-	-	-	-	5	56	11	131	-	-	11	123
1.0,-0.5	1	2	36	1	18	2	37	4	18	1	-	2	-
	2	-	-	-	-	-	-	-	-	-	-	2	-
	3	-	-	21	231	21	150	21	246	21	-	4	-
-1.2-0.4	1	1	9	2	53	1	10	2	53	1	-	1	-
	2	1	10	-	-	1	10	-	-	1	-	1	-
	3	1	10	2	21	1	13	3	47	1	-	1	-
0.0,8.0	1	3	63	-	-	3	63	-	-	3	-	2	60
	2	4	47	1	41	4	48	1	41	2	-	1	45
	3	10	56	7	57	3	37	7	69	21	-	5	63

The first number indicates a number of iterations, the second one - a computational time / in sec/.

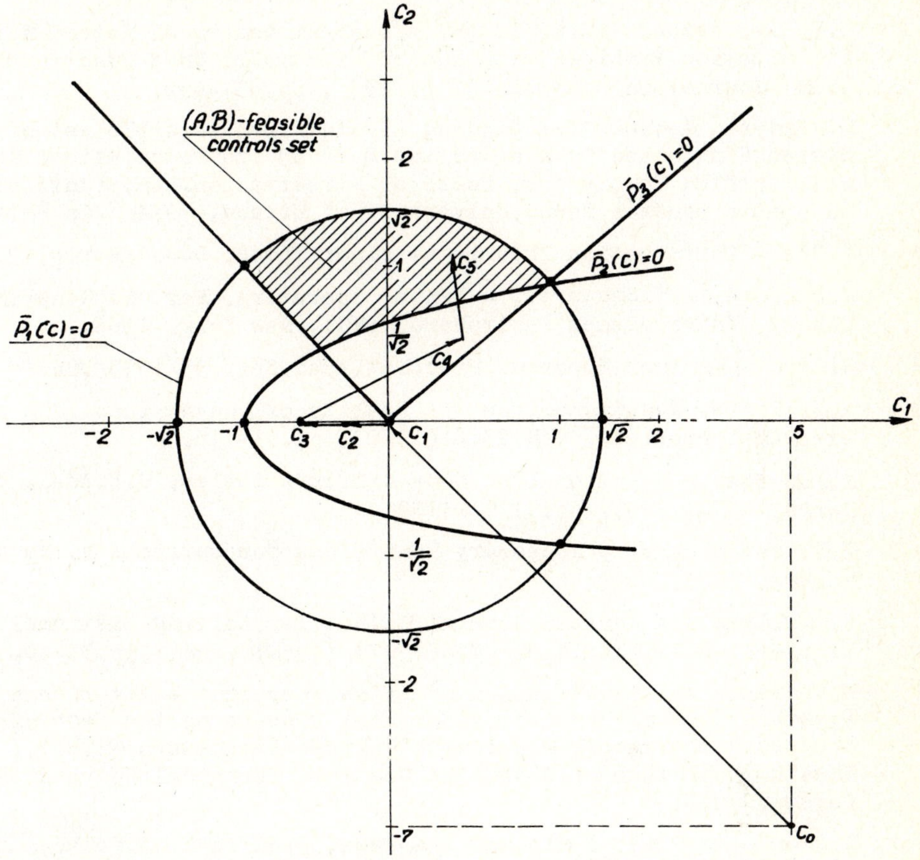
c_2
c_1
(A,B)-feasible controls set
$\bar{P}_1(c)=0$
$\bar{P}_2(c)=0$
$\bar{P}_3(c)=0$
C_0
C_1
C_2
C_3
C_4
C_5
-2
$-\sqrt{2}$
-1
1
$\sqrt{2}$
2
5
$\frac{1}{\sqrt{2}}$
$-\frac{1}{\sqrt{2}}$
$-\sqrt{2}$
-2
-7

Fig. 1

REFERENCES

1 A. Auslender,-Problémes de Minimax via l'Analiyse Convexe et les Inégalités Variationnelles: Théorie et Algorithmes, Springer-Verlag, Berlin, Heidelberg, New-York, 1972.

2 D.P. Bertsekas, S.K.Mitter, A.Descent Numerical Method for Optimization Problems with Nondifferentiable Cost Functionals, SIAM Journal on Control, 11 (1973) , pp.637-652.

3 M. Brdyś - Methods for Finding a Point from Feasible set of the System Controlled On-Line on the Base of its Mathematical Model with Shifted Parameters, Doctoral Dissertation, Institute of Automatic Control Techn.University of Warsaw, 1974 /in Polish/.

4 J.Céa - Optimization Théorie et algorithmes, Dunod,Paris,1971.

5 N.Dunford, J.T.Schwartz, - Linear operators, Part I: General Theory, Interscience Publishers, INC., New York, 1966.

6 K. Fan - Minimax Theorems.Proc.Nat.Acad.Sci.,39/1953/,pp.42-47.

7 A.D. Ioffe, W.M.Tichomirow - A Duality of convex functions and exstremal problems, UMN 23,6 (1968), pp.51-116.

8 J.J.Moreau - Functionelles sous-differentiables, C.R.Acad. Sci. Paris, 257 (1963), pp.4117 -4119.

9 B.N. Pshenichnyi - Necessary Conditions for Optimality, Nauka Moskwa, 1969.

10 B.T.Polyak - A Certain General Method for Solving Exstremal Problems, Dokl.Akad.Nauk SSSR, 174 (1967) , no.1,pp.33-36.

11 H.Radwan - An Investigation of Efficiency of the Algorithms of Feasible Control Generation for Real Process on the Base of Process Mathematical Model with Unknown Parameters, M.Sci., Thesis, Institute of Automatic Control, Technical University of Warsaw, 1974.

12 M.Sion - On General Minimax Theorems, Pacific J.of Math., 8/ 1958/, pp.171-176.

13 M.Valadier - Sous-Différentiels d'une Borne Supérieure at d'Une Somme Continue de Functions Convexes, C.R. Acad.Sci. Paris 268, pp.39-42, 1969.

INFINITE DIMENSIONAL ESTIMATION THEORY APPLIED TO A WATER POLLUTION PROBLEM

Ruth F. Curtain
Control Theory Centre, University of Warwick, Coventry CV4 7AL, U.K.

1. INTRODUCTION

There is now a fairly complete theory for filtering, prediction and smoothing for linear infinite dimensional systems where one assumes a Gaussian white noise type disturbance in the system model (see [1], [2], [3], [5]). Recently in the finite dimensional stochastic control literature, there has been interest in problems involving jump processes (see [9]) as these have applications to a wide range of problems. For example in [8], Kwakernaak applies the filtering theory for linear systems excited by Poisson white noise with noisy observations corrupted by Gaussian white noise to solve a river pollution problem. In fact the model of the river is distributed and so he uses a finite dimensional approximation. Here we show how the same problem can be solved using an infinite dimensional model by appropriately modelling the Poisson type noise as an infinite dimensional stochastic process and applying the recent theory of estimation for linear systems disturbed by general noise processes in [7]. The necessary mathematical background of stochastic differential equations in Hilbert space is outlined in § 2 and the results on infinite dimensional estimation theory for linear systems corrupted by Poisson type noise processes are summarized in § 3. In the final section § 4, the model for river pollution is discussed in detail and comparisons are made with the model of Kwakernaak in [8].

2. STOCHASTIC DIFFERENTIAL EQUATIONS IN HILBERT SPACES

The theory we summarize here is a special case of the theory for stochastic equations with general white noise disturbance [4].

Let H, K be real separable Hilbert spaces, (Ω, β, μ) a complete probability space and $T = [0,T]$, a real finite time interval.

Then we shall use the following definitions

Definition 2.1

An H-valued random variable is a map $u: \Omega \to H$ which is measurable with respect to the μ measure.

If $u \in L_1(\Omega, \mu; H)$, we define its expectation

$$E\{u\} = \int_{\Omega} u \, d\mu$$

If $u \in L_2(\Omega, \mu; H)$, we define its covariance operator

$$\mathrm{Cov}(u) = E\{(u - E\{u\}) \circ (u - E\{u\})\}$$

where $u_0 u \in \mathcal{L}(H)$ is self adjoint nuclear operator given by

$$(u_0 u)\, h = u \langle u, h \rangle \text{ for all } h \in H$$

Definition 2.2

An H-valued stochastic process is a map $u(\cdot,\cdot) : T \times \Omega \to H$ which is measurable on $T \times \Omega$ using the Lebesgue measure on T.

Consider the following estimation problem:
Estimate the random variable $x \in L_2(\Omega, \mu; H)$ from the random variable $y \in L_2(\Omega, \mu; K)$. Let $\tilde{L}_2(K, \sigma; H)$ be the closed subspace of $L_2(\Omega, \mu; H)$ which is isometric to $L_2(K, \sigma; H)$, where (K, σ) is the probability spaced induced by y. Then we shall be concerned with two types of estimates.

Definition 2.3

The best global estimate $\hat{x} = E\{x \mid y\}$ of x from y is the projection of x on $\tilde{L}_2(K, \sigma; H)$.

The best linear estimate $\bar{x}$ of x from y is $\bar{x} = \Lambda_0 y$, where $\Lambda_0 \in \mathcal{L}(K, H)$ minimizes $E\{\|x - \Lambda y\|^2\}$ over all $\Lambda \in \mathcal{L}(K, H)$. &

If x and y are Gaussian, then the linear and global estimates are identical.

The following class of stochastic process is used for our application model and is a special case of the orthogonal increments process introduced in [4].

Definition 2.4

An H-valued compound Poisson process $\{q(t);\ t \in T\}$ is such that

$$q(t) = \sum_{i=0}^{\infty} q_i(t)\, e_i \tag{2.1}$$

where $\{e_i\}$ is a complete orthonormal basis for H and where $\{q_i(t);\ t \in T\}$ is a real compound Poisson process such that

$$\begin{cases} E\{q_i(t) - q_i(s)\} = \mu_i (t-s) \\ E\{(\bar{q}_i(t_1) - \bar{q}_i(s_1))(\bar{q}_j(t_2) - \bar{q}_j(s_2))\} = 0 \quad ; \ 0 \le s_1 < t_1 \le s_2 < t_2 \le T \\ E\{(\bar{q}_i(t) - \bar{q}_i(s))(\bar{q}_j(t) - \bar{q}_j(s))\} = \lambda_{ij}(t-s) \quad ; \ 0 \le s \le t \le T \end{cases} \tag{2.2}$$

where $\bar{q}_i(t) = q_i(t) - \mu_i t$ and $\sum_{i=0}^{\infty} \mu_i < \infty$

For convenience we rewrite (2.2) as

$$\text{(2.3)} \quad \begin{cases} E\{q(t)\} = t \sum_{i=0}^{\infty} \mu_i e_i \\ Cov(\bar{q}(t) - \bar{q}(s)) = \Lambda (t-s) \quad ; \quad 0 \le s \le t \le T. \end{cases}$$

where $\bar{q}(t) = q(t) - t \sum_{i=0}^{\infty} \mu_i e_i$

and Λ is the covariance operator of $q(t)$ and is such that

$$\text{(2.4)} \quad \Lambda e_i = \sum_{j=0}^{\infty} \lambda_{ij} e_j$$

We remark that $\lambda^2_{ij} \le \lambda_i \lambda_j \quad (\lambda_{ii} = \lambda_i)$ and

$$E\{\|\bar{q}(t) - \bar{q}(s)\|^2\} = \text{trace}\,\Lambda\,(t-s) = \sum_{i=0}^{\infty} \lambda_i (t-s)$$

We now define a stochastic integral for these processes.

<u>Definition 2.5</u>

Let $q(t)$ be as in definition 2.4 and suppose $\Phi \in B_2(T; \mathcal{L}(H,K))$, the class of strongly-measurable $\mathcal{L}(H,H)$ -valued functions with $\int_T \|\Phi\|^2 ds < \infty$. Then we define the stochastic integral

$$\int_0^t \Phi(s)\, dq(s) = \sum_{i=0}^{\infty} \int_0^t \Phi(s)\, d\bar{q}_i(s) + \sum_{i=0}^{\infty} \mu_i \int_0^t \Phi(s) e_i\, ds$$

$\int_0^t \Phi(s)\, dq(s)$ is a well defined H -valued stochastic process with the properties

$$\text{(2.5)} \quad E\left\{ \int_0^t \Phi(s)\, dq(s) \right\} = \sum_{i=0}^{\infty} \mu_i \int_0^t \Phi(s) e_i\, ds$$

$$\text{(2.6)} \quad E\left\{ \left\| \int_0^t \Phi(s)\, dq(s) \right\|^2 \right\} \le \text{trace}\,\Lambda \int_0^t \|\Phi(s)\|^2 ds$$

We shall be concerned with the following stochastic evolution equation

$$\text{(2.7)} \quad \begin{cases} du(t) = Au(t)dt + B\, dq(t) \\ u(0) = u_0 \end{cases}$$

where A is the infinitesimal generator of an analytic semigroup J_t on H, $B \in \mathcal{L}(H)$, $q(t)$ is as in definition 2.4, $u_0 \in L_2(\Omega, \mu; H)$ and $u(t)$ is an -valued stochastic process.

By a solution of (2.7) we shall mean the following:

Definition 2.6

(2.7) has a strong solution u, if $u \in C(T; L_2(\Omega,\mu; H))$, $u(t) \in \mathcal{D}(A(t))$ w.p. and $u(t)$ satisfies (2.8) almost everywhere on $T \times \Omega$. u is unique if whenever u_1 and u_2 are strong solutions,

$$\mu \{ \omega : \sup_{t \in T} \| u_1(t) - u_2(t) \| = 0 \} = 1$$

Theorem 2.1

(2.7) has the unique solution

$$u(t) = \mathcal{J}_t u_0 + \int_0^t \mathcal{J}_{t-s} B dq(s)$$

provided the following extra assumptions are satisfied.

(2.8) $$\sum_{i=0}^{\infty} \lambda_i \int_0^t \| A \mathcal{J}_t B e_i \|^2 ds < \infty$$

(2.9) $$\sum_{i=0}^{\infty} \mu_i \int_0^t \| A \mathcal{J}_t B e_i \| ds < \infty$$

3. ESTIMATION THEOREY FOR INFINITE DIMENSIONAL SYSTEMS WITH POISSON TYPE DISTURBANCE

We summarize the results of [7], specialized to time invariant systems corrupted by Poisson type noise disturbance.

With H, T and $(\Omega, \mathcal{P}, \mu)$ as in § 2, consider the following abstract system and observation models.

(3.1) $$u(t) = \mathcal{J}_t u_0 + \int_0^t \mathcal{J}_{t-s} B dq(s)$$

(3.2) $$z(t) = \int_0^t C u(s) ds + w(t)$$

where $\mathcal{J}_t$ is an analytic semigroup on H with generator A, $B \in \mathcal{L}(H)$, $q(t)$ is as in definition 2.4, $u_0 \in L_2(\Omega, \mu; H)$, $C \in \mathcal{L}(H, R^k)$ and W is a k-dimensional Wiener process with incremental covariance matrix the identitiy.

The estimation problem is to find the best linear unbiased estimate $\hat{u}(t|t_0)$ of the state $u(t)$ at time t, based on the observations $z(s)$; $0 \le s \le t$.

Under the above assumptions, the optimal filter $\hat{u}(t) = \hat{u}(t|t)$ is given by

(3.3) $$\hat{u}(t) = v(t) + \int_0^t \mathcal{Y}(t,s)P(s)C^*dz(s)$$

where $$v(t) = \sum_{i=0}^{\infty} \mu_i \int_0^t \mathcal{Y}(t,s)Be_i ds$$

and $\mathcal{Y}(t,s)$ is the unique solution of

(3.4) $$\mathcal{Y}(t,s)x = \mathcal{J}_{t-s}x - \int_s^t \mathcal{J}_{t-\rho}P(\rho)C^*C\mathcal{Y}(\rho,s)x\,d\rho$$

and $P(t)$ is the unique solution of the Riccati equation in the class of self adjoint weakly continuous operator-valued functions with $\langle x, P(t)y\rangle$ absolutely continuous for $x, y \in \mathcal{D}(A^*)$.

(3.5) $$\begin{cases} \frac{d}{dt}\langle P(t)x,y\rangle - \langle P(t)x, A^*y\rangle - \langle A^*x, P(t)y\rangle \\ \qquad + \langle P(t)C^*CP(t)x,y\rangle = \langle B\Lambda B^*x,y\rangle \quad \text{a.e.} \\ P(0) = P_0 \qquad ; \qquad x,y \in \mathcal{D}(A^*). \end{cases}$$

Moreover, $$P(t) = \mathrm{Cov}(u(t) - \hat{u}(t)).$$

The optimal smoother $\hat{u}(t|t_0)$, where $t_0 > t$ is given by

(3.6) $$\hat{u}(t|t_0) = \hat{u}(t) + P(t)\lambda(t)$$

where $$\lambda(t) = \int_t^{t_0} \mathcal{Y}^*(s,t)C^*(dz(s) - C\hat{u}(s)ds)$$

The optimal predictor $\hat{u}(t|t_0)$, where $t > t_0$ is given by

(3.7) $$\hat{u}(t|t_0) = \sum_{i=0}^{\infty} \mu_i \int_0^t \mathcal{J}_{t-s}Be_i ds + \mathcal{J}_{t-t_0}\hat{u}(t_0)$$

To obtain differential equations for the estimators we need to impose the following extra assumptions

(3.8) $$\sum_{i=0}^{\infty} \lambda_i \int_0^t \| A\mathcal{J}_{t-s}Be_i\| ds < \infty$$

(3.9) $$\sum_{i=0}^{\infty} \lambda_i \| A\mathcal{J}_t P_0 e_i\| < \infty$$

Then writing $\hat{u}(t) = v(t) + y(t)$, $y(t)$ is the unique solution of the stochastic evolution equation

(3.10) $$\begin{cases} dy(t) = (A - P(t)C^*C)\,y(t)\,dt + P(t)C^*C\,u(t)\,dt + P(t)C^*dw(t) \\ y(0) = 0 \end{cases}$$

and $V(t)$ is the unique solution of the deterministic differential equation

(3.11) $$\begin{cases} \dot{v}(t) = (A - P(t)C^*C)\,v(t) + \sum_{i=0}^{\infty} \mu_i B e_i \\ V(0) = 0 \end{cases}$$

For the smoothing problem $t_0 > t$, writing

$\hat{u}(t|t_0) = V(t) + y(t, t_0)$, we have that $y(t, t_0)$ is the unique strong solution of the stochastic evolution equation

(3.12) $$\begin{cases} dy(t,t_0) = (A - P(t)C^*C)\,y(t,t_0)\,dt + P(t)C^*C\,\hat{u}(t)\,dt + B\Lambda B^*\lambda(t)\,dt \\ y(t_0,t_0) = y(t_0) \end{cases}$$

For the prediction problem $t > t_0$, writing

$$\hat{u}(t|t_0) = \bar{u}(t) + y(t,t_0)$$

where $\bar{u}(t) = E\{u(t)\}$, we have that $y(t, t_0)$ is the unique strong solution of the stochastic evolution equation

(3.13) $$\begin{cases} dy(t,t_0) = A\,y(t,t_0)\,dt \\ y(t,t_0) = y(t_0) \end{cases}$$

and $\bar{u}(t)$ is the unique solution of

(3.14) $$\begin{cases} \dot{\bar{u}}(t) = A\,\bar{u}(t) + \sum_{i=0}^{\infty} \mu_i B e_i \\ \bar{u}(0) = 0 \end{cases}$$

4. POLLUTION MODEL

The problem considered by Kwakernaak in [8] is an environmental problem of river pollution. For his model of the pollution by chemical wastes he assumes that the number of deposits in a section of the river of infinitesimal length dx (x being the distance coordinate along the river) behave according to a Poisson process with rate parameter $\lambda(x)dx$ where $\lambda(x)$ is a given function. He assumes that the number of deposits in non-overlapping sections are independent processes and that the amounts of chemical deposited each time at location x are independent stochastic variables with given distribution H_x and characteristic function ϕ_x. The time evolution of the concentration of the chemical at location x at time t, $y(t,x)$ is supposed to be given by

$$(4.1)\qquad \frac{\partial y(t,x)}{\partial t} = D\frac{\partial^2 y(t,x)}{\partial x^2} - V\frac{\partial y(t,x)}{\partial x} + \xi(t,x) \quad ; \ 0 \le x \le l.$$

where D is the dispersion coefficient, V the water velocity and $\xi(t,x)$ is the rate of increase of the concentration at (t,x) due to the deposits of chemical wastes described above.

Along the river are a finite number of measuring stations which continuously measure the observed local concentrations of the chemical. Kwakernaak solves the problem of reconstructing the times, locations and amounts of deposits from the measured data by approximating (4.1) by a finite dimensional model and applying the finite dimensional filtering theory using a martingale approach.

Here we consider the problem of estimating the concentration of the chemical at any point along the river at any time in the past or future, based on continuous noisy measurements in time at a finite number of measuring stations along the river. Our approach is quite different in that we retain the infinite dimensional character of the problem throughout and show how the final equations can be solved by eigenfunction approximations.

4.1 Characterization of the pollution process

In order that (4.1) describes a stochastic differential equation it is necessary to define $\xi(t,x)$ as some suitable distributed stochastic process. Following the discussion in §2, we try a model of the following type

$$(4.2)\qquad \begin{aligned} dy(t) &= A_0\, y(t)\,dt + dq(t) \\ y(0) &= y_0 \end{aligned}$$

where $y(t) \in H = L_2(0,l)$ for each $t \in [0,T]$ models the amount of pollutant dumped at (t,x) and A_0 is the diffusion operator defined by

$$(4.3)\qquad \begin{cases} A_0 y = D\dfrac{\partial^2 y}{\partial x^2} - V\dfrac{\partial y}{\partial x} \\ \mathcal{D}(A_0) = \left\{ y \in H : \dfrac{\partial y}{\partial x}, \dfrac{\partial^2 y}{\partial x^2} \in H \text{ and } y = 0 \text{ at } x = 0, l \right\} \end{cases}$$

In order to apply the theory of stochastic differential equations on a Hilbert space from §2, we need to establish that $q(t)$ is a compound Poisson process according to definition 2.4.

In [8], Kwakernaak defines $q(t)$ formally in terms of a random functional, that is

$F_{q_t} : H \to L_2(\Omega \times T; R)$ is defined by

$$F_{q_t}(u) = \langle q(t), u \rangle_H \tag{4.4}$$

From his assumptions on the polluting process, he deduces that for each $u \in H$, $F_{q_t}(u)$ is a real valued compound Poisson process with rate parameter $\bar{\lambda} = \int_0^l \lambda(x)dx$ and jumps with characteristic function $1/\bar{\lambda} \int_0^l \lambda(x)\phi_x(\gamma\, u(x))\, dx$. $F_{q_t}(u)$ has characteristic function

$$E\{e^{i\gamma F_q(u)}\} = \exp\{t \int_0^l \lambda(x)[\phi_x(\gamma u(x)) - 1]\, dx\} \tag{4.5}$$

We now show that provided $\lambda(x)$ and H_x are suitably chosen $F_{q_t}(\cdot)$ induces an H-valued stochastic process $q(t)$ defined by (4.4)

Lemma 4.1

Suppose $\{H_x;\ 0 \le x \le l\}$ is a real stochastic process with characteristic function $\phi_x(\gamma)$ and $\lambda(x)$ is a real function on $(0,l)$.

Let $H = L_2(0,l)$ and $\{e_k\}_{k=0}^{\infty}$ be the following orthonormal bases for H

$$e_k(x) = \sqrt{\tfrac{2}{l}}\, \sin\left(\frac{k\pi x}{l} + \epsilon_k\right) \quad ; \tan \epsilon_k = -\frac{2\pi kD}{lV} \quad ; k = 1,2,.. \tag{4.6}$$

Under the following assumptions on λ and H_x

$$(\lambda(x)E\{H_x\})^2,\ \lambda(x)E\{H_x^2\} \in L_2(0,l) \tag{4.7}$$
$$(\lambda(x)\, E\{H_x\})^2 = \sum_{k=1}^{\infty} \beta_k\, e_{2k}(x)$$
$$\lambda(x)E\{H^2x\} = \sum_{k=1}^{\infty} \gamma_k\, e_{2k}(x)$$

$q(t)$ induced by (4.4) is a well-defined H-valued stochastic process $\in L_2(\Omega \times T; H)$; $q(t) = \sum_{k \ge 1}^{\infty} q_k(t)\, e_k$, where $q_k(t)$ is a real compound Poisson process with characteristic function

$$\exp\{t \int_0^l \lambda(x)[\phi_x(\gamma e_k(x)) - 1]\, dx\}$$

If we further assume that

(4.8) $$E\{(q_K(t)-E\{q_K(t)\})(q_i(t)-E\{q_i(t)\})\} = \lambda_{ij}\, t$$

then $q(t)$ is compound Poisson process according to definition 2.4 with

(4.9) $$\begin{cases} E\{q(t)\} = t\sum_{K=0}^{\infty} \mu_K e_K \\ Cov(q(t)) = t\,\Lambda \end{cases}$$

where Λ is a nuclear, self adjoint operator on H, given by

$$\Lambda e_K = \sum_{i=1}^{\infty} \lambda_{Ki} e_i \quad ; \quad \text{trace } \Lambda = \sum_{i=1}^{\infty} \lambda_{ii} < \infty .$$

and

(4.10) $$\begin{cases} \mu_K = \int_0^{\ell} \lambda(x) e_K(x) E\{H_x\} dx \\ \lambda_K = \lambda_{KK} = \int_0^{\ell} \lambda(x) e_K^2(x) E\{H_x^2\} dx \\ \lambda_{ij}^2 \le \lambda_i \lambda_j \end{cases}$$

Proof

From (4.4), $q(t) = \sum_{K=0}^{\infty} q_K(t) e_K$; where $q_K(t) = Fq_t(e_K)$.

So $q_K(t)$ is a real compound Poisson process and has the characteristic function
$\exp\{t\int_0^{\ell}\lambda(x)[\phi_x(\gamma e_K(x))-1]dx\}$ and hence its moments are

(4.11) $$E\{q_K(t)\} = t\int_0^{\ell}\lambda(x)e_K(x)E\{H_x\}dx = \mu_K t \quad \text{, say}$$

(4.12) $$E\{q_K^2(t)\} - E\{q_K(t)\}^2 = t\int_0^{\ell}\lambda(x)e_K^2(x)E\{H_x^2\}dx = \lambda_K t \quad \text{, say}$$

and $q(t) \in L_2(\Omega\times T; H)$ provided $\sum_{K=0}^{\infty}\mu_K < \infty$ and $\sum_{K=1}^{\infty}\lambda_K < \infty$

From (4.6), it may be verified that

$$\int_0^{\ell} e_{2K}(x)\, e_i^2(x)\, dx = 0 \qquad \text{for} \qquad i \ne K$$

and $\int_0^{\ell} e_{2K}(x)\, e_K^2(x)\, dx = O(\frac{1}{K})$

So $$\sum_{i=1}^{\infty}\lambda_i = \sum_{i=1}^{\infty}\gamma_i \langle e_{2i}, e_i^2\rangle = \sum_{i=1}^{\infty}\gamma_i O(\tfrac{1}{i}) < \infty .$$

and similarly $\sum_{i=1}^{\infty}\mu_i < \infty$. (Note that $\mu_0 = 0 = \lambda_0$).

So under assumption (4.7), $q(t) \in L_\nu(\Omega \times T; H)$

By construction $q_k(t)$ has independent increments and assumption (4.8) ensures $q(t)$ satsfies definition 2.4,(4.9) and (4.10) follow from (4.11), (4.12).

Remark

(4.7)(b) is necessary, since $\int_0^\ell e_k(x)\, e_k^2(x)\, dx \not\to 0$ as $k \to \infty$ for odd k.

4.2 State and observation models

In the filtering theory, both A and its adjoint appear in the equations and since A_0 is not self adjoint this makes an eigenfunction expansion approach difficult. So instead we transform the original model (4.2) to the following:

$$
(4.13) \qquad \begin{cases} du(t) = Au(t)\,dt + B\,dq(t) \\ u(0) = u_0 \end{cases}
$$

where $y(t,x) = e^{ax} u(t,x)$; $y_0 = e^{ax} u_0$; $a = V/2D$.

A is a self adjoint operator on H given by

$$
(4.14) \qquad Ah = \frac{\partial^2 h}{\partial x^2} - a^2 h
$$

$$
\mathcal{D}(A) = \{ h \in H : \tfrac{\partial h}{\partial x}, \tfrac{\partial^2 h}{\partial x^2} \in H \text{ and } ah(x) + \tfrac{\partial h}{\partial x} = 0 \text{ at } x = 0, \ell \}
$$

and $B \in \mathcal{L}(H)$ is given by

$$
(4.15) \qquad (Bh)(x) = e^{-ax} h(x) \qquad \text{for} \quad h \in H
$$

We suppose that the initial state u_0 has zero expectation and covariance operator P_0 given by

$$
(4.16) \qquad P_0 e_k = \alpha_k e_k \quad ; \quad \sum_{k=1}^{\infty} \alpha_k < \infty \quad ; \alpha_0 = 0.
$$

A has the eigenfunctions $\{e_i\}$ given by (4.6) and generates the analytic semigroup J_t given by

$$
(4.17) \qquad (J_t h)(x) = \sum_{n=1}^{\infty} \langle h, e_n \rangle e^{(a^2 - n^2/\ell^2)t} e_n(x) \qquad \text{for} \quad h \in H
$$

(4.13) has the well-defined mild solution

(4.18) $$u(t) = J_t u_0 + \int_0^t J_{t-s} B\, dq(s)$$

which we can take as our state model for the evolution of the concentration of pollutant in the river.

Now we show that (4.18) is a strong solution of (4.13) (see definition 2.6).

From theorem 2.1, we need to verify

(4.19) $$\sum_{i=1}^{\infty} \lambda_i \int_0^t \| A J_{t-s} B e_i \|^2 ds < \infty$$

(4.20) $$\sum_{i=1}^{\infty} \mu_i \int_0^t \| A J_{t-s} B e_i \| ds < \infty$$

Writing $B e_i = \sum_{j=0}^{\infty} \mu_{ij} e_j$, we find

$$A J_{t-s} B e_i = -\frac{\pi^2}{\ell^2} \sum_{j=1}^{\infty} \mu_{ij} j^2 e^{(a^2 - j^2/\ell^2)(t-s)} e_j$$

and

$$\int_0^t \| A J_{t-s} B e_i \|^2 ds \leq \text{const.} \sum_{j=0}^{\infty} \frac{j^4 \mu_{ij}^2}{(j^2 - a^2 \ell^2)}$$

By direct calculation, we find that for $i \neq j$

(4.21) $$\mu_{ij} = \int_0^\ell e^{-ax} e_i(x) e_j(x) dx$$

$$= \frac{\text{const. } ij\,(3a^2\ell^2 - i^2 - j^2)}{\sqrt{\pi^2 j^2 + a^2 \ell^2}\, \sqrt{\pi^2 i^2 + a^2 \ell^2}\, (a^2\ell^2 + \pi^2(i+j)^2)(a^2\ell^2 + \pi^2(i-j)^2)}$$

and since $\sum_{i=1}^{\infty} \lambda_i < \infty$, we have

$$\sum_{i=1}^{\infty} \sum_{j=1}^{\infty} \frac{j^4 \lambda_i \mu_{ij}^2}{(j^2 - a^2 \ell^2)} < \infty$$

and (4.19) is satisfied.

(4.20)is similarly verified.

For our observation process we take

(4.22) $$dz(t) = C u(t) dt + dw(t)$$

where $w(t)$ is a k dimensional Wiener process with covariance matrix the identity and $C \in \mathcal{L}(H, R^k)$ is given by

(4.23) $$(Cu)_j = \frac{1}{2\epsilon} \int_{x_j - \epsilon}^{x_j + \epsilon} u(x)\, dx \qquad \text{for small} \quad \epsilon > 0$$

This approximates point observations at the fixed locations $x_1, \ldots, x_k$.

4.3 Solution to the estimation problem

Our model in §4.2 satisfies all the assumptions of the theory of § 3 and so there exists a unique optimal estimator, given by (3.3) - (3.7).

In order to obtain computable solutions we first obtain recursive equations for our estimates and the covariance of the error process $P(t)$. From § 3, $P(t)$ is the unique solution of the differential Riccati equation

$$(4.24)\quad \begin{cases} \frac{d}{dt}\langle P(t)f,h\rangle - \langle P(t)f, Ah\rangle - \langle Af, P(t)h\rangle + \langle P(t)C^*Cf,h\rangle \\ \qquad = \langle B\Lambda B^* f, h\rangle \quad a.e. \\ P(0) = P_0 \quad \text{for} \quad f,h \in \mathcal{D}(A) \end{cases}$$

(note that).

We shall try for a solution $P(t)$ of the form

$$(4.25)\quad P(t)h = \sum_{i,j=1}^{\infty}\sum^{\infty} p_{ij}(t)\, e_i(x)\langle e_j, h\rangle \qquad ; \; p_{ij}(t) = p_{ji}(t).$$

Substituting (4.24) into (4.23) and equating coefficients of $h = \sum_{i=1}^{\infty} h_i e_i$, $f = \sum_{i=1}^{\infty} f_i e_i$ we obtain

$$(4.26)\quad \begin{cases} \frac{d}{dt} p_{ij}(t) + \frac{\pi^2}{\ell^2}(i^2+j^2)\,p_{ij}(t) + \sum_{n,m=1}^{\infty}\sum^{\infty} p_{ni}(t)p_{mj}(t)\,A_{mn} = \rho_{ij} \\ p_{ij}(0) = \delta_{ij}\,\alpha_i \end{cases}$$

where $\rho_{ij} = \sum_{k=1}^{\infty}\sum_{n=1}^{\infty} \lambda_{kn}\,\mu_{in}\,\mu_{jk}$ (see 4.21)

and $A_{mn} = \sum_{r=1}^{K} a_{mr}a_{nr}$

$$a_{sr} = \frac{\sqrt{2\ell}}{s\pi\epsilon}\sin\frac{s\pi\epsilon}{\ell}\sin\left(\frac{s\pi x_r}{\ell} + \epsilon_s\right) = (Ce_s)_r$$

We now verify that the additional assumptions (3.8) and (3.9) are satisfied and hence $\hat{u}(t/t_0)$ may be expressed in differential form.

From § 3, we must verify the following

$$(4.28)\quad \sum_{i=1}^{\infty} \lambda_i \int_0^t \| AJ_s e_i \| ds < \infty$$

$$(4.29)\quad \sum_{i=1}^{\infty} \lambda_i \| A J_t P_0 e_i \| < \infty \qquad \text{for} \quad t>0$$

Now $$A J_t P_0 e_i = -\alpha_i \frac{i^2\pi^2}{\ell^2} e^{(a^2 - i^2/\ell^2)t} e_i$$

and so $\| A J_t P_0 e_i \| \leq$ constant for $t>0$

and $\sum_{i=1}^{\infty} \lambda_i < \infty$, ensures that (4.29) holds.

$$\int_0^t \| A J_s e_i \| ds = \left| \frac{i^2\pi^2}{(l^2a^2-i^2)} \left(1 - e^{(a^2-i^2/l^2)t}\right) \right|$$

$$\leq \quad \text{const.}$$

and so again (4.28) is satisfied since $\sum_{i=1}^{\infty} \lambda_i < \infty$.

So $\hat{u}(t|t_0)$ may be expressed in differential form. For the optimal filter let us try the following expansion:

$$\hat{u}(t) = \sum_{i=1}^{\infty} v_i(t) e_i + \sum_{i=1}^{\infty} \beta_i(t) e_i \tag{4.30}$$

where $v(t) = \sum_{i=1}^{\infty} v_i(t) e_i$ is deterministic. Substituting in (3.10) and (3.11) we obtain the following equations for $\beta_i(t)$ and $v_i(t)$

$$\begin{cases} d\beta_i(t) = -\frac{i^2\pi^2}{l^2} \beta_i(t)\,dt + \sum_{n=1}^{\infty}\sum_{r=1}^{\infty} A_{nr}\beta_r(t) p_{in}(t)\,dt \\ \qquad\qquad + \sum_{n=1}^{\infty} p_{ni}(t) \sum_{r=1}^{k} a_{nr}\,dz_r(t) \\ \beta_i(0) = 0 \end{cases} \tag{4.31}$$

$$\begin{cases} \dot{v}_i(t) = -\frac{i^2\pi^2}{l^2} v_i(t) + \sum_{n=1}^{\infty}\sum_{r=1}^{\infty} A_{nr} v_r(t) p_{in}(t) + \sum_{r=1}^{\infty} \mu_r \mu_{ri} \\ v_i(0) = 0 \end{cases} \tag{4.32}$$

So the filtering problem may be solved recursively from the infinite systems , (4.26), (4.31), (4.32) for $p_{ij}(t)$, $\hat{u}_i(t)$, $v_i(t)$ respectively, using the usual truncation methods.

For the prediction problem $t > t_0$ writing

$$\hat{u}(t|t_0) = \bar{u}(t) + y(t,t_0)$$

$$= \sum_{i=1}^{\infty} \gamma_i(t) e_i + \sum_{i=1}^{\infty} y_i(t,t_0) e_i$$

and substituting in (3.13) and (3.14) we obtain the following equations for $\gamma_i(t)$, $y_i(t,t_0)$.

$$\begin{cases} dy_i(t,t_0) = -\frac{\pi^2}{l^2} i^2 y_i(t,t_0)\,dt \\ y_i(t_0,t_0) = \beta_i(t_0) \end{cases} \tag{4.33}$$

(4.34) $$\begin{cases} \dot{\gamma}_i(t) = -\frac{\pi^2 i^2}{\ell^2}\gamma_i(t) + \sum_{k=1}^{\infty}\mu_k\mu_{ik} \\ \gamma_i(0) = 0 \end{cases}$$

Similarly for the smoothing problem $t_0 > t$, writing

$$\hat{u}(t/t_0) = v(t) + \sum_{i=1}^{\infty} y_i(t, t_0)$$

you may obtain equations for $y_i(t, t_0)$.

We remark that $\beta_i(t)$, $y_i(t, t_0)$ and $x_i(t, t_0)$ are unique strong solutions of stochastic evolution equations, whereas $v_i(t)$ and $\gamma_i(t)$ are deterministi functions.

Returning to the original problem of estimating the amount of concentration $\hat{y}(t/t_0; x)$ of pollutant at (t, x) based on measurements $y(x_1, s)$, .. $y(x_k, s); 0 \le s \le t_0$, we have

$$\hat{y}(t/t_0; x) = e^{ax}\,\hat{u}(t/t_0)(x)$$

Finally we note that the estimators $\hat{u}(t/t_0)$ obtained are the best linear estimates and not the best global estimates, because the noise process $q(t)$ is Poisson-like. In the Gaussian case the best linear estimates are also the best global estimates (see § 2). The advantage with working with linear estimates is that you obtain Kalman-Bucy type recursive equations.

In his finite dimensional approximation Kwakernaak found the best global estimate and so obtained an infinite set of filtering equations which he then proceeded to develop approximation algorithms for. In his conclusions he notes that little if any improvement was obtained over the Kalman filter and this is not surprising since the Kalman filter approach obtains the best linear estimate to a high degree of accuracy; whereas the global estimate was obtained using approximating algorithms, which will incur errors.

In our infinite dimensional approach we again obtain Kalman Bucy type equations, but an infinite system this time. However they can be conveniently solved by an eigenfunction approach.

REFERENCES

1. BENSOUSSAN, A. "Filtrage Optimal des Systèmes Linéaires", Dunod, 1971.

2. CURTAIN, R.F. "Infinite Dimensional Filtering", SIAM J. Control, 1975.

3. CURTAIN, R.F. "A Survey of Infinite Dimensional Filtering", SIAM Review, 17, 1975.

4. CURTAIN, R.F. "Stochastic Evolution Equations with General White Noise Disturbance", Control Theory Centre Report No. 41, University of Warwick, 1975.

5. CURTAIN, R.F. "Infinite Dimensional Estimation Theory for Linear Systems", Control Theory Centre Report No. 38, University of Warwick, 1975.

6. CURTAIN, R.F. & PRITCHARD, A.J. "The Infinite Dimensional Riccati Equation for Systems Defined by Evolution Operators", 1975 (to appear in SIAM J. Control).

7. CURTAIN, R.F. "Estimation Theory for Abstract Evolution Equations Excited by General White Noise Processes", Control Theory Centre Report No. 40, University of Warwick, 1975.

8. KWAKERNAAK, H. "Filtering for Systems Excited by Poisson White Noise", Int. Symposium on Control Theory, Numerical Methods & Computer Systems Modelling, 1974. (Lecutre Notes in Economics & Math. Systems, 107, Springer Verlag.)

9. VARAIYA, P. "Filtering and Control of Jump Processes", Ibid.

NUMERICAL SOLUTION OF THE OPERATOR RICCATI EQUATION FOR THE FILTERING OF LINEAR STOCHASTIC HEREDITARY DIFFERENTIAL SYSTEMS*

M.C. Delfour

Centre de Recherches Mathématiques
Université de Montréal, C.P. 6128,
Montréal, P.Q. H3C 3J7 Canada.

1. Introduction.

The filtering problem for hereditary systems has been considered by a number of authors. To the author's knowledge, the first paper in this field is the one of H. KWAKERNAAK [1], where simultaneously the smoothing and filtering problems for linear differential systems with multiple constant time delays are studied. Other more recent papers by A. LINDQUIST [1], A. BENSOUSSAN [2], BENSOUSSAN-DELFOUR-MITTER [1], MITTER-VINTER [1], R. CURTAIN [1] and R. KWONG [1] have also discussed the theory of this problem and extended the well-known duality theorem of KALMAN-BUCY in various forms.

In this paper we put our hereditary system in state form and use the work of A. BENSOUSSAN [1]. This leads to the study of the dual optimal control problem. It allows us to obtain the existence of the covariance operator $\Pi(t)$ and to study its properties without deriving the Riccati differential operator equation. One major difficulty is to make sense of that equation without adding any extra hypotheses on the matrices defining the original systems. For instance R. VINTER [1] and MITTER-VINTER [1] have shown that the intersection over the time t of the domains of a certain unbounded operator $\tilde{A}(t)^*$ (cf. eq. (4.23)) is generally not dense in the product space $X \times L^2(-a,0;X)$. Also the equation for the map r (cf. section 4.2) which appears in the decoupling of the optimality system (4.6)-(4.9) (cf. Theorem 4.1) does not belong to the class of hereditary systems; it could be interpreted as a special type of transport equation. But the above mentionned equations are not required to use J.C. NÉDELEC [1]'s method (see also A. BENSOUSSAN [1]). As a result we obtain a numerical scheme to compute the covariance operator for which we have convergence proofs. A similar approach has been successfully used by M.C. DELFOUR [1],[2],[3] and [4] to numerically solve the Riccati differential equation describing the evolution of the feedback operator for the linear quadratic optimal control problem.

Through several simple examples the numerical results cast some light on the nature of the operator $\Pi(t)$. In particular delays seem to create discontinuities in the derivative of the map $\alpha \mapsto \Pi_{01}(t,\alpha)$ and the covariance operator for this problem does not seem to have the same properties as the feedback operator in M.C. DELFOUR [1],[2],[3],[4].

---

* This research was supported in part by a "Subvention F.C.A.C. du Ministère de l'Éducation du Québec" and by National Research Council (Canada) Grant A8730 at the Centre de Recherches Mathématiques, Université de Montréal, Montréal, Québec, Canada.

Notation. Let $\mathbb{R}$ be the field of all real numbers. Let $\mathbb{R}^n$ be the Euclidean real Hilbert space of finite dimension n ($n\geq 1$, an integer). Given two real Hilbert spaces X and Y we denote by $\mathcal{L}(X,Y)$ the real Banach space of all continuous linear maps $L:X \to Y$ endowed with the natural norm $\|L\|$. The adjoint of L in $\mathcal{L}(X,Y)$ will be denoted by $L^* \in \mathcal{L}(Y,X)$. When X=Y, we write $\mathcal{L}(X,X)$ and the identity in $\mathcal{L}(X)$ is denoted by I_X. An element L of $\mathcal{L}(X)$ is said to be self adjoint (resp. positive or ≥ 0) when $L^*=L$ (resp. for all x in X, the inner product of Lx and x in X is positive or zero).

Given F a closed convex subset of $\mathbb{R}^n$ and E a real Banach space, we denote by $\mathcal{L}^p(F;E)$ the real vector space of all m-measurable (m, the Lebesgue measure on F) maps $F \to E$ which are p-integrable ($1\leq p<\infty$) or essentially bounded ($p=\infty$). We denote by $L^p(F;E)$ the natural real Banach space associated with $\mathcal{L}^p(F;E)$ and by $\| \; \|_p$ its natural norm. When F is an interval we use the notation $L^p(a,b;E)$ where a and b are the end points of the interval. Given t_0 in $\mathbb{R}$ and $t_0 < t_1 \leq +\infty$, we denote by $H^1(t_0,t_1;E)$ the Sobolev space of all maps x in $L^2(t_0,t_1;E)$ with a distributional derivative Dx in $L^2(t_0,t_1;E)$. Let I(a,b) denote the interval $]-\infty,+\infty[\cap [a,b]$. C(a,b;E) will be the real Banach space of all bounded continuous maps $I(a,b) \to E$ endowed with the natural sup-norm. Given two reals $t_0 < t_1$ in $\mathbb{R}$ we define

$$\mathcal{P}(t_0,t_1) = \{(t,s) \in [t_0,t_1] \times [t_0,t_1]: t \geq s\}.$$

Finally we shall denote by $L^2_{loc}(0,\infty;E)$ and $H^1_{loc}(0,\infty;E)$ the Fréchet spaces of all maps $[0,\infty[\to E$, the restriction of which to each compact interval of the form [0,T] belongs to $L^2(0,T;E)$ and $H^1(0,T;E)$.

2. System description and formulation of the problem.

2.1. Deterministic features.

Let $X = \mathbb{R}^n$, $U = \mathbb{R}^m$, $Z = \mathbb{R}^k$ for some positive non-zero integers n, m and k. Let $(\cdot,\cdot)$ and $|\cdot|$ denote the inner product and norm in X. Let $(\cdot,\cdot)_U$ (resp. $(\cdot,\cdot)_Z$) and $|\cdot|_U$ (resp. $|\cdot|_Z$) denote the inner product and norm in U (resp. Z). We are given an integer $N \geq 1$, real numbers $0 < a < +\infty$, $0 < T < +\infty$, $-a = \theta_N < \ldots < \theta_{i+1} < \theta_i < \ldots < \theta_0 = 0$. The product space $H = X \times L^2(-a,0;X)$ is endowed with inner product and norm

$$(2.1) \qquad ((h,k)) = (h^0,k^0) + \int_{-a}^{0} (h^1(\theta),k^1(\theta))\,d\theta, \quad \|h\| = ((h,h))^{\frac{1}{2}}.$$

Let $A_i:[0,\infty[\to \mathcal{L}(X)$ $(i=0,\ldots,N)$ and $B:[0,\infty[\to \mathcal{L}(U,X)$ be measurable and bounded maps on compact intervals. Let $A_{01}:[0,\infty[\times [-a,0] \to \mathcal{L}(X)$ be also a measurable and bounded map on sets of the form $[0,t] \times [-a,0]$ for each $t\geq 0$. For f in $L^2_{loc}(0,\infty,X)$ and ξ in $L^2_{loc}(0,\infty;U)$, we consider the hereditary differential system (HDS)

$$(2.2) \qquad \begin{cases} \dfrac{dx}{dt}(t) = \displaystyle\sum_{i=0}^{N} A_i(t) \begin{cases} x(t+\theta_i), & t+\theta_i \geq 0 \\ h^1(t+\theta_i), & \text{otherwise} \end{cases} + \int_{-a}^{0} A_{01}(t,\theta) \begin{cases} x(t+\theta), & t+\theta \geq 0 \\ h^1(t+\theta), & \text{otherwise} \end{cases} d\theta \\ \qquad + B(t)\xi(t) + f(t) \quad \text{in} \quad]0,T[\\ x(0) = h^0, \; h = (h^0,h^1) \in H. \end{cases}$$

It is easily shown that for given h, ξ and f equation (2.2) has a unique solution x in $H^1_{loc}(0,\infty;X)$ and that the map $(h,\xi) \mapsto x: H \times L^2_{loc}(0,\infty;U) \to H^1_{loc}(0,\infty;X)$ is affine and continuous. When h belongs to the subspace

$$(2.3) \qquad V = \{(h(0),h) : h \in H^1(-a,0;X)\}$$

of H, equation (2.1) can be made equivalent to the following partial differential equation (PDE) by introducing the function $y(t,\theta) = x(t+\theta)$ if $t+\theta \geq 0$ and $h(t+\theta)$ if $t+\theta < 0$:

PDE $\qquad \frac{\partial}{\partial t}y(t,\theta) = \frac{\partial}{\partial \theta}y(t,\theta) \quad \text{in }]0,\infty[\times]-a,0[$

BOUNDARY CONDITION $\qquad \frac{\partial}{\partial t}y(t,0) = \sum_{i=0}^{N} A_i(t)y(t,\theta_i)+B(t)\xi(t)+\int_{-a}^{0} A_{01}(t,\theta)y(t,\theta)d\theta+f(t) \quad \text{in }]0,\infty[$

INITIAL CONDITION $\qquad y(0,\theta) = h(\theta) \quad \text{in } [-a,0].$

The above formulation very naturally leads to an operational differential equation . To see this we define the state of system (2.2) as an element $\tilde{x}(t)$ of H,

$$(2.4) \qquad \tilde{x}(t)^0 = x(t), \ \tilde{x}(t)^1(\theta) = \begin{Bmatrix} x(t+\theta) \ , & -t<\theta\leq 0 \\ h^1(t+\theta), & -a\leq\theta\leq -t \end{Bmatrix}.$$

We introduce the continuous linear operators $\tilde{A}_0(t):V \to X$, $\tilde{A}_1(t):V \to L^2(-a,0;X)$, $\tilde{A}(t):V \to H$, $\tilde{B}(t):U \to H$ and the vector $\tilde{f}(t)$ in H:

$$(2.5) \qquad \tilde{A}_0(t)h = \sum_{i=0}^{N} A_i(t)h(\theta_i)+\int_{-a}^{0} A_{01}(t,\theta)h(\theta)d\theta, (\tilde{A}_1(t)h)(\theta) = \frac{dh}{d\theta}(\theta)$$

$$(2.6) \qquad \tilde{A}(t)h = (\tilde{A}_0(t)h,\tilde{A}_1(t)h), \ \tilde{B}(t)w = (B(t)w,0), \ \tilde{f}(t) = (f(t),0).$$

The proofs of the following two theorems can be found in M.C. DELFOUR [5].

<u>Theorem 2.1</u>. For a given T>0 and all h in V, f in $L^2_{loc}(0,\infty;X)$ and ξ in $L^2_{loc}(0,\infty;U)$, $\tilde{x}$ is the unique solution in $W(0,T) = \{z \in L^2(0,T;V):Dz \in L^2(0,T;H\}$ (Dz denotes the distributional derivative of z) of the equation

$$(2.7) \qquad \frac{dz}{dt}(t) = \tilde{A}(t)z(t) + \tilde{B}(t)\xi(t) + \tilde{f}(t) \quad \text{in }]0,T[, \ z(0) = h,$$

and the map $(h,\xi) \mapsto \tilde{x}:V \times L^2(0,T;X) \to W(0,T)$ is affine and continuous (V is endowed with the H^1-topology and W(0,T) with the norm $\|z\|_{W(0,T)}=[\|z\|^2_{L^2(0,T;V)}+\|Dz\|^2_{L^2(0,T;H)}]^{\frac{1}{2}}$). By density there exists a lifting of this map to a continuous affine map $H \times L^2(0,T;U) \to C(0,T;H)$. Moreover there exists an evolution operator $\tilde{\Phi}:\mathcal{P}(0,T) \to \mathcal{L}(H)$ such that (i) $\forall$ h $\in$ H, $(t,s) \mapsto \tilde{\Phi}(t,s)h$ is continuous, (ii) $\forall$ (r,s), $0\leq r\leq s\leq t\leq T$, $\tilde{\Phi}(t,r) = \tilde{\Phi}(t,s)\tilde{\Phi}(s,r)$, and (iii) the solution of (2.7) can be written

$$(2.8) \qquad \tilde{x}(t) = \tilde{\Phi}(t,0)h + \int_0^t \tilde{\Phi}(t,r)[\tilde{B}(r)\xi(r)+\tilde{f}(r)]dr.$$

We shall also need the <u>adjoint state</u>. Let H' (resp. V') be the topological dual of H (resp. V). We identify elements of H and H' and denote by $\Lambda:V \to H$ and $\Lambda^*:H \to V'$

the continuous dense injections as in the theory of operational differential equations (cf. LIONS-MAGENES [1] and J.L. LIONS [1]).

Theorem 2.2. Given k in H and g in $L^2(0,T;H)$ the equations

$$\frac{dz}{dt}(t) + \tilde{A}(t)^*z(t) + \Lambda^*g(t) = 0 \quad \text{in} \quad]0,T[, \; z(T) = k, \tag{2.9}$$

has a unique solution $p(\cdot;k,g)$ in the space

$$W^*(0,T) = \{z \in C(0,T;H) : Dz \in L^2(0,T;V')\}. \tag{2.10}$$

The map $(k,g) \mapsto p(\cdot;k,g): H \times L^2(0,T;H) \to W^*(0,T)$ is linear and continuous ($W^*(0,T)$ is endowed with the norm $\|z\|_{W^*(0,T)} = \|z\|_{C(0,T;H)} + \|Dz\|_{L^2(0,T;V')}$). Moreover

$$p(t;k,g) = \tilde{\Phi}(T,t)^*k + \int_t^T \tilde{\Phi}(r,t)^*g(r)dr. \quad \square \tag{2.11}$$

2.2. Stochastic features.

We now consider a noisy initial condition, that is

$$x(0) = h^0 + \zeta^0, \; x(\theta) = h^1(\theta) + \zeta^1(\theta), \; -a \le \theta < 0, \tag{2.12}$$

where $\zeta = (\zeta^0, \zeta^1)$ belong to H. From now on ξ and ζ will be the noise at the input and the noise in the initial condition, respectively. We shall also assume an observation of the form

$$z(t) = C(t)x(t) + \eta(t), \tag{2.13}$$

where $C: [0,T] \to \mathcal{L}(X,Z)$ is measurable and bounded on compact intervals and η represents the error in measurement. As in A. BENSOUSSAN [1] $\{\zeta^0, \zeta^1, \xi, \eta\}$ will be modelled as a Gaussian linear random functional on the Hilbert space $\Phi = H \times L^2(0,T;U) \times L^2(0,T;Z)$ with zero mean and covariance operator

$$\Xi = \begin{bmatrix} P_0 & 0 & 0 & 0 \\ 0 & P_1(\theta) & 0 & 0 \\ 0 & 0 & Q(t) & 0 \\ 0 & 0 & 0 & R(t) \end{bmatrix}. \tag{2.14}$$

It will be convenient to introduce the covariance operator P in $\mathcal{L}(H)$ defined as follows

$$(\!(Ph, \bar{h})\!) = (P_0 h^0, \bar{h}^0) + \int_{-a}^{0} (P_1(\theta)h^1(\theta), \bar{h}^1(\theta))d\theta. \tag{2.15}$$

In view of the continuity of the map $(h,\xi) \mapsto x$ and the properties of the image of a linear random functional under an affine continuous map, we can look at $x(t)$ as a Gaussian linear random functional on X (for any t), where the mean of $x(t)$, $\bar{x}(t)$, is a solution of (2.2) with $\xi = 0$ and $\zeta = 0$. But it is easy to check that the mean of $\tilde{x}(t)$, $\bar{\tilde{x}}(t)$, is obtained from the mean of $x(t)$ and the mean, $\bar{h}^1(\theta)$, of $h^1(\theta)$ as follows:

$$\bar{\tilde{x}}(t)^0 = \bar{x}(t), \; \bar{\tilde{x}}(t)^1(\theta) = \begin{cases} \bar{x}(t+\theta), & t+\theta \ge 0 \\ \bar{h}^1(t+\theta), & \text{otherwise} \end{cases}. \tag{2.16}$$

As a result $\bar{\tilde{x}}(t)$ is a solution of the state equation (2.7) with $\xi = 0$ and $\zeta = 0$ and the covariance operator $\Gamma(t)$ of $\tilde{x}(t)$ is a "weak solution" of the equation

$$(2.17)\qquad \frac{d\Gamma}{dt}(t) = \tilde{A}(t)\Gamma(t) + \Gamma(t)\tilde{A}(t)^* + \tilde{B}(t)Q(t)\tilde{B}(t)^*,\ \Gamma(0) = P.$$

2.3. Formulation of the problem.

For each T we want to determine the best estimator of the linear random functional $\tilde{x}(T)$ with respect to the linear random functional $z(s)$, $0 \le s \le T$. It is a linear random functional $\hat{\tilde{x}}(T)$ which can be obtained (see A. BENSOUSSAN [1]) through the following control problem. We start with the deterministic system (2.7) with the initial condition (2.17) at time 0 and consider the variables ξ and $\zeta = (\zeta^0,\zeta^1)$ as control variables. We want to minimize the cost function for a given h.

$$(2.18)\qquad J_T(\xi,\zeta,h) = (\!(P^{-1}\zeta,\zeta)\!) + \int_0^T (Q(t)^{-1}\xi(t),\xi(t))_U dt + \int_0^T (R(t)^{-1}(z(t)-C(t)x(t), z(t)-C(t)x(t))_Z dt$$

(provided that P, Q(t) and R(t) be invertible almost everywhere).

3. Solution of the optimal control problem (2.7)-(2.17)-(2.18).

It will be technically advantageous to work in the state space. For this purpose we redefine the cost function (2.18) in terms of the state

$$(3.1)\qquad \frac{d\tilde{x}}{dt}(t) = \tilde{A}(t)\tilde{x}(t) + \tilde{B}(t)\xi(t) + \tilde{f}(t) \quad \text{in }]0,T[,\ \tilde{x}(0) = h + \zeta$$

$$(3.2)\qquad J_T(\xi,\zeta) = (\!(P^{-1}\zeta,\zeta)\!) + \int_0^T (Q(t)^{-1}\xi(t),\xi(t))_U dt + \int_0^T (R(t)^{-1}(z(t)-\tilde{C}(t)\tilde{x}(t)), z(t)-\tilde{C}(t)\tilde{x}(t))_Z dt$$

where $\tilde{C}(t)h = C(t)h^0$. It will be convenient to introduce the variables y and $\bar{y}$:

$$(3.3)\qquad \frac{dy}{dt}(t) = \tilde{A}(t)y(t) + \tilde{B}(t)\xi(t) \quad \text{in }]0,T[,\ y(0) = \zeta,$$

$$(3.4)\qquad \frac{d\bar{y}}{dt}(t) = \tilde{A}(t)\bar{y}(t) + \tilde{f}(t) \quad \text{in }]0,T[,\ \bar{y}(0) = h.$$

We notice that $\tilde{x}(t) = y(t) + \bar{y}(t)$ and rewrite the cost function (3.2) in terms of y

$$(3.5)\quad J_T(\xi,\zeta) = (\!(P^{-1}\zeta,\zeta)\!) + \int_0^T (Q(t)^{-1}\xi(t),\xi(t))_U dt + \int_0^T (R(t)^{-1}\tilde{C}(t)y(t), \tilde{C}(t)y(t))_Z dt$$
$$-2\int_0^T (R(t)^{-1}[z(t)-\tilde{C}(t)\bar{y}(t)], \tilde{C}(t)y(t))_Z dt + \text{terms independent of } \xi \text{ and } \zeta.$$

The pair $(\hat{\xi},\hat{\zeta})$ which minimizes the cost function over all (ξ,ζ) in $L^2(0,T;U) \times H$ is characterized by

$$(3.6)\quad (\!(P^{-1}\hat{\zeta},\zeta)\!) + \int_0^T (Q(t)^{-1}\hat{\xi}(t),\xi(t))_U dt + \int_0^T (R(t)^{-1}\tilde{C}(t)\hat{y}(t), \tilde{C}(t)y(t))_Z dt$$
$$= \int_0^T (\tilde{C}(t)\hat{y}(t), R(t)^{-1}[z(t)-\tilde{C}(t)\bar{y}(t)])_Z dt \quad \forall\ \xi,\zeta.$$

By introducing the adjoint system

$$(3.7)\quad \begin{cases} \frac{d\hat{p}}{dt}(t) + \tilde{A}(t)^*\hat{p}(t) + \Lambda^*\tilde{C}(t)^*R(t)^{-1}[\tilde{C}(t)\hat{y}(t)-(z(t)-\tilde{C}(t)\bar{y}(t))] = 0 \quad \text{in} \quad]0,T[\\ \hat{p}(T) = 0, \end{cases}$$

we obtain that $(\hat{\xi},\hat{\zeta})$ are characterized by

$$(3.8)\qquad \hat{\zeta} = -P\hat{p}(0), \quad \hat{\xi}(t) = -Q(t)\tilde{B}(t)^*\hat{p}(t).$$

By substituting $\hat{\zeta}$ and $\hat{\xi}$ in (3.3) we obtain

$$(3.9)\qquad \frac{d\hat{y}}{dt}(t) = \tilde{A}(t)\hat{y}(t) - \tilde{B}(t)Q(t)\tilde{B}(t)^*\hat{p}(t) \quad \text{in} \quad]0,T[,\ \hat{y}(0) = -P\hat{p}(0),$$

and system (3.9)-(3.8) is called the <u>optimality system</u>. The optimal $\hat{\hat{x}}$ corresponding to $(\hat{\zeta},\hat{\xi})$ is given by $\hat{\hat{x}}(t) = \hat{y}(t) + \bar{y}(t)$.

4. <u>Dual optimal control problem.</u>

We introduce the maps $g:[0,T] \to H$ and $N:[0,T] \to \mathcal{L}(H)$,

$$(4.1)\qquad g(t) = \tilde{C}(t)^*R(t)^{-1}[\tilde{C}(t)\bar{y}(t)-z(t)],\ N(t) = \tilde{B}(t)Q(t)\tilde{B}(t)^*,$$

and we consider the following control system and its associated cost function

$$(4.2)\qquad \frac{dp}{dt}(t) + \tilde{A}(t)^*p(t) + \Lambda^*[\tilde{C}(t)^*w(t)+g(t)] = 0 \quad \text{in} \quad]0,T[,\ p(T) = k$$

$$(4.3)\qquad J^*(w,k) = (\!(Pp(0),p(0))\!) + \int_0^T [(\!(N(t)p(t),p(t))\!)+ (R(t)w(t),w(t))_Z]dt,$$

where k and w belong to H and $L^2(0,T;Z)$, respectively. Given k there exists a unique $\hat{w}$ in $L^2(0,T;Z)$ which minimizes $J^*(w,k)$ over all w in $L^2(0,T;Z)$. The minimizing $\hat{w}$ is characterized by Euler's equation:

$$(4.4)\qquad (\!(P\hat{p}(0),q(0))\!) + \int_0^T [(\!(N(t)\hat{p}(t),q(t))\!)+ (R(t)\hat{w}(t),w(t))_Z]dt = 0,\ \forall\ w,$$

where

$$(4.5)\qquad \frac{dq}{dt}(t) + \tilde{A}(t)^*q(t) + \Lambda^*\tilde{C}(t)^*w(t) = 0 \quad \text{in} \quad]0,T[,\ q(T) = 0.$$

If we introduce the dual system of system (4.2)

$$(4.6)^*\qquad \frac{d\hat{y}}{dt}(t) = \tilde{A}(t)\hat{y}(t) - N(t)\hat{p}(t) \quad \text{in} \quad]0,T[,\ \hat{y}(0) = -P\hat{p}(0),$$

identity (4.4) reduces to

$$(4.7)\qquad \forall\ m \in L^2(0,T;Z),\ \int_0^T (R(t)\hat{w}(t)-\tilde{C}(t)\hat{y}(t),w(t))_Z dt = 0.$$

Since R(t) is invertible (4.7) is equivalent to

$$(4.8)\qquad \hat{w}(t) = R(t)^{-1}\tilde{C}(t)\hat{y}(t), \text{ a.e. in } [0,T].$$

If we now substitute identity (4.8) into equation (4.2) we obtain

$$(4.9)\qquad \frac{d\hat{p}}{dt}(t) + \tilde{A}(t)^*\hat{p}(t) + \Lambda^*[\tilde{C}(t)^*R(t)^{-1}\tilde{C}(t)\hat{y}(t)+g(t)] = 0 \quad \text{in} \quad]0,T[,\ \hat{p}(T) = k.$$

---

* This equation must be interpreted in an appropriate weak sense, namely,

$$\hat{y}(t) = -\tilde{\Phi}(t,0)P\hat{p}(0) - \int_0^t \tilde{\Phi}(t,r)N(r)\hat{p}(r)dr \quad \text{in} \quad [0,T].$$

Equations (4.6)-(4.9) form the optimality system of problem (4.2)-(4.3). The reader will notice that system (4.6)-(4.9) with k=0 is identical to the optimality system (3.7)-(3.9) with the significant difference that (4.6)-(4.9) was derived without the hypothesis that P and Q(t) be invertible.

Remark. In the formulation of the dual optimal control problem $\tilde{C}$ can be any strongly measurable map $[0,T] \to \mathcal{L}(H,Z)$ which is bounded in [0,T]. It is not necessary to restrict ourselves to $\tilde{C}$'s of the form $\tilde{C}(t)h = C(t)h^0$ for some $C:[0,T] \to \mathcal{L}(X,Z)$.

We shall now proceed to the decoupling of optimality system (4.6)-(4.9) and to the study of the decoupling operator as in J.L. LIONS [2] and A. BENSOUSSAN [1].

In the sequel we shall use the notation $M(t) = \tilde{C}(t)^*R(t)^{-1}\tilde{C}(t)$.

Theorem 4.1. Let $\hat{p}$ and $\hat{y}$ be the solution of system (4.6)-(4.9). Then there exists a family of linear operators $\Pi(t):H \to H$ and a family of elements r(t) in H, $0 \le t \le T$, such that

$$\hat{y}(t) = -\Pi(t)\hat{p}(t) + r(t), \quad \text{in} \quad [0,T]. \tag{4.10}$$

$\Pi(t)$ and r(t) are obtained in the following manner: (i) we solve the system

$$\begin{cases} \frac{d\beta}{dt}(t) = \tilde{A}(t)\beta(t) - N(t)\gamma(t) \quad \text{in} \quad]0,s[,\ \beta(0) = -P\gamma(0) \\ \frac{d\gamma}{dt}(t) + \tilde{A}(t)^*\gamma(t) + \Lambda^*M(t)\beta(t) = 0 \quad \text{in} \quad]0,s[,\ \gamma(s) = k \end{cases} \tag{4.11}$$

and $\Pi(s)k = -\beta(s)$; (ii) we solve the system

$$\begin{cases} \frac{d\eta}{dt}(t) = \tilde{A}(t)\eta(t) - N(t)\chi(t) \quad \text{in} \quad]0,s[,\ \eta(0) = -P\chi(0) \\ \frac{d\chi}{dt}(t) + \tilde{A}(t)^*\chi(t) + \Lambda^*[M(t)\eta(t)+g(t)] = 0 \quad \text{in} \quad]0,s[,\ \chi(s) = 0 \end{cases} \tag{4.12}$$

and $r(s) = \eta(s)$. □

4.1. Study of the operator $\Pi(t)$.

In order to study the operator $\Pi(t)$ we make use of Theorem 4.1 and consider system (4.2) with g=0 and the cost function (4.3) in a time interval [0,s] for some s in]0,T].

Theorem 4.2. (i) If we denote by γ (resp. $\bar{\gamma}$) the solution of equation

$$\frac{d\gamma}{dt}(t) + [\tilde{A}(t)^*-\Lambda^*M(t)\Pi(t)]\gamma(t) = 0 \quad \text{in} \quad]0,s[,\ \gamma(s) = k \ (\text{resp. } \bar{k}), \tag{4.13}$$

then

$$(\!(\Pi(s)k,\bar{k})\!) = (\!(P\gamma(0),\bar{\gamma}(0))\!) + \int_0^s (\!([N(r)+\Pi(r)M(r)\Pi(r)]\gamma(r),\bar{\gamma}(r))\!)dr \tag{4.14}$$

and in particular if $\hat{w}$ is the optimal control corresponding to k

$$J_s^*(\hat{w},k) = (\!(\Pi(s)k,k)\!). \tag{4.15}$$

(ii) The operator $\Pi(s)$ is a self adjoint element of $\mathcal{L}(H)$, there exists a constant $c > 0$ (independent of s and h) such that

$$\forall\, s,\ \forall\, k,\ \|\Pi(s)k\| \le c\|k\|, \tag{4.16}$$

and the map $s \mapsto \Pi(s):[0,T] \to \mathcal{L}(H)$ is weakly continuous (hence strongly measurable and bounded). $\Pi(s)$ can be decomposed in a unique way into a matrix of operators

$$\begin{bmatrix}\Pi_{00}(s), & \Pi_{01}(s)\\ \Pi_{10}(s), & \Pi_{11}(s)\end{bmatrix} \begin{matrix}\Pi_{00}(s) \in \mathcal{L}(X), \ \Pi_{01}(s) \in \mathcal{L}(L^2(-a,0;X),X)\\ \Pi_{10}(s) \in \mathcal{L}(X,L^2(-a,0;X)), \ \Pi_{11}(s) \in \mathcal{L}(L^2(-a,0;X)).\end{matrix} \tag{4.17}$$

Moreover

$$\Pi_{00}(s)^* = \Pi_{00}(s) \geq 0, \ \Pi_{01}(s) = \Pi_{10}(s)^*, \ \Pi_{11}(s)^* = \Pi_{11}(s) \geq 0, \tag{4.18}$$

and

$$J_s^*(\hat{m},k) = (\Pi_{00}(s),k^0,k^0) + 2(\Pi_{01}(s)k^1,k^0) + (\Pi_{11}(s)k^1,k^1)_2. \tag{4.19}$$

(iii) The equation

$$y_s(t) = \tilde{\Phi}(t,s)h - \int_s^t \tilde{\Phi}(t,r)\Pi(r)M(r)y_s(r)dr \quad \text{in} \quad [s,T] \tag{4.20}$$

has a unique solution y_s in $C(s,T;H)$ which generates an evolution operator $\Lambda(t,s)$ defined as $\Lambda(t,s)h = y_s(t)$ with the following properties:

a) $\forall\ (t,s) \in \mathcal{P}(0,T),\ \Lambda(t,s) \in \mathcal{L}(H)$;

b) $\forall\ 0 \leq s \leq r \leq t \leq T,\ \Lambda(t,s) = \Lambda(t,r)\Lambda(r,s)$;

c) $\forall\ h \in H,\ (t,s) \mapsto \Lambda(t,s)h:\mathcal{P}(0,T) \to H$ is continuous.

Equation (4.14) can now be rewritten in the form

$$((\Pi(s)k,\bar{k})) = ((P\Lambda(s,0)^*k,\Lambda(s,0)^*\bar{k})) + \int_0^s (([N(r)+\Pi(r)M(r)\Pi(r)]\Lambda(s,r)^*k,\Lambda(s,r)\bar{k}))dr. \ \square \tag{4.21}$$

Equation (4.21) is the "integral form" of the desired Riccati operator differential equation. It is identical to the one in MITTER-VINTER [1]. Formally one should obtain in $]0,T[$ an equation of the form

$$\frac{d\Pi}{dt}(t) = \Pi(t)\tilde{A}(t)^* + \tilde{A}(t)\Pi(t) - \Pi(t)M(t)\Pi(t) + N(t), \ \Pi(0) = P. \tag{4.22}$$

However it is not easy to interpret this equation properly since $\tilde{A}(t)$ and $\tilde{A}(t)^*$ are unbounded operators which depend on time. In particular

$$\bigcap_{0\leq t\leq T} \{k \in H:\tilde{A}(t)^*k \in H\} \tag{4.23}$$

is not necessarily dense in H (cf. R. VINTER [1] and MITTER-VINTER [1]).

<u>Remark</u>. Notice that the equation

$$\frac{dp}{dt}(t) + \tilde{A}(t)^*p(t) - \Lambda^*M(t)\Pi(t)p(t) = 0 \quad \text{in} \quad]0,T[, \ p(T) = k$$

is perfectly legitimate and that its solution is $p(t) = \Lambda(T,t)^*k$.

4.2. <u>Study of the function r(t)</u>.

In order to study the function r we consider the problem (4.2)-(4.3) with k=0. By Theorem 4.1 we know that

$$r(t) = \hat{y}(t) + \Pi(t)\hat{p}(t). \tag{4.24}$$

Straightforward computations using equations (4.6), (4.9) and (4.21) will show that

$$\begin{cases}\frac{dr}{dt}(t) = [\tilde{A}(t)-\Pi(t)M(t)\Lambda]r(t) - \Pi(t)g(t), \quad \text{in} \quad]0,T[,\\ r(0) = 0,\end{cases} \tag{4.25}$$

where the above equation is to be interpreted as

$$(4.26) \qquad r(t) = -\int_0^t \Lambda(t,s)\Pi(s)g(s)ds \quad \text{in} \quad [0,T]$$

or

$$(4.27) \qquad r(t) = -\int_0^T \tilde{\Phi}(t,s)\Pi(s)[M(s)r(s)+g(s)]ds \quad \text{in} \quad [0,T].$$

5. Approximation of the dual system.

In this section we shall exploit earlier results of M. DELFOUR [1],[4] on the standard optimal control problem. We assume that we can find non zero positive integers $M, L, L_0, \ldots, L_N$ and a discretization step $\delta > 0$ such that $T = M\delta$, $a = L\delta$, $\theta_i = -L_i\delta$, $i=0,\ldots,N$.

5.1. Approximation of initial data.

We approximate the product space $H = X \times L^2(-b,0;X)$ by the finite dimensional space $H^\delta = X^{L+1}$ endowed with inner products

$$(5.1) \qquad (\underline{h},\underline{k})_\delta = (h_0,k_0) + \delta \sum_{\ell=-L}^{-1} (h_\ell,k_\ell), (\underline{h},\underline{k})_L = \sum_{\ell=-L}^{0} (h_\ell,k_\ell).$$

We introduce the maps

$$(5.2) \qquad h = (h^0,h^1) \mapsto r_\delta(h) = (h^0,h^1_{-1},\ldots,h^1_{-L}) : H \to H^\delta,$$

$$(5.3) \qquad \underline{h} = (h_0,h_{-1},\ldots,h_{-L}) \mapsto q_\delta(\underline{h}) = (h_0, \sum_{\ell=-L}^{-1} h_\ell \chi_\ell) : H^\delta \to H,$$

where χ_ℓ is the characteristic function of $[\ell\delta,(\ell+1)\delta[$ and

$$(5.4) \qquad h^1_\ell = \frac{1}{\delta} \int_{\ell\delta}^{(\ell+1)\delta} h^1(\theta)d\theta, \; -L \le \ell \le -1.$$

It is readily seen that $\|q_\delta r_\delta(h)\| \le \|h\|$. We shall also need the transformation

$$i_\delta : H^\delta \to H^\delta, \; [i_\delta(\underline{h})]_0 = h_0, \; [i_\delta(\underline{h})]_\ell = \delta^{\frac{1}{2}} h_\ell, \; \ell=-L,\ldots,-1.$$

If we introduce the map $I_\delta = i_\delta^2$, we notice that for all $\underline{h}$ and $\underline{k}$

$$(\underline{h},\underline{k})_\delta = (i_\delta \underline{h}, i_\delta \underline{k})_L = (i_\delta^2 \underline{h},\underline{k})_L = (I_\delta \underline{h},\underline{k})_L.$$

5.2. Approximation of the differential equation.

We associate with A_i a family of matrices and with f an element $\underline{f}$ in X^M

$$(5.5) \qquad A_i^m = \frac{1}{\delta} \int_{m\delta}^{(m+1)\delta} A_i(t)dt, \; (i=0,\ldots,N), \; f^m = \frac{1}{\delta} \int_{m\delta}^{(m+1)\delta} f(t)dt, \; m=0,\ldots,M-1.$$

With A_{01} we associate the family of matrices

$$(5.6) \qquad \begin{cases} A_{01}^{m,0} = \dfrac{1}{\delta^2} \displaystyle\int_{m\delta}^{(m+1)\delta} dt \int_{-(t-m\delta)}^{0} d\theta \, A_{01}(t,\theta), A_{01}^{m,-L} = \dfrac{1}{\delta^2} \displaystyle\int_{m\delta}^{(m+1)\delta} dt \int_{-b}^{-a+(m+1)\delta-t} d\theta \, A_{01}(t,\theta) \\ A_{01}^{m,\ell} = \dfrac{1}{\delta^2} \displaystyle\int_{m\delta}^{(m+1)\delta} dt \int_{(m+\ell)\delta-t}^{(m+\ell+1)\delta-t} d\theta \, A_{01}(t,\theta), \; \ell=-1,\ldots,-(L-1), \end{cases}$$

We construct a finite dimensional approximation $U^\delta = U^M$ to the space of control maps $L^2(0,T;U)$. We define the maps

(5.7) $$\xi \mapsto r_\delta(\xi) = (\xi_0,\dots,\xi_{M-1}):L^2(0,T;U) \to U^M,$$

(5.8) $$\xi_m = \frac{1}{\delta}\int_{m\delta}^{(m+1)\delta} \xi(t)dt,\ m=0,\dots,M-1,$$

(5.9) $$\xi = (\xi_0,\dots,\xi_{M-1}) \mapsto q_\delta(\xi) = \sum_{m=0}^{M-1} \xi_m\chi_m : U^M \to L^2(0,T;U),$$

where χ_m is the characteristic function of $[m\delta,(m+1)\delta[$. It is reasily seen that $\|q_\delta \mathbf{r}_\delta(\xi)\|_2 \le \|\xi\|_2$. It will be clear from the context whether q_δ and r_δ are associated with $L^2(0,T;U)$ or with the product space $H = X \times L^2(-b,0;X)$. Finally we associate with B the following family of matrices

(5.10) $$B^m = \frac{1}{\delta}\int_{m\delta}^{(m+1)\delta} B(t)dt,\ m=0,\dots,M-1.$$

We now associate with $\underline{h}$ in H^δ, $\underline{\xi} = (\xi_0,\dots,\xi_{M-1})$ and $\underline{f} = (f_0,\dots,f_{M-1})$ the following numerical scheme

(5.11) $$\begin{cases} x_{m+1}-x_m=\delta\left[\sum_{i=0}^{N} A_i^m \begin{cases} x_{m-L_i}, & m-L_i\ge 0 \\ h_{m-L_i}, & m-L_i<0 \end{cases} + \sum_{\ell=-L}^{0} \delta\, A_{01}^{m,\ell} \begin{cases} x_{m+\ell}, & m+\ell\ge 0 \\ h_{m+\ell}, & m+\ell<0 \end{cases} + B^m\xi_m + f^m\right], \\ \qquad m=0,\dots,M-1, \\ x_0 = h_0. \end{cases}$$

The following propositions summarize the results we shall need.

<u>Proposition 5.1.</u> (Stability). We denote by $(x_0^\delta,\dots,x_{M-1}^\delta)$ the unique solution of (5.11) corresponding to $\underline{h} = r_\delta(h)$, $\underline{\xi} = r_\delta(\xi)$ and $\underline{f} = r_\delta(f)$ for h in H, $\boldsymbol{\xi}$ in $L^2(0,T;U)$ and f in $L^2(0,T;X)$. We define the maps

(5.12) $$x^\delta(t) = \sum_{m=0}^{M-1} x_m^\delta\chi_m(t),\ x^\delta(T) = x_M^\delta,\quad Dx^\delta(t) = \sum_{m=0}^{M-1} \frac{x_{m+1}^\delta - x_m^\delta}{\delta}\chi_m(t),$$

where χ_m denotes the characteristic function of $[m\delta,(m+1)\delta[$. As δ goes to zero there exists a constant $c > 0$ (independent of h, f, v and δ) such that

(5.13) $$\max\{|x_m^\delta|:m=0,\dots,M\} + \|x^\delta\|_\infty + \|Dx^\delta\|_2 \le c[\|h\|+\|f\|_2+\|\boldsymbol{\xi}\|_2].\ \square$$

<u>Proposition 5.2.</u> (Convergence). Fix h in H, f in $L^2(0,T;X)$, ξ in $L^2(0,T;U)$. As δ goes to zero with $M\delta = T$

(5.14) $$\max\{|x_m^\delta - x(m\delta)|:0\le m\le M\} + \|x^\delta - x\|_2 + \|Dx^\delta - Dx\|_2$$

converges to zero, where x is the solution in $W^{1,2}(0,T;X)$ of equation (2.2). $\square$

<u>Corollary</u>. Assume that $A_1,\dots,A_N$ and B are constant matrices and that A_{01} is identically zero. As δ goes to zero there exists a constant $c > 0$ (independent of δ, h, v and f) such that

(5.15) $$\|x-x^{\delta}\|_2 + \max\{|x(m\delta)-x_m^{\delta}|:0\le m\le M\} \le c\delta\|Dx\|_2. \quad \square$$

5.3. Approximation of the differential equation for the state.

We now introduce an explicit finite difference scheme to approximate the differential equation for the state. Given $\underline{h}$ in H^{δ}, we want to determine $\{x_{m,n}:0\le m\le M, -L\le n\le 0\}$ from the following set of equations

(5.16) $$x_{m+1,0}-x_{m,0} = \delta[\sum_{i=0}^{N} A_i^m x_{m,-L_i} + \delta\sum_{\ell=-L}^{0} A_{01}^{m,\ell} x_{m,\ell} + B^m \xi_m + f^m]. \quad m=0,\dots,M-1,$$

(5.17) $$x_{0,\ell} = h_\ell,\ \ell = -L,\dots,0,\quad x_{m+1,\ell} = x_{m,\ell+1},\quad m=0,\dots,M-1,\quad \ell=-L,\dots,-1.$$

We define an $(L+1)n \times (L+1)n$ matrix $\tilde{A}^m$ and an $(L+1)n \times m$ matrix $\tilde{B}^m$

$$[\tilde{A}^m\underline{h}]_0 = \sum_{i=0}^{N} A_i^m h_{-L_i} + \delta\sum_{\ell=-L}^{0} A_{01}^{m,\ell} h_\ell,\quad [\tilde{A}^m\underline{h}]_\ell = \frac{1}{\delta}[h_{\ell+1}-h_\ell],\ \ell=-L,\dots,-1,$$

$$[\tilde{B}^m w]_0 = B^m w,\quad [\tilde{B}^m w]_\ell = 0,\quad \ell=-L,\dots,-1,$$

and the vectors $\tilde{x}_m$ and $\tilde{f}^m$ in H^{δ}

(5.18) $$\tilde{x}_m = (x_{m,0},x_{m,-1},\dots,x_{m,-L}),\quad \tilde{f}^m = (f^m,0,\dots,0).$$

Equations (5.16) and (5.17) can now be rewritten in the more compact form

(5.19) $$\tilde{x}_{m+1} - \tilde{x}_m = \delta[\tilde{A}^m\tilde{x}_m + \tilde{B}^m \xi_m + \tilde{f}^m],\quad m=0,\dots,M-1,\quad \tilde{x}_0 = \underline{h}.$$

We notice that equation (5.19) has been constructed in such a way that $x_{m,\ell}$ remains constant along the characteristics of the differential equation for the state

$$x_{m+1,\ell} = x_{m,\ell+1},\quad m=0,\dots,M-1,\quad \ell=-L,\dots,-1.$$

As a result there exists $\underline{y} = (y_0,\dots,y_M)$ such that $x_{m,\ell} = y_{m+\ell}$, $m+\ell \ge 0$ and it is easy to see that $(y_0,\dots,y_M)$ is the solution of (5.11) with initial condition $\underline{h}$. Hence the scheme (5.19) has a unique solution.

Proposition 5.4. Let $(\tilde{x}_0,\dots,\tilde{x}_M)$ be the solution of scheme (5.19) for $\underline{h} = r_\delta(h)$, $\underline{\xi} = r_\delta(\xi)$ and $\underline{f} = r_\delta(f)$ for some h in H, ξ in $L^2(0,T;U)$ and f in $L^2(0,T;X)$. We define the map $\tilde{x}^{\delta}:[0,T] \to H$ as follows

(5.20) $$\begin{cases} \tilde{x}^{\delta}(t)^0 = \sum_{m=0}^{M-1} x_{m,0}\chi_m(t), t \in [0,T[,\ \tilde{x}^{\delta}(T) = q_\delta(\tilde{x}_M), \\ \tilde{x}^{\delta}(t)^1(\theta) = \sum_{m=0}^{M-1}\sum_{n=-L}^{-1} [x_{m,n}\chi_{m,n}^{\ell}(t,\theta) + x_{m,n+1}\chi_{m,n}^{u}(t,\theta)],\ (t,\theta) \in [0,T[\times [-a,0], \end{cases}$$

where χ_m is the characteristic function of $[m\delta,(m+1)\delta[$, $\chi_{m,n}^{\ell}$ is the characteristic function of

(5.21) $$\{(t,\theta) \in [m\delta,(m+1)\delta[\times [n\delta,(n+1)\delta[: t + \theta < (m+n+1)\delta\},$$

and $\chi_{m,n}^{u}$ the characteristic function of

(5.22) $$\{(t,\theta) \in [m\delta,(m+1)\delta[\times [n\delta,(n+1)\delta[: (m+n+1)\delta \le t+\theta\}.$$

(i) (Stability). There exists a constant $c > 0$, independent of δ, h, ξ and f, such that for all h in H, f in $L^2(0,T;X)$ and ξ in $L^2(0,T;U)$

(5.23) $$\max\{\|q_\delta(\tilde{x}_m)\|: m=0,\dots,M\} + \|\tilde{x}^\delta\|_\infty \le c[\|h\|+\|f\|_2+\|\xi\|_2].$$

(ii) (Convergence). Fix h, f and ξ. As δ goes to zero with $M\delta = T$

(5.24) $$\max\{\|q_\delta(\tilde{x}_m)-\tilde{x}(m\delta)\|: m=0,\dots,M\} \to 0,$$

and $\tilde{x}^\delta$ converges to $\tilde{x}$ in $L^\infty(0,T;H)$, where $\tilde{x}$ denotes the solution of equation (2.7).□

5.4. Approximation of the differential equation for the adjoint state.

In this section we introduce an approximation of the adjoint state equation (2.9). Consider the following scheme

(5.25) $$p_{m+1} - p_m + \delta[I_\delta^{-1}(\tilde{A}^m)^* I_\delta p_{m+1} + g^m] = 0, \quad m=0,\dots,M-1, \text{ in } H^\delta, \quad p_M = \underline{k} \text{ in } H^\delta$$

where $\underline{k} = r_\delta(k)$ and $\underline{g} = (g^0,\dots,g^{M-1})$ is constructed from g in $L^2(0,T;H)$

(5.26) $$g^m = \frac{1}{\delta}\int_{m\delta}^{(m+1)\delta} r_\delta(g(t))dt, \quad m=0,\dots,M-1.$$

Proposition 5.5. Let $(p_0,\dots,p_M)$ be the solution of the scheme (5.25) with final datum $\underline{k}$ for some k in H and $\underline{g} = (g^0,g^1,\dots,g^{M-1})$ for some g in $L^2(0,T;H)$. Let the approximation $p^\delta:[0,T] \to H$ be defined as

(5.27) $$p^\delta(t) = \sum_{m=0}^{M-1} q_\delta(p_{m+1})\chi_m(t), \quad 0 \le t < T, \quad p^\delta(T) = q_\delta(p_M).$$

(i) (Stability). There exists a constant $c > 0$ (independent of δ, h and g) such that

(5.28) $$\max\{\|q_\delta(p_m)\|: m=0,\dots,M\} + \|p^\delta\|_\infty \le c[\|k\|+\|g\|_2].$$

(ii) (Convergence). Let p denote the solution of equation (2.9) in the space $W^*(0,T)$. Then for all h in H

$$\max\{|((q_\delta(p_m),h))-((p(m\delta),h))|: m=0,\dots,M\} \to 0 \text{ as } \delta \to 0 \text{ with } M\delta = T$$

and the map $t \mapsto ((h,p^\delta(t)))$ converges to the map $t \mapsto ((h,p(t)))$ in $L^\infty(0,T;\mathbb{R})$. □

Corollary. Assume that the matrices $A_1,\dots,A_N$ are constant, that A_{01} is identically zero and that $k = (k^0,0)$ and $g(t) = (g^0(t),0)$. As δ goes to zero there exists a constant $c' > 0$ (independent of k^0, g^0 and δ) such that

(5.29) $$\|p-p^\delta\|_2 + \max\{\|p(m\delta)-q_\delta(p_m)\|: 0\le m\le M\} \le c'\delta[|k^0|+\|g^0\|_2]. \quad □$$

6. Approximation of the dual optimal control problem.

We now construct an approximation to the dual optimal control problem of section 2.4. We start with the approximation (5.25) to system (4.2):

(6.1) $$\begin{cases} p_{m+1}-p_m+\delta[I_\delta^{-1}(\tilde{A}^m)^* I_\delta p_{m+1}+(\tilde{\underline{C}}^m)^* w_m+g^m] = 0, \quad m=0,\dots,M-1, \text{ in } H^\delta \\ p_M=\underline{k}=r_\delta(k) \text{ for some k in H}, \ \underline{w}=(w_0,\dots,w_{M-1})=r_\delta(w) \text{ for } w \in L^2(0,T;Z), \end{cases}$$

where

$$(6.2) \qquad g^m = \frac{1}{\delta}\int_{m\delta}^{(m+1)\delta} r_\delta[g(t)]dt, \quad \underline{\tilde{C}}^m = \frac{1}{\delta}\int_{m\delta}^{(m+1)\delta} \tilde{C}(t)q_\delta dt, \quad m = 0,\dots,M-1.$$

We associate with system (6.1) the following approximation of the cost function

$$(6.3) \qquad J_\delta^*(\underline{w},\underline{k}) = (\underline{P}p_0,p_0)_L + \delta\sum_{m=0}^{M-1}[(\underline{N}^m p_{m+1},p_{m+1})_L + (\underline{R}^m w_m,w_m)_Z],$$

where

$$(6.4) \qquad \begin{cases} \underline{P} = q_\delta^* P q_\delta,\ \underline{N}^m = \underline{\tilde{B}}^m \underline{Q}^m(\underline{\tilde{B}}^m)^*,\ \underline{R}^m = \frac{1}{\delta}\int_{m\delta}^{(m+1)\delta} R(t)dt, \\ \underline{Q}^m = \frac{1}{\delta}\int_{m\delta}^{(m+1)\delta} Q(t)dt,\ \underline{\tilde{B}}^m = \frac{1}{\delta}\int_{m\delta}^{(m+1)\delta} r_\delta(\tilde{B}(t))dt. \end{cases}, \quad m = 0,\dots,M-1,$$

The approximate optimal control problem consists in minimizing $J_\delta^*(\underline{w},\underline{k})$ over all $\underline{w}$ in Z^δ:

$$\mathrm{Inf}\{J_\delta^*(\underline{w},\underline{k}) : \underline{w} \in Z^\delta\}.$$

Lemma 6.1. Given $\underline{k}$ in H^δ, the approximate optimal control problem (6.1)-(6.3) has a unique solution $\hat{\underline{w}}$ in U^δ. This solution is completely characterized by the optimality system:

$$(6.5) \qquad \hat{p}_{m+1}-\hat{p}_m+\delta[I_\delta^{-1}(\tilde{A}^m)^* I_\delta \hat{p}_{m+1}+(\tilde{C}^m)^*\hat{w}_m+g^m] = 0,\ 0 \le m \le M-1,\ \hat{p}_M = \underline{k},$$

$$(6.6) \qquad \hat{y}_{m+1}-\hat{y}_m = \delta[\tilde{A}^m\hat{y}_m - I_\delta^{-1}\underline{N}^m\hat{p}_{m+1}],\ 0 \le m \le M-1,\ \hat{y}_0 = I_\delta^{-1}\underline{P}\hat{p}_0,$$

$$(6.7) \qquad \hat{w}_m = (\underline{R}^m)^{-1}\underline{\tilde{C}}^m I_\delta \hat{y}_m, \quad 0 \le m \le M-1. \quad \square$$

Proposition 6.2. (i) Given h in H, for each $\delta > 0$ the approximate optimal control problem with initial condition $\underline{h} = r_\delta(h)$ has a unique solution $\hat{\underline{w}}$ in U^δ. As δ goes to zero, $q_\delta(\hat{\underline{w}})$ converges to $\hat{w}$ in $L^2(0,T;Z)$, where $\hat{w}$ is the optimal control in the minimization problem (4.2)-(4.3). (ii) We define the maps p^δ and $y^\delta : [0,T] \to H$

$$(6.8) \qquad p^\delta(t) = \sum_{m=0}^{M-1} q_\delta(\hat{p}_m)\chi_m(t),\ 0 \le t < T,\ p^\delta(T) = q_\delta(\hat{p}_M),$$

$$(6.9) \qquad y^\delta(t) = \sum_{m=0}^{M-1} q_\delta(\hat{y}_m)\chi_m(t),\ 0 \le t < T,\ y^\delta(T) = q_\delta(\hat{y}_M).$$

As δ goes to zero $J_\delta(\hat{\underline{w}},r_\delta(h))$ converges to $J(\hat{w},h)$,

$$(6.10) \qquad \begin{cases} \forall\, s \in [0,T],\ q_\delta(\hat{p}_m) \to \hat{p}(s) \text{ in H weak with } m\delta = s \\ \forall\, h \in H, \text{ the map } t \mapsto ((p^\delta(t),h)) \text{ converges to} \\ \qquad \text{the map } t \mapsto ((\hat{p}(t),h)) \text{ in } L^\infty(0,T;\mathbb{R}), \end{cases}$$

$$(6.11) \qquad \begin{cases} \forall\, s \in [0,T],\ q_\delta(\hat{y}_m) \to \hat{y}(s) \text{ in H weak with } m\delta = s \\ \forall\, h \in H, \text{ the map } t \mapsto ((y^\delta(t),h)) \text{ converges to} \\ \qquad \text{the map } t \mapsto ((\hat{y}(t),h)) \text{ in } L^\infty(0,T;\mathbb{R}), \end{cases}$$

where $\hat{p}$ and $\hat{y}$ are the solutions of system (4.6)-(4.9). □

Proposition 6.3. Let the sequences $\{\hat{p}_m\}$ and $\{\hat{y}_m\}$ be the solutions of the optimality equations (6.5) to (6.7). There exists a family of matrices $\{\Pi_m^\delta : m = 0,\dots,M\}$ in $\mathcal{L}(H^\delta)$ and a family of elements $\{\rho_m : m = 0,\dots,M\}$ in H^δ such that

(6.12) $$I_\delta \hat{y}_m = -\Pi_m^\delta \hat{p}_m + \rho_m, \quad m = 0,\dots,M.$$

Moreover Π_m^δ and ρ_m are obtained in the following manner: (i) we solve the system

(6.13) $$\gamma_{m+1}-\gamma_m+\delta[I_\delta^{-1}(\tilde{A}^m)^* I_\delta \gamma_{m+1}+\underline{M}^m I_\delta \beta_m], \ 0 \le m \le r-1, \ \gamma_r = \underline{k},$$

(6.14) $$\beta_{m+1}-\beta_m = \delta[\tilde{A}^m \beta_m - I_\delta^{-1}\underline{N}^m \gamma_{m+1}], \ 0 \le m \le r-1, \ \beta_0 = I_\delta^{-1}\underline{P}\gamma_0,$$

and $\Pi_r^\delta \underline{k} = -I_\delta \beta_r$ (where $\underline{M}^m = (\underline{\tilde{C}}^m)^*(\underline{R}^m)^{-1}\underline{\tilde{C}}^m$); (ii) we solve the system

(6.15) $$\xi_{m+1}-\xi_m+\delta[I_\delta^{-1}(\tilde{A}^m)^* I_\delta \xi_{m+1}+\underline{M}^m I_\delta \eta_m + g^m] = 0, \ 0 \le m \le r-1, \ \xi_r = \underline{k},$$

(6.16) $$\eta_{m+1}-\eta_m = \delta[\tilde{A}^m \eta_m - I_\delta^{-1}\underline{N}^m \xi_{m+1}], \ 0 \le m \le r-1, \ \eta_0 = I_\delta^{-1}\underline{P}\xi_0$$

and $\rho_r = I_\delta \eta_r$. □

7. Approximation of the equations for Π and ρ.

To study the family of operators $\Pi(s)$, $0 < s \le T$, we have considered the optimal control problem in the interval $[0,s]$ with $g=0$. By analogy we fix an integer r, $0 < r \le M$, and consider the system

(7.1) $$p_{m+1}-p_m+\delta[I_\delta^{-1}(\tilde{A}^m)^* I_\delta p_{m+1}+(\tilde{C}^m)^* w_m] = 0, \ m = 0,\dots,M-1, \ p_M = \underline{k},$$

and the optimal control problem for the cost function $J^*_{\delta,r}(\underline{w},\underline{k})$ (that is, (6.3) on $[0,r]$) associated with (7.1).

Proposition 7.1. We fix an integer $r > 0$. (i) If $\hat{\underline{w}}$ is the minimizing control

(7.2) $$J^*_{\delta,r}(\hat{\underline{w}},\underline{k}) = (\Pi_r^\delta \underline{k},\underline{k})_L, \ \Pi_r^\delta = (\Pi_r^\delta)^* \ge 0,$$

where * denotes the adjoint in H^δ with respect to the inner product $(\ ,\ )_L$, (ii) As δ goes to zero, there exists a constant $c > 0$ such that

(7.3) $$\forall\ k \in H, \ (\Pi_r^\delta r_\delta(k), r_\delta(k))_L \le c\|k\|^2.$$

(iii) If γ_m, β_m, $m = 0,\dots,M$, are the solutions of (6.13)-(6.14), then $\Pi_m^\delta \gamma_m = -I_\delta \beta_m$, $r \le m \le M$. □

Theorem 7.2. We define

(7.4) $$P_m = r_\delta^* \Pi_m^\delta r_\delta, \ 0 \le m \le M.$$

We introduce the map $\Pi^\delta : [0,T] \to \mathcal{L}(H)$

(7.5) $$\Pi^\delta(t) = \sum_{m=0}^{M-1} P_m \chi_m(t) \text{ in } [0,T[, \ \Pi^\delta(T) = P_M.$$

(i) (Stability). As δ goes to zero, there exists a constant $c > 0$ (independent of δ) such that

(7.6) $$\max\{\|P_m\|_{\mathcal{L}(H)} : m = 0,\dots,M\} \le c.$$

(ii) (Convergence). For fixed s in [0,T] and all h and k in H

(7.7) $$(\!(P_m h,k)\!) \to (\!(\Pi(s)h,k)\!) \text{ as } \delta \to 0 \text{ with } m\delta = s. \quad \square$$

Corollary. We can easily verify that

(7.8) $$[P_m]_{00} = [\Pi_m^\delta]_{00},\ [P_m]_{01}h^1 = \int_{-a}^{0} \sum_{n=-L}^{-1} \delta^{-1}[\Pi_m^\delta]_{0n}\chi_n(\theta)h^1(\theta)d\theta$$

(7.9) $$([P_m]_{11}h^1)(\alpha) = \int_{-a}^{0} \sum_{\ell=-L}^{-1} \sum_{n=-L}^{-1} \delta^{-2}[\Pi_m^\delta]_{\ell n}\chi_\ell(\alpha)\chi_n(\theta)h^1(\theta)d\theta.$$

For fixed s in [0,T], as δ goes to zero with $m\delta = s$

(7.10) $$[P_m]_{00} \to [\Pi(s)]_{00} \text{ in } \mathcal{L}(X),\ \forall\, h^1 \in L^2(-a,0;X),\ [P_m]_{01}h^1 \to [\Pi(s)]_{01}h^1 \text{ in } X,$$

(7.11) $$\forall\, h^1 \in L^2(-a,0;X),\ [P_m]_{11}h^1 \to [\Pi(s)]_{11}h^1 \text{ in } L^2(-a,0;X) \text{ weak}$$

and the norms of $[P_m]_{00}$, $[P_m]_{01}$ and $[P_m]_{11}$ are uniformly bounded. $\square$

Theorem 7.3. For δ small enough the family Π_m^δ, $0 \le m \le M$, as defined by equations (6.13)-(6.14) of Proposition 6.3 is the solution of the following set of equations:

(7.12) $$\begin{cases} \Pi_{m+1} = \delta \underline{N}^m + (I+\delta I_\delta \tilde{A}^m I_\delta^{-1})\Pi_m(I+\delta \underline{M}^m \Pi_m)^{-1}(I+\delta I_\delta \tilde{A}^m I_\delta^{-1})^*, & 0 < m \le M, \\ \Pi_0 = \underline{P}, \text{ where I is the identity matrix in } \mathcal{L}(H^\delta). & \square \end{cases}$$

To study the family of vectors $\rho(s)$, $0 < s \le T$, we have considered the optimal control problem in the interval [0,s] with k=0. By analogy we can fix an integer r, $0 < r \le M$, and consider the system (6.1) with $\underline{k}=0$ and the optimal control problem for the associated cost function (6.3). We will obtain a set of equations for ρ_m.

8. Numerical examples.

In this section we consider a number of examples which will illustrate the behaviour of the map $\alpha \mapsto \Pi_{01}(t,\alpha)$ as a function of the time t in [0,T]. In all examples a=1, T=2, $X = U = Z = \mathbb{R}$ and the observation equation is $z(t) = x(t)+\eta(t)$ with $R(t) = 1$, $0 \le t \le 2$. All equations are to be interpreted in an appropriate way as stochastic differential equations.

Example 1. This example has an analytical solution. Consider

(8.1) $$\dot{x}(t) = x(t-1),\ 0 \le t \le 2,\ x(0) = h^0+\zeta^0,\ x(\theta) = h^1(\theta),\ -1 \le \theta < 0.$$

Let $c = P_0^{-1}$. It can be shown that

(8.2) $$\Pi_{00}(s) = \begin{Bmatrix} (s+c)^{-1}, & 0 \le s \le 1 \\ s^2(1+c+\frac{s^3-1}{3})^{-1}, & 1 \le s \le 2 \end{Bmatrix}$$

$$(8.3)\qquad \Pi_{01}(s,\theta) = \begin{cases} (s+c)^{-1}\chi_{[-s,0]}(\theta) & , 0\le s\le 1 \\ s(1+c+\frac{s^3-1}{3})^{-1}\begin{Bmatrix} s+\theta, & 1-s\le\theta\le 0 \\ 1, & \text{otherwise}\end{Bmatrix}, & 1\le s\le 2 \end{cases}.$$

The results appear in Figures 1a and 1b for $P_0 = 1$.

Example 2. We now consider a system with two delays

$$(8.4)\qquad \dot{x}(t) = x(t-\tfrac{1}{2})+x(t-1),\ 0\le t\le 2,\ x(0) = h^0+\zeta^0,\ x(\theta) = h^1(\theta),\ -1\le\theta<0.$$

The results appear in Figure 2 for $P_0 = 1$.

Example 3. We allow input noise in (8.4)

$$(8.5)\qquad \dot{x}(t) = x(t-\tfrac{1}{2})+x(t-1)+\xi(t),\ 0\le t\le 2,\ x(0) = h^0+\zeta^0,\ x(\theta) = h^1(\theta),\ -1\le\theta<0.$$

The results appear in Figure 3 for $P_0 = 1$ and $Q(t) = 1$, $0\le t\le 2$.

Example 4. We consider system (8.5) but without noise in the initial condition, $x(0) = h^0$. The results are shown in Figure 4.

Example 5. We consider the system

$$(8.6)\qquad \dot{x}(t) = A_1(t)x(t-1),\ 0\le t\le 2,\ x(0) = h^0+\zeta^0,\ x(\theta) = h^1(\theta),\ -1\le\theta<0.$$

The results appear in Figures 1a and 5 for $P_0 = 1$ and

$$(8.7)\qquad A_1(t) = \begin{Bmatrix} 1, & n/10 \le t < (n+1)/10, & n \text{ even} \\ 0, & n/10 \le t < (n+1)/10, & n \text{ odd}\end{Bmatrix}.$$

Example 6. It is similar to Example 5 but without noise in the initial condition and with input noise:

$$(8.8)\qquad \dot{x}(t) = A_1(t)x(t-1)+\xi(t),\ 0\le t\le 2,\ x(0) = h^0,\ x(\theta) = h^1(\theta),\ -1\le\theta<0.$$

The results appear in Figure 6 for $Q(t) = 1$, $0\le t\le 2$, and A_1 as in (8.7).

References.

A. BENSOUSSAN [1], Filtrage optimal des systèmes linéaires, Dunod, Paris, 1971.
[2], Filtrage des systèmes linéaires avec retard, IRIA Report INF 7118/71027; Oct. 1971.

A. BENSOUSSAN, M.C. DELFOUR and S.K. MITTER [1], Optimal filtering for linear stochastic hereditary differential systems, Proc. 1972 IEEE Conference on Decision and Control, New Orleans, 378-380.

R.T. CURTAIN [1], A Kalman-Bucy theory for affine hereditary differential equations, in Control theory, numerical methods and computer systems modelling, eds. A. Bensoussan and J.L. Lions, Springer-Verlag, New York 1975, 22-43.

M.C. DELFOUR [1], Solution numérique de l'équation différentielle de Riccati rencontrée en théorie de la commande optimale des systèmes héréditaires linéaires, in Control theory, numerical methods and computer systems modelling, eds. A. Bensoussan and J.L. Lions, Springer-Verlag, New York 1975, 362-383.
[2], Numerical solution of the optimal control problem for linear hereditary differential systems with a linear-quadratic cost function and approximation of the Riccati differential equation, CRM-Report 408, Université de Montréal, Montréal, Canada 1974.
[3], Numerical solution of the operational Riccati differential equation in the optimal control theory of linear hereditary differential systems with a linear-quadratic cost function, Proc. 1974 IEEE Conference on Decision and Control, Phoenix, 784-790.

[4], The linear quadratic optimal control problem for hereditary differential systems: theory and numerical solution, to appear in Journal of Applied Mathematics and Optimization.

[5], State theory of linear hereditary differential systems, Centre de Recherches Mathématiques Report CRM-395, Université de Montréal, Montréal, Canada.

H. KWAKERNAAK [1], Optimal filtering in linear systems with time delay, IEEE Trans. Automatic Control 12 (1967), 169-173.

R. KWONG [1], Doctoral dissertation, Massachusetts Institute of Technology, 1975.

A. LINDQUIST [1], A theorem on duality between estimation and control for linear stochastic system with time delay, J. Math. Anal. Appl. 37 (1972), 516-536.

J.L. LIONS [1], Equations différentielles opérationnelles, Springer-Verlag, Berlin 1961.

[2], Optimal control of systems governed by partial differential equations, Springer-Verlag, New York, 1971.

J.L. LIONS and E. MAGENES [1], Problèmes aux limites non homogènes, Vol. 1,2,3, Paris, Dunod 1968.

S.K. MITTER and R. VINTER [1], Filtering for linear stochastic hereditary differential systems, in Control theory, numerical methods and computer systems modelling, eds. A. Bensoussan and J.L. Lions, Springer-Verlag, New York 1975, 1-21.

J.C. NÉDELEC [1], Schémas d'approximation pour des équations intégro-différentielles de Riccati, Thèse de doctorat d'état, Paris, 1970.

M. SENDAULA [1], On the duality between estimation and control for linear hereditary differential systems, Internal Report, Makerere University, Kampala, Uganda.

R. VINTER [1], On the evolution of the state of linear differential delay equations in M^2: properties of the generator, Report ESL-R-541, Electronic Systems Laboratory, M.I.T., Cambridge, Mass., U.S.A..

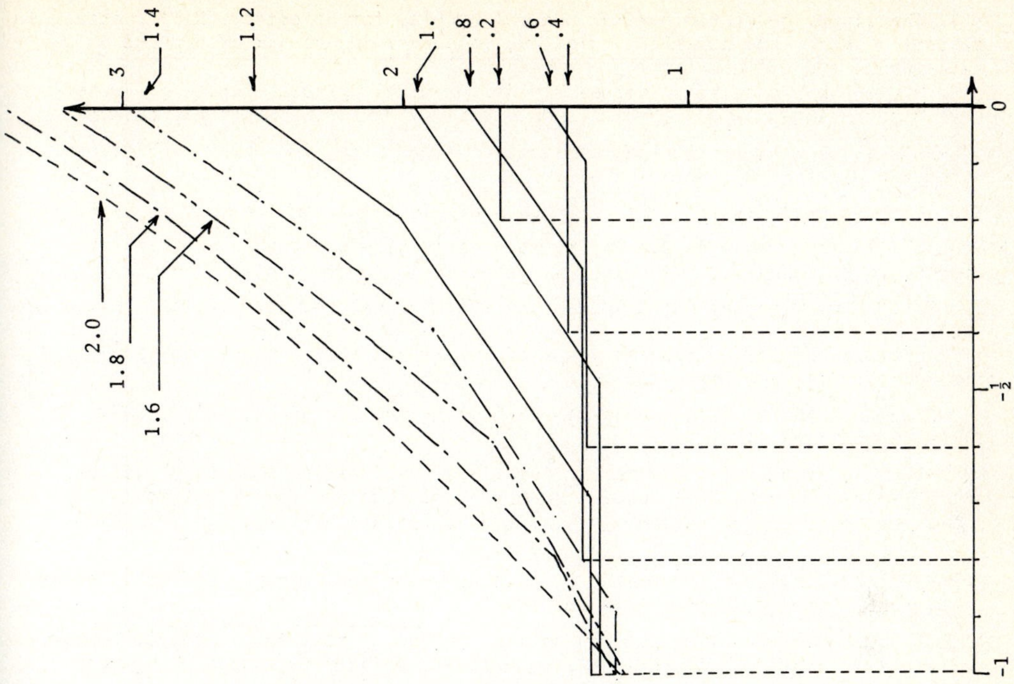

Figure 2.

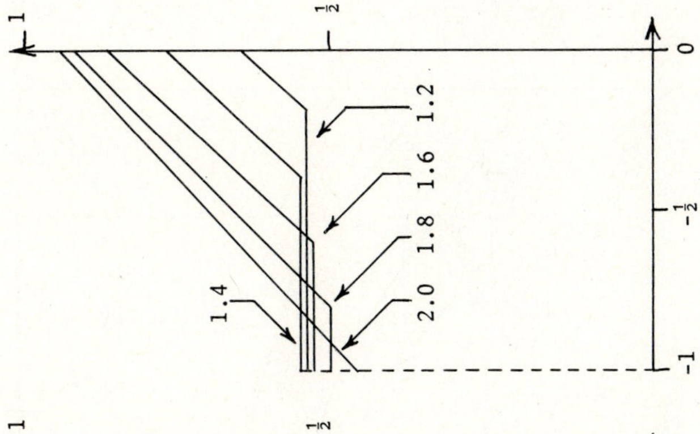

Figure 1b.

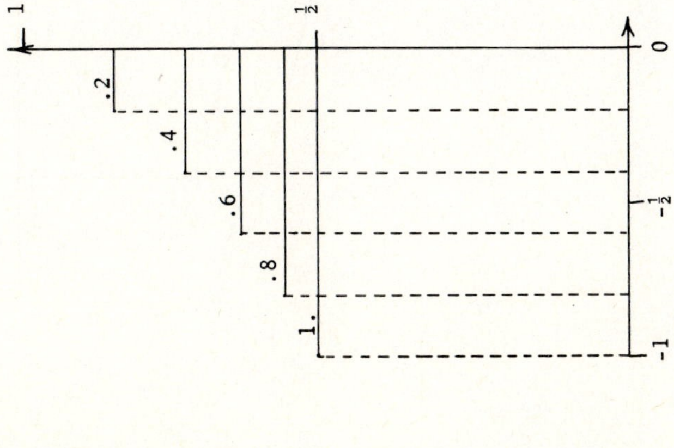

Figure 1a.

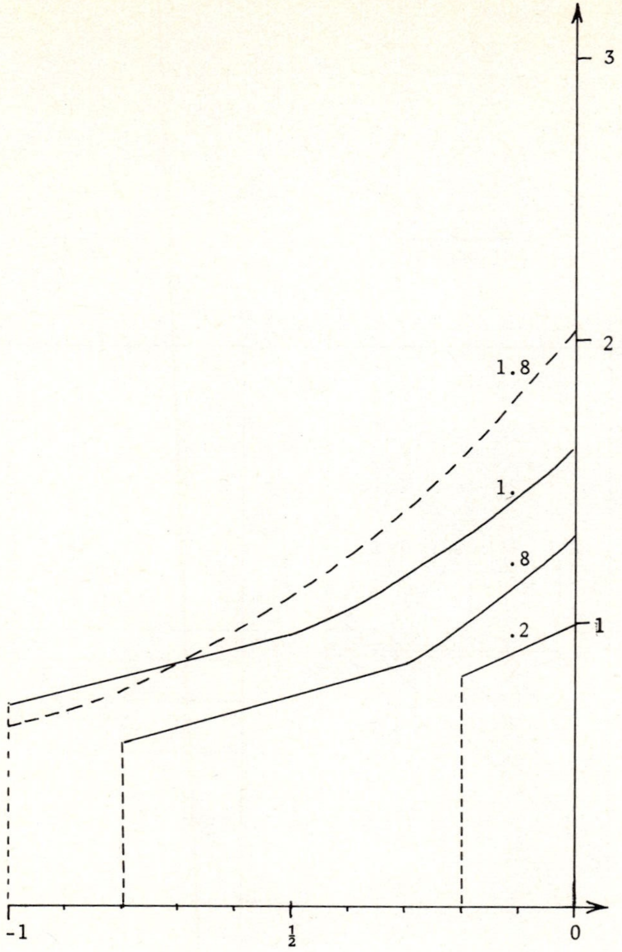

Figure 3.

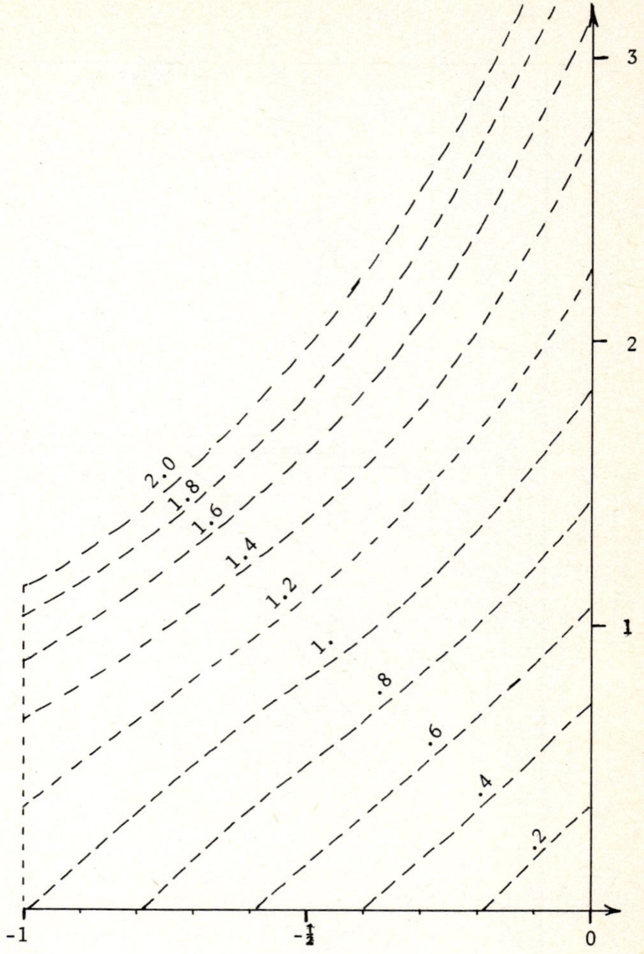

Figure 4.

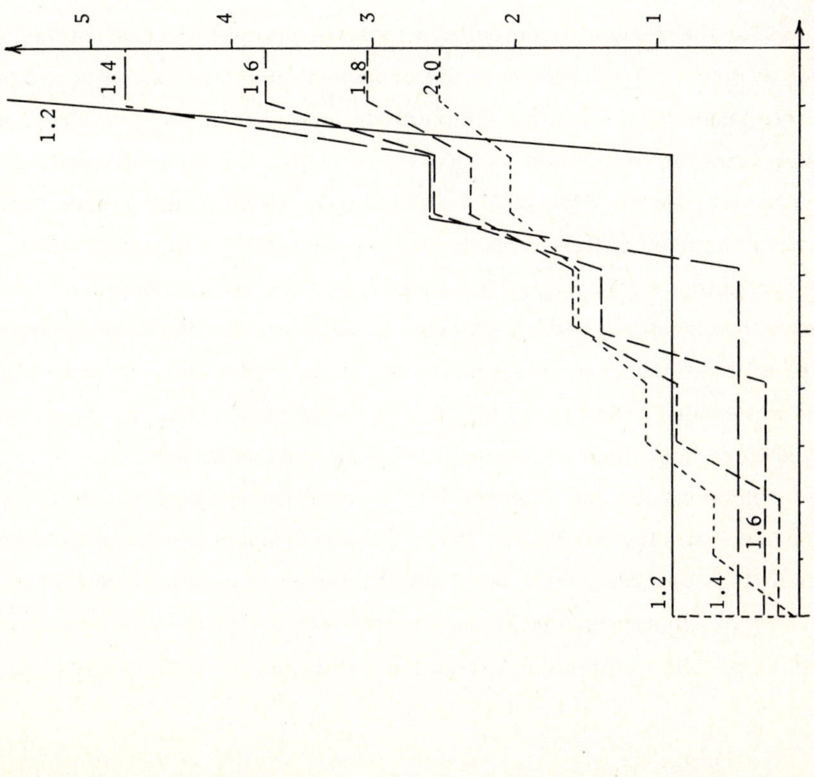

Figure 5.

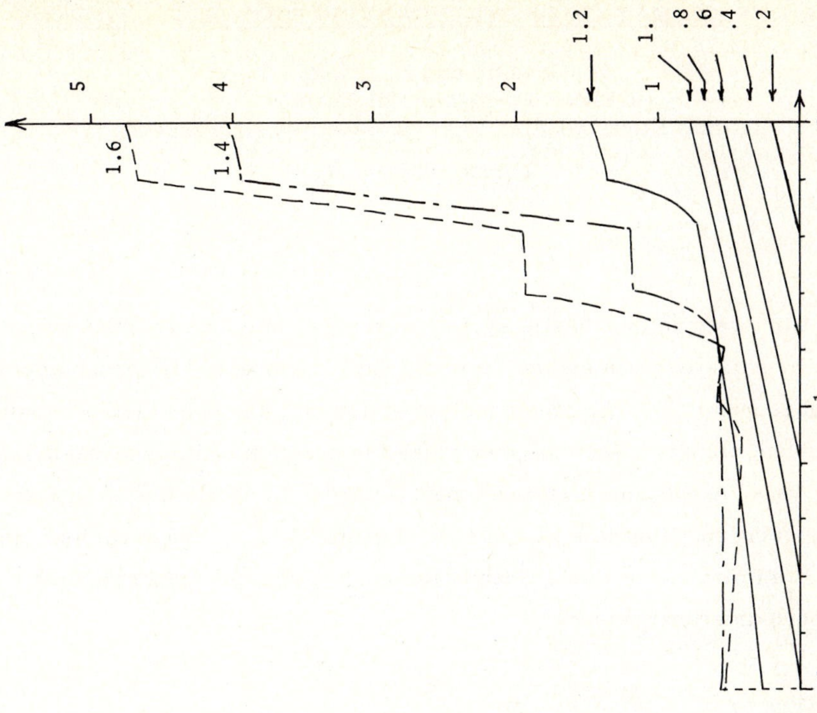

Figure 6.

ON THE APPROXIMATION OF TIME-VARYING STOCHASTIC SYSTEMS

R. Genesio and R. Pomé
CENS - Politecnico di Torino
Istituto Elettrotecnico Nazionale Galileo Ferraris

Torino, Italy

Abstract

A linear time-varying stochastic system described in terms of input-output data corrupted by noise is given and an optimal, time-invariant, low-order approximating model is required. After the problem statement, the paper introduces an input-independent criterion and then considers the problem of its evaluation from the available data. A procedure is developed in order to obtain in closed form the upper bound, corresponding to a given level of probability, of the error functional. Finally, the minimization of this quantity leads to the optimal model parameters and to the approximation measure.

1. - Introduction

The problem of modelling a high-order linear differential system by means of a low-order reduced model has been recently studied by several authors both for the analysis and for the design of control systems. (A complete set of references on this subject is given in [1].) Most of the proposed techniques attempt to approximate typical responses of the system (impulse or step responses, transfer function, z-transfer function, time-moments, etc.) in terms of the corresponding responses of the model (see for example [2] [3] [4]). An alternate approach is given by the uniform approximation methods which essentially minimize a worst case error (see for example [5] [6] [7]). Usually the proposed procedures concern linear time-invariant systems and a relatively small attention has been devoted to the case where a linear time-varying system has to be modelled by a linear time-invariant one of reduced order [5] [8] [9]. In these references the given system is assumed to be described in exact form by its state equations.

This paper considers the approximation by a linear time-invariant model of low-order of a time-varying stochastic linear system from input-output data corrupted by noise. This problem, which is of some interest in communication field [10] [11] [12] [13], is approached from the point of view of the uniform approximation. After the problem statement a modelling criterion is defined and a method is given in order to determine the optimal model from the real data which describe the system.

2. - Problem statement

Consider the linear, time-varying, dynamic system described by the relation

$$y(t) = \int_0^{\infty} g(t, s)\, u(t-s)\, ds \tag{2.1}$$

where u and y represent input and output respectively. The system is stochastic since g is a real-valued stochastic process.

In the measurement situation the input system u is exactly determinable while the observed output is given by

$$z(t) = y(t) + d(t) \tag{2.2}$$

i. e. the true output of the system is corrupted by an additive random noise d.

It is desired to determine a linear time-invariant model of low-order described by the input-output relation

$$\hat{y}(t) = \int_0^{\infty} \hat{g}(s)\, u(t-s)\, ds \tag{2.3}$$

(where $\hat{g}$ is a deterministic function) which gives the "best" approximation of the above system according to a certain criterion (see section 3).

To this end some assumptions are made:

a) on the system

g is a real-valued, Gaussian, stochastic process for which the following conditions hold:

$$g(t, s) = 0 \qquad \text{for} \qquad s < 0 \tag{2.4}$$

except on a set of sample functions of g of probability zero;

$$E \iint_{\Lambda} g(t, s)^2\, dtds < \infty \tag{2.5}$$

(E means expected value) for any bounded measurable set Λ in the plane;

$$E\, g(t, s) = g_o(s) \qquad \text{for} \qquad s \geq 0 \tag{2.6}$$

and

$$E\left[g(t, s)\, g(t', s')\right] = \varphi_g(t-t'; s, s') + g_o(s)\, g_o(s') \tag{2.7}$$

The impulse response g_o is unknown. It is stable, sufficiently smooth and a time

T_s is known at which $g_o(t)$ can be considered negligible. The system correlation function $\varphi_g(t; s, s')$ is known.

b) on the model

The model impulse response $\hat{g}$ is defined by a low-dimension parameter vector p, i.e. $\hat{g}(s) = \hat{g}\,[s(p)]$.

c) on the noise

The noise d is a zero-mean, Gaussian, stationary, stochastic process having a known correlation function $\varphi_d(t)$.

d) on the measurements

Samples u_M of the input and z_M of the corresponding measurable output of the system are given, observed on a time interval of suitable length T_M.

3.- Approximation criterion

Let $\mathcal{U}$ be the set of the inputs u of system and model and let $\mathcal{U}$ be formed by the square integrable functions on a certain time interval $[O, T_A]$. According to the system assumptions also the corresponding outputs y of the system are square integrable (with probability one) and therefore two norms can be introduced

$$\|u\|^2 = \int_O^{T_A} u^2(t)\, dt \tag{3.1}$$

$$\|\hat{y} - y\|^2 = E \int_O^{T_A} [\hat{y}(t) - y(t)]^2\, dt \tag{3.2}$$

Now, for every u, it is well known that [14]

$$\|\hat{y} - y\|^2 \leq K^2\, \|u\|^2 \tag{3.3}$$

where K^2 represents the Hilbert-Schmidt norm of the difference between model and system, i.e.

$$K^2 = E \int_O^{T_A} \int_O^{t} \left\{ \hat{g}\,[s(p)] - g(t, s) \right\}^2 dtds \tag{3.4}$$

Observe that K^2 represents an upper bound for the usual norm of the difference between model and system induced by definitions (3.1) and (3.2).

At this point, according to relation (3.3), an input independent approximation criterion is introduced and the optimal time-invariant model of low order for the given system is defined as that which corresponds to the value p_o of p minimizing the quantity (3.4).

Finally, in order to carry out this operation it can be remarked that the value p_o to be determined also corresponds to the minimum of the expression

$$\int_O^{T_A} (T_A - s) \left\{ \hat{g} \, [s(p)] - g_o(s) \right\}^2 ds \qquad (3.5)$$

which can be easily obtained from (3.4) and which differs from this of a known value depending on the system correlation function φ_g (see section 2).

4. - Model determination

The evaluation of the index (3.4) (or (3.5)) and then of the optimal model is not directly possible since $\hat{g}$ can be easily computed for any fixed p but the impulse response g_o is unknown. Therefore, this function must be estimated from the available input-output data represented by samples u_M and z_M (see section 2).

According to (2.1) the relation (2.2) can be expressed in the form

$$z(t) = \int_o^{\infty} g_o(s)\, u(t-s) ds + \int_o^{\infty} [g(t, s) - g_o(s)]\, u(t-s) ds + d(t) \qquad (4.1)$$

and letting

$$y_v(t) = \int_o^{\infty} [g(t, s) - g_o(s)]\; u(t-s)\, ds \qquad (4.2)$$

as

$$z(t) = \int_o^{\infty} g_o(s)\, u(t-s) ds + y_v(t) + d(t) \qquad (4.3)$$

Now, by choosing a suitable sampling time T_c such that the functions g_o (see section 2) and u_M can be sufficiently approximated from their samples and by introducing the following matrix notations (the symbol ^{T} means transpose)

$$Z^T = [z_M(O) \quad z_M(T_c) \ldots\ldots\ldots z_M[(m-1)T_c]] \qquad (4.4)$$

$$G_o^T = [g_o(O) \quad g_o(T_c) \ldots\ldots\ldots g_o[(n-1)T_c]]\, \gamma \qquad (4.5)$$

$$Y_v = [y_v(O) \quad y_v(T_c) \ldots\ldots\ldots y_v[(m-1)T_c]] \qquad (4.6)$$

$$D = [d(O) \quad d(T_c) \ldots\ldots\ldots d[(m-1)T_c]] \qquad (4.7)$$

$$U = \begin{bmatrix} u_M(O) & O & \cdot & O \\ u_M(T_c) & u_M(O) & \cdot & O \\ \cdot & \cdot & \cdot & \cdot \\ u_M[(n-1)T_c] & u_M[(n-2)T_c] & \cdot & u_M(O) \\ \cdot & \cdot & \cdot & \cdot \\ u_M[(m-1)T_c] & u_M[(m-2)T_c] & \cdot & u_M[(m-n)T_c] \end{bmatrix} \qquad (4.8)$$

the relation (4. 3), which corresponds to the observed data, becomes in discrete form

$$Z = U\, G_O + Y_v + D \qquad (4.9)$$

Of course, γ represents a known normalizing factor ,

$$n = T_S/T_c - 1, \quad m = T_M/T_c - 1 \qquad (4.10)$$

(see section 2). It is assumed $m \geq n$ and $u_M(O) \neq O$. The minimum variance unbiased estimate (Gauss-Markov) of the unknown vector G_O is [15]

$$\overline{G}_O = (U^T C^{-1} U)^{-1}\, U^T C^{-1} Z \qquad (4.11)$$

where G is the covariance matrix of Y_v+D, that is, in the case of g and d not correlated

$$C = E\,[Y_v Y_v^T] + E\,[D D^T] \qquad (4.12)$$

According to the assumptions of section 2, this matrix can be evaluated from the knowledge of u_M and φ_g for the first term and from the knowledge of φ_d in a very direct form for the second term.

The estimate error

$$\varepsilon = \overline{G}_O - G_O \qquad (4.13)$$

is a Gaussian random vector having zero mean value and covariance matrix

$$Q = (U^T\, C^{-1}\, U)^{-1} \qquad (4.14)$$

The computation of the optimal model is obviously based on the knowledge of the estimate $\overline{G}_O$.

In fact, coming back to relation (3. 5) and letting

$$\hat{G}^T = [\hat{g}(O)\ \hat{g}(T_c) \ldots\ldots\ldots\ \hat{g}\,[(n-1)T_c]]\,\gamma \qquad (4.15)$$

this approximation index can be written as

$$[\hat{G}(p) - G_O]^T\ W[\hat{G}(p) - G_O] \qquad (4.16)$$

where W is a diagonal n x n positive definite weighting matrix depending on the weighting function (T_A - s) of (3.5).

According to (4.13), if one defines

$$e(p) = \hat{G}(p) - \bar{G}_O \tag{4.17}$$

(4.16) becomes

$$[e(p) + \varepsilon]^T W [e(p) + \varepsilon] \tag{4.18}$$

which has to be minimized with respect to p.

Indeed the expression (4.18) is a random variable with known statistics. Therefore it seems to be reasonable, by following a worst case criterion, to assume that the optimal model which is possible to determine is sought by minimizing (always with respect to p)

$$\max_{\varepsilon} \; (e + \varepsilon)^T W (e + \varepsilon) \tag{4.19}$$

where the vector ε lies in the confidence region which corresponds to a given level of probability P. This region (see above) is obviously defined by the relation

$$\varepsilon^T Q^{-1} \varepsilon \leq F(P) \tag{4.20}$$

where F is a known function of P.

The problem of finding the quantity (4.19) with the constraint (4.20) is now taken into account.

Let be

$$\varepsilon = Q^{1/2} \alpha \quad , \quad e = Q^{1/2} a \tag{4.21}$$

By substitution in (4.20) and (4.21) it is obtained

$$\max_{\alpha} \; (a + \alpha) Q^{1/2} W Q^{1/2} (a + \alpha) \tag{4.22}$$

$$\alpha^T \alpha \leq F(P) \tag{4.23}$$

and introducing the lagrangian multiplier μ the necessary condition for determining (4.22) is written as

$$\alpha = - [Q^{1/2} W Q^{1/2} - \mu I]^{-1} Q^{1/2} W Q^{1/2} e \tag{4.24}$$

The use of (4.24) in (4.23) in order to derive μ yields

$$a^T (Q^{1/2} W Q^{1/2}) [Q^{1/2} W Q^{1/2} - \mu I]^{-2} (Q^{1/2} W Q^{1/2}) a \leq F(P) \tag{4.25}$$

The first member of this equation is a quadratic form which is defined by a matrix whose eigenvectors are the same [16] of the matrix $(Q^{1/2} W Q^{1/2})$. Therefore, the diagonalization of this quadratic form is possible by introducing the transformation

$$b = M^{-1} a \tag{4.26}$$

where M is the normalized modal matrix of $(Q^{1/2} W Q^{1/2})$ whose eigenvalues are indicated by λ_i. Then (4.25) can be rewritten in the form

$$\sum_{i=1}^{n} b_i^2 \frac{\lambda_i^2}{(\lambda_i - \mu)^2} = F \tag{4.27}$$

where in terms of the original variable it is

$$b = M^{-1} Q^{-1/2} e \tag{4.28}$$

and where it is taken into account that the solution of the problem belongs to the boundary of the region defined by (4.20).

Equation (4.27) has 2n solution for μ. What is, among these, the value μ_o which allows to determine the quantity (4.19)? A sufficient condition can be easily derived from the second derivative of the functional (modified with the introduction of the lagrangian multiplier). So, the matrix

$$[Q^{1/2} W Q^{1/2} - \mu I] \tag{4.29}$$

must be definite negative and hence the relation

$$\mu > \lambda_i \, , \quad i = 1, 2 \, \ldots\ldots \, n \tag{4.30}$$

must be satisfied. Therefore, according to the form of (4.27), μ_o is usually the only solution of (4.27) greater than the maximum eigenvalue of $(Q^{1/2} W Q^{1/2})$. Observe that singular cases can arise when the condition

$$\sum_{i=1}^{n} b_i^2 \frac{\lambda_i^2}{(\lambda_i - \mu)^2} \leq F \tag{4.31}$$

holds, due to the fact that the coefficient b_i corresponding to the maximum λ_i is equal to zero. The problem can still be solved but multiple solutions can happen [17].

From μ_o it is possible to calculate the expression (4.19). In fact, by use of (4.24), expression (4.22) becomes

$$\begin{aligned} a^T [I - Q^{1/2} W Q^{1/2} (Q^{1/2} W Q^{1/2} - \mu_o I)^{-1} \, Q^{1/2} W Q^{1/2} \\ [I - Q^{1/2} W Q^{1/2} (Q^{1/2} W Q^{1/2}) - \mu_o I)^{-1}] a \end{aligned} \tag{4.32}$$

and this quadratic form can be diagonalized by introducing the transformation (4.26).

Finally, the quantity (4.19) is written in the form

$$\sum_{i=1}^{n} b_i^2 \frac{\lambda_i \mu_o^2}{(\lambda_i - \mu_o)^2} \tag{4.33}$$

and this represents the required result.

At this point the whole procedure of approximation can be summarized. From the real data u_M and z_M, the matrices U and Z are constructed and the estimate $\overline{G}_O$ of the unknown vector G_O is evaluated according to (4.11). Then, using eigenvalues and eigenvector of $Q^{1/2} W Q^{1/2}$ the solution μ_o of equation (4.27) is found. Finally, the optimal approximating model is derived as that corresponding to the value of p which gives (see (4.33))

$$\min_{p} \sum_{i=1}^{n} b_i^2 \frac{\lambda_i \mu_o^2}{(\lambda_i - \mu_o)^2} \tag{4.34}$$

Remark that from (4.34) and through (3.5) the value $K^2(p_o)$ of (3.4) corresponding to p_o can be easily computed and therefore the method furnishes, together with the optimal model, also a measure of the reached approximation. It is possible to conclude that, for any square integrable signal u on $[O, T_A]$, on the basis of system knowledge which is possible to derive from the real data u_M and z_M, there is at least a given probability P that the condition

$$E\int_O^{T_A} [\hat{y}_o(t) - y(t)]^2\, dt \leq K^2(p_o) \int_O^{T_A} u^2(t)\, dt \tag{4.35}$$

holds ($\hat{y}_o$ represents the optimal model output corresponding to u).

5. - Conclusions

The problem of modelling a linear, time-varying, stochastic system described by input-output data corrupted by noise has been considered. The assumption is made that the approximating model is time-invariant and has a simple structure described in terms of a parameter vector. An approximation criterion is introduced following a min-max approach and the optimal model is defined as that which minimizes the Hilbert-Schmidt norm of the difference between model and system. The evaluation of this performance index is not directly possible from the available data on the system: therefore a procedure is given in order to obtain an estimate of such an index and in order to determine its upper bound corresponding to a given level of probability. Finally, the minimization of this quantity, which has to be carried out by a numerical iterative procedure, leads to the optimal approximating model and furnishes a measure of the obtained approximation which results, of course, input-independent.

References

[1] R. Genesio and M. Milanese: "A note on derivation and use of reduced order models" - IEEE Trans. Automat. Contr., vol. AC-21, Feb. 1976.

[2] C.F. Chen and L.S. Shieh: "A novel approach to linear model simplification" - Int. J. Contr., vol. 8, pp. 561-570, 1968.

[3] M.R. Chidambara: "Two simple techniques for the simplification of large dynamic systems" - Proc. JACC '69, pp. 669-674, 1969.

[4] T.C. Hsia: "On the simplification of linear systems" - IEEE Trans. Autom. Contr., vol; AC-17, pp. 372-374, 1972.

[5] H. Heffes and P.E. Sarachik: "Uniform approximation of linear systems" - Bell Syst. Tech. J., vol. 48, pp. 209-231, 1969.

[6] M. Milanese and A. Negro: "Uniform approximation of systems. A Banach space approach" - J.O.T.A., vol. 12, pp. 203-217, 1973.

[7] R.J.P. De Figuereido, A. Caprihan and A.N. Netrevali: "On optimal modeling of systems" - J.O.T.A., vol. 11, pp. 68-83, 1973.

[8] G.J. Bierman: "Weighted least squares stationary approximations to linear systems" - IEEE Trans. Automat. Contr., vol. AC-17, pp. 232-234, 1972.

[9] H. Nosrati and H.E. Meadows: "Modeling of linear time-varying systems by linear time-invariant systems of lower order" - IEEE Trans. Automat. Contr., vol. AC-18, pp. 50-52, 1973.

[10] W.L. Root: "On the measurement and use of time-varying communication channels" - Inform. and Contr., vol. 8, pp. 390-422, 1965.

[11] I. Bar-David: "Estimation of linear weighting functions in Gaussian noise" - IEEE Trans. Inform. Theory, vol. IT-14, pp. 288-293, 1968.

[12] P.A. Bello: "Measurement of random time-variant linear channels" - IEEE Trans. Inform. Theory, vol. IT-15, pp. 469-475, 1969.

[13] E. Mosca: "A deterministic approach to a class of nonparametric system identification problems" - IEEE Trans. Inform. Theory, vol. IT-17, pp. 686-696, 1971.

[14] N.I. Akhiezer and I.M. Glazman: "Theory of Linear Operators in Hilbert Space", vol. I, F. Ungar, New York, 1966.

[15] R. Deutsch: "Estimation Theory", Prentice-Hall, Englewood Cliffs, 1965.

[16] F.R. Gantmacher: "The Theory of Matrices", vol. I, Chelsea, New York, 1959.

[17] R. Genesio and R. Pomé: "Identification of reduced models from noisy data" - Int. J. Contr., vol. 21, pp. 203-211, 1975.

STABILIZING CONTROL FOR LINEAR SYSTEMS WITH BOUNDED PARAMETER AND INPUT UNCERTAINTY

S. Gutman*
NASA
Ames, California

and

G. Leitmann
University of California
Berkeley, California

ABSTRACT

We consider dynamical systems with norm-bounded uncertainty in (i) the system parameters (model uncertainty) or in (ii) the input (disturbance).

For case (i), the nominal (null uncertainty) system is linear with constant matrices. Such systems with norm-bounded control as well as with a control penalty are treated. However, in the former the treatment is restricted to single input systems in companion form, and in the latter to second order systems. For case (ii), the system is linear with time-varying matrices and norm-bounded control.

Using some results from the theories of differential games and general dynamical systems, we deduce feedback controls which render the origin uniformly asymptotically stable in the large for all admissible parameter uncertainties or input disturbances; these may be both time and state dependent.

The application of the theory is illustrated by examples.

*Research Associate, National Research Council.

1. INTRODUCTION

The problem of designing a feedback control for uncertain systems has been discussed in a series of articles, [1-7]. In principle, we distinguish among three types of uncertainties:

a) Uncertainty in the <u>model</u> (parameter)

b) Uncertainty in the <u>input</u> (disturbance)

c) Uncertainty in the <u>state</u> (measurement)

Here we deal only with the first two types of uncertainty, model and input uncertainties for linear† systems.

To motivate the discussion, consider an aircraft maneuvering at a high angle of attack. It is possible to describe the dynamical behavior by a set of nonlinear differential equations such that the "nominal" part is a set of linear differential equations. Often two difficulties arise:

1) The nonlinear characteristics of the parameters are known but it is impossible to find a "best" controller for achieving desired specifications, e.g. controlling the system asymptotically to rest.

2) Because of lack of experimental data, there is incomplete information about the parameter characteristics, except that their value belong to known sets.

In both cases we approach the difficulty by allowing for the "worst" nonlinear characteristics with respect to an appropriate performance index and for that nonlinearity we seek the "best" controller. This "worst case" philosophy does not imply that the "worst" situation will occur, but rather that a controller capable of achieving the desired end under the "worst" of circumstances will also do so under more favorable ones, and hence under all allowable ones.

The theory of two-person zero-sum games is employed to generate "worst case" controllers. Towards this end, an appropriate performance index is stipulated; it is to be maximized by the uncertainty and minimized by the controller, respectively. If a saddlepoint strategy pair exists, then the controller assures himself a cost (in terms of the assumed performance index) that is no greater than the saddlepoint one, no matter what the strategy of the disturbance.

2. MODEL UNCERTAINTY WITH CONTROL PENALTY

2.1 <u>Problem Statement</u>

Here we treat a class of second order dynamical systems with parameter uncertainty. Consider

†That is, when the "nominal" system (namely, the system without uncertainty) is linear.

$$\dot{x}(t) = [A_o + \sum_{i=1}^{p} A_i\, v_i(t)]\, x(t) + B\, v_{p+1}(t)\, u(t) \tag{1}$$

$$x(t_o) = x_o\ ,\quad t \in [t_o, t_1]$$

where

$x(t) \in R^2$ is the state of the system at time t ;

A_i , $i = 0, 1, \ldots, p$, are constant 2×2 matrices, each containing a single non-zero element;

B is a constant $2 \times m$ matrix;

$v_i(t)$, $i = 1, \ldots, p$, with $|v_i(t)| \leqslant 1$, and

$v_{p+1}(t)$ with $v_{p+1}(t) \in [1,q]$, q = constant > 1 ,

are values of parameter uncertainty at time t ;

$u(t) \in R^m$ is the value of the control at time t .

We are interested in the asymptotic behavior of the system (1) under all possible uncertainties

$v(t) = (v_1(t), v_2(t), \ldots, v_{p+1}(t))'$, $t \in [t_o, \infty)$.

Since we are concerned with the asymptotic stability of the origin $x = \{0\}$, we introduce a measure of deviation from that state subject to a control penalty. That is, we introduce the performance index

$$J = \int_{t_o}^{t_1} [x'(t)\, Q\, x(t) + u'(t)\, R\, u(t)]\, dt \tag{2}$$

where

Q is a constant positive semidefinite symmetric 2×2 matrix;

R is a constant positive definite symmetric $m \times m$ matrix;

and consider the differential game with state equation (1) and cost (2) .

That is, we seek a saddlepoint $(p^*(\cdot), e^*(\cdot))$ in a given class of strategies

$p(\cdot) : R^2 \times R^1 \to R^m$, $e(\cdot) : R^2 \times R^1 \to R^{p+1}$

such that

$u(t) = p(x(t),t)$, $v(t) = e(x(t),t)$

Note that we are looking for a feedback control $p^*(\cdot)$ while admitting an uncertainty $e(\cdot)$ that may depend on state and time.

Having found a saddlepoint candidate $(p^*(\cdot), e^*(\cdot))$, we inquire then under what conditions the feedback control $p^*(\cdot)$ renders $x = \{0\}$ uniformly asymptotically stable in the large (in the sense of Lyapunov) against every allowable uncertainty $e(\cdot)$.

2.2 Stability

Before discussing the asymptotic behavior of the system, we invoke necessary conditions for a saddlepoint candidate $(p^*(\cdot), e^*(\cdot))$, e.g. [8-9]. These conditions lead us to consider the following procedure, [10]:

Step 1. For each possible combination

$\{v_i^* : v_i^* = 1 \text{ or } -1 , i = 1, \ldots, p\}$

Compute

$$A = A_o + \sum_{i=1}^{p} A_i v_i^* \tag{3}$$

Let $A_{(k)}$ denote the value of A corresponding to the k-th possible combination.

Step 2. Using (3), compute $P_{(k)}$, the solution of

$$PA + A'P - PBR^{-1} B'P + Q = 0 \tag{4}$$

corresponding to the k-th possible combination of the v_i^*.

Step 3. Define $\sigma_i^k(\cdot) : R^2 \rightarrow R^1$ by

$$\sigma_i^k(x) \triangleq x' [P_{(k)} A_i + A_i' P_{(k)}] x \tag{5}$$

Step 4. Define a decomposition of R^2 by the lines given by

$$\sigma_i^k(x) = 0 \tag{6}$$

$i = 1, 2, \ldots, p, \quad k = 1, 2, \ldots, 2^p$

and designate the decomposition by $D \triangleq \{X_1, X_2, \ldots, X_\ell\}$.

The X_i are the open subsets of the decomposition, where $X_i \cap X_j = \emptyset$ for $i \neq j$ and $R^2 = \bigcup_{i=1}^{\ell} \bar{X}_i$.

Step 5. Determine a control candidate $p^*(\cdot)$ by the following algorithm:

Algorithm 1:

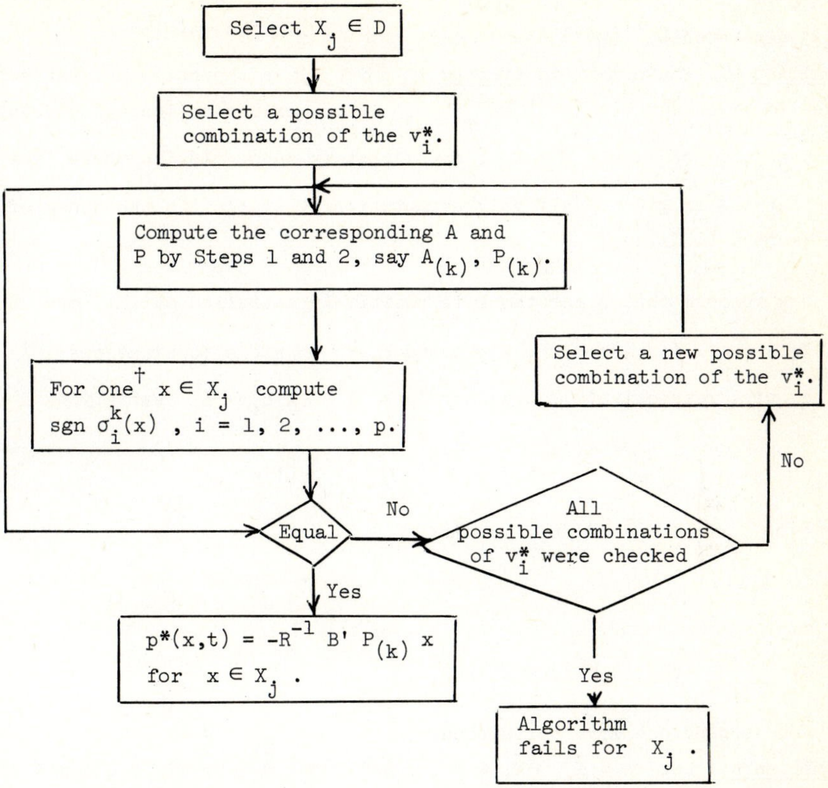

Definition 1. Algorithm 1 is said to be positively satisfied on X_j iff there is at least one possible combination of the v_i^* such that the equality test is answered in the affirmative.

Remark 1. If there is a region of R^2 on which the equality test is met by more than one possible combination of the v_i^*, then one may be able to introduce an altered decomposition of R^2, on each of whose members $p^*(\cdot)$ takes on values corresponding to one of the combinations meeting the sign test; for instance, see Example 1. If this has been done, we still denote the members of the decomposition by X_i, $i \in \{1, 2, \ldots, \ell\}$.

[†]Since $\sigma_i^k(\cdot)$ is continuous on R^2, the $\operatorname{sgn} \sigma_i^k(x)$, $i = 1, 2, \ldots, p$, remain constant on X_j .

Before continuing, we introduce some definitions.

Definition 2. The set $Z \subset R^2$ is positively invariant with respect to $(p(\cdot), e(\cdot))$ iff $x_o \in Z \Rightarrow x(t) \in \bar{Z} \ \forall t \in [t_o, \infty)$, where $x(\cdot) : [t_o, \infty) \to R^2$ is a solution of (1) generated by $(p(\cdot), e(\cdot))$ and $x(t_o) = x_o$.

Definition 3. The origin is eventually uniformly asymptotically stable in the large iff given any $(x_o, t_o) \in R^2 \times R^1_+$ there exists at least one solution $x(\cdot) : [t_o, \infty) \to R^2$, $x(t_o) = x_o$, and for every such solution there is a $T \geq t_o$ such that the origin is uniformly asymptotically stable (in the sense of Lyapunov) with respect to $x(\cdot)\,|_{[T,\infty)}$.

Now suppose that Algorithm 1 is positively satisfied on X_i and $X_j \in D^\dagger$, $\bar{X}_i \cap \bar{X}_j \neq \phi$, $i \neq j$, for the k_i-th and k_j-th possible combinations of the v_i^*, respectively. Consider $\tilde{p}(\cdot) : R^2 \times R^1_+ \to R^m$ satisfying

$$\tilde{p}(x,t) \begin{cases} = -R^{-1} B' P_{(k_i)} x & \forall (x,t) \in X_i \times R^1_+ \\ = -R^{-1} B' P_{(k_j)} x & \forall (x,t) \in X_j \times R^1_+ \\ \in \{-R^{-1} B'[\alpha P_{(k_i)} + (1-\alpha) P_{(k_j)}]\, x : \alpha \in [0,1]\} & \\ \qquad \forall (x,t) \in \bar{X}_j \cap \bar{X}_j \times R^1_+ . & \end{cases} \tag{7}$$

Now we need one more definition.

Definition 4. Let $\hat{x} \in \bar{X}_i \cap \bar{X}_j \neq \phi$, $i \neq j$, $\Rightarrow \hat{x} \in \{x : d'x = 0, x \in R^2\}$. Let $h_i(x,t)$ and $h_j(x,t)$ be the r. h. s. of (1) corresponding to $(\tilde{p}(\cdot), e(\cdot))$ on $\bar{X}_i$ and $\bar{X}_j$, respectively. The boundary $\bar{X}_i \cap \bar{X}_j$ is attractive iff $\forall\, (\tilde{p}(\cdot), e(\cdot))$ and $\forall\, \hat{x} \in \bar{X}_i \cap \bar{X}_j$ there is a ball $B(\hat{x})$ in R^2 with center at $\hat{x}$ such that

$$d'h_i(x,t) \geq 0 \quad \forall (x,t) \in B(\hat{x}) \cap \bar{X}_i \times R^1_+ , \text{ and}$$

$$d'h_j(x,t) \leq 0 \quad \forall (x,t) \in B(\hat{x}) \cap \bar{X}_j \times R^1_+ ,$$

where d points into X_j. The boundary is a transition one iff $\forall (\tilde{p}(\cdot), e(\cdot))$

$$d'h_i(x,t) > 0 \text{ and } d'h_j(x,t) > 0 \text{ or}$$

$$d'h_i(x,t) < 0 \text{ and } d'h_j(x,t) < 0$$

for all $(x,t) \in \bar{X}_i \cap \bar{X}_j \times R^1_+$.

---

†Decomposition D may be an altered decomposition; see Remark 1.

Consider the following assumptions.

Assumptions 1.

(i) Admissible uncertainty $e(\cdot) : R^2 \times R^1_+ \to R^{p+1}$ is continuous on R^2 and piecewise continuous on any compact subinterval of R^1_+.

(ii) Algorithm 1 is positively satisfied on every $X_j \in D$.

(iii) Triple $\{C, A, B\}$, where $Q = C'C$ and A is given by (3), is completely controllable and observable for all possible combinations of the v_i^*.

(iv) Every boundary $\bar{X}_i \cap \bar{X}_j \neq \phi$, $i \neq j$, is either an attractive or a transition one.

(v) Decomposition D is such that there exists at least one $k \in \{1, 2, \ldots, \ell\}$ such that, given $\tilde{p}(\cdot)$, X_k is positively invariant with respect to $(\tilde{p}(\cdot), e(\cdot))$ for all admissible $e(\cdot)$.

(vi) If a solution $x(\cdot) : [t_o, \infty) \to R^2$ generated by $(\tilde{p}(\cdot), e(\cdot))$ reaches an attractive boundary at $x(T)$, then the origin is uniformly asymptotically stable with respect to $x(\cdot)|_{[T,\infty)}$; see Remark 3.

Now we are ready to state a stability theorem.

Theorem 1. Consider system (1). If Assumptions 1 are met there exists a feedback control $\tilde{p}(\cdot)$ satisfying (7) such that the origin is eventually uniformly asymptotically stable in the large for all admissible uncertainties $e(\cdot)$.

Proof. Since $\tilde{p}(\cdot)$ is discontinuous and hence considered not unique, (1) becomes a generalized dynamical system, [11-16],

$$\dot{x}(t) \in C(x(t),t) \tag{8}$$

where the set valued function $C(\cdot)$ is given by

$$C(x,t) = \{[A_o + \sum_{i=1}^{p} A_i \, e_i(x,t)]x + B \, e_{p+1}(x,t) \, u : u = \tilde{p}(x,t)\}$$

We show first that, given any $(x_o, t_o) \in R^2 \times R^1_+$, there exists at least one solution of (8) and that such a solution can be continued on any compact subset of $R^2 \times R^1_+$.† This can be done by showing, [10],

(i) $C(x,t)$ is convex for all $(x,t) \in R^2 \times R^1_+$.

(ii) $C(x,t)$ is compact on any compact subset of $R^2 \times R^1_+$.

(iii) $C(\cdot)$ is upper semicontinuous on $R^2 \times R^1_+$.

(iv) Every member of $C(x,t)$ satisfies a linear growth condition.

To prove the eventual uniform asymptotic stability of the origin, we show that the origin is eventually uniformly asymptotically stable with respect to every solution, and, as indicated above, at least one solution exists and is continuable for every initial point (x_o, t_o).

† At points of discontinuity of $e(x,\cdot)$, solutions can be joined in the usual way.

First consider any $X_j \in D$. By Assumption 1 (ii), Algorithm 1 is positively satisfied; suppose this is accomplished by the k-th possible combination of the v_i^* . Let $V_k(\cdot) : X_j \to R^1$ be given by

$$V_k(x) = x' P_{(k)} x \tag{9}$$

where $P = P_{(k)}$ is the solution of

$$P A_{(k)} + A'_{(k)} P - P B R^{-1} B' P + Q = 0 \tag{10}$$

with

$$A_{(k)} = A_o + \sum_{i=1}^{p} A_i \operatorname{sgn} \sigma_i^k (x)$$

$$\sigma_i^k(x) = x'[P A_i + A_i' P] x$$

In view of (iii) of Assumptions 1, $P_{(k)}$ is positive definite and symmetric, [17].

Next we show that $V_k \circ x(t)$ decreases along a solution $x(\cdot)$ of (3) generated by $(\tilde{p}(\cdot), e(\cdot))$ for all $x(t) \in X_j$. For all $x(t) \in X_j$

$$W(t) = \operatorname{grad} V_k(x(t)) \dot{x}(t)$$

$$= 2 x'(t) P_{(k)} [(A_o + \sum_{i=1}^{p} A_i v_i(t) x(t) + B v_{p+1}(t) \tilde{p}(x(t),t)]$$

where

$$v_i(t) = e_i(x(t),t)$$

$$\tilde{p}(x(t),t) = - R^{-1} B' P_{(k)} x(t)$$

However, since $|v_i(t)| \leqslant 1, i \in \{1, 2, \ldots, p\}$,

$$\sigma_i^k (x(t)) \operatorname{sgn} \sigma_i^k (x(t)) \geqslant v_i(t) \sigma_i^k(x(t)) \tag{11}$$

so that

$$W(t) \leqslant x'(t) [P_{(k)} A_{(k)} + A'_{(k)} P_{(k)}] x(t)$$

$$- 2 x'(t) [P_{(k)} B R^{-1} B' P_{(k)}] x(t) v_{p+1}(t)$$

Since $P_{(k)} B R^{-1} B' P_{(k)}$ is positive semidefinite and $v_{p+1}(t) \in [1, q]$,

$$W(t) \leqslant x'(t) [P_{(k)} A_{(k)} + A'_{(k)} P_{(k)} - P_{(k)} B R^{-1} B' P_{(k)}] x(t)$$

$$- x'(t) [P_{(k)} B R^{-1} B' P_{(k)}] x(t)$$

In view of (10) we have

$$W(t) \leqslant - x'(t) Q x(t) - x'(t) [P_{(k)} B R^{-1} B' P_{(k)}] x(t) \tag{12}$$

Now we have two possibilities:

(i) $x(t) \in X_j$, $t \in [t', t'']$, and "Nature" does not use her "optimal" strategy. Then (11), and hence (12), is a strict inequality; thus, $W(t) < 0$.

(ii) $x(t) \in X_j$, $v_i(t) = \operatorname{sgn} \sigma_i^k(x(t))$, $i = 1, 2, \ldots, p$, $v_{p+1}(t) = 1$,

$t \in [t', t'']$. Then the system is linear with constant coefficients (linear, time-invariant). Further more,

$$- W(t) = x'(t)\, Q\, x(t) + x'(t)\, P_{(k)}\, B\, R^{-1}\, B' P_{(k)}\, x(t)$$

$$= x'(t)\, C'C\, x(t) + u'(t)\, R\, u(t) \equiv 0 \tag{13}$$

on $[t', t'']$. Since both terms in (13) are non-negative

$x'(t)\, C'C\, x(t) \equiv 0$, $u'(t)\, R\, u(t) \equiv 0$.

Since R is positive definite, $u(t) \equiv 0$, and the system is

$\dot{x}(t) = A_{(k)}\, x(t)$, $t \in [t', t'']$.

But, since $\{C, A_{(k)}, B\}$ is assumed to be observable, $x'(t)\, C'C\, x(t) \equiv 0$ cannot occur and so neither can $W(t) \equiv 0$. We conclude that $V_{(k)} \circ x(t)$ decreases along a solution $x(\cdot)$ for all t such that $x(t) \in X_j \in D$.

Finally we note:

a) If a solution $x(\cdot) : [t_o, \infty) \rightarrow R^2$ remains in an $X_j \in D$ for all $t \in [T, \infty)$, $T \geq t_o$, the origin is eventually uniformly asymptotically stable with respect to $x(\cdot)$ since the requirements for Lyapunov stability are met with respect to $x(\cdot)|_{[T, \infty)}$

b) If a solution leaves an $X_j \in D$ it cannot return to it by Assumptions 1 (iv) and (v). Since the decomposition D is finite, a solution must remain in some X_j (case a)), or enter an attractive boundary, or reach an invariant set $X_k \in D$. If it enters an attractive boundary, Assumption 1 (vi) assures eventual uniform asymptotic stability. If it enters an invariant X_k , it must remain in X_k (case a)) or reach an attractive boundary. In either case, eventual uniform asymptotic stability is assured, since X_k exists by Assumption 1(v).

Remarks

2. Assumptions 1 are sufficient but not necessary to assure that $\tilde{p}(\cdot)$ is stabilizing, [10].

3. Assumptions 1 (iv) - (vi) depend on the properties of boundaries $\bar{X}_i \cap \bar{X}_j$. For some cases, for instance single input systems in companion form, these properties are readily checked, [10].

4. Feedback control $\tilde{p}(\cdot)$ is defined almost everywhere on $R^n \times R^1_+$. Due to a real controller's delay in switching, chattering across an attractive boundary occurs, [11 - 13]; see also Example 1.

2.3 Example 1

Here we consider a simple example to illustrate the theory developed in Section 2.2, namely, a second order system with a single input and a single uncertainty:

$$\dot{x}_1(t) = x_2(t) \tag{14}$$

$$\dot{x}_2(t) = 1.6\ v(t)\ x_2(t) + u(t)$$

with uncertainty $v(t) \in [-1,1]$, and control penalty matrix $R = 1$. Furthermore, let matrix

$$Q = \begin{bmatrix} 1 & 0 \\ 0 & 0 \end{bmatrix}$$

With the system so specified, we have only two possible combinations for v^*. These, together with the pertinent $P_{(k)}$, $\sigma^k(x)$ and $\tilde{p}(x,t)$ are listed below. The decomposition induced by $\sigma^k(x) = 0$, $k = 1, 2$, is shown in Fig. 1.

$v^* = 1$ $\qquad\qquad$ $v^* = -1$

$$P_{(1)} = \begin{bmatrix} 2.1 & 1 \\ 1 & 3.7 \end{bmatrix} \qquad P_{(2)} = \begin{bmatrix} 2.1 & 1 \\ 1 & 0.5 \end{bmatrix}$$

$$\sigma^1(x) = x_2\ (3.2\ x_1 + 12\ x_2) \qquad \sigma^2 = x_2\ (3.2\ x_1 + 1.7\ x_2)$$

$$\tilde{p}(x,t) = -x_1 - 3.7\ x_2 \qquad \tilde{p}(x,t) = -x_1 - 0.5\ x_2$$

The algorithm is positively satisfied on each member of the decomposition induced by $\sigma^k(x) = 0$, $k = 1, 2$; the corresponding switching functions are indicated on Fig. 1. As can be seen, on two members of the decomposition, the algorithm is positively satisfied with both possible combinations. Furthermore, both combinations satisfy the algorithm positively on two pairs of adjacent members of the decomposition. Thus, these adjacent members can be combined into a single one; e.g., the ones for $\operatorname{sgn} \sigma^2 = -1$. Recalling that one assumption underlying Theorem 2 requires that every boundary of the decomposition be either attractive or a transition one, we verify readily that the boundaries given by $x_2 = 0$ are transition ones; however, the boundaries given by $3.2\ x_1 + 12\ x_2 = 0$ are neither. Thus, we alter the decomposition by rotating this line until we obtain boundaries satisfying the above assumption, in this case attractivity. The final decomposition is shown in Fig. 2.

If we denote system (14) by

$$\dot{x}(t) = A\ x(t) + b\ u(t)$$

then, upon setting $u(t) = \tilde{p}(x(t),t)$ and $v(t) = \operatorname{sgn} \sigma^k(x(t))$ for $k = 1, 2$, we get†

$$\dot{x}(t) = A_{CL}\ x(t)$$

where matrix

$$A_{CL} = \begin{bmatrix} 0 & 1 \\ -1 & -2.1 \end{bmatrix}$$

Line AOA contains one eigenvector of A_{CL}.

†For special features of single input systems in companion form see [10].

Finally, Fig. 3 shows some typical solution curves of system (14) subject to a parameter uncertainty that is a random piecewise constant function of time. Note that the solution curves reach the attractive boundary $\bar{X}_2 \cap \bar{X}_3$ and then move along it towards the origin.

3. INPUT DISTURBANCE WITH BOUNDED CONTROL

3.1 Problem Statement

Now we treat a class of dynamical systems with input disturbance. Consider

$$\dot{x}(t) = A(t)\, x(t) + B(t)\, u(t) + B(t)\, v(t) \tag{15}$$

$$x(t_o) = x_o \ , \ t \in [t_o, t_1]$$

where

$x(t) \in R^n$ is the state of the system at time t ;

$A(\cdot)$ is an $n \times n$ matrix, continuous on R^1 ;

$B(\cdot)$ is an $n \times m$ matrix, continuous on R^1 ;

$u(t) \in U = \{u \in R^m : \| u \| \leqslant \rho_u = \text{constant} \in (0,\infty)\}$ is the control;

$v(t) \in V = \{v \in R^m : \| v \| \leqslant \rho_v = \text{constant} \in (0,\infty)\}$ is the disturbance.

Since we are again concerned with the asymptotic stability of the origin $x = \{0\}$, we introduce a measure of deviation

$$J = \int_{t_o}^{t_1} x'(t)\, Q(t)\, x(t)\, dt \tag{16}$$

where $Q(\cdot)$ is a symmetric $n \times n$ matrix, continuous on R^1 , and consider the differential game with state equation (15) and cost (16). That is, we seek a saddlepoint $(p^*(\cdot), e^*(\cdot))$ in a given class of strategies

$$p(\cdot) : R^n \times R^1 \to R^m \ , \quad e(\cdot) : R^n \times R^1 \to R^m$$

such that

$$u(t) = p(x(t), t) \quad , \quad v(t) = e(x(t), t).$$

Again, we look for a feedback control $p^*(\cdot)$ while admitting a disturbance that may depend on state and time.

Having found a saddlepoint $(p^*(\cdot), e^*(\cdot))$, we inquire under what conditions feedback control $p^*(\cdot)$ renders $x = \{0\}$ uniformly asymptotically stable in the large (in the sense of Lyapunov) <u>against every allowable disturbance</u> $e(\cdot)$.

3.2 Saddlepoint Strategy

On invoking necessary conditions for a saddlepoint, e.g. [8-9], and then sufficient conditions, e.g. [18-19], we find the following saddlepoint for the case $\rho_u = \rho_v = \rho$:

$$p^*(x,t) = - e^*(x,t) = \begin{cases} - \dfrac{B'(t)P(t)x}{\| B'(t)P(t)x \|}\, \rho & \forall (x,t) \notin N \\ \text{any admissible value} & \forall (x,t) \in N \end{cases} \tag{17}$$

where

$$N = \{(x,t) \in R^n \times R^1 : B'(t)\, P(t)\, x = 0\}$$

and matrix $P(\cdot)$ is the solution of

$$\dot{P}(t) + P(t)\, A(t) + A'(t)\, P(t) + Q(t) = 0 \qquad (18)$$

$$P(t_1) = 0$$

The details of the derivation can be found in Chapter 2 of [10].

3.3 Stability

Having deduced a saddlepoint, we ask now whether the controller's saddlepoint strategy $p^*(\cdot)$ results in asymptotic stability of the origin against any allowable disturbance strategy $e(\cdot)$. Thus, consider a feedback control $\tilde{p}(\cdot) : R^n \times R^1_+ \to U$ given by

$$\tilde{p}(x,t) = \begin{cases} -\dfrac{B'(t)P(t)x}{\|B'(t)P(t)x\|}\, \rho_u & \forall (x,t) \notin N \\ u \in U = \{u \in R^m : \|u\| \leq \rho_u\} & \forall (x,t) \in N \end{cases} \qquad (19)$$

with, [20],

$$P(t) = \int_t^{\infty} \Phi'(\tau - t)\, Q(\tau)\, \Phi(\tau - t)\, d\tau \qquad (20)$$

which is a particular solution of (18), where $\Phi(\cdot)$ is the transition matrix of $\dot{x}(t) = A(t)\, x(t)$.

Consider the following assumptions.

Assumptions 2.

(i) Admissible disturbance $e(\cdot) : R^n \times R^1_+ \to V \subset R^m$ is continuous on R^n and piecewise continuous on any compact subinterval of R^1_+.

(ii) $\exists\, c_1, c_2 \in (0, \infty)$ such that $\|A(t)\| \leq c_1$, $\|B(t)\| \leq c_2$ $\forall t \in R^1_+$.

(iii) $Q(t)$ is positive definite (symmetric); that is, $\exists\, c_3, c_4 \in (0, \infty)$, $c_3 \leq c_4$, such that $c_3 I \leq Q(t) \leq c_4 I$ $\forall t \in R^1_+$.

(iv) $A(t)$ is uniformly asymptotically stable.

(v) $\rho_u \geq \rho_v$.

Now we are ready to state a stability theorem.

Theorem 2. Consider system (15). If Assumptions 2 are met there exists a feedback control $\tilde{p}(\cdot)$ satisfying (19) such that the origin is uniformly asymptotically stable in the large (Lyapunov) for all admissible disturbances $e(\cdot)$.

Proof. Since $\tilde{p}(\cdot)$ is discontinuous and hence considered not unique, (15) becomes a generalized dynamical system, [11-16],

$$\dot{x}(t) \in C(x(t), t) \qquad (21)$$

where the set valued function $C(\cdot)$ is given by

$$C(x,t) = \{A(t)x + B(t)u + B(t)e(x,t) : u = \tilde{p}(x,t)\}$$

As in the proof of Theorem 1, it can again be shown, [10], that, given any $(x_o, t_o) \in R^n \times R^1_+$, there exists at least one solution of (21) and that such a solution can be continued on any compact subset of $R^n \times R^1_+$.

To demonstrate the uniform asymptotic stability of the origin, we consider the function $V(\cdot) : R^n \times R^1_+ \to R^1$ given by

$$V(x,t) = x'P(t)x \tag{22}$$

where $P(t)$ is defined by (20).

Since $A(t)$ is uniformly asymptotically stable and $Q(t)$ is positive definite according to (iii) and (iv) of Assumptions 2, matrix $P(t)$ is positive definite, [20]. In particular, there exist $c_5, c_6 \in (0, \infty)$, $c_6 \geq c_5$, such that $c_5\|x\|^2 \leq V(x,t) \leq c_6\|x\|^2 \quad \forall (x,t) \in R^n \times R^1_+$.

Thus, $V(\cdot)$ is a Lyapunov function candidate.

Finally, we observe that $V \circ x(t)$ decreases along a solution $x(\cdot)$ of (21) generated by $(\tilde{p}(\cdot), e(\cdot))$. Namely, for $(x(t), t) \notin N$,

$$\begin{aligned} W(t) &= \text{grad}_x V(x(t), t)\dot{x}(t) + \frac{\partial V(x(t),t)}{\partial t} \\ &= 2x'(t)P(t)[A(t)\, x(t) - B(t)\, \frac{B'(t)P(t)x(t)}{\|B'(t)P(t)x(t)\|}\, \rho_u + B(t)e(x(t),t)] \\ &\qquad + x'(t)\dot{P}(t)x(t) \\ &= x'(t)[\dot{P}(t) + P(t)A(t) + A'(t)P(t)]\, x(t) \\ &\qquad -2\rho_u\|B'(t)P(t)x(t)\| + 2x'(t)P(t)B(t)e(x(t),t) \\ &= -x'(t)Q(t)x(t) - 2\rho_u\|B'(t)P(t)x(t)\| + 2x'(t)P(t)B(t)e(x(t),t) \\ &\leq -\, x'(t)Q(t)x(t) - 2(\rho_u - \rho_v)\|B'(t)P(t)x(t)\| \\ &< 0 \quad \forall\, \rho_u \geq \rho_v . \end{aligned}$$

For $x(t) \in N$ but $x(t) \neq 0$,

$W(t) = -\, x'(t)Q(t)x(t) < 0$.

This concludes the proof.

Theorem 2 has an immediate corollary.

<u>Corollary 1</u>. The average measure of deviation from the origin along a solution $x(\cdot) : [t_o, \infty] \to R^n$, $x(t_o) = x_o$, generated by $(\tilde{p}(\cdot), e(\cdot))$ is

$$\int_{t_o}^{\infty} x'(t)Q(t)x(t)dt \leq x_o'\, P(t_o)x_o .$$

<u>Proof</u>. In view of (ii) - (iv) of Assumptions 2, $P(t)$ is bounded on R^1_+, [20]. Thus, the result follows upon integration of $W(t)$.

<u>Remarks</u>

5. If matrices $A(t)$ and $Q(t)$ are constant and $t_1 \to \infty$, then $P(\cdot)$ is the constant matrix solution of the Lyapunov equation, [20],

$$PA + A'P + Q = 0 \tag{23}$$

6. Chattering across the singular manifold N is possible, [11-13].

7. In the scalar input case, the control $\tilde{p}(x,t)$ is bang-bang.
8. If the matrix A is not stable but $\{A,B\}$ is stabilizable, Theorem 2 is applicable,[21].
9. The results of this section, in particular Theorem 2, remain unaltered if input matrix B is state and time-dependent; i.e., $B(\cdot)$ may be continuous on $R^n \times R^1$.
10. For state-independent input matrix $B(\cdot)$, control $\tilde{p}(x,t)$ is only output-dependent for outputs $y = C(t)x$ where $C(t) = B'(t)P(t)$ depends on $Q(\cdot)$.

3.11 Example 2

As an example illustrating the theory of Section 3.2 consider the third order system

$$\dot{x}(t) = A\,x(t) + B\,u(t) + B\,v(t)$$

where

$$A = \begin{bmatrix} 0 & 1 & 0 \\ 0 & 0 & 1 \\ -1 & -3 & -2.9 \end{bmatrix} \qquad B = \begin{bmatrix} 0 & 0 \\ 0 & 1 \\ 1 & 0 \end{bmatrix}, \qquad Q = 2I$$

The solution of (15) is

$$P = \begin{bmatrix} 4.6 & 3.8 & 1 \\ 3.8 & 6.3 & 1.6 \\ 1 & 1.6 & 0.9 \end{bmatrix}$$

and

$$B'Px = \begin{bmatrix} x_1 + 1.6\,x_2 + 0.9\,x_3 \\ 3.8\,x_1 + 6.3\,x_2 + 1.6\,x_3 \end{bmatrix}$$

Note that N is of dimension $n - 2 = 1$.

Figure 4 shows the response of this system under a random piecewise constant disturbance and a control given by (19).

3.5 Example 3

Finally, as another illustration of the stabilization of a system with input disturbance consider the second order single input system

$$\dot{x}(t) = A\,x(t) + b\,u(t) + b\,v(t) \tag{25}$$

$$A = \begin{bmatrix} 0 & 1 \\ -22 & -24 \end{bmatrix}, \qquad b = \begin{bmatrix} 0 \\ 22 \end{bmatrix}, \qquad Q = \begin{bmatrix} 1 & 0 \\ 0 & \frac{23}{24} \end{bmatrix}$$

Following Section 3.2, it is readily shown that

$$\tilde{p}(x,t) = -\mathrm{sgn}(x_1 + x_2) \quad \text{for} \quad x_1 + x_2 \neq 0 \tag{26}$$

Here, the singular manifold N is of dimension $n - 1 = 1$, and chattering occurs due to delay in digital computation.

Figures 5 and 6 show the system's response under four types of disturbance -- constant, sinusoidal, random piecewise constant, and "worst" -- and control (26). For comparison, Figure 7 shows the analog computer solution for zero as well as sinusoidal disturbance. As expected, the analog solution is smoother than the digital computer one (of the discretized system); the response slides along N rather than chattering across it.

4. MODEL UNCERTAINTY WITH BOUNDED CONTROL

4.1 Problem Statement

Now we return to a class of model uncertainty problems. Here we treat n-th order single input systems in companion form. Consider

$$\dot{x}(t) = [A_o + \sum_{i=1}^{p} A_i v_i(t)]\, x(t) + b\, u(t) \tag{27}$$

$$x(t_o) = x_o \ , \quad t \in [t_o, t_1]$$

where

$x(t) \in R^n$ is the state of the system at time t;

A_o is a constant $n \times n$ matrix of the form

$$A_0 = \begin{bmatrix} 0 & 1 & 0 & \cdots & 0 \\ 0 & 0 & 1 & & 0 \\ \vdots & & & \ddots & \vdots \\ 0 & 0 & 0 & & 1 \\ -\alpha_1 & -\alpha_2 & -\alpha_3 & \cdots & -\alpha_n \end{bmatrix}$$

with α_1 = constant, $i = 1, 2, \ldots, n$;

A_i , $i = 1, 2, \ldots, p$, are constant $n \times n$ matrices of form

$$A_i = \begin{bmatrix} & \mathbf{0} & \\ \hdashline 0 \ \cdots & a_i & \cdots \ 0 \end{bmatrix}$$

with a_i = constant ≥ 0 , $b = [0 \ . \ . \ . \ 0 \ \ 1]' \in R^n$;

$v_i(t)$, $i = 1, 2, \ldots, p$, with $|v_i(t)| \leq 1$, are values of parameter uncertainty at time t;

$u(t) \in R^1$, with $|u(t)| \leq \rho_u \in (0, \infty)$, is the value of control at time t .

Again, we are interested in the asymptotic behavior of system (27) under all possible parameter uncertainties. Towards that end we introduce a performance index

$$J = \int_{t_o}^{t_1} x'(t)\, Q\, x(t)dt \tag{28}$$

where Q is a constant positive definite symmetric n X n matrix.

4.2 Stability

Before proceeding we note that the system may be converted into an equivalent input disturbance one:

$$\dot{x}(t) = A_o x(t) + b\, u(t) + b\, \tilde{v}(t) \tag{29}$$

with

$$\tilde{v}(t) = c'(t)x(t)$$

where

$$c'(t) = [a_1 v_1(t) \quad a_2 v_2(t) \; . \; . \; . \; a_n v_n(t)] \in R^n \text{ , and}$$

$$|\tilde{v}(t)| \leqslant \|c(t)\| \, \|x(t)\| \leqslant \left(\sum_{i=1}^{p} a_i^2\right)^{\frac{1}{2}} \|x(t)\|$$

Thus, we allow disturbances subject to

$$|\tilde{v}(t)| \leqslant \rho_v = \left(\sum_{i=1}^{p} a_i^2\right)^{\frac{1}{2}} \|x(t)\| \tag{30}$$

We see now that the equivalent input disturbance problem is of the type treated in Section 3, with the sole exception of the state dependence of the disturbance constraint.

Upon applying necessary conditions for a saddlepoint $(p^*(\cdot), e^*(\cdot))$, with

$$\rho_u = \rho_v = \left(\sum_{i=1}^{p} a_i^2\right)^{\frac{1}{2}} \|x\| = \rho(\|x\|) \tag{31}$$

one finds the results of Section 3 unchanged. Hence, we can state a stability theorem for the equivalent input disturbance problem.

Theorem 3. Consider system (29). If Assumptions 2(i) and (iv) are met there exists a feedback control $\tilde{p}(\cdot) : R^n \times R^1_+ \to R^1$ satisfying

$$\tilde{p}(x,t) = \begin{cases} -\dfrac{b'Px}{|b'Px|}\,\rho(\|x\|) & \forall (x,t) \in \{(x,t) : b'Px \neq 0\} \\ \text{an admissible value} & \forall (x,t) \in \{(x,t) : b'Px = 0\} \end{cases}$$

where

$$PA_o + A'_o P + Q = 0$$

such that the origin is uniformly asymptotically stable in the large for all admissible disturbances $e(\cdot)$.

Remarks

11. If, in addition to the parameter uncertainties, there is also an input disturbance, say w with $|w| \leqslant \rho_w$, then $\tilde{p}(\cdot)$ is stabilizing provided $\rho(\|x\|) = \rho_w + \left(\sum_{i=1}^{p} a_i^2\right)^{\frac{1}{2}} \|x\|$.

12. The results are readily extended to the case of time-varying matrix A_o by means of Section 3,[10] .

4.3 Example 4

To illustrate the preceding results let us consider a second order system (27) with

$$A_o = \begin{bmatrix} 0 & 1 \\ 0 & 0 \end{bmatrix} , \quad A_1 = \begin{bmatrix} 0 & 0 \\ 1 & 0 \end{bmatrix} , \quad A_2 = \begin{bmatrix} 0 & 0 \\ 0 & 3 \end{bmatrix}$$

and

$$Q = \begin{bmatrix} 1 & 0 \\ 0 & 1 \end{bmatrix}$$

Of course, here

$$\rho(\|x\|) = \|x\| \sqrt{10}$$

Then it is readily shown that

$$\tilde{p}(x,t) = -(x_1 + 2.4\, x_2) - \rho(\|x\|)\, \mathrm{sgn}\,(x_1 + 2.4\, x_2)$$

for all $(x,t) \notin \{(x,t) : x_1 + 2.4\, x_2 = 0\}$.

Here, matrix A_o is not stable but $\{A_o, b\}$ is stabilizable by linear feedback; this accounts for the first term in the expression for $\tilde{p}(\cdot)$; see Remark 8.

Finally, the digital computer response of the system under the indicated parameter uncertainty and control $\tilde{p}(\cdot)$ is shown in Figure 8.

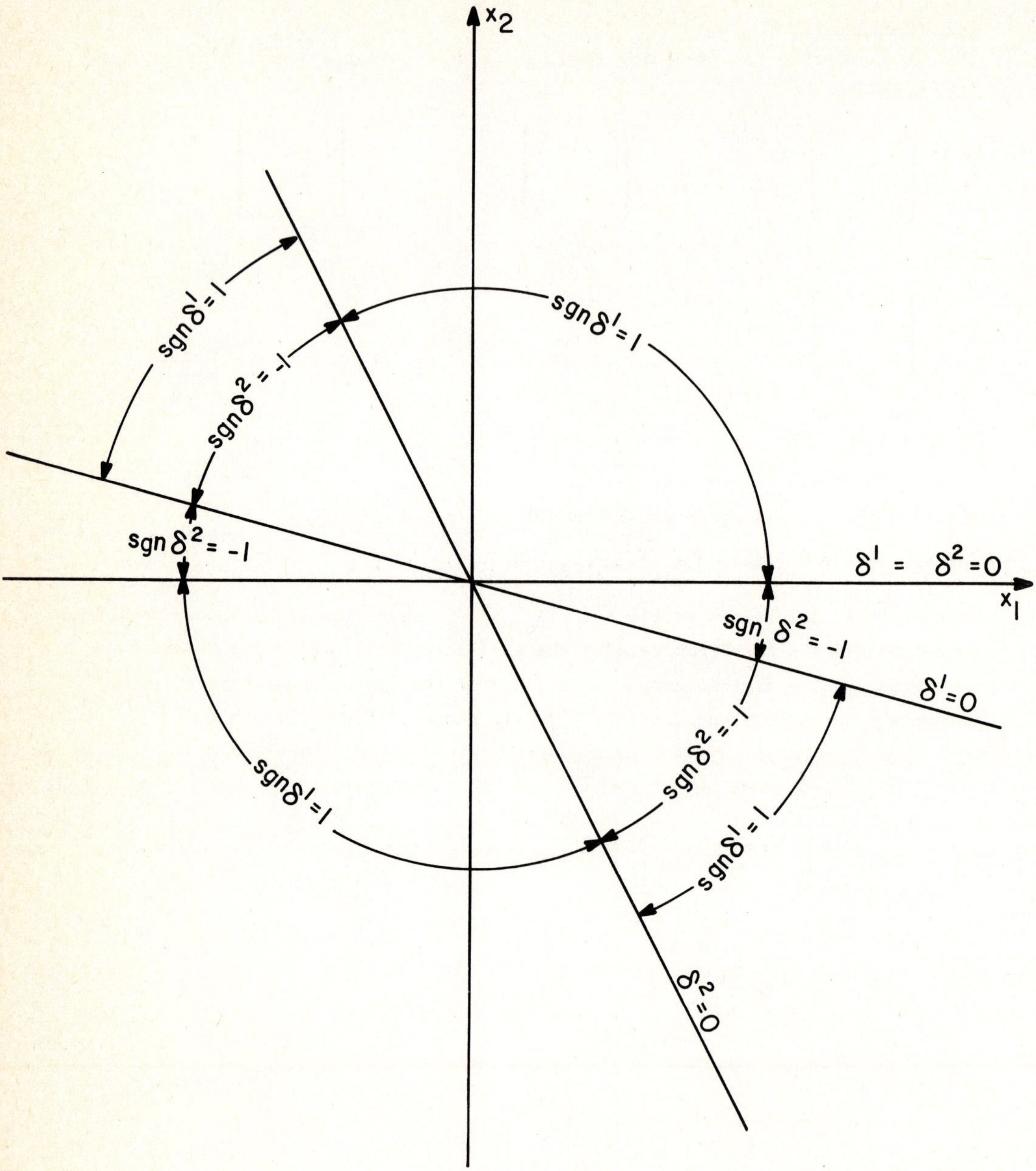

FIGURE 1, EXAMPLE 1

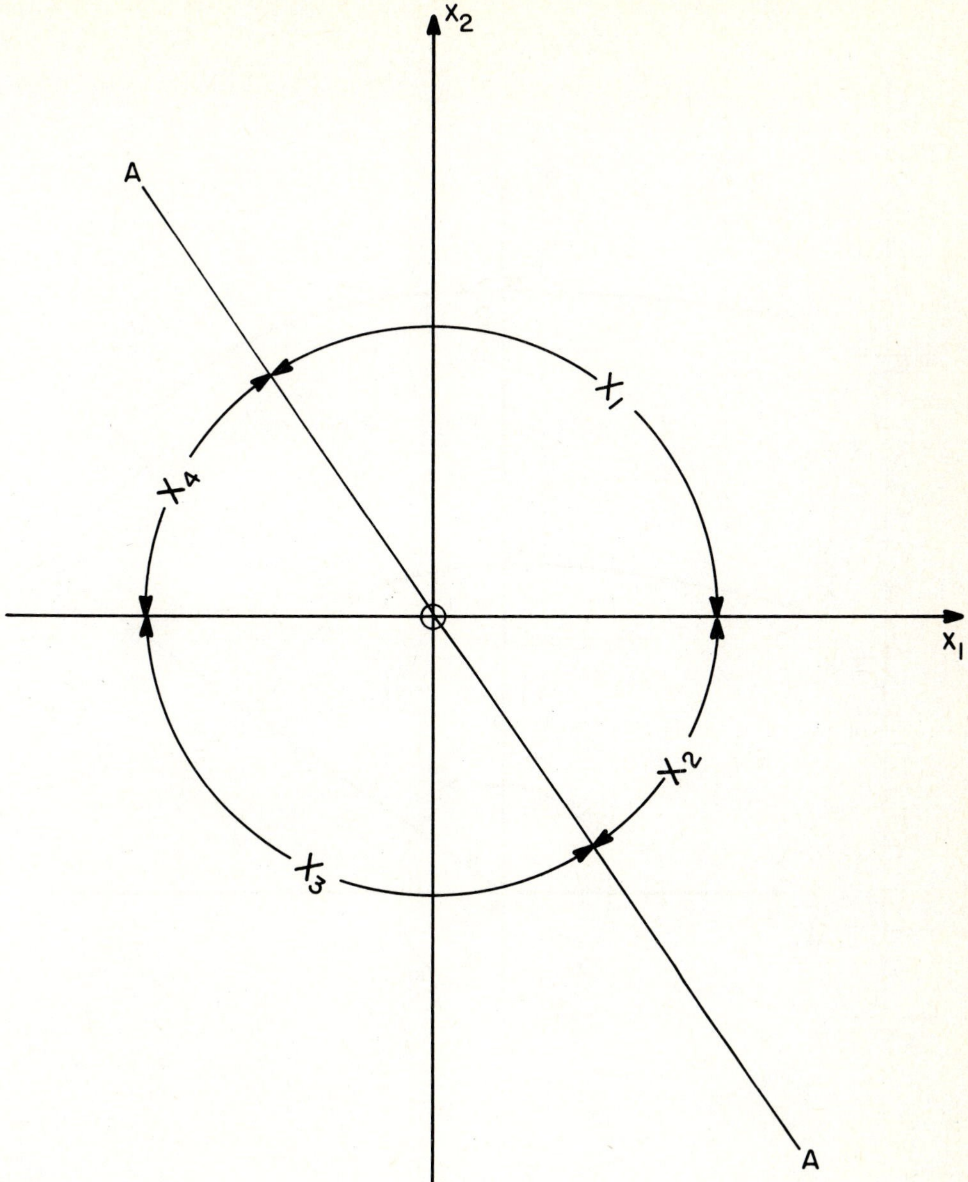

FIGURE 2, EXAMPLE 1

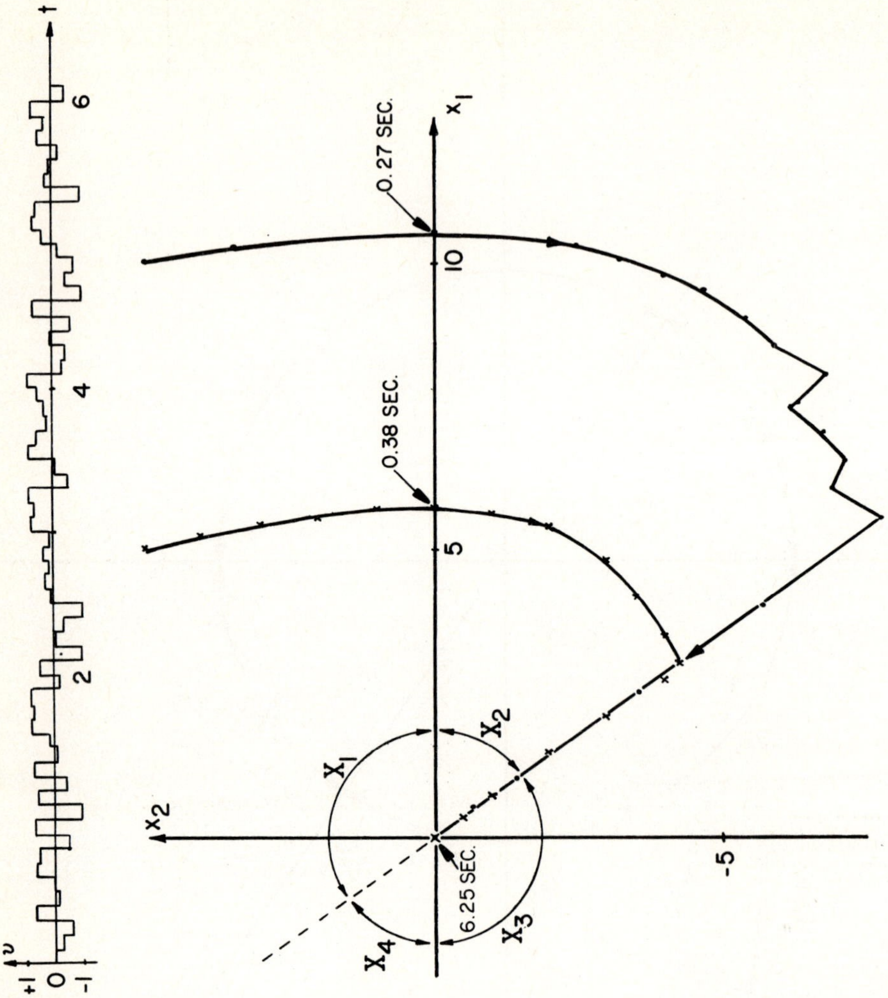

FIGURE 3, EXAMPLE 1

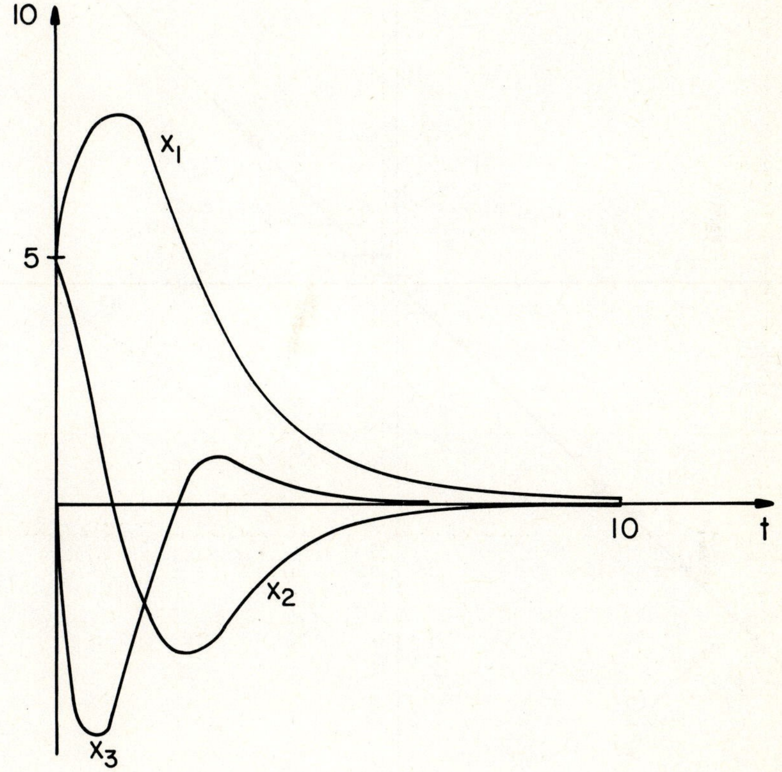

FIGURE 4, EXAMPLE 2

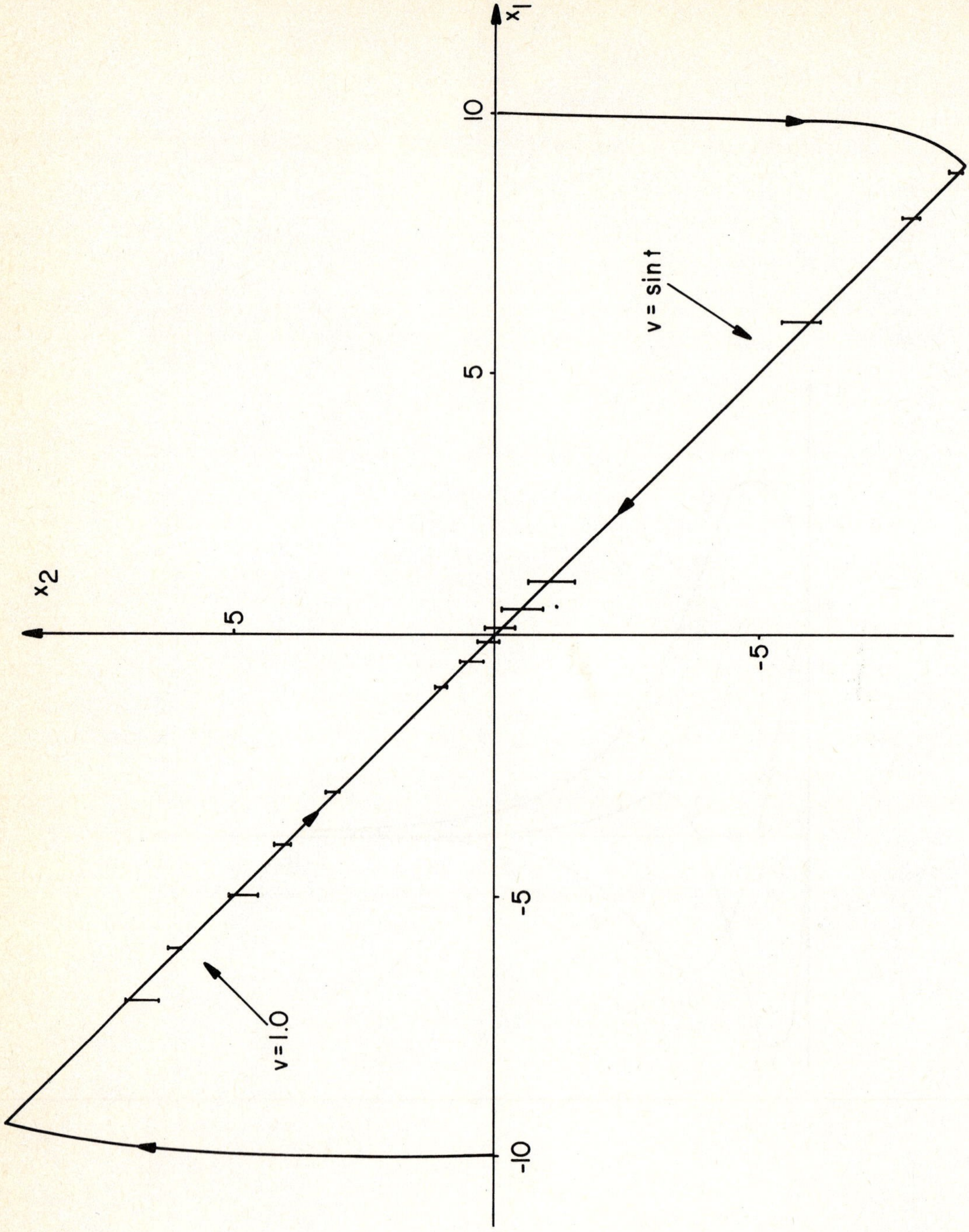

FIGURE 5, EXAMPLE 3

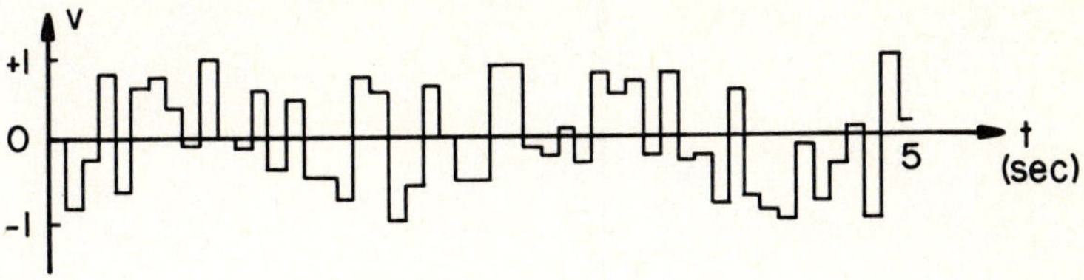

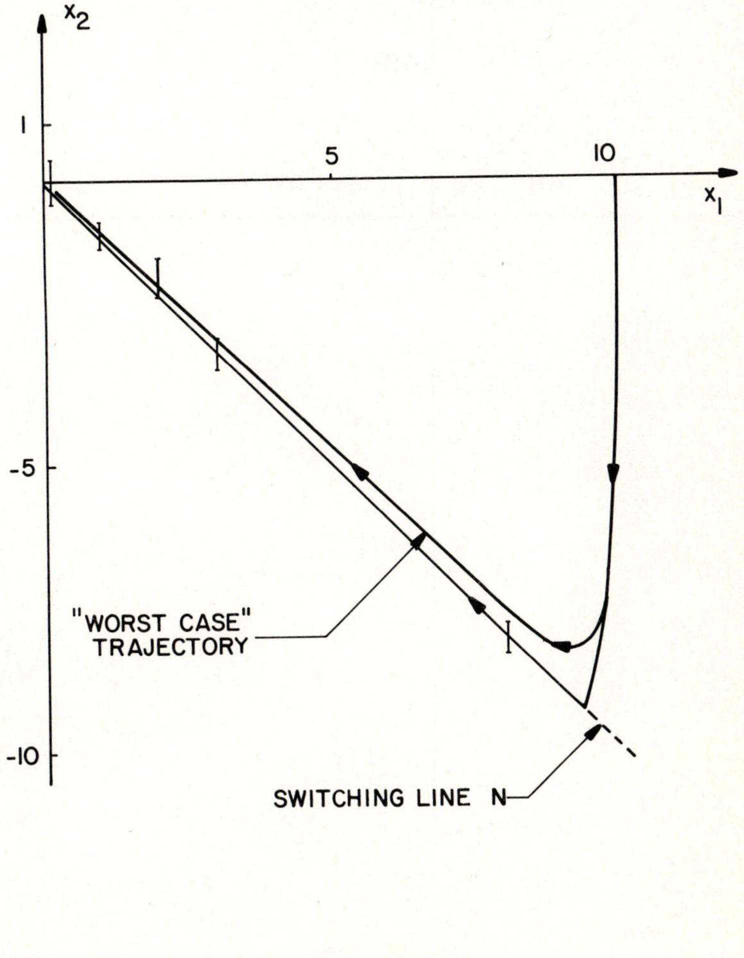

FIGURE 6, EXAMPLE 3

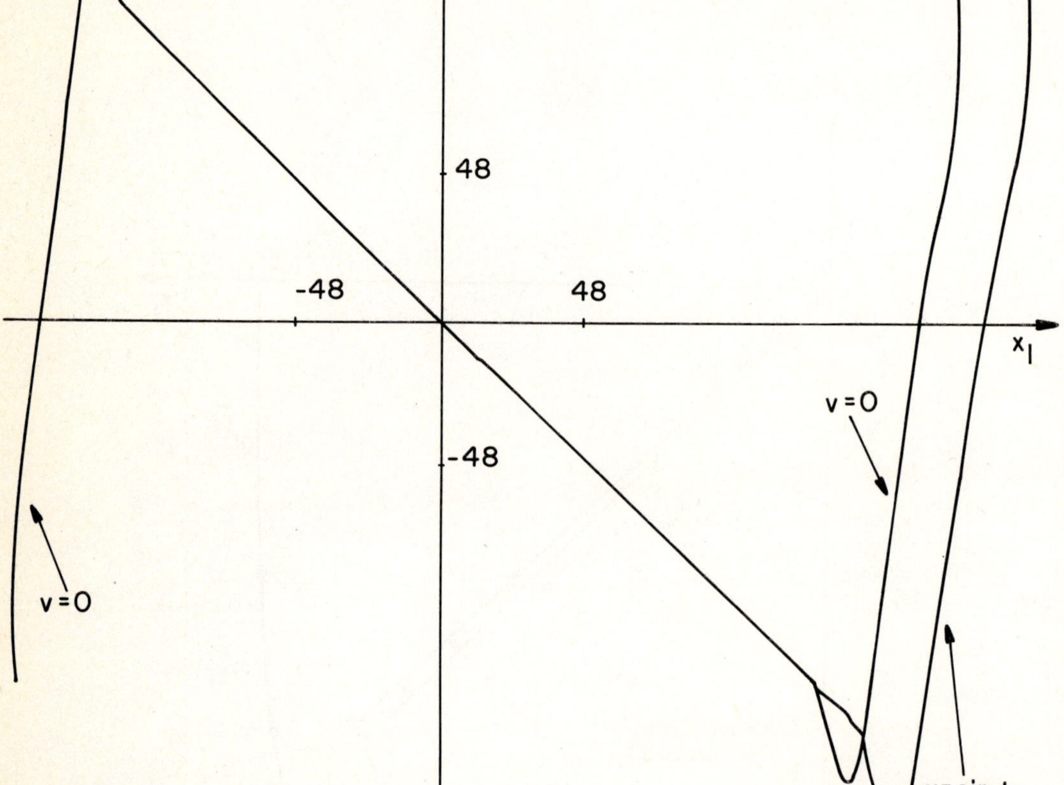

FIGURE 7, EXAMPLE 3

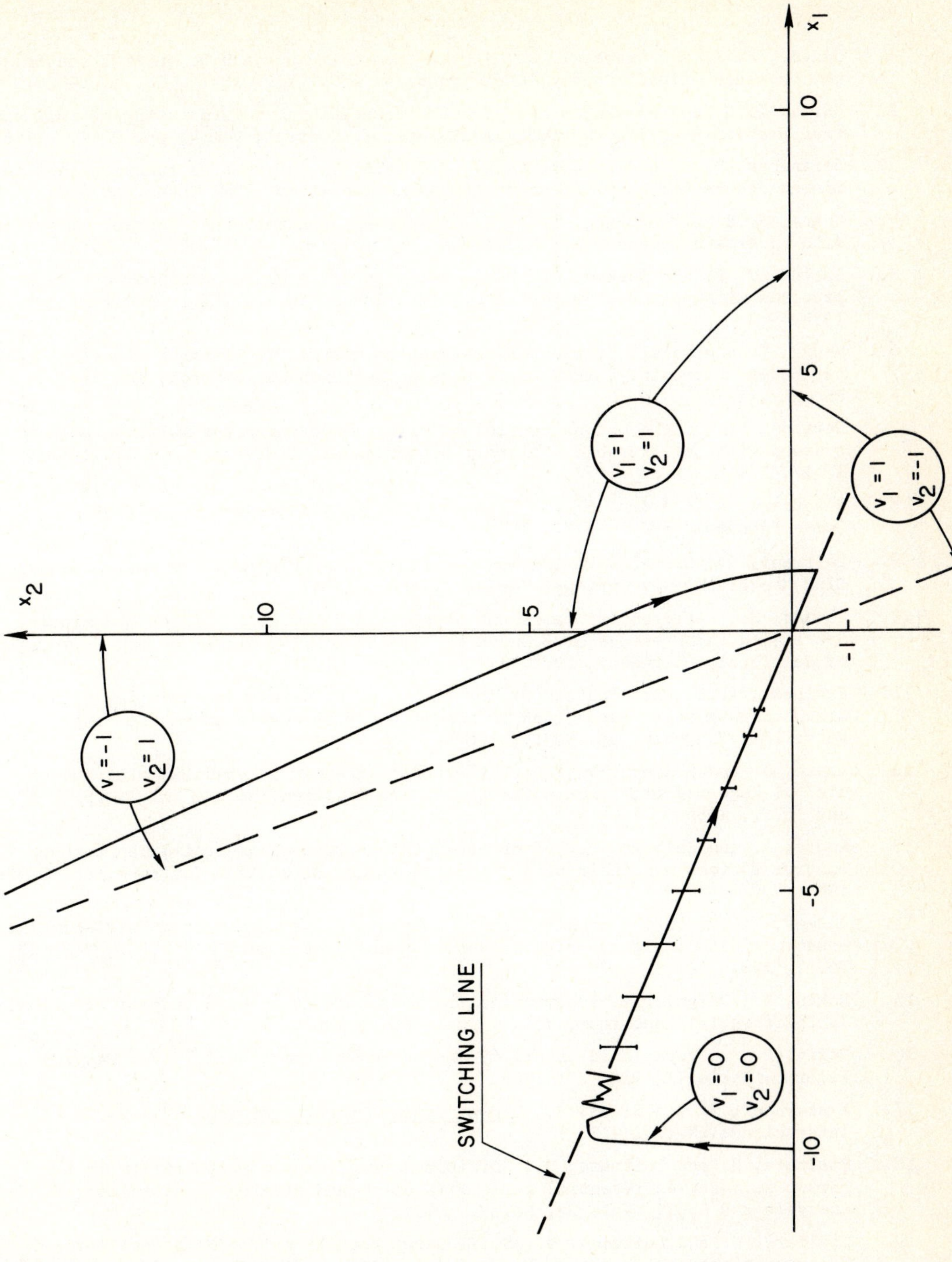

FIGURE 8, EXAMPLE 4

REFERENCES

1. Ragade, R. K. and Sarma, I. G., A game-theoretic approach to optimal control in the presence of uncertainty, IEEE Trans. on A.C., Vol. AC-12, No. 4, 1967.

2. Sarma, I. G. and Ragade, R. K., Some considerations in formulating optimal control problems as differential games, Intl. J. Control, Vol 4, pp. 265f., 1966.

3. Bertsekas, D. P. and Rhodes, I. B., Sufficiently informative functions and the minmax feedback control of uncertain dynamic systems, IEEE Trans. on A.C.

4. Chang, S. S. L. and Peng, T. K. C., Adaptive guaranteed cost control of systems with uncertain parameters, IEEE Trans. on A.C., Vol. AC-17, No. 4, 1972.

5. Speyer, J. L. and Shaked, U., Minimax design for a class of linear quadratic problems with parameter uncertainty, IEEE Trans. on A.C., Vol. AC-19, No. 2, 1974.

6. Menga, G. and Dorato, P., Observer-feedback design for linear systems with large parameter uncertainty, IEEE Conference on Decision and Control, pp. 872f., Phoenix, 1974.

7. Davison, E. J., The output control of linear time invariant multivariable systems with unmeasurable arbitrary disturbances, IEEE Trans. on A.C., Vol. AC-17, No. 5, 1972.

8. Blaquière, A., Gérard, F. and Leitmann, G., Quantitative and Qualitative Games, Academic Press, N.Y., 1969.

9. Leitmann, G., Cooperative and Noncooperative Many Player Differential Games, CISM Monograph 190, Springer Verlag, Vienna, 1974.

10. Gutman, S., Differential Games and Asymptotic Behavior of Linear Dynamical Systems in the Presence of Bounded Uncertainty, Ph.D. dissertation, University of California, Berkeley, 1975.

11. Filippov, A. G., Application of the theory of differential equations with discontinuous right-hand sides to non-linear problems in automatic control, First IFAC Congress, pp. 923f., 1960.

12. André, J. and Seibert, P., Über stückweise lineare Differential-gleichungen, die bei Regelungsproblemen auftreten, I and II, Arch. Math., Vol 7, pp. 148f. and 157f., 1956.

13. André, J. and Seibert, P., After end-point motions of general discontinuous control systems and their stability properties, First IFAC Congress, pp. 919f., 1960.

14. Alimov, Y. I., On the application of Lyapunov's direct method to differential equations with ambiguous right sides, Automation and Remote Control, Vol. 22, No. 7, 1961

15. Roxin, E., On generalized dynamical systems defined by a contingent equation, J. Differential Equations, Vol. 1, pp. 188f., 1965.

16. Roxin, E., On asymptotic stability in control systems, Rend. Circ. Mat. di Palermo, Serie II, Tomo XV, 1966.

17. Kwakernaak, H. and Sivan, R., Linear Optimal Control Systems, Wiley-Interscience, N.Y., 1972.

18. Stalford, H. and Leitmann, G., Sufficient conditions for optimality in two-person zero-sum differential games with state and strategy constraints, J. Math. Analysis and Appl., Vol. 33, No. 3, 1971.

19. Stalford, H. and Leitmann, G., Sufficiency conditions for Nash equilibria in N-person differential games, in Topics in Differential Games, ed. A. Blaquière, pp. 345f., North-Holland, Amsterdam, 1973.

20. Kalman, R. E. and Bertram, J. E., Control system analysis and design via the "second method" of Lyapunov I, J. Basic Engin., ASME Trans., Vol. 82, No. 2, 1960.

21. Gutman, S. and Leitmann, G., On class of linear differential games, J. of Optimization Theory and Application, to appear.

APPLICATION OF THE OPTIMAL CONTROL THEORY WITH DISTRIBUTED PARAMETERS ON A SEARCHING PROBLEM

Olavi Hellman
University of Turku
Turku, Finland

1§ The free motion of the target

We shall assume that the target is moving randomly in such a way that $u(x,t)$ the probability density of the location of the target at time t, satisfies - if no search is going on - the parabolic equation

$$(1) \qquad \frac{\partial u}{\partial t} = \sum_{i=1}^{n} \frac{\partial}{\partial x_i} \left[\sum_{j=1}^{n} a_{ij}(x,t) \frac{\partial u}{\partial x_j} - b_i(x,t) u \right]$$

valid for $(x,t) \in Q_T = \{(x,t) : x \in \Omega \subset R^n,\ t \in (0,T)\}$,
the initial condition

$$(2) \qquad u(x,0) = \psi_0(x),$$

and the boundary condition

$$(3) \qquad \sum_{i=1}^{n} n_i(x,t) \left[\sum_{j=1}^{n} a_{ij}(x,t) \frac{\partial u}{\partial x_j} - b_i(x,t) u \right] \Big|_{S_T} = 0,$$

where $S_T = \{(x,t) : x \in S, t \in [0,T]\}$ and where $n_i(x,t)$ is the normal to S_T at point (x,t). The initial-and boundary value problem (1)...(3) has a unique solution such that $u(x,t) \geq 0$ for all $x \in \Omega$ and $t \geq 0$ and such that $\int_\Omega u(x,t)dx = 1$ for all $t \geq 0$. The functions $a_{ij}(x,t)$, $b_i(x,t)$, and $\psi_0(x)$ may be of a fairly general nature. Equation (1) is the most general parabolic equation of second order which has these properties.

2§ Mathematical formulation of a search

The search for a target is usually carried out by some system of searching equipment which is distributed in Ω and which in general changes as a function of time. - We shall assume that the searching system may be represented by a function $\lambda(x,t)$ with the following properties:

i) $\lambda(x,t) \geq 0$ for $(x,t) \in \Omega \times [0,T]$

ii) $\int_\Omega \lambda(x,t)dx = L_0 = const.$ for all $t \in [0,T]$

iii) $\lambda(x,t)\Delta t + o(\Delta t)$ is the probability that the target will be located, during time $(t, t+\Delta t)$, given that it is at $x \in \Omega$ (for other ways of defining a search see for instance refs. [3] and [4])

Within the theory of generalized parabolic equations applied here, $\lambda(x,t)$ must be a measurable function in $\Omega \times [0,T]$.

3§ Equation for the a posteriori probability density of the location of the target

With the search going on, the probability density of the location of the target will no longer change according to equation (1). Let $\hat{u}(x,t;\lambda)$ denote the said probability density, given that search during time $[0,t]$ has not been successful. Then [5]

$$(4) \qquad \frac{\partial \hat{u}}{\partial t} = \mathcal{L}(\hat{u}) + \hat{u}\left[\int_\Omega \hat{u}(\bar{x},t;\lambda)\lambda(\bar{x},t)d\bar{x} - \lambda(x,t)\right],$$

where we wrote more briefly

$$(5) \qquad \mathcal{L}(u) = \sum_{i=1}^{n} \frac{\partial}{\partial x_i}\left[\sum_{j=1}^{n} a_{ij}(x,t)\frac{\partial u}{\partial x_j} - b_i(x,t)u\right].$$

Equation (4) may be replaced, through the substitution

$$(6) \qquad \hat{u}(x,t;\lambda) = y(x,t;\lambda)\exp\left(\int_0^t d\tau \int_\Omega u(\bar{x},\tau;\lambda)\lambda(\bar{x},\tau)d\bar{x}\right)$$

by the simpler parabolic equation

$$(7) \qquad \frac{\partial y}{\partial t} = \mathcal{L}(y) - \lambda y$$

while the conditions (2) and (3) become, respectively,

$$(8) \qquad y(x,0;\lambda) = \psi_0(x)$$

and

$$(9) \qquad \sum_{i=1}^{n} n_i(x,t)\left[\sum_{j=1}^{n} a_{ij}(x,t)\frac{\partial y}{\partial x_j} - b_i(x,t)y\right]\Big|_{S_T} = 0$$

The probability that the target will be detected, during time $[0,T]$, as a result of the search is [2]

$$(10) \qquad P(T,\lambda) = 1 - \int_\Omega y(x,T;\lambda)dx$$

4§ The optimal search

This far nothing has been said about the choise of search density $\lambda(x,t)$. It is natural to try to find a $\lambda(x,t)$ which maximizes $P(T,\lambda)$, the probability of detection of the target during time $[0,T]$. One has now the following optimal control problem with distributed variables:

$$(11)\qquad \int_\Omega y(x,T;\lambda)dx = Min!$$

$$(12)\qquad \frac{\partial y(x,t;\lambda)}{\partial t} = \mathcal{L}(y(x,t;\lambda)) - \lambda(x,t)y(x,t;\lambda)$$

$$(13)\qquad y(x,0;\lambda) = \Psi_0(x)$$

and

$$(14)\qquad \Sigma_{i=1}^n n_i(x,t)\Big[\Sigma_{j=1}^n a_{ij}(x,t)\frac{\partial y(x,t;\lambda)}{\partial x_j} - b_i(x,t)y(x,t;\lambda)\Big]\Big/_{S_T} = 0.$$

The condition of optimality for $\lambda(x,t)$ will now be obtained by following closely the approach by Lions [6]. Let $\lambda_0(x,t)$ be the optimal search density. Then

$$(15)\qquad \int_\Omega [y(x,T;\lambda) - y(x,T;\lambda_0)]dx \geq 0,$$

for all $\lambda(x,t)$. Condition (15) may now be expressed (cf. ref. [7] sec. 16.2.1) in the form

$$(16)\qquad \int_0^T d\tau\int_\Omega p(x,t;\lambda_0)(\lambda_0(x,t) - \lambda(x,t))y(x,t;\lambda)\,dx\,d\tau \geq 0,$$

where $p(x,t;\lambda_0)$, the adjoint of $y(x,t;\lambda_0)$, is the solution of the boundary value problem

$$(17)\qquad \frac{\partial p(x,t;\lambda)}{\partial t} = -\Sigma_{i,j=1}^n \frac{\partial}{\partial x_i}\Big(a_{ij}(x,t)\frac{\partial p(x,t;\lambda)}{\partial x_j}\Big) + \Sigma_{i=1}^n b_i(x,t)\frac{\partial p(x,t;\lambda)}{\partial x_i} + \lambda(x,t)p(x,t;\lambda) \quad \text{for } (x,t)\in Q_T,$$

$$(18)\qquad p(x,T;\lambda) = 1$$

$$(19)\qquad \Sigma_{ij=1}^n n_i(x,t)a_{ij}(x,t)\frac{\partial p(x,t;\lambda)}{\partial x_j}\Big|_{S_T} = 0.$$

It is a well known fact of the theory of search, that there will be a region $A(t)\subset\Omega$ such that $\lambda(x,t)>0$ for all $x\in A(t)$ and such that $\lambda(x,t)\equiv 0$ for $x\in\Omega\setminus A(t)$. Furthermore, $\lambda(x,t)$ will very likely be discontinuous across the border of $A(t)$. - By using essentially the technique of [3], one now obtains, from condition (16), the following condition of optimality:

(20) $$p(x,t;\lambda)\, y(x,t;\lambda) = \Lambda(t), \quad \text{for } x \in A(t)$$

and

(21) $$p(x,t;\lambda)\, y(x,t;\lambda) \leq \Lambda(t), \quad \text{for } x \in \Omega \setminus A(t).$$

Here $\Lambda(t)$ is a function wich will remain to be determined later.

5§ Application of the condition of optimality

It may be shown, through a fairly straightforward calculation, that equations (12), (17), and condition (20) imply, for $(x,t) \in A(t)$, that

(22) $$\sum\nolimits'^{n}_{i,j=1} \frac{\partial}{\partial x_i}\left[a_{ij}(x,t)\frac{\partial z}{\partial x_j}\right] = -\frac{1}{2}\frac{d}{dt}(\ln \Lambda(t)) + \sum_{i=1}^{n} \frac{\partial b_i(x,t)}{\partial x_i}$$

where

(23) $$z(x,t) = \ln y(x,t;\lambda).$$

It is of interest that equation (22) contains no derivatives with respect to time.

The optimal $\lambda_0(x,t)$ is finally obtained from (12) as follows:

(24) $$\lambda_0(x,t) = \sum_{i,j=1}^{n} a_{ij}(x,t)\frac{\partial z}{\partial x_i}\frac{\partial z}{\partial x_j} - \sum_{i=1}^{n} b_i(x,t)\frac{\partial z}{\partial x_i} - \frac{\partial z}{\partial t} - \frac{1}{2}\frac{d}{dt}(\ln \Lambda(t)).$$

6§ An example

In order to illustrate the way the theory given in previous sections would be applied in practical cases, let us consider the perhaps simplest case, the one dimensional diffuse motion of the target on the real axis. Therefore, now,

(25) $$\mathcal{L}(u) = a^2 u_{xx}(x,t).$$

Furthermore, let $u(x,0) = f(x)$, where $f(-x) = f(x)$. The region $A(t)$ of sec. 5 becomes now the interval $[-x(t), x(t)]$, where $x(t)$ is an unknown function of time which remains to be determined later. We have now for $x \in [-x(t), x(t)]$, the equation

(26) $$(\ln y(x,t))_{xx} = -2C_0(t)$$

obtained form equation (22), where we wrote more briefly

$$C_0(t) = \frac{1}{4a^2}\frac{d}{dt}(\ln \Lambda(t)).$$

Equation (26) is solved at once to give, for $x \in [-x(t), x(t)]$,

(27) $$y_1(x,t) = \exp\left(-C_0(t)x^2 + C(t)\right),$$

where $C(t)$ is an unknown function of time. Solution (27) satisfies the obvious requirement that $y_1(-x,t) = y_1(x,t)$. Outside interval $[-x(t), x(t)]$ function $y(x,t)$ satisfies equation

(28) $$y_t(x,t) = a^2 y_{xx}(x,t),$$

while $y(x,t)$ and $y_x(x,t)$ must be continuous at points $-x(t)$ and $x(t)$. Our problem of optimizing the search has now been reduced to an initial- and boundary value problem with a moving boundary. - We shall now proceed in the well known manner. First let

(29) $$w(x,t) = y(x,t) - \bar{y}(x,t)$$

where

$$\bar{y}(x,t) = (4\pi a^2 t)^{-\frac{1}{2}} \int_{-\infty}^{+\infty} f(\xi) \exp\left\{-\frac{(x-\xi)^2}{4a^2 t}\right\} d\xi,$$

so that

(30) $$w_t(x,t) = a^2 w_{xx}(x,t)$$

(31) $$w(x,0) = 0$$

(32) $$w(\pm x(t),t) = y_1(\pm x(t),t) - \bar{y}(\pm x(t),t)$$

and

(33) $$w_x(\pm x(t),t) = y_{1_x}(\pm x(t),t) - \bar{y}_x(\pm x(t),t)$$

Next we substitute, for $x \geq x(t)$ and for $x \leq -x(t)$

$$w(x,t) = (4a^2\pi)^{-\frac{1}{2}} \int_0^t \frac{K(t')}{\sqrt{t-t'}} \left[\exp\left(-\frac{(x-x(t'))^2}{4a^2(t-t')}\right) + \right.$$

(34) $$\left. + \exp\left(-\frac{(x+x(t'))^2}{4a^2(t-t')}\right)\right] dt'$$

Expression (34) satisfies equation (30) and condition (31), but contains the unknown functions $x(t)$ and $K(t')$. For their determination one has conditions (32) and (33). Function $y_1(x,t)$ contains, however, the unknown functions $C_0(t)$ and $C(t)$ so that two more conditions will be needed. One obtains now, from equation (24), that

$$\lambda(x,t) = (a^2 C_0^2(t) + \dot{C}_0(t))x^2 - (a^2 C_0(t) + \dot{C}(t))$$

It will be required that

$$\lambda(x,t) \geq 0 \tag{35}$$

Since finally $\int_{-x(t)}^{+x(t)} \lambda(x,t)\,dx = L_0$

we obtain the condition that

$$\frac{2}{3}(a^2 C_0^2(t) + \dot{C}_0(t))x^3(t) - 2(a^2 C_0(t) + \dot{C}(t))x(t) = L_0. \tag{36}$$

Condition (35) can be satisfied by $C_0(t) \equiv 0$, provided that $\dot{C}(t) \leq 0$, in which case condition (36) reduces to

$$-2x(t)\dot{C}(t) = L_0$$

It is easy to see, that the determination of unknown functions $x(t)$, $C(t)$, and $K(t)$ still is a fairly complicated problem. By assuming a^2 of (25) to be small, perturbation calculations associated with an asymptotic treatment of function $w(x,t)$, enable one to obtain the unknown functions as power series of a^2. This seems to indicate that the assumption $C_0(t) \equiv 0$ leads to a function $y(x,t)$ which satisfies our initial- and boundary value problem. Since the said problem has a unique solution, our solution would be the solution of the problem. We shall not reproduce the lengthy perturbation calculations here.

The case $a^2 = 0$, that of a stationary target, is well known [3] However, since a^2 appears as the coefficient of the second partial derivative u_{xx}, the limiting process $a^2 \to 0$ would not be a trivial one. When considering the case $a^2 = 0$ one has to give up condition (33). The situation is very much related by nature to the corresponding situation in the theory of fluid flow in the limit of a vanishing viscosity.

References

[1] O.A.Ladyzenskaya, V.A. Solonnikov, and N.N.Uraltseva, Linear and quasi-linear parabolic equations, AMS,Trans.Math.Monogr. 23 (1968)

[2] O.Hellman, On the optimal search for a randomly moving target. SIAM J.Appl.Math.22,545 - 552

[3] B.O.Koopman, Search and screening, OEG Rep. 56, Washington D.C.,(1946)

[4] L.O.Saretsalo, On stochastic models of search for stationary and moving objects, Publication of the institute for applied mathematics, University of Turku, Finland, (1971) (Dissertation).

[5] O.Hellman, On the effect of search upon the probability distribution of a target whose motion is a diffusion process, Ann.Math.Statitst. 41 (1970), pp. 1717 - 1724.

[6] J.L.Lions, Controle optimal de systems gouvernes par des equations aux derivees partielles, Dunod Gauthier-Villars, Paris (1968).

HIRIART-URRUTY Jean-Baptiste
Département de Mathématiques Appliquées
Université de Clermont
Boîte Postale n° 45
63170 AUBIERE

ABOUT PROPERTIES OF THE MEAN VALUE FUNCTIONAL AND OF THE CONTINUOUS INFIMAL CONVOLUTION IN STOCHASTIC CONVEX ANALYSIS

Abstract : In stochastic convex programming numerous examples are to be found where the cost functional to be minimized is of the form of a mean value functional $Ef(x) = \int_\Omega f(x, \omega)\, dP(\omega)$ where $x \in \mathbb{R}^n$ and ω is an uncertain quantity-element of a probability space. The problem of minimizing Ef is a deterministic problem related to the stochastic convex program. To be able to apply the methods of convex optimization and the theorems of convex analysis, it is important to know the properties of Ef, both topological and algebraic. The aim of this paper is to determine the main properties and characteristics of the mean value functional Ef resulting from these corresponding to the functions $f(., \omega)$. By the conjugacy operation, the mean value functional is closely related to the continuous infimal convolution of which we shall also give some properties. Finally the different results obtained are applied to stochastic optimization problems.

I - Notations and terminology

We shall adopt throughout the notations and terminology of P.J. Laurent ([3]) for convex analysis and those of R.T. Rockafellar ([7]) for measurable multivalued mappings.Let $(\Omega, \mathcal{A}, P)$ be a complete probability space, a.s. meaning "almost surely". $\mathcal{F}(\mathbb{R}^n \times \Omega, \overline{\mathbb{R}})$ will denote the set of functions defined on $\mathbb{R}^n \times \Omega$ and taking their values in $\overline{\mathbb{R}}$. In what follows C will usually designate a measurable multivalued mapping defined on Ω. We recall that if $C(\omega)$ is a nonempty closed convex set containing no whole lines, the measurability of C is equivalent to the mesurability of the support functions $\chi^*_{C(.)}(x')$ for every x' ([7] Corollary 3.2). We shall say that C is a CK-valued mapping if $C(\omega)$ is a.s. a nonempty compact convex set of $\mathbb{R}^n$.

As for the integrability of a measurable multivalued mapping, we shall recall the following fundamental definition and properties.

I.1. Definition : Let C be a measurable multivalued mapping defined on Ω. By *mathematical expectation* of C, we shall mean the set (it may be empty) denoted E (C) and defined by :

$$E(C) = \{E(X) \;/\; X \text{ integrable selector of } C\}$$

where an integrable selector of C is an integrable function $X : \Omega \to \mathbb{R}^n$ such that $X(\omega) \in C(\omega)$ a.s.

I.2. Definition : Let C be a CK-valued mapping ; it is said that C is P-integrable if and only if C is measurable and the random variable $||C||$ is integrable, where $||C||$ is defined by :

$$\forall \omega \in \Omega \qquad ||C||(\omega) = ||C(\omega)|| = \operatorname{Sup}\{||x|| \;/\; x \in C(\omega)\}$$

I.3. Fundamental property ([11])

If C is a CK-valued mapping, P-integrable, then E (C) is a nonempty compact convex set characterized by the support functions as follows :

$$(1) \qquad \forall x \qquad \chi^*_{E(C)}(x) = E\{\chi^*_{C(\omega)}(x)\}$$

In this case, the integrability of C is equivalent to the integrability of the functions $\chi^*_{C(.)}(x)$ for every x.

Remarks : I.3.1. If C is a measurable multivalued mapping such that $C(\omega)$ is a.s. a closed convex set, then C is said quasi P-integrable if there exists an integrable selector of C. In this case, $\chi^*_{C(.)}(x)$ is a quasi-integrable function and the relation (1) remains true ([12]).

I.3.2. In a wider sense, if C is such that : $\exists\, A,\ P(A) > 0$ $\omega \in A \Longrightarrow C(\omega) = \emptyset$, then E (C) is empty.

I.3.3. If C is a CK-valued mapping, P-integrable, we have the inequality between norms :

$$||E(C)|| \leq E(||C||)$$

Our purpose in this paper is to examine the properties of the mean value functional and of the continuous infimal convolution. Let us recall here their definitions as well as those of integrands.

I.4. Definitions ([7]). By integrand we mean a function $f : \mathbb{R}^n \times \Omega \to \mathbb{R} \cup \{+\infty\}$. If $f(.,\omega)$ is proper, l.s.c. for every ω and f is $\mathcal{B}_n \otimes \mathcal{A}$-measurable where $\mathcal{B}_n$ denotes the σ-algebra of borelian subsets of $\mathbb{R}^n$, f is said to be a normal integrand. In a wider sense, by normal integrand, we shall also mean f such that $f(.,\omega) \in \Gamma_0(\mathbb{R}^n)$ almost surely. If, moreover, $f(.,\omega)$ is convex for every ω f is called convex.

Normality ensures in particular that for every random variable X, the function $\omega \to f(X(\omega),\omega)$ is measurable.

I.5. Mean value functional : Let $f \in \mathcal{F}(\mathbb{R}^n \times \Omega, \overline{\mathbb{R}})$ be such that, for every x, $f(x,.)$ is measurable. The mean value functional Ef of the collection $\{f(.,\omega)\}_{\omega \in \Omega}$ is defined by :

$$\forall x \qquad Ef(x) = \int_\Omega f(x,\omega)^+ dP(\omega) \dotplus \int_\Omega f(x,\omega)^- dP(\omega)$$

Of course, for any measurable function $g : \Omega \to \mathbb{R}^+ \cup \{+\infty\}$ such that, if we denote $\Omega_\infty(f) = \{\omega \mid f(\omega) = +\infty\}$, we have $P(\Omega_\infty(f)) > 0$, then $Ef = +\infty$.

I.6. Continuous infimal convolution ([2], [12]). Let f be a normal integrand on $\mathbb{R}^n \times \Omega$. By continuous infimal convolution of the family $\{f(.,\omega)\}_{\omega \in \Omega}$ relating to the probability measure P, we mean the functional denoted by $F = \oint_\Omega f(.,\omega)\, dP(\omega)$ and defined by :

$$\forall x \qquad F(x) = \inf_{X \in \mathcal{L}^1(x)} \left\{ \int_\Omega f(X(\omega),\omega)^+ dP(\omega) \dotplus \int_\Omega f(X(\omega),\omega)^- dP(\omega) \right\}$$

where $\mathcal{L}^1(x) = \{X \in \mathcal{L}^1 / E(X) = x\}$.

II - Properties of the mean value functional Ef.

A certain class of stochastic optimization problems are characterized by a cost functional of the form Ef (x). It is therefore interesting to know how the properties of the functions f (., ω) are transmitted to the function Ef. We shall examine such topological properties as l.s.c. continuity and determine such convex characteristics as recession function, θ-subdifferential ...

II.1. L.s.c. continuity of Ef.

II.1. Theorem : *Let $f \in \mathcal{F}(\mathbb{R}^n \times \Omega, \overline{\mathbb{R}})$ be such that : $f(x, .)$ is measurable for every x and a.s. $f(., \ )$ is l.s.c. If $f^*(0, .)^+$ is integrable, then Ef is a l.s.c. function taking its values in $\mathbb{R} \cup \{+\infty\}$.*

Proof : The inequality $f(x, .) \geq -f^*(0, .)$ implies that for every x $f(x, .)$ is quasi-integrable and $\forall x$ $Ef(x) > -\infty$. Suppose that Ef is a proper function (if Ef is identically $+\infty$, Ef is l.s.c).

Let $\{x_n\}_{n \in \mathbb{N}}$ a sequence converging to x ; a.s. f (., ω) is l.s.c., then :

$$\text{a.s.} \qquad f(x, \omega) \leq \varliminf_{n\to\infty} f(x_n, \omega)$$

According to Fatou's lemma ([5]), $E(\varliminf_{n\to\infty} f(x_n, \omega)) \leq \varliminf_{n\to\infty} E f(x_n, \omega)$

$$\Longrightarrow \quad EF(x) \leq \varliminf_{n\to\infty} Ef(x_n, \omega).$$ Hence the l.s.c. of Ef

II.2. Recession function of Ef.

II.2.1. Theorem : *Let $f \in \mathcal{F}(\mathbb{R}^n \times \Omega, \overline{\mathbb{R}})$ be such that : $f(x, .)$ is measurable for every x and a.s. $f(., \omega) \in \Gamma_0(\mathbb{R}^n)$. If moreover Ef is l.s.c. and proper, then the recession function $(Ef)_\infty$ of Ef is given by :*

$$(Ef)_\infty = E(f_\infty).$$

Proof : Ef being proper, there exists x_0 such that $f(x_0, .)$ is integrable : thus a.s. $x_0 \in \text{dom}\, f(., \omega)$

a.s. $f(., \omega) \in \Gamma_0(\mathbb{R}^n)$ and the recession function $f_\infty(., \omega)$ is given by the formula : ([3] Proposition 6.8.3).

$$\forall x \qquad f_\infty(x,\omega) = \operatorname*{Sup}_{\lambda>0} \frac{f(x_o+\lambda x,\omega) - f(x_o,\omega)}{\lambda} = \lim_{\lambda\to\infty} \frac{f(x_o+\lambda x,\omega) - f(x_o,\omega)}{\lambda}$$

This formula is only valid for functions of $\Gamma_o(\mathbb{R}^n)$. Ef is supposed proper and l.s.c., it is also convex. Like previously

$$\forall x \qquad (Ef)_\infty(x) = \lim_{\lambda\to\infty} \frac{Ef(x_o+\lambda x) - Ef(x_o)}{\lambda}$$

Let $\{\lambda_n\}_{n\in\mathbb{N}}$ an increasing sequence of real numbers such that : $\lambda_n > 0$ and $\lim_{n\to\infty} \lambda_n = +\infty$. We take :

$$f_n(x,\omega) = \frac{f(x_o+\lambda_n x,\omega) - f(x_o,\omega)}{\lambda_n}$$

$\{f_n(x,.)\}_{n\in\mathbb{N}}$ is an increasing sequence of quasi-integrable functions ($f_n(x,.)^-$ being integrable) and $\lim_{n\to\infty} f_n(x,.) = f_\infty(x,.)$

It follows that $f_\infty(x,.)$ is quasi-integrable and according to the Beppo-Levi monotone convergence theorem, $Ef_n \to E(f_\infty)$.

II.2.2. Remarks

a) Theorem II.1 gives conditions for Ef being l.s.c.

b) The l.s.c. of Ef was proved in the convex case in [14]. Moreover, R.J.B. Wets ([15]) has proved the inf-compactness of Ef when $f(.,\omega)$ is inf-compact. Using the result of theorem II.2.1 and making a proof similar to that of Wets, we prove more generally that : *if $f(.,\omega)$ is a.s. inf-compact for a slope $X_o(\omega)$ where X_o is an integrable random variable, then Ef is inf-compact for the slope $E(X_o)$.*

II.3. Conjugate of Ef

The computation of the conjugate of Ef will show us the connection between the mean value functional and the continuous infimal convolution. This will allow us to deduce the properties of one through the properties of the other.

II.3.1. Theorem : *Let f be a convex integrand such that $f(x,.)$ is integrable for every x. Then*

$$(Ef)^* = \oint_\Omega f^*(.,\omega)\, dP(\omega)$$

Proof : We shall use mainly the theorem given by Valadier ([12] Theorem 7) determing the conjugate of the continuous infimal convolution.

"Let f be a normal convex integrand such that, for every x' , f^* (x', .) is integrable. Then the continuous infimal convolution $F = \oint_\Omega f(., \omega)\, dP(\omega)$ is exact (i.e. $\forall x, \exists X \in \mathcal{L}^1$ such that $E(X) = x$ and $F(x) = \int_\Omega f(X(\omega), \omega)\, dP(\omega)$).
Moreover $F \in \Gamma_o(\mathbb{R}^n)$ and the conjugate F^* is given by : $F^*(x') = \int_\Omega f^*(x',\omega)\, dP(\omega)$".

We apply this theorem to the normal convex integrand f^* (it will be shown in the proof of theorem II.4.2. that f^* is a normal convex integrand). We have $f(., \omega) = f^{**}(., \omega)$ and $(Ef)^{**} = Ef$. The conjugate of $\oint_\Omega f^*(., \omega)\, dP(\omega)$ is Ef ; hence the result.

II.3.2.Remarks : Let f be anormal convex integrand such that f^* (x', .) is integrable for every x'. Then the continuous infimal convolution F of the family $\{f(., \omega)\}_{\omega \in \Omega}$ is a *co-finite* convex function ([8] p. 116) and $F_\infty = \chi_{\{o\}}$.

Moreover, it is deduced an interesting result of continuity of F : *let X_o be an integrable random variable, if $f(., \omega)$ is finite and continuous at $X_o(\omega)$ a.s., then F is finite and continuous at $E(X_o)$.* This arises from the relation between continuity of f and inf-compactness of f^* ([3] theorem 6.3.9) and from Remark II.2.2. If f takes particular forms, specially with random matrix, the equality of theorem II.3.1 implies interesting results.

II.3.3. Proposition : *Let f be a normal convex integrand. We suppose that $Ef \in \Gamma_o(\mathbb{R}^n)$ and that $f_\infty(x, .)$ is integrable for every x. Then, the multivalued mapping $\omega \to \overline{dom}\, f^*(., \omega)$ is a CK-valued mapping, P-integrable and :*

$$E\{\overline{dom}\, f^*(., \omega)\} = \overline{dom}\ \oint_\Omega f^*(., \omega)\, dP(\omega)$$

Proof : Let $x_o \in$ dom Ef ; $x_o \in$ dom f (., ω) a.s. We have :

$$\forall x, \quad \forall z \qquad f(z, \omega) \leqslant f(x, \omega) + f_\infty(z-x, \omega) \qquad ([8] \text{ Corollary } 8.5.1)$$

This implies that f (x, .) is integrable for every x. Moreover, $f(., \omega) \in \Gamma_o(\mathbb{R}^n)$ and $f_\infty(., \omega)$ is a function of $\Gamma_o(\mathbb{R}^n)$ such that :

$$f_\infty(x,\omega) = \chi^*_{\overline{\text{dom}}\, f^*(.,\omega)}(x) \qquad ([3] \text{ theorem } 6.8.5)$$

Likewise : $(Ef)_\infty(x) = \chi^*_{\overline{\text{dom}}(Ef)^*}(x)$. f^* is a normal convex integrand ([6] lemma 5) and the multivalued mapping $\omega \to$ epi $f^*(., \omega)$ is measurable ([7] theorem 4) ; hence the mesurability of the mapping $\omega \to$ dom $f^*(.,\omega)$ = proj (epi $f^*(., \omega)$. The multivalued mapping $\omega \to \overline{\text{dom}}\, f^*(., \omega)$ is a CK-valued $\mathbb{R}^n$ mapping, P-integrable because the support function $f_\infty(x, .)$ is integrable for every x. Then,

$$\chi^*_{E\{\overline{\text{dom}}\, f^*(.,\omega)\}} = E(\chi^*_{\overline{\text{dom}}\, f^*(.,\omega)})$$

It arises from the equality $(Ef)_\infty = E(f_\infty)$ that $\overline{\text{dom}}\,(Ef)^* = E\{\overline{\text{dom}}\, f^*(., \omega)\}$ and the result is deduced from the expression $(Ef)^* = \oint_\Omega f^*(., \omega)\, dP(\omega)$.

II.3.4 Remark : Under the assumptions of the previous theorem, $f(., \omega)$ is Lipchitzian with coefficient $\alpha(\omega)$ and α is an integrable random variable. Likewise, Ef is Lipchitzian with coefficient A and we have : $A \leqslant E(\alpha)$.

Indeed $\alpha(\omega)$ is given by $||\overline{\text{dom}}\, f^*(., \omega)|| = \text{Sup}\,\{||x^*|| \,/\, x^* \in \text{dom } f^*(., \omega)\}$ ([8] Corollary 13.3.3). Similarly $A = ||\overline{\text{dom}}\,(Ef)^*|| = ||E\{\overline{\text{dom}}\, f^*(., \omega)\}||$

We have seen that $||E\{\text{dom } f^*(., \omega)\}|| \leqslant E(||\overline{\text{dom}}\, f^*(., \omega)||)$ (Remark I.3.3) ; hence the result.

II.4. Determination of the θ-subdifferential of Ef

Convex analysis ([8]) and algorithms for minimizing convex functions ([4]) have brought out the importance of the θ-subdifferential of a convex function ; in this paragraph, we intend to determine the θ-subdifferential of the mean value functional Ef.

II.4.1. Definition : Let $f \in \mathcal{F}(\mathbb{R}^n, \overline{\mathbb{R}})$ be a functional finite at x_0. A vector x^* is called an θ-subgradient of f et x_0 (where $\theta \geqslant 0$) if

$$\forall x \qquad f(x) \geqslant f(x_0) + \langle x^*, x - x_0 \rangle - \theta$$

The set of all θ-subgradients is denoted by $\partial_\theta f(x_0)$ and is called θ-subdifferential of f at x_0.

II.4.2. Theorem : *If f is a convex integrand on $\mathbb{R}^n \times \Omega$ such that $f(x, .)$ is measurable for every x and such that*

$$(H) \qquad \forall x \quad \int_\Omega |f(x, \omega)| \, dP(\omega) < +\infty$$

Then $\forall \theta \geq 0 \quad \partial_\theta \; Ef(x) = \bigcup_{\substack{\theta \in \mathcal{L}^1(\theta) \\ \theta(\omega) \geq 0}} E \; \{\partial_{\theta(\omega)} f(x, \omega)\}$

$$\text{where } \mathcal{L}^1(\theta) = \{\theta \in \mathcal{L}^1 \;/\; E(\theta) = \theta\}$$

Proof : Let us take a sequence $\{z_p\}$ satisfying the following condition : for every x, x is the barycenter of a finite number of z_p (take $z_p \in \mathbb{Z}^n$ for example) ; $f(., \omega)$ being convex functions, we can deduce that : $\exists N \in \mathcal{C}$, $P(N) = 0$ such that : $\forall \omega \notin N$, $f(x, \omega) \in \mathbb{R} \quad \forall x$ ([12]) proof of the lemma 6). Thus there are no major drawbacks in supposing f real-valued. Let $\theta \in \mathcal{L}^1(\theta)$, $\theta(\omega) \geq 0$. We have :

$$\partial_{\theta(\omega)} f(x, \omega) = \{x^* \;/\; f^*(x^*, \omega) - \langle x, x^* \rangle \leq \theta(\omega) - f(x, \omega)\}$$

f is a normal convex integrand ([8]), f^* is also one ([8] lemma 5), as is g^* defined by : $g^*(x^*, \omega) = f^*(x^*, \omega) - \langle x, x^* \rangle$. Moreover, the function $\omega \to \theta(\omega) - f(x, \omega)$ is measurable. Consequently, for every x, the multivalued mapping $\omega \to \partial_{\theta(\omega)} f(x, \omega)$ is measurable ([7] Corollary 4.3).

Let X^* be a measurable selection of this multivalued mapping. By definition, we have :

$$\forall y \qquad f(x+y, \omega) \geq f(x, \omega) + \langle X^*(\omega), y \rangle - \theta(\omega)$$

$$f(x-y, \omega) \geq f(x, \omega) - \langle X^*(\omega), y \rangle - \theta(\omega)$$

$$\Longrightarrow \forall y \qquad -f(x-y, \omega) + f(x, \omega) - \theta(\omega) \leq \langle X^*(\omega), y \rangle$$

$$\leq f(x+y, \omega) - f(x, \omega) + \theta(\omega)$$

Then, for every y, $\langle X^*(.), y \rangle$ is an integrable function. Every selection X^* of the multivalued mapping being integrable, the multivalued mapping $\omega \to \partial_{\theta(\omega)} f(x, \omega)$ is P-integrable.

By definition of the θ-subdifferential, we have :

$$\forall y \qquad f(y, \omega) \geq f(x, \omega) + \langle X^*(\omega), y-x \rangle - \theta(\omega)$$

$$\Longrightarrow \forall y \qquad Ef(y) \geq Ef(x) + \langle E(X^*), y-x \rangle - \theta$$

$$\Longrightarrow \qquad E(X^*) \in \partial_\theta \; Ef(x)$$

Therefore, according to the definition of $E\ \{\partial_{\theta(\omega)}\ f\ (x, \omega)\}$, we deduce that :

$$E\ \{\partial_{\theta(\omega)}\ f\ (x, \omega)\} \subset \partial_\theta\ Ef\ (x)$$

$$\Longrightarrow \bigcup_{\substack{\theta \in \mathcal{L}^1(\theta) \\ \theta(\omega) \geq 0}} E\{\ \partial_{\theta(\omega)}\ f\ (x, \omega)\} \subset \partial_\theta\ \ Ef\ (x)$$

Conversely, let $x^* \in \partial_\theta$ Ef (x). We can also write :

$$\partial_\theta\ \ Ef\ (x) = \{x^* \ /\ (Ef)^*\ (x^*) + Ef\ (x) - < x, x^* > \leqslant \theta\}$$

From the theorem II.3.1, there exists $X^* \in \mathcal{L}^1$ such that :

$$E\ (X^*) = x^* \text{ and } (Ef)^*\ (x^*) = \int_\Omega f^*\ (X^*\ (\omega), \omega)\ dP\ (\omega)$$

Let $\theta(\omega) = f^*\ (X^*\ (\omega), \omega) + f\ (x, \omega) - < x, X^*\ (\omega) >$. Obviously we have :

$$\theta(\omega) \geqslant 0, \quad \theta \in \mathcal{L}^1, \ E\ (\theta) \leqslant \theta \quad \text{and } X^*\ (\omega) \in \partial_{\theta\ (\omega)}\ f\ (x, \omega)$$

By definition of the expectation, $x^* \in E\ \{\ \partial_{\theta(\omega)}\ f\ (x, \omega)\}$; hence the result.

II.4.3. Remark : in the previous theorem, making $\theta = 0$, we obtain the formula for $\partial_0\ Ef = \partial\ Ef$.

Then the formula of theorem II.4.2 becomes :

$$\partial(Ef) = E\ (\partial f) \qquad \text{(denoted symbolically)}$$

Thus we again find the known formula of the subdifferential of the mean value functional ([1] ; [9] p. 62)

II.4.4. Application : extremums of the expectation E (C)

For a convex compact K, let K' be the extremum in the direction x^*, that is to say:

$$K'_{x^*} = \{x \in K\ /\ < x, x^* > = \text{Sup}\ (< z, x^* > /\ z \in K)\} \qquad ([13]\ \text{p. } 10)$$

Let us consider a CK-valued mapping, P integrable. For each x^* , we denote by C'_{x^*} the multivalued mapping defined as following : $\forall\ \omega \in \Omega$

$C'_{x^*}\ (\omega) = [C\ (\omega)]'_{x^*}$. C'_{x^*} is also a CK-valued mapping.

II.4.4. Proposition : For each x^*, the multivalued mapping C'_{x^*} is P-integrable and we have : $E\ (C'_{x^*}) = [E\ (C)]'_{x^*}$; that is to say : *the extremum of the expectation in the direction x^* is the expectation of the extremum in the same direction.*

Proof : We have the following equivalence ([8] Corollary 23.5.3) : for a non-empty closed convex set K, $K'_{x^*} = \partial \chi^*_K (x^*)$

Then, $C'_{x^*}(\omega) = \partial \chi^*_{C(\omega)}(x^*)$ and $[E(C)]'_{x^*} = \partial \chi^*_{E(C)}(x^*)$

From the fundamental property I.3 and the formula II.4.3., it is deduced that :

$$E(C'_{x^*}) = [E(C)]'_{x^*}$$

II.4.5. Remark : for a non-empty closed convex set C, it is said that $x \in C$ is "exposed in the direction x^*" if χ^*_C is differentiable at x^* and if $\partial \chi^*_C(x^*) = \{x\}$. It follows from the previous proposition that :

x is an exposed point of $E(C)$ in the direction x^* if and only if $x = E(X)$ where $X(\omega)$ is a.s. an exposed point of $C(\omega)$ in the direction x^*.

II.5. θ-directional derivative of Ef

In minimization methods, when using certain methods of descent, we replace the directional derivative $f'(x ; d)$ by an approximation $f'_\theta(x ; d)$ which is the θ-directional derivative. We consider here the functional Ef and an explicit characterization is given for the θ-directional derivative of Ef.

II.5.1. Definition : Let $f \in \Gamma_0(\mathbb{R}^n)$, finite at x and $\theta \geq 0$. The θ-directional derivative of f at x with respect to a vector d is defined by :

$$f'_\theta(x ; d) = \text{Sup}\,\{ < c, d > / c \in \partial_\theta f(x)\}$$

II.5.2. Theorem : *Under the assumption of theorem II.4.2., we have :*

$$\forall \theta \geq 0 \qquad (Ef)'_\theta(x ; d) = \sup_{\substack{\theta \in \mathcal{L}^1(\theta) \\ \theta(\omega) \geq 0}} E\,\{f'_{\theta(\omega)}(x ; d, \omega)\}$$

Proof : According to the previous definition, for any $\theta \in \mathcal{L}^1(\theta)$, we write :

$$(Ef)'_\theta(x ; d) = \chi^*_{\partial_\theta Ef(x)} \quad \text{and} \quad f'_{\theta(\omega)}(x ; d, \omega) = \chi^*_{\partial_{\theta(\omega)} f(x,\omega)}(d)$$

From the fundamental property I.3 ;

$$\chi^*_{E\{\partial_{\theta(\omega)} f(x,\omega)\}} = E\{\chi^*_{\partial_{\theta(\omega)}} f(x,\omega)\}$$

The functional Ef is finite and continuous at x ; ∂_θ Ef (x) is a nonempty compact convex set characterized by the formula of Theorem II.4.2. Then, according to the lemma 16.5.1 of [8], it is deduced that :

$$\chi^*_{\partial_\theta} Ef(x) = \sup_{\substack{\theta \in \mathcal{L}^1(\theta) \\ \theta(\omega) \geq 0}} E\{\chi^*_{\partial_{\theta}(\omega)} f(x,\omega)\}$$

II.5.3. <u>Remark</u> : Like in the case of the previous theorem ; if we take $\theta = 0$, the directional derivative of Ef at x with respect to the vector d can be expressed as following :

$$(Ef)'(x\,;\,d) = E\,\{f'(x\,;\,d,\omega)\}$$

In the present case, f (., ω) being a.s. finite and continuous at x, there is no discrepancy between f' (x ; d) as usually defined and f'_θ (x ; d) for θ = 0. Likewise Ef is finite and continuous at x ; so, (Ef)' (x ; d) = $\chi^*_{\partial Ef(x)}$.

II.5.4. <u>Necessary and sufficient optimality condition for a class of stochastic optimization problems</u>

Let us consider stochastic programming models consisting of two-stage formulations ([14]). A first stage problem is that in which an optimization problem is performed without having the prior knowledge of the random outcomes.
After the random outcomes have been observed, the inaccuracies occured are compensated in another optimization problem : second stage program. Generally, the functional to be minimized is of the form Ef, x may be subject to certain constraints :

$$\begin{cases} x \in C \\ f_i(x) \leq 0 \quad i = 1, \ldots, m \end{cases}$$ where C is a convex set and f_i are convex

This deterministic program can be written as :

($\mathcal{P}$) Find $\overline{x}$ such that : $\overline{x} \in Q$ and Ef ($\overline{x}$) = Inf {Ef (x) / x ∈ Q}.

II.5.4. Proposition : Let f a convex integrand a $\mathbb{R}^n \times \Omega$ such that f (x, .) is integrable for every x, Q a convex set. Then a necessary and sufficient condition for $\bar{x}$ being a solution of $(\mathcal{P})$ is that :

$$\forall x \in Q \qquad E\{f'(\bar{x}\,;\, x-\bar{x}, \omega)\} \geq 0$$

Proof : This arises from the usual optimality conditions in convex programming and from the remark II.5.3. concerning the directional derivative of Ef.

III - Properties of the continuous infimal convolution

The continuous infimal convolution F is closely related to the average functional Ef by the conjugacy operation ; for the operations concerning the measurable multivalued mappings, we remark that ; on the one hand : $\chi^*_{E(C)} = E(\chi^*_{C(\omega)})$ under certain assumptions ; on the other hand, more generally : $\chi_{E(C)} = \oint_\Omega \chi_{C(\omega)}\, dP(\omega)$. This last equality brings out the fact that $F = \oint_\Omega f(., \omega)\, dP(\omega)$ may not be l.s.c. for a normal convex integrand f. Let f_1 and $f_2 \in \mathcal{F}(\mathbb{R}^n, \overline{\mathbb{R}})$ (not identically equal to $+\infty$) ; if the infimal convolution $f_1 \nabla f_2$ is exact in $x = x_1 + x_2$, we have : $\partial(f_1 \nabla f_2)(x) = \partial f_1(x_1) \cap \partial f_2(x_2)$ ([3] Proposition 6.6.4). Our purpose is to obtain a similar formula for the subdifferential of the continuous infimal convolution. For that end, we shall define the continuous intersection of a family of sets indexed by Ω.

III.1. Definition : Let A be a multivalued mapping defined on Ω ; we shall call *continuous intersection* of the family $\{A(\omega)\}_{\omega \in \Omega}$, the set denoted by A^* and defined by :

$$A^* = \bigcup_{N \in \mathcal{N}} \Big(\bigcap_{\omega \in \Omega \setminus N} A(\omega) \Big)$$

where $\mathcal{N}$ indicates the family of P-null sets of $\mathcal{A}$.

Remark : It is indeed a generalization to the continuous case of the intersecting operation. For example, we know that if we consider two functions f_1 et f_2 we obtain epi $[\sup(f_1, f_2)]$ = epi $f_1 \cap$ epi f_2. Similarly, let $f \in \mathcal{F}(\mathbb{R}^n \times \Omega, \overline{\mathbb{R}})$ such that for each x, f (x, .) is measurable. Let Φ defined by :

$$\Phi(x) = \text{ess. sup. } f(x, \omega) = \text{Inf } \{\alpha \ / \ f(x, \omega) \leq \alpha \quad \text{a.s.}\}$$

Then, it is easy to see that : epi $\Phi = [\text{epi } f(., \omega)]^*$

III.2. Properties

a) If we modify A on a P-null set, we do not alter A^*.

b) If A is a.s. convex-valued (resp. closed valued, compact valued) mapping, then A^* is convex (resp. closed, compact)

c) The indicator function of A^* is : $\chi_{A^*} = E(\chi_{A(\omega)})$

Proofs :

a) Let A and B two multivalued mappings such that if we denote by N_o the set $\{\omega \in \Omega \ / \ A(\omega) \neq B(\omega)\}$ we have $P(N_o) = 0$

Let $x \in A^*$; $\exists N \in \mathcal{N}$ such that : $\forall \omega \in \Omega \setminus N \quad x \in A(\omega)$; this implies that : $\forall \omega \in \Omega \setminus N \cup N_o \quad x \in B(\omega)$; thus $x \in B^*$. Conversely, a similar proof shows us that $B^* \subset A^*$.

b) According to the property a) we may suppose the required hypothesis assumed for every $\omega \in \Omega$.

Convexity of A^* : let $x, y \in A^*$; $\exists N_x, N_y \in \mathcal{N}$ such that : $x \in A(\omega) \ \forall \omega \in \Omega \setminus N_x$; $x \in B(\omega) \ \forall \omega \in \Omega \setminus Ny$. Taking $\lambda \in [0, 1]$, $A(\omega)$ being convex, we have : $\lambda x + (1-\lambda) y \in A(\omega) \forall \omega \in \Omega \setminus N_x \cup N_y$. So, $\lambda x + (1-\lambda) y \in A^*$.

Closedness of A^* : let $\{x_n\}_{n \in \mathbb{N}}$ a convergent sequence of A^*. For each $n \in \mathbb{N}$, there exists $N_n \in \mathcal{N}$ such that : $x_n \in A(\omega) \quad \forall \omega \in \Omega \setminus N_n$. But :

$$\bigcup_{n \in \mathbb{N}} \Big(\bigcap_{\omega \in \Omega \setminus N_n} A(\omega) \Big) \subset \bigcap_{\omega \in \Omega \setminus \bigcup_{n \in \mathbb{N}} N_n} A(\omega)$$

$$\Longrightarrow \quad \{x_n\}_{n \in \mathbb{N}} \quad \bigcap_{\omega \in \Omega \setminus \bigcup_{n \in \mathbb{N}} N_n} A(\omega)$$ which is closed and included in A^*. Thus, $\lim_{n \to \infty} x_n \in A^*$ and A^* is a closed set.

Compactness of A^* : the same inclusion as previously shows us that A^* is compact when $A(\omega)$ is compact.

c) Indicator function of A^* : $E(\chi_{A(\omega)}(x)) = 0 \Longleftrightarrow \exists N \in \mathcal{N}\ \forall \omega \in \Omega \setminus N \quad \chi_{A(\omega)}(x)=0$

that is to say : $E(\chi_{A(\omega)}(x)) = 0 \Longleftrightarrow x \in A^*$.

III.3. Subdifferential of the continuous infimal convolution F

III.3.1. Theorem : *Let f be a normal convex integrand such that $f^*(x^*, .)$ is integrable for every x^*. Let $x_0 \in \operatorname{dom} F$ and X_0 a random variable whose expectation is x_0 and giving the exactness of the continuous infimal convolution at x_0. If we denote by D_{X_0} the multivalued mapping : $\omega \to \partial f(X_0(\omega), \omega)$, then*

$$\partial F(x_0) = (D_{X_0})^*$$

Proof : Let $x^* \in \partial F(x_0)$; $x^* \in \partial F(x_0) \Longleftrightarrow F(x_0) + F^*(x^*) = \langle x_0, x^* \rangle$.

Moreover x_0 giving the exactness of the continuous infimal convolution F at x_0, we have :

$$E(X_0) = x_0 \quad \text{and} \quad F(x_0) = \int_\Omega f(X_0(\omega), \omega)\, dP(\omega)$$

By Theorem 7 of [12] , $F^* = Ef^*$ and

$$\int_\Omega f(X_0(\omega), \omega)\, dP(\omega) + \int_\Omega f^*(x^*, \omega)\, dP(\omega) = \int_\Omega \langle X_0(\omega), x^* \rangle\, dP(\omega)$$

We always have : $f(X_0(\omega), \omega) + f^*(x^*, \omega) \geq \langle X_0(\omega), x^* \rangle$. The equality between integrals implies then : a.s. $f(X_0(\omega), \omega) + f^*(x^*, \omega) = \langle X_0(\omega), x^* \rangle$, that is to say : a.s. $x^* \in \partial f(X_0(\omega), \omega)$. Consequently $\partial F(x_0) \subset (D_{X_0})^*$.

Conversely, let x^* belonging to $\partial f(X_0(\omega), \omega)$ a.s. :

$$f(X_0(\omega), \omega) + f^*(x^*, \omega) = \langle X_0(\omega), x^* \rangle \text{ a.s.}$$

$$\Longrightarrow F(x_0) + F^*(x^*) = \langle x_0, x^* \rangle \text{ that is to say : } x^* \in \partial F(x_0).$$

Thus $\partial F(x_0) = \bigcup_{N \in \mathcal{N}} \Big(\bigcap_{\omega \in \Omega \setminus N} \partial f(X_0(\omega), \omega) \Big)$ Q.E.D.

III.3.2. Application : normal cone to E(C)

We shall apply the previous result to determining the normal cone to E (C). For a convex set K, the normal cone to K at x_0 is denoted by $N_K(x_0)$ and is defined by :

$$N_K(x_0) = \{x^* / \forall x \in K \quad \langle x^*, x - x_0 \rangle \leq 0\}$$

III.3.2. Proposition : Let C be a CK-valued mapping, P-integrable ; x_o belonging to E(C). If we denote by X_o a random variable such that $E(X_o) = x_o$, $X_o(\omega) \in C(\omega)$ a.s. and by N_{X_o} the multivalued mapping : $\omega \to N_{C(\omega)}(X_o(\omega))$, then we have :

$$N_{E(C)}(x_o) = (N_{X_o})^*$$

Proof : Let f be defined by $f(x, \omega) = \chi_{C(\omega)}(x)$; the properties of C imply that f is a normal convex integrand ([7]). The normal cone is related to the indicator function by the following equality

$$N_{C(\omega)}(x) = \partial\chi_{C(\omega)}(x) \qquad ([8] \text{ Page } 215)$$

It is enough then to apply Theorem III.3.1. whose assumptions are satisfied, bearing in mind that $\chi_{E(C)} = \oint_\Omega \chi_{C(\omega)}\, dP(\omega)$.

III.4. Directional derivative of the continuous infimal convolution

III.4.1. Lemma : Let A be a measurable multivalued mapping defined on Ω such that $A(\omega)$ is a.s. a nonempty closed convex set. Then the support function of A^* is given by :

$$\chi^*_{A^*} = \overline{\oint_\Omega \chi^*_{A(\omega)}\, dP(\omega)}$$

Proof : Let $\Phi = \oint_\Omega \chi^*_{A(\omega)}\, dP(\omega)$. According to the definition of the continuous infimal convolution, Φ is convex and positively homogeneous. The l.s.c. regularization $\overline{\Phi}$ of Φ is the support function of a certain closed convex set C, namely :

$$C = \{x^* \ / \ \forall x \quad < x, x^* > \leq \Phi(x)\} \qquad ([8] \text{ Corollary } 13.2.1)$$

Let $X \in \mathcal{L}^1(x)$; $\forall y \in A^* \quad < X(\omega), y > \leq \chi^*_{A(\omega)}(X(\omega))$ a.s.

$$\Longrightarrow \quad \forall y \in A^* \quad < x, y > \leq \Phi(x)$$

$$\Longrightarrow \quad \chi^*_{A^*} \leq \overline{\Phi} \quad , \text{ thus } A^* \subset C$$

Conversely, let $x^* \in C$. For every $A \in \mathcal{A}$ such that $P(A) > 0$, we define the random variable X_A by :

$$X_A(\omega) = \begin{cases} \dfrac{x}{P(A)} & \text{if } \omega \in A \\ 0 & \text{elsewhere} \end{cases}$$

$E(X_A) = x$ and according to the definitions of Φ and C, we deduce that :

$$(R) \quad \forall x, \forall A \in \mathcal{A} \quad P(A) > 0 \qquad \langle x, x^* \rangle \leq \frac{1}{P(A)} \int_A \chi^*_{A(\omega)} (x) \, dP(\omega)$$

This inequality implies that :

$$\text{a.s.} \quad \forall x \qquad \langle x, x^* \rangle \leq \chi^*_{A(\omega)} (x)$$

Suppose that this last inequality is not satisfied, there is $A \in \mathcal{A}$, $P(A) > 0$ and $\varepsilon > 0$ such that :

$$\forall \omega \in A \qquad \langle x, x^* \rangle \geq \chi^*_{A(\omega)} (x) + \varepsilon$$

$$\Longrightarrow \qquad \langle x, x^* \rangle \geq \frac{1}{P(A)} \int_A \chi^*_{A(\omega)} (x) + \varepsilon$$

and this is in contradiction with the inequality (R).

III.4.2. Remark : The result of the previous lemma is a generalization to the continuous case of the following formula : if $A_1, \ldots, A_m$ are non-empty closed convex sets of $\mathbb{R}^n$,

$$\chi^*_{\bigcap_{i=1}^m A_i} = \overline{\mathop{\nabla}_{i=1}^{m} \chi^*_{A_i}}$$

III.4.3. Theorem : *Let f be a normal convex integrand such that $f^*(x^*, .)$ is integrable for every x^*. Let x_0 such that $\partial F(x_0) \neq \emptyset$, X_0 a random variable whose expectation is x_0 and giving the exactness of the continuous infimal convolution at x_0. Then, the l.s.c. regularization of the directional derivative F' is given by :*

$$\overline{F'}(x_0, .) = \overline{\oint_\Omega \overline{f'}(X_0(\omega), .) \, dP(\omega)}$$

Proof : We suppose that $\partial F(x_0) \neq \emptyset$. According to Theorem III.3.1., $\partial f(X_0(\omega), \omega)$ is a.s. a non empty closed convex set. The multivalued mapping $D_{X_0} : \omega \to \partial f(X_0(\omega), \omega)$ is measurable : consequently, the support function of D_{X_0} , i.e. the function g defined by : $g(d, \omega) = \overline{f'}(X_0(\omega), d)$ is a normal convex integrand. We apply then the previous lemma and conclude with the equality : $\chi^*_{\partial F(x_0)} = \overline{F'}(x_0, .)$ ([3] Theorem 6.4.8).

Remark : the formula of Theorem III.4.3. generalizes the corresponding formula for two convex functions g_1 et g_2 : let g_1, g_2 proper convex functions and $g = g_1 \nabla g_2$. If the infimal convolution g is exact in $x_0 = x_0^1 + x_0^2$ and if $\partial g(x_0) \neq 0$, then :

$$\overline{g'}(x_0, .) = \overline{\overline{g'}(x_0^1, .) \nabla \overline{g'}(x_0^2, .)}$$

IV. Applications

A - We first apply the obtained results to determining some convex characteristics of the mean value functional Ef and continuous infimal convolution such as : level sets, l.s.c. conical hull, gauge functional of the polar of the expectation E(C)...

1 - Level sets of the continuous infimal convolution F

For a convex function g, it is interesting to be able to determine the λ-level set of g, i.e. $g^{\leqslant}(\lambda) = \{x \,/\, g(x) \leqslant \lambda\}$. For the continuous infimal convolution $F = \oint_\Omega f(., \omega)\, dP(\omega)$, our purpose is to determine $F^{\leqslant}(\lambda)$ by expressing it with the level sets of $f(., \omega)$.

Let $T_f(\omega)$ the projection of epi $f(., \omega)$ on $\mathbb{R}$, that is to say : $\{\lambda \,/\, \exists x \; f(x, \omega) \leqslant \lambda\}$. Likewise, $\tau_F = \mathrm{proj}_{\mathbb{R}}(\mathrm{epi}F)$

1.1. Theorem : *Let f be a normal convex integrand such that $f^*(x^*, .)$ is integrable for every x^*. Then :*

$$\forall \lambda \in \tau_F \qquad F^{\leqslant}(\lambda) = \bigcup_{\substack{\Lambda \in \mathcal{L}^1(\lambda) \\ \Lambda(\omega) \in T_f(\omega) \text{ a.s.}}} E\{f(., \omega)^{\leqslant}(\Lambda(\omega))\}$$

Proof : We have $T_f(\omega) = [m(\omega), +\infty[$ and $\tau_F = [M, +\infty[$ where we denote by $m(\omega) = \inf\{f(x, \omega) \,/\, x \in \mathbb{R}^n\}$ and $M = \inf\{F(x) \,/\, x \in \mathbb{R}^n\}$. According to the equality $F^* = E(f^*)$ ([12] Theorem 7), it is easy to remark that $M = E(m)$. Moreover, it is clear that T_f is a measurable multivalued mapping such that, a.s. $T_f(\omega)$ is a non-empty closed set. Let $\lambda \in \tau_F$; $\exists \Lambda \in \mathcal{L}^1(\lambda)$ such that : a.s. $\Lambda(\omega) \in T_f(\omega)$. Let us consider the multivalued mapping : $\omega \to f(., \omega)^{\leqslant}(\Lambda(\omega))$. The conjugate $f^*(., \omega)$ is a.s. finite and continuous at 0 (proof of theorem II.4.2) and $\partial_\theta f^*(0, \omega)$ is a.s. a nonempty compact convex set characterized by :

$$\partial_\theta \ f^* (0, \omega) = \{x \ / \ f (x, \omega) \leq \theta - f^* (0, \omega)\}$$

Let $\theta(\omega) = \Lambda(\omega) + f^*(0, \omega)$; $\theta(\omega) \geq 0$ a.s. because $\Lambda(\omega) \in T_f(\omega)$ a.s. The random variable $\Lambda + f^*(0, .)$ is integrable and we can deduce like in the proof of Theorem II.4.2 that the multivalued mapping $\omega \to \partial_{\theta(\omega)} f^*(0, \omega)$ is P-integrable. On the other hand ; we have : $\forall \lambda \in \tau_F \quad F^{\leq}(\lambda) = \partial_{\lambda + E(f^*)(o)} E(f^*)(o)$

Then it is enough to apply theorem II.4.2 to obtain that :

$$F^{\leq}(\lambda) = \bigcup_{\substack{\Lambda \in \mathcal{L}^1(\lambda) \\ \Lambda(\omega) + f^*(0,\omega) \geq 0}} E\{f(., \omega)^{\leq} (\Lambda(\omega))\}$$

a.s. Hence the result.

2 - Level sets of the mean value functional Ef.

2.1. Theorem : *Let $f \in \mathcal{F}(\mathbb{R}^n \times \Omega, \overline{\mathbb{R}})$ be such that $f(x, .)$ is measurable for every x and Ef the mean value functional. If, for each random variable Λ, we denote by S_Λ the multivalued mapping : $\omega \to f(., \omega)^{\leq}(\Lambda(\omega))$, then :*

$$(Ef)^{\leq}(\lambda) = \bigcup_{\Lambda \in \mathcal{L}^1(\lambda)} (S_\Lambda)^* \qquad ((S_\Lambda)^* \text{is defined in III.1})$$

Proof : It is trivial to see that : $\forall \Lambda \in \mathcal{L}^1(\lambda) \ (S_\Lambda)^* \subset Ef^{\leq}(\lambda)$. Conversely, let $x \in Ef^{\leq}(\lambda)$ and define Λ_o by

$\Lambda_o(\omega) = f(x, \omega) - Ef(x) + \lambda$ a.s. Then $\Lambda_o \in \mathcal{L}^1(\lambda)$ and $x \in (S_{\Lambda_o})^*$.

This theorem is proved to show the analogy between the expressions of the level sets of the mean value functional and continuous infimal convolution.

3 - L.s.c. conical hull of Ef

For $g \in \mathcal{F}(\mathbb{R}^n, \overline{\mathbb{R}})$, we denote by g^c the l.s.c. conical hull of g ([3] definition 6.8.6).

3.1. Theorem : *Let f be a convex integrand such that, for every x, $f(x, .)$ is integrable. We suppose that $Ef(0) \geq 0$. Then :*

$$(Ef)^c \neq -\infty \quad \text{and} \quad (Ef)^c = \sup_{\substack{\Lambda \in \mathcal{L}^1(0) \\ f(0,\omega)+\Lambda(\omega) \geq 0 \ \text{a.s.}}} \left\{ \int_\Omega [f(.,\omega) + \Lambda(\omega)]^c \, dP(\omega) \right\}$$

Proof : Ef is real-valued ; $\partial Ef(0) \neq \emptyset$. Let $x^* \in \partial Ef(0)$, we have :

$\forall x \quad Ef(x) \geq Ef(o) + \langle x^*, x \rangle$; this implies that $(Ef)^*(x^*) \leq 0$. Moreover $(Ef)^c$ is the support function of C, where $C = \{x^* / (Ef)^*(x^*) \leq 0\}$. ([3] Theorem 6.8.7). Therefore $C \neq \emptyset$ and $(Ef)^c \not\equiv -\infty$. $(Ef)^* = \oint_\Omega f^*(., \omega)\, dP(\omega)$, then, according to Theorem 1.1.

$$C = \bigcup_{\substack{\Lambda \in \mathcal{L}^1(0) \\ f(0,\omega)+\Lambda(\omega)\geq 0 \text{ a.s.}}} E\{f^*(.,\omega)^{\leq}(\Lambda(\omega))\}$$

The conjugate of $f + \Lambda$ being $f^* - \Lambda$, $[f(., \omega) + \Lambda(\omega)]^c$ is the support function of the set $f^*(., \omega)^{\leq}(\Lambda(\omega))$.

$f(0, .) + \Lambda$ is integrable and the multivalued mapping

$\omega \rightarrow f^*(., \omega)^{\leq}(\Lambda(\omega)) = \partial_{f(0,\omega)+\Lambda(\omega)} f(0, \omega)$ is P-integrable. Hence :

$$\chi_C^* = \operatorname*{Sup}_{\Lambda} \chi^*_{E\{f^*(.,\omega)^{\leq}(\Lambda(\omega))\}} = \operatorname*{Sup}_{\Lambda} \int_\Omega [f(., \omega) + \Lambda(\omega)]^c\, dP(\omega)$$

4 - L.s.c. conical hull of F

4.1. Theorem : *Let f a normal convex integrand such that $f^*(x^*, .)$ is integrable for every x^* and $F = \oint_\Omega f(., \omega)\, dP(\omega)$. The l.s.c. conical hull of the continuous infimal convolution F is given by :*

$$F^c = \operatorname*{Sup}_{\substack{\Lambda \in \mathcal{L}^1(0) \\ f(0,\omega)+\Lambda(\omega)>0 \text{ a.s.}}} \overline{\oint [f(.,\omega) + \Lambda(\omega)]^c\, dP(\omega)}$$

Proof : For a function $g \in \Gamma(\mathbb{R}^n)$, it is easy to state the following equivalence :

$$g_c \not\equiv -\infty \iff g(0) > 0$$

Let $\Lambda \in \mathcal{L}^1(0)$ such that $f(0, \omega) + \Lambda(\omega) > 0$ a.s. . $[f(., \omega) + \Lambda(\omega)]^c$ which is the support function of the multivalued mapping $S_\Lambda : \omega \rightarrow f^*(.,\omega)^{\leq}(\Lambda(\omega))$ is a normal convex integrand. If $D = \{x / F^*(x) \leq 0\}$, according to Theorem 2.1, we have :

$$\chi_D^* = \operatorname*{Sup}_{\Lambda \in \mathcal{L}^1(0)} \chi^*_{(S_\Lambda)^*}$$

According to lemma III.4.1. $\chi^*_{(S_\Lambda)^*} = \overline{\oint \chi^*_{S_{\Lambda(\omega)}}\, dP(\omega)}$. Hence the formulation of F^c.

5 - Polar of E (C). Gauge function of $[E(C)]^\circ$

5.1. Theorem : *Let C be a measurable multivalued mapping such that, for every* ω, *$C(\omega)$ is a closed convex set containing 0. Then the gauge function* $\gamma_{[E(C)]^\circ}$ *and the polar* $[E(C)]^\circ$ *are given by :*

a) $\gamma_{[E(C)]^\circ} = E(\gamma_{C^\circ})$

b) $[E(C)]^\circ = \bigcup_{\Lambda \in \mathcal{L}^1(1)} [\Lambda C^\circ]^*$, *ΛC° indicating the multivalued mapping* $\omega \to \Lambda(\omega) C^\circ(\omega)$.

Proof : C being measurable, the multivalued mapping C° : $\omega \to [C(\omega)]^\circ$ is also measurable ([7] Corollary 3.5).

a) E(C) is a convex set containing 0 and : $\gamma_{[E(C)]^\circ} = \chi^*_{E(C)}$ ([8] Theorem 14.5)
Moreover, $0 \in C(\omega)$ and C is a quasi P-integrable multivalued mapping. Therefore, according to Remark I.3.1.

$$\chi^*_{E(C)} = \int_\Omega \chi^*_{C(\omega)} \, dP(\omega) = \int_\Omega \gamma_{C^\circ(\omega)} \, dP(\omega)$$

b) $\gamma_{[E(C)]^\circ}$ is a l.s.c. and positively homogeneous function such that :

$$\forall \lambda > 0 \quad \gamma^{\leq}_{[E(C)]^\circ}(\lambda) = \lambda [E(C)]^\circ \quad ([8] \text{ Corollary } 9.7.1)$$

Thus $[E(C)]^\circ = \gamma^{\leq}_{[E(C)]^\circ}(1) = \bigcup_{\Lambda \in \mathcal{L}^1(1)} (S_\Lambda)^*$ with $S_\Lambda(\omega) = \gamma^{\leq}_{C^\circ(\omega)}(\Lambda(\omega)) = \Lambda(\omega) C^\circ(\omega)$

IV.1.6. Remark : If, furthermore $C(\omega)$ is a cone, we have : $\chi_{C(\omega)} = \gamma_{C(\omega)}$ and $\chi_{[E(C)]^\circ} = E(\chi_{C^\circ})$. So, $[E(C)]^\circ = (C^\circ)^*$.

B - Optimization problems

In this second part of applications, we consider different optimization problems where the function to be minimized and the constraints may depend on the random outcome ω.

6 - A first minimization problem

Let f be a real-valued convex function and C a CK-valued mapping which is supposed P-integrable. Let us consider the following problems :

$$(\mathcal{P}_\omega) \qquad \varphi(x,\omega) = \operatorname{Inf}\{f(x-y) \,/\, y \in C(\omega)\}$$

$$(\mathcal{P}) \qquad \Phi(x) = \operatorname{Inf}\{f(x-y) \,/\, y \in E(C)\}$$

The relation between φ and Φ is given by the proposition below :

6.1. Proposition : $\Phi = \oint_\Omega \varphi(.,\omega)\, dP(\omega)$

Proof :

a) f being convex, the definition of φ shows us that φ is a real-valued convex function. C is a measurable multivalued mapping such that $C(\omega)$ is nonempty and closed. Then, there exists a countable collection of measurable functions $\{f_i\}_{i\in D}$ such that :

$$C(\omega) = \overline{\{f_i(\omega) \,/\, i \in D\}} \qquad ([7] \text{ Theorem } 1)$$

Then $\varphi(x,\omega) = \inf_{i\in D} f(x - f_i(\omega))$ is measurable ; φ is then a normal convex integrand. Let $S(x,\omega) = \{(y \in C(\omega) \,/\, f(x-y) = \varphi(x,\omega)\}$. For every x, $S(x,.)$ is a CK-valued measurable multivalued mapping. According to the theorem of Kuratowski - Ryll-Nardzewski ([7] Corollary 1.1), there exists a measurable selector of $S(x,.)$ i.e. a measurable function Y such that : $Y(\omega) \in S(x,\omega)$ a.s. Then, a.s. $f(x-Y(\omega)) = \varphi(x,\omega)$ and for every random variable X such that $E(X) = x$, we have :

$$\int_\Omega f(X(\omega) - Y(\omega))\, dP(\omega) \geq f(x - E(Y)) \geq \inf\{f(x-y) \,/\, y \in E(C)\}$$

$$\Longrightarrow \oint_\Omega \varphi(.,\omega)\, dP(\omega) \geq \Phi$$

b) Conversely, let $\overline{Y} \in E(C)$ such that $f(x-\overline{Y}) = \Phi(x)$. Denote by Y a measurable selector of C such that $E(Y) = \overline{Y}$. Let $X(\omega) = x + Y(\omega) - \overline{Y}$; we deduce that :

$$E(X) = x \text{ and } \varphi(X(\omega),\omega) \leq f(X(\omega) - Y(\omega)) = f(x - \overline{Y})$$

$$\Longrightarrow \int_\Omega \varphi(X(\omega),\omega)\, dP(\omega) \leq f(x-\overline{Y})$$

Consequently : $\Phi \geq \oint_\Omega \varphi(.,\omega)\, dP(\omega)$. We also conclude that the continuous infimal convolution is exact.

Application

Let $f(x) = ||x||$; then $\varphi(x, \omega) = d(x, C(\omega))$. According to the previous result ; $d(., E(C)) = \oint_\Omega d(., C(\omega))\, dP(\omega)$.

7 - Minimization of the continuous infimal convolution

Let f a normal integrand and F the continuous infimal convolution of the family $\{f(., \omega)\}_{\omega \in \Omega}$ relating to the probability measure P. We consider the following optimization problems :

$(\mathcal{P}_\omega)$ Find $\bar{x}$ such that : $f(\bar{x}, \omega) = Inf \left\{ f(x, \omega) \,/\, x \in \mathbb{R}^n \right\}$

$(\mathcal{P})$ Find $\tilde{x}$ such that : $F(\tilde{x}) = Inf \left\{ F(x) \,/\, x \in \mathbb{R}^n \right\}$

We call $S(\omega)$ and $\sum$ the solution sets of respectively $(\mathcal{P}_\omega)$ and $(\mathcal{P})$. The following theorem allow us to compare $S(\omega)$ and $\sum$ as well as the optimal values.

7.1. Theorem : *Let f be a normal integrand on $\mathbb{R}^n \times \Omega$. We denote by $m(\omega) = Inf \{ f(x, \omega) \,/\, x \in \mathbb{R}^n \}$ and $M = inf \{ F(x) \,/\, x \in \mathbb{R}^n \}$. Then*

a) If there exists a measurable selector for the multivalued mapping S, we have :

$$M = E(m).$$

b) Moreover, if f is a normal convex integrand such that $f(x', .)$ is integrable for every x', S is a CK-valued mapping, P-integrable and $E(S) = \sum$. In other words :

$$\tilde{x} \text{ solution of } (\mathcal{P}) \iff \tilde{x} = E(X) \text{ where } X(\omega) \text{ is a.s. solution of } (\mathcal{P}_\omega).$$

Proof :

a) Let X_o a measurable selector of S and $E(X_o) = x_o$. From the definition of F, $F(x_o) \leq \int_\Omega f(X_o(\omega), \omega)\, dP(\omega) = \int_\Omega m(\omega)\, dP(\omega)$; thus $M \leq E(m)$. Conversely, $\forall x, \forall X \in \mathcal{L}^1(x) \int_\Omega f(X(\omega), \omega)\, dP(\omega) \geq \int_\Omega m(\omega)\, dP(\omega)$ because a.s. $f(X(\omega), \omega) \geq m(\omega)$. Therefore : $\forall x \quad F(x) \geq E(m)$ and $M \geq E(m)$

b) We have already seen that F and a.s. $f(., \omega)$ are inf-compact functions. $f(., \omega) \in \Gamma_0(\mathbb{R}^n)$; the solution set of $(\mathcal{P}_\omega)$ is $\partial f^*(0, \omega)$ ([8] Theorem 27.1) Likewise, the solution set $\sum$ of $(\mathcal{P})$ is $\partial F^*(0) = \partial E(f^*)(0)$. According to Theorem II.4.1., we have $\partial E(f^*)(0) = E(\partial f^*(0, \omega))$. Q.E.D.

7.2. Remarks :

a) More generally, for a normal integrand f, we have only the following inequality : $E(m) \leqslant M$ in $\overline{\mathbb{R}}$. The equality $M = E(m)$ appears in a different form and in the convex case in [10] (Proposition 1).

b) Under the assumptions b) of Theorem 7.1., concerning the uniqueness of solutions of $(\mathcal{P})$ and $(\mathcal{P}_\omega)$, we may assert that :

$$\sum = \{\bar{x}\} \iff S(\omega) = \{\bar{X}(\omega)\} \text{ a.s. with } E(\bar{X}) = \bar{x}$$

The previous theorem may be extended when considering approximated optimization problems. Let $(\mathcal{P}^\alpha_\omega)$ and $(\mathcal{P}^\theta)$ the following approximated minimization problems.

$(\mathcal{P}^\alpha_\omega)$ Find $\bar{x}$ such that : $m(\omega) \leqslant f(\bar{x}, \omega) \leqslant m(\omega) + \alpha$

$(\mathcal{P}^\theta)$ Find $\tilde{x}$ such that : $M \leqslant F(\tilde{x}) \leqslant M + \theta$.

7.3. Theorem : *Let f be a normal convex integrand such that $f^*(x', .)$ is integrable for every x'. Then, for each $\theta \geqslant 0$,*

$$\tilde{x} \textit{ solution of } (\mathcal{P}^\theta) \iff \begin{cases} \exists \bar{X},\ E(\bar{X}) = \tilde{x} \\ \exists \theta,\ \theta(\omega) \geqslant 0 \qquad E(\theta) = \theta \\ \textit{such that : a.s. } \bar{X}(\omega) \textit{ is solution of } \mathcal{P}^{\theta(\omega)}_\omega \end{cases}$$

Proof : According to the definition of the θ-subdifferential, we have :

$$\bar{X}(\omega) \text{ solution of } (\mathcal{P}^\alpha_\omega) \iff 0 \in \partial_\alpha f(\bar{X}(\omega), \omega) \iff \bar{X}(\omega) \in \partial_\alpha f^*(0,\omega)$$

Likewise : $\tilde{x}$ solution of $(\mathcal{P}^\theta) \iff \tilde{x} \in \partial_\theta E(f^*)(0)$ because $F^* = E(f^*)$. It is enough to apply Theorem II.4.1. to determine $\partial_\theta E(f^*)(0)$

8 - Minimization of a normal convex integrand on a random closed convex set.

Let f be a normal convex integrand and C a measurable multivalued mapping such That C (ω) is a nonempty closed convex set (in a stochastic linear program, C (ω) may be π (ω) = $\{x \,/\, A(\omega)\, x \leqslant b(\omega)\}$ where A and b are measurable). Let :

$$\tilde{f}(x, \omega) = f(x, \omega) + \chi_{C(\omega)}(x)$$

8.1. Theorem : *Let f be a normal convex integrand such that for every x' $f^*(x', .)$ is integrable, C a nonempty closed convex valued measurable mapping. Moreover, assume that* $\tilde{F} = \oint_\Omega \tilde{f}(., \omega)\, dP(\omega)$ *is not identically equal to + ∞. Then* $\tilde{f}$ *is a normal convex integrand and if we denote by* S_C^1 *the set of integrable selectors of C, we have :*

$$\tilde{M} = \operatorname{Inf}\{\tilde{F}(x)/x \in \mathbb{R}^n\} = \operatorname*{Inf}_{X \in S_C^1} \int_\Omega f(X(\omega),\omega)\, dP(\omega) = \int_\Omega \operatorname{Inf}\{f(x,\omega)/x \in C(\omega)\}\, dP(\omega)$$

Proof : The assumption $\tilde{F} \not\equiv +\infty$ implies that : $\exists x_0$ such that $\tilde{F}(x_0) < +\infty$. Then, there exists $X_0 \in \mathcal{L}^1$ such that $\tilde{f}(X_0(.), .)^+$ is integrable.

Necessarily : a.s. $X_0(\omega) \in C(\omega)$. So, S_C^1 is a nonempty set.

Let $\Omega_0 = \{\omega \in \Omega / \exists x \in \mathbb{R}^n \;\; f(x,\omega) + \chi_{C(\omega)}(x) < +\infty\}$. We have shown that $P(\Omega_0) = 1$.

Thus, $\tilde{f}$ a normal convex integrand ([7] Corollary 4.2).

Moreover :

$$f(., \omega) \leqslant \tilde{f}(., \omega) \Longrightarrow \tilde{f}^*(., \omega) \leqslant f^*(., \omega)$$

$$\forall x' \quad f^*(x', \omega) \geqslant \langle X_0(\omega), x' \rangle - f(X_0(\omega), \omega)$$

$$\Longrightarrow X_0(\omega), x' \rangle - f(X_0(\omega), \omega) \leqslant \tilde{f}^*(x', \omega) \leqslant f^*(x', \omega)$$

These inequalities imply that for every x', $\tilde{f}^*(x', .)$ is integrable. According to results of Theorem 7.1., we obtain :

$$M = \operatorname{Inf}\{\tilde{F}(x) / x \in \mathbb{R}^n\} = \int_\Omega \operatorname{Inf} f(x, \omega) / x \in C(\omega)\}\, dP(\omega)$$

On the other hand, it arises from the definition of $\tilde{F}$ that :

$$\tilde{M} = \operatorname*{Inf}_{X \in S_C^1} \int_\Omega f(X(\omega), \omega)\, dP(\omega)$$

8.2. Remark : The required assumptions for f and C imply that $\tilde{F} \in \Gamma_0(\mathbb{R}^n)$. Moreover, it is obvious that dom $\tilde{F} \subset E(C)$. The problem we have dealt with is quite different from the minimization of $F = \oint_\Omega f(., \omega)\, dP(\omega)$ on E (C).

9 - Minimization of the continuous infimal convolution on compact convex set Q

9.1. Theorem : *Let f be a normal convex integrand such that $f^*(x', .)$ is integrable for every x' and C a CK-valued mapping P-integrable. We set :*

$$F = \oint_{\Omega} f(., \omega)\, dP(\omega) \qquad \overline{\alpha} = Min\left\{F(x) \;/\; x \in E(C)\right\}$$

For every $u \in \mathcal{L}^1(0)$, $\alpha_u(\omega) = Inf\left\{f(x+u(\omega), \omega) \;/\; x \in C(\omega)\right\}$.

Then :

$$\overline{\alpha} = Min\left\{\int_{\Omega} \alpha_u(\omega)\, dP(\omega) \;/\; u \in \mathcal{L}^1(0)\right\}$$

Proof : Denote $f(., \omega) \nabla \chi_{-C(\omega)}$ by $g(., \omega)$. g is obviously a normal convex integrand. For each $u \in \mathcal{L}^1(0)$, $\alpha_u(\omega)$ may be written : $\alpha_u(\omega) = g(u(\omega), \omega)$. Likewise :

$\overline{\alpha} = [F \nabla \chi_{-E(C)}](0)$. The functions $f^*(x', .)$ and $\chi^*_{C(.)}(x')$ are integrable for every x', thus $F \nabla \chi_{-E(C)} = \oint_{\Omega} g(., \omega)\, dP(\omega)$ and the continuous infimal convolution is exact. That is to say :

$$F \nabla \chi_{-E(C)}(0) = \underset{u \in \mathcal{L}^1(0)}{Min} \int_{\Omega} g(u(\omega), \omega)\, dP(\omega)$$

So, these exists $\overline{u} \in \mathcal{L}^1(0)$ such that : $\overline{\alpha} = \int_{\Omega} \alpha_{\overline{u}}(\omega)\, dP(\omega)$. Hence, the result :

If previously we take $C(\omega) = Q$ where Q is a nonempty convex set, the previous theorem takes the following form :

9.2. Corollary : Let g be a normal convex integrand such that for every x' $f^*(x', .)$ is integrable. Q is a nonempty compact convex set. Let :

$$(\mathcal{P}) \qquad \overline{\alpha} = Inf\left\{F(x) \;/\; x \in Q\right\}$$

$(\mathcal{P}_\varepsilon)$ the perturbed problems :

$$(\mathcal{P}_\varepsilon) \qquad \alpha_\varepsilon(\omega) = Inf\left\{f(x, \omega) \;/\; x \in Q + \varepsilon(\omega)\right\}$$

Then : $\exists\, \varepsilon \in \mathcal{L}^1(0)$ such that $\overline{\alpha} = \int_{\Omega} \alpha_\varepsilon(\omega)\, dP(\omega)$

Proof : We take in the previous theorem $\varepsilon(\omega) = -\overline{u}(\omega)$.

REFERENCES

[1] D.P. BERTSEKAS, Stochastic optimization problems with non differentiable cost functionals. Journal of Optimization Theory and Applications. 1973, Vol. 12, n° 2, p. 218-231.

[2] A.D. IOFFE - V.M. TIHOMIROV, Dualité des fonctions convexes et problèmes d'extrêmum. Uspehi Mat. N. 23-6 (1968) p. 51-116.

[3] P.J. LAURENT, Approximation et optimisation, Hermann, Paris 1972.

[4] C. LEMARECHAL, Sur les méthodes de gradient conjugué appliquées aux critères non différentiables.C.R. Acad. Sc. Paris, t. 278 (1974).

[5] J. NEVEU, Bases mathématiques du Calcul des Probabilités. Masson, Paris 1970.

[6] R.T. ROCKAFELLAR, Integrals which are convex functionals. Pacific Journal of Mathematics 1968. Vol. 24, n° 3, p. 525-539.

[7] R.T. ROCKAFELLAR, Measurable dependance of convex sets and functions on parameters. Journal of Mathematical Analysis and applications, 1969, 28, p. 4-25.

[8] R.T. ROCKAFELLAR, Convex analysis. Princeton 1970.

[9] R.T. ROCKAFELLAR, Conjugate duality and optimization, Lecture Notes, Série n° 16, SIAM Publications, 1974.

[10] R.T. ROCKAFELLAR, R.J.B. WETS, Stochastic convex programming : Basic Duality. Pacific Journal of Mathematics (to appear)

[11] M. VALADIER, Multiapplications mesurables à valeurs convexes compactes, Journal de Mathématiques Pures et Appliquées 50, 1971, p. 265-297.

[12] M. VALADIER, Intégration d'ensembles convexes fermés, notamment d'épigraphes. Rev. Franç. Inf. Rech. Op. 4 (1970) p. 57-73.

[13] B. VAN CUTSEM, Eléments aléatoires à valeurs convexes compactes. Thèse Université de Grenoble, 1971.

[14] D.W. WALKUP - R.J.B. WETS, Stochastic programs with recourse II : on the continuity of the objective. Siam J. Appl. Math. 1969, Vol. 17, n° 1, p. 98-103.

[15] R.J.B. WETS, On inf-compact mathematical programs, Proceedings 5th I.F.I.P. Conference on optimization, Springer Verlag, (1974).

EVOLUTION OF SOME PROBLEMS OF STOCHASTIC CONTROL WHEN THE DISCOUNT VANISHES

J.M. LASRY

CNRS et Université de Paris IX Dauphine Mathématiques de la Décision, 75775 Paris

H. HOWARD [7] , D. BLACKWELL [3] , and other authors have studied the evolution of the Markovian problems of control (P_τ) involving discounted cost functions, when the discount parameter $\tau<1$ tends to 1. The main result is that there is a limit problem (P_1) such that the optimal policy P_1 of (P_1) is the limit of optimal policies p_τ of the problems (P_τ) $(\tau<1)$.

These results concern general Markovian processes with discrete time $t\in N$, but might perhaps be extended to general Markovian processes with continuous time $t\in [0,+\infty[$.

In this article we study the case in which the Markovian processes are diffusions ξ , which are solutions of a differential stochastic equation : $d\xi = a(t)dt + dW$, where W is a brownian motion and where the control is the non-anticipative stochastic function a.

For this class of Markovian processes the results of convergence when the discount goes to one are mainly the same. The difference is that in the case of diffusion processes, the general frame of BELMAN-HALMITON-JACOBI theory [2] leads to an elliptic quasi-linear partial differential equation : H. FLEMING [4,5,...] has shown that this equation has a regular solution from which the optimal control can be computed. We shall recall (theorem 1) H. FLEMING's result which was not stated exactly in the same context; then give a similar analytical treatment for the limit (i.e : no-discounted) problem (theorem 2); and finaly give the convergence results (theorem 3). The proofs of these three theorems are in [8,9] . Here we shall emphasize on the analytic treatment of the "ergodic" (limit) problem and give a proof of theorem 2 different from that of [8,9] .

In R^n , n=1, such results have been found by P. MANDL [11] using ordinary differential equations methods.

Discussions with A. BENSOUSSAN and L. TARTAR were helpful for this work.

§ 1. RESULTS

a) We note f a twice continuously differentiable function from $R^n \times R^n$ into R such that

(1) $f(x+m,y) = f(x,y) \qquad \forall m \in N ; \forall x,y \; R^n$

(2) $\exists \alpha > 0$ such that for all $x,y,z \in R^n$

$$\sum_{i,j=1}^{n} \frac{\partial^2}{\partial y_i \partial y_j} f(x,y) z_i z_j \geqslant \alpha z^2$$

(3) $\exists c > 0$ such that for all $x,y \; R^n$:

$$\left\| \frac{\partial}{\partial x} f(x,y) \right\| + \left\| \frac{\partial}{\partial y} f(x,y) \right\| \leqslant cf(x,y)$$

Convexity hypothesis (1) is natural if one wants the existence of a regular control (see cost function below). Periodicity hypothesis (2) is a way to introduce the compacity (i.e: $R^n/\mathbb{Z}^n$ is compact) we need for "ergodic" results. Growth hypothesis(3) is not very strong : For example $f(x,y) = \exp(x+y)$ satisfy (3). In application it can be useful to replace f by f+ constant (which does not change the control problem below) in order to verify (3).

We note $f^*(x,.)$ the conjugate function of $f(x,.)$, i.e. (ROCKAFELLAR [13]):

(4) $f^*(x,p) = \text{Sup}\{py - f(x,y) \mid y \in R^n\}$

One can verify that $f^*: R^n \times R^n \to R$ is twice continuously differentiable, and that

(5) $\varphi(x,p) = \frac{\partial}{\partial p} f^*(x,p)$

is the point where the function $y \to py - f(x,y)$ is minimum.

b) We denote W a standart brownian motion; we note E the expectation. Let A be the set of stochastic functions $a : R_+ \to R^n$ which are bounded non anticipative functionnals of W. For all $x \in R^n$, $s > 0$ and $a \in A$, define $I(s,x,a)$ and $\lambda(x,a)$ by :

(6) $$I(s,x,a) = E\int_0^{+\infty} f(\xi(t),a(t))e^{-st}dt$$

(7) $$\lambda(x,a) = \lim_{T \to +\infty} . \inf E\frac{1}{T}\int_0^T f(\xi(t), a(t))dt$$

where ξ is the stochastic process define by

(8) $\xi(o)=x \quad , \quad d\xi = a(t)dt + \sqrt{2}\, dW$

($\sqrt{2}$ in order to simplify formulas).

Let H be the set of all bounded lipschitzian functions $h : R^n \to R^n$. For all $h \in H$, $x \in R^n$, $s > o$, let

(9) $$J(s,x,h) = E\int_0^{+\infty} e^{-st} f(\eta(t),a(t))dt$$

(10) $$\mu(x,h) = \lim_{T \to +\infty} E\frac{1}{T}\int_0^T f(\eta(t), a(t))dt$$

where η is defined by the following stochastic differential equation :

(11) $\eta(o) = x$, $d\eta = h(\eta)\,dt + \sqrt{2}\,dw$

and where $a \in A$ is defined by

(12) $a(t) = h(\ (t))$

Take $x_o \in R^n$ and $h_o \in H$.

Let η_o be defined by (11); then let $a_o \in A$ be defined by (12). Then we can compute ξ_o by (8) and we find $\xi_o = \eta_o$ by (11). So we have

(13) $I(x_o,s,a_o) = J(x_o,s,h_o)$ and $\lambda(x_o,a_o) = \mu(x_o,h_o)$

Elements of H are called closed loop controls and elements of A are called open loop controls. By (13), it seems that closed loop controls are less efficient than the open ones. Precisely we have :

(14) $\inf_{h \in H} J(x,s,h) \geqslant \inf_{a \in A} I(x,s,a)$

(15) $\inf_{h \in H} \mu(x,h) \quad \inf_{a \in A} \lambda(x,\ \ a)$

Actually theorem 1 and 2 below will imply that there is equality in (14) and (15), and that the optimal control is a closed loop control.

THEOREM 1 Let $s > 0$. The function u_s defined by

(16) $u_s(x) = \inf\{I(x,s,a) \mid a \in A\ ,\ \ x \in R^n\}$ is three time continuously differentiable, periodic (i-e : $u_s(x+m) = u_s(x)$, $\forall x \in R^n$, $\forall m \in \mathbb{Z}^n$)

This function is the unique periodic solution of the following PDE :

(17) $-\Delta u_s = s\, u_s(x) + f^*(x, -\nabla u_s\ (x)) = 0$, $\forall x \in R^n$

(where : Δ = Laplacian, ∇ = gradient)

Let $h_s \in H$ be defined by (see(5)) :

(18) $h_s(x) = \varphi(x, -\nabla u_s(x))$, $\forall x \in R^n$

Then we have

(19) $u_s(x) = J(x,s,h_s) = \inf\{J(x,s,h) \mid h \in H\}$

which means in particular that h_s is the optimal control in the problem

(20) $\underset{h \in H}{\text{Minimize}}\ J(x,s,h)$

THEOREM 2. The function λ defined by

(21) $\lambda(x) = \inf\{\lambda(x,a) \mid a \in A\}$, $\forall x \in R^n$

(22) is constant : $\lambda(x) = \lambda_o$, $\forall x \in R^n$.

There exists a function $v_o : R^n \longrightarrow R$ periodic, i.e:

$$v_o(x+m) = v_o(x)\ ,\quad \forall m \in \mathbb{Z}^n\ ,\ \forall x \in R^n,$$

three time continuously differentiable, such that

(23) $-\Delta v_o(x) + \lambda_o + f^*(x, -\nabla v_o(x)) = 0$, $\forall x \in R^n$

Let $h_o \in H$ be defined by (see 15)):

(24) $h_o(x) = \varphi(x, -\nabla v_o(x))$, $\forall x \in R^n$

we have, for all $x \in R^n$:

(25) $\lambda_o = \mu(x, h_o) = \inf\{ \mu(x,h) \mid h \in L\}$

Which means in particular that h_o is the solution of the following problem of stochastic control

(26) $\underset{h \in H}{\text{Minimize}} \quad \mu(x,h)$

Moreover, if (λ', h') is another solution of (23) then $\lambda' = \lambda_o$ and $h' = h_o$ + constant.

THEOREM 3 . Let u_s, h_s, λ_o, v_o be defined as in theorem 1 and 2. Then when $s > o$ converges to o, we have

(27) $|||\, s u_s - \lambda_o |||\longrightarrow 0$

(28) $|||\nabla u_s - \nabla v_o||| \longrightarrow 0$

(29) $|||\, h_s - h_o ||| \longrightarrow 0$

(30) with $|||g||| = \sup\{ \|g(x)\| + \|\nabla g(x)\| \mid x \in R^n\}$

§ 2. PROOF OF THEOREM 2

Proofs of theorem 1 and 3, and of unicity in theorem 2 can be found in [8, 9].

Let $C^{k,\alpha}(R^n)$ be the space of functions $u : R^n \longrightarrow R$ which are k-time differentiable with α-Holder-continuous derivatives, (see [10]). Let $C^{k,\alpha}_{\cdot}$ be the subspace of $C^{k,\alpha}(R^n)$ of functions u such that

(1) $u(x+m) = u(x) \qquad \forall x \in R^n, \forall m \in \mathbb{Z}^n$

(2) $u(0) = 0$

From the classical results about Laplacian (Schauder, see [10] , Agmon-Douglis-Niremberg [1]), one can deduce the following lemma

Lemma 1 If $g \in C^o_{\cdot}$ there exists a unique $u \in C^1_{\cdot}$ and a unique $\lambda_o \in R$ such that (in a weak sens):

(3) $-\Delta u + \lambda_o = g$ in R^n

Moreover : $u \in C^{1,\alpha}_{\cdot}$ for all $\alpha \in]0,1[$, and there exists a constant C_α , not depending on g, such that

(4) $\|u\|_{1,\alpha} \leqslant C_\alpha \|g\|_o$

(where $\| \|_{k,\alpha}$ the norm in $C^{k,\alpha}$)

If $g \in C^{k,\alpha}_{\cdot}$ then $u \in C^{k+2,\alpha}_{\cdot}$ and

(5) $\| u \|_{k+2,\alpha} \leqslant b_\alpha \| g \|_{k,\alpha}$

where b_α is a constant which does not depend on g

Définition 2 : According to lemma 1 on can define on operator K from $C^1_{\cdot}$ into itself by $v = K(u)$ iff there exists $\lambda_o \in R$ such that

(6) $-\Delta v + \lambda_o = -f^*(.,-\nabla u(.))$

Remark that (6) and periodicity imply that

(7) $\lambda_o = -\int_{[0,1]^n} f^*(x,-\nabla u(x))dx$

Lemma 3 : K is continuous and compact.

Proof : For $r \geqslant 0$, let

(8) $\theta(r) = \sup \{ \| f^*(x,y) \| \mid x,y \in R^n , \|y\| \leq r \}$

Then by (4) and (6) we have

(9) $\|u\|_1 \leqslant r \Rightarrow \|K_u\|_{1,\alpha} \leqslant C_\alpha \theta(r)$

As the injection $C^{1,\alpha}_{\cdot} \to C^1$ is compact this prove that K is compact.

Now let $\|u_n - u\|_1 \to 0$ and let $v_n = K_{u_n}$. By compacity there exists $v \in C^1$ and a subsequence v_{n_k} such that $\|v_{n_k} - v\|_1 \to 0$.

From $-\Delta v_n + \lambda_n = -f^*(., -\nabla u_n(.))$, from the continuity of f^* and from (7), we deduce that $-\Delta v + \lambda = -f^*(., u(.))$, with $\lambda = \lim \lambda_{n_k}$

So v = Ku. This shows the continuity of K.

Lemma 4(*). There exists a constant $c_3 \in R$ such that if $u \in C^1_{\cdot}$, if $\alpha \in [0,1]$, and if $u = \alpha Ku$, then $\|u\|_1 \leqslant c_3$.

proof Let $u \in C^1$, $u = \alpha Ku$, $0 < \alpha \leqslant 1$ ($\alpha = 0$ is trivial). Then by definition 2, we have

(10) $-\Delta u + \alpha\lambda = -\alpha f^*(x,-\nabla u(x))$, $\forall x \in R^n$

Let us first show that $u \in C^3_{\cdot}$, by a standart a posteriori regularity argument. As $u \in C^1_{\cdot}$, we see that $f^*(.,-\nabla u(.))$ belongs to $C^0_{\cdot}$. From lemma 1 and (10) we deduce that $u \in C^{1,\alpha}_{\cdot}$. Now we see that $f^*(.,-\nabla u(.))$ belongs to $C^{0,\alpha}_{\cdot}$. Again from (10) and lemma 1, this implies $u \in C^{2,\alpha}_{\cdot}$. Now we see that $f^*(.,-\nabla u(.)) \in C^{1,\alpha}_{\cdot}$, and again by (10) and lemma 1 we have $u \in C^{3,\alpha}_{\cdot}$. That should be enough.

As u is continuous and periodic there exists $x_o \in R^n$ such that $u(x_o) = \max u$. As u is twice continuously differentiable, this implies $\nabla u(x_o) = o$ and $\Delta u(x_o) \leqslant o$. So from (10) we get :

$$\alpha\lambda \leqslant f^*(x_o,o)$$

By the same argument on the minimum of u we find

(11) $|\lambda| \leqslant c_1 = \sup \{ \|f^*(x,0)\| \mid x \in R^n \}$

(*)Remark The proof of lemma 4 given here was kindly commuticated by L. TARTAR. The initial probabilistic proof can be found in [8,9] .

Now let y belongs to R^n, and $\|y\| = 1$. Put

(12) $$v(x) = y.\nabla u(x), \quad \forall x \in R^n$$

By derivation we deduce from (10) that

(13) $$-\Delta v(x) = -g_1(x,-\nabla u(x)) + g_2(x,-\nabla u(x)).\nabla u(x)$$

where, for all x,p in R^n:

(14) $$g_1(x,p) = y.\frac{\partial}{\partial x} f^*(x,p) \text{ and } g_2(x,p) = \frac{\partial}{\partial p} f^*(x,p)$$

Let $w = c(u-\min u + c_1)-v$, where c_1 come from (11) and c come from (3) of § 1. Then from (10) and (13) one deduces

(15) $$-\Delta w + \alpha w - \alpha a.\nabla w = b \quad \text{with}$$

(16) $$\begin{cases} a(x) = g_2(x,-\nabla u(x)) \text{ and} \\ b(x) = c\alpha\left[-\lambda + c_1 - f^*(x,-\nabla u(x) + u(x) - \min u \right. \\ \qquad \left. - a(x).\nabla u(x)\right] - \alpha\left[v(x) - g_1(x,-\nabla u(x))\right] \end{cases}$$

From the definition § 1 (4) of f^* one can deduce the usual relations of convex analysis(see Rockafellar [13]).

(17) $$\begin{aligned} f(x,y) &= \sup\{y.p - f^*(x,p) \mid p \in R^n\} \\ &= y.\bar{p} - f^*(x,\bar{p}) \end{aligned}$$

where $\bar{p}$ si defined by one of the equivalent following relations.

(18) $$y = \frac{\partial}{\partial p} f^*(x,p) \Leftrightarrow p = \frac{\partial}{\partial y} f(x,y)$$

One can also show that

(19) $$\frac{\partial}{\partial x} f(x,y) = -\frac{\partial}{\partial x} f^*(x,p), \text{ when y,p satisfy (18)}$$

Then from (17,18,19) and from § 1.3one deduces

(20) $$C\left[p.\frac{\partial}{\partial p} f^*(x,p) - f^*(x,p)\right] \geqslant \left\|\frac{\partial}{\partial x} f^*(x,p)\right\| + \|p\|$$

From (20) with $p = -\nabla u(x)$, from the definition of a (see(16)), from $\|v(x)\| = \|y.\nabla u(x)\| \leqslant u(x)$ (because $\|y\| = 1$), and from the definition (14) of g, one deduces :

(21) $$c\left[-a(x).\nabla u(x) - f^*(x,-\nabla u(x))\right] - v(x) + g_1(x,-\nabla u(x)) \geqslant 0$$

From $u(x) - \min u \geqslant 0$, from $c_1 - \lambda \geqslant 0$ (see (11)), from (16,21) we deduce

(22) $$b(x) \geqslant 0, \forall x \in R^n$$

As w is continuous and periodic there exists $x_o \in R^n$ such that $w(x_o) = \min w$. As w is twice continuously differentiable, one has $\nabla w(x_o)=0$ and $\Delta w(x_o) \geqslant o$. So from (15) and (22) one gets $\alpha w(x_o) \geqslant b(x_o) \geqslant o$. So $w \geqslant \min w = w(x_o) \geqslant 0$. From the definitions of v and w one deduces

(23) $$y.\nabla u(x) \leqslant C\left[u(x) - \min u + c_1\right], \forall x \in R^n$$

But (23) is true for all y such that $\|y\| = 1$. So we have

(24) $$\|\nabla u(x)\| \leqslant c\left[u(x) - \min u + c_1\right], \forall x \in R^n$$

Let $x_o \in R^n$ be such that $u(x_o) = \min u$. From (24) we get by Gronwall lemma :

(25) $\quad u(x) - \min u + c_1 \leqslant c_1 \exp\,(c\,\|x - x_o\|)$

Periodicity allows us to suppose $\|x-x_o\| \leqslant \sqrt{n}$. So from (24) and (25) we deduce

(26) $\quad \|\nabla u(x)\| \leqslant c_2 = c\,c_1 \exp\,(c\sqrt{n})$

As $u \in C^1_\bullet$ imply $u(0) = 0$, one deduces $\|u\|_1 \leqslant c_3$ from (26) with c_3 depending on c_2.

Lemma 5. There exists a function $v_o \in C^3_\bullet$ and a real λ_o such that

(27) $\quad -\Delta v_o(x) + \lambda_o + f^*(x,-\nabla v_o(x)) = 0, \quad \forall x \in R^n$

Proof Thanks to lemma 3 and 4, and to the Schaefer fixed point theorem (Schafer [14]) there exists $v_o \in C^1_\bullet$ such that $v_o = K\,v_o$. So by an argument given at the beginning of the proof of lemma 4, one has $v_o \in C^3_\bullet$ and (27).

Lemma 6. Let $_o$ R and v_o C^3 satisfy (27). Then

(28) $\quad v_o(x) \leqslant \lambda(x,a) \quad$ for all $a \in A$

(29) $\quad v_o(x) \leqslant \mu(x,h) \quad$ for all $h \in H$

(30) $\quad v_o(x) = \mu(x,h_o) \quad$ where h_o is defined by

$h_o(x) = g_2\,(x,-\nabla u(x)) \qquad \forall x \in R^n \qquad$ (see(14))

Proof Let us compute $E\,(v(\xi(t))$ where ξ is defined by : $\xi(0) = x$, $d\xi = a(t) + \sqrt{2}\,dW$

By Ito's lemma (Gihman - Skorohod [6]) we have

(31) $\quad E(v(\xi(t))) - E(v(\xi(0))) =$

$$E\int_0^T \Delta\,v(\xi(t)) + a(t).\nabla v(\xi(t))dt + \sqrt{2}\int_0^T \nabla v(\xi(t))dW$$

As $t \longrightarrow \nabla v(\xi(t))$ is non-anticipative we have $E\int_0^T \nabla v(\xi(t))dW = 0$. So we deduce from (17,27,31) that

(31) $\quad E(v(\xi(t))) - v(x) \geqslant E\int_0^T \left[\,\lambda_o - f(\xi(t),a(t))\right]\,dt$

Divide each side of (31) by T, and let $T \rightarrow +\infty$. We find (28) and from the definition (§1,7) of λ. From (28) and from (§1,13) we deduce (29). If $h = h_o$ we deduce from (17) that there is equality in (31), which leads to (30).

REFERENCES

(1) S. AGMON, A. DOUGLIS, L. NIREMBERG

Estimates near the boundary for solutions of elliptic partial differential equations satisfying general boundary conditions, I, Comm. pure and appl. Math., t.12, 1959, p. 623 - 727.

(2) R. BELMAN,

Dynamic programming, Princeton, 1957

(3) D. BLACKWELL,

Discret dynamic programming,
Ann. Math. Stat., t.33 - p. 719 à 726, 1962

(4) W.H. FLEMING,

The Cauchy problem for a non linear first order partial differential equation, J. of diff. eq., t.5 - 1969.

(5) W.H. FLEMING,

Some markovian optimization problems, J. of Math. Mech., 12, n°1 - 1963, p. 131-140.

(6) I. GIHMAN A.V. SKOROHOD,

Stochastic differential equations, Springer Verlag , 1972.

(7) R. HOWARD,

Dynamic programming and Markov processes, Technology Presse and Wiley, N.Y..

(8) J.M. LASRY,

Comptes rendus du congrès de contrôle optimal de Juin 1974 à l'I.R.I.A., lecture notes - 1975, econ. math. systems., n° 107.

(9) J.M. LASRY,

Thèse, Université de Paris IX Dauphine, secrétariat de Mathématiques de la Décision, 75775 PARIS CEDEX 16.

(10) C.A. LADYZENSKAJA et N.N. URAL'CEVA,

Equations aux dérivées partielles de type elliptique, Nanka, Moscou 1964 ; traduction française : Dunod - 1968.

(11) P. MANDL,

Analytic Treatment of one dimensionnal Markov Processes, ch. VI, Springer Verlag, N.Y. - 1968.

(12) H. MOULIN,

Prolongements des jeux à deux personnes de somme nulle, Thèse, S.M.F., 1976.

(13) T. ROCKAFFELAR,

Convex Analysis
Princeton University Press - 1970.

(14) H. SCHAEFFER,

Uber die Methode der a priori Schranken, Math. ann., t.279, 1955, p. 415-416.

THE EFFECT ON OPTIMAL CONSUMPTION OF INCREASED UNCERTAINTY IN LABOR INCOME IN THE MULTIPERIOD CASE*

Bruce L. Miller†

WESTERN MANAGEMENT SCIENCE INSTITUTE

University of California, Los Angeles

ABSTRACT

We consider a multiperiod, additive utility, optimal consumption model with a riskless investment and a stochastic labor income. The main result is that for utility functions belonging to the set F, consumption decreases when we go from any sequence of distribution functions representing labor income to a more risky sequence. It is shown that a concave utility function belongs to F if and only if its first derivative exists everywhere and is convex.

*This research was supported by the National Science Foundation under ENG 74-13494.

†I thank Nils Hakansson and Steve Lippman for contributing several suggestions.

I. Introduction

The impact on consumption of increased uncertainty in future labor or capital income has been examined by a number of authors in the last ten years. As illustrated by Sandmo [23], the answers one gets are different in the two cases of uncertain labor income and uncertain capital income. Therefore, in order to separate these effects, the models with random labor income generally have one non-risky investment opportunity, and those with random investment opportunities have a deterministic (or zero) labor income. The model in this paper conforms to the above dichotomy. The only exception to that rule seems to be in Section 8 of Merton [14] which treats the case of a nondecreasing Poisson income stream, an exponential utility function, and two investment opportunities, one riskless and the other described by Brownian motion.

Three relatively early papers which examine the random labor income case are the two-period models of Leland [11], Sandmo [23] and Drèze and Modigliani [5]. Their problem is: given the first period labor income y_1 and the distribution function Y_2 of the labor income in period two, choose consumption c_1 in period 1, $0 \leq c_1 \leq y_1$, so as to maximize $EU(c_1,(1+r)(y_1-c_1) + Y_2)$.

Assuming the utilities are additive, $(U = u_1 + u_2)$, Leland [11, eq. (25)] concludes that concavity and a positive third derivative imply that there is a decrease in consumption when going from the deterministic income case to the random income case with the same mean and an infinitesimal random element (such that a second-order Taylor approximation is valid). Sandmo compares parameterized versions of Y_2 of the form

$\alpha Y_2 + (1-\alpha)E(Y_2)$, $0 \le \alpha \le k$, where k is such that the income remains nonnegative, and he demonstrates that c_1 is a decreasing function of α (and of risk) when U has decreasing temporal risk aversion. His results imply that in the case of a concave additive utility function, c_1 is a decreasing function of α when the third derivative is positive. In [5] Drèze and Modigliani look at the income and substitution effects of increased risk in labor income.

The model that we will be working with is an infinite horizon additive utility model which the author used in [15]. There the main qualitative result was that for isoelastic utility functions, consumption decreases when we go from the deterministic labor income case to the random labor income case with the same mean. In this paper we will show that for utility functions belonging to the set F (defined in Section 3), consumption decreases when we go from any sequence of distribution functions representing labor income to a more risky sequence where we are using increased risk in the sense of Rothschild and Stiglitz [20,22]. In Section 3 we show that a concave utility function belongs to F if and only if its first derivative exists everywhere and is convex. Therefore if a concave utility function is thrice differentiable, then it belongs to F if and only if its third derivative is nonnegative. It is an easy exercise to verify that the isoelastic utility functions, $\frac{1}{\gamma} c^{\gamma}$, $\gamma < 1$, $\gamma \neq 0$, belong to F.

In the random capital income case with an isoelastic utility function, the effect on consumption of increased risk in the return on capital is different depending on whether $\gamma < 0$ or $\gamma > 0$ (Rothschild and Stiglitz [21, Section 3]). Therefore we get the differing conclusions in the random labor income and random capital income models with an infinite horizon that Sandmo observed in his two-period model. In [4], p. 354, Diamond and Stiglitz have further analyzed and clarified the effect of increased risk in the random capital income case with an isoelastic utility function using the concept of mean utility preserving increase in risk.

In view of the importance of a nonnegative third derivative as exhibited in Leland [11] and Mirman [16, Appendix], it is not surprising that the third derivative is also the key condition in the model considered here. Its import is made all the more plausible when we recall the certainty-equivalence results of Theil [25], Simon [24], and recently Duchan [6]. Essentially, their results state that with a quadratic utility function (third derivative zero) and linear state equation with an additive random disturbance, the decisions are unaltered if the random elements are replaced by their means.

In Section 2 we consider the special case of an n period model with a quadratic utility function. In addition to showing that the higher derivatives of the optimal return function may not exist, this model serves as an example where the qualitative results of a two-period model do not hold for a multiperiod model. The general model is introduced in Section 3, and, using some earlier results of Miller [15], the main result is established.

II. A Quadratic Utility Multiperiod Model

As described in the introduction, the effect of uncertainty on consumption has been investigated in both two-period models and multiperiod models. The difference between analyzing the two-period and the multiperiod models is that the multiperiod models require an induction step on some property of the optimal return functions of dynamic programming. This induction step has been carried out in the uncertain capital income case by Fama [7] with the property of concavity (he does not assume additive utilities), by Neave [18] for decreasing absolute risk aversion, and by both Hakansson [8] and Mossin [17] for isoelastic functions. However, Hakansson [9] has given examples where the induction property does not hold and this section provides another. At the same time we show that the third derivative of the optimal return function may not exist even if that of the utility function does. This motivates the definition of the set of functions F in Section 3.

In order to describe the multiperiod quadratic utility model we define:

- x_j: the nonnegative amount of capital at the beginning of the period when j periods remain.
- R_j: the nonnegative labor income received at the end of the period when j periods remain. We assume that the R_j are bounded by K, and allow no borrowing against future labor income. We let $\overline{R}_j$ stand for the mean of R_j.
- $u(c_j)$: the utility from consumption of c_j when j periods remain. We assume $u(c_j) = c_j + bc_j^2$ where $b < 0$. We also assume that the point where u is decreasing, $-1/2b$, is very large relative to K and the problem parameters.
- $r-1$: the rate of interest. In this model we set $r = 1$.

Our objective is to determine the decision rule which maximizes

$$E\left(\sum_{j=1}^{N} u(c_j)\right) .$$

Here, as elsewhere, E stands for expected value.

If we let $V(x,j)$ be the optimal return function when capital equals x and j periods remain, then the terminal condition is $V(x,0) = 0$. Since we consume all the capital when there is one period to go as long as u is increasing, $V(x,1) = x + bx^2$, $0 \le x \le -1/2b$.

For $0 \le x \le -1/2b - K$, and $j = 2$,

$$V(x,2) = \max_{0 \le c \le x} E(c + bc^2 + V(x-c+R_2,1))$$

$$= \max_{0 \le c \le x} E(c + bc^2 + (x-c+R_2) + b(x-c+R_2)^2) .$$

By concavity, the optimality condition is

$$E(1 + 2bc - 1 - 2b(x-c+R_2)) = 0 ,$$

or

$$c = \frac{x + \bar{R}_2}{2}, \tag{1}$$

subject to $0 \le c \le x$. Since $x, \bar{R}_2 \ge 0$, the inequality $c \ge 0$ is always satisfied. If $x \ge \bar{R}_2$, then the optimal decision is given by (1). If $x < \bar{R}_2$ the best we can do is set $c = x$.

Therefore for the two-period additive model we have the following result: the optimal decision depends only on labor income through its mean.

In order to evaluate $V(x,2)$ we substitute the optimal decision to obtain

$$V(x,2) = K_0 + a_0 x + b_0 x^2 \qquad 0 \le x < \bar{R}_2$$

$$= K_1 + a_1 x + b_1 x^2 \qquad \bar{R}_2 \le x \le -1/2b - K$$

where $a_0 = 1$, $b_0 = b$, $a_1 = 1 + b\bar{R}_2$, $b_1 = b/2$.

Thus $V(\cdot,2)$ is described by two quadratic functions. Its third derivative is zero except at $x = \overline{R}_2$ where it is undefined and where the second derivative is discontinuous.

For $0 \leq x \leq -1/2b - 2K$, and $j = 3$,

$$V(x,3) = \max_{0 \leq c \leq x} E(c + bc^2 + V(x-c+R_3,2))$$

so that the optimality condition is

$$1 + 2bc - \int_{y:x-c+y < \overline{R}_2} (a_0 + 2b_0(x-c+y))dF(y) - \int_{y:x-c+y \geq \overline{R}_2} (a_1 + 2b_1(x-c+y))dF(y) = 0$$

where F is the distribution function of R_3. Let $x = 10$, $\overline{R}_2 = 10.01$, $R_3 = 10$ with probability one in case (a), and $R_3 = 6, 14$ with probability one half each in case (b). In case (a), $c = (10 + 10)/2 = 10$. In case (b), $c = (30 + 12 + 14 + 10.01)/7 = 9.43$. Therefore for the multiperiod model it is no longer true that the optimal decision only depends on labor income through its mean.

One might wonder how this example could be consistent with the certainty-equivalence results of [6], [24] and [25] which apply in the multiperiod case. The answer is that those results require that there be no constraints on the decision variables. It is precisely the constraint $c \leq x$ which causes $V(x,2)$ to be described by two quadratic functions.

III. The Model and the Main Results

Except for a more general class of utility functions, the model we consider is the same as that presented by Miller in [15], so that we will limit ourselves to the bare essentials and refer the reader to [15] for discussion of the model. Unlike the example of Section 2, the periods are numbered chronologically and some borrowing is permitted against future labor income. Consider

(x,j): the state of the system where x represents the capital at the beginning of period j.

$r-1$: the rate of interest for both lending and borrowing where $r > 1$.

Y_j: the nonnegative random income received at the end of period j. It is convenient to divide Y_j into certain and uncertain parts by $Y_j = y_j + R_j$ where $y_j = \sup\{h: F_j(h) = 0\}$ and $F_j(\cdot)$ is the distribution function of Y_j. We also assume that the Y_j are independent, but not identically distributed, that the means of R_j are uniformly bounded, and that $\Sigma_{i=1}^{\infty} r^{-i} y_i < \infty$. It is significant that we do not assume that the Y_j are identically distributed, for otherwise the optimal decision in period j would depend on the value of x_j and not on j.

D_j: the amount of debt allowed in period j equals $\Sigma_{i=1}^{\infty} r^{-i} y_{j+i-1}$. D_j is finite by our assumption above concerning the y_j. Thus we allow the individual to borrow against certain future income and x_j can take on values in $[-D_j, \infty)$.

c_j: the consumption in period j. We require that $0 \leq c_j \leq x_j + D_j$.

$U(c_1, c_2, c_3 \ldots)$: the utility function for all feasible $c_1, c_2 \ldots$, equals $\Sigma_{i=1}^{\infty} \alpha^{i-1} u(c_i)$, where α is a discount factor between 0 and 1. We will restrict our attention to $u \in F$ where F is defined below

Definition. A concave function $g: X \to R$, with the convex set $X \subset R$, belongs to F if for every set of λ_i, Δ_i, $i = 1, \ldots, n$, satisfying $\lambda_i \geq 0$, $\Sigma \lambda_i = 1$, and $\Sigma \lambda_i \Delta_i = 0$,

$$g(x_1) - \sum_{i=1}^{n} \lambda_i g(x_1 + \Delta_i) \geq g(x_2) - \sum_{i=1}^{n} \lambda_i g(x_2 + \Delta_i) \tag{2}$$

where $x_2 \geq x_1$, x_2, x_1 are in the interior of X.

In this paper X will be $[0,\infty)$ or $(0,\infty)$ if g is a utility function, and $[-D_j,\infty)$ if g is related to the optimal return function of period j.

The decision making takes place as follows. In period 1 the individual has x_1 units of capital. He decides to consume c_1, where $0 \leq c_1 \leq x_1 + D_1$, and he receives a utility $u(c_1)$. The resulting capital (or debt) grows to $r(x_1-c_1)$ and a random income Y_1 is received at the end of period 1 so that x_2 equals $r(x_1-c_1) + Y_1$. In general starting from state (x,j) the new state is given by

$$T(x,j) = (r(x-c) + Y_j,\ j + 1) \ . \tag{3}$$

By a policy δ we mean a decision rule that specifies the amount $c_j = \delta(x,j)$ that we consume given that we are in state (x,j). We let $f_\delta(x,j)$ be the expected value of U when using an admissible policy δ and starting from state (x,j), and define $f(x,j) = \sup_\delta f_\delta(x,j)$.

A policy δ^* is said to be optimal if $f_{\delta^*} = f$. The functional equation of dynamic programming is

$$f(x,j) = \sup_{0 \leq c_j \leq x_j + D_j} (u(c_j) + \alpha E f(T(x,j))) \ . \tag{4}$$

Some useful notation is

$$h((x,j),c,v) = u(c) + \alpha E v(T(x,j))$$

$$(Av)(x,j) = \sup_{0 \leq c \leq x + D_j} h((x,j),c,v) \ .$$

Thus equation (4) can be written as

$$f(x,j) = Af(x,j) \tag{5}$$

so that the problem of finding solutions to (4) is then equivalent to the problem of finding fixed points of A.

An interpretation that can be given to the function $h((x,j),c,v)$ is that it represents the expected return in a one-period model where the state is (x,j), the decision is c, and v is the terminal reward function. In turn, $(Av)(x,j)$ represents the expected return in the same situation when an optimal decision is made. Often the v chosen will be the optimal return function.

Let v be fixed, and for a given state (x,j) let $c^*(x,j)$ be the (feasible) value of c which maximizes $h((x,j),c,v)$. If both v and u are concave functions, then it is known (and also very easy to prove) that both

$$c^*(x,j) \quad \text{and} \quad x-c^*(x,j) \quad \text{are nondecreasing functions of } x\ . \tag{6}$$

In the event that there is more than one optimal decision we let $c^*(x,j)$ be the smallest such decision.

In this paper we will assume that we are only considering utility functions such that a unique finite-valued f satisfying (5) exists. In [15] this question was examined in detail for the isoelastic functions. For example, with the log utility function it was shown that a unique finite valued f satisfying (5) exists if we restrict $\varepsilon(x_j+D_j) \le c_j \le (1-\varepsilon)(x_j+D_j)$ for any fixed $\varepsilon > 0$. In order to go to the case here of $0 \le c_j \le (x_j+D_j)$ one needs to go through the exercise of showing the nonoptimality (with respect to f) of the newly admissible c_j. The difficulty is that the basic papers of discrete dynamic programming, Blackwell

[1] and Denardo [3], require that the reward function be bounded over all admissible states and decisions, an assumption which is not satisfied by any unbounded u. Only recently have techniques been developed which get away from this restriction (Lippman [12,13] and Harrison [10]). Fortunately, there is no difficulty whatsoever in the finite period case, so our results apply without qualification for all $u \in F$.

Theorem 1. A concave function $g: X \to R$, where the convex set $X \subset R$, belongs to F if and only if the first derivative of g, g', exists everywhere on the interior of X and is convex. From Rockafeller [19, Theorems 23.1, 24.1, 24.2 and Corollary 24.2.1.] the convexity of g' implies that

(a) the right hand and left hand derivatives of g', g''_+ and g''_-, exist everywhere on the interior of X, are increasing, and satisfy $g''_+ \geq g''_-$.

(b) for any $x, y \in X$,

$$g'(y) - g'(x) = \int_x^y g''_+(t)\,dt = \int_x^y g''_-(t)\,dt \ .$$

Proof. We first establish the "if" part of the theorem by showing that if the first derivative of g exists everywhere on the interior of X and is convex, then g belongs to F. We must show that (2) holds which we rewrite as

$$\Sigma\lambda_i [g(x_1) - g(x_1+\Delta_i) - g(x_2) + g(x_2+\Delta_i)] \geq 0 \ .$$

For any i,

$$(g(x_1) - g(x_1+\Delta_i) - g(x_2) + g(x_2+\Delta_i))$$

$$= -\int_{x_1}^{x_1+\Delta_i} g'(y)\,dy + \int_{x_2}^{x_2+\Delta_i} g'(y)\,dy$$

$$= -\int_{x_1}^{x_1+\Delta_i} [g'(x_1) + \int_{x_1}^{y} g''_+(z)\,dz]\,dy$$

$$+ \int_{x_2}^{x_2+\Delta_i} [g'(x_2) + \int_{x_2}^{y} g''_+(z)\,dz]\,dy \ ,$$

using (b) of Theorem 1,

$$\geq -\Delta_i [g'(x_1) - g'(x_2)] \tag{7}$$

since g''_+ is increasing. To see this inequality, observe that $g''_+(z_2) \geq g''_+(z_1)$ where z_2 is the same distance above x_2 that z_1 is above x_1 when $\Delta_i > 0$. If $\Delta_i < 0$ then $y < x_1$ or $y < x_2$ as the case may be, and $g''_+(z_2) \geq g''_+(z_1)$ where z_2 is the same distance below x_2 that z_1 is below x_1. Therefore

$$\Sigma\lambda_i(g(x_1) - g(x_1 + \Delta_i) - g(x_2) + g(x_2 + \Delta_i))$$

$$\geq -\Sigma\lambda_i\Delta_i[g'(x_1) - g'(x_2)] = 0 .$$

We begin the proof of the "only if" part by establishing that g' exists everywhere on the interior of X. Assume the contrary, that is for some x, the derivative does not exist. Since g is concave, both the right and left hand derivatives exist at x and we must have $g'_+(x) - g'_-(x) = k < 0$, or

$$\lim_{y \downarrow 0} \frac{g(x+y) - g(x) - g(x) + g(x-y)}{y} = k < 0 .$$

Since $g \in F$ we observe from (2) that $(-g(x) + g(x+y)/2 + g(x-y)/2)$ is an increasing function of x. Consequently the derivative cannot exist for any $x' < x$ which is inconsistent with the concavity of g.

It remains to show that for any $x_2 > x_1$ in the interior of X and $0 < \lambda < 1$,

$$-g'(\lambda x_1 + (1-\lambda)x_2) + \lambda g'(x_1) + (1-\lambda)g'(x_2) \geq 0 .$$

The derivatives equal the right hand derivatives so that the left hand side equals

$$\lim_{y \downarrow 0} \frac{1}{y}[-g(\lambda x_1 + (1-\lambda)x_2 + y) + g(\lambda x_1 + (1-\lambda)x_2) + \lambda g(x_1+y) - \lambda g(x_1) + (1-\lambda)g(x_2+y) - (1-\lambda)g(x_2)] .$$

The term in brackets is nonnegative, since by (2)

$$g(\lambda x_1 + (1-\lambda)x_2) - \lambda g(x_1) - (1-\lambda)g(x_2) \geq g(\lambda x_1 + (1-\lambda)x_2 + y) - \lambda g(x_1+y) - (1-\lambda)g(x_2+y) .$$

Q.E.D.

We note that F is large enough to include utility functions u whose absolute risk aversion, -u"/u', is decreasing. This is true since u' is decreasing by concavity and therefore we must have -u" decreasing.

We also want the result that if $g \in F$ and Z is any random variable with zero expectation such that $Eg(x_1+Z)$ and $Eg(x_2+Z)$ exist (are finite), then

$$E[g(x_1) - g(x_1+Z) - g(x_2) + g(x_2+Z)] \geq 0, \text{ when } x_2 > x_1 . \tag{8}$$

This follows from the definition of F if Z is a simple function. In order to go from simple functions to random variables we apply the same method of proof as that in Chung [2], Theorem 9.1.4. The result is also true if we replace x_1 and x_2 by $x_1 + X$ and $x_2 + X$ where X is a random variable, and if Z is a random variable such that $E(Z|X = x) = 0$ for all x. Again assuming that all expectations are defined we get (by conditioning on X = x)

$$E[g(x_1+X) - g(x_1+X+Z) - g(x_2+X) + g(x_2+X+Z)] \geq 0 . \tag{9}$$

If X is a nonnegative random variable with a finite mean, then $v(x) = Eg(x+X)$ also exists and is concave. If we let Z be the discrete random variable $P(Z = \Delta_i) = \lambda_i$, where Δ_i, λ_i, $i = 1,\ldots,n$, have the properties of those same terms in the definition of F, and be independent of X, then (9) shows that $v \in F$.

Next, we examine the idea of increasing risk as defined by Rothschild and Stiglitz [20,22]. There they establish the equivalence of three measures of risk when comparing two random variables. The definition most useful for our purposes is that Y is more risky than X if and only if

$$Y \underset{d}{=} X + Z$$

where $\underset{d}{=}$ means "has the same distribution as" and Z is a random variable such that $E(Z|X = x) = 0$ for all x. Clearly from (9) we have that if Y is more risky than X (and all expectations exist) then $g \in F$ and $x_2 > x_1$ imply that

$$E[g(x_1+X) - g(x_1+Y) - g(x_2+X) + g(x_2+Y)] \geq 0 . \tag{10}$$

The proof of Theorem 2 starts with a lemma which establishes the induction step of the kind mentioned at the beginning of Section 2 for property of belonging to the set F.

Lemma 1. If $u \in F$, then $f \in F$.

Proof. Our starting point is (5) which states that f is a fixed point of A. We have assumed that we are only considering utility functions such that a unique finite valued f satisfying (5) exists. It remains to show that the range of A is contained in F, since this will imply that the fixed point of A is in F.

It is known (for a proof in this particular case, see Miller [15]) that if g is concave then Ag is concave. We need to show that if $g(x,j) \in F$ (and hence is concave) then the concave function Ag satisfies

$$Ag(x_1,j) - \sum_{i=1}^{n} \lambda_i Ag(x_1+\Delta_i,j) - Ag(x_2,j) + \sum_{i=1}^{n} \lambda_i Ag(x_2+\Delta_i,j) \geq 0 . \tag{11}$$

Let c_1^i, $i = 1,\ldots,n$, be the optimal decisions (with respect to g) for the states $x_1 + \Delta_i$ and c_2 and c_1 be the optimal decisions for the states x_2 and x_1 respectively. From (6) we know that $c_2 \geq c_1$ and $x_2-c_2 \geq x_1-c_1$. Let c_2^i, $i = 1,\ldots,n$, be the decisions associated with the state $x_2 + \Delta_i$, and be given by $c_2^i = c_2-c_1+c_1^i$. They are feasible $(0 \leq c_2^i \leq x_2 + D_j)$ since the c_1^i are feasible, $c_2 \geq c_1$, and $x_2-x_1 \geq c_2-c_1$.

We have that

$$Ag(x_1+\Delta_i,j) = u(c_1^i) + \alpha Eg(x_1 + \Delta_i - c_1^i + Y_j,\ j+1) \ ,$$

and similar equations hold for x_1 and x_2. Since the c_2^i may not be optimal

$$Ag(x_2+\Delta_i,j) \geq u(c_2^i) + \alpha Eg(x_2 + \Delta_i - c_2^i + Y_j,\ j+1) \ .$$

Let $\Delta_1^i = c_1^i-c_1 = c_2^i-c_2$. By (7)

$$u(c_1) - u(c_1^i) - u(c_2) + u(c_2^i) \geq -\Delta_1^i(u'(c_1) - u'(c_2)) \ ,$$

since $c_2 \geq c_1$ and $u \in F$. By the development after equation (9), the function $v(x) = \alpha Eg(x+Y_j,j+1)$ belongs to F. Therefore by (7)

$$(v(x_1-c_1) - v(x_1+\Delta_i-c_1^i) - v(x_2-c_2) + v(x_2+\Delta_i-c_2^i))$$
$$\geq -(\Delta_i-\Delta_1^i)[v'(x_1-c_1) - v'(x_2-c_2)] \ ,$$

since $v \in F$ and $x_2-c_2 \geq x_1-c_1$, and $(x_1+\Delta_i-c_1^i) - (x_1-c_1) = (x_2+\Delta_i-c_2^i) - (x_2-c_2) = \Delta_i-\Delta_1^i$. Combining the above equalities and inequalities we have that the left hand side of (11) is greater than or equal to

$$-\Sigma\lambda_i[\Delta_1^i(u'(c_1) - u'(c_2)) + (\Delta_i-\Delta_1^i)(v'(x_1-c_1) - v'(x_2-c_2))] \ . \tag{12}$$

If both c_1 and c_2 are interior points of their respective constraint sets, $0 \leq c_1 \leq x_1 + D_j$ and $0 \leq c_2 \leq x_2 + D_j$, then $u'(c_1) = v'(x_1-c_1)$ and $u'(c_2) = v'(x_2-c_2)$, by the optimality of c_1 and c_2 and the fact that the derivatives of u and v exist everywhere in the interior. In this case (12) equals $-\Sigma\lambda_i\Delta_i[u'(c_1) - u'(c_2)] = 0$.

We will consider the boundary cases of $c_1 = 0$ or $c_1 = x_1+D_j$ and $c_2 = 0$ or $c_2 = x_2+D_j$ by giving the proofs for the cases $c_1 = 0$ and $c_2 = 0$ only. A similar situation arises in the proof of Theorem 2, and there we give the proofs of the cases $c_1 = x_1+D_j$ and $c_2 = x_2+D_j$ only.

One possibility at the boundary is c_1 at a boundary, say $c_1 = 0$, but c_2 is not. In order to apply (12) where c_1 is a left end point we need to verify that

$$u(x) - u(x+\Delta_i) \geq -\int_x^{x+\Delta_i} [u'_+(x) + \int_x^y u''_+(z)]dz \tag{13}$$

for $x = c_1$. Since u is nondecreasing $u(c_1) \leq \lim_{x\downarrow c_1} u(x)$. If $u(c_1) < \lim_{x\downarrow c_1} u(x)$, then $u'_+(c_1) = +\infty$ and (13) holds. If $u(c_1) = \lim_{x\downarrow c_1} u(x)$, then (13) holds for $x = c_1$ since $u'_+(c_1) \geq \lim_{x\downarrow c_1} u'_+(x) = \lim_{x\downarrow c_1} u'(x)$ (Rockafeller [19, Theorem 24.1]), and (13) holds (with equality) for all $x > c_1$. Returning to the main argument, the optimality of c_1 implies that $u'_+(c_1) \leq v'(x_1-c_1)$ and clearly $\Delta_1^i \geq 0$ for all i. Therefore (12) is greater than or equal to

$$-\Sigma\lambda_i[\Delta_1^i v'(x_1-c_1) + (\Delta_i-\Delta_1^i)v'(x_1-c_1)] - \Sigma\lambda_i\Delta_i(-u'(c_2)) = 0 .$$

The other possibility at the boundary is c_2 at a boundary, say $c_2 = 0$. By (6) $c_1 = 0$, and therefore $c_1^i = c_2^i$. Then (11) becomes

$$v(x_1) - \Sigma\lambda_i v(x_1+\Delta_i-c_1^i) - v(x_2) + \Sigma\lambda_i v(x_2+\Delta_i-c_2^i) .$$

This quantity is nonnegative since $v(x_1) - \Sigma\lambda_i v(x_1+\Delta_i) - v(x_2) + \Sigma\lambda_i v(x_2+\Delta_i)$ ≥ 0 by (2) since $v \in F$, and $v(x_1+\Delta_i) - v(x_1+\Delta_i-c_1^i) - v(x_2+\Delta_i) + v(x_2+\Delta_i-c_2^i)$ ≥ 0 for all i since v is concave. Q.E.D.

Theorem 2. Let $u \in F$ and $X_1, X_2, \ldots$ be a sequence of random variables describing labor income (case a), and $Y_1, Y_2, \ldots$ be a second sequence of random variables describing labor income (case b). If for each i, Y_i is riskier than X_i, then the optimal amount to consume as a function of the state (x,j) in case a is greater than the optimal amount to consume in case b.

Verifying the hypothesis of the following lemma leads directly to a proof of Theorem 2.

Lemma 2. Let f_X be the optimal return function in case a and f_Y be the optimal return function in case b. If $d(x,j) = f_Y(x,j) - f_X(x,j)$ is a nondecreasing function of x then the conclusion of Theorem 2 holds.

Proof. By Lemma 1 we know that $f_X, f_Y \in F$. Let c^* be the optimal decision for state (x,j) with the optimal return function f_X. For $c > c^*$,

$$\begin{aligned} h((x,j),c^*,f_Y) - h((x,j),c,f_Y) &= u(c^*) \\ &+ \alpha E f_X(r(x-c^*) + Y_j, j+1) + \alpha E d(r(x-c^*) + Y_j, j+1) \\ &- u(c) - \alpha E f_X(r(x-c) + Y_j, j+1) - \alpha E d(r(x-c) + Y_j, j+1) \, . \end{aligned}$$

Since d is nondecreasing and $c > c^*$ we have that the right hand side is

$$\begin{aligned} &\geq u(c^*) + \alpha E f_X(r(x-c^*) + Y_j, j+1) \\ &\quad - u(c) - \alpha E f_X(r(x-c) + Y_j, j+1) \\ &\geq u(c^*) + \alpha E f_X(r(x-c^*) + X_j, j+1) \\ &\quad - u(c) - \alpha E f_X(r(x-c) + X_j, j+1) \end{aligned}$$

by (10) since $r(x-c^*) > r(x-c)$ and $f_X \in F$,

$$= h((x,j),c^*,f_X) - h((x,j),c,f_X) \geq 0, \text{ since } c^* \text{ is optimal.}$$

Therefore the optimal amount to consume in case b is less than or equal to c^*. Recall that in case of ties (which could not happen with strict concavity) we pick the smallest c.

The following lemma from [15] is needed to establish that d is nondecreasing.

Lemma 3. Consider the model in the case where $Y_1, Y_2, \ldots$ are the random variables describing labor income (case b). Let $v \in F$ and suppose that v satisfies Condition A below. Then $f_Y(x,j) - v(x,j)$ is a nondecreasing function of x.

Condition A. Given any two states x_1, j and x_2, j, $x_2 > x_1$, and decision c_1 for (x_1,j), there is a feasible decision c_2 for (x_2,j) such that

(a) $x_2 - c_2 \geq x_1 - c_1$

(b) $(v(x_1,j) - h((x_1,j),c_1,v) - v(x_2,j) + h((x_2,j),c_2,v)) \geq 0$.

Equation (b) by itself is a necessary and sufficient condition that $Av-v$ be a nondecreasing function. The proof of the lemma consists of verifying an induction hypothesis in order to show that $A^n v-v$ is nondecreasing, and using the fact that $f_Y = \lim_{n\to\infty} A^n v$. By $A^n v$ we mean A applied n times to v. $A^2 v = A(Av)$.

Proof of Theorem 2. By Lemma 2 and Lemma 3 we need to show that Condition A holds where we let $v = f_X$. Let c_1^* and c_2^* be the optimal decisions for states (x_1,j) and (x_2,j) with the optimal return function f_X. Given a c_1 we set $c_2 = c_1 + c_2^* - c_1^*$.

Since c_1^* and c_2^* satisfy (6), $c_2^* \geq c_1^*$, and $x_2-x_1 \geq c_2^*-c_1^*$. Thus c_2 is feasible since c_1 is feasible, and $c_2-c_1 = c_2^*-c_1^* \leq x_2-x_1$ and (a) of Condition A holds.

Recalling that $f_X = Af_X$, the left hand side of (b) in Lemma 3 equals

$$\begin{aligned}
&(u(c_1^*) + \alpha Ef_X(x_1 - c_1^* + X_j, j+1) - u(c_1) - \alpha Ef_X(x_1 - c_1 + Y_j, j+1) \\
&\quad - u(c_2^*) - \alpha Ef_X(x_2 - c_2^* + X_j, j+1) + u(c_2) + \alpha Ef_X(x_2 - c_2 + Y_j, j+1)) \\
&\geq u(c_1^*) + \alpha Ef_X(x_1 - c_1^* + X_j, j+1) - u(c_1) - \alpha Ef_X(x_1 - c_1 + X_j, j+1) \\
&\quad - u(c_2^*) - \alpha Ef_X(x_2 - c_2^* + X_j, j+1) + u(c_2) + \ Ef_X(x_2 - c_2 + X_j, j+1)
\end{aligned} \tag{14}$$

by (10) since $x_2-c_2 \geq x_1-c_1$ and $f_X \in F$.

Now let $\Delta = c_2-c_2^* = c_1-c_1^*$, and we will show that (14) is nonnegative. By (7) $(u(c_1^*) - u(c_1) - u(c_2^*) + u(c_2)) \geq -\Delta(u'(c_1^*) - u'(c_2^*))$, since $u \in F$ and $c_2^* \geq c_1^*$. As in the proof of Lemma 1 we let $v(x) = \alpha Eg(x+X_j,j)$. Then $v \in F$ and $(v(x_1-c_1^*) - v(x_1-c_1) - v(x_2-c_2^*) + v(x_2-c_2)) \geq \Delta(v'(x_1-c_1^*) - v'(x_2-c_2^*))$. Therefore (14) is greater than or equal to

$$\Delta(-u'(c_1^*) + v'(x_1-c_1^*) + u'(c_2^*) - v'(x_2-c_2^*)) \ . \tag{15}$$

If both c_1^* and c_2^* are interior points of their constraint sets, then $u'(c_1^*) = v'(x_1-c_1^*)$, $u'(c_2^*) = v'(x_2-c_2^*)$ and (15) equals zero. If c_1^* is at a boundary, say $c_1^* = x_1 + D_j$, and c_2^* is not, then $u'(c_1^*) \geq v'_+(x_1-c_1^*)$ and $u'(c_2^*) = v'(x_2-c_2^*)$. Since Δ must be nonpositive in this case, (15) will be nonnegative. Here we use (13) as applied to v. If c_2^* is at a boundary, say $c_2^* = x_2 + D_j$, then c_1^* must equal $x_1 + D_j$, and $v(x_2-c_2^*) = v(x_1-c_1^*)$ and $v(x_2-c_2) = v(x_1-c_1)$. Then (14) becomes $(u(c_1^*) - u(c_1) - u(c_2^*) + u(c_2))$ which is nonnegative by the concavity of u. Q.E.D.

References

1. Blackwell, D., "Discounted Dynamic Programming," Ann. Math. Stat., 36 (1965), 226-235.

2. Chung, K., A Course in Probability Theory, Harcourt, Brace and World, New York, 1968.

3. Denardo, E., "Contraction Mappings in the Theory Underlying Dynamic Programming," SIAM Review, 9 (1967), 165-177.

4. Diamond, P. and J. Stiglitz, "Increases in Risk and in Risk Aversion," J. of Economic Theory, 8 (1974), 337-360.

5. Drèze, J. and F. Modigliani, "Consumption Decisions Under Uncertainty," J. of Economic Theory, 5 (1972), 308-335.

6. Duchan, A., "A Clarification and a New Proof of the Certainty Equivalence Theorem," International Economic Review, 15 (1974), 216-224.

7. Fama, E., "Multiperiod Consumption-Investment Decisions," American Economic Review, 60 (1970), 163-174.

8. Hakansson, N., "Optimal Investment and Consumption Strategies Under Risk for a Class of Utility Functions," Econometrica, 38 (1970), 587-607.

9. Hakansson, N., "On Optimal Myopic Portfolio Policies, With and Without Serial Correlation Yields," J. of Business, 44 (1971), 324-334.

10. Harrison, M., "Discrete Dynamic Programming With Unbounded Rewards," Ann. Math. Stat., 43 (1972), 636-644.

11. Leland, H., "Saving and Uncertainty: The Precautionary Demand for Saving," Quarterly J. of Economics, 82 (1968), 465-473.

12. Lippman, S., "Semi-Markov Decision Processes With Unbounded Rewards," Management Science, 19 (1973), 717-731.

13. Lippman, S., "On Dynamic Programming With Unbounded Rewards," Working Paper No. 212, Western Management Science Institute, University of California, Los Angeles, November 1973.

14. Merton, R., "Optimal Consumption and Portfolio Rules in a Continuous Time Model," J. of Econ. Theory, 3 (1971), 373-413.

15. Miller, B., "Optimal Consumption with a Stochastic Income Stream," Econometrica, 42 (1974), 253-266.

16. Mirman, L., "Uncertainty and Optimal Consumption Decisions," Econometrica, 39 (1971), 179-185.

17. Mossin, J., "Optimal Multiperiod Portfolio Policies," J. of Business, 41 (1968), 215-229.

18. Neave, E., "Multiperiod Consumption-Investment Decisions and Risk Preferences," J. of Econ. Theory, 3 (1971), 40-53.

19. Rockafeller, R.T., Convex Analysis, Princeton University Press, Princeton, N.J., 1970.

20. Rothschild, M. and J. Stiglitz, "Increasing Risk: I, A Definition," J. of Econ. Theory, 2 (1970), 225-243.

21. Rothschild, M. and J. Stiglitz, "Increasing Risk: II, Its Economic Consequences," J. of Econ. Theory, 3 (1971), 66-84.

22. Rothschild, M. and J. Stiglitz, "Addendum to 'Increasing Risk: I, A Definition,'," J. of Econ. Theory, 5 (1972), 306.

23. Sandmo, A., "The Effect of Uncertainty on Saving Decisions," The Review of Economic Studies, 37 (1970), 353-360.

24. Simon, H., "Dynamic Programming Under Uncertainty with a Quadratic Criterion Function," Econometrica, 24 (1956), 74-81.

25. Theil, H., "A Note on Certainty Equivalence in Dynamic Programming," Econometrica, 25 (1957), 346-349.

NONLINEAR OPTIMAL STOCHASTIC CONTROL

- SOME APPROXIMATIONS WHEN THE NOISE IS SMALL

by J.D. Perkins, Department of Chemical Engineering, University of Cambridge

and R.W.H. Sargent, Department of Chemical Engineering and Chemical Technology, Imperial College, London.

1. Introduction

We consider the design of optimal controllers for the following nonlinear stochastic system:

System S1

The evolution of the "state" is described by

$$\dot{x}(t) = f(t,x(t), u(t), \xi(t)) , t\epsilon[0,t_f], \tag{1}$$

$$\text{with measurements} \quad y(t) = g(t,x(t), \eta(t)) , \tag{2}$$

$$\text{controls} \quad u(t) = h(t,I(t)) \epsilon U , \tag{3}$$

$$\text{and performance index} \quad J = E\{c^T x(t_f)\}. \tag{4}$$

Here $x(t)\epsilon E^n$ is an n-vector of "state" variables, and the initial "state" $x(0)\epsilon X(0)$ with probability one, where $X(0)\subset E^n$ is a given convex, bounded set. The system noise, $\xi(t)\epsilon V$ w.p.1., and the measurement noise, $\eta(t)\epsilon W$ w.p.1., are both vector stochastic processes and $V\subset E^r$, $W\subset E^q$ are given convex, bounded sets of possible values. $U\subseteq E^m$ is a given set of admissible values for the controls $u(t)$, which are vector-valued functions $h(t,I(t))$ of the set of information $I(t)$ available to the controller at time t; $I(t)$ is a specified subset of the a priori information $I(0)$ and the measurements $y(t)$, $t\epsilon[0,t_f]$. In the expression for the performance index, $c\epsilon E^n$ is a given vector, and the expectation is taken over all possible initial conditions $x(0)$ and all possible realizations of the noise processes $\xi(t)$ and $\eta(t)$.
The control design problem is to find the function $h(t,I(t))$, $t\epsilon[0,t_f]$ which minimizes the performance index J for specified information sets $I(t)$, $t\epsilon[0,t_f]$.
Following Joffe and Sargent[1], we appeal to the theory of ordinary differential equations with Lebesgue-measurable input functions in order to obtain a set of assumptions which guarantee the existence and uniqueness of solutions to equation (1) for given initial condition $x(0)$, noise realization $\xi(t)$ and admissible control $u(t)$, $t\epsilon[0,t_f]$:-

(i) Each admissible control $u(t)$ and each noise realization $\xi(t)$ is a measurable function of t on the interval $[0,t_f]$.

(ii) For each possible set of values x, u, ξ, the function $f(t, x, u, \xi)$ is a measurable function of t on the interval $[0,t_f]$.

(iii) For each $t\epsilon[0,t_f]$, $f(t, x, u, \xi)$ is continuous in x, u, and ξ for all possible values of these variables.

(iv) There exists a function $S(t)$, summable on $[0,t_f]$, and a function $\phi(z)$, positive and continuous for $z \geqslant 0$ but not summable on $[0,\infty)$, such that

$$||f(t,x,u,\xi)|| \leqslant S(t)/\phi(||x||)$$

for each possible t, x, u, and ξ.

(v) There exists a function $M(t)$, summable on $[0,t_f]$, such that

$$|(x-x')^T(f(t,x,u,\xi) - f(t,x',u,\xi))| \leqslant M(t).||x-x'||^2$$

for all possible t, x, x', u and ξ.

These conditions also ensure that the set $X(t_f)$ of states reachable w.p.1. with some possible initial condition, control and noise realization, is a bounded set.

We note that by requiring $\xi(t)$ to be measurable in t we exclude the possibility that $\xi(t)$, and hence also $x(t)$, are Markov processes. Thus $x(t)$ does not represent a state in the stochastic sense, but for a given realization of the system noise it has all the properties required of a state vector for a deterministic system.

It is well known that systems with more general performance indices than (4) can often be put in the form of S1 by adjoining extra state variables with appropriate definitions. These systems can therefore be treated, provided that the adjoined variables and functions also satisfy the above conditions.

A variety of control problems can be posed for S1 by suitable choice of the information sets $I(t)$. For example this set may consist of just the current set of measurements $y(t)$, or perhaps past measurements at a sequence of times or over a specified time interval. Feed-forward control policies are obtained if $I(t) = I(0)$, $t\in[0,t_f]$.

Joffe and Sargent[1] considered the cases of feed-forward control, and feed-back control based on continuous exact measurement of the state $x(t)$. They also considered sub-optimal controls based on the functional form:

$$u(t) = h(t,x(t),\alpha(t)) \tag{5}$$

where $h(t,x,\alpha)$ is a given heuristically derived function, and $\alpha(t)$ is a vector of parameters which are chosen optimally.

Their approach was to assume that the noise $\xi(t)$ and the initial state $x(0)$ satisfy certain smallness conditions, and then to derive a deterministic model which predicts $E\{x(t)\}$ for a specified control policy. In this paper we shall extend their treatment to deal with more general feed-back controls.

2. A Modelling Theorem

Suppose that we have a system as described in S1, and a mathematical model of this system which enables us to predict the performance index for any initial condition and any control law of the form of equation (3). More precisely, if $J_s(h)$ is the value of the system performance index, using the control function $h(t,I(t))$ with specified $I(t)$, $t\in[0,t_f]$ and initial condition $x(0)$, and $J_m(h)$ is the value of the model performance index using the same control function, information sets and initial condition, then there exists a $\delta \geq 0$ such that

$$|J_s(h) - J_m(h)| \leq \delta \tag{6}$$

for all possible $h(\cdot,\cdot)$, $I(t)$ and $x(0)$.

Now suppose that we have solved the control design problem for the system for a given specification of the $I(t)$, obtaining the optimal system control function $h_s(t,I(t))$ with corresponding performance index $J_s(h_s)$. Similarly, we solve the design problem for the model with the same specification of the $I(t)$, obtaining the optimal model control function $h_m(t,I(t))$ with performance index $J_m(h_m)$. Then we have the following theorem:

Theorem 1

If the system and model satisfy condition (6), and if $h_s(t,I(t))$ and $h_m(t,I(t))$ are the optimal system and model control laws respectively for specified information sets $I(t)$, $t \in [0,t_f]$, then

$$|J_s(h_m) - J_s(h_s)| \leq 2\delta, \quad |J_m(h_m) - J_s(h_s)| \leq 3\delta \tag{7}$$

Proof

From condition (6) we have

$$|J_s(h_m) - J_m(h_m)| \leq \delta, \quad |J_s(h_s) - J_m(h_s)| \leq \delta \tag{8}$$

It follows from the triangle inequality that

$$|J_s(h_m) - J_m(h_m) + J_m(h_s) - J_s(h_s)| \leq 2\delta$$

But from the optimality conditions $J_s(h_s) \leq J_s(h_m)$ and $J_m(h_m) \leq J_m(h_s)$, and hence

$$|J_s(h_m) - J_s(h_s)| \leq 2\delta, \quad |J_m(h_s) - J_m(h_m)| \leq 2\delta$$

To obtain the second inequality in (7), we use the first part and condition (8):

$$\begin{aligned}|J_m(h_m) - J_s(h_s)| &= |J_m(h_m) - J_s(h_m) + J_s(h_m) - J_s(h_s)| \\ &\leq |J_m(h_m) - J_s(h_m)| + |J_s(h_m) - J_s(h_s)| \leq 3\delta\end{aligned}$$

Q.E.D.

This theorem tells us that the model optimal control law applied to the system will produce a performance index within $O[\delta]$ of the true optimal result, and that the optimal performance index $J_m(h_m)$ computed from the model is an $O[\delta]$ approximation to the true optimal value.

It therefore remains to find good manageable models for nonlinear stochastic systems, and since J is an expected value it is worth looking for deterministic models for its prediction.

3. Deterministic Models for Small-Noise Stochastic Systems

If, in addition to the conditions listed in Section 1, the second partial derivatives of $f(t,x,u,\xi)$ and $g(t,x,\eta)$ with respect to x, ξ, η exist, and are Lipschitz continuous on $X(t_f)\times V$ and $X(t_f)\times W$ respectively, then it is possible to use Taylor expansions to obtain the following models:

Model M1

$$\dot{x}^*(t) = f(t,x^*(t),u(t),\xi^*(t)) \ , \ t\in[0,t_f] \tag{9}$$

$$y^*(t) = g(t,x^*(t),\eta^*(t)) \tag{10}$$

$$J^* = c^T x^*(t_f) \tag{11}$$

$$\text{where } \xi^*(t) = E\{\xi(t)\} \ , \ \eta^*(t) = E\{\eta(t)\} \tag{12}$$

Model M2

$$\dot{x}^*(t) = f(t,x^*(t),u(t),\xi^*(t))+z(t) \ , \ t\in[0,t_f] \tag{13}$$

$$y^*(t) = g(t,x^*(t),\eta^*(t))+\zeta(t) \tag{14}$$

$$J^* = c^T x^*(t_f) \tag{15}$$

where [†]

$$z^k(t) = \tfrac{1}{2} \text{ trace } [f^k_{xx}P(t)+f^k_{\xi\xi}Q(t)] \tag{16}$$

$$\zeta^k(t) = \tfrac{1}{2} \text{ trace } [g^k_{xx}P(t)+g^k_{\eta\eta}R(t)] \tag{17}$$

$$\dot{P}(t) = f_x P + P f_x^T \tag{18}$$

$$\left.\begin{aligned} Q(t) &= E\{\delta\xi(t).\delta\xi^T(t)] \\ R(t) &= E\{\delta\eta(t).\delta\eta^T(t)] \\ \xi^*(t) &= E\{\xi(t)\} \qquad \delta\xi(t) = \xi(t)-\xi^*(t) \\ \eta^*(t) &= E\{\eta(t)\} \qquad \delta\eta(t) = \eta(t)-\eta^*(t) \end{aligned}\right) \tag{19}$$

We then have the following basic approximation theorem for Model M2:

Theorem 2

Suppose that, in addition to the assumptions stated previously for system S1, the following inequalities hold for some finite, positive constants A, B, C, ε, and each $t'\in[0,t_f]$:

$$\begin{aligned}
&\int_0^{t_f} E\{||\delta\xi(t)||^3\}dt \leqslant t_f\varepsilon^3 && E\{||\delta x(0)||^3\} \leqslant A^3\varepsilon^3 \\
&\int_0^{t_f} E\{||\delta\xi(t)||^2||\delta\xi(t')||\}dt \leqslant t_f\varepsilon^3 && E\{||\delta x(0)||^2||\delta\xi(t')||\} \leqslant A^2\varepsilon^3 \\
&\int_0^{t_f} E\{||\delta\xi(t)||.||\delta\xi(t')||^2\}dt \leqslant t_f\varepsilon^3 && E\{||\delta x(0)||.||\delta\xi(t')||^2\} \leqslant A\varepsilon^3 \\
&\int_0^{t_f} ||E\{\delta\xi(t).\delta\xi^T(t')\}||dt \leqslant t_f\varepsilon^3 && ||E\{\delta x(0).\delta\xi^T(t')\}|| \leqslant A\varepsilon^3 \\
&E\{||\delta\eta(t)||^3\} \leqslant B^3\varepsilon^3 && E\{||\delta x(t)||.||\delta\eta(t)||^2\} \leqslant AB^2\varepsilon^3 \\
&||E\{\delta x(t).\delta\eta^T(t)\}|| \leqslant AB\varepsilon^3 && E\{||\delta x(t)||^2||\delta\eta(t)||\} \leqslant A^2B\varepsilon^3
\end{aligned} \tag{20}$$

$$||x^*(0)-E\{x(0)\}|| \leqslant C\varepsilon^3 \quad , \quad ||P(0)-E\{\delta x(0).\delta x^T(0)\}|| \leqslant C\varepsilon^3 \tag{21}$$

where $\delta x(t) = x(t) - E\{x(t)\}$.

Then, for each admissible control $u(t)$, $t\in[0,t_f]$ and given initial conditions $x^*(0)$, $P(0)$, there exists a unique solution to equations (13) - (19) of Model M2, and this

---

† Notation Subscripts denote differentiation with respect to the variables concerned, and superscripts denote elements of vectors. Arguments of functions, where omitted, are $(t,x^*(t),u(t),\xi^*(t))$.

solution satisfies the inequalities:

$$||x^*(t)-E\{x(t)\}|| \leqslant K_1\varepsilon^3 \ , \ ||P(t)-E\{\delta x(t).\delta x^T(t)\}|| \leqslant K_2\varepsilon^3 \qquad (22)$$
$$||y^*(t)-E\{y(t)\}|| \leqslant K_3\varepsilon^3 \ , \ |J^*-J| \leqslant K_1\varepsilon^3||c||$$

where K_1, K_2 and K_3 are finite positive numbers.

Joffe and Sargent(1) give the proof of this theorem without the results on $y^*(t)$, and on the assumption that C = 0. However the required extension of their proof is immediate and will not be given here. It is useful to note at this point that if C = 0 then K_1 and K_2 are both directly proportional to t_f.

We also note that if X(0) consists of a single point there is no uncertainty in the initial state and its covariance matrix is zero. We may therefore set $x^*(0) = x(0)$ and $P(0) = 0$, so that C can be taken as zero; it also follows from (18) that $P(t) = 0$ for all $t\epsilon[0,t_f]$, thereby removing a matrix differential equation from the model and simplifying $z(t)$.

The same proof shows that Model M1 provides $O[\varepsilon^2]$ bounds for the quantities in (22), and again it is not necessary to generate $P(t)$ in order to obtain $x^*(t)$.

4. Optimal Controller Design

If the information sets $I(t)$, $t\epsilon(0,t_f]$ are empty the control is simply a function of time, and Theorem 1 applies with optimal control policies $h_s(t)$ and $h_m(t)$ for the system and model respectively. Model M2 satisfies condition (6) with $\delta = K_1\varepsilon^3||c||$, and it follows that the deterministic optimal control policy for Model M2 is within $O[\varepsilon^3]$ of the true optimal policy for System S1. Similarly Model M1 approximates the true policy with an error of $O[\varepsilon^2]$.

Theorem 2 does not apply as it stands to general feed-back control laws as given by equation (3). However, if the information sets $I(t)$ are written in terms of the measurements $y(t)$, which are in turn expressed in terms of $x(t)$ through equation (2), then substitution of the resulting control law into equation (1) yields an equation of the same form as equation (1) with a given control policy $u(t)$ substituted, and Theorem 1 then yields the general result that the optimal model control law is a suitable approximation to the true stochastic optimal law for any feed-back control law as defined in equation (3).

We note that the small-noise and smoothness assumptions of Theorem 2 must now apply to the substituted form of equation (1), and hence to the combined stochastic processes $\xi(t)$ and $\eta(t)$. Because of the conditions already imposed on $g(t,x,\eta)$, the smoothness conditions will be satisfied if $f(t,x,u,\xi)$ is twice differentiable in u, and $h(t,I(t))$ is twice differentiable in its arguments, and all these second derivatives are Lipschitz continuous in the attainable domain of their arguments. Of course these strong smoothness assumptions may be unduly restrictive, and exclude control laws giving much more favourable performance. A further difficulty arises if the information set $I(t)$ involves information at times other than the current time, for the composite substituted equation then describes a generalized dynamical system falling outside the scope of the theory of differential equations used in Theorem 2. It turns out that these difficulties can be overcome, and a more general form of Theorem 2 obtained(2), as will be shown in a separate paper. For the present discussion however, we shall restrict ourselves to the subclass of smooth control functions satisfying the above conditions, with the information sets $I(t)$, $t\epsilon(0,t_f]$ either empty or consisting only of current measurements $y(t)$.

4.1 Problems with Known Initial State

In an important class of problems the initial state $x(0)$ is known to good accuracy - for example in problems of optimum start-up of a process, or change of regime from one steady state to another, or where the state variables can all be measured with negligible error.

In such cases the smallness conditions concerning the initial state are satisfied with $A = 0$, and we may set $x^*(0) = x(0)$, $P(0) = 0$ in Model M2, satisfying (21) with $C = 0$. The model then generates second-order estimates of expected quantities conditioned on the known initial state, and Theorems 1 and 2 show that the optimal model control law is a second-order approximation to the stochastic optimal control for the system for any feed-back control law.

However, we note that Model M2 is a deterministic system with given initial conditions, which generates for itself second-order approximations to the measurements $y(t)$ for all $t\epsilon[0,t_f]$. Thus the optimal feed-forward policy based on the given initial state will be within a margin of $0[\varepsilon^3]$ of the optimal result for any feed-back control law, and it follows that little advantage is to be gained from the use of measurements in a feed-back controller in such a situation.

On the other hand, we noted in Section 3 that K_1, and hence the error, is directly proportional to t_f, so that the uncertainty grows with time. If the time interval of operation is large it may therefore be worth utilizing the extra information gained from measurements to compensate for this growth of uncertainty, and this brings us to the case where the initial state is no longer determined to high precision.

4.2 Problems with Unknown Initial State

We now consider the situation where we know only the expected value of the initial state with its corresponding covariance matrix, conditioned on the initial information available $I(0)$.

If we set $x^*(0) = E\{x(0)|I(0)\}$ and $P(0) = E\{\delta x(0)\delta x^T(0)|I(0)\}$, then Model M2 generates approximations to the corresponding conditional expectations, and we obtain the same result for the optimal model and system control laws for a given specification of the information sets $I(t)$. The design of the optimal feed-back control for the deterministic model is still much simpler than for the original stochastic system, but it is nevertheless a substantial problem, and it is worth seeking further insight.

If the information sets $I(t)$, $t\epsilon(0,t_f]$ are empty, then the model control is simply a feed-forward control policy, and the optimum policy for the given initial state is easily obtained by standard methods. With non-empty information sets it is clear that the optimal feed-back control must implicitly make use of the additional information to improve the estimate of the state, and it is of interest to examine the conjecture that an approximate form of Wonham's Separation Theorem [3] will apply to the nonlinear stochastic system. This would imply that the optimal controller can be split into an estimator, producing an estimate $x^*(t)$ of the state from the measurements, and an optimal feed-back controller based on continuous observation of the state, which uses $x^*(t)$ in place of the actual state.

In this connection we have the following theorem:

Theorem 3

The performance of System S1 is compared under the two control laws

$$u(t) = h(t,x(t)) \tag{23a}$$

$$\text{and} \qquad u(t) = h(t,x^*(t)) \tag{23b}$$

giving rise to states $x_a(t)$ and $x_b(t)$ respectively, where $x^*(t)\in X(t_f)$ and $h(t,x)$ are measurable functions of t on $[0,t_f]$, and $h(t,x)$ is Lipschitz continuous in x on $X(t_f)$. Then if the corresponding performance indices are J_a and J_b we have

$$\begin{aligned} E\{||x_a(t)-x_b(t)||\} &\leqslant K\int_0^t E\{||x_b(\tau)-x^*(\tau)||\}d\tau \\ |J_a-J_b| &\leqslant K||c||\int_0^{t_f} E\{||x_b(\tau)-x^*(\tau)||\}d\tau \end{aligned} \tag{24}$$

for all $t\in[0,t_f]$ and some $K\in[0,\infty)$.

Proof

We compare performances with a given noise realization and initial state:

$$\begin{aligned} \dot{x}_a(t) &= f(t,x_a(t),h(t,x_a(t)),\xi(t)) \quad , \quad x_a(0) = x_o \\ \dot{x}_b(t) &= f(t,x_b(t),h(t,x^*(t)),\xi(t)) \quad , \quad x_b(0) = x_o \end{aligned} \tag{25}$$

From the properties of $f(t,x,u,\xi)$ and $h(t,x)$ we have w.p.1.

$$\begin{aligned} ||f(t,x_a,u,\xi)-f(t,x_b,u,\xi)|| &\leqslant L_x||x_a-x_b||, \\ ||f(t,x,u_a,\xi)-f(t,x,u_b,\xi)|| &\leqslant L_u||u_a-u_b||, \\ ||h(t,x_a)-h(t,x_b)|| &\leqslant L_h||x_a-x_b||, \end{aligned} \tag{26}$$

for some finite, positive constants L_x, L_u, L_h. Then from (25) and (26):

$$\begin{aligned} \frac{d}{dt}||x_a(t)-x_b(t)|| &\leqslant ||f(t,x_a(t),h(t,x_a(t)),\xi(t))-f(t,x_b(t),h(t,x_b(t)),\xi(t))|| \\ &+ ||f(t,x_b(t),h(t,x_b(t)),\xi(t))-f(t,x_b(t),h(t,x^*(t)),\xi(t))|| \\ &\leqslant (L_x+L_uL_h)||x_a(t)-x_b(t)||+L_uL_h||x_b(t)-x^*(t)|| \end{aligned} \tag{27}$$

Taking expectations of (27), noting that $||x_a(0)-x_b(0)|| = 0$, and applying Halkin's "Generalized Gronwall Lemma"[1], we obtain the first relation in (24) with

$$K = L_uL_h\exp(L_x+L_uL_h)t_f < \infty$$

and the second relation follows immediately from the definition of J.

Q.E.D.

Of course, Models M1 and M2 can themselves be used as estimators, and with additional smoothness conditions we can, for example, prove the following:

Theorem 4

The performance of System S1 is compared under the control laws given in (23a) and (23b), where $h(t,x)$ is measurable in t and has continuous derivatives with respect to x on $[0,t_f]\times X(t_f)$, and $x^*(t)$ is the solution for Model M1 with $u(t)$ given by (23b) and initial condition $x^*(0) = E\{x(0)|I(0)\}$. Then with the small-noise conditions of (20):

$$||E\{x_a(t)-x_b(t)\}|| \leqslant K_1\int_0^t ||E\{x_b(\tau)-x^*(\tau)\}||d\tau + K_2\varepsilon^2 \tag{28}$$

for some finite K_1 and K_2, where all expectations are conditioned on the initial information $I(0)$.

Proof

$$\text{Write}\quad \delta x_a(t)=x_a(t)-x^*(t),\quad \delta x_b(t)=x_b(t)-x^*(t)$$
$$\delta\xi(t)=\xi(t)-\xi^*(t),\quad \xi^*(t) = E\{\xi(t)|I(0)\}$$

Using (26) with a similar Lipschitz condition for ξ:

$$||f(t,x,u,\xi)-f(t,x,u,\xi^*)|| \leqslant L_\xi||\xi-\xi^*||, \tag{26'}$$

we obtain from (9) and (25)

$$\frac{d}{dt}||\delta x_a(t)|| \leqslant (L_x+L_uL_h)||\delta x_a(t)|| + L_\xi||\delta\xi(t)||,$$

$$\frac{d}{dt}||\delta x_b(t)|| \leqslant L_x||\delta x_b(t)|| + L_\xi||\delta\xi(t)||,$$

and using the Generalized Gronwall Lemma these yield

$$\begin{aligned}||\delta x_a(t)|| &\leqslant \{||\delta x_a(0)|| + L_\xi\int_0^t ||\delta\xi(t)||dt\}.\exp(L_x+L_uL_h)t,\\ ||\delta x_b(t)|| &\leqslant \{||\delta x_b(0)|| + L_\xi\int_0^t ||\delta\xi(t)||dt\}.\exp(L_xt).\end{aligned} \tag{29}$$

Taking expectations of (29) and using (20) we obtain

$$\begin{aligned}E\{||\delta x_a(t)||\} &\leqslant K_a\varepsilon\ , \quad K_a = (A+L_\xi t).\exp(L_x+L_uL_h)t,\\ E\{||\delta x_b(t)||\} &\leqslant K_b\varepsilon\ , \quad K_b = (A+L_\xi t).\end{aligned} \tag{30}$$

Also from (20), (29) and (30), using the Hölder inequality for both expectations and integrals, we can obtain

$$\begin{aligned}&E\{||\delta x_a(t)||^2\} \leqslant K_a^2\varepsilon^3\ , \quad E\{||\delta x_b(t)||^2\} \leqslant K_b^2\varepsilon^2\\ &\int_0^{t_f}E\{||\delta x_a(t)||.||\delta\xi(t)||\}dt \leqslant K_a\varepsilon^2\ , \quad \int_0^{t_f}E\{||\delta x_b(t)||.||\delta\xi(t)||\}dt \leqslant K_b\varepsilon^2\end{aligned} \tag{31}$$

Now from (25), expanding about the values t, $x^*(t)$, $\xi(t)$ we have

$$\dot{x}_a(t) = f + (f_x+f_uh_x).\delta x_a(t) + v_a(t)\ , \quad ||v_a|| \leqslant L_a||\delta x_a(t)||^2$$
$$\dot{x}_b(t) = f + f_x.\delta x_b(t) + v_b(t)\ , \quad ||v_b|| \leqslant L_b||\delta x_b(t)||^2$$

for some finite, positive L_a, L_b. But we may now expand the derivatives f_x, f_u about the value $\xi^*(t)$, and since the second derivatives $f_{x\xi}$, $f_{u\xi}$ are bounded there exists an $L < \infty$ such that

$$\frac{d}{dt}\{x_a(t) - x_b(t)\} = (f_x^*+f_u^*h_x^*)\{x_a(t) - x_b(t)\} + f_u^*h_x^*\delta x_b(t) + w(t) \tag{32}$$

where f_x^*, f_u^*, h_x^* are evaluated for t, $x^*(t)$, $\xi^*(t)$ - and hence are deterministic - and $w(t)$ satisfies:

$$||w(t)|| \leqslant L\{||\delta x_a(t)||^2+||\delta x_b(t)||^2+||\delta x_a(t)||.||\delta\xi(t)||+||\delta x_b(t)||.||\delta\xi(t)||\} \quad (33)$$

Taking expectations and then norms of (32) leads to

$$\frac{d}{dt}||E\{x_a(t)-x_b(t)\}|| \leqslant (L_x+L_uL_h)||E\{x_a(t)-x_b(t)\}||+L_uL_h||E\{\delta x_b(t)\}||+E\{||w(t)||\} \quad (34)$$

and applying the Generalized Gronwall Lemma and using (33) and (31) finally yields (28) with $K_1 = L_uL_h\exp(L_x+L_uL_h)t$ and $K_2 = \{K_a+K_b+K_a^2+K_b^2\}.L\exp(L_x+L_uL_h)t_f$

Q.E.D.

From the analogue of Theorem 2 for Model M1 we note further that the integral on the right-hand side of (28) is $O[\varepsilon^2]$. A similar proof will show that Model M2 provides an $O[\varepsilon^3]$ error for $||E\{x_a(t)-x_b(t)\}||$ and hence for the difference in performance indices.

Theorem 3 is of interest since it does not depend on the form of the estimator and shows that the loss in performance of the "separated" controller depends only on the integral error of the estimate, allowing for example a poor estimate provided that the time over which it is used is short. Note however that the error also depends on the magnitude of the noise $\xi(t)$ since $x_b(\tau)$ is the actual state, and in fact, Theorem 3 places quite strong conditions on both noise and estimate to ensure a small loss in performance.

Condition (28) of Theorem 4 is weaker, but the small-noise conditions (20) in essence require that the uncertainty in the estimate of the initial state is small. This brings us back to the situation considered in Section 4.1, and indeed, since Models M1 and M2 as given in Section 3 simply predict states from the initial state estimate, without utilizing information from measurements, it is not surprising that we reach the same conclusion. However Theorem 4 is also of use if the state estimate is updated. Suppose, for example, that the state can be measured with negligible error at a sequence of times $t_o = 0, t_1, t_2, \ldots$ and Model M1 is used for prediction of the state within these intervals, with $x^*(t_k) = x_b(t_k)$ as initial condition. Then we may set $A = 0$ and note from (30) that K_a and K_b are $O[t]$. If the integral conditions in (20) apply over each subinterval $[t_k,t_{k+1}]$, more careful integration of (34) then shows that $||E\{x_a(t_f)-x_b(t_f)\}||$ is $O[\varepsilon^2 t_f.\Delta t]$, where Δt is the duration of the largest interval between measurements. Hence, as would be expected, the loss in performance can be reduced by increasing the frequency of measurement.

If only noisy measurements are available, as in (2), then Model M1 or M2 can be used as a predictor in an estimator which makes use of these measurements. For example, if measurements $y(t_k)$, $k = 0, 1, 2, \ldots$ are available, then an estimator based on Model M2 is as follows:

(i) Given estimates $x^*(t_{k-1}|I(t_{k-1}))$, $P(t_{k-1}|I(t_{k-1}))$, these are used as initial conditions with equations (13) - (19) to generate predictions $x^*(t_k|I(t_{k-1}))$, $P(t_k|I(t_{k-1}))$.

(ii) These predictions are updated using the measurements at t_l by the equations:

$$K(t_k) = P(t_k|I(t_{k-1})).(g_x^*)^T[g_x^*P(t_k|I(t_{k-1}))(g_x^*)^T + R(t_k)]^{-1}$$

$$P(t_k|I(t_k)) = P(t_k|I(t_{k-1})) - K(t_k)g_x^*P(t_k|I(t_{k-1})) \tag{35}$$

$$x^*(t_k|I(t_k)) = x^*(t_k|I(t_{k-1})) - K(t_k)(y(t_k) - y^*(t_k))$$

where g_x^* is the partial derivative with respect to x of the right-hand side of equation (14); in the iterated form of the estimator it is the arithmetic mean of the values at t_{k-1} and t_k.

The proof of Theorem 4 makes no use of the nature of the conditioning in the expectations, and hence the method of proof remains valid for the above estimator, the update of (35) serving to reduce the magnitude of the error due to the integral in (28). It is important to note that the update also reduces the variance of the state, and hence the variance of the performance index, so that feed-back increases the probability of the achieved performance being close to the expected optimal value.[2]

The analogy with the extended Kalman filter[4] is obvious, and a similar estimator and corresponding theorem can be developed for the case of continuous measurements. The small-noise assumptions concerning the initial state are still restrictive, and in essence are made necessary by the fact that comparison is made between use of the estimator and continuous exact measurement of the state. What is really required is a comparison with the true optimal controller based on the same information sets, but since all the past information is available to the controller this inevitably leads us to consideration of a generalized dynamical system, which is outside the scope of the present paper.

5. Conclusions

Stochastic control problems for a rather general class of nonlinear systems have been considered in this paper. Deterministic models for the prediction of expected performance have been presented, and it has been shown that optimal controllers designed using the models are good approximations to the true optimal controllers for the stochastic system. Some results have been presented on the use of an estimator coupled with an optimal controller based on continuous observation of the state.

The problems associated with the non-Markov nature of the system, and relaxation of the smoothness conditions imposed on the state and control laws will be the subject of a separate paper.

Acknowledgments

One of the authors (J. D. Perkins) wishes to thank the Salters' Company for the award of a Scholarship.

References

1. Joffe, B. L., and R. W. H. Sargent, "A Naïve Approach to the Optimal Control of Nonlinear Stochastic Systems", in "Recent Mathematical Developments in Control", D. J. Bell (Ed.), Academic Press (1973)

2. Perkins, J. D., "Optimal Stochastic Control of Continuous Processes - Some Approximations Applicable to Chemical Engineering Problems", Ph.D. Thesis, London (1976)

3. Wonham, W. M., "On the Separation Theorem of Stochastic Control", SIAM J. Control,

6 (2), 312 (1968)

4. Jazwinski, A. H., "Stochastic Processes and Filtering Theory", (Academic Press, New York, 1970)

ASYMPTOTIC BEHAVIOR OF POSTERIOR DISTRIBUTIONS FOR RANDOM PROCESSES UNDER INCORRECT MODELS

Keigo Yamada
Japan UNIVAC Research Institute
Tokyo DIA Building, 1-28-38
Shinkawa, Chuo-ku, Tokyo 104 Japan

Abstract

In this paper, the asymptotic behavior of posterior distributions on parameters contained in random processes is examined when the specified model for the densities is not necessarily correct. Uniform convergence of likelihood functions in some way is shown to be a sufficient condition for the posterior distributions to be asymptotically confined to a set (Theorem 1). For ergodic stationary Markov processes uniform convergence of likelihood functions is established by the ergodic theorem for Banach-valued stationary processes (Proposition 1). A sufficient condition for the uniform convergence is also shown for general random processes (Proposition 2). These results are used to analyze the asymptotic behavior of posterior distributions on parameters contained in linear systems under incorrect models (Example 1 and 2).

1. INTRODUCTION. Let $\{X_n\}$, $n=1, 2, \cdots$ be a family of random variables defined on a probability space $(\Omega, \mathcal{B}, P)$. A model is given which specifies that the joint density of random variables $X_1, \cdots, X_n$ is one of the densities $f_n(x_1, x_2, \cdots, x_n \mid \theta)$, where the indexing parameter θ takes its values in the parameter space Θ, assumed to be a compact metric space. π denotes a prior distribution on $(\Theta, \mathcal{B}(\Theta))$, where $\mathcal{B}(\Theta)$ is the Borel σ-field of Θ, and π_n denotes the corresponding posterior distribution of the parameter given $X_1, X_2, \cdots, X_n$. Thus, for any $A \in \mathcal{B}(\Theta)$,

$$(1) \quad \pi_n A = \int_A f_n(X_1, \cdots, X_n \mid \theta)\, d\pi(\theta) \Big/ \int_\Theta f_n(X_1, \cdots, X_n \mid \theta)\, d\pi(\theta).$$

In this paper we study the asymptotic behavior of the sequence $\{\pi_n\}$ under the situation that the joint density of $\{X_n\}$ need not correspond to any of the densities in the specified model. Such an analysis was done by Berk [1] when $\{X_n\}$ are identically and independently distributed $(i.i.d.)$.

It is, however, desirable to do the same kind of analysis for more general cases since most of the stochastic processes we encounter in practical problems are not $i.i.d.$.

As was shown in Berk [1], when the process $\{X_n\}$ is $i.i.d.$, uniform convergence of the likelihood functions $f_n(X_1, \cdots, X_n \mid \theta)$ in some way ensures that the posterior distribution for the parameter θ is asymptotically confined to a set (which is called the asymptotic carrier by Berk). In Theorem 1, it is shown that the same thing is true when $\{X_n\}$ are not necessarily $i.i.d.$

In general, it is impossible to determine the asymptotic carrier since the true density for the observed process is not known. We can, however, analyze to some extent the asymptotic behavior of posterior distributions under a misspecified (incorrect) model by investigating the property of the asymptotic carrier.

In section 3, the uniform convergence of likelihood functions is established for ergodic stationary Markov processes using the ergodic theorem in Banach space, and an example of the analysis is given. A sufficient condition for the uniform convergence is given for general processes in Section 4. These results are then applied to the analysis of the asymptotic behavior of posterior distributions on parameters involved in multi-input, multi-output linear systems when the model is incorrect.

2. CONVERGENCE OF POSTERIOR DISTRIBUTIONS.

We assume the following:

(A1) For any n and $\theta \in \Theta$, $f_n(x_1, \cdots, x_n \mid \theta)$ is jointly Borel-measurable.

(A2) $f_n(X_1, \cdots, X_n \mid \theta) > 0$ with probability one.

(A3) For any nonempty open set $A \in \mathcal{B}(\Theta)$, $\pi(A) > 0$.

As was indicated in Introduction, the following theorem states that, if the likelihood functions $f_n(X_1, \cdots, X_n \mid \theta)$ converges uniformly in θ in some way, then the posterior distribution $\{\pi_n\}$ defined in (1) is asymptotically confiend to a set.

Theorem 1. Assume (A1 - 3). Suppose that, for a continuous function $\eta(\theta)$ defined on Θ,

(2) $$\sup_{\theta \in \Theta} \left| \frac{1}{n} \log f_n(X_1, \cdots, X_n \mid \theta) - \eta(\theta) \right| \to 0 \quad \text{in probability,}$$

then, for any open set $A \in \mathcal{B}(\Theta)$ which contains the asymptotic carrier A_0,

(3) $$\pi_n A \to 1 \quad \text{in probability}$$

where A_0 is defined as $A_0 = \{\theta \,;\, \eta^* = \eta(\theta),\ \eta^* = \sup_{\theta \in \Theta} \eta(\theta)\}$

Remark 1. Since $\eta(\cdot)$ is continuous on the compact set Θ, η^* is finite and A_0 is not empty.

Proof. The proof follows the method given by Berk [1]. It is sufficient to prove

$$L_n A = \frac{\pi_n A^c}{\pi_n A} \to 0 \quad \text{in probability}$$

where A^c is complement of the set A.

We should note that $\pi_n A > 0$ for all n with probability one because of (A2 - 3).

Now

$$L_nA=\int_{A^c} f_n(X_1,\cdots,X_n\mid\theta)\,d\pi(\theta)\Big/\int_A f_n(X_1,\cdots,X_n\mid\theta)\,d\pi(\theta)$$

$$=\int_{A^c}(\exp\frac{1}{n}\log f_n(X_1,\cdots,X_n\mid\theta))^n d\pi(\theta)\Big/\int_A(\exp\frac{1}{n}\log f_n(X_1,\cdots,X_n\mid\theta))^n d\pi(\theta).$$

We shall show that, for any $A\in\mathcal{B}(\Theta)$,

$$(4)\qquad (\int_A(\exp\frac{1}{n}\log f_n(X_1,\cdots,X_n\mid\theta))^n d\pi(\theta))^{\frac{1}{n}}\to\sup_{\theta\in A}\exp\eta(\theta)$$

in probability.

By the condition (2) in the theorem,

$$(5)\qquad \sup_{\theta\in A}\left|\frac{1}{n}\log f_n(X_1,\cdots,X_n\mid\theta)-\eta(\theta)\right|\to 0$$

in probability.

Hence we have

$$(6)\qquad \sup_{\theta\in A}\left|\exp\frac{1}{n}\log f_n(X_1,\cdots,X_n\mid\theta)-\exp(\theta)\right|\to 0$$

in probability.

In fact, defining $F_n(\omega;\theta)$ by

$$F_n(\omega;\theta)=\frac{1}{n}\log f_n(X_1,\cdots,X_n\mid\theta),$$

$$(7)\qquad \begin{aligned}&\sup_{\theta\in A}\left|\exp F_n(\omega;\theta)-\exp\eta(\theta)\right|\\ &\leq\sup_{\theta\in A}|F_n(\omega;\theta)-\eta(\theta)|\exp\{|\eta(\theta)|+|F_n(\omega;\theta)-\eta(\theta)|\}\\ &\leq\sup_{\theta\in A}|F_n(\omega;\theta)-\eta(\theta)|\cdot\exp\{\sup_{\theta\in A}|\eta(\theta)|+\sup_{\theta\in A}|F_n(\omega;\theta)-\eta(\theta)|\}\end{aligned}$$

Since $\exp(\cdot)$ is continuous, using (5) and Theorem 6 in 3, II in Gihman - Skorohod [2],

$$\exp\{\sup_{\theta\in A}|F_n(\omega;\theta)-\eta(\theta)|\}\to 1$$

in probability.

Hence again, by noting $\exp\{\sup_{\theta\in A}|\eta(\theta)|\}<\infty$ the last term in (7) converges to 0 in probability and (6) follows.

Now by Minkowski's inequality

$$\left|(\int_A(\exp F_n(\omega;\theta))^n d\pi(\theta))^{\frac{1}{n}}-(\int_A(\exp\eta(\theta)^n d\pi(\theta))^{\frac{1}{n}}\right|$$

$$\leq(\int_A|\exp F_n(\omega;\theta)-\exp\eta(\theta)|^n d\pi(\theta))^{\frac{1}{n}}$$

$$\leq \sup_{\theta \in \Theta} |\exp F_n(\omega;\theta) - \exp\eta(\theta)| \to 0$$

in probability.

On the other hand

$$\left(\int_A (\exp\eta(\theta))^n d\pi(\theta)\right)^{\frac{1}{n}} \to \sup_{\theta \in A} \exp\eta(\theta)$$

(see Yoshida ([8], Theorem 1, 3, I, P. 34)). Combining these results, (4) follows.

Now

$$\lim_{n\to\infty} L_n A = \lim_{n\to\infty} \left[\frac{\sup_{\theta \in A^c} \exp\eta(\theta)}{\sup_{\theta \in A} \exp\eta(\theta)}\right]^n$$

in probability.

By noting that A^c is compact, $\eta(\cdot)$ is continuous and that $A \supset A_0$,

$$0 \leq \sup_{\theta \in A^c} \exp\eta(\theta) / \sup_{\theta \in A} \exp\eta(\theta) < 1$$

and hence

$$L_n A \to 0 \qquad \text{in probability.} \qquad \text{Q.E.D.}$$

Remark 1. The above proof shows that, if the convergence in (2) holds with probability one, then the convergence in (3) holds with probability one.

Remark 2. Suppose

$$\frac{1}{n} \mathrm{E} \log f_n(X_1, \cdots, X_n \mid \theta) \to \eta(\theta)$$

for each θ. Let $f_n(X_1, \cdots, X_n \mid \theta_0)$ be the true density of $\{X_n\}$. Then we have $\eta(\theta_0) \geq \eta(\theta)$. In fact,

$$\begin{aligned}\eta(\theta_0) - \eta(\theta) &= \lim_{n\to\infty} \frac{1}{n} (\mathrm{E} \log f_n(X_1, \cdots, X_n \mid \theta_0) \\ &\quad - \mathrm{E} \log f_n(X_1, \cdots, X_n \mid \theta)) \\ &= \lim_{n\to\infty} \frac{1}{n} \mathrm{E} \log \frac{f_n(X_1, \cdots, X_n \mid \theta_0)}{f_n(X_1, \cdots, X_n \mid \theta)} \geq 0.\end{aligned}$$

It is well known (Kullback [3]) that

$$\mathrm{E} \log \frac{f_n(X_1, \cdots, X_n \mid \theta_0)}{f_n(X_1, \cdots, X_n \mid \theta)} \geq 0$$

Theorem 1 shows that the asymptotic behavior of the posterior distributions under a specified model can be analyzed by using $\eta(\theta)$ once likelihood functions converges uniformly to $\eta(\theta)$ in a manner defined in (2). For *i.i.d.* random variables $\{X_n\}$, Berk [1] established the condition (2) by using the strong law of large numbers for Banach-valued *i.i.d.* random variables. By the similar idea, we can show that the condition (2) holds for ergodic stationary Markov processes by the ergodic theorem for Banach-valued stationary processes. This will be done in the next section.

3. THE CASE OF ERGODIC STATIONARY MARKOV PROCESSES. In this section we treat the case where the process $\{X_n\}$, $n=1, 2, \cdots$ is an ergodic stationary Markov process. We shall show that, under a specified model described soon, the condition (2) in Theorem 1 is satisfied for this class of stochastic processes.

Let $f(y|x,\theta_0)$ be the transition probability density of the process $\{X_n\}$ characterized by a parameter θ_0. $f(x|\theta_0)$ denotes the density of the random variable X_1. Then, given a parameter set Θ which is a compact metric space,

$$f_n(X_1,\cdots,X_n|\theta)=f(X_1|\theta)\prod_{i=2}^{n} f(X_i|X_{i-1},\theta),\, n=1,2,\cdots$$

is the likelihood functions of $\{X_n\}$ defined on the parameter set Θ. We shall adopt the functions $f_n(x_1,\cdots,x_n|\theta), n=1,2,\cdots$ as a model for the densities of the process $\{X_n\}$. The following assumptions are made:

(B1) $f(\cdot,\cdot|\cdot)$ is jointly measurable and, for each fixed (x, y), $f(y|x,\cdot)$ is continuous. $f(\cdot|\theta)$ is measurable for each $\theta\in\Theta$.

(B2) There exists a measurable function $K(y|x)$ such that

$$E\,K(X_2|X_1)<\infty \qquad \text{and} \qquad |\log f(y|x,\theta)|\leq K(y|x).$$

Let us define

$$\eta(\theta)=E\log f(X_2|X_1,\theta) \tag{8}$$
$$=\int\log f(y|x,\theta)\cdot f(y|x,\theta_0)\cdot f(x|\theta_0)\,dy\,dx.$$

Note that under (B1 - 2) $E\log f(X_2|X_1,\cdot)$ exists in the sense of Bochner's integral. Then we have

Proposition 1. Under the assumption (B1) and (B2),

$$\sup_{\theta\in\Theta}\left|\frac{1}{n}\log f_n(X_1,\cdots,X_n|\theta)-\eta(\theta)\right|\to 0$$

with probability one.

Proof. We use the ergodic theorem for Banach-valued stationary processes. Let $C(\Theta)$

be the space of all continuous functions defined on θ with the sup-norm. Since

$$\sup_{\theta \in \Theta} \left| \frac{1}{n} f(X_1 \mid \theta) \right| \to 0 \qquad \text{with probability one,}$$

it suffices to prove

$$\text{(9)} \qquad \sup_{\theta \in \Theta} \left| \frac{1}{n} \sum_{i=2}^{n+1} \log f(X_i \mid X_{i-1}, \theta) \right| \to 0$$

with probability one.

By the assumption (B1) and separability of the space $C(\Theta)$ with the aid of Lemma 2.2.1 of Padgett-Taylor [4], $Y_i = \log f(X_{i+1} \mid X_i, \cdot)$, $i=1, \ldots$ are random variables in $C(\Theta)$. Hence if we can show that the process $\{Y_n\}$, $n=1, 2, \cdots$ is an ergodic stationary process in Banach space $C(\Theta)$, then the ergodic theorem for Banach-valued stationary processes (see Parthasarathy [5]) asserts (9) and the conclusion follows.

To show the stationarity of the process $\{Y_n\}$, it is sufficient to prove, for example,

$$\text{(10)} \qquad P\{\omega ; (Y_1(\omega), \cdots, Y_k(\omega)) \in A\} = P\{\omega ; (Y_2(\omega), \cdots, Y_{k+1}(\omega)) \in A\}$$

for any k and $A \in \mathcal{B}(C^k(\Theta))$ where $C^k(\Theta)$ is the product space of k copies of $C(\Theta)$ This can be done by the same method as in Lemma 2.3.4 of Padget-Taylor [4].

The set

$$U = \{\{x \in C^k(\Theta), f(x) < b\} : f \in C^k(\Theta)^* \text{ and } b \in R^1\},$$

where $C^k(\Theta)^*$ is the dual space of $C^k(\Theta)$, is a family of unicity for the Borel-field $\mathcal{B}(C^k(\Theta))$ (P. 25 of Padget-Taylor [4]), and it suffices to show that (10) holds for any $A \in U$. Now,

for $B = \{x \in C^k(\Theta), f(x) < b\} \in U$,

$$P\{\omega ; (Y_1, \cdots, Y_k) \in B\} = P\{\omega ; f(Y_1, \cdots, Y_k) < b\}.$$

Since $Y_i = \log f(X_{i+1} \mid X_i, \cdot)$, $f(Y_1, \cdots, Y_k)$ is a function of $X_1, \cdots, X_{k+1}$, i.e.,

$$f(Y_1, \cdots, Y_k) = g(X_1, \cdots, X_{k+1}).$$

g is a composite function of g_1 and g_2 where

$$g_1 : (x_1, \cdots, x_{k+1}) \in R^{k+1} \to (f(x_2 \mid x_1, \cdot), \cdots, f(x_{k+1} \mid x_k, \cdot)) \in C^k(\Theta)$$

$$g_2 : \omega \in \Omega \to (X_1(\omega), \cdots, X_{k+1}(\omega)) \in R^{k+1}$$

The assumption (B1) and separability of $C^k(\Theta)$ ensure measurability of g_1, and hence g is measurable. With this fact and stationarity of the process $\{X_n\}$, $g(X_1,\cdots,X_{k+1})$ and $g(X_2,\cdots,X_{k+2})$ have the same distribution. Hence

$$P\{\omega\,;\,f(Y_1,\cdots,Y_k)<b\}$$

$$=P\{\omega\,;\,g(X_1,\cdots,X_{k+1})<b\}$$

$$=P\{\omega\,;\,g(X_2,\cdots,X_{k+2})<b\}$$

$$=P\{\omega\,;\,f(Y_2,\cdots,Y_{k+1})<b\}$$

$$=P\{\omega\,;\,(Y_2,\cdots,Y_{k+1})\in B\}.$$

This establishes (10) and, hence, stationarity of $\{Y_n\}$.

Finally

ergodicity of $\{Y_n\}$ is also proved by showing that for every $A\in\mathcal{B}(C^{k+1}(\Theta))$, $k=1,2,\cdots$,

$$\text{(11)}\qquad \frac{1}{N}\sum_{n=1}^{N}\chi_A(Y_n,\cdots,Y_{n+k})\to P(\omega\,;\,(Y_1,\cdots,Y_{k+1})\in A)$$

with probability one where χ_A is the characteristic function of the set A. Since $\chi_A(Y_n,\cdots,Y_{n+k})$ is a function of $\{X_n\}$, *i.e.*,

$$\chi_A(Y_n,\cdots,Y_{n+k})=g(X_n,\cdots,X_{n+k},X_{n+k+1}),$$

just as before $g(\cdot)$ is a measurable function. Hence the process $\{Z_n\}$, where $Z_n=g(X_n,\cdots,X_{n+k+1})$, is ergodic and stationary, and by the ergodic theorem

$$\frac{1}{N}\sum_{n=1}^{N}\chi_A(Y_n,\cdots,Y_{n+k})=\frac{1}{N}\sum_{n=1}^{N}g(X_n,\cdots,X_{n+k+1})$$

$$\to E\,g(X_1,\cdots,X_{k+2})$$

$$=E\chi_A(Y_1,\cdots,Y_{k+1})$$

$$=P\{\omega\,;Y_1,\cdots,Y_{k+1})\in A\}.$$

This shows the relation (11). Q.E.D.

Remark 3. From Remark 2 we have $\eta(\theta_0)\geqq\eta(\theta)$. If the following condition

$$(*):\qquad f(y\,|\,x,\theta_1)\neq f(y\,|\,x,\theta_2)\ a.e.\quad \theta_1\neq\theta_2$$

and $f(y\,|\,x,\theta_0)$ and $f(x\,|\,\theta_0)$ are both positive, holds then $\eta(\theta_0)=\eta(\theta)$ implies $\theta_0=\theta$.

Example 1. Let us consider a first order ergodic stationary Markov process $\{X(n)\}$ given by

$$X(n)=\Phi_0 X(n-1)+G_0 w(n-1),\ n=\cdots,-1,0,1,\cdots$$

where $X(n)\in R^d$ and $\{w(n)\}$ is a k-dimensional vector valued $i.i.d.$ random sequence with normal distribution $N(O,I_k)$, I_k identity matrix. Φ_0 and G_0 are unknown $d\times d$ and $d\times k$ matrices respectively, and we assume that absolute values of all eigenvalues of Φ_0 lie in a unit circle. We investigate the asymptotic behavior of posterior distributions of the parameter $\theta=(\Phi,G)$ on a compact set Θ which need not contain the true parameter $\theta_0=(\Phi_0,G_0)$ The transition density of the process $\{X(n)\}$ is given by

$$f(y|x,\theta_0)=((2\pi)^d|G_0G_0'|)^{-\frac{1}{2}}\exp\{-\frac{1}{2}(y-\Phi_0 x)'(G_0G_0')^{-1}(y-\Phi_0 x)\}$$

where we assumed nonsingularity of the matrix G_0G_0'. The density $f(x|\theta_0)$ of $X(n)$ is normal $N(O,\Gamma)$ where Γ is given by

$$\Gamma=G_0G_0'+\Phi_0G_0G_0'\Phi_0'+\Phi_0^2\ G_0G_0'(\Phi_0)^2+\cdots.$$

Now the function $\eta(\theta)$ defined by (8) is

$$\begin{aligned}\eta(\theta)&=-\frac{1}{2}\log(2\pi)^d|GG'|\\&\quad-\frac{1}{2}\int(y-\Phi x)'(GG')^{-1}(y-\Phi x)f(y|x;\theta_0)f(x|\theta_0)\,dydx\\&=-\frac{1}{2}\log(2\pi)^d|GG'|-\frac{1}{2}\mathrm{trace}[(GG')^{-1}G_0G_0']\\&\quad-\frac{1}{2}\int x'(\Phi_0-\Phi)'(GG')^{-1}(\Phi_0-\Phi)xf(x|\theta_0)\,dx\\&=-\frac{1}{2}\log(2\pi)^d|GG'|-\frac{1}{2}\mathrm{trace}[(GG')^{-1}G_0G_0']\\&\quad-\frac{1}{2}\mathrm{trace}(\Phi_0-\Phi)'(GG')^{-1}(\Phi_0-\Phi)\Gamma\end{aligned}\tag{12}$$

where we are assuming nonsingularity of GG' for all $\theta=(\Phi,G)\in\Theta$.
As we have shown in Remark 2

$$\begin{aligned}\eta(\theta_0)&=-\frac{1}{2}\log(2\pi)^d|G_0G_0'|-\frac{1}{2}d\\&\geq\eta(\theta),\ \forall\theta\in\Theta.\end{aligned}$$

The asymptotic carrier A_0 for the parameter set Θ can be calculated by using (12) and this enables us various kinds of analysis for the asymptotic behavior of posterior distributions. For example, let the parameter set Θ be such that $\Theta=\{(\Phi,G_1),$

$\Phi \in \Theta_1$, $G_1 \neq G_0\}$ where Θ_1 is a compact set regarding the parameter Φ and contains Φ_0. For this case the asymptotic carrier A_0 contains only one point (Φ_0, G_1) since $\mathrm{trace}(\Phi_0-\Phi)'(GG')^{-1}(\Phi_0-\Phi)\Gamma=0$ if and only if $\Phi=\Phi_0$. Hence even if the specified model does not include the true density, *i.e.*, $G_1 \neq G_0$, as far as the parameter Φ is concerned the posterior distributions on Φ converge to the true point Φ_0.

4. A SUFFICIENT CONDITION FOR GENERAL CASES. Returning to the general case, let $\{X_n\}, n=1,\cdots,$ be a random sequence. Given a joint density model $f_n(x_1,\cdots,x_n|\theta)$, $\theta\in\Theta$, $n=1,2,\cdots$ for the process $\{X_n\}$, we have the following proposition regarding the uniform convergence of likelihood functions.

Proposition 2. Assume that

(i) $$\mathrm{Var}\left(\frac{1}{n}\log f_n(X_1,\cdots,X_n|\theta)\right)\to 0$$

uniformly in θ

(ii) $$E\left(\frac{1}{n}\log f_n(X_1,\cdots,X_n|\theta)\right)\to\eta(\theta)$$

uniformly in θ, then the condition (2) in Theorem 1 is satisfied

Proof. Application of Chebyshev's inequality easily shows the result. In fact, for an arbitrary number $c>0$,

$$\begin{aligned}&P\{\omega\,;\sup_{\theta\in\Theta}|\frac{1}{n}\log f_n(X_1,\cdots,X_n|\theta)-\eta(\theta)|>c\}\\ &\leq P\{\omega\,;\sup_{\theta\in\Theta}|\frac{1}{n}\log f_n(X_1,\cdots,X_n|\theta)-\frac{1}{n}E\log f_n(X_1,\cdots,X_n|\theta)|>\frac{c}{2}\}\\ &+P\{\omega\,;\sup_{\theta\in\Theta}|\frac{1}{n}\exists\log f_n(X_1,\cdots,X_n|\theta)-\eta(\theta)|>\frac{c}{2}\}\end{aligned}$$

By Chebyshev's inequality

$$\begin{aligned}&\leq\frac{4}{c^2}E\{\sup_{\theta\in\Theta}|\frac{1}{n}\log f_n(X_1,\cdots,X_n|\theta)-\frac{1}{n}E\log f_n(X_1,\cdots,X_n|\theta)|\}^2\\ &+\frac{4}{c^2}\{\sup_{\theta\in\Theta}|\frac{1}{n}E\log f_n(X_1,\cdots,X_n|\theta)-\eta(\theta)|\}^2\end{aligned}$$

By Fatou's lemma, the first term in the above equation is less than

$$\frac{4}{c^2}\{\sup_{\theta\in\Theta}E|\frac{1}{n}\log f_n(X_1,\cdots,X_n|\theta)-\frac{1}{n}E\log f_n(X_1,\cdots,X_n|\theta)|\}^2$$

Hence the conclusion follows. Q.E.D.

Remark 4. (a) In Proposition 2 if $\eta(\theta)$ is continuous, then we can apply Theorem 1.
(b) When the process $\{X_n\}$ is an independent (but not necessarily identically

distributed) sequence, the density model is given by

$$f_n(x_1,\cdots,x_n\mid\theta)=f_1(x_1\mid\theta)\cdots f_n(x_n\mid\theta),\quad \theta\in\Theta$$

where $f_i(x_i\mid\theta)$ is the density model for random variable X_i.
In this case the assumption in Proposition 2 takes the form:

(i) $\frac{1}{n^2}\sum_{i=1}^{n}$ Var log $f_i(X_i\mid\theta)\to 0$ uniformly in θ.

(ii) $\frac{1}{n}\sum_{i=1}^{n}$ E log $f_i(X_i\mid\theta)\to\eta(\theta)$ uniformly in θ.

Example 2. Let us consider the following linear system:

$$X_{n+1}=\Phi X_n+GU_n,\quad X_1 \text{ ; given}$$

$$Y_n=HX_n+\xi_n$$

where $X_n\in R^d, Y_n, \xi_n\in R^l, U_n\in R^m$ and matrices Φ, G, H · have appropriate dimensions. $\{U_n\}$ is a given control sequence and we assume that $\{\xi_n\}$ is an independent and normally distributed sequence with mean zero and covariance matrix Γ. The unknown parameter θ consists of Φ, G, H and Γ. The process $\{Y_n\}$ is clearly independent but not identically distributed. Φ_0, G_0, H_0 and Γ_0 denote the true parameter. Then Y_n has a normal distribution $N(H_0X_n(\theta_0),\Gamma_0)$ where $X_n(\theta_0)$ is the state vector corresponding to the true parameter θ_0, and the density model is given by

$$f_n(Y_1,\cdots,Y_n\mid\theta)=\prod_{i=1}^{n} f_i(Y_i\mid\theta)$$

where

$$f_i(Y_i\mid\theta)=((2\pi)^l|\Gamma|)^{-\frac{1}{2}}\exp\{-\frac{1}{2}(Y_i-HX_i(\theta))'\Gamma^{-1}(Y_i-HX_i(\theta))\}$$

and $X_i(\theta)$ is the state vector corresponding to the parameter θ.
We assume the following:
(C1) Φ_0 and $\Phi\in\Theta$ are stable matrices, *i.e.*, absolute values of all eigenvalues of these matrices lie in a unit circle.
(C2) The control sequence is uniformly bounded, *i.e.*, $|U_u|\le K$. and, it has the property such that

$$\frac{1}{N}\Sigma(H_0X_n(\theta_0)-HX_n(\theta))(H_0X_n(\theta)-HX_n(\theta))'$$

converges to a function of θ uniformly in θ.

Then conditions (i) and (ii) in Proposition 2 are satisfied. To show this, first we note that the state vector $X_n(\theta)$ is uniformly bounded, *i.e.*, $|X_n(\theta)| \leq M$ for all n and $\theta \in \Theta \cup \{\theta_0\}$. In fact, since

$$X_n(\theta) = GU_{n-1} + \Phi GU_{n-2} + \Phi^2 GU_{n-3} + \cdots + \Phi^{n-1} GU_0 + \Phi^n X_0 ,$$

$$|X_n(\theta)| \leq \|G\| \, |U_{n-1}| + \|\Phi\| \, \|G\| \, |U_{n-2}| + \cdots + \|\Phi\|^{n-1} \|G\| \, |U_0| + \|\Phi\|^n |X_0|$$

where the matrix norm $\|A\|$ is defined by $\|A\| = \sup_{|x| \leq 1} |Ax|$.

The right hand side of the above inequality is less than

$$
\begin{aligned}
&K \|G\| (1 + \|\Phi\| + \|\Phi\|^2 + \cdots \quad) + |X_0| \\
&= K \|G\| \frac{1}{1 - \|\Phi\|} + |X_0| \leq M
\end{aligned}
\tag{13}
$$

Note that by (C1) $\|\Phi\| < 1$. The last inequality in (13) is due to the compactness of the set Θ, and the uniform boundedness of $X_n(\theta)$ follows. Since $\{Y_n\}$ are independent, according to Remark 4 (b), let us calculate $\operatorname{Var} \log f_n(Y_n | \theta)$. Since

$$
\begin{aligned}
\log f_n(Y_n | \theta) = {} & \log((2\pi)^l |\Gamma|)^{-\frac{1}{2}} - \frac{1}{2} (H_0 X_n(\theta_0) - HX_n(\theta))' \Gamma^{-1} (H_0 X_n(\theta_0) - HX_n(\theta)) \\
& - \xi_n' \Gamma^{-1} (H_0 X_n(\theta_0) - HX_n(\theta)) \\
& - \frac{1}{2} \xi_n' \Gamma^{-1} \xi_n ,
\end{aligned}
$$

we have

$$
\begin{aligned}
\operatorname{Var} \log f_n(Y_n | \theta) = {} & \mathrm{E} (\xi_n' \Gamma^{-1} (H_0 X_n(\theta_0) - HX_n(\theta)) + \frac{1}{2} \xi_n' \Gamma^{-1} \xi_n \\
& - \frac{1}{2} \mathrm{E} \xi_n' \Gamma^{-1} \xi_n)^2 \\
\leq {} & 2 \mathrm{E} (\xi_n' \Gamma^{-1} (H_0 X_n(\theta_0) - HX_n(\theta)))^2 + 2 \operatorname{Var} (\frac{1}{2} \xi_n' \Gamma^{-1} \xi_n) \\
\leq {} & M
\end{aligned}
$$

where M does not depend on θ.

The last inequality comes from finiteness of the moment of ξ_n and uniform boundedness of $H_0 X_n(\theta_0) - HX_n(\theta)$. Now the condition (i) in Remark (b) is easily checked by noting that

$$
\begin{aligned}
\operatorname{Var} (\frac{1}{2} \log f_n(Y_1, \cdots, Y_n | \theta)) &= \frac{1}{n^2} \sum_{i=1}^{n} \operatorname{Var} \log f_i(Y_i | \theta) \\
&\leq \frac{M}{n} \to 0 \qquad \text{uniformly in } \theta .
\end{aligned}
$$

Similarly, by the condition (C2),

$$\frac{1}{n}\sum_{i=1}^{n} \mathrm{E} \log f_n(Y_n \mid \theta) \to \lim_{n\to\infty}\frac{1}{n}\sum_{i=1}^{n} \mathrm{E} \log f_i(Y_i \mid \theta) = \eta(\theta)$$

uniformly in θ.

Thus by Proposition 2 we have

$$\sup_{\theta\in\Theta} \left| \frac{1}{n} \log f_n(Y_1, \cdots, Y_n \mid \theta) - \eta(\theta) \right| \to 0$$

in probability and $\eta(\theta)$ is given by

$$\eta(\theta) = \log((2\pi)^{l} |\Gamma|)^{-\frac{1}{2}} - \frac{1}{2} \lim_{N\to\infty} \frac{1}{N} \sum_{i=1}^{n} (H_0 X_n(\theta_0) - HX_n(\theta))' \Gamma^{-1}$$

$$(H_0 X_n(\theta_0) - HX_n(\theta)) - \frac{1}{2} \operatorname{trace} \Gamma^{-1}\Gamma_0$$

As was done in Example 1 in Section 3, we shall investigate the asymptotic behavior of posterior distributions on (Φ, G, H) under the condition that Γ is arbitrarily fixed, *i. e.* $\Gamma = \Gamma_1$. Let the parameter set Θ be such that $\Theta = \Gamma_1 \times \Theta_1$ where Θ_1 is a parameter set regarding (Θ, G, H) and we shall assume that Θ_1 contains the true parameter(Φ_0, G_0, H_0). Clearly $\eta(\theta_1) = \sup_{\theta\in\Theta} \eta(\theta)$ where $\theta_1 = (\Gamma_1, \Phi_0, G_0, H_0)$ and hence the asymptotic carrier A_0 contains θ_1. But in this case $\eta(\theta_1) = \eta(\theta)$ does not imply $\theta_1 = \theta$ since, for all non-singular matrix T, $\theta = (\Gamma_1, T\Phi_0 T^{-1}, TG_0, H_0 T^{-1})$ gives the same values to $\eta(\theta)$. When the input U_n and the output Y_n are both one dimensional, a necessary and sufficient condition on the input sequence $\{U_n\}$ for the asymptotic carrier A_0 to contain only one point, *i. e.* $\theta_1 = (\Gamma_1, \Phi_0, G_0, H_0)$ is known (Aoki and Yue [6]) under the condition that (Φ, H) has a cannonical observable form and (Φ, G) is a controllable pair.

We shall consider more general cases. To simplify the analysis, we assume (C3): The control sequence $\{U_n\}$ is a uniformly bounded *i.i.d* random process and

$$\mathrm{E}\, U_i U_j' = \begin{cases} W & i=j \,; W: \text{ positive definite} \\ 0 & i=j \end{cases}$$

Furthermore, since we are only concerned with the asymptotic behavior of posterior distributions, we assume that the time index n of $\{Y_n\}$ tends to infinite past, *i. e.*, $n = \cdots, -1, 0, 1 \cdots$.

Then since

$$X_n = GU_n + \Phi GU_{u-1} + \Phi^2 GU_{n-2} + \cdots$$

and $\{U_n\}$ is a uniformly bounded *i.i.d* sequence, by (C1) $\{X_n\}$ is an ergodic stationary process and so is $\{Y_n\}$.

Now $\{Y_n\}$ are no more independent, but since $\xi_n = Y_n - H_0 X_n(\theta_0)$ is an independent and normally distributed sequence, the likelihood function $f_n(Y_1, \cdots, Y_n \mid \theta)$ of $Y_1, \cdots, Y_n$ is given by

$$(14)\quad f_n(Y_1, \cdots, Y_n \mid \theta) = \prod_{i=1}^{n} ((2\pi)^{l} |\Gamma|)^{-\frac{1}{2}} \exp\{-\frac{1}{2}(Y_i - HX_i(\theta))' \Gamma^{-1}(Y_i - HX_i(\theta))\},$$

and we shall assume that this function $f_n(Y_1, \cdots, Y_n \mid \theta)$ is to be the density model for $\{Y_n\}$. Let us investigate the asymptotic behavior of $(1/n)\log f_n(Y_1, \cdots, Y_n \mid \theta)$ directly without using Proposition 2. Since

$$\frac{1}{n}\log f_n(Y_1, \cdots, Y_n \mid \theta) = \log((2\pi)^{l}|\Gamma|)^{-\frac{1}{2}} - \frac{1}{2}\frac{1}{n}\sum_{i=1}^{n}(H_0X_i(\theta_0) - HX_i(\theta))'\Gamma^{-1}(H_0X_i(\theta_0) - HX_i(\theta)) - \frac{1}{n}\sum_{i=1}^{n}\xi_i'\Gamma^{-1}(H_0X_i(\theta_0) - HX_i(\theta)) - \frac{1}{2}\frac{1}{n}\sum_{i=1}^{n}\xi_i'\Gamma^{-1}\xi_i,$$

we shall apply the ergodic theorem to each term of the above equation. Noting that by the same discussion as in Section 2 $\{X_n(\theta)\}$ is an ergodic stationary process in Banach space $C(\Theta)$, by the ergodic theorem in Banach space we have

$$\sup_{\theta\in\Theta}\left|\frac{1}{n}\sum_{i=1}^{n}\xi_n'\Gamma^{-1}(H_0X_n(\theta_0) - HX_n(\theta))\right| \to 0$$

with probability one.
Here we used the fact that ξ_n and $H_0X_n(\theta_0) - HX_n(\theta)$ are independent.
Similarly

$$\sup_{\theta\in\Theta}\left|\frac{1}{n}\sum_{i=1}^{n}\xi_i'\Gamma^{-1}\xi_i - \mathrm{trace}\,\Gamma^{-1}\Gamma_0\right| \to 0$$

with probability one,
and

$$\sup_{\theta\in\Theta}\left|\frac{1}{n}\sum_{i=1}^{n}(H_0X_i(\theta_0) - HX_i(\theta))'\Gamma^{-1}(H_0X_i(\theta_0) - HX_i(\theta)) - E(H_0X_1(\theta_0) - HX_1(\theta))'\Gamma^{-1}(H_0X_1(\theta_0) - HX_1(\theta))\right| \to 0$$

with probability one.
Hence we have

$$\sup_{\theta\in\Theta}\left|\frac{1}{n}\log f_n(Y_1, \cdots, Y_n \mid \theta) - \eta(\theta)\right| \to 0 \quad \text{with probability one}$$

where

$$\eta(\theta)=\log((2\pi)^{l}|\Gamma|)^{-\frac{1}{2}}-\frac{1}{n}\ \mathrm{trace}\,\Gamma^{-1}\Gamma_0$$

$$-\frac{1}{2}\,\mathrm{E}\,(H_0X_1(\theta_0)-HX_1(\theta))'\Gamma^{-1}(H_0X_1(\theta_0)-HX_1(\theta))$$

Let us calculate

$$\mathrm{E}\,(H_0X_1(\theta_0)-HX_1(\theta))'\Gamma^{-1}(H_0X_1(\theta_0)-HX_1(\theta))$$

$$=\mathrm{trace}\,\Gamma^{-1}\mathrm{E}(H_0X_1(\theta_0)-HX_1(\theta))(H_0X_1(\theta_0)-HX_1(\theta))'.$$

Since

$$HX_1(\theta)=HGU_0+H\Phi GU_{-1}+H\Phi^2GU_{-2}+\cdots$$

and by (C3), we have

$$\mathrm{E}\,(H_0X_1(\theta_0)-HX_1(\theta))(H_0X_1(\theta_0)-HX_1(\theta))'$$
$$=\sum_{i=0}^{\infty}(H_0\Phi_0^iG_0-H\Phi^iG)W(H_0\Phi_0^iG_0-H\Phi^iG)'$$

Hence

$$\eta(\theta)=\log((2\pi)^{l}|\Gamma|)^{-\frac{1}{2}}-\frac{1}{2}\ \mathrm{trace}\,\Gamma^{-1}\Gamma^0 \tag{15}$$
$$-\frac{1}{2}\,\mathrm{trace}\,\Gamma^{-1}\sum_{i=0}^{\infty}(H_0\Phi_0^iG_0-H\Phi^iG)W(H_0\Phi_0^iG_0-H\Phi^iG)'.$$

Let $\eta(\theta_1)=\eta(\theta)$ where $\theta_1=(\Gamma_1,\Theta_0,G_0,H_0)$. Then

$$H_0\Phi_0^iG_0=H\Phi^iG \qquad \text{for all} \quad i\geqq 0.$$

Thus under the assumption

(C4): For all $\theta\in\Theta\cup\{\theta_0\}$ the system

$$\begin{cases} X_n=\Phi X_{n-1}+GU_{n-1} \\ Y_n=HX_n \end{cases}$$

is a minimal realization,
there exists a non-singular matrix T such that

$$\Phi=T\Phi_0T^{-1},\ G=TG_0,\ H=H_0T^{-1} \tag{16}$$

(See Brockett [7]).
As we have already shown, if the parameter (Φ, G, H) is completely unknown, $\eta(\theta_0)=\eta(\theta)$ does not imply $\theta_0=\theta$. Motivated by this fact, we shall consider the case where some of elements of (Φ_0, G_0, H_0) are known a priori so that the following condition (C5) holds:

(C5): $\theta\in\Theta$, which satisfies (16), is equal to θ_1.

Then under this assumption $\eta(\theta_1)=\eta(\theta)$ implies $\theta_1=\theta$.
We summarize here the obtained result.

Proposition 3. Consider the linear system:

$$X_n=\Phi X_{n-1}+GU_n$$

$$Y_n=HX_n+\xi_n \qquad n=\cdots,-1,0,1,\cdots$$

Under the condition (C1) and (C3),

$$\sup_{\theta\in\Theta}\left|\frac{1}{n}\log f_n(Y_1,\cdots,Y_n|\theta)-\eta(\theta)\right|\to 0$$

with probability one
where $f_n(Y_1\cdots,Y_n|\theta)$ and $\eta(\theta)$ are given in (14) and (15).
For a case where $\Theta=\Gamma_1\times\Theta_1$ (defined earlier), under further assumptions (C4) and (C5), $\eta(\theta_1)=\eta(\theta)$ implies $\theta_1=\theta$ and hence the asymptotic carrier A_0 contains only one point $\theta_1=(\Gamma_1, \Phi_0, G_0, H_0)$

References

[1] R.H. Berk, Limiting behavior of posterior distributions when the model is incorrect, Ann. Math. Statist., 37 (1966), 51-58

[2] I.I. Gikhman and A.V. Skorphod, "Introduction to the Theory of Random Processes," W.B. Sounders Co., Philadelphia (1969)

[3] S. Kullback, "Information and Statistics," Wiley, New York (1959)

[4] W.J. Padgett and R.L. Taylar, "Laws of Large Numbers for Normed Linear Spaces and Certain Fréchet Spaces," Lecture Notes in Mathematics, Springer-Verlag, Berlin (1973)

[5] K.R. Parthasarathy, "Probability Measures on Metric Spaces," Academic Press, New York (1967)

[6] M. Aoki and P.C. Yue, On certain convergence questions in system identification, SIAM. J. Control, 8 (1970)

[7] R.W. Brockett, "Finite Dimensional Linear Systems," Wiley, New York (1970)

[8] K. Yoshida, "Functional Analysis," Academic Press, New York (1968)

LIST OF AUTHORS

*paper not received

This series aims to report new developments in computer science research and teaching – quickly, informally and at a high level. The type of material considered for publication includes:

1. Preliminary drafts of original papers and monographs
2. Lectures on a new field, or presenting a new angle on a classical field
3. Seminar work-outs
4. Reports of meetings, provided they are
   a) of exceptional interest and
   b) devoted to a single topic.

Texts which are out of print but still in demand may also be considered if they fall within these categories.

The timeliness of a manuscript is more important than its form, which may be unfinished or tentative. Thus, in some instances, proofs may be merely outlined and results presented which have been or will later be published elsewhere. If possible, a subject index should be included. Publication of Lecture Notes is intended as a service to the international computer science community, in that a commercial publisher, Springer-Verlag, can offer a wider distribution to documents which would otherwise have a restricted readership. Once published and copyrighted, they can be documented in the scientific literature.

Manuscripts

Manuscripts should comprise not less than 100 pages.

They are reproduced by a photographic process and therefore must be typed with extreme care. Symbols not on the typewriter should be inserted by hand in indelible black ink. Corrections to the typescript should be made by pasting the amended text over the old one, or by obliterating errors with white correcting fluid. Authors receive 75 free copies and are free to use the material in other publications. The typescript is reduced slightly in size during reproduction; best results will not be obtained unless the text on any one page is kept within the overall limit of 18 x 26.5 cm (7 x 10½ inches). The publishers will be pleased to supply on request special stationery with the typing area outlined.

Manuscripts in English, German or French should be sent to Prof. G. Goos, Institut für Informatik, Universität Karlsruhe, 75 Karlsruhe/Germany, Zirkel 2, Prof. J. Hartmanis, Cornell University, Dept. of Computer-Science, Ithaca, NY/USA 14850, or directly to Springer-Verlag Heidelberg.

Springer-Verlag, D-1000 Berlin 33, Heidelberger Platz 3
Springer-Verlag, D-6900 Heidelberg 1, Neuenheimer Landstraße 28–30
Springer-Verlag, 175 Fifth Avenue, New York, NY 10010/USA

ISBN 3-540-**07623**-9
ISBN 0-387-**07623**-9